THE ROUTLEDGE INTERNATIONAL HANDBOOK OF PRACTICE-BASED RESEARCH

The Routledge International Handbook of Practice-Based Research presents a cohesive framework with which to conduct practice-based research or to support, manage and supervise practice-based researchers. It has been written with an inclusive approach, with the intention of presenting deep and meaningful knowledge for the benefit of all readers.

This handbook has been designed to present specific detail of practice-based research by outlining its shared traits with all forms of research and to highlight its core distinguishing features into a cohesive, principled and methodical approach. To this end, the handbook is presented in five sections: 1. Practice-Based Research, 2. Knowledge, 3. Method, 4. The Practice-Based PhD and 5. Practitioner Voices. Each section begins with a leading chapter that outlines each of the distinct areas as they relate to practice-based research. This is followed by a series of contributing chapters that discuss pertinent themes in more detail.

Practitioners from a broad range of backgrounds will find these chapters helpful:

- research students or final year graduates will be introduced to the principled nature of practice-based research
- PhD researchers embarking on a research project or in the flow of research will find this guidance supportive
- professionals such as designers, makers, engineers, artists and creative technologists wishing to strengthen their research into their practice will be guided through the principled and focused nature of practice-based research
- supervisors, managers and policy makers will benefit from the potential and rigour of practice-based researchers in the pursuit of new knowledge.

Craig Vear is Research Professor at De Montfort University, where he is a director of the Creative AI and Robotics lab in the Institute of Creative Technologies. His research is naturally hybrid as he draws together the fields of music, digital performance, creative technologies, Artificial Intelligence, creativity, gaming and robotics.

THE ROUTLEDGE INTERNATIONAL HANDBOOK OF PRACTICE-BASED RESEARCH

Edited by Craig Vear

Consultant Editors

Linda Candy and Ernest Edmonds

Routledge
Taylor & Francis Group

LONDON AND NEW YORK

Cover image: "Busy Mind" (2021) by Serena Menezes Poltronieri.
Crayon drawing retraced in Adobe Illustrator. © Serena Menezes Poltronieri

First published 2022
by Routledge
4 Park Square, Milton Park, Abingdon, Oxon OX14 4RN

and by Routledge
605 Third Avenue, New York, NY 10158

Routledge is an imprint of the Taylor & Francis Group, an informa business

British Library Cataloguing-in-Publication Data
Names: Vear, Craig, editor.
Title: The Routledge international handbook of practice-based research /
edited by Craig Vear ; consultant editors, Linda Candy and Ernest
Edmonds.
Description: Abingdon, Oxon ; New York, NY : Routledge, 2022. |
Includes bibliographical references and index. | Summary: "This
Handbook provides readers with an overview of the field of Practice-Based
Research (PBR): different approaches, disciplines that frequently employ
PBR, methodologies and creative outputs. It gives guidance and exemplars
of the way PBR is carried out, including both within academic study and
through 'practitioner voice' case studies and examples. It takes a wide-
ranging view of the field, including both the more established use of PBR
in the creative arts, as well as fields in which PBR is emerging, such as
Education, and digital art and technology"—Provided by publisher.
Identifiers: LCCN 2021036458 (print) | LCCN 2021036459
(ebook) | ISBN 9780367341435 (hardback) | ISBN 9781032182209
(paperback) | ISBN 9780429324154 (ebook)
Subjects: LCSH: Research—Methodology. | Social sciences—
Research—Methodology. | Evidence-based medicine. |
Evidence-based design. | Practice (Philosophy)
Classification: LCC Q180.55.M4 R68 2022 (print) |
LCC Q180.55.M4 (ebook) | DDC 001.4/2—dc23
LC record available at https://lccn.loc.gov/2021036458
LC ebook record available at https://lccn.loc.gov/2021036459

Library of Congress Cataloging-in-Publication Data
catalog record for this book has been requested

ISBN: 978-0-367-34143-5 (hbk)
ISBN: 978-1-032-18220-9 (pbk)
ISBN: 978-0-429-32415-4 (ebk)

DOI: 10.4324/9780429324154

Typeset in Bembo
by Apex CoVantage, LLC

Dedicated to all the inspiring practitioner researchers who we have worked with over many years.

CONTENTS

Contents

Contents

FIGURES

Chapter 1.6

Chapter 1.8

Part II – Knowledge

Chapter 2.2

Chapter 2.3

Chapter 2.4

Chapter 2.5

Chapter 2.7

Chapter 2.8

Part III – Method

Chapter 3.1

Chapter 3.2

Chapter 3.3

Chapter 3.10

Part IV – The Practice-Based PhD

Chapter 4.2

Chapter 4.3

Chapter 4.4

Chapter 4.5

Part V – Practitioner Voices

Chapter 5.2

Chapter 5.3

Chapter 5.4

Chapter 5.5

Chapter 5.6

Chapter 5.7

Chapter 5.8

Chapter 5.9

Chapter 5.10

Chapter 5.11

Chapter 5.12

Chapter 5.13

Chapter 5.14

Chapter 5.15

Chapter 5.17

TABLES

Chapter 4.4

Chapter 4.6

Section 5 – Practitioner Voices

Chapter 5.14

Chapter 5.16

AUTHOR BIOGRAPHIES

Alice Charlotte Bell is Senior Lecturer in Fine Art at the University of Lincoln. Educated at Goldsmiths College and the Slade School of Fine Art, gaining a BA First Class Hons, Fine Art, 1997. I achieved an MA in Creative Technologies with Distinction, 2014, and a PhD in Participatory Practice-Based Research, 2021, both De Montfort University (DMU), IOCT. I have further postgraduate qualifications in Integrative Arts Psychotherapy and Psychosynthesis Coaching and a professional background in factual arts broadcasting and higher education teaching. A mother to a blended family of five boys, I combine my professional and personal experience with the artistic, educational and psychological.

Stacie Lee Bennett-Worth is an artist-researcher and PhD candidate at De Montfort University. Her PhD focuses on the role of creative digital technologies in post-16 performing arts education. Stacie was research associate on the Leverhulme Trust-funded project 'Physical Actor Training: An Online A–Z' at the University of Kent with Professor Paul Allain, which was published by Methuen Drama Bloomsbury in 2018. In 2019, she co-edited the *Theatre Dance and Performance Training* special issue on Digital Training and became a trustee at the Jasmin Vardimon Education Company. In 2021, Stacie became a selected beneficiary of the Mercury Creatives Programme.

Giovanni Biglino is a biomedical engineer. He studied at Imperial College London and obtained his PhD in cardiovascular mechanics from the Brunel Institute of Bioengineering. He carried out research at Great Ormond Street Hospital for Children and University College London, focusing on congenital heart disease. Now he is Senior Lecturer in Biostatistics at the Bristol Medical School. He has studied biostatistics at Harvard Medical School and started to enthusiastically explore the world of narrative medicine at Columbia University. His research is very collaborative, involving cardiologists, surgeons, imagers, psychologists and artists. He has a natural curiosity for and propensity toward interdisciplinary collaborations and strongly believes in the importance of engaging and involving patients in medical research.

Graeme Brooker is Professor and Head of Interiors at The Royal College of Art, London. He has published numerous books on many aspects of the interior, including the recent publications *Brinkworth: So Good So Far* (Lund Humphreys 2019), *Adaptations* (Bloomsbury 2016) and *Key Interiors Since 1900* (Laurence King 2013). He has co-authored/edited ten books on the interior, including the highly acclaimed *Re-readings*, (RIBA 2005, Volume 2,2018). In 2022, he will publish *50|50*

Words for Reuse (Canalside Press) and *The Story of the Interior* (Phaidon), a re-treading of Gombrich's classic, but re-telling histories of inside spaces.

Oliver Bown is Associate Professor and Co-Director of the Interactive Media Lab at the Faculty of Arts, Architecture and Design at the University of New South Wales, in Sydney, Australia. He is a researcher and maker working with creative technologies, with a highly diverse academic background spanning social anthropology, evolutionary and adaptive systems, music informatics and interaction design, with a parallel career in electronic music and digital art spanning over 15 years. He is interested in how artists, designers and musicians can use advanced computing technologies to produce complex creative works. His current active research areas include media multiplicites, musical metacreation, the theories and methodologies of computational creativity, new interfaces for musical expression and multi-agent models of social creativity.

Øyvind Brandtsegg is a composer and performer working in the fields of computer improvisation and sound installations. He has a deep interest in developing new instruments and audio processing methods for artistic purposes, and he has contributed novel extensions to granular synthesis, feedback systems and live convolution techniques. Brandtsegg has participated on more than 25 music albums in a variety of genres. As of 2010 he, is a professor of music technology at NTNU, Trondheim, Norway.

Linda Candy is a researcher with a deep and enduring interest in how creative practitioners think, act and make artefacts. For over 30 years, she has studied the work of artists, designers, engineers and technologists with a particular focus on how digital systems amplify creativity. Her books documenting this research include *Explorations in Art and Technology* (2002, 2018), *Interactive Experience in the Digital Age: Evaluating New Art Practice* (2014) and *Interacting: Art, Research and the Creative Practitioner* (2011) and *The Creative Reflective Practitioner* (2020). She is currently a visiting professor at the University of Technology Sydney and co-director of ArtworksrActive (ArA), Sheffield, England. http://www.lindacandy.com

Maria Chatzichristodoulou (aka Maria X) is Professor of Theatre, Performance and Technology and Associate Dean of Research Business and Innovation at Kingston School of Art, Kingston University, since 2020. She is a scholar and curator working in the field of digital performance and arts who has published extensively on the subject, contributed to numerous conferences and symposia and curated festivals and events. Maria was Lead Academic for LSBU on the ERDF-funded project ACE IT (Accelerating the Creative Economy through Immersive Technologies, total value: £1.4 million, 2019–2020) and PI on an Innovate UK and AHRC funded KTP project with Footprint Scenery (total value: £277K, 2018–19). She is Editor-in-Chief of the *International Journal of Performance Arts and Digital Media* (*IJPADM*) (Taylor & Francis).

Benjamin Carey is a Sydney-based composer, improviser and educator. He makes electronic music using the modular synthesiser, develops interactive music software and creates audio-visual works. Ben's research and practice is concerned with musical interactivity, generativity and the delicate dance between human and machine agencies in composition and performance. Ben has released five albums, including TRACK – Sumn Conduit (2020, self-released), ANTIMATTER (2019, Hospital Hill Records) and *derivations | human-machine improvisations* (2014, self-released). Ben completed a PhD in interactive musical composition at the University of Technology Sydney in 2016 and is currently Lecturer in Composition and Music Technology at the Sydney Conservatorium of Music, University of Sydney. bencarey.net

Cesar & Lois imagines hybrid biological and digital technologies. The collective includes artists **Lucy HG Solomon** (California) and **Cesar Baio** (Brazil). Lucy HG Solomon is an associate professor in the Department of Art, Media and Design at California State University San Marcos (CSUSM), where she leads the DaTA Lab. Cesar Baio is an associate professor of Art and Technology at Universidade Estadual de Campinas (UNICAMP) and the director of ACTlab. Cesar & Lois received the 2018 Lumen Prize in AI and were selected for Singapore's Global Digital Art Prize biennial in 2019 and the Aesthetica Prize Future Now anthology in 2021.

Amy Yo Sue Chen is a PhD candidate working with Dr William Odom in the School of Interactive Arts and Technology at Simon Fraser University in Vancouver, Canada. Her research focuses on slow interaction design, human–computer interaction and experiential quality, exploring how people's lived experience with their personal digital possessions could be enriched through a temporal lens in the creations of design artifacts. She enjoys soldering electronics and crafting silicone with 3D-printed moulds in her research lab – Everyday Design Studio.

Brigid Mary Costello is an interactive artist and researcher who lectures in interaction design at the University of New South Wales, Australia. Brigid's research focuses on playful interaction design, user experience, evaluation methods for interactive experience and video games. Brigid also creates interactive artworks, games and designs. Her recent research explores rhythmic experience within digital environments and improvisational play within serious games.

Balandino Di Donato is Lecturer in Interactive Audio at Edinburgh Napier University. He holds a PhD in designing embodied human–computer interactions in music performance from Royal Birmingham Conservatoire. He covered the roles of Lecturer in Creative Computing at University of Leicester and Research Associate at Goldsmiths. Previously, he worked at Integra Lab and at Centro Ricerche Musicali di Roma as a software developer ad artistic assistant. Alongside his academic career, Balandino is a sound artist. He realised award-winning sound art installations (Biennale of Contemporary Art and Design 2018, Biennale ArteScienza 2019) and performed in international conferences (ICMC, AudioMostly, NPAPW, etc).

Ernest Edmonds has supervised and examined practice-based PhDs since the 1970s. In 2017 he received the SIGGRAPH Distinguished Artist Award for Lifetime Achievement in Digital Art and the SIGCHI Lifetime Achievement Award for the Practice of Computer Human Interaction. Recent retrospective exhibitions include ones at Microsoft Research, Beijing and De Montfort University, Leicester. His most recent book, with Margaret Boden, is *From Fingers to Digits: An Artificial Aesthetic* (MIT). His work is described by Francesca Franco in *Generative Systems Art: The Work of Ernest Edmonds* (Routledge). He is currently Emeritus Professor at De Montfort University.

Kerry Francksen is an experienced educator, independent researcher and digital dance artist with over two decades of experience in the university sector and within the creative industries. Her artistic work has been screened and performed nationally and internationally, and she is also published in books and in esteemed journals, such as *Leonardo Transactions*. Kerry is Senior Fellow of the HEA and holds an MA in Dance Video Installation Art and a PhD in Performance and Digital Art. She also serves on the editorial board of the IJPADM and is a co-founder of the cross-sector digital arts organisation DAPPER.

Iain Findlay-Walsh is an experimental music producer, sound artist and researcher whose work combines field recording and studio production with autoethnographic methods to research personal listening. He makes sound installations, multichannel audio works, text scores and spatial audio pieces for headphones. Scholarly research engages with auditory reception and perception, spatiality and virtuality in record production, and sonic and textual autoethnography among other things. Iain is Lecturer in Music at the University of Glasgow and co-director of the Immersive Experiences ArtsLAB on VR/XR research in the Arts and Humanities.

Petra Gemeinboeck is an artist and researcher exploring questions of embodiment, performativity and agency in human–machine relations. Petra is Associate Professor at Swinburne University, AU, and ARC Future Fellow (2021–2025) and also leads an artistic FWF-funded research project at the University of Applied Arts Vienna. Her artworks have been shown internationally, including at the Museum of Contemporary Art (MCA, Chicago, USA); National Art Museum China (Beijing, CN); NTT InterCommunication Center (Tokyo, JP); and Ars Electronica Center (Linz, AT).

Falk Hübner is Professor of Artistic Connective Practices at Fontys University of the Arts in Tilburg, the Netherlands. With a background as composer, theatre maker, researcher and educator, he is active in a huge diversity of collaborations within and outside of the arts. His research focuses on the social–societal potential of artistic research, research methodologies and the relation of the arts and art education in relation to society. In 2019–2021, Falk conducted post-doctoral research at HKU University of the Arts on research methodology and ethics.

Janis Jefferies is Emeritus Professor of Visual Arts, Goldsmiths, University of London. She established the Goldsmiths Digital Studios and associated arts and technology research programmes in computing in 2005. Her areas of expertise lie at the intersection of arts and technology across textiles, performance, sound, design, publishing and curating. Since 2005, she has collaborated with Professor Barbara Layne, subTela Lab, Concordia University, Montreal on four Canada Council-funded research projects. Jefferies was the first International Creative Thinking Fellow, University of Auckland (2014), and has been an advisory board member of CHAT, Hong Kong (2012–2019) and Hangzhou Triennial, China (2012–2020). http://research.gold.ac.uk/view/goldsmiths/Jefferies=3AJanis_K=2E=3A=3A.html

Pearl John graduated from the Royal College of Art in 1992 with an MA in Holography and obtained a PhD in practice-based research from DMU in 2019. She is a visiting arts researcher in the Holographic Research Group at De Montfort University (DMU) and the Public Engagement Leader in Physics and Astronomy at the University of Southampton. John's art work aims to humanise technology using holography and lenticular imaging. John's science communication engages different communities creatively with photonics – the science of light – making holograms and running a travelling laser light show.

Andrew Johnston is a research professor and co-director of the Creativity and Cognition Studios at the University of Technology Sydney (UTS). He is also Research and Course Director of the UTS Animal Logic Academy, a unique, professionally equipped studio focusing on the creative application and design of digital technologies. His work focuses on the design of systems that support experimental, exploratory approaches to interaction and visualisation and the experiences and practices of the people who use them.

Gail Kenning is an artist, designer and interdisciplinary researcher working at the intersection of art, health and technology. She engages in participatory practices, co-design and co-creation exploring inclusivity and diversity in art and design. The impact of creative engagement and creative practice is a central concern to her research and art and design practice. Gail is an interdisciplinary research fellow at University of New South Wales, Sydney, Australia.

Sofie Layton's artistic practice is an investigation of illness and how it is experienced and understood and brings a new visual narrative and knowledge to the clinical setting. She received an arts award from the Wellcome Trust for *Under the Microscope* 2015, which explored rare disease at Great Ormond Street Hospital London, and *The Hearts of the Matter* 2016–2018, a national touring project which explored the medical and poetic heart – funded by Wellcome, ACE England and the Blavatnik Family Foundation. She is in receipt of LAHP funding and is currently pursuing a PhD at the Royal College of Art.

Anna Ledgard, a producer, project manager, researcher, trainer and professional development leader, has 30 years of experience shaping collaborative arts practice with hospitals, communities, schools and cultural organisations, most recently in NHS intensive care or end of life settings. Her role is to pay attention, to listen and to weave the web of trusting relationships which makes deep artistic collaboration across disciplines possible. Anna is an end of life doula trainer with Living Well Dying Well training (www.lwdwtraining.uk), building understanding and confidence in talking about and planning for death and dying. www.annaledgard.com

Bruce Mackh is an artist, faculty member, administrator and author of *Pivoting Your Instruction* (Routledge 2021), *Higher Education by Design* (Routledge 2018) and *Surveying the Landscape: Arts Integration at Research Universities* (University of Michigan, 2015). He is Associate Vice-President of Teaching and Learning at Chadron State College (Chadron, Nebraska).

Jon McCormack works at the nexus of technology and creativity, his research spanning design, media art and computer science. He is currently an Australian Research Council Future Fellow, research professor and the founder and director of SensiLab, a creative technologies research space based at Monash University in Melbourne, Australia. SensiLab brings different disciplines together to explore the untapped potential of technology.

Debbie Michaels has a background as an art psychotherapy practitioner and educator, with research interests in the application of psychoanalytic ideas outside the consulting room and in art as a reflexive space for learning. She has contributed to psychoanalytic and art psychotherapy literature and, at the time of writing, is completing a practice-based PhD in the fine arts at Sheffield Hallam University. www.debbiemichaels.co.uk

Jonathan Michaels is Honorary Professor of Clinical Decision Science at the School of Health and Related Research, University of Sheffield. He was previously Professor of Vascular Surgery and Honorary Consultant Vascular Surgeon in Sheffield. He has undertaken wide-ranging research, including randomised controlled trials, quantitative and qualitative research in healthcare and practice-based research in fine art, and has particular interests in decision science, healthcare policy and research methodology. https://www.sheffield.ac.uk/scharr/people/staff/jonathan-michaels; Twitter handle: @JonM_ScHARR

Judith Mottram is Professor of Visual Arts at Lancaster University. Her research interests include colour, drawing and pattern and the inter-relationships between subject knowledge, creativity, research and practice. She is on the 2021 UK Research Excellence Framework (REF) sub-panel for Art and Design: History, Practice and Theory. She is Fellow of the Design Research Society and on the editorial boards of the *Journal of Visual Arts Practice* (JVAP), the *Journal of Textile Design Research and Practice* and the *Journal of the International Colour Association* (JAIC).

Lizzie Muller is a curator and researcher specialising in audience experience and interdisciplinary collaboration. She researches the future of museums as sites of knowledge production and the relationship between curatorial practice and changing disciplinary structures. Her international exhibitions celebrate the intersection of art, science and technology. Lizzie is Chair of the Sydney Culture Network Education, Research and Innovation committee and co-convener with Keir Winesmith of the bi-monthly Sydney Culture Data Salon.

Alexandra Murray-Leslie is a performer and educator. She is Professor of Digital Performance at Trondheim Academy of Fine Art, Norwegian University of Science and Technology, and is the co-founder and member of art band Chicks on Speed.

Corey Mwamba, born and based in Derby, is a critically acclaimed vibraphonist, as well as a researcher, live music promoter and award-winning radio presenter. Corey Mwamba ceased all live public performance on 23 March 2019, as a protest against the experiences of racism and objectification in the British jazz+ scenes.

Kristina Niedderer (PhD, MA [RCA]) is Professor of Design at Manchester Metropolitan University. She was originally apprenticed and worked as a goldsmith and silversmith in Germany. She then trained as a designer and design researcher in the United Kingdom. Niedderer's research focuses on the role of design to engender mindful interaction and behaviour change for health and sustainability. She led the European project MinD 'Designing for People with Dementia' (2016–2020, MSCA GA No. 691001) as well as the IDoService project (2020–2022, MSCA GA No. 895620). She also has an interest in craft and design research and knowledge frameworks and methodologies. www.niedderer.org

William Odom is Assistant Professor in the School of Interactive Arts and Technology at Simon Fraser University in Vancouver, Canada, where he is co-director of the Everyday Design Studio. He leads a range of projects in slow interaction design, the growing digitization of people's possessions and methods for developing the practice of research-through-design. He holds a PhD in Human-Computer Interaction from Carnegie Mellon University and was previously a Fulbright Scholar in Australia, a Banting Fellow in Canada and a Design United Research Fellow in the Netherlands.

Garth Paine is a composer, scholar and acoustic ecologist. As Professor of Digital Sound and Interactive Media at Arizona State University, he crosses art–science boundaries with his community work on environmental listening and creative place-making. His research drives towards new approaches to acoustic ecology and the exploration of sound as our lived context, including the application of VR in health. Paine is a visiting researcher/artist in residence at IRCAM (Centre Pompidou), where in 2018–19 he produced the 50-minute media opera *Future Perfect*, examining our future impacted by climate change. He co-directs the Acoustic Ecology Lab (AELab@ASU).

Mike Phillips is Professor of Interdisciplinary Arts at University of Plymouth and Director of Research at i-DAT.org, an Open Research Lab for playful experimentation with creative technology. His R&D orbits a portfolio of practice-based projects that explore the ubiquity of data 'harvested' from an instrumentalised world and its potential as a material for revealing things that lie outside our normal frames of reference – things so far away, so close, so massive, so small and so ad infinitum.

Fabrizio Augusto Poltronieri is an academic and artist who explores the relationship between technology and deep-rooted philosophical concepts, such as chance. His current artwork involves Artificial Intelligence, applying machine and deep learning techniques to create and design narratives, moving images and objects. He is a self-taught programmer who started to code during his childhood. His first degree was in maths. He holds a PhD in Semiotics (Pontifical Catholic University of São Paulo – PUC/SP). Poltronieri is Associate Professor at the IOCT (Institute of Creative Technologies), De Montfort University, Leicester, UK, supervising PhD students and teaching creative code in the Creative Technologies MA.

Gavin Sade is Associate Dean, Teaching and Learning in the Creative Industries Faculty at Queensland University of Technology QUT. He is responsible for leading and managing the Faculty's teaching and learning strategy, curriculum innovation and learning design (including online), accreditation and program quality. His portfolio also includes the Bachelor of Creative Industries (BCI) program, Work Integrated Learning (WIL) and Student Support. In 2019, Gavin led QUT's involvement with World Science Festival Brisbane 2019.

Rob Saunders is Associate Professor in the Leiden Institute of Advanced Computer Science (LIACS). His research explores the role of intrinsic motivation, emergent languages and physical embodiment in the computational modelling of creative processes, individuals and societies. His collaborative robotic art practice provides a platform for knowledge mobilisation by materially engaging audiences in questions of machine creativity. He is a founding member of the Association of Computational Creativity.

Stephen Scrivener studied Fine Art at De Montfort University and the Slade School of Fine Art, London, where he began to use the computer as a means of art production. He subsequently completed a PhD in computer science and researched the design and development of interactive systems for artists and designers and how such systems are used. He moved back into art and design in 1992, and his research has since focused on the theory and practice of what is often called practice-based research, which he combines with art practice. He is currently Emeritus Professor at Chelsea School of Art.

Jennifer Seevinck is an artist, researcher and educator. In her work she draws on scientific and poetic interpretations of the world, qualitative research into audience experience, human–computer interaction design methods and digital technology to create aesthetic experiences – provoking new and surprising understandings to expand the interaction design discipline and propose new ways of designing for human experience with technology. An Associate Professor at the Queensland University of Technology, Australia, Jen has participated in artist residencies, created award-winning work for international exhibitions and published on practice-based research processes and insights, including the research monograph *Emergence in Interactive Art* (Springer 2017). www.smArtnoise.net

Becky Shaw makes live, collaborative artworks that explore and animate institutions and infrastructures. Works have been commissioned by organisations including Grizedale Arts Cumbria, City of Calgary Water Services, New Art Gallery Walsall, Culture Consortium London and Amstelveen Art Incentive Prize NL. Becky was co-investigator on the AHRC-funded project 'Odd', exploring children's experiences of school. Becky's 1998 practice-based doctorate explored collaboration with palliative care patients. Between 2000 and 2006, Becky co-directed Static, Liverpool, an architecture and art organization. Becky is Reader in Fine Art at Sheffield Hallam University and leads the Art, Design and Media Research Centre doctoral programme.

Hanna Slättne is a Swedish freelance dramaturg, researcher, theatre-maker and facilitator based in UK/Ireland. She works across live performance, XR and immersive audio as a maker and dramaturg and facilitates interdisciplinary and international artists development programs. In 2016 she received the Kenneth Tynan Award for Excellence in Dramaturgy and in 2017 an Elliot Hayes Award special commendation for her work on the immersive audio play *Reassembled Slightly Askew* by Shannon Yee. She is a co-founder of the Dramaturgs' Network. www.hannaslattne.com

Sophy Smith is Professor of Creative Technologies Practice and Director of the Institute of Creative Technologies at De Montfort University, Leicester (UK). As a practice-based researcher, Sophy's expertise extends across inter/multi/transdisciplinary and cross-sector collaboration within performance. She also co-leads the University's Doctoral Training in Practice-Based Research and manages the PBR PhD. www.pbrcookbook.com

Deborah Turnbull Tillman is Lecturer in Media Arts and Curatorial at the University of NSW, Australia. She exhibits and publishes extensively from her independent curatorial research platform, New Media Curation, where she is Creative Director. Deborah served as the Design and Technology Curator (2012–14) at the Museum of Applied Arts and Sciences (Powerhouse Museum), where she initially curated Beta_Space (2007–11). For more information on Deborah or her creative research practice, please visit: https://research.unsw.edu.au/people/dr-deborah-jane-turnbull-tillman | www.newmediacuration.com/new.

Sian Vaughan is Reader in Research Practice at Birmingham School of Art, Birmingham City University, in the United Kingdom. Broadly, her research interests concern the pedagogies that underpin research in art and design and the modalities of interpretation and mediation of public engagement with contemporary art. Her art research focuses on artistic practices with archives, history and institutions with a particular focus on creative research methods as knowledge generation. Her educational research is focused on the practices and pedagogies of doctoral education and particularly in how these are responding to creative practice in research.

Craig Vear is Professor of Digital Performance (Music) at De Montfort University, where he is a director of the Creative AI and Robotics lab in the Institute of Creative Technologies. His research is naturally hybrid as he draws together the fields of music, digital performance, creative technologies, Artificial Intelligence, creativity, gaming, mixed reality and robotics. He has been engaged in practice-based research with emerging technologies for nearly three decades. His recent monograph, *The Digital Score: Creativity, Musicianship and Innovation*, was published by Routledge in 2019, and he is Series Editor of Springer's Cultural Computing Series. In 2021 he was awarded a €2 million ERC Consolidator Grant to continue to develop his Digital Score research.

Marloeke van der Vlugt is a Dutch cross-disciplinary artist-researcher with a background in performance and scenography. Her work has been exhibited and performed in multiple gallery shows and theatres. In 2015 her book, *Performance as Interface | Interface as Performance*, was released. She is a lecturer at the HKU University of the Arts Utrecht and a researcher at the Professorship Performative Processes. In 2018 she started her PhD project, which enquires into artistic strategies that activate the sense of touch.

PREFACE

For as long as I can remember, I have enjoyed the spirit of adventure and play that my creative practice has offered to me. Spanning over four decades, my creative music practice has remained a central part of my life, initially starting as a hobby and a curiosity, quickly turning into professional activity. Following scholarships to music schools, degree studies and a successful career as a musician, my experimental practices started to gain some interest and funding that enabled further experimental and adventurous play in the pursuit of new insights and experience. Through this period, I explored my practice through making, coding and performing and from different perspectives. I sought to understand and develop the specific skills and technical procedures of my experimental practice, the ways and methods of carrying out or applying these skills through making and doing, new ideas and models of practice and their effect upon the audience and participants and the shifts in my understanding and conceptualisation as a practitioner. In short, you could say that I explored new ways of knowing that were:[1]

- embedded *in* practice
- embodied *with* practice
- enacted *through* practice
- extended *by* practice

Throughout this period and even to this day, my practice feels like a portal through which I visit another dimension. On a poetic level this feels like something akin to the Stargate at the centre of the 1990s movie of the same name. My world-of-concern would shift, and I would too in order to attune myself to my new environment. I would enter a flow and be completely absorbed by this activity, or at least most of me would. The more attuned I became to this world-of-concern, the more natural it felt, and over time I became less focused on the physical activity and more interested in exploring the unique facets of each portal I entered into. You could say that I became naturalised, and the more natural it became the more processing power I had at my disposal to dig deeper into the nuance and extraordinary elements of this dimension. Furthermore, each evocation of this world-of-concern brought with it unique opportunities to explore and to expand my sense of play, and further my personal knowing of my practice. But I must confess, that in retrospect this knowing was only personal to me.

I considered myself to be a researcher in this dimension. I felt that I was exploring new ways of knowing and understanding the practice and from different perspectives. I also felt that the term *practice-based research* perfectly described my activity as my explorations were *based inside* my practice, not outside. I also shared my new insights and knowing with others through performance, CDs, talks and lectures. But what I didn't fully grasp then, as I do now, is that my pursuits for new insights were really only a matter of personal development, and so really didn't constitute research in the formal sense.

A critical point occurred when I was creative artist-in-residence with the British Antarctic Survey (2003–4), spending three months embedded with the scientific team in Antarctica. It was here that I had my first-hand experience of formal, academic research and I was surprised at how close it was to my own experimental adventures in my practice. They entered their world-of-concern for a brief period of time and maximised each opportunity, came prepared for surprises and had an open approach and pioneering can-do attitude. Because of this, there was creativity in their actions and research, there was reflection and curiosity, and a spirit of adventure than I recognised. But with one big difference: a pursuit for new knowledge that a) dealt with a gap and b) benefited others. It was following this residency that I decided to formalise my experimental practice into a PhD, and to switch the focus of my experiments towards knowledge production for the betterment of the field and others in and around it.

I must confess that this switch was an easy one to make: instead of being a curious creative practitioner and developing work through informed artistic intuition, I would flip those instincts around to form a question. Once I had the start of a question, I could then verify it by looking further into the field to see what else existed, and how my contribution might enhance or contribute towards better knowledge. At the heart of this process were the '4 Es'[2] listed earlier: what can I as a professional creative practitioner bring to the field, and from which perspective can I advance the field?

Having now been inside academia for several decades, I can honestly say that the spirit of adventure in my practice still remains. I still look forward to entering into the dimension of my practice and exploring it further. The systems and rigour that I put in place are different to those from when I started my explorations, but not wildly different. However, the focus is strongly towards the pursuit of new knowledge. This outlook I now use to guide the way I structure the practice-based research of others, whether it is through mentoring early career researchers, PhD supervision, or as director of PhD and doctoral training programmes.

Almost five years ago, I was approached by Linda Candy and Ernest Edmonds about whether I would be interested in being editor for this handbook. Routledge had reached out to Linda with the proposition, as both she and Ernest are pioneers of practice-based research, having been central to its inception in the university system since the 1960s. They have contributed an immense amount of significant writing and thinking about the rigorous nature of it as a mode of research and, through discussions it was clear that we three shared a core philosophy about the nature of practice-based research. I agreed to take it on, with one caveat: that Linda and Ernest operate as consultant editors, and that I may stand on their shoulders throughout the process. They agreed – to my relief – and have been instrumental throughout each phase of the handbook development.

From the outset I was determined to make this handbook a community affair, rather than merely my take on the form, and for it to read like a cohesive volume, rather than a disparate collection of chapters. It is worth outlining the process of developing the handbook and how these values guided the process.

Phase 1 – initial workshop

In June 2018, Linda, Ernest and I organised an initial workshop at De Montfort University with the aim of laying down the groundwork for the theoretical construct of the handbook and to start the process of community building. We invited 20 academics from around the world who we felt would contribute to this discussion, and who would eventually write chapters for the handbook. I also invited a handful of PhD candidates from the university as respondents; their role was to take notes on the topics that mattered to them and to respond to each panel from their perspective. The academics were invited to present a paper or provocation in advance of the workshop, with the intention that these would be read in advance and provide the impetus for the roundtable discussions. The workshops were split into three sections: a) Theory: Definitions, Scope; b) Methodologies: Approaches, Methods; c) The Academy: Frameworks, Regulations, Guidance. The workshop was warmly supported by those involved, and we moved forward with developing the synopsis of the handbook. Surprisingly, a third value was added: to make the handbook 'pan-disciplinary', by which I mean that we sought to present the handbook as a framework across the many disciplines that utilise practice as a mode of research. A proposal and synopsis were outlined and sent to Routledge, from which the handbook was commissioned and contracts signed.

Phase 2 – author invites and extended abstracts

Following the workshop, Linda, Ernest and I compiled a list of established and emerging practice-based researchers mindful of achieving a relative balance of geographic location, discipline spread and experience. Each person was asked to present a 1000-word abstract on a subject of their choosing or aligned to the Routledge synopsis. The purpose of this extended abstract was to get the authors to think more clearly about their topic and to start the process of formulating a cohesive argument and evaluation. Once collated, we assessed the spread of topics and identified key areas that were missing. We then reached out to additional authors for extended abstracts on the missing themes.

Phase 3 – preliminary final drafts and feedback

Following the request for abstracts to be completed by 1 July 2019, I asked for '1st phase drafts' to be submitted to me by 3 February 2020. These were then categorised into sections (now the four that remain in this handbook), and each chapter was then scrutinised carefully by all three editors and general feedback was issued in advance of July 2020 deadline for delivery of their final versions. The purpose of this process was to review the sections for the following:

- evaluate what is missing from the section topic as a whole and draft a list of areas that need to be addressed in the leading chapters
- find any overlap between the section chapters, and seek out points for consolidation
- start the process of understanding what and how we collectively deal with the notion of a 'pan-disciplinary' approach.

Phase 4 – online workshop/colloquia

All '1st phase final drafts' were returned to me in advance of the September 2020 workshop. These were then re-assessed and categorised into sections (some had changed focus). They were then distributed among the section authors. The pre-workshop reading, and also the workshop itself, operated as part of a community peer review.

Two significant things happened during this period that changed the world and also changed the focus of the third value and strengthened the others: COVID and *Black Lives Matter*. The latter highlighted the importance of focusing on inclusivity rather than formulating some meta-formulae for 'pan-disciplinarity' (a concept that I found difficult to explain when asked). As such, the aim of this peer-review process and the workshop was to seek to find (as much) unity between chapters and across the book and promote an inclusive approach that opens our discourse and experiences out for a broad range of readers and disciplines. The COVID pandemic and extended periods of lockdown highlighted to me the importance of family and local community, and although the authors of this handbook were spread across the globe, we were a community (or at least a small section of a global practice-based research community). As such, the objectives of the workshop were to bring positive suggestions, guidance and connections through peer review of each other's writings, to find connections between author's experiences and writings and to share underlying values and principles for the benefit of fellow authors, the handbook and a broad range of readers.

Phase 5 – final drafts and practitioner voices

Following a thoroughly enjoyable and unifying workshop, the authors were tasked with bringing their chapter over the finishing line ready for editing and production for print. To help them with this process, I collated and distributed the individual peer reviews of each chapter, and we presented them with the following (temporary) definitions as a 'lodestone' with which to focus their thoughts or to make a challenge:

> Simply put, practice-based research is a research investigation based in, with or through practice leading to new knowledge. From this perspective the base-line definitions are:
>
> **Practice** is *doing* something that extends beyond everyday thinking into actions that may lead to new outcomes; the making, modifying or designing of objects, events or processes. It is the use of ideas, beliefs or methods, as opposed to investigating theories relating to them. Note the contrast with 'theory' and the emphasis on the use or application of ideas and methods towards some end.
>
> **Research** is, in one well-honed definition, an original, significant, rigorous study generating new insights that are effectively shared in retainable material. Research must be disseminated, original and contextualised. It is important to distinguish between research as a *public* activity that results in generally available outcomes and *personal* research, which is private and need not generate insights that are new in the world.
>
> **Practice-based research** is a principled and original investigation undertaken in order to gain new knowledge by means of practice and the outcomes of that practice. A basic principle of practice-based research is that not only is practice embedded in the research process, but research questions arise from the process of practice, crucially placing the practitioner researcher at the centre of the research. Practice-based research *involves practice* and *is research*, as defined earlier.

At the same time as the peer reviews were sent out, Linda, Ernest and I created a wish list of practitioner researchers who we would invite to author 3000-word 'Practitioner's Voice' chapters. These would centre on a practice-based project of theirs through either PhD, post-doc fellowships or professional practice. The focus was to be a discursive reflection on the processes and methods of the project, and to highlight the lessons learnt for the betterment of others. Equally, the three values of community, cohesiveness and inclusivity were shared among the authors.

All 50+ chapters were returned in the first three months of 2021. A massive editorial process then ensued with me, in close consultation with Linda and Ernest, editing each chapter several times. At the same time, we co-wrote the leading chapters for each section. This added further to the sense of cohesiveness of the handbook, with the leading chapters presenting the 'nuts and bolts' of the section themes and framing the contributing chapters of each section within the general discourse.

Even though it was a huge process, it was enjoyable, illuminating and educational. It was invigorating entering into others' worlds-of-concerns and research knowledge. And a privilege to do so. I feel that this handbook represents the values that were present through the editorial journey, and more so, represent a comprehensive insight into the rigorous world of practice-based research. I believe that we have achieved a community of knowledge that puts forth a strong and convincing argument of the best practices in practice-based research. I believe that the handbook is cohesive, that it speaks across the chapters and sections, and even if authors don't necessarily agree, their underlying values and principles form a cohesive whole. Finally, and most definitely not least of all, the chapters as a whole represent an inclusive and diverse set of perspectives that speak outwardly to the readers with knowledge and experience of exceptional value and true worth.

Craig Vear

Notes

1 The 4 Es are adapted from Menary, R. A. (2010). Introduction to the Special Issue on 4E Cognition. *Phenomenology and the Cognitive Sciences*, 9, 459–463.
2 Ibid.

ACKNOWLEDGEMENTS

My sincerest thanks go to Linda Candy and Ernest Edmonds whose trust, support, encouragement, wisdom, knowledge, experience, kindness and friendship have been of great sustenance to me through the process of editing this handbook. I simply couldn't have done this without their presence and support and have thoroughly enjoyed looking out over their vast wisdom while standing on their shoulders.

Thanks also to Hannah Shakespeare, the Senior Commissioning Editor of Research Methods at Routledge, for commissioning this handbook and for her continued support through the process. It has been a wonderful journey together. Also, thanks to Matt Bickerton for his support through the editing and production process.

This handbook needed to be multi-voiced yet cohesive; a field such as practice-based research cannot be written from a single perspective, nor as a disparate collective. I extend my sincerest thanks to all the authors of this handbook that have journeyed through the process with me and supported the shared values of community, cohesiveness and inclusivity. Your contributions are truly valuable, and an inspiration.

Special thanks, too, to all the students and PhD researchers that I have had the good fortune to work alongside, whose enquiring minds have always challenged me to be at my best. And to my colleagues, especially in the Institute of Creative Technologies who make research and academia playful, fun, supportive, and enriching . . . this is what a research institute should be.

And finally, but by no means least, my family who are so kind and encouraging, and whose love nurtures my playful pursuits through experimental research.

INTRODUCTION TO THE HANDBOOK

Linda Candy, Ernest Edmonds and Craig Vear

Introduction

This handbook has been written for anyone interested in practice-based research. You may be a post-graduate research student or a supervisor or adviser of practice-based research. You may be conducting your own practice-based research as an academic or as a professional, for the first time, or for the umpteenth time. You may be working as a research educator or manager in a university, industrial setting or as a freelance professional. This handbook is designed to be especially helpful to those practitioners who seek to strengthen their studies of their practice, or who wish to contribute new knowledge to their field through their practice.

This handbook presents a cohesive framework with which to conduct practice-based research or to support, manage and supervise practice-based researchers. The goal is, in the words of Tim Ingold, 'to provide us with facts *about* the world [of practice-based research] as to enable us to be taught *by* it'.[1] It has been written with an inclusive approach, with the intention of presenting deep and meaningful knowledge for the benefit of all readers. To this end, this handbook presents a robust way to navigate practice-based research developed by those who have expertise. This unique perspective is hard to get at from the outside alone or by theoretical observations of what seems to be going on inside practice. For that reason, we have included chapters written by highly experienced practitioner researchers from across the world who provide an insider's perspective on doing practice-based research. They come from disciplines spanning art, music, dance, design, software development, clinical medicine, curation, digital art, creative AI, interior design and robotics, to name but a few.

Practitioners from a broad range of backgrounds will find these chapters helpful:

- Research students or final year graduates will be introduced to the principled nature of practice-based research
- PhD researchers embarking on a research project or in the flow of research will find this guidance supportive
- Professionals such as designers, makers, engineers, artists and creative technologists wishing to strengthen their research into their practice will be guided through the principled and focused nature of practice-based research
- Supervisors, managers and policy makers will benefit from the potential and rigour of practice-based researchers in the pursuit of new knowledge

DOI: 10.4324/9780429324154-1

From the outset, it is of paramount importance to understand that practice and research are interdependent and complementary processes in practice-based research. The attraction of this form of research for practitioners is that by connecting closely to existing practice, it provides a means of exploration that extends that work in a personal sense as well as contributing to broader considerations such as the production of new forms of knowledge. Furthermore, it gives researchers access to new knowledge arising from practice conducted in the real world.

As a relatively new development in the landscape of research disciplines, practice-based research has arisen where existing approaches have not served the needs of practice sufficiently well. Its origins as an academic research form can be traced by to the 1960s and 1970s and has since grown as a significant and valid approach to research and knowledge generation. Through this period, practitioners in a wide variety of fields have sought better ways to explore the kinds of questions that arise from the problems they encounter during the thinking, acting, appraising and reflecting on activities that make up the essential elements of their practices.

Throughout this handbook we define practice-based research as: *a principled approach to research by means of practice in which the research and the practice operate as interdependent and complementary processes leading to new and original forms of knowledge.*

This definition highlights the four main principles that:

1   Practice and research are complementary but distinctive
2   The research is based within a world-of-concern defined by practice
3   The practitioner researcher is at the centre of the research
4   The research aim is to generate new knowledge

While practice-based research shares many traits with all forms of research, some core features distinguish it. These are: the centrality of practice to the research, the role of artefacts in research and the forms of knowledge that arise from it. As a methodology, practice-based research is characterised by practitioner researcher-designed strategies that determine the methods, tools and techniques to be used that draw upon or observe practice, such as the documentation of reflection.

This handbook has been designed to bring specific detail to the shared traits and to highlight the core distinguishing features into a cohesive, principled and methodical approach. To this end, the handbook is presented in five parts: Part I, 'Practice-Based Research'; Part II, 'Knowledge; Part III, 'Method'; Part IV, 'The Practice-Based PhD'; and Part V, 'Practitioner Voices'. Each section begins with a leading chapter that outlines each of the distinct areas as they relate to practice-based research. This is followed by a series of contributing chapters that discuss pertinent themes in more detail.

Part I: Practice-Based Research begins by introducing the principled nature of practice-based research. The chapter outlines the defining features of practice-based research, including the centrality of practice, the role of artefacts in research and the forms of knowledge that arise from it. It then traces some of the origins of practice-based research and how it has emerged in some practice-oriented disciplines such as healthcare, education and software engineering and the creative arts and design. This discussion is then extended into practice-based interdisciplinary research initiatives. The next part highlights the importance of community and how practice-based researchers can enable and advance community building and innovation, research training and cultural change.

The contributing chapters that follow on immediately discuss in detail topics such as:

•   The origins of practice-based research in institutional and regulatory changes
•   Fostering practice-based research through places and spaces for interdisciplinary relationships

- How interdisciplinary activities happen within an individual practitioner's research practice
- How crossing and blurring boundaries can be a way to reach fresh perspectives and innovative outcomes
- Studio and public spaces for making and evaluating the production of new knowledge
- How practitioner-focused organisations can strengthen practice-based communities
- The benefits of crossing disciplines to generate new knowledge through practice
- Developing the language of research from the practitioner's lexicon fosters social identity
- Community knowledge building is a contribution to practice-based discourse
- Fostering community, place and purpose through shared interests and practices
- The criteria for promoting cultural shifts in doctoral practice-based research

Part II: Knowledge introduces the nature of knowledge and defines it in research terms specifically to practice-based research. The starting point is a discussion about the meaning of the term and distinguishing between *knowledge* that is visible in the world and tacit *knowing* that is invisible but fundamental to how practitioners do their practice. Through this, the introduction outlines the difference between knowledge that is explicit and available for communication across time, and knowing that is something tacit and implicit to the practitioner alone. This distinction is crucial to understand when approaching practice-based research with a principled and focused perspective. From this, five principles of new knowledge are characterised as *original*, *validated*, *contextualised*, *shareable* and *retainable*. The chapter then considers the key questions of knowledge generation in practice-based research, such as the role of practice in generating new knowledge and the relationship of the artefact to knowledge. This part highlights how artefacts are typically central to practice and can be equally central to research in the practice-based context. Following on, it discusses how a special aspect of practice-based research is that the delivered outcomes often have to include artefacts, or documentation of them, together with the written description.

The contributing chapters that follow discuss many different categories of knowledge, each of which relates to a different aspect of practice or formalising the nature of knowing, providing a broad picture of what knowledge is and how it is advanced in practice-based research. Among the many diverse aspects of this topic, they explore issues such as:

- How practitioners develop theories and new knowledge through practice-based research
- Frameworks as support for thinking and making
- How methodological knowledge arises from practice-based research
- The value of tacit knowing to expertise and connoisseurship
- New directions in creative practice leading to unexpected and stimulating outcomes
- Empirical frameworks for generating new forms of embodied knowledge
- Philosophy, science and theory in new knowledge
- The artefact and the generation of new knowledge
- Advances in art practice are different to knowledge from research

Part III: Method describes the methodologies, methods and systematic approaches to conducting practice-based research. The opening chapter defines methodology and methods in the context of practice-based research and explains why this distinction matters. Very often the terms are used interchangeably, but for the practitioner researcher it is important to differentiate between them. Simply put, methodology provides a principled framework that supports the design of the research inquiry; methods are the means of data collection, the evaluation of results and the reflections on the processes and outcomes leading to new insights, artefacts and knowledge. The second part of this introductory chapter describes key methods adopted for practice-based research, specifically reflective practices and evidence-based studies.

The contributing chapters that follow focus in different ways on key features of practice-based research methodology and its many kinds of methods. This includes models and frameworks for designing practice-based research projects, how questions emerge through thinking, making, appraising and reflecting and the role and value of practice-based research to practice, among other topics such as:

- Proposed models for designing a practice-based research project
- Identifying questions through making, appraising and reflecting
- Frameworks articulating and sharing the practice-based process
- The role and value of practice-based research to practice
- Evidence-based and reflective practice methods with examples
- Methodologies as novel contributions to research

Part IV: The Practice-Based PhD considers practice-based research from the practical perspective. The aim of this section is to provide guidance to all practice-based researchers, whether as PhD candidates and advisers, or those planning written proposals for funding, or academic teachers designing new curricula for practice-based research in any discipline. The first chapter introduces the PhD and the specific nature of a practice-based research PhD. It then discusses a range of topics that parallel the journey that a practitioner researcher will take from inception through to final dissemination or examination of the thesis. This includes how to develop a systematic approach to designing the research methodology. Following this, it then highlights and introduces the challenges of a practice-based PhD and the role of the supervisor followed by guidance on the purpose of a literature review, the structure and purpose of a thesis, and the examination. Overall, this section describes the core elements and activities of the practice-based PhD, drawing on the material presented in the earlier sections of the handbook. In effect, it summarises the key issues as they apply to PhD study. The emphasis is on the distinguishing features of the practice-based PhD as well as on the particular challenges. The contributing chapters following this leading chapter elaborate on a number of specific issues:

- The role of play in practice-based research
- The role of improvisation in practice-based research
- Autoethnography as a reflexive method
- A dramaturgical approach to practice-based research
- The relationship of the researcher to the organisation
- Community-building for practice-based researchers
- Research ethics for practice-based studies
- Structuring and delivering practice-based research results

Part V: Practitioner Voices, the final part of this handbook, presents a series of practitioner perspectives from a range of fields and disciplines. The authors have written a short piece that centres on a practice-based research project they have undertaken, whether part of a PhD or another kind of research project. The focus is on their understanding of the work as it relates to both practice and research and presents lessons learnt in a way that speaks outwardly to the diverse readership of the handbook. To do this, they concentrate on personal reflections of the experiences and what they learnt through doing their projects. As such, they are not specific to a single case study in a single field but speak outwardly to all practitioners. Some decided to focus on the challenges they faced, or the pitfalls and problems, conflicts and solutions they needed to negotiate. Others discuss outcomes such as frameworks and models that contribute to the kind of knowledge or methods that other practitioner researchers might be able to use to advance their practice-based research. Others present discursive reflections on the processes and methods of their projects, including an account of how any creative works were made and appraised or evaluated. Overall, these chapters present a rich insight into the processes and research practice of a diverse range of practice-based researchers in order to bring back into the practice the knowledge that has been discussed in the previous sections of handbook.

All the contributing chapters to this handbook were rigorously reviewed before being accepted for inclusion. Throughout the process, we have maintained certain values that we hope you will feel when you read these pages. Of primary importance is that the knowledge discussed in each chapter is precise in its argumentation and supported by relevant evidence and examples as appropriate. Second, it was important that the different perspectives were capable of speaking from direct individual experience within the confines of a specific field, in a way as to benefit many more readers than the individual. Third, that the handbook as a whole has a cohesive quality to it, with chapters speaking across the pages to each other. We hope that this approach strengthens your research and supports your practice through a principled and focused process. Above all, we care about the quality of practice-based research and for the wonderful new knowledge that can emerge from inside practice. This is the true goal of this handbook. To all our readers, we wish a creative, productive and fulfilling time as you carry out your practice-based research.

Reference

Ingold, T. (2013). *Making*. London: Routledge.

Chapter summaries

Part I: Practice-Based Research

Chapter 1.1 – Practice-Based Research

Linda Candy, Ernest Edmonds, Craig Vear

Practice-based research is a principled approach to research by means of practice, in which the research and the practice operate as interdependent and complementary processes leading to new and original forms of knowledge. This chapter outlines the defining features of practice-based research as: the centrality of practice, the role of artefacts in research and the forms of knowledge that arise from it. We then consider some of the origins of practice-based research and how it has emerged in some practice-oriented disciplines and in interdisciplinary research initiatives that have fostered the growth of community. Communities of practice-based research are dedicated to enabling and advancing community building and innovation, research training and cultural change. Throughout, we highlight areas of significance that are explored in contributor chapters that follow in the first section of the handbook.

Chapter 1.2 – Interdisciplinary perspectives on practice-based research

Jonathan Michaels

This chapter is about the many different ways in which the term 'practice-based research' is used and understood, both across and within a variety of disciplines. I draw upon my own research experience in a number of fields, and my exploration of the published literature, to consider the features of practice-based research in its numerous embodiments. I argue that, while certain features are more prevalent in practice-based research, there is no single set of methodological, paradigmatic, episte-mological, or other criteria that is sufficiently specific to differentiate it from practice and/or other forms of research, in all instances. I go on to suggest that a possible thread that links the disparate uses of the term relates to a desire to break down some of the value systems that privilege certain method-ologies, professions, academic disciplines, or other subgroups in our epistemic endeavours. I further suggest that interdisciplinary approaches, which accept and explore the tensions that arise from such differences in perspective, can enhance understanding and become a rich source of new insights.

Chapter 1.3 – *The academisation of creativity and the morphogenesis of the practice-based researcher*

Mike Phillips

This chapter explores the origins of creative practice-based research, its roots in the Art College sys-tem, and its struggle to emerge imago like in the University system. It reflects on the 'academisation' of art and design and the slow and relentless 'metricisation' of culture and creativity, both within the context of research and the broader cultural landscape. As such, some of these characteristics have a strong UK focus, but the reader may recognise similar tropes in their recent cultural history. There is hope in the emergence of a new creature born in vitro to embrace the challenges of this new context, as an interdisciplinary practitioner with a transdisciplinary agenda. Much of this refers to the kinds of interdisciplinary work that emerged through a history of 20th-century artists engaging with digital technologies and subsequent entanglement with the broader science communities. As such, there is a focus on examples drawn from the work of Roy Ascott and the Planetary Collegium, a nomadic practice-based PhD programme with its origins in 1960s networked enlightenment. There is consideration of the ingredients necessary to support interdisciplinary practice-based research for the creative practitioner, in terms of the necessary international networks, the kinds of event/space that provide a provocative environment to nurture such work, and a couple of examples of the instruments and artefacts that open up new creative relationships.

Chapter 1.4 – *The studio and living laboratory models for practice-based research*

Linda Candy and Ernest Edmonds

This chapter presents organisational models of interdisciplinary practice-based research designed for studio and public spaces in the field of art and technology. It charts the development of the 'studio as laboratory' approach to practice and research and its later association with a 'living laboratory' in a public museum. The foundational ideas arose at the first Creativity and Cognition conference, a forum for exchange between artists, technologists, scientists, engineers, historians and others, dur-ing which meetings outside the main conference programme took place. From this the Creativity and Cognition Research Studios (C&CRS) was created at Loughborough University and later re-established at the University of Technology Sydney as the Creativity and Cognition Studios (CCS). In its new location, CCS continued the work initiated in C&CRS but with an enlarged and stronger PhD programme. Shortly after the inception of CCS, Beta_Space was established with the Powerhouse Museum (PHM), now the Museum for Applied Arts and Sciences (MAAS). The partnership between CCS and PHM constituted a new model for digital art practice that, in paral-lel, provided a vehicle for research. From the history of establishing such centres of practice-based research, strategies for running practice-based PhD programmes that include support for interdisci-plinary community building are described.

Chapter 1.5 – Practice-based research at SensiLab

Jon McCormack, Alon Ilsar, Tom Chandler, Mike Yeates, Elliott Wilson,
Camilo Cruz Gambardella, Nina Rajcic, Maria Teresa Llano,
and Sojung Bahng

This chapter discusses our approach to research at SensiLab, a creative technologies research laboratory at Monash University in Melbourne, Australia. Spanning two Faculties within the university structure (Information Technology and Art Design and Architecture), the lab's mission is to understand the innovative creative applications and undiscovered opportunities of technology. Our research is based in practice, but we extend the traditional concerns of creative practices to other disciplines, particularly those related to engineering and technology. After introducing basic concepts around knowledge and its communication, we discuss the specific methodologies and the culture that support our research practice. We illustrate this by describing some example projects created in the lab and reflect on the changing nature of practice-based research in a technological setting.

Chapter 1.6 – Working the space: augmenting training for practice-based research

Becky Shaw

The worlds of practice and academic research produce powerful cultural identities: for this reason, practice-based researchers must work hard to construct a space for themselves within these differing forces. The need to 'work the *space*' in this way is a creative and constructive part of practice-based research but often is not recognised or catered for in university doctoral training. In this chapter, I explore some of the national requirements for doctoral training in the UK. These indicate that cohort development is desirable for all PhD candidates, but the delivery of doctoral training in UK universities sometimes offers a version of universal researcher identity that excludes practice-based researchers. Drawing on work with art and design PhD candidates in a post-92 UK university, I offer alternative ways to design group research training which build researchers' understanding and construction of their intellectual and cultural research *space* without eschewing engagement with more generic notions of research training. These approaches draw on historic cultures of organising in art and design, employ critical and creative practices and challenge how we might see training, skills and being a researcher. From this I suggest a manifesto for how practice-based research training can happen that might also be applied beyond art and design.

Chapter 1.7 – Understanding doctoral communities in practice-based research

Sian Vaughan

Practice can be the mode, method, tool, object, subject and/or embodiment of research in a doctorate. This changes the traditional assumptions of a written text as the thesis and of doctoral education as being designed to support the production of text; how can institutions develop doctoral provision that supports practice-based research? To address this question, this chapter draws on the findings from semi-structured interviews with centre directors and coordinators of doctoral education in a number of institutions worldwide that represent a diversity of approaches to doctoral education in the broad fields of art, design and performance. As there is a growing emphasis on cohort-based approaches to doctoral education, particularly in the UK, the chapter includes an exemplar of how a multi-institution Doctoral Training Partnership (DTP) has supported practice-based research. My focus is on lived-experience of supporting doctoral researchers engaging in practice-based research.

Chapter 1.8 – Research doctorates in the arts: a perspective from Goldsmiths

Janis Jefferies

In this chapter, I discuss the advent of arts research in the context of the Goldsmiths University of London doctoral programmes which I led over many years. It explores the experiences, research narratives and journeys of interdisciplinary practitioner researchers who successfully completed PhDs between 2009 and 2018. It includes commentaries and reflections by the researchers together with their supervisor, drawing on the transcriptions of a roundtable discussion, online meetings, submitted summaries of practice and email exchanges. Individual and collaborative interdisciplinary practice research projects are presented, providing future practice-based researchers with examples of different methodological approaches. Challenges and opportunities for practice research are discussed in terms of the underpinning of principles and values as well as writing, dissemination and communication.

Chapter 1.9 – The PhD in visual arts practice in the USA: beyond Elkins' Artists with PhDs

Bruce Mackh

Art is a productive site for practice-based research, as evidenced by the gradual proliferation of doctoral programs in art worldwide. In the US, however, the pinnacle of academic achievement in arts practice remains the Master of Fine Arts, with few options for doctoral study in art despite increasing acceptance of this degree elsewhere. Lacking a common phrase for referring to doctoral-caliber study involving the practice of visual art, the term 'visual arts practice' is employed to distinguish between the norms of study in entirely academic fields such as art history, theory, and criticism on the one hand and the hands-on work of the studio on the other. The plural 'arts' rather than 'art' encompasses the many and varied forms of visual media, thus remaining inclusive rather than perpetuating longstanding exclusionary practices associated with terms such as 'studio art' or 'fine arts'. Although this chapter deals specifically with the PhD degree in visual arts practice, the 7 *Key Criteria for Success* and other discussions could apply to many fields where practice and research intersect, especially where stratification exists between 'thinkers' in a field and 'makers' or 'doers' engaged in the professional practice of that field. This chapter would be of interest to educators or academic administrators of doctoral programs involving practice-based research or to individuals concerned with graduate study in, through, with, or about visual arts practice. Its primary focus is the PhD in visual arts practice as it exists in the US, although many exemplars are well established across the globe.

Chapter 1.10 – The relationship between practice and research

Gavin Sade

This chapter discusses how communities of practice-based researchers have developed methodological approaches which mobilise their practice within the context of research. To do so, the chapter draws on theories of practice and how this has shaped approaches to practice-based research. It then considers the way some communities of practice-based researchers have considered it to be a new *paradigm* and highlights different approaches to research in comparison to other fields of research and within practice-based research. This highlights key concerns practice-based researchers address as they move between different practices of research within academic settings and those of a specific professional field or discipline. These concerns are focused on the way practitioners position themselves and their practices as research, and how they conceptualise non-traditional research outcomes as expressions of new knowledge.

Part II: Knowledge

Chapter 2.1 – Knowledge

Linda Candy, Ernest Edmonds, Craig Vear

In this chapter, we discuss the nature of knowledge, define it in research terms and show how it relates to practice-based research. We open with a discussion about the meaning of the term and distinguish between knowledge that is visible in the world and tacit knowing that is invisible but fundamental to how practitioners do their practice. Through this, we outline the difference between knowledge which is explicit and available for communication across time, and knowing as something hidden inside the practitioner. We then discuss five principles of new knowledge as original, validated, contextualised, shareable and retainable. Key questions are examined, including the role of practice in generating new knowledge and the relationship of the artefact to knowledge. Artefacts are typically central to practice and can be equally central to research in the practice-based context. One aspect of practice-based research that is special is that the delivered outcomes often have to include artefacts, or documentation of them, together with the written description. The chapters that follow in this section relate to different aspects of practice-based knowledge or formalising the nature of knowing, providing a broad picture of what knowledge is and how it is advanced in practice-based research.

Chapter 2.2 – Theory as an active agent in practice-based knowledge development

Linda Candy

This chapter is about the nature of theory and how a new kind of theory is used and evolved through practice-based research. I call this *theory as active agent*, by which I mean that existing theory plays a dynamic role in the research process that can lead to the emergence of new theory. This theory may take the form of a practitioner 'framework' which operates as an instrument for thinking and, where applicable, making something new. I begin with a discussion of theory as used in different contexts and then discuss how theory operates as an active agent in practitioner research. Expanding on this process, I describe a trajectory model of theory in practice that represents how practice-based knowledge emerges from research. The trajectory model was derived by observing and analysing practice-based research practitioners in order to describe common and differentiating features. What began as a researcher's method for generalising about the practice-based research process has proven to be useful to practitioners. Examples of practitioner theory as 'frameworks' that have been developed from practice through research are described.

Chapter 2.3 – Mapping practitioner knowledge: a framework for identifying new knowledge through practice-based research

Craig Vear

This chapter outlines a basic mapping framework for targeting or identifying new knowledge through the process of practice-based research. Part 1 presents a common problem experienced by practice-based researchers, that of the difficulty of consolidating the nature of the process of research with the nature of practice. To deal with this, I introduce Bergson's notion of *analysis* and *intuition* as defining two modes of knowledge: *outside-looking-in* and *inside-looking-out*. I present a way of understanding practice-based research as a consolidation of these two types. Central to understanding this consolidation of practice and research is a visual metaphor, which I describe and then offer an exemplar of how new knowledge in practice-based research can be identified through its use. Part 2 of this chapter describes the mapping framework and the underlying theories that support it. Part 3 presents several examples of mapping in use. This mapping framework will be of use to PhD students, first-time practice-based researchers and supervisors in developing an understanding of the specific nature of their research projects, and the development of a more systematic and targeted focus to investigating new knowledge in practice-based research.

Chapter 2.4 – Mapping the nature of knowledge in creative and practice-based research

Kristina Niedderer

This chapter presents a reflection on the nature and role of knowledge in creative and practice-based research (PBR). It begins with a brief historical overview of the development of PBR, especially in the creative disciplines, how the need for PBR arose, and the challenges associated with it. The discussion highlights how the requirement for explicit or propositional knowledge in research required new practice-based methods and approaches to be introduced to enable the use and recognition of knowledge rooted in practice, called tacit knowledge. This leads to an examination of tacit knowledge in relation to skills and expertise, using two examples, to better understand the role of tacit knowledge in the act of knowledge creation. The discussion of examples reveals further dimensions of knowledge, which necessitate a review of the foundations of traditional understandings of knowledge from philosophical perspectives. The discussion proposes a new framework to relate the different types of knowledge. The conclusion is that there is no issue, in principle, with including tacit knowledge in research, but there remains a tension between the exclusive use of tacit knowledge and requirements of justification in research as well as practical application.

Chapter 2.5 – Un-knowing: a strategy for forging new directions and innovative works through experiential materiality

Garth Paine

Artists work in systematic and deeply considered ways, but equally they may utilise a strategy I call *un-knowing*: a state of productive allowing, whereby they diverge from familiar paths and practices in a search for new directions that can lead to unexpected and stimulating outcomes informed by the experiential dimensions of the material. In this chapter, I present my ideas on how exploration, interrogation and invention in the arts can lead to a certain aspect of practitioner knowing that is communicated through the individual's unique creative process and the artistic works that emerge.

I propose that the process of *un-knowing*, of eschewing known techniques, workflows and approaches in favour of intuitive exploration, can be a valid investigatory method in creative practice-based research. I provide several historical examples and demonstrate the proposition through my own creative practice and research.

Chapter 2.6 – Appreciative systems in doing and supervising curatorial practice-based research

Lizzie Muller

As a creative practitioner, how do you judge whether what you have done is good? This question of judgement is central to the production of new knowledge through reflective practice. In this chapter, I develop the concept of *appreciative system* – an individual and collectively held lens that determines what is thinkable and doable – as integral to this process. Through the example of my own experience of supervising a curatorial research candidate, I argue that paying close attention to appreciative systems provides access to deep seams of knowledge in practice-based research degrees, and the means for researchers and supervisors to collaboratively articulate and consciously expand that knowledge. I describe how the concept of appreciative systems can help curators recognise, articulate and interrogate their own authorial agency as a crucial 'through-line' that provides structure and coherence to doctoral projects.

Chapter 2.7 – The art object does not embody a form of knowledge revisited

Stephen Scrivener

This is a slightly revised version of a paper that was first published in 2002, with the addition of an epilogue. In the chapter proper, I acknowledge the turn of the academic artworld towards research, understood as an investigation undertaken to acquire new knowledge, and also its desire to place making and the products of making at the heart of the enterprise, i.e., practice-based research. I make my position on the subject clear but assert that the proper goal of visual arts research is visual art. I then go on to explore the question of whether artworks can be understood as conveyors of knowledge, i.e., true, justified, belief. Although I do not exclude the possibility that artworks convey knowledge, I discuss a number of reasons why it is not straightforward to make them do so, including whether such 'artworks' would be considered as art. Hence, not only do I doubt that artworks can convey knowledge, I also doubt the value of those that can be understood to do so, because I value them as thought provoking rather thought appeasing artefacts. Hence, I conclude by defining arts research as original creation undertaken in order to generate novel apprehension, i.e., the unfamiliar, the unknown, or a state of affairs that confounds one's current knowledge. In the epilogue to the chapter proper, I reflect on its origins and ambition after the passage of 19 years.

Chapter 2.8 – Research, shared knowledge and the artefact

Ernest Edmonds

The chapter reviews the use of the word 'knowledge' in the context of this handbook, including a brief discussion of significant other usages. The review is predicated on an understanding of the use of the word 'research', standing for formal, or academic, research. The core discussion is about forms of knowledge from practice that are shared as an outcome of research. New knowledge generated from practice is discussed and a number of famous examples are presented in order to illustrate what new knowledge can look like in practice-based research even though these examples do not come from research. The theory of research knowledge that arises from practice is then reviewed.

Part III: Method

Chapter 3.1 – Method

Linda Candy, Ernest Edmonds, Craig Vear

This chapter discusses what we mean by methodology and methods in the context of practice-based research and explains why this distinction matters. Very often the terms are used interchangeably, but for the practitioner researcher, particularly one undertaking a PhD programme of research, it is important to differentiate between them. In simple terms, a method is a process, a technique or a tool that is used in a research investigation. Practitioners coming to practice-based research are already well equipped with methods for practice which will be specific to the field they inhabit, and those methods will continue into the research. Here we consider method that can apply across any field or discipline, and only in the sense of having a research function. In any given research project, there will usually be several methods depending on the nature of the study, the questions to be addressed, the aims and procedures directed towards achieving certain results, the gaps in knowledge to be filled: in other words, the overall approach. We refer to the totality of the approach as the methodology, which literally means a *study of methods* undertaken in order to decide on which methods are most appropriate for the envisaged investigation. In a specific research context, the results of that study can be described as a *system of methods* comprising the research strategy and the selection of procedures, tools and techniques to be used. To illustrate what this means, we describe some of the methods adopted for practice-based PhD research programmes where the use of reflective practices and evidence-based studies frequently occur. In the chapters that follow, features of practice-based research methodology and its many kinds of methods are discussed, including models and frameworks for designing a practice-based research projects, how questions emerge through thinking, making, appraising and reflecting and the role and value of practice-based research to practice among other things.

Chapter 3.2 – The **Common Ground** *model for practice-based research design*

Falk Hübner

This chapter proposes a model for the design of practice-based research. This model is equally useful in peer-feedback, supervision and teaching contexts. Based on the notion of putting the practitioners and their practice in the centre of research, it offers a methodological framework that departs from Henk Borgdorff's notion of 'methodological pluralism'. Through this, it argues for a flexible approach that does not predominantly build existing frameworks but rather seeks to create a bespoke design based on the specificity of the project and context at hand. The design of a research strategy is thereby regarded as a creative process, with a strong emphasis on the work of emergence, in both the design process and while carrying out the research. The *Common Ground* model of practice-based research design is applicable in diverse fields, contexts and disciplines. It offers a common ground that enables researchers of various kinds and disciplines to articulate, communicate and share their methods and methodological choices. Underpinning this model is the notion of flexibility, which builds on the understanding of networks as the two main layers of the practice-based research model:

the first one is concerned with the designing methods as concrete research actions, while the second one is on the level of the overall research strategy. Both layers act as flexible networks in themselves, as well as being intertwined. Together, these layers build the basis, as well as the counterpoint, for emergence during the research process. The definitions of research, practice, methods and methodologies are shared with those defined in this handbook.

Chapter 3.3 – Finding the groove: the rhythms of practice-based research

Brigid Mary Costello

The rhythms of the process of doing research and their impact on researcher and research outcomes are the focus of this chapter. After defining rhythm, I explore research rhythms in terms of speed, intensity, synchrony, predictability, and transitions. Lastly, I consider the overall rhythmic pattern of practice-based research and how these patterns might be composed and orchestrated when planning a research timeline. Insights are drawn from my own experience as a practice-based researcher and from a wide range of practice disciplines, including music, dance, design, and architecture. No matter the discipline, or whether novice or experienced, paying attention to the rhythms of a research process is a perspective valuable for planning ahead, maintaining momentum, and reflecting on the past. A focus on rhythm can also help researchers identify obstacles and understand how to resolve them.

Chapter 3.4 – Practice-based research in the visual arts: exploring the systems of practice and the practices of research

Judith Mottram

This chapter explores the context for practice-based research in the visual arts in relation to cultural, social, and economic frameworks that recognise both innovation and continuity. In order to determine appropriate methods for practice-based research in this space, it is suggested that clarity on what constitutes the parameters of contemporary creative practice gives us a basis for exploring and possibly expanding our presumptions. I note that our methods of practice are no longer restricted to just those of the studio and that it is a rather narrow conception for practice-based research and the production of new knowledge. Drawing on my experience of research, practice as a painter, research degree supervision, academic management, art gallery administration, and exhibition organisation, I argue that there are questions yet to be explored about what the objectives of practice-based research in the visual arts might be and the methods of how it might be undertaken.

Chapter 3.5 – Crafting temporality in design: introducing a designer-researcher approach through the creation of Chronoscope

Amy Yo Sue Chen and William Odom

This chapter describes the generative role that a practice-based research inquiry can play in incorporating temporality into the design of longer-term human–technology relations. We draw attention to how first-person experiences of time, over time in the design process, are of great importance when designing slow technologies. A practice-based research approach enables creative practitioners to attend to how temporality is shaped and manifested through the crafting of a design artifact which represents a key benefit of this approach and a key area for future research. Through an example of crafting temporality into *Chronoscope*, we introduce our designer-researcher approach to reveal how first-person perspectives among our design team could benefit and support practice-based research. Four main points are described to position our designer-researcher approach through articulating the motivation and design decisions that follow the goal of embodying temporality across the *Chronoscope* design process. Finally, we highlight how *Chronoscope* exemplifies a designer-researcher approach to craft slowness and temporality into practices and provide four lessons for future design-oriented researchers and practitioners to apply similar concepts and perspectives to a broader range of disciplines. Our experience of practice-based research is from an academic perspective in North America. Our thinking aligns with the principles of practice-based research stated in the definitions of this handbook.

Chapter 3.6 – *Thinking together through practice and research: collaborations across living and non-living systems*

Lucy HG Solomon and Cesar Baio (AKA Cesar & Lois)

In this chapter, the art collective *Cesar & Lois* delves into the nuts and bolts of the process of making our experimental artworks. This process is rooted in non-discipline-focused research, collaboration, and an interrogatory approach. We take the reader on a journey through our practice-based research, which entails a post-anthropocentric perspective and multispecies dialogues, as well as networked thinking and collaboration as propulsion for our work as researcher-artists. Along the way, we hope to engage creative practitioners from different areas to reflect on multidisciplinary practice as knowledge production that goes beyond specific fields and becomes a place for experimentation where questions are identified, hypotheses are tested and ideas materialise as prototypes. This chapter introduces the reader to the specifics of our work and excavates some of the sources of our methods. It is not intended as a step-by-step guide for every practitioner, but rather a set of considerations within an interdisciplinary and collaborative art practice and as an example of reflective creativity.

Chapter 3.7 – *Site: an inventories approach to practice-led research*

Graeme Brooker

This chapter charts the exploration of approaches to site in practice-led research, with a particular focus on the field of interiors, yet with the view to determining affinities with other creative practices. Practice-led is used to describe the forms of knowledge generation utilised by designers, researchers and academics in this field, using the numerous ways in which research is undertaken when working directly with found materials: primarily existing buildings. It describes practice-led approaches that are gleaned from the many actions of handling materials and their subsequent documentation and reuse. These are described as performative practices, ones that designate 'the found', that is, site-specific elements, as the material for the generation of new insights.

Chapter 3.8 – *Reflective practice variants and the creative practitioner*

Linda Candy

This chapter is about the kinds of reflection found in the practice of creative practitioners and what we can learn from this. It begins by revisiting Donald Schön's original definition of reflective practice, which was derived from studies of professional practitioners. It then describes several variants of reflective practice that extend the original definitions of reflection in and on action in the context of creative practice in music composition, visual and sculptural art, design and public art installations This reveals a more nuanced picture of how practitioners reflect as they explore ideas and experiment with materials in the creation of artworks. The understandings that have emerged from this research have implications for applying reflective practice as a method in practice-based research, the principles of which are described in the introductory chapter of this handbook. The chapter concludes with a brief account of how we can learn to be more reflective following the example of the creative practitioners. Overall, it summarises key findings and observations that are described in detail in the book *The Creative Reflective Practitioner.*

Chapter 3.9 – Reflection in practice: inter-disciplinary arts collaborations in medical settings

Anna Ledgard, Sofie Layton, and Giovanni Biglino

This chapter will discuss the role of reflection in practice in four interdisciplinary arts collaborations in medical contexts involving artists, clinicians, carers, and patients and resulting in high profile public presentations. It is written by a biomedical research scientist, an artist, and a producer who were central to the projects described. The chapter examines how reflection is employed in different ways by different players at four stages of project evolution:

1   Project conception (how to initiate and resource artistic collaboration in sensitive medical settings).
2   Arts engagement (creative engagement with patients and clinicians in workshops and conversations at the hospital bedside).
3   Re-presentation (translation of images, metaphors and material from the engagement processes into material for public presentation).
4   Presentation to the public (sharing the artistic work, inviting further reflection on others' lived experiences).

The case studies are positioned within participatory arts practice and the chapter draws out key principles which underpin a reflective practice approach to arts engagement. By articulating these principles, readers will be better equipped to advocate for cultures of reflection as central to creative collaborations on ethical as well as practical grounds. It is hoped that the authors' experience of reflective practice in grass-roots project conception and delivery can offer practical suggestions to inform future practitioners working in arts and health. To this end, the chapter explores how creative processes can present the human narrative alongside medical science, acknowledging the emotions as well as the clinical facts. Theories of 'situated learning', or learning embedded in social processes and physical contexts, and understanding gained through experience and participation in activities in relationships with others can be applied to much of this work.

Chapter 3.10 – Making reflection-in-action happen: methods for perceptual emergence

Jennifer Seevinck

When a reflective practitioner converts an existing phenomenon into a new way of seeing that phenomenon, we call this *reframing*. Reframing is an emergent process. This chapter synthesises emergence theory with Schön's concepts from reflection-in-action to aid the practitioner researcher in generating new insights within, and across, the dimensions of practice and research. I draw on my prior work in *perceptual emergence* to demonstrate a cycle for *emergent reflection-in-action*. Emergence is broadly understood as when something new or unexpected arises out of a given set of conditions.

The chapter focuses on mechanisms for *making emergence happen within reflection-in-action* and a set of emergent techniques that can be used as *moves* to generate reframing are proposed. My experience as a creative practitioner, researcher and a supervisor of Higher Degree Researchers informs a discussion of how this has been implemented, providing examples of *emergent reframing* in practice-based research efforts that range from ideation to thematic analysis, with mechanisms from Gestalts to data displays.

Part IV – The Practice-Based PhD

Chapter 4.1 – The Practice-Based PhD

Linda Candy, Ernest Edmonds, Craig Vear

In this chapter, we look at practice-based research from the practical perspective. Our aim is to provide practical guidance to all practice-based researchers, whether as PhD candidates and advisers or those planning written proposals for funding, or academic teachers designing new curricula for practice-based research in any discipline. We start by introducing the PhD and the specific nature of a practice-based research PhD. The subsequent topics parallel the journey that a practitioner researcher will take from inception through to final dissemination or examination of the thesis. This includes how to develop a systematic approach to designing the research methodology. We describe the challenges of a practice-based PhD and the role of the supervisor followed by guidance on the purpose of a literature review, the structure and purpose of a thesis, and the examination. Throughout this chapter, we introduce the contributor's chapters that form the rest of the section by way of highlighting areas of significance that are the focus of these chapters.

Chapter 4.2 – A play space for practice-based PhD research

Sophy Smith

This chapter offers a practical approach to support the role of play within knowledge construction through doctoral practice. In developing this approach, I draw on psychology and learning theory – primarily Peter Gray's *five characteristics of play*, Sandseter's figurative summary of the phenomenological structure of risky play and Kolb's updated Experiential Learning Theory. The chapter draws on my experience as programme leader for a Practice-Based Research Doctoral Training Programme at De Montfort University (2012–present) which runs across various disciplines, including computational intelligence, creative technologies, fine art, digital art and holography. While this chapter relates primarily to a doctoral training programme, the lessons learnt relating to the importance of play within practice extend beyond PhD training to practice-based researchers more generally. My experience of practice-based research is from an academic perspective in the UK, and as a professional practitioner in music composition and interdisciplinary performance practice. The principles of practice-based research that shape this chapter are reflective of those defined in this handbook. This chapter builds on a previous chapter written by the author for *The Power of Play*.

Chapter 4.3 – The sound of my hands typing: autoethnography as reflexive method in practice-based research

Iain Findlay-Walsh

Through this chapter, I present both an overview and an example of autoethnography, a research process and method that uses critically engaged, autobiographical writing as the basis for sociocultural inquiry, and that may be useful to practice-based researchers. My aims in writing this are threefold: 1. to introduce autoethnography as an approach that comprises certain methods, forms and objectives that in turn afford certain possibilities; 2. to consider and discuss its usefulness and relevance to practice-based researchers in general, and specifically to those working the arts, drawing upon my own work as a sound artist and researcher; and 3. to generate an example of this method 'in action' by telling a story of doing *this* research, the writing that you are now reading, a stop-start process unfolding over a nine month period, and embedded within the particular, personal, subjective, embodied, emotional, local and specifically challenging context of working, parenting and living with my partner and young children during COVID-19 and under lockdown.

Chapter 4.4 – Navigating the unknown: a dramaturgical approach

Hanna Slättne

This chapter provides some tools and methods based on my personal methodology as a working dramaturg within performance practice developed over 20 years as a full-time dramaturg. My experiences underpinning these tools and methods come predominantly from the field of new writing, dance and devising for the stage working in the UK and Ireland. In this chapter, I will share two approaches relevant to the practitioner researcher of interrogating their ideas and their own process through dramaturgical questions. The chapter will look at some tension that might arise in the transition from practice to practice-research.

Chapter 4.5 – The practice of practice-based research: challenges and strategies

Andrew Johnston

This chapter is about the realities of practice-based research (PBR) from the perspective of recent PhD graduates at the University of Technology Sydney. Drawing on a series of interviews with graduates and experienced PBR supervisors, I identify a number of challenges that are commonly encountered and strategies that have been applied to address them. Challenges include difficulty in managing the scope of projects and effectively documenting creative work and the tension between the need to produce creative work (the 'practice') and the time required to generate meaningful theoretical insights (the 'research'). Strategies that PBR PhDs and their supervisors have applied to address these challenges include collaboration, establishing and maintaining clear reflective strategies and practices, and ensuring the program of research is flexible enough to accommodate artistic exploration.

Chapter 4.6 – Community-building for practice-based doctoral researchers: mapping key dimensions for creating flexible frameworks

Sian Vaughan

Undertaking a practice-based doctorate brings new challenges. Being part of a community can make navigating the challenges not only easier but also more rewarding. This chapter maps the dimensions and elements to consider in building and supporting community for practice-based doctoral researchers. These dimensions are not exclusive facets of community, but should be conceived of as entangled and fluid. As practice can be the mode, method, tool, object, subject and/or embodiment of research in a doctorate, so too can institutional contexts and individual research projects differ.

Each thematic section includes some key questions that offer important points for consideration in supporting practice-based doctoral communities in different institutional contexts. I also offer suggestions for practice-based doctoral researchers as to how they might individually engage with and benefit from particular types of community-building activities. The relative importance and priority attached to each suggestion here will depend on the possibilities and particularities of your own context.

Chapter 4.7 – Strategies for supporting PhD practice-based research: the CTx ecosystem

Craig Vear, Sophy Smith, Stacie Lee Bennett-Worth

This chapter outlines the strategies implemented by the Institute of Creative Technologies (IOCT) at De Montfort University in the UK to support their doctoral candidates. IOCT considers these strategies from a holistic perspective, viewing how they can operate as an ecosystem of doctoral support. The CTx ecosystem is a model of how to design a support strategy that can be applied in any PBR context or, alternatively, adapted to different institutional contexts where necessary. In part 1 of this chapter, we present the underlying principles that guide the support of doctoral candidates in the IOCT. Part 2 then describes each component in the ecosystem and offers an evaluation from the student body. The final section concludes how the ecosystem interlinks and what the benefits are. Additionally, an appendix contains several of the resources that are offered to the doctoral candidates. These resources are intended for post-graduate practitioner researchers and their supervisors so they can guide and strengthen each individual's programme of work.

Chapter 4.8 – Ethics through an empathetic lens: a human-centred approach to ethics in practice-based research

Falk Hübner

This chapter is concerned with ethics in practice-based research through an empathetic lens. It focuses on identifying with others involved in research by seeing and understanding their involvement from their perspective and building an ethical approach from such inter-relationships. This chapter explores basic ethical principles that are relevant for researchers in general and discusses potential dilemmas for practice-based researchers from such a lens. Ethics and research ethics play a role in every research project, whether practice-based or not. A key point is that acting ethically does not equate to acting lawfully, despite obvious overlaps. Often, general rules are guidance only, and much is left to the judgement, behaviour and integrity of the researcher. Through this chapter, the reader–practitioner–researcher should

- Obtain an idea and awareness of the ethical issues that may become relevant in her respective field of work and research.
- Get a sense of the different angles and perspectives on ethics in the often interdisciplinary field of practice-based research.
- Be able to develop an idea on the various steps in professional work and research from the angle of ethics.
- Be ready to break down general ethical principles into concrete, step-by-step decisions.
- Understand that such concrete judgements need to be part of an overarching position that informs these decisions.

Chapter 4.9 – The practice-based PhD: some practical considerations

Ernest Edmonds

This chapter is primarily intended for practice-based PhD researchers. In the first section, it proposes actions and issues that should be addressed at the beginning of a project. The next section discusses the various text sections that might form part of the thesis as well as the supplementary material that is often included. Then there is a section that gives a few tips about checking the final text of the PhD before submitting it for examination. Finally, an outline thesis structure and a possible ethics approval procedure are provided.

Part V – Practitioner Voices

Chapter 5.1 – Practitioner Voices

Craig Vear

Chapter 5.2 – A new framework for enabling deep relational encounter through participatory practice-based research

Practitioner's Voice: Alice Charlotte Bell

This chapter will discuss the generation of my new Participatory Practice-Based Research (PartPb) framework through which deep relational encounters between practitioners and participants can be enabled. It provides a scaffold for generating opportunities for sustained one-to-one encounters between creative practitioners operating in performance, arts and health contexts with participant-subjects. As part of my practice-based research creative strategy, I first introduce the methods included in my overall research design, interwoven with psychological, phenomenological and maternal theories tested in-action. This is then followed by an overview of my PartPb project *Transformational Encounters: Touch, Traction, Transform (TETTT)*. I discuss the co-formation of my PartPb framework, which includes an outer PbR scaffold and inner Gestalt psychotherapeutic core, in interplay with *TETTT*. This chapter incorporates a discussion of the ethical and artistic challenges faced in the realisation of my PartPb research and signposts the new knowledge and insights gained.

Chapter 5.3 – Risk, creative spaces and creative identity in creative technologies research

Practitioner's Voice: Oliver Bown

In this chapter, I will draw on over ten years of practice-based research that applies emerging technologies, algorithms and code as a creative medium in the areas of music performance, music composition and media art installation. I will consider the competing demands of making things work technically and artistically, and the nature of collaborative work. I will consider my own recent research and that of a current PhD student as case studies in how this works in an academic context. Two concepts will structure the discussion. The first considers the question of time commitment and risk when undertaking specific activities, in relation to expected outcomes. I will look at how a practitioner handles the risk of an idea not working out at all and the more pragmatic risk of unexpected barriers to success. The second considers how much effort goes into setting up a creative space, through prior technical work and design thinking. I will look at examples of building creative freedom into a tool.

Chapter 5.4 – FEEDBACK: vibrotactile materials informing artistic practice

Practitioner's Voice: Øyvind Brandtsegg and Alexandra Murray-Leslie

The cross-cutting collaborative project *FEEDBACK: Vibrotactile Materials Informing Artistic Practice* searches for transformative artistic potential through the experience of co-making new musical instruments. We aim to explore vibrational modes in materials and an embodied interaction with these objects. The concept of feedback across nature and culture inspires and pervades our art and engineering collaboration to explore the inherent resonant properties, metaphorical readings and new aesthetics of these materials. Co-creating new dialogic spaces between practitioners' diverse and at times opposite processes inform polymorphic art practices in asynchronous ways. Our work is based on research into engineering natural and artificial materials which are developed in tandem with musical and technological devices. Custom programming is a key area of experimentation, providing an applied context for a feedback loop with dance, music and performance art. This prospective contribution presents the project and reflects on aesthetic and practical lessons learned through this multidisciplinary collaboration.

Chapter 5.5 – Co-evolving research and practice – _derivations and the performer-developer

Practitioner's Voice: Benjamin Carey

This chapter discusses the approach taken to practice-based research in my doctoral project entitled *_derivations and the Performer-Developer: Co-Evolving Digital Artefacts and Human-Machine Performance Practices*. The project involved the development of both a novel software artefact and an associated performance practice in the area of interactive musical performance. This chapter details the approach to reflective practice taken in this creative-production research project, reflects upon the benefits and challenges of this methodological approach and outlines the theoretical findings that emerged from cycles of practice and reflection in the research process.

Chapter 5.6 – Publishing practice research: reflections of an editor

Practitioner's Voice: Maria Chatzichristodoulou

This chapter reflects on the publication of Practice Research from Chatzichristodoulou's perspective as Editor-in-Chief of the *International Journal of Performance Arts and Digital Media* (*IJPADM*). Though several publications over the last 15 years have reflected on Practice Research from the perspective of the practitioner researcher, there has hardly been any discussion around publishing (or effectively sharing) Practice Research. The chapter thus offers a different perspective to the Practice Research debate, which can be relevant to researchers themselves as well as other academic journals, publishers and research publication outlets. The discussion is framed by Chatzichristodoulou's own experience as a practitioner researcher.

Chapter 5.7 – From a PhD to assisting BioMusic research

Practitioner's Voice: Balandino Di Donato

This chapter describes the author's practice and experience in transitioning from being a PhD in Music Technology candidate to a research assistant on a large-scale funded project. Both research works were delivered in a multidisciplinary environment and through a continuous feedback loop between academics, musicians and industry partners. The practice of these works was driven by computer science approaches applied in a musical context. Challenges imposed by the research and the timeline were significant, but at the same time, they fostered musical creativity and contribution to knowledge.

Different aspects of practice-based research in this setting will be described relying on the author's observations and experience.

Chapter 5.8 – The curious nature of negotiating studio-based practice in PhD research: intimate bodies and technologies

Practitioner's Voice: Kerry Francksen

This short retrospective considers some of the important activities involved in negotiating practice in PhD research. By reflecting on my investigations into the embodied and somatic practices of moving in media-rich environments, I focus on my project *Intimate Bodies and Technologies: A Concept for Live-Digital Dancing (Intimate Bodies)* as a specific case in point (see Figure 5.8.1 as an example of the work). In my retrospective, I contemplate the often complex, changeable, and multi-layered processes of negotiating practice-based research, and discuss key topics such as methodology and knowledge as practice.

Specifically, I reflect on some of the practicalities of exploring practice via a studio-based investigation and highlight a number of key discoveries that were encountered during this process. While the subject of *Intimate Bodies* is particular to the areas of dance performance and digital media, the methods and strategies developed are also applicable across a range of practical subjects. For example, the interconnections between practice and theory, as they played out in the studio context, highlighted some key landmark activities that necessitated a changed approach. This helped to define the emerging thesis and ultimately enabled me to explore the production of knowledge via practice. As such, some of the discoveries made pose interesting questions for the practice-based researcher.

Chapter 5.9 – Encounters at the fringe: a relational approach to human-robot interaction

Practitioner's Voice: Petra Gemeinboeck and Rob Saunders

Our collaborative practice aims to contribute to the public imaginary of our sociotechnical future by exploring how machines creatively and aesthetically participate in social encounters. In this chapter, we begin by briefly describing our long-term art–science collaboration before introducing our current Machine Movement Lab project. Methodological development and knowledge production are deeply entangled in our practice. We discuss the underlying core concepts that are constitutive to and materially enacted by our methodology of embodied, performative inquiry, as well as the new insights and understandings they produce.

Chapter 5.10 – *The impact of public engagement with research on a holographic practice-based study*

Practitioner's Voice: Pearl John

This chapter will introduce my field of practice as an artist working with holography and lenticular imaging and provide an overview of my research project. Drawing from my experience and my own development as a researcher, I will discuss lessons learnt during my research journey, aiming to answer the questions:

- In what ways is the *Vitae* Researcher Development Framework (RDF) beneficial to the research process?
- And how could supervisors make better use of the framework to assist researchers with their development?

Chapter 5.11 – *Project-based participatory practice and research: reflections on being 'in the field'*

Practitioner's Voice: Gail Kenning

This chapter focuses on an arts health research project. The aims of the project were to understand how older adults felt about where they lived, use arts-based approaches for data collection, and create a digital artwork for exhibition in the community. The project reiterated the need for flexibility, investment of time prior to the engagement, and reciprocity when working on socially engaged or collaborative projects. It shows how art and design engagement approaches can become methods for use in research opening up new possibilities and suggests my practice as an artist and designer prepared me for engaging in this research.

Chapter 5.12 – *Bearing witness – the artist within the medical landscape: reflections on a participatory and personal research by practice*

Practitioner's Voice: Sofie Layton

In this chapter I will discuss my practice as an artist, which is rooted in participatory arts within the medical context, and I will reflect on the key research elements within my practice. These include the participatory process, collaboration, and the artistic filtering of the narratives and metaphors which emerge through the participatory research process as a final artwork. I also draw on my own lived experience as part of my practice. My work is interdisciplinary and explores how by working with different specialists new methods of research can be developed. For me, within the arts, and arts/health, and scientific fields of transdisciplinary research, the partnership with the institution and the development of individual relationships, which may become a collaboration, are essential aspects of good research practice.

Chapter 5.13 – *Organisational encounters and speculative weavings: questioning a body of material*

Practitioner's Voice: Debbie Michaels

This is a personal reflection on an interdisciplinary project in which I transpose approaches from psychoanalysis and art psychotherapy to the fine arts in an experiment with method. Conceptualised as a 'speculative weaving' in three transpositions, my research follows the intertwining dialogues and entanglements that emerge as I traverse institutional boundaries in healthcare and academia, exploring practices of 'making' as a means of enquiry into organisational processes. Working *with* and *through* these practices, I come to understand the potential contribution that artistic processes and strategies may offer as part of a reflexive research approach, and the learning that can arise through a clash of disciplinary perspectives. New understanding emerges *in/through* the moving, (re) assembling, handling, (re)configuring, and sharing of diverse practices and material, the interweaving of dialogues, and the negotiation of tensions and resistances encountered at the borders between domains.

Chapter 5.14 – improvising as practice/research method

Practitioner's Voice: Corey Mwamba

I work as a musician, using vibraphone and audio software in both improvised and prepared settings. My research interest engages with the relationship between the vibraphone and the improviser, and how that relationship projects the idea of a personal sound in jazz and related forms. I talk about using improvising as a method for investigation in practice-based research, and how the practitioner researcher can use improvising to form new research questions and insights.

Chapter 5.15 – Dreaming of utopian cities: art, technology, Creative AI, and new knowledge

Practitioner's Voice: Fabrizio Augusto Poltronieri

In this chapter, I discuss some of the questions that intrigue me as researcher and artist working with practice-based research. These questions arise from my experience in combining intuition with technology, philosophy, art, and science. I believe the most significant contribution of this chapter is my pragmatic, *in vivo* experience, which showed me the importance of working with prototypes and small-scale projects using a methodology developed mainly from art. In the end, I discuss a practical project involving Creative AI, *ArchXtonic*, which uses the methodology described in the initial sections of the chapter.

Chapter 5.16 – Curating interactive art as a practice-based researcher: an enquiry into the role of autoethnography and reflective practice

Practitioner's Voice: Deborah Turnbull Tillman

This chapter investigates the role of reflective practice and autoethnography in curatorial practice. It examines two case studies, the first focusing on the methods articulated in a curatorial case study as part of a practice-based PhD on curating interactive art; the second focusing on an enquiry into methods acquired academically being useful professionally. The first, titled *ISEA2015: disruption*, was the central study in a series of three, resulting in one of three sets of criteria for a *multiple voices* approach to curating. This is particularly fitting for understanding language and organising criteria around experience-based over objects-based artworks and events. The voices captured in the resulting criteria were a) the audience's (through survey), b) the independent producer-as-curator (my own, through autoethnography and reflective practice), and c) a wider audience of curators working across art, science, and technology with interactive, engaging, or media based-artists and their work (through in-situ and focused interviews using conversation analysis). This chapter both situates my practice in the PhD case study and expands on the criteria in a new case study that uses method b) utilising interviews, autoethnography, and reflective practice to understand if knowledge gained within a PhD is useful in professional settings that support contemporary curatorial practice.

Chapter 5.17 – Please touch!

Practitioner's Voice: Marloeke van der Vlugt

In part one of this chapter, I commence by positioning my artistic PhD project in its field of practice, before describing and presenting the variety of methods that I deploy to research, develop and document the questions that I am concerned with. In part two, I zoom in on the case study, *Thresholds of Touch*, a performative experiment based on an interdisciplinary collaboration between a composer/ researcher, a sociologist and an artist/researcher (myself). I share how we set up a collaborative methodology between social science and artistic research, and what it contributed to researching *touch* from my perspective on practice-based research. The power relations between disciplines, methods and forms of expression/ knowledge will be traced and discussed. Finally, in the conclusion I reflect on the research outcomes and speculate on how different documentation strategies would have foregrounded other experiences, insights and/or knowledge.

Note

1 Ingold (2013, p. 9). Original emphasis.

PART I

Practice-Based Research

1.1

PRACTICE-BASED RESEARCH

Linda Candy, Ernest Edmonds and Craig Vear

Introduction

This chapter presents the central theme of this handbook – practice-based research, a principled approach to research by means of practice. As a relatively new development in the landscape of research disciplines, practice-based research has arisen from those domains where existing approaches have not served the needs of practice sufficiently well. Practitioners in a wide variety of fields have sought better ways to explore the kinds of questions that arise from the problems they encounter during thinking, acting, appraising and reflecting on activities that make up the essential elements of their practices. While it shares some core features with all forms of research, there are three aspects which distinguish it: the centrality of practice to the research, the role of artefacts in research and the forms of knowledge that arise from it.

A search for ways to bring practice and research together in a quest for knowledge that is relevant and effective in practice is one of the key motivating factors for the rise of practice-based research. Practice-based research emerged in the context of 20th-century, post-war political and cultural change and the growth of professional practice in fields such as clinical medicine, healthcare, nursing, education and software engineering. There are also organisational and cultural drivers that have played key roles, not least of which has been changes in the way academies institute regulatory structures for the validation of research degrees. The chapter introduces these motivating factors and also discusses the impetus given to inter-disciplinary exchanges for stimulating new thinking and new knowledge. We then describe examples of inter-disciplinary endeavours that aim to advance this form of research through community building and innovation, research training and cultural change. Throughout the discussion, we introduce the contributor chapters in this section that highlight areas of significance.

What is practice-based research?

Practice-based research is a principled approach to research by means of practice in which the research and the practice operate as interdependent and complementary processes leading to new and original forms of knowledge. By 'practice', we mean taking purposeful actions within a specific context, typically in a creative or professional way: the making, modifying or designing of objects, events or processes. As a methodology, practice-based research is characterised by strategies designed

 DOI: 10.4324/9780429324154-3

by the practitioner researcher that determine the methods, tools and techniques to be used to draw upon or observe practice, such as the documentation of reflection. The attraction of this form of research for practitioners is that, by connecting closely to existing practice, it provides a means of exploration for extending that work in a personal sense whilst contributing to broader considerations, such as the production of new forms and new knowledge. Furthermore, it gives researchers access to new knowledge arising from within practice conducted in the real world.

Practice-based research has been described variously as practice research, practice as research, practice-led research, research-led practice, evidence-based practice, research-through practice, practice-related research, artistic practice, professional research.For the purposes of this handbook, we have chosen to use the term *practice-based research* as the clearest definition and most frequently used label for this research activity.

Practice-based research encapsulates four main principles:

1   Practice and research are complementary but distinctive
2   The research is based within a world-of-concern defined by practice
3   The practitioner researcher is at the centre of the research
4   The research aim is to generate new knowledge

A basic principle of practice-based research is that practice and research operate in tandem, as related activities but with distinctive attributes. However, because practice is central to research activities, this means that research questions, for example, often arise from the process of practice, and the answers may lead to the enhancement of that practice. This may take the form of new practice methods or innovative products, such as artefacts or performances. The production of new artefacts, and the documentation of the context and process that gave rise to them, is frequently a distinctive feature of practice-based research.

What is research?

It is helpful to take a step back and look at the meaning of research in general terms. Research aims to seek new knowledge through a principled and original investigation. This aim is defined by research councils from around the globe: for example, the Australian Research Council defines research as the 'creation of new knowledge and/or the use of existing knowledge in a new and creative way so as to generate new concepts, methodologies, inventions and understandings'.[1] In the UK, the Research Excellence Framework 2021 defines research as 'a process of investigation leading to new insights, effectively shared'.[2] Definitions are also written into federal documents and legislation; for example, the definition given by the European Joint Quality Initiative as part of the 'Dublin descriptors' defines research as:

> The word 'research' is used to cover a wide variety of activities, with the context often related to a field of study; the term is used here to represent a careful study or investigation based on a systematic understanding and critical awareness of knowledge. The word is used in an inclusive way to accommodate the range of activities that support original and innovative work in the whole range of academic, professional and technological fields, including the humanities, and traditional, performing, and other creative arts. It is not used in any limited or restricted sense or relating solely to a traditional 'scientific method'.[3]

The production of new knowledge that is communicable to a wider community is an essential component of any research activity, and practice-based research is no exception. This is in contrast to private research, creative exploration, or personal development, which are sometimes called research by practitioners in many fields, especially those undertaking the kind of leading-edge work that requires a search for information about advanced technologies or materials, for example. This kind of research may lead the individual to produce original work, but if they are the only beneficiary and the outcomes remain private to the individual concerned, they are unlikely to have an impact on the wider world. Knowledge from research that is private and inaccessible to others is not the concern of this handbook.

Defining features of practice-based research

There are three features that distinguish practice-based research from other forms of research:

- The centrality of practice
- The role of artefacts
- The forms of knowledge

The centrality of practice

Practice is doing something that extends beyond everyday thinking into actions that may lead to new outcomes: for example, the making, modifying or designing of objects, events or processes. It is the use of ideas, beliefs or methods, as opposed to investigating theories relating to them. Note the contrast with 'theory' and the emphasis on the use or application of ideas and methods towards some end. Practice involves taking these ideas further by realising them in some way. By extension, creative practice combines the act of creating something novel with the necessary processes and techniques belonging to a given field, whether art, music, design, engineering or science. Practice that is creative is characterised by not only a focus on creating something new but also the way that the making process itself leads to a transformation of the ideas – which in turn leads to new works. Practice-based research is situated in the world-of-concern defined by the practice usually 'in the field', that is in a real-world context with real world outcomes. Practice-based research will generate knowledge through the core involvement of the practitioner researcher conducting practice that includes making, appraising and reflecting on artefacts of many different kinds.

The role of artefacts

Artefacts is a term used to refer to outcomes from practice of many forms. They include objects such as artwork, musical compositions, dance performances, engineered bridges, software code and, sometimes, partly ephemeral experiences, such as interactive installations. The practitioner researcher may use artefacts themselves, or the process of developing an artefact, as the object of study. Creating artefacts is a core activity in many forms of practice-based research, particularly in the field of creative arts. A given artefact will give rise to different experiences in different people in different contexts. In the arts, for example, this transmission of experience contrasts with the direct communication of knowledge. However, the artefacts generated within a practice-based research programme may be at the core of the 'new knowledge' generated by the research, in which case the clarity with which that knowledge is communicated directly through the artefact is a key question. If we accept that the artefact can, in some sense, represent new knowledge, the problem of sharing what that knowledge is implies a need for a parallel means of communication that illuminates the knowledge – in effect, a linguistic one. The textual element is vital in completing the contribution to knowledge that the artefact may, in its core, represent. In a practice-based PhD submission, for example, the thesis will often comprise a written text together with supplementary material in other forms, such as film, artwork or the documentation of a dance performance.

The place of artefacts in practice-based research and their relationship to new knowledge is discussed in the lead chapter of Part II, 'Knowledge'.

The forms of knowledge

The aim of practice-based research, as with any other mode of research, is to seek new knowledge where none exists in the particular field or discipline in which it is undertaken. Once this knowledge has been acquired, it is communicated in such a way that it can be shared in an archivable form for the benefit of others. The knowledge from research typically meets these criteria:

- *Original*: it is new in the world
- *Validated*: there are identifiable reasons to believe that it is true
- *Contextualised*: intended beneficiaries can understand its relevancy and reasons why it is new
- *Shareable*: others can benefit from it and understand it within its context
- *Retainable*: in archives in sustainable material of any form

These criteria are elaborated in the chapter on 'Knowledge' in Part II.

Practice-based research knowledge is expected to meet these criteria, as does any other form of research knowledge. The new knowledge can take the form of new facts, principles, frameworks, taxonomies or models. However, in practice-based research, it can also be a new form of artefact or practitioner knowledge generated by making the tacit explicit. The knowledge encompasses the theoretical and practical aspects of a subject as well as the awareness and familiarity gained through experience of both the theoretical and practical aspects. Knowledge can advance the individual's practice into new areas of innovation, or it can change the way a discipline conducts practice. Knowledge can move the practice forward into new areas of innovation and understanding, or it can bring small, incremental nuance to the practice. It is important to understand the difference between *knowing* and *knowledge* and the various forms that these take with practice-based research. Simply put, *knowing* is felt within and can be demonstrated by actions, *knowledge* is externalised, validated and can be archived. Research is concerned with knowledge, but in practice-based research, turning knowing into knowledge can be a significant activity. This is discussed in more detail in the lead chapter of Part II on 'Knowledge'.

The disciplines of practice-based research

This section is on the beginnings of practice-based research and the disciplines that have embraced it. It does not pretend to cover the full story – far from it. That is a book yet to be written. Instead, we give a brief overview of those disciplines where practice-based research, sometimes referred to as evidence-based practice or practice-based evidence, is carried out. These are principally, but not exclusively, the fields of clinical medicine, healthcare, nursing, education, software engineering, art and design. This is followed by a selective account of the way practice-based research began to emerge in the 1970s and how the influence of new definitions of research and regulatory changes provided a spur in the creative arts in academic institutions to develop ways of validating practice and the role of artefacts in the generation of new knowledge. Finally, we describe the impetus given to inter-disciplinary exchanges for stimulating new thinking and new knowledge, prompted by initiatives that brought artists together with scientists in *crossing disciplines*.

Practice-based research in healthcare, education and software engineering

In clinical practices and nursing, practice-based research can be traced back to the early 1970s[4] as a response by practicing clinicians to the reliance on evidence-based medicine as the dominant mode for knowledge development.[5] Clinicians were responding to a growing concern that academic settings for clinical research through evidence-based processes was setting up a two-tier system for standards and development of care: the well-funded 'academic' research hospital and the practicing clinician in an office practice. The role that practicing physicians can play in the development of their field was recognised in a landmark report by Sir James McKenzie,[6] in which he stated that 'future clinical research would be conducted with the cooperation of ambulatory patients in the doctor's office, the health centre, the clinic, the outpatient department, and the home'. By the 1980s, medical schools and university departments supported practice-based training and research, especially in internal medicine and paediatrics, as well as for primary care physicians.[7]

In his chapter, 'Interdisciplinary Perspectives on Practice-Based Research',[8] Jonathan Michaels discusses practice-based research in surgical and medical studies and contrasts them with his current research in art. He argues that the 'nature and purpose of practice in different fields of endeavour is diverse', and 'real or perceived dichotomies are often drawn between aspects of practice and research' for a broad range of disciplines and fields. However, the boundaries of definitions and practice are 'blurred or overlapping', and core aspects of these are activities can be understood through questions of 'intention, generalisability, sharing, experimentation and the temporal and spatial relationship between practice and research'.[9] His experiences in clinical practice and research across healthcare, decision science and art[10] are examples of cross-disciplinary activity in one individual, in which research skills and methods are transferrable.

Practice-based research in nursing, understood as evidence-based practice, has emerged as a strong practitioner-led field of enquiry. The term 'evidence-based practice' has been used to counter the predominance of the technical rationalist approach of evidence-based medicine.[11] Originally concerned with 'improving patient outcomes through sound evidence',[12] using reflective and analytical processes through practice, it 'evolved from being strictly clinically based to incorporate a more holistic approach that appropriately reflects the entirety of nursing research and practice'.[13] This field is recognised world-wide and 'continues to advance and change along with the nursing discipline'.[14]

In education, practice-based research is an established community across the world. The critical and professional development of teachers and practice-based teaching research can be said to have clear origins in two foundational academic concepts: *action research* and *reflective practice*.[15] Both of these concepts can be traced to the 1930s–1940s as methods for developing critical practice in teaching, but it was not until Stenhouse's (1975) notion of the teacher-as-researcher that the two came most compellingly into relationship and educational action research as a process, which held at its centre different kinds of reflection, began to be reformulated in Britain.[16]

Schön, in the mid-1980s, went on to develop the rigor behind the notion of 'teacher-as-researcher' through his theory of the 'reflective practitioner' by 'extending Dewey's (1933) foundational ideas on reflection through observing how practitioners think in action'. This went on to 'form the core professional artistry of the reflective practitioner'.

There are a growing number of post-graduate courses and research projects across the world that have been developed for many years and show strong support for reflective practice in developing educational practice. Many of these are 'based on a more integrative model of thinking, feeling and acting, which elaborates Schön's notions of reflection in and on action'. Ultimately this field of research builds a

> self-aware, self-reflexive teaching population, capable of producing the highest quality learning situations for pupils, is a laudable and necessary aim in a world characterised by social fragmentation, increasing economic competition and personal turbulence.[17]

Areas such as software engineering are recognising the 'potential to improve software products or development processes' through practice-based research.[18] In situations where, for example, practitioners have to make informed decisions about whether to adopt a new technology, it can be difficult because there is little collective knowledge beyond personal opinion about significant issues such as 'suitability, limits, qualities, costs, and inherent risks'.[19] Although termed 'evidence-based' in line with clinical practice research, the concern is with progressing and advancing practitioner knowledge for the field based on sound research processes, and for software engineering research to be a mechanism to support and improve the technology adoption decisions of the practitioners.

Practice-based research in the creative arts

Practice-based research, in the academic context, started to emerge as a research strategy in the 1970s and early 1980s. It was particularly noticeable in the UK, with a first generation of pioneering artists and designers exploring and developing their creative practices through higher degrees. UK research degrees in creative practice were largely the reserve of the polytechnic institutions[20] at that time, regulated by the Council for National Academic Awards (CNAA),[21] which validated their degrees. An exception was the independent Royal College of Art, London, where Professor Bruce Archer was particularly influential in developing models of art and design research. Archer's founding paper set the debate going in the 1970s.[22] He argued for the inclusion of design, alongside science and the humanities, in education (including research education). Archer went on to head the RCA's Department of Design Research, overseeing practice related to PhDs in design and art.

A significant point for the polytechnics was that CNAA regulations came to differ from many universities in allowing supplementary material, other than the text, to be submitted for examination. Carole Gray summarised the early days of practice-based research degrees in the UK in this way:

> it is possible to identify examples of 'pioneers' who used their own practice as a vehicle for inquiry. Andrew Stonyer's PhD completed in 1978: 'The development of kinetic sculpture by the utilization of solar energy' – demonstrates the beginnings of inquiry through practice.[23]

In the following ten years, at least a dozen more PhDs and MPhils were completed, all involving the development of some 'experimental', creative practice. Some examples are:

1980	Raz: Fashion and Textiles
	Connor, Newling (MPhil): Fine Art
1981	Saleh: Graphic Design
1982	Cooper: Graphic Design
	Scrivener: Computer-Aided Graphic Design
	Goodwin: Painting
	Newton (MPhil): Fine Art/Computing
1983	Tebby: Sculpture

Several of these (Stonyer, Scrivener and Tebby) were undertaken at Leicester Polytechnic, now De Montfort University, and provide interesting early cases. For example, Susan Tebby presented a written thesis together with an exhibition of the work that her practice had generated. The external examiner both inspected the exhibition and studied the text. Of course, the exhibition could not be archived in the library, but a volume of 35 mm slides documenting it was lodged there as part of the permanently available thesis. When a multidisciplinary team, led by Ernest Edmonds and including Stephen Scrivener, moved from Leicester Polytechnic to Loughborough University in 1985, one of the first achievements was to successfully propose that the higher degree regulations should be changed to allow supplementary material to be submitted along the CNAA lines, which was then added to the university's existing regulations. It is significant that the main funding body for arts research in the UK, the AHRC, has taken practice-based research (using the term practice-led research) into its main stream of work for some time. In fact, it has stopped running some specific practice-based programmes because the work is now covered within the standard schemes.[24]

As Mike Phillips explains in his Chapte, 'The Academisation of Creativity and the Morphogenesis of the Practice-Based Researcher',[25] the role of Roy Ascott was important in promoting practice-based research in the UK. Ascott initiated the Planetary Collegium, a networked culture that essentially existed as a network of people supporting each other's research through regular meetings and events. Phillips wrote:

> Things got interesting when Roy Ascott joined the Interactive Media Subject Group which I ran in 1997 to establish STAR (Science Technology and Art Research) alongside the Centre for Advanced Inquiry in the Interactive Arts . . . CAiiA-STAR constituted a trans-institutional research platform which was superseded in 2003 by the formation of the Planetary Collegium . . . primarily a practice-based PhD programme.[26]

More broadly, Phillips looks at the way the creative arts, originating in art colleges once independent but now swept into the university system, were affected by the sector's requirements for measuring research. He sees room for optimism in the rise of inter-disciplinary practitioners engaging with digital technologies and becoming involved with the broader science communities exemplified by the Planetary Collegium and its practice-based PhD programme. He considers how support for inter-disciplinary practice-based research practitioners is happening through international networks, events and spaces that provide places where new creative relationships are forged.

The origins of practice-based research are not limited to the UK, although it was early in the game, and other countries are building significant traditions in the area. Although the terms of reference used may not be the same, related developments are occurring in other countries, such as Brazil and Sweden.[27] The Norwegian Artistic Fellowship programme, described by Kjørup,[28] is another example of funding for research involving practice.[29] After a hesitant start, there are now a number of universities in the USA that offer practice-based PhDs, notably at Washington University, where the Centre for Digital Arts and Experimental Media (DXARTS) offers a structured PhD programme that includes a practice-based component. The primary focus of DXARTS is to create opportunities for artists to discover and document new knowledge, and creating new art is at the centre of all activities in the programme.[30]

In the English-speaking world, the UK, Australia and New Zealand lead the way in the development of structures for practice-based research as post-graduate research awards such as PhDs, particularly in the art, design and digital media domains where the creative artefact assumes a central role. In other countries, including Sweden and the USA, initiatives in practice-based research have similar characteristics, although the organisational structures and regulatory frameworks are fewer in number. An important influential factor shaping the way these initiatives take root and grow is a country's university system and its regulatory standards, which affect the take-up and expansion of such initiatives.

Without a suitable regulatory framework, a legitimate role for the designing and making of artefacts leading to post-graduate research awards is hard to justify. Opportunities for including artefacts in formal research remain limited on a world-wide scale, and within the context of academic research, the changes to existing rules that permit artefacts, artworks, exhibitions and performances to be included in a PhD submission are found in a relatively small number of universities. Where a PhD is undertaken, university rules have a significant impact on the research process. In Australia, the Australian Research Council (ARC) has been funding research in creative practice and has entered a partnership with the Australian Council in which collaborative art/science projects are funded jointly. The ARC includes the possibility of artefact 'inventions' in its definition of research as the 'creation of new knowledge and/or the use of existing knowledge in a new and creative way so as to generate new concepts, methodologies, inventions and understandings'.[31] Similarly, in the UK, the Arts and Humanities Research Council, in its definition of research, states that 'creative output can be produced, or practice undertaken, as an integral part of a research process'.[32]

Throughout the developments described earlier, a number of texts and books were written that documented the beginnings of a varied practice-based discourse, sometimes at variance with itself. Nevertheless, these writings were important markers in the evolution of the field, and they remain valuable to the practitioner researcher seeking understanding through examples of what has preceded their own work.

Crossing disciplines

Practice-based research has been interpreted and enacted differently across different disciplines as each area developed its own modus operandi in entirely different conditions and from various historical origins. In design and performing arts, it has been a real challenge to have practice-based research acknowledged as a principled and rigorous activity. For many years, research in this area has privileged historical and theoretical study over practice in research, and the practitioner's voice is not always heard. In surgery, by contrast, the activity of practice-based research is simply referred to as 'research'. The label *practice-based* in this context seems tautological, as research in surgery is predominantly based on the practice of surgery and conducted by surgeons where collaborative teamwork is the norm.

Inter-disciplinary work is a distinctive feature of many practice-based activities across different fields. In order to move away from the constraints of tightly enforced norms of their formative disciplines, practitioners have sometimes found ways forward by moving across to other disciplines. One of the unique aspects of practice-based research is the way different disciplines can come together through practice to benefit all the fields involved. As demonstrated in several chapters in Part I, cross-disciplinary practice-based research can bring forth interesting new knowledge and surprising insights.

There are several chapters that highlight the benefits of crossing disciplines, for example, Marloeke van der Vlugt's[33] chapter 'Please Touch!' in Part V advocates an 'experimental method' with the intention of 'blurring the boundaries between artistic and sociological research'.

There are other ways to cross disciplines and, as initiatives in art and science have shown, this is by now an established way to instigate fresh perspectives and produce innovative outcomes. A number of funded research initiatives have been taken in which inter-disciplinary collaboration between, for example, science and art have been facilitated. Two examples are SciArt in the UK and Synapse[34] in Australia. In the latter, the Australia Council for the Arts and the Australian Research Council jointly fund artist and scientist collaborations. In such cases, although the normal outcomes including learned papers are expected, artefacts that are exhibited are also seen as legitimate and valued contributions from the research.

The SciArt programme,[35] begun in the 1990s, was a major funded programme in inter-disciplinary art and science which acted as a catalyst for change.[36] This was important in demonstrating the potential benefits for inter-disciplinary practice that could be combined with research. Inter-disciplinary initiatives have the potential to change the way practitioners think about how they work. Often, the programme introduced artists to the idea of research as an element of practice. Hence, it is interesting to know that the practice was changed as a result. It was primarily the artists who were making artefacts within the programmes and the incorporation of that making into research that initiated the changes. Moreover, they seeded the growth of a community of practitioners working across art, science and technology, some important examples of which are described in the chapters by Candy, McCormack, Shaw, Phillips and Jefferies in Part I of the handbook.

Communities of practice

As a relatively new area of research compared to established disciplines, support communities were initiated in different disciplines and fields. The initiatives varied across the broad field of practice-based research, but a common feature was the recognition that for innovative research practices to gain traction, there needed to be strong and enabling communities of interest. This presented practitioner researchers with an opportunity to build environments in which communities of common interest developed supportive structures for these new ventures in research and practice.

Community building

Linda Candy and Ernest Edmonds, in their chapter 'The Studio and Living Laboratory Models for Practice-Based Research',[37] describe an initiative that began in the 1990s and exemplifies the inter-disciplinary character of many forms of practice-based research. Creativity and Cognition centres were models of studio and public spaces for inter-disciplinary practice-based research in the field of art and technology.[38] The 'studio as laboratory' was established in a university computer science department which offered a home to artists experimenting with new forms of digital technology.

The concept of an inter-disciplinary space for art and technology practice-based research was later extended to include a 'living laboratory' in a public museum where outcomes from practice could be explored in a public setting. This was a crucial development in supporting their community and encouraging rigour and innovation in practice. They discuss various organisational methods that were used to strengthen the group. This led to a strong community of practice among the researchers, which was 'underpinned by a principled approach to research that continues to this day and is reflected in several contributions by practitioner researchers in this handbook'.[39]

Jon McCormack et al.'s chapter, 'Practice-Based Research at SensiLab',[40] outlines a multidisciplinary research culture at the creative technologies research laboratory, SensiLab,[41] as a way of discussing the benefits of crossing disciplines. Based on expertise in creative technologies, such as media arts, virtual world-making, computational creativity and machine learning, they have experienced practice-based research as a

> mode of enquiry that seeks to generate new knowledge through practice, i.e. by making, doing, building, experimenting, creating and experiencing, so it may often involve the use of *knowing-how* types of knowledge, gained through physical exploration and direct action.[42]

SensiLab's mode of research incorporates research practices outside the art and design disciplines and purposefully seeks research that crosses disciplines, including areas such as artificial intelligence, human–computer interaction, interactive media and games, simulation and cultural heritage, reflecting contemporary globalised concerns and industrial foci. McCormack et al. state that:

> Universities are no longer the exclusive innovators of knowledge, as global technology companies, for example, employ more PhD researchers than many universities. Rising

superpower nations, such as China, are expected in the next decade to graduate more students annually than the entire population of the United States.[43]

SensiLab's aim is to 'bring a successful research methodology in the creative arts to a broader group of researchers: those who can understand their research as practice, even if it doesn't fit into a traditional creative arts stereotype'. To this end, they have developed a strategy that includes 'the physical workplace design and the infrastructure it contains; articulating our goals and values in an open and clear way' and conducting 'on-going critique and reflection on our processes and outcomes'.[44]

In a quite different perspective on inter-disciplinarity community building, Becky Shaw's chapter, 'Working the *Space*: Augmenting Training for Practice-Based Research',[45] argues that the 'practice-based researcher is located between the professional world of practice and the academic model of research'. She illustrates how the

> worlds of practice and academic research produce powerful cultural identities: for this reason, practice-based researchers must work hard to construct a space for themselves within these differing forces.[46]

She sees the need to 'work the *space*' as a 'creative and constructive' part of practice-based research, something often not catered for in university doctoral support. The '*space*' relates to a group identify, a 'cultural social *space*' that is within the physical place of a university and is a product of social relationships, formed through interaction.

Sian Vaughan's chapter, 'Understanding Doctoral Communities in Practice-Based Research',[47] highlights the need to build community, arguing that a 'contribution to knowledge is a contribution to discourse and a community engaging in that discourse'.[48] A core part of practice-based researcher development is learning how to 'join and belong to those discourse communities as they are learning how to undertake research'. Vaughan quotes Mantai in stressing that doctoral researchers need 'a growing sense of belonging to a scholarly, academic, or research community. This sense of community is a well-established requirement for the doctoral student's scholarly development'.[49]

Vaughan goes on to show how building a community for practice-based research can 'provide the sense of belonging supportive of doctoral becoming, enabling productive conversations that aid the navigation of methodological, epistemological and ontological threshold crossing'. She stresses the importance of 'community-building events and spaces for conversations to explore issues, rather than training to provide answers and solutions'. For Vaughan, and the others who contributed to her chapter, they have witnessed how such a community can 'enable safe spaces to vent as well as to share experiences', thereby supporting a feeling of being part of a community that can 'provide reassurance as well as opportunities to test out ideas and learn from others'.[50]

Community innovation

It is sometimes necessary within the fixity of institutional support systems to radically rethink and design new structures to support practice-based research. One such initiative was the Goldsmiths Digital Studios, an inter-disciplinary centre for post-graduate studies based between the Departments of Computer Science and Art. The Digital Studios was dedicated to multidisciplinary research and practice across arts, design, technologies and cultural studies[51] and was led by Janis Jefferies, Professor of Visual Arts and Robert Zimmer, Professor of Computing. The goal was to bring artistic and scientific research together and to approach inter-disciplinary research with a sense of openness and inclusivity; but also to challenge assumptions based in established disciplinary frameworks and to engage with the different vocabularies across the disciplines. This led to a sense of community, place, identity and purpose among the PhD candidates and researchers involved in the Digital Studios, the benefits of which are discussed in Janis Jefferies's chapter, 'Research Doctorates in the Arts: A Goldsmiths Perspective'.[52]

Innovative and radical approaches to forming a sense of community in practice-based research are also discussed in Mike Phillips's chapter.[53] In a similar strategy to that of the Digital Studios, and SensiLab as described by McCormack at al., an inter-disciplinary community was forged between creative-based disciplines and computer science. Another strategy that Phillips discusses is the development of i-DAT.org in 1998 as a way to circumvent the strict structure in place at his host university. Phillips sees i-DAT.org as operating as a collective of 'digital artists' that sponsor 'collaborations with other areas of the University, such as robotics, earth sciences and engineering'. This in turn has generated a sense of collegiate research that engages with people, communities and institutions through collaborative and participatory design methods.[54]

Criteria for cultural change

In the USA, practice-based research is establishing itself in discreet pockets of academia. PhDs in music, creative technologies, design and performance, for example, are now being regarded as the terminal qualification, with long-established centres of excellence supporting systematic and rigorous doctoral research. Once the USA was the 'only country in the world to offer a doctorate for studio artists';[55] however, as Bruce Mackh reports in his chapter, 'The PhD in Visual Arts Practice in the USA: Beyond Elkins' *Artists With PhDs*',[56] 'the pinnacle of academic achievement in arts practice remains the Master of Fine Arts', with 'few options for doctoral study in art despite increasing acceptance of this degree elsewhere'.

Based on his reflections of James Elkins's publication *Artists With PhDs: On the New Doctoral Degree in Studio Art*[57] Mackh re-examines the impact of this situation on the research culture of an institution. Although focused on visual arts communities, these findings present a stark warning, and solutions to any research community, where 'stratification exists between "thinkers" in a field and "makers" or "doers" engaged in the professional practice of that field'.[58] This, he observes, can lead to a divide between those with and without formal research qualifications, such as a PhD, and the way research is viewed and supported in faculties that are predominated with the practitioner qualification of the MFA.

To counter this divide, Mackh presents '7 Key Criteria for Success' that could be implemented to support a cultural shift in a faculty wishing to support doctoral practice-based research. These are:

• Supporting academics in their development of research skills and qualifications
• Developing program structures that align to graduate school requirements
• Investing in faculty members that are research focused and research qualified (e.g. have earned a PhD)
• Evaluating how the local and regional culture can support the research community
• Ensuring there are resources to launch and sustain a PhD program
• Supporting the 'buy-in' from all faculty members to support the program and researchers
• Ensuring the program is attractive enough for successful enrolment.

Practice-based research training

Vaughan, Shaw, Mackh and Phillips highlight the need for specific training for practice-based research beyond what is generally offered by graduate schools and doctoral colleges. Shaw argues that 'the problem is not necessarily the idea of generic training *per se*, as there is value in striving to teach research skills or issues that all might have in common', but that the focus of generic training can be biased towards the type of research that generally ends up as a text-only thesis. She argues that trainers should not pretend that one particular world of academia is a universal given.

Shaw, Vaughan, Smith, Candy and Edmonds, as well as Vear et al., present solutions to support how the 'rules of the game' of practice-based research are exemplified within a specific community. These include cohort-designed and -led activities, community-driven focus groups, community building, playful and creative ways of understanding research and practice, embedding professionals into research groups, peer teaching and peer learning, and the development of formal training programs and PhD programmes designed to 'address the gap between the policies of doctoral training and the experiences of practice-based research PhD candidates'. The benefit of initiatives such as these is that

> a formal programme makes this area of research visible, distinct, shareable, open to improvement and formally recognised by the university. It also performs a task of enculturation in which university needs and practitioner worlds are welded into a new form.[59]

Pearl John's *Practitioner's Voice* chapter *The impact of public engagement with research on a holographic practice-based study*[60] presents the *Vitae* Researcher Development Framework (RDF) and proposes that it is beneficial to the research process. She shows how practice-based communities, including supervisors, could use this framework to assist their development. *Vitae* is a charitable organisation that describes itself as 'the global leader in supporting the professional development of researchers, experienced in working with institutions as they strive for research excellence, innovation and impact'.[61] The RDF framework identifies four domains:

1 *Knowledge and intellectual abilities*: the knowledge, intellectual abilities and techniques to do research
2 *Personal effectiveness*: this includes the personal qualities and approach needed to be an effective researcher
3 *Research governance and organisation*: knowledge of the professional standards and requirements to do research
4 *Engagement, influence and impact*: includes the knowledge and skills to work with others and ensure the wider impact of research

For Johns, the final domain, *Engagement*, can be a particularly fruitful way of developing a deeper understanding of the reach of the research, creating shared experiences across their community and contributing to sense of well-being. Public engagement, she proposes, 'is a valuable pathway to impact for researchers, and can be used to provide impact case studies', that ultimately provided her with 'valuable feedback which helped to develop my work including the crafting of research questions and methods'.[62]

Topics on practice-based research in the following chapters

- Origins of practice-based research in institutional and regulatory changes
- Fostering practice-based research through places and spaces for inter-disciplinary relationships
- Across disciplines, there are variations of how practice-based research is defined
- Inter-disciplinary activities happen within an individual practitioner's research practice
- Crossing and blurring boundaries can be a way to reach fresh perspectives and innovative outcomes
- Studio and public spaces for making and evaluating the production of new knowledge

- Practitioner-focused organisations can strengthen practice-based communities
- Benefits of crossing disciplines to generate new knowledge through practice
- Developing the language of research from the practitioner's lexicon fosters social identity
- Community knowledge building is a contribution to practice-based discourse
- Sense of belonging gives reassurance and opportunities to learn from others
- Fostering community, place and purpose through shared interests and practices
- Criteria for promoting cultural shifts in doctoral practice-based research
- How practice-based researchers use research frameworks to guide the processes

Notes

1 www.arc.gov.au/sites/default/files/minisite/static/10419/ERA2015/intro-3_define-research. html#:~:text=For%20the%20purposes%20of%20ERA,%2C%20methodologies%2C%20inventions%20 and%20understandings.
2 UK Research Excellence Framework 2021. www.ref.ac.uk/.
3 Dublin descriptors: contextualised: intended beneficiaries can understand its relevancy and reasons why it is new.
4 Wood et al. (1986, p. 359).
5 Ibid.
6 Discussed in Mair (1973).
7 Ibid., p. 359.
8 Michaels (2022).
9 Ibid.
10 Ibid.
11 McIntosh (2010, p. 23).
12 Mackey and Bassendowski (2017).
13 Ibid.
14 Ibid.
15 Leitch and Day (2000).
16 Ibid.
17 Ibid.
18 Jarząbek et al. (2020).
19 Dybå et al. (2005).
20 Polytechnics were tertiary institutions in England, Wales and Northern Ireland offering undergraduate and post-graduate degrees. In 1992, they became independent universities awarding their own degrees.
21 Council for National Academic Awards: https://discovery.nationalarchives.gov.uk/details/r/C73.
22 Archer (2005).
23 Gray (1998).
24 https://webarchive.nationalarchives.gov.uk/20200923112848/https://ahrc.ukri.org/funding/apply-for-funding/archived-opportunities/researchgrantspracticeledandapplied/ (accessed June 2, 2021).
25 Phillips (2022).
26 Ibid.
27 Buchler et al. (2009).
28 Kjørup (2011).
29 Ibid., p. 26.
30 The DXARTS PhD. www.washington.edu/dxarts/academics_phd_requirements.php.
31 www.arc.gov.au/sites/default/files/minisite/static/10419/ERA2015/intro-3_define-research. html#:~:text=For%20the%20purposes%20of%20ERA,%2C%20methodologies%2C%20inventions%20 and%20understandings.
32 AHRC. https://ahrc.ukri.org/funding/research/researchfundingguide/introduction/definitionofresearch/.
33 Vlugt (2022).
34 Synapse: Art Science Collaborations: www.synapse.net.au/.
35 Set up by the Wellcome Trust in 1996, SciArt was run by a consortium between 1999 and 2002 of the Arts Council of England, the British Council, the Calouste Gulbenkian Foundation, the National Endowment for Science, Technology and the Arts (NESTA) and the Wellcome Trust. From 2002, the programme was run independently by Wellcome.

36 Glinkowski and Bamford (2009).
37 Candy and Edmonds (2022).
38 Ibid.
39 Ibid.
40 McCormack et al. (2022).
41 SensiLab is a technology-driven, design-focused research lab based at Monash University in Melbourne, Australia. https://sensilab.monash.edu/.
42 McCormack et al. (2022).
43 Ibid.
44 Ibid.
45 Shaw (2022).
46 Ibid.
47 Vaughan (2022)
48 Ibid.
49 Mantai (2019, p. 368).
50 Vaughan (2022).
51 Jefferies (2022).
52 Ibid.
53 Phillips (2022).
54 Ibid.
55 Schwarzenbach and Hackett (2017).
56 Mackh (2022).
57 James Elkins published *Artists with PhDs: On the New Doctoral Degree in Studio Art* (2009).
58 Mackh.
59 Ibid.
60 John (2022).
61 www.vitae.ac.uk/researchers-professional-development/about-the-vitae-researcher-development-framework.
62 John (2022).

Bibliography

Archer, B. (2005). The Three Rs. In B. Archer, K. Baynes, & P. Roberts (Eds.), *A Framework for Design and Design Education. A Reader Containing Papers from 1970s and 80s.* Warwickshire: The Design and Technology Association.

Buchler, D., Biggs, M., Sandin, G., & Ståhl, L. H. (2009). Architectural Design and the Problem of Practice-Based Research. *Cadernos de Pós-Graduação em Arquitetura e Urbanismo*, 8(2). https://uhra.herts.ac.uk/handle/2299/7470.

Candy, L., & Edmonds, E. (2022). The Studio and Living Laboratory Models for Practice-Based Research. In C. Vear (Ed.), *The Routledge International Handbook of Practice-Based Research.* London and New York: Routledge.

Dewey, J. (1933). *How We Think: A Restatement of the Relation of Reflective Thinking to the Educative Process.* Boston, MA: D.C. Heath Co Publishers.

Dybå, T., Kitchenham, B. A., & Jørgensen, M. (2005). Evidence-Based Software Engineering for Practitioners. *IEEE Software*, 22(1), 58–65.

Elkins, J. (Ed.) (2009). *Artists with PhDs: On the New Doctoral Degree in Studio Art.* Washington, DC: New Academia Publishing, 1st ed.

Glinkowski, P., & Bamford, A. (2009). *Insight and Exchange: An Evaluation of the Wellcome Trust's Sciart Programme.* London: Wellcome Trust.

Gray, C. (1998). *Inquiry Through Practice: Developing Appropriate Research Strategies.* Keynote speech in Proceedings of No Guru, No Method? Discussions on Art and Design Research, University of Art and Design, UIAH, Helsinki, Finland, 82–95.

Jarząbek, S., Poniszewska-Marańda, A., & Madeyski, L. (Eds.). (2020). *Integrating Research and Practice in Software Engineering.* Germany: Springer.

Jefferies, J. (2022). Research Doctorates in the Arts: A Goldsmiths Perspective. In C. Vear (Ed.), *The Routledge International Handbook of Practice-Based Research.* London and New York: Routledge.

John, P. (2022). Using the Vitae framework to Guide Supervisors and Well-Being. In C. Vear (Ed.), *The Routledge International Handbook of Practice-Based Research.* London and New York: Routledge.

Kjørup, S. (2011). Pleading for Artistic Plurality: Artistic and other Kinds of Research. In M. Biggs & H. Karlsson (Eds.), *The Routledge Companion to Research in the Arts*. London and New York: Routledge, Chapter 2.

Leitch, R., & Day, C. (2000). Action Research and Reflective Practice: Towards a Holistic View. *Educational Action Research*, 8(1), 179–193.

Mackey, A., & Bassendowski, S. (2017). The History of Evidence-Based Practice in Nursing Education and Practice. *Journal of Professional Nursing*, 33(1), 51–55.

Mackh, B. (2022). The PhD in Visual Arts Practice in the USA: Beyond Elkins' Artists with PhD. In C. Vear (Ed.), *The Routledge International Handbook of Practice-Based Research*. London and New York: Routledge.

Mair, A. (1973). *Sir James McKenzie, MD, 1853–1925: General Practitioner*. Edinburgh and London: Churchill-Livingstone.

Mantai, L. (2019). A Source of Sanity: The Role of Social Support for Doctoral Candidates' Belonging and Becoming. *International Journal of Doctoral Studies*, *14*, 367–382.

McCormack, J. et al. (2022). Practice-Based Research at SensiLab. In C. Vear (Ed.), *The Routledge International Handbook of Practice-Based Research*. London and New York: Routledge.

McIntosh, P. (2010). *Action Research and Reflective Practice: Creative and Visual Methods to Facilitate Reflection and Learning*. London: Routledge.

Michaels, J. (2022). Interdisciplinary Perspectives on Practice-Based Research. In C. Vear (Ed.), *The Routledge International Handbook of Practice-Based Research*. London and New York: Routledge.

Phillips, M. (2022). The Academisation of Creativity and the Morphogenesis of the Practice-Based Researcher. In C. Vear (Ed.), *The Routledge International Handbook of Practice-Based Research*. London and New York: Routledge.

Schwarzenbach, J., & Hackett, P. (2017). *Transatlantic Reflections on the Practice-Based PhD in Fine Art*. London: Routledge.

Shaw, B. (2022). Working the *Space*: Augmenting Training for Practice-Based Research. In C. Vear (Ed.), *The Routledge International Handbook of Practice-Based Research*. London and New York: Routledge.

Stenhouse, L. (1975). *An Introduction to Curriculum Research and Development*. London: Heinemann.

Vaughan, S. (2022). Understanding Doctoral Communities in Practice-Based Research. In C. Vear (Ed.), *The Routledge International Handbook of Practice-Based Research*. London and New York: Routledge.

Vlugt, M. van der. (2022). Practitioner Voice 'Please Touch! In C. Vear (Ed.), *The Routledge International Handbook of Practice-Based Research*. London and New York: Routledge.

Wood, M., Mayo, F., & Marsland, D. (1986). Practice-Based Recording as an Epidemiological Tool. *Annual Review of Public Health*, 7(1), 357–389.

1.2

INTERDISCIPLINARY PERSPECTIVES ON PRACTICE-BASED RESEARCH

Jonathan Michaels

Introduction

In this chapter, I consider of some of the issues around the relationships between practice, research and knowledge, as they are understood in different disciplines, particularly considering the differences and intersections between them. In this process, I refer to examples from my personal experience in clinical healthcare practice and in undertaking, teaching and evaluating research across a range of subject and methodological areas related to healthcare, decision science and art. In doing so, I identify:

- Philosophical or practical differences in the ways in which such relationships are understood in the various disciplines
- Where such distinctions appear to arise primarily from alternative ways of framing comparable activities
- How and why the label 'practice-based research' may be applied in particular circumstances.

I highlight a number of areas for debate. These include how the nature and purpose of practice in different fields of endeavour is diverse, and how real or perceived dichotomies are often drawn between aspects of practice and research. On close examination of some such distinctions, I find that the boundaries are blurred or overlapping. Aspects that I explore in this chapter include questions of intention, generalisability, sharing, experimentation and the temporal and spatial relationship between practice and research. In addressing each of these, I consider aspects of paradigm, methodology, framing and knowledge production.

I attempt to unpack some of the difficulties and distinctions that create confusion in definitions of practice-based research across different fields of activity. I argue that, to be recognised as a distinct entity, practice-based research must be differentiated from both a form of practice that is not considered to be research and from other forms of research that are not deemed to be 'practice-based'. Such distinctions may have very different implications for research and practice in widely disparate disciplines. My analysis suggests that the differing embodiments of practice-based research do not share a specific set of features relating to the subject, location, process or outputs of research. Rather, they may reflect a challenge to the values attributed to these features by the academic establishment in particular, and wider society in general.

DOI: 10.4324/9780429324154-4   42

My background is that of an academic vascular surgeon and health services researcher, combining my clinical work with mixed methods and interdisciplinary healthcare research.[1] Some years ago, I retired from clinical practice and, alongside continuing some healthcare research, developed my interest in three-dimensional art, studying it for the first time as an academic subject and exploring computer-generated and multi-media installation work. This has been a challenging journey. My personal experience of art has been that it is a source of insight and knowledge, provoking contemplation and reflection on unfamiliar themes, leading me to explore and reframe my prior understandings. However, I have found that the underlying theory and implementation of research processes are often presented very differently to those that I had previously encountered. Both within and across disciplines, I have found different perceptions of the relevance, framing and value of particular research methods, objectives and outcomes.

Exploring these differing perspectives, through reference to my own experience and a review of published material, I argue that there are no specific identifiable criteria that can be used to provide clear demarcation between practice, practice-based research and other forms of research. I conclude that a common thread relates to value systems that privilege certain methodologies, professions, academic disciplines or other subgroups, in our epistemic endeavours. I also suggest that the investigation of these different perspectives can, in itself, be a source of new insights.

I begin by considering a number of aspects of research and practice, where specific characteristics that constitute practice-based research have been suggested. I argue that these do not form clear boundaries and the emphasis and understanding may vary between the different contexts in which the term is used. Specific topics that I consider in relation to these potential tensions are:

- Discipline and location[2]
- Intention and generalisability[3]
- Discipline-specific terminology[4]
- Research paradigms[5]
- Research methodology[6]
- Practice as research[7]
- Practice-based knowledge[8]

Discipline and location

Searches for terms that encompass practice-based research, in a large bibliographic database that includes scientific, arts and humanities publications, shows that the term first started to be used in a significant number of published papers in the 1990s, and its use has increased steadily over the past 30 years (Figure 1.2.1). Considering the different fields in which practice-based research has become prominent reveals that the greatest and earliest use of the term was in various fields of medicine and healthcare professions, with increasing use in education, creative arts and humanities in more recent years.

When I qualified in medicine in 1980, the scientific approach to medicine was in its infancy. Randomised controlled trials were rapidly increasing in popularity and becoming the accepted method of researching medical treatments, and the systematic use of research evidence to inform practice was being widely promoted.[9] In the earliest uses in the healthcare field, practice-based research could be seen as a reaction to the rise of 'evidence-based medicine' and a hierarchy of evidence that placed emphasis on the rigorously controlled experimental studies carried out in large academic institutions.[10] Part of the response to this was a view among some practicing clinicians that research evidence was becoming the province of 'ivory towers academics'.[11] Furthermore, it was seen as divorced from real-world practice, leading some clinicians to form networks to carry

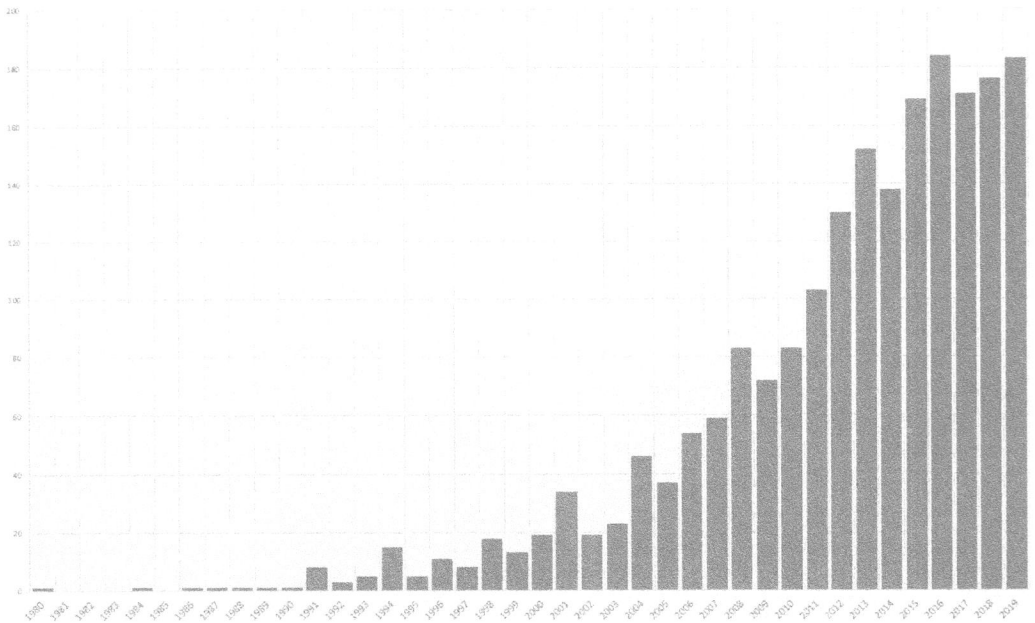

Figure 1.2.1 Number of publications identified in 'Web of Science' search with terms identified as related to practice-based research in the title, abstract or keywords by year of publication.

out research in practice settings.[12] Over the years, the term practice-based research has been associated with the development of practice-based doctorates. Fields in which such doctorates have subsequently become adopted include creative practice,[13] professions allied to health such as nursing,[14] occupational therapy[15] and psychotherapy,[16] as well as social work,[17] education[18] and business management.[19]

In 2011, the UK Council for Graduate Education commissioned a report to inform the debate relating to the development of professional doctorates that highlighted some of the confusion that exists between the requirements for conventional PhDs, practice-based PhDs in the creative arts and practice-based or professional doctorates in a variety of other professional fields.[20] The picture has become increasingly confused with the wide variety of routes to doctoral qualifications that blur the boundaries between practice-based, research-based and taught professional qualifications.[21] Thus, in the context of healthcare, and more recently in education and other professional fields, the emphasis in many publications is not on any specific features of research methodology or the underlying paradigm, but appears to be primarily a response to concerns that much research funding and effort has been devoted to the interests of a relatively small group of leading academics. While sometimes challenging conventional research methodology,[22] practice-based research was targeted mainly at ensuring direct applicability to practice, addressing ideas and questions that arose in practice[23] and locating the research in a practice setting.[24]

In the context of creative arts, much of the literature addresses the recognition of practice-based research as a distinct entity, focusing on doctoral research. Regulations frequently identify a separate doctoral route, aimed at subjects including art, drama, music and English, in which the practice, artefacts or documentation of that practice are an assessed component of the final submission, alongside a reduced written component.[25] In contrast, the documentation of practice, outcomes of practice and experimental results may be included as appendices in many other subjects.

In summary, the location of research would not appear to be a defining feature of practice-based research, with some using the term to emphasise the value of research carried out in a practice setting, while others emphasise the value of practice outputs and/or documentation in research carried out in academic establishments.

Intention and generalisability

Activities undertaken in many fields may be commonly described under the umbrella term of research. For example, a geologist may arrange test-drills to establish the underlying rock composition, a physician may undertake blood tests and radiological investigations to aid a diagnosis, an architect may carry out a site survey before planning a building, a journalist will investigate their subject matter for an article, a ceramicist might produce line-blends to find a required glaze or an artist may talk about 'researching' a location for a site-specific installation. While terms such as 'research', test, investigation or experiment may be applied to many of these activities, they may also be considered part of routine practice in the relevant disciplines.

Some differentiation may be related to the extent to which such 'research' may be generalised. For the most part, when such activities are carried out as a part of 'usual practice', they are aimed at addressing a specific situation rather than adding to a shared and generalisable body of knowledge. The physician investigates to improve their care of an individual patient, the journalist investigates to write a specific story, the artist researches to develop a particular artwork. However, such boundaries are blurred. The practitioner, faced with the findings of such activities, may be led to challenge received wisdom or formulate new theories.

This highlights the issues of generalisability and intention. In many fields, the intention of practice is to achieve a specific objective, whether it is the treatment of a health condition, the design of a building or the attainment of an educational outcome. In doing so, the practitioner uses an existing knowledge base and their experiential knowledge to the best of their ability. They may require some investigative techniques to achieve this, but there is no intention to generate new or generalisable knowledge.

Thus, a potential distinction may be drawn between knowledge that comes about as a by-product of 'routine' practice, and that where practice is undertaken, and potentially manipulated, with the express intention of generating knowledge. While this is more clearly seen in the difference between 'discovery' and 'experimentation', the boundary is not clear-cut. Experiential knowledge inevitably occurs through a cycle of repeated practice and observation of the results of that practice. Additional insights may be the result of subsequent reflection.

Within this broad categorisation, there are subtle nuances: as discussed later in relation to methodology, routine practice in many fields includes a component that promotes reflection, data aggregation, audit or review intended to advance the knowledge in that domain. There are also fields of activity where the primary purpose of 'practice' would appear to be the generation of new knowledge. It is hard to think of the practice of a particle physicist, an astronomer, an archaeologist or an anthropologist as being anything other than research.

Discipline-specific terminology

I suggest that it is fundamental that practice is inseparable from research or knowledge generation in a particular field; the knowledge that underpins any specific discipline must relate to the practice of that discipline. There can be no research without practice, and even claims to new knowledge that are founded on theoretical arguments can only be substantiated or validated through practice. Furthermore, the potential value of new knowledge would seem inextricably linked to any impact that it might have on future practice. This raises the question of whether practice-based research is a unique and separate category of research, or simply a particular way in which some forms of research are framed in the language of specific disciplines.

This point may be illustrated by comparing the role of practice in the development of a novel product or procedure in two different fields. A surgeon, after managing the same condition on many occasions, may develop a new surgical procedure that can be used to treat the condition. Similarly, the artist or musician may develop an original artwork, performance or composition that they consider, by some criteria, to be successful. Without further supporting evidence, or some additional research and development, it seems unlikely that either the description of a novel surgical procedure or the creative output of the artist or musician would be considered to be a 'significant contribution to knowledge'.

Were the surgeon to go on to repeat the new procedure, identifying generalisable findings about the suitability and advantages of the new treatment, or collecting information about outcomes or patients' responses, then this would be likely to be considered to be research that might lead to a contribution to knowledge. Similarly, were the artist or musician to go on to repeat or review the work, undertake performances or exhibitions and share generalisable conclusions about the aspects that constituted 'success' or to evaluate audience responses in some way, then this might well be considered to be practice-based research. It is difficult to see a clear conceptual distinction between the role of the novel surgical procedure and the novel artwork, or their documentation, in relation to a contribution to knowledge.

Early in my training as a vascular surgeon, I undertook a period of research and completed a thesis on the subject of laser angioplasty.[26] At the time, there were rapid developments in medical lasers and my research considered their use in unblocking arteries to improve circulation to the legs. Balloons were being used increasingly to open up narrowed arteries without the need for a major operation, but it was often difficult to get them into place in a completely blocked artery. Lasers offered the potential to open up a passage for the balloon device, leading to a research question that arose from a significant problem in practice. My research involved laboratory experiments, followed by the use of lasers in clinical practice to open up such blocked arteries.[27] Part of the definition of practice-based research used in this handbook describes 'a principled and original investigation undertaken in order to gain new knowledge by means of practice and the outcomes of that practice'. Thus, despite never being described as such, it seems to me that my research meets this definition of practice-based research, as might many other examples of clinical research in healthcare.

There are, however, clear regulatory differences at those universities that offer a distinct route to a practice-based doctorate. The main difference is that the practice, or documentation of that practice, may be included as part of the assessed contribution to knowledge. This raises the question of the basis upon which an artwork, composition or performance can be assessed as a contribution to knowledge. While they appear to be illustrative of a thesis, or to represent a source of the data that underpins the contribution to knowledge, this seems little different from the examples or documentation of practice that accompany a thesis, often as appendices, in fields where there is no route to a practice-based doctorate.

In contrast, the 'professional doctorates' offered for a variety of professional subjects in many universities around the world[28] frequently have a significant 'practice-based' research component. These may use a variety of quantitative and qualitative methodologies,[29] but are also likely to have a significant taught component,[30] and the practice element may be assessed as advanced professional practice, rather than as a 'contribution to knowledge'.[31] In conclusion, the regulations for the award of practice-based and professional doctorates suggest that various disciplines have very different approaches to the relationship between research and practice in higher education.

Research paradigms

It has been suggested that practice-based research in the creative arts is based upon a separate and distinct research paradigm.[32] However, across the broad range of professional subjects in which the term practice-based research is used, it is clear that a range of research paradigms are in play.[33] The Oxford English Dictionary defines a paradigm as 'a world view underlying the theories and methodology of a particular scientific subject.'[34] In contrast, Guba, in his book *The Paradigm Dialog* (1990), states:

> I will use the term in this chapter only in its most common or generic sense; a basic set of beliefs that guides action, whether of the everyday garden variety or action taken in connection with a disciplined inquiry.[35]

He goes on to define four paradigms: positivism, post-positivism, critical theory and constructivism, in which particular ontological, epistemological and methodological questions are answered based upon an underlying set of beliefs.

Gray and Malins build upon Guba's summary of these paradigms in asking what constitutes an 'artistic' or 'designerly' paradigm of inquiry.[36] They suggest that

> from these basic philosophical positions, it is clear that researchers have been characteristically eclectic, diverse and creative in the methodologies they have adopted. When necessary, they have drawn on positivist experimental methodologies, constructivist interpretation and reflection, and invented hybrid methodologies involving a synthesis of many diverse research methods and techniques. So a characteristic of 'artistic' methodology is a pluralist approach using a multi-method technique, tailored to the individual project.

From my perspective, this is a bit of a magpie approach, taking what is attractive from other paradigms, which seems at odds with the concept of paradigms as a 'basic set of beliefs'. The categorisation of specific disciplines into Guba's four categories is contrary to my experience across several disciplines. It is frequently the case that my research has involved a variety of methodologies that appear to stem from different paradigms. Mixed methods research, using a combination of quantitative and qualitative methodologies, has become common in healthcare[37] and other fields,[38] although some have expressed concerns about this development.[39] The growth of requirements for patient and public involvement in all aspects of the research process,[40] despite its limitations,[41] would appear to be a recognition of the situatedness of knowledge and the need to involve differing perspectives in its generation and interpretation.

Guba suggests that 'as a constructivist I can confidently assert that none of these four is the paradigm of choice. Each is an alternative that deserves, on its merits (and I have no doubt that all are meritorious), to be considered'.[42] However, if one sees a paradigm as being a reflection of a 'world view' or an underlying belief system, then it is difficult to see how an individual can reconcile these apparently contradictory paradigms. Furthermore, Guba subsequently states 'accommodation between paradigms is impossible . . . we are led to vastly diverse, disparate, and totally antithetical ends'.[43]

An alternative way of thinking, to which I would subscribe, is to see the different paradigms of inquiry as theoretical models of the principles underlying knowledge production more generally. As George Box said, 'All models are wrong, but some are useful'.[44] As a model, there is no requirement that an individual prescribes to a particular belief system or a single paradigm but, as in the description by Gray and Malins of artistic research, they are free to pick and choose the model that best fits a particular circumstance. Thus, I would not state, as Guba does, that 'I am a constructivist', but rather that there are situations in which I find a constructivist model of knowledge production to be useful.

In recent years, the main focus of my research has been on the difficult issue of the ways in which health services are organised. This has involved multiple research methods ranging from the statistical analysis of huge and complex national datasets to individual interviews with service users to understand their priorities and preferences in respect to service delivery.[45] Over the years I have carried out research with others from a range of specialities, including healthcare professionals, basic scientists, economists, philosophers, ethicists and artists. My experience has been that researchers are often very flexible and open to a variety of research models that are appropriate to differing circumstances, particularly when working across disciplines.

In conclusion, it is evident that a variety of research paradigms are in play across the professional and creative fields in which the term practice-based research is used. It would thus appear that there is no specific paradigm that can be considered necessary or sufficient to characterise practice-based research.

Research methodologies

As discussed in the previous section, doctorates in practice-based research make use of many different research strategies, so it is difficult to define a specific methodology that is characteristic of practice-based research. However, if there is a contender that seems most associated with practice-based research, it seems to me that this is in the area of reflective/reflexive practice[46] or action research. These seem to be examples of a range of conceptually similar descriptions of research methodology that are represented as cyclical processes. There are at least three basic elements that, depending on the model, may be split into other components. These three primary components are:

- Doing, acting, practicing, making etc. in which some action is taken that is part of professional practice in a specific discipline
- Experiencing, observing, recording, documenting etc. in which some form of data is identified or derived from the previous activity
- Reflection, conceptualisation, analysis etc. in which the experience or observations are processed and analysed in some way to derive insights, understanding or new knowledge

These basic structures are seen within Schön's reflective practice,[47] Gibbs reflective cycle,[48] Kolb's experiential learning cycle,[49] Greenaway's active reviewing cycle,[50] Argyris's double loop learning[51] and the various models of action research.[52] In some respects, these generalised components that define the cyclical process might be applied to any research method. But there are conceptual, practical and ethical differences between a research process in which the primary aim of the 'action' or practice is the generation of knowledge or the testing of a generalisable hypothesis, and one in which the purpose of practice remains the unique circumstances of the specific case or situation under consideration, whether this is a patient, a client, an artwork or design.

This difference is well illustrated in healthcare, where a clear distinction is made between 'audit' and 'research'.[53] Multi-disciplinary audit or 'mortality and morbidity' meetings have been a common part of medical practice for many years, although the term 'reflective practice' has only recently entered the vocabulary.[54] These are intended to be processes to review and learn from experience and I have little doubt that these processes can lead to new knowledge and understanding. However, considerable emphasis is placed on differentiating between audit and research in the healthcare field, perhaps largely due to the ethical implications of labelling practice as research. Thus, the principal difference would appear to be that audit does not involve 'experimentation': the manipulation of practice in order to meet research objectives.[55]

As with many issues in this area, things are never black and white, and the distinction between experimental research and practice is not so clear-cut. Experimentation can be a fundamental aspect of practice in many fields: the clinician may undertake a therapeutic trial, the psychotherapist may experiment with different interpretations, the artist or architect may experiment with sketches and models, all as part of their professional practice. However, while such experiments may lead to insights, perhaps resonating with other experience and leading to generalised and generalisable knowledge, the prime motivation for such experiments is the matter in hand, the patient, client or design that presents an issue to be addressed.

These may be related to the different forms of practical experimentation that Schön describes as taking place in professional practice: exploratory, move-testing and hypothesis testing.[56] These together form part of the process that he categorises as 'reflection-in-action', as opposed to 'reflection-on-action'.[57] In Schön's model of reflective practice, which distinguishes 'reflection-in-action' from 'reflection-on-action', although there may be temporal differences, both are presented as orderly cycles of discrete events in which practice is followed by reflection. The more formalised collective processes of audit in healthcare, or the reflection that takes place in clinical supervision, may look back on past events in order to inform future practice, which may be characterised as reflection-*on*-action. In contrast, reflection-*in*-action may need to occur rapidly in the heat of practice, as in Schön's example of the skilled tennis player who takes a moment to plan their next shot.[58]

In Linda Candy's chapter in this handbook, 'Reflective Practice Variants and the Creative Practitioner', she extends Schön's model based upon her work with creative practitioners[59] and categorises four kinds of reflection in the creative process:

- Reflection-for-action
- Reflection-in-the-making-moment
- Reflection-on-surprise
- Reflection-at-a-distance

Although this was developed in respect to reflection in creative practice, much of it resonates with my personal experience in vascular surgery. Reflection-for-action and reflection-at-a-distance are reminiscent, respectively, of the processes of multi-disciplinary team meetings to discuss and plan treatment, and the mortality and morbidity meetings, at which we would reflect upon cases with poor outcomes, to see what could be learnt. Reflection-in-the-making-moment seems analogous to the many decisions that are required in the course of a surgical procedure, which call upon experience to solve any problems that may be encountered.

In surgical practice, reflection-on-surprise may take place in the moment, such as when opening an abdomen reveals an unexpected problem, or at a distance, when something may be learnt from reflection upon unexpected outcomes. This does, however, highlight a possible distinction between such reflection in creative and professional practice. As Candy highlights, many creative practitioners deliberately seek surprise and unpredictability and may be free to follow and develop upon an unexpected event in their creative process. In contrast, the professional practitioner is very much 'ends-driven', in that their actions are directed at achieving a desired outcome, and they must work within strict regulatory and professional boundaries.

Each of these forms of reflection may be recognisable in professional practice. For example, as a vascular surgeon, there were many instances when I would carry out an operation, opening an abdomen in an emergency, in the expectation of finding a particular cause of bleeding, only to be surprised to find a different cause. Reflection-in-the-making-moment may result in an immediate re-thinking of the plans for the current operation. Reflection-on-surprise may cause me to re-examine the basis of my diagnosis and consider whether the event may have been predictable. Reflection-at-a-distance may take place when I discuss the case with colleagues at a morbidity and mortality meeting, to see what lessons may be learnt for future cases.

In summary, although there are particular reflective research processes that would appear to be more prominent in the literature on practice-based research methodology, these are not exclusive to practice-based research or necessary to characterise such research.

Practice as research

Another possibility that may distinguish the role of reflection in practice-based research is that there is a further category of reflection-*through*-action, perhaps analogous to Frayling's research *through* art.[60] For the individual creative practitioner, it may be the case that practice, experience and reflection are occurring simultaneously. It may also follow that these maybe haphazard and unpredictable and that part of the mechanism of reflection is directly through their creative activities, performance or manipulation of materials. This suggests a potential difference between the role of practice in creative arts, as opposed to other disciplines, where the emphasis is on professional practice. However, it could be argued that the trial and error and haptic feedback through which a physical therapist might develop new insights, or the testing of interpretations made by a psychoanalyst, offer a similar potential for knowledge generation through practice.

The numerous roles for practice may help to explain the apparent difference that I perceive in its status as an aspect of research in various disciplines. Many artists and performers see their creative practice as a way of exploring the world through models and metaphors,[61] potentially leading to new understandings, reminiscent of Alva Noë's description of art as a 'strange tool'.[62] This may also help to explain the difficulty in differentiating artistic practice from practice-based research.[63] Artistic practice takes many forms and may have a variety of objectives. If, under certain circumstances, artistic activity is instantiated as a reflective process leading to new insights or understanding, then it becomes difficult to distinguish this on a conceptual basis from other forms of research, other than through the way that it is formalised, articulated or shared.[64]

Practice-based knowledge

Another possible distinction between practice-based research and other forms of research is the potential for a distinct form of 'practice-based knowledge' that is the subject of practice-based research. This idea seems to be consistent with the suggestions of a number of those writing about practice-based research in the creative arts. For example, Nelson refers to 'know-what' as distinct from 'know-how' and 'know-that',[65] Sullivan distinguishes between knowledge and 'understanding' or 'creative insights'[66] and others have described practice-based research as producing embodied, tacit, haptic or experiential knowledge.[67] Such definitions of knowledge are also relevant to other disciplines, where observation of a skilled professional may clearly demonstrate their personal knowledge, despite the fact that they may be unaware of or unable to articulate it.

Defining knowledge is no easy task, and we use the term in many different ways in different circumstances.[68] What is understood as knowledge in terms of doctoral research outputs may be very different from the way that we talk about 'knowing something' in everyday conversation. I may claim to 'know' how to carry out complex surgical procedures, who won the 1966 World Cup, what foods I like and the difference between right and wrong. The classical philosophical view of knowledge as 'justified true belief'[69] would seem contrary to most current thinking and common usage; 'truth' and 'belief' are terms more likely to be heard in religious and political circles than in academic departments. In philosophy, the field of epistemology is seen as distinct from moral philosophy, which deals with issues of value. However, a more constructivist approach breaks down this clear distinction with 'facts' considered to be value-laden and values being open to scientific investigation. In health services research, there is an expanding area of 'utility analysis' concerned with investigating values and preferences in order to inform decision-making.[70]

In the context of practice-based research, I suggest that it is helpful to distinguish between data and knowledge and between personal and shared knowledge. Research methods are sometimes equated with the processes for observing or generating data. Whether such data is the numerical output from a physics experiment, transcripts from semi-structured interviews or a series of experimental artworks, for these to contribute to knowledge there must be a process of analysis, interpretation or meaning making. Every object, experience and process may act as a source for data, sometimes easily observed, such as physical form or temporal and spatial relationships, sometimes requiring more in-depth investigation, such as chemical properties or emotional effects.

An artwork may be a rich source of data about the work itself, the production method, artistic process or subject matter, from which the artist or others who encounter the work may derive meanings or generate knowledge. In some cases, the data may reflect the intention of the maker, aiming to convey a particular meaning or provoke consideration of a specific subject, such as raising awareness of a social or political issue. Often, the meaning that is made may be unintended or unexpected. A fashion student may use artwork as material for a dissertation about international comparisons in eighteenth-century costume, or a doctor may make a spot diagnosis on the basis of reviewing a historic portrait, but in neither case can the artist or artworks be considered to possess the knowledge, only the data upon which the knowledge is founded.

Although this distinction between the elements of data and knowledge may be of little importance in everyday practice, I suggest that they may take on particular significance when considering the place of artefacts or documentation of practice as aspects of the contribution to knowledge in academic research.[71]

For me, knowledge is a fundamentally human process that involves the development of a set of models and metaphors.[72] Through these we can understand the way that the world in which we live operates, and how we interact with it and with the others who inhabit it. These models are generalised and generalisable representations of our personal reality that may arise through conscious and unconscious processes. They are informed by data from many sources: our own instinct, experience or observation and data, with or without proposed interpretations, that are communicated to us by others. As models, to build on George Box's premise, they are always simplified, inaccurate, incomplete and provisional, but may be potentially useful. However, they may also be misleading, biased, misused and the source of epistemic injustices.[73]

When I claim to 'know what I like', I am reporting that I have an internal 'model' of things that I find pleasurable. Suggesting that I like chocolate and dislike tomatoes is an aspect of this generalised model. This is not fully accurate or consistent, as I know of tomato-containing foods that I enjoy and can imagine chocolate products (not many) that I would dislike, and yet the model may be 'good-enough' to help inform my decision-making when faced with a restaurant menu. When I state that I 'know' a person, I have an internal model that allows me to recognise that person's face from visual data, their voice from auditory data and, perhaps, to predict (with a varying degree of accuracy), their behaviour in response to certain circumstances.

In both of these examples, the 'knowledge' is tacit or personal, in the terms described by Polanyi,[74] in that I have come to know it without a conscious rational process and am unable to articulate it fully. However, there are different processes at work – facial recognition is a complex task of pattern recognition. Although I may be unable to articulate my method of facial recognition, it is a cognitive task based upon observable data. In the early days of computers, it was generally considered that they would be unable to undertake complex tasks of pattern recognition, such as facial recognition, voice recognition and deciphering of spoken or written language. This has proved to be untrue, not because we have been able to understand and codify our methods of facial recognition, but because we have been able to develop machine learning algorithms that may allow computers to simulate the way in which we develop our own models of the world.

Even if I am unable to articulate my process of facial recognition in words, if I were a good portrait artist or working with a photo-fit, I may be able to articulate it in a way that allows me to communicate this knowledge and come to shared understanding with others. In contrast, my food preferences are personal experiences, which I cannot communicate in a way that I would expect to lead to a shared understanding of what constitutes a pleasurable taste. I may be able to describe these through language or illustration and potentially influence the preferences of others, perhaps even using artistic methods that might convey the experience in a more accessible way than a literal description.[75]

'Know-how' or practical knowledge is another form of personal knowledge which may be difficult to communicate. I may be profoundly affected by an artistic work or performance with which I find resonances. Observing a skilled professional at work may cause me to marvel at the skills and knowledge that they demonstrate through their actions. However, this may be dependent upon the context in which the work is encountered and my prior experience: to describe this as 'shared knowledge' seems quite a stretch and would place all such activities, performances and products as forms of knowledge sharing. An artist, performer or professional practitioner may be communicating their know-how through such works but, as an audience, an encounter with the work does not allow you to acquire that know-how. Thus, in summary, I would suggest that there is no specific form of practice-based knowledge that is exclusive to practice-based research, but that in discussing knowledge in all contexts, it may be helpful to differentiate between data and knowledge and between the demonstration and sharing of personal knowledge.

Interdisciplinarity

Over 60 years ago, C. P. Snow published his original article entitled 'Two Cultures',[76] lamenting the divisions that had arisen between the sciences and humanities. He related this to the British educational system in which, as he described it, 'intense specialisation, like nothing else on earth, is dictated by the Oxford and Cambridge scholarship examinations'.[77] Since that time, far from breaking down the barriers, I have observed increasing sub-specialisation in the academic world, with academic and commercial competition producing a culture of highly specialised academics working in separate silos.[78] This becomes evident in a failure to accept knowledge from external sources'[79] distorted patterns of citation[80] and may contribute to a failure of collaboration across disciplines.[81]

Experimental research is often designed to isolate an underlying mechanism or create carefully controlled conditions that allow the effects of a specific intervention to be measured without confounding factors, as in the randomised controlled trial that is often seen as the 'gold standard' in healthcare.[82] However, professionals working outside academia are called on to deal with a messy world, where clients, patients, students or customers present with complex requirements that frequently cross professional and disciplinary boundaries. At least a part of the impetus for practice-based research would appear to be a view that academic research was becoming too specialist and divorced from the concerns of practicing professionals.[83]

Practice-based research in the creative arts has been described as inherently interdisciplinary in that it 'draws on combinations of different methods, approaches and subjects, and invariably involves a number of different perspectives that the artist/researcher has to combine'.[84] In reviewing a number of leading texts on practice-based research in the arts, Cazeaux suggests that the general thesis in them all is that it is this inherent interdisciplinarity that provides the opportunity for practice-based research in the arts to generate new knowledge.[85]

The terms that are used to describe the interactions between disciplines vary, but a distinction is drawn between multidisciplinarity, interdisciplinarity, and transdisciplinarity. Rosenfield characterises these respectively as 'working in parallel or sequentially', 'working jointly but from a disciplinary-specific basis' and 'working together from a shared conceptual framework'.[86] Choi and

Pak describe these approaches in simple terms as 'additive', 'interactive' or 'holistic'.[87] A discipline is held together by a shared epistemology that defines the assumptions about the nature of knowledge and the ways in which it is generated.[88] In contrast, professions draw on many disciplines, often requiring a grounding in several fields across sciences, social sciences and humanities or collaborating with other professionals with different disciplinary backgrounds.

It is, perhaps, the cross-disciplinary nature of practice-based research that provides the opportunity for a particular type of knowledge generation. Cazeaux suggests that it is when the perception of a situation from two different disciplinary perspectives collides that the opportunity for new understanding arises, through not just 'seeing things differently' but also introducing new concepts to make sense of the transition.[89] Choi and Pak introduce the idea of 'inter-discipline distance', suggesting that the greater the epistemological dissonance between disciplines, the more likely it is that multiple disciplinary thinking will lead to new insights for complex problems.[90]

This principle can be illustrated in an example of my own practice-based research. When I moved into an academic art setting, at the same time as continuing my own research within healthcare, I found that there was both a dissonance and a cross-fertilisation of ideas. My experience of computer modelling informed my artistic practice and reflection on my creative processes raised issues that resonated with, and informed, my understanding of healthcare issues. An example was my software implementation of a Sol LeWitt wall drawing, which resulted in a reflective process that informed my understanding of hierarchies of evidence and the handling of uncertainty in healthcare.[91]

Conclusion

Considering the characteristics of practice-based research in the context of the relationship between research and practice in different disciplines highlights a number of issues. I have argued that there is no single set of criteria that can be identified as necessary or sufficient to characterise practice-based research. The relationships between practice-based research and practice, on one hand, and between practice-based research and 'traditional' research, on the other, are fuzzy, with blurred boundaries in many respects.

One aspect that is most evident is the political dimension, with practice-based research appearing to arise as a challenge to established research practices in different fields. However, even in this, the drivers appear to be diametrically opposed. In the case of some professions, such as those related to healthcare and education, the motivation for practice-based research appears to have come from a desire to re-establish the professional practice setting as the site for research that is based upon, relevant to, and implemented in professional practice, rather than meeting the aims and needs of the academic institutions.

In contrast, in the creative arts the motivation appears to come from a desire to give practice-based research an academic legitimacy and establish a scholastic pedigree within the higher educational setting for subjects that have previously been under-valued. Thus, it is not the epistemological issues that come to the fore but rather questions of value, relating to the esteem in which different methods of research and modes of knowing are held by the academic and professional communities and wider society. There are exceptions to this, as there are some creative disciplines with a long pedigree of academic research.[92] However, many of the creative arts are relatively new to the postgraduate academic environment. Changes such as the incorporation of polytechnics and art schools into the university system in the UK were followed in the 1990s by the introduction of a Research Assessment Exercise that created funding incentives for particular types of research outputs in creative disciplines. Similar trends have been seen in other countries, changing the landscape of academia.[93]

In the final analysis, the thread that binds together the various embodiments of practice-based research may be not a shared paradigmatic, methodological or epistemological underpinning but rather a response to the epistemic injustice inherent in a value system that privileges the testimony and knowledge of the sub-specialist academic expert.[94] For example, the creative practitioner, despite specialist expertise in their chosen medium, may work across disciplines, approaching other fields as a 'non-expert' and offer insights from an alternative perspective to the academic establishment. The practicing professional may generate knowledge through a pragmatic and holistic approach that is at odds with the exclusivity, reductionism and purism of academia.

Over 30 years ago, Vandana Shiva used the term 'epistemological violence'[95] to describe the potential injustices that may be introduced between the scientific expert and non-expert:

> Here violence is inflicted on the subject socially through the sharp divide between the expert and the non-expert – a divide which converts the vast majority of non-experts into non-knowers even in those areas of life in which the responsibility of practice and action rests with them.
>
> But even the expert is not spared: fragmentation of knowledge converts the expert into a non-knower in fields of knowledge other than his or her specialization.[96]

It is, perhaps, an increasing recognition of the potential negative consequences of these growing divides – between rigid disciplinary boundaries, between expert and non-expert, and between researcher and practitioner – that is the driving force behind the emergence of practice-based research.

Lessons to be learnt

This chapter is a personal reflection on practice-based research from my experience of professional and academic practice within the healthcare sector and, to a lesser extent, in academic art. However, there are some potential lessons that may be learnt through reflecting on these issues.

The first, which will be familiar to anyone who has worked across disciplinary boundaries, is that it cannot be assumed that the use of similar terminology to describe an academic practice implies a shared understanding, similar motivation or a common theoretical framework underpinning this practice.

Another important lesson relates to the way that interdisciplinarity has the potential to enhance and contribute to the generation of new knowledge. When two or more disciplines come together to provide different perspectives on a subject, it should not be with the aim of reaching a consensus or accommodation between them. There may be little to be learnt if this simply results in the dominance of one or other viewpoint or a weak generalisation that avoids dissent. More may be learnt from a synthesis of the different perspectives that can accept and accommodate multiple viewpoints.

Perhaps the greatest opportunities for the generation of new and sometimes unexpected insights arise through these clashes of perspective and the negotiation of positions. The ability to jointly undertake a careful forensic examination of, and to reflect on, these tensions and differences in perception may create an environment that is most conducive the furthering of knowledge. This potential to learn from careful consideration of the implications of differences in perspective is not restricted to distinct disciplines. It may apply equally to the benefits of scrutinising the gulfs that can develop between academics and practitioners, or any other distinct factions, within a discipline.

Notes

1 My main research interests have related to health service configuration, medical decision-making and research methodology. Through this, I have been involved in many mixed methods studies that have taken interdisciplinary approaches, supervised research students in various fields and chaired prioritisation and review panels for allocating research funding.

2 For example, the development of practice-based research networks in community care has been seen as an attempt to relocate research away from academic tertiary referral centres Westfall et al. (2006, pp. 8–14).

3 For example, see the AHRC definition of research at: AHRC (2020, p. 10).

4 One example of this is the distinction that is drawn between research and audit in healthcare discussed later; see: Closs and Cheater (1996, pp. 249–256).

5 For example, McIntosh uses the term 'practice-based evidence' as a counter to a 'deity of technical rationalism' McIntosh (2010, p. 23).

6 Practice-based research has been described by some as restricted to particular methods, for example to exclude randomised experimental studies, e.g. Epstein and Blumenfield (2001, pp. 17–18)

7 Nelson, for example, explores practice itself as a form of research that generates experiential knowledge: Nelson (2006, pp. 105–116)

8 For example, see Borgdorff (2011).

9 In 1971, Archie Cochrane had published Cochrane (1972), which had been hugely influential in the development of evidence-based medicine.

10 The earliest use of an evidence hierarchy is generally thought to be from a Canadian Task Force report (e.g. Hill et al. 1979), but there has subsequently been considerable work in developing such hierarchies in numerous areas; for example, see the Centre for Evidence Based Medicine at: https://www.cebm.ox.ac.uk/resources/levels-of-evidence/ocebm-levels-of-evidence (accessed October 8, 2021).

11 The idea of academics being in 'ivory towers' and divorced from the realities that practitioners face in the workplace would appear to be at the root of many of the origins of practice-based research. For an interesting discussion of the origin of the term, see Shapin (2012).

12 Wood et al. (1986).

13 The role of the doctorate in creative practices is a fairly recent innovation that partly stems from the move of creative subjects into the university academic setting. For example, see Frayling (1993) and Elkins (2009).

14 Edwards (2009).

15 Morley and Petty (2010).

16 McBeath et al. (2018).

17 Hothersall (2018).

18 Armsby et al. (2017).

19 Clegg et al. (2018), Morley and Banerjee (2013).

20 Fell et al. (2011).

21 For example, for a discussion of the variety of routes available for a doctorate relating to nursing, see Edwards (2009). A more general discussion of the inconsistencies in such qualifications can be found in Robinson (2018).

22 Giddings (2006).

23 Edwards (2009).

24 For example, similar concerns have been expressed in public health: Margaret A. Potter et al. (2006); and in education: Eunsook Hong and Lonnie Rowell (2018).

25 For example, see Katy Macloed and Lin Hildridge (2004).

26 Michaels (1989).

27 Michaels et al. (1989).

28 Felly Chiteng Kot and Darwin D. Hendel (2012).

29 Robin Mellors-Bourne et al. (2016).

30 Armsby et al. (2017).

31 Martin Eubank and Mark Forshaw (2018).

32 James Haywood Rolling (2015).

33 Richard Winter et al. (2000).

34 Oxford English Dictionary, *Paradigm, N.* https://en.oxforddictionaries.com/definition/paradigm (accessed October 8, 2021).

35 Egon G. Guba (1990).

36 Carole Gray and Julian Malins (2004, p. 20).

37 Louise Doyle et al. (2009).

38 Charles Teddlie and Abbas Tashakkori (2011).

39 Giddings (2006).

40 For example, INVOLVE was set up in 1996 and has been funded by the NHS and National Institute for Health Research to encourage patient and public involvement in NHS research. For further details, see their website at: www.invo.org.uk (accessed October 8, 2021).

41 Carole Mockford et al. (2012).

42 Guba (1990, p. 27).

43 Ibid., p. 81.

44 This is a frequently cited quotation in statistics and decision science attributed to the statistician George Box, who used it as a chapter title in George E. P. Box, *Robustness in the Strategy of Scientific Model Building* (Madison: Wisconsin University, Mathematics Research Center: Defense Technical Information Center, 1979).

45 Jonathan Michaels et al., (2021).

46 Editor note: a deep discussion on the merits or reflective practice in practice-based research can be read in Chapter 3.8 of this volume by Linda Candy's 'Reflective Practice Variants and the Creative Practitioner'.

47 Donald A. Schön (1991).

48 G. Gibbs (1988), Reprint, Kitchen (1999).

49 David A. Kolb (2014).

50 Roger Greenaway (2004).

51 Chris Argyris (1977).

52 For example, see http://cei.ust.hk/teaching-resources/action-research (accessed October 8, 2021) for an introduction to action research.

53 Closs and Cheater (1996).

54 Beverley Taylor (2010).

55 Closs and Cheater (1996, pp. 251–252).

56 Schön (1991, pp. 145–146).

57 Ibid., p. 61.

58 Ibid., p. 279.

59 See Chapter 3 in Linda Candy, *The Creative Reflective Practitioner: Research Through Making and Practice* (London: Routledge, 2019).

60 Frayling (1993).

61 Annette Arlander (2010, pp. 315–332).

62 Noë (2015) states: 'Works of art are strange tools. Technology is not just something we use or apply to achieve a goal, although this is right to a first approximation; technologies organize our lives in ways that make it impossible to conceive of our lives in their absence; they make us what we are. Art, really, is an engagement with the ways in which our practices, techniques, and technologies, organize us and it is, finally, a way to understand that organization and, inevitably, to reorganize ourselves'.

63 H. Borgdorff (2007).

64 Kathrin Busch (2009).

65 Robin Nelson (2013, p. 44).

66 Graeme Sullivan (2010, p. 96).

67 For a detailed discussion of this area, see Borgdorff (2011).

68 Editor note: Ernest Edmonds's chapter, 'Research, Shared Knowledge and the Artefact', discusses this argument in more detail.

69 The idea of 'justified, true belief' as a definition for knowledge stretches back to Plato. For further details see, for example, Ichikawa and Steup (2014).

70 Torrance (2006).

71 For example, Büchler writes 'the artefacts produced should have an essential role in the conduct of the research, and as a result that the research could not be conducted or communicated without them' Büchler, Biggs, and Ståhl (2011, pp. 318–327).

72 For an excellent discussion of the way that metaphors are fundamental to our thinking, see George Lakoff and Mark Johnson (2003).

73 Miranda Fricker (2007).

74 Michael Polanyi (1998).

75 For example, try 'Chocolate Cake' in Rosen and Blake (2011).

76 C. P. Snow (1956, pp. 413–414). These ideas were subsequently explored further in his 1959 Rede Lecture.

77 Snow (1959) at 20.

78 Hershey H Friedman and Linda Weiser Friedman (2018).

79 David Antons and Frank T. Piller (2015).

80 Malcolm Tight (2014).

81 Arthur M. Feldman (2008).

82 Laura E. Bothwell et al. (2016).

83 L. W. Green (2008, pp. i20–4).

84 Jane Messer (2012), partially quoting from Sullivan (2010).

85 Clive Cazeaux (2008).
86 Patricia L. Rosenfield (1992).
87 Bernard C. K. Choi and Anita W. P. Pak (2006).
88 Bernard C. K. Choi and Anita W. P. Pak (2008).
89 Cazeaux (2008, p. 128).
90 Choi and Pak (2008, p. E43).
91 A paper regarding this was presented at the Electronic Visualisation and the Arts Conference 2017. See Jonathan Michaels (2017)
92 For example, music has a history in leading universities, extending back to the Renaissance, and arguments about the extent to which academia should imitate other subjects or retain a unique curriculum were happening in the mid-nineteenth century (Golding Rosemary 2016).
93 Danny Butt (2017).
94 Fricker (2007, p. 20).
95 Shiva used the term epistemological violence in the title of a book chapter (Vandana Shiva, 'Reductionist Science as Epistemological Violence', in Ashis Nandy (ed.), *Science, Hegemony and Violence: A Requiem for Modernity* (Oxford University Press, 1988), 232–56), which was an edited version of a previous paper (Vandana Shiva, 'The Violence of Reductionist Science', *Alternatives* 12(2) (1987), 243–261).
96 Shiva, *The Violence of Reductionist Science, Shiva, Reductionist Science as Epistemological Violence,* pp. 243–244.

Bibliography

AHRC. (2020). *Research Funding Guide (Version 5.1).* https://ahrc.ukri.org/documents/guides/research-funding-guide1/ (accessed October 8, 2021).

Antons, D., & Piller, F. T. (2015). Opening the Black Box of 'Not Invented Here': Attitudes, Decision Biases, and Behavioral Consequences. *Academy of Management Perspectives,* 29(2), 193–217.

Argyris, C. (1977). Double Loop Learning in Organizations. *Harvard Business Review,* 55(5), 115–125.

Arlander, A. (2010). Characteristics of Visual and Performing Arts. In *The Routledge Companion to Research in the Arts.* London: Routledge, 315–332.

Armsby, P., Costley, C., & Cranfield, S. (2017). The Design of Doctorate Curricula for Practising Professionals. *Studies in Higher Education,* 43(12), 2226–2237.

Borgdorff, H. (2007). The Debate on Research in the Arts. *Dutch Journal of Music Theory,* 12(1).

Borgdorff, H. (2011). The Production of Knowledge in Artistic Research. In H. Karlsson & M. Biggs (Eds.), *The Routledge Companion to Research in the Arts.* London: Routledge.

Bothwell, L. E., et al. (2016). Assessing the Gold Standard – Lessons from the History of RCTs. *The New England Journal of Medicine,* 374(22), 2175–2181.

Box, G. E. P. (1979). *Robustness in the Strategy of Scientific Model Building.* Madison: Wisconsin University-Madison Mathematics Research Center, Defense Technical Information Center.

Büchler, D., Biggs, M. A. R., & Ståhl, L. H. (2011). A Critical Mapping of Practice-Based Research as Evidenced by Swedish Architectural Theses. *International Journal of Art & Design Education,* 30(2), 318–327.

Busch, K. (2009). Artistic Research and the Poetics of Knowledge. *Art & Research: A Journal of Ideas, Contexts and Methods,* 2(2), 1–7.

Butt, D. (2017). Artistic Research: Defining the Field. In D. Butt (Ed.), *Artistic Research in the Future Academy.* London: Intellect Ltd.

Candy, L. (2019). *The Creative Reflective Practitioner: Research Through Making and Practice* London: Routledge, 1st ed.

Cazeaux, C. (2008). Inherently Interdisciplinary: Four Perspectives on Practice-Based Research. *Journal of Visual Art Practice,* 7(2), 107–132.

Choi, B. C. K., & Pak, A. W. P. (2006). Multidisciplinarity, Interdisciplinarity and Transdisciplinarity in Health Research, Services, Education and Policy: 1. Definitions, Objectives, and Evidence of Effectiveness. *Clinical and Investigative Medicine: Medecine Clinique et Experimentale,* 29(6), 351–364.

Choi, B. C. K., & Pak, A. W. P. (2008). Multidisciplinarity, Interdisciplinarity, and Transdisciplinarity in Health Research, Services, Education and Policy: 3. Discipline, Inter-Discipline Distance, and Selection of Discipline. *Clinical and Investigative Medicine = Médecine clinique et expérimentale,* 31(1), E41–E48.

Clegg, S., et al. (2018). Practices, Projects and Portfolios: Current Research Trends and New Directions. *International Journal of Project Management,* 36(5), 762–772.

Closs, S. J., & Cheater, F. M. (1996). Audit or Research – What Is the Difference? *Journal of Clinical Nursing,* 5(4), 249–256.

Cochrane, A. L. (1972). *Effectiveness and Efficiency: Random Reflections on Health Services*. S.l.: The Nuffield Provincial Hospitals Trust.

Dictionary, Oxford English. *Paradigm, N.* https://en.oxforddictionaries.com/definition/paradigm (accessed October 8, 2021).

Doyle, L., Brady, A. M., & Byrne, G. (2009). An Overview of Mixed Methods Research. *Journal of Research in Nursing*, 14(2), 175–185.

Edwards, S. (2009). A Professional Practice-Based Doctorate: Developing Advanced Nursing Practice. *Nurse Education Today*, 29(1), 1–4.

Elkins, J. (2009). *Artists with PhDs: On the New Doctoral Degree in Studio Art*. Washington, DC: New Academia.

Epstein, I., & Blumenfield, S. (2001). *Clinical Data-Mining in Practice-Based Research: Social Work in Hospital Settings*. London: Routledge, 33.

Eubank, M., & Forshaw, M. (2018). Professional Doctorates for Practitioner Psychologists: Understanding the Territory and Its Impact on Programme Development. *Studies in Continuing Education*, 41(2), 141–156.

Feldman, A. M. (2008). Does Academic Culture Support Translational Research? *Clinical and Translational Science*, 1(2), 87–88.

Fell, T., Flint, K., & Haines, I. (2011). *Professional Doctorates in the UK*. London: UK Council for Graduate Education Lichfield.

Frayling, C. (1993). Research in Art and Design. *Royal College of Art Research Papers*, 1(1).

Fricker, M. (2007). *Epistemic Injustice: Power and the Ethics of Knowing*. Oxford: Oxford University Press.

Friedman, H. H., & Friedman, L. W. (2018). Does Growing the Number of Academic Departments Improve the Quality of Higher Education? *Psychosociological Issues in Human Resource Management*, 6(1), 96–114.

Gibbs, G. (1988). The Reflective Cycle (Reprinted in Kitchen, S. (1999). An Appraisal of Methods of Reflection and Clinical Supervision. *British Journal of Theatre Nursing*, 9(7), 313–317).

Giddings, L. S. (2006). Mixed-Methods Research: Positivism Dressed in Drag? *Journal of Research in Nursing*, 11(3), 195–203.

Gray, C., & Malins, J. (2004). *Visualising Research: A Guide for Postgraduate Students in Art and Design*. Aldershot: Ashgate.

Green, L. W. (2008). Making Research Relevant: If It Is an Evidence-Based Practice, Where's the Practice-Based Evidence? *Family Practice*, 25(Suppl 1), i20.

Greenaway, R. (2004). Reviewing for Development. *Active Reviewing Tips*, 7.

Guba, E. G. (1990). *The Paradigm Dialog: Conference Entitled 'Alternative Paradigms': Papers*. Newbury Park, CA: Sage Publications.

Hill, N., Frappier-Davignon, L., & Morrison, B. (1979). The Periodic Health Examination. *Canadian Medical Association Journal*, 121, 1193–1254.

Hong, E., & Rowell, L. (2018). Challenging Knowledge Monopoly in Education in the U.S. Through Democratizing Knowledge Production and Dissemination. *Educational Action Research*, 27(1), 125–143.

Hothersall, S. J. (2018). Epistemology and Social Work: Enhancing the Integration of Theory, Practice and Research Through Philosophical Pragmatism. *European Journal of Social Work*, 22(5), 860–870.

Ichikawa, J. J., & Steup, M. (2014). The Analysis of Knowledge. In *The Stanford Encyclopedia of Philosophy*. http://plato.stanford.edu/archives/spr2014/entries/knowledge-analysis/ (accessed October 8, 2021).

Kolb, D. A. (2014). *Experiential Learning: Experience as the Source of Learning and Development*. Upper Saddle River, NJ: FT Press.

Kot, F. C., & Hendel, D. D. (2012). Emergence and Growth of Professional Doctorates in the United States, United Kingdom, Canada and Australia: A Comparative Analysis. *Studies in Higher Education*, 37(3), 345–364.

Lakoff, G., & Johnson, M. (2003). *Metaphors We Live by*. Chicago, IL and London: University of Chicago Press.

Macloed, K., & Hildridge, L. (2004). The Doctorate in Fine Art: The Importance of Exemplars to the Research Culture. *International Journal of Art: Design Education*, 23(2), 155168.

McBeath, A., Du Plock, S., & Bager-Charleson, S. (2018). Therapists Have a Lot to Add to the Field of Research, but Many Don't Make It There: A Narrative Thematic Inquiry into Counsellors' and Psychotherapists' Embodied Engagement with Research. *Language and Psychoanalysis*, 7(1), 1–18.

McIntosh, P. (2010). *Action Research and Reflective Practice: Creative and Visual Methods to Facilitate Reflection and Learning*. London: Routledge.

Mellors-Bourne, R., Robinson, C., & Metcalfe, J. (2016). *Provision of Professional Doctorates in English HE Institutions*. https://dera.ioe.ac.uk//25165/ (accessed October 8, 2021).

Messer, J. (2012). Practicing Interdisciplinarity. *TEXT*, 14(Special Issue, Beyond Practice-Led Research).

Michaels, J. A. (1989). *Laser Angioplasty: Evaluation of Devices for Clinical Use in the Peripheral Circulation*. Cambridge: University of Cambridge.

Michaels, J. A. (2017). Drawing out Ideas: Computer Models, Artworks and the Generation of Knowledge. *Proceedings of the Conference on Electronic Visualisation and the Arts*, 227–234.

Michaels, J. A., et al. (1989). Laser Angioplasty with a Pulsed NdYAG Laser: Early Clinical Experience. *British Journal of Surgery*, 76(9), 921–924.

Michaels, J A., et al., (2021) 'Configuration of Vascular Services : A Multiple Methods Research Programme', *Programme Grants for Applied Research*, 9(5), 1–150, 2021/10/08

Mockford, C., et al. (2012). The Impact of Patient and Public Involvement on UK NHS Health Care: A Systematic Review. *International Journal for Quality in Health Care*, 24(1), 28–38.

Morley, C., & Banerjee, S. (2013). Professional Doctorates in Management: Toward a Practice-Based Approach to Doctoral Education. *Academy of Management Learning & Education*, 12(2), 173–193.

Morley, M., & Petty, N. J. (2010). Professional Doctorate: Combining Professional Practice with Scholarly Inquiry. *British Journal of Occupational Therapy*, 73(4), 186–188.

Nelson, R. (2006). Practice-as-Research and the Problem of Knowledge. *Performance Research*, 11(4), 105–116.

Nelson, R. (2013). *Practice as Research in the Arts: Principles, Protocols, Pedagogies, Resistances*. London: Palgrave Macmillan.

Noë, A. (2015). *Strange Tools: Art and Human Nature*. New York: Hill and Wang, a Division of Farrar, Straus and Giroux, 1st ed., xiii, 285.

Polanyi, M. (1998). *Personal Knowledge Towards a Post-Critical Philosophy*. London: Routledge.

Potter, M. A., et al. (2006). Demonstrating Excellence in Practice-Based Research for Public Health. *Public Health Reports*, 121(1), 1–16.

Robinson, C. (2018). The Landscape of Professional Doctorate Provision in English Higher Education Institutions: Inconsistencies, Tensions and Unsustainability. *London Review of Education*, 16(1), 90–103.

Rolling, J. H. (2015). A Paradigm Analysis of Arts-Based Research and Implications for Education. *Studies in Art Education*, 51(2), 102–114.

Rosemary, G. (2016). Seeking a Philosophy of Music in Higher Education: The Case of Mid-Nineteenth Century Edinburgh. *Philosophy of Music Education Review*, 24(2).

Rosen, M., & Blake, Q. (2011). *Quick, Let's Get out of Here*. London and New York: Penguin.

Rosenfield, P. L. (1992). The Potential of Transdiciplinary Research for Sustaining and Extending Linkages Between the Health and Social Sciences. *Social Science & Medicine*, 35(11), 1343–1357.

Schön, D. A. (1991). *The Reflective Practitioner: How Professionals Think in Action*. Aldershot: Ashgate.

Shapin, S. (2012). The Ivory Tower: The History of a Figure of Speech and Its Cultural Uses. *The British Journal for the History of Science*, 45(1), 1–27.

Snow, C. P. (1956). The Two Cultures. *New Statesman and Nation*, 413–414, October 6.

Snow, C. P. (1959). *The Two Cultures and the Scientific Revolution: The Rede Lecture 1959*. Cambridge: Cambridge University Press, 14, 62.

Sullivan, G. (2010). *Art Practice as Research: Inquiry in Visual Arts*. Los Angeles: Sage.

Taylor, B. (2010). *Reflective Practice for Healthcare Professionals: A Practical Guide*. London: McGraw-Hill Education.

Teddlie, C., & Tashakkori, A. (2011). Mixed Methods Research. *The Sage Handbook of Qualitative Research*, 285–300.

Tight, M. (2014). Working in Separate Silos? What Citation Patterns Reveal About Higher Education Research Internationally. *Higher Education*, 68(3), 379–395.

Torrance, G. W. (2006). Utility Measurement in Healthcare. *Pharmacoeconomics*, 24(11), 1069–1078.

Westfall, J. M., et al. (2006). Community-Based Participatory Research in Practice-Based Research Networks. *Annals of Family Medicine*, 4(1), 8–14.

Winter, R., Griffiths, M., & Green, K. (2000). The 'Academic' Qualities of Practice: What Are the Criteria for a Practice-Based PhD? *Studies in Higher Education*, 25(1), 25–37.

Wood, M., Mayo, F., & Marsland, D. (1986). Practice-Based Recording as an Epidemiological Tool. *Annual Review of Public Health*, 7(1), 357–389.

1.3

THE ACADEMISATION OF CREATIVITY AND THE MORPHOGENESIS OF THE PRACTICE-BASED RESEARCHER

Mike Phillips

A bloody replacement

E. B. White's[2] observations on the similarity of dissecting humour and a frog can equally be applied to creative practice. But one could possibly argue that the mutilated remains that litter galleries and conferences serve culture less well than the anatomical knowledge gained by scientists. Ever heard someone try to explain the Dead Parrot sketch? Not perform, embody, or live it, but just tell it, maybe with the addition of a funny voice here and there? While the level of sureality may be doubled by shifting the perspective to a third-person account, the level of humour is exponentially decreased. Describing and analysing the sketch has a slug-like quality which confirms Mr. Praline's punch line: 'Well it's hardly a bloody replacement, is it!'[3]

And so, with the academisation of creativity, the consumption of the Art College by the multiheaded hydra of the University system has seen its transmogrification into something rich and strange is referenced. It is not a replacement but rather a sea-change we are suffering, bones to coral and eyes to pearls,[4] a process that has happened so slowly, albeit accelerating over the last 35 years, that it is hard to remember what it was like before, or that we were experiencing the change, or that we understand the thing it has been changed into.

This is not a cynical critique or romantic memento of the end of an era, mourning the demise of the post-war period of creative force that was the Art College, something Brian Eno termed 'a fertile ecology of an education system'.[5] Instead, it hopes to find useful remains and potential in order to stitch them back together into something new. Understanding this sea-change will hopefully kill two parrots with one stone: the nature and value of the PhD, and the nature and value of creative practice; or at least cast some light on the murky turbulence in the fathoms below the surface of the global concern to better understand practice-based research in Higher Education (HE).

This chapter also explores research-based practice that has emerged and flourished through this transformation. The potential for rich interdisciplinary interactions enabled by the slow collision of disciplinary silos that make up HE institutions is explored, in particular through an entanglement with digital practices which have the potential to operate as a 'Rosetta Stone' to unlock the utility of transdisciplinarity.[6] At the very least, these practices generate the kind of friction necessary for disciplinary transformation. It does not attempt to unpick the easy rhetoric that surrounds 'cross', 'inter', and 'trans', but rather highlights the potential/difficulties of these marriages of inconvenience.

DOI: 10.4324/9780429324154-5   60

Even within the blurry boundaries of a single discipline, there is often discord. Earth Sciences engage in a fractious debate around the value, nature, and meaning of 'data'. To the instrumentalists that measure the world, data is something clean and uncorrupted by human hands. For them, the data collected by people, through unmediated observation, citizen science processes, and the scouring of historic archives, is described as 'dirty' and therefore fallacious. Likewise, never put an experimental physicist and a theoretical physicist in the same room, unless you wish to redefine the meaning of the Big Bang. The gap between these boundaries is (potentially) artist/designer shaped.

Never eat anything bigger than your head

The emergence of practice-based/led/driven/informed . . . research was forced by the process of academisation of art and design. This was made manifest in the UK following the slow absorption into polytechnics through the 1970s, and in 1992, by the University system gobbling up Art Colleges whole. As a consequence, both swallower and swallowee have suffered from indigestion; universities still have that surprised look a python has when trying to swallow a deer twice its diameter, and the partially digested remains of Art College struggle to remember what they once were: places where 'you didn't have to cohere, it was the incoherence that was exciting and fruitful'.[7]

So much of this practice-based making culture has been lost, but what remains is a genetic propensity for disruption that once defined the function of Art and Design. The impact of a sense of creative practice on the University and the various disciplines it contains should not be underestimated. Where creative practice contaminates and infects other disciplines, the transformation can be significant, way beyond the simple illustration of science of an antiquated Art/Sci agenda. The disruptive quality of art and design practice, as it struggles to define itself as research, is in turn influencing the identity of research practices in other disciplines. It should not be seen as osmosis – the movement of something from one density to another until an equilibrium is reached – but as a vivisection, with the car crash of animal metaphors earlier more reminiscent of an episode of Doctor Moreau's *Love Island*.[8]

Consequently, practice-based research is a blasphemy of body parts, a tensegrity held together by forces which are at times contradictory and confused, but when seen in transition, they have a strange power to challenge convention, breach disciplinary boundaries, and ignite new knowledge. It is a body of knowledge and experience in a state of transformation, cultivating emergent methods, enabling an evolution of form – from solid to immaterial, object to process, and script to algorithm. Qualities identified by Roy Ascott in this prophetic quote from 1968:

> When art is a form of behaviour, software predominates over hardware in the creative sphere. Process replaces product in importance, just as system supersedes structure.
> Consider the art object in its total process: a behavourable in its history, a futurible in its structure, a trigger in its effect.[9]

Ascott's influence on practice-based PhDs will be discussed later in this chapter, but for me, when I first encountered this manifesto, somewhere in the early 1980s in the library of Exeter College of Art and Design, came the realisation that the Art College was doomed. More significant than the impact of the then Conservative Prime Minister, Margaret Thatcher's reign was the sense that the incredible pedagogic, haptic, hallucinatory, behaviourally disruptive experiential education the Art College provided was being undermined by its worship of objects. Ascott was here providing a eulogy for the *objet d'art* and the value systems, markets, inequalities, arrogance, and institutions that glorify them.

So, it is taking a little longer than expected, but the quote still holds true when applied to the broader cultural sector that the majority of practice impacts. For instance, on a visit to the *Future*

History v1.0 exhibition,[10] celebrating the influence of Ascott's work, Darren Henley, CEO of Arts Council England, whimsically mused that he wished he had used the statement as a rubric for the last round of National Portfolio Organisation awards. He hints at a fundamental problem the culture/creative sector has with a sense of value, something which is intrinsically linked with the way creative and cultural disciplines in HE understand and represent themselves. What is cultural value, and how do you measure it? Is footfall synonymous with quality?

If we are slaughtering or vivisecting the things we once held sacred, what meaningful methods and tools are there for measuring *intrinsic* as opposed to *instrumental* value in the new things that emerge? The arts sector has for years suffered from an imposed need to measure its value in terms of the yard sticks provided by the empirical methods identified in the HM Treasury's *The Green Book*,[11] the UK government's official guide to cost–benefit analysis. Essentially designed to measure return on investment through economic and social impact, they have been deployed by economists in areas such as health and the environment, where value sits outside of normal economic markets and transactions and has been shown to have success, if sometimes generating their own controversies. Arts Council England's drive to engage with audience metrics overlapped with the push to explore the challenges in understanding cultural value by the UK's Arts and Humanities Research Council through the Cultural Value Project. This initiative had two main objectives:

> The first was to identify the various components that make up cultural value. And the second was to consider and develop the methodologies and the evidence that might be used to evaluate these components of cultural value.[12]

Among the broader concerns and reticence to adopt a reductionist approach to value, the deficit in meaningful methods and tools for data collection across the cultural sector, combined with the sector's limited acceptance of effective digital processes, was identified. This also exposed the 'divide' between the motivations and requirements of the funder and the funded and highlighted an 'innovation problem' across the cultural sector in respect of technology adoption. The concern for the cultural sector is that artistic output is viewed only in economic terms, using things that are easy to measure and ignoring those that are not. This presents a particular difficulty for the cultural sector, where value is considered to be intrinsic and hard to quantify. Inevitably, the low-hanging fruit is audiences rather than art works and artists. These technological innovations have inevitably challenged traditional discursive methods of measuring cultural impact and outcomes and highlighted capacity issues to undertake the necessary measurements and data collection across the sector. The resulting metrics frameworks deployed by Arts Council England, such as the Audience Agency[13] and the Audience Impact and Insight Toolkit[14] – much loathed by arts organisations – focus on the audience and not artistic and creative practice itself (and in these pandemic days one must wonder what there is now to measure).

This process of dissection and disassembly of practice neuters it, rendering it neither a behaviourable, a futurible, or a trigger. The things that can be funded through this self-fulfilling algorithmic prophecy are those which comfortably sit within the cells in rows and columns of the spreadsheet, usually things of size and quantity. This critically impacts smaller arts organisations, which, one could argue, interface more directly with communities of emerging artists and creative practitioners, those more likely to engage with practice-based research.

Creative practice-based research suffers a similar plight as the broader cultural economy through the metricisation of academia.[15] It is increasingly apparent that the way things are measured ultimately defines them, not least the impact of measuring things that are easy-to-measure on the future development of artistic and cultural forms. If the educational framework, inception, commissioning, development, curation, and dissemination of artistic and cultural works are defined by short-term measures, the chains of influence and value will be diminished. Some things are hard to quantify.

For UK academics, this is played out through the Research Excellence Framework[16] (REF), a national five-yearly cycle (give or take), performance-based research funding evaluation mechanism. Here metrics have moulded the shape of research in many institutions, and practice-based research often struggles to tick the relevant boxes, often creating confusion and uncertainty in researchers, institutional REF coordinators, and higher up the hierarchy, let alone the broader REF panel review community. The problems of metricisation are eloquently articulated in *The Metric Tide*,[17] and while the REF has been tweaked and massaged to better embrace practice-based research, the struggle will always be quantity versus quality and instrumental verses intrinsic. In this context, how will new forms be measured if their intrinsic qualities require a different tape measure? The behaviourables and futuribles that could emerge from the friction generated by the collision of disciplines may simply not be recognised by the orthodoxy, and the survival of the metricised fittest will artificially shape our cultural outputs.

In parallel, the nature and shape of knowledge itself is shifting, and increasingly this is being recognised by institutions, such as the British Library, which is actively developing its EThOS web-service to capture new knowledge emerging from practice-based PhDs in the form of 'multimedia' and non-text theses.[18] However, this necessity to capture and disseminate new knowledge within the sciences in a close to real-time way, such as through temporal and interactive forms, has led to the construction of platforms such as Figshare.com, which challenge the established publishing models and powerbases.

The mechanisms for the dissemination of new knowledge generated by practice-based research are more often than not the same spaces used for other cultural endeavours, galleries, theatres, performances, video, *Fulldome* productions, apps, software, and online environments. In many cases, they are real-time experiences which only need an archive for storing the detritus, flotsam, and jetsam of the research manifest in the work. Possibly, as these real-time spaces emerge and converge, there is a new space for the creative practitioner, where one ends and the other begins, or somewhere in between. As the boundaries blur between the processes of publishing, archiving, and other forms of creation and dissemination, the necessity of distinguishing between what is practice and what is research raises its chimeric head. Wherever this distinction lies, it should not be defined by reductionist metrics means but should rather remain a negotiation and an open conversation.

How long is the piece of string that can measure this space of practice? The tropes of practice can be identified as the *things* that are produced, the spaces that they are produced in, such as the studio or lab, and the transactions that take place before and after the thing enters the world. These transactions might be conversations, exchanges of knowledge and skills, and the subsequent deployment and manifestation in other spaces, such as galleries or publications, which create longitudinal chains of value and influence. These tropes formed the very foundations of the Art College; the very fabric of these institutions was formulated around the replication of the artist's studio, if maybe radically reduced to a shared 8 x 4 white emulsioned chipboard panel. As these spaces morphed, compressed, and squeezed into university buildings, one is reminded of Derek Zoolander's plea: 'How can we be expected to teach children to learn how to read if they can't even fit inside the building?'[19]

The imitation game

For me, back in the mid-eighties, the career pathway for the artist was clearly defined during my undergraduate Fine Art course. To adhere to the grand historic traditions of an emergent practitioner, I would leave as a fully formed artist, live and work in an attic, catch tuberculosis, die, and then become famous. The horrible irony of this statement is that the example of Donald Rodney's *Psalms*, discussed later, tragically followed this trajectory, and his work is now being recovered as the subject of many academic research, curatorial, and conservation initiatives. Produced as a product of interdisciplinary and collaborative research by the artist, roboticist, and myself, the work itself is now the subject of research in curation and conservation. The further implications of this are discussed in the section on instruments/artefacts.

The trajectory of most other academic disciplines requires a slightly lengthier process. To be called an engineer requires the successful completion of an undergraduate, master's, PhD, postdoc, and then maybe, and only maybe, can you call yourself an engineer. A pattern replicated in most disciplines creates a shared understanding of what rigours are expected when referring to the word 'research'.

In many ways, creative practice is still claustrophobically hobbled by processes that are market driven, such as the parasitic relationship of galleries and museums or the client-driven nature of design problem solving. In more traditional insular areas of the creative domain, this pathway may still have some nostalgic attraction, but, certainly in the more digitally contaminated areas of creative practice, portholes to other disciplines, industries, and cross-sector applications have opened. Practice-based research has the potential to permeate disciplinary membranes; its agility, (im)materiality, and critical sense of audience enables rich conversations across disciplines, not just in the service of some public understanding of science agenda and a cultivation of new relationships and new knowledge.

This chapter loosely bases its timeline for the emergence of practice-based research on the parallel evolution of digital practices and processes, as evidenced in the educational initiatives of Ascott from the 1960s, Rodney artwork in the 1990s, and contemporary imaging technologies that found their transdisciplinary foundation in the early part of this century. The pivot point identified earlier in the early 1990s was the absorption of the Art College into the larger beast of the University and the subsequent problems of mastication and digestion. The advent of digital practices and platforms in the 1990s catalysed interdisciplinary conversations and connectivity which stretched the topology of the University sector and challenged disciplinary silos. The slippage of Human-Computer Interaction (HCI) into Interaction Design and New Media Art, the convergence of Biology, Psychology, and Computing gave birth to modern Neuroscience, Artificial Intelligence, and Robotics, and the harvesting and analysis of data is challenging traditional hypothesis–driven scientific methods in multiple disciplines are all emergent properties of this period.

While mourning the loss of the Art School, its disappearance through assimilation also framed a new potential for interdisciplinary collaboration. A cultural philosophy dominated by the hegemony of the eye and a reliance on instruments with lenses and celluloid suddenly had access to new tools and methods for understanding the world beyond the visible and imaginary. The invisible and the obscured due to scale and distance (infinitely big or nanoscopically small) became accessible through access to trans-scalar instruments, such as microscopy departments with scanning electron and atomic force microscopes, or chemistry and physics departments to reveal more things in heaven and earth than dreamt of in these ocular philosophies. To quote the TV character Father Ted: 'These are small . . . but the ones out there are far away. Small . . . far away'.[20]

It was not an easy transition, and in many ways the Humanities are still catching up with this potential. I have fond memories of an interdisciplinary workshop organised by Masoud Yasdani, who later founded the Intellect Press,[21] at Exeter University circa 1990, organised by the Computing Department with the ambition to explore collaborative opportunities with the Language Department. Computing saw immense potential and great excitement in code and programming as a new Modern Language, with the Language Department only able to contribute a request for a better version of Microsoft Word. The depression and disappointment at this lack of vision was palpable.

Regardless, during this period of assimilation, this potential received backing from the Wellcome Trust in collaborative consortium, which comprised Arts Council England, the Scottish Arts Council, the British Council, the Calouste Gulbenkian Foundation, and the National Endowment for Science, Technology and the Arts, to establish the SciArt programme between 1996 and 2006:

> Whether it was a matter of timing or incentive, during the decade of Sciart there was
> a cultural shift, especially within the arts, towards more interdisciplinary practice. The

burgeoning academic research culture of the visual arts in particular benefited from having Sciart as a context within which artists could develop projects.[22]

Building on the Wellcome Trust's biomedical agenda, the programme expanded to engage with other issues and was informed by C. P. Snows' 1959 Rede Lecture, *The Two Cultures*, in which he argues for the importance of

> closing the gap between our cultures is a necessity in the most abstract intellectual sense, as well as in the most practical. When those two senses have grown apart, then no society is going to be able to think with wisdom.[23]

This ambition shares the same motivation of small and large initiatives such as the Leonardo Journal,[24] the Swiss Artists in Labs Program, founded in 2003 by Jill Scott,[25] and more recent projects such as the UCLA Art|Sci Center[26] and the Gluon Scientists in Residency program collaboration with BOZAR Lab[27] – to name but a few in a massive emergent field of global organisations which embrace transdisciplinary collaboration. While many of these projects repeat the mistakes of their predecessors, such as getting the balance of power between the artist and the scientist and skills out of kilter, illustrating or imitating science or even that one needs the other, there is an inevitable vertical influence down through the education system and spillage into the wider cultural landscape. It is in this collapsing of disciplines that the spark of new practice-based research can be kindled, nestling in interdisciplinary conversations and emerging through transdisciplinary transformations, something more than the sum of its body parts.

In vitro

Untimely ripped from the nurturing environment of the Art College to be transplanted into the alien varsity, this bricolage practice-based creature struggled to build a new *umwelt* with conditions and parameters conducive to inter/transdisciplinary practice-based research in vitro. In this context, this section explores not what practice-based research is but, instead, the delicate ecology that allows it to thrive. It identifies the network, event/space, and instruments/artefacts and draws on an analysis of my engagement with PhD supervision, which began in 1993 in the School of Computing at the University of Plymouth, where fledgling digital art practice encountered potential impacts beyond their creative pragmatic application. In the time when new media was the big idea, in the early days that marked the death of HCI and the birth of interaction design, the School of Computing was keen to engage with the new (business) opportunities of multimedia, where the database collided with publishing and TV production.

Things got interesting when Roy Ascott joined the Interactive Media Subject Group which I ran in 1997 to establish STAR (Science Technology and Art Research) alongside the Centre for Advanced Inquiry in the Interactive Arts (CAiiA) in Caerleon (later to become part of the University of Wales, Newport). CAiiA-STAR constituted a trans-institutional research platform which was superseded in 2003 by the formation of the Planetary Collegium[28] when Ascott exited the University of Wales for the University of Plymouth. Planetary Collegium was primarily a practice-based PhD programme, a form still very much in its infancy, but a form which seamlessly integrated with the profile of a computing PhD. There was a clear mapping of the practice of coding and engineering of robotic research onto the creative practice model that had already emerged through the Planetary Collegium doctoral candidates' research activities. The science and technology model of a PhD appeared to be more accepting and compatible for creative practice-based research than the Humanities model of a PhD. A fundamental aspect of this relationship was an appreciation of network concepts. While the Humanities tend to fetishise objects and events (although both are also celebrated), the culture of computation was already adjusted to embracing the asynchronous network, and this was very much a part of the Planetary Collegium ethos: it was nomadic and constantly in a state of transition:

> Such is the case with Ascott's own theorization in 1966 of interdisciplinary collaborations over computer networks, a concept that became the central focus of his theory and practice in 1980, and subsequently has been popularized through web-based multimedia at the turn of the century.[29]

Ascott's radical educational principles established through the formation of the 'Ground Course' at the Ealing and Ipswich Art Schools in the 1960s and played out to controversial effect in Ontario College of Art, Toronto, in 1971, when 'the doors of perception were not only flung open but wrenched off their hinges'.[30] On this Kiendl writes:

> he was hired at the Ontario College of Art and he came in with a plan that broke down departmental boundaries. He eliminated painting and sculpture, and he redid the school's approach to pedagogy to look more at structure and communication. For him, professors were a data bank of ideas and all of the students could go to those primary sources for direction rather than be specifically filed into sculpture or ceramics. He also released a 33-⅓ RPM record as the curriculum.[31]

Ascott's creative practice was syncretic with his pedagogic strategies, which drew on this inspiration to enhance creative behaviour through processes based on the principles of cybernetics. By founding the Ground Course at Ealing College of Art in 1961, he created a platform for a radical cybernetic approach to learning through participatory, interactive, and disruptive strategies that locate the viewer/learner as an active participant. A peculiarity or feature of Ascott's practice and his influence on practice-based PhDs is that the 'things' he is most known for are ephemeral, temporal, and transitionary: things with the least physical trace. The recalibration of human behaviour enacted through the Ground Course leaves few physical relics, other than influential behaviours of the likes of Brian Eno. As with the later telematic projects, such as which predicted the activities of the Planetary Collegium, the behaviourables and futuribles define strategy of dissolving the materiality of the University itself into a mobile networked community of ideas and practice.

Network

These activities were synthesised into the nomadic model of the Planetary Collegium PhD Programme, which ran in Plymouth until 2020. It spawned networked 'nodes' in Milan (M-Node at Nuova Accademia di Belle Arti, NABA, Milan, Italy, with Professor Francesco Monico), Zurich (Z-Node in the Hochschule fuer Gestaltung und Kunst, Zurich (HGKZ), with Dr Jill Scott), and smaller nodes in Kefalonia (i-Node: (Ionian Node), Ionian Centre for Arts and Culture, Kefalonia, Greece, with Dr Katerina Karoussos) and Lucerne (NGL Node, Neue Galerie Luzern, Swiss Biennial on Science, Technics + Aesthetics, Lucerne, Switzerland, with Dr René Stettler). And it lives on in the DeTao-Node, Detao Masters Academy, Shanghai Institute of Visual Arts (SIVA) public university, Shanghai, China. Each node attracted its own cohort of doctoral candidates, flavoured by the local supervisory teams but all operating from a rubric of nomadic and practice-based research.

The model for this networked platform is mapped out in a manifesto paper published in 1994. *The Planetary Collegium: Electronic Art and Education in the Post-Biological Era* declared:

> We're not talking about a few curriculum changes here. We're not talking about the gradual replacement of some of the library stacks with a few computers. We are talking about the total dissolution, disintegration, and dispersal of Higher Education. From real estate to cyber estate. The university is becoming the intervarsity.[32]

In it, he describes a collegial model for 'non-hierarchical, non-linear, and intrinsically interactive; a gathering together, a connecting, an integration of people and ideas'. It is with regret that, in these pandemic times, where real estate hangs around the necks of Higher Education institutions at the cost of embracing a networked asynchronous model for learning by engaging the telematic, this manifesto was not more widely embraced.

Fundamentally practice-based, it attracted some of the leading contemporary Media Arts practitioners; with over 80 PhD completions, these individuals now play significant roles in defining the future of practice-based research in a plethora of international institutions. One can easily point to Dr Jill Scott and her Artists in Labs programme; Victoria Vesna at UCLA and Director of the Art|Sci Center at the School of the Arts and California NanoSystems Institute; Donna J. Cox as the Director for Visualization and Experimental Technologies at National Center for Supercomputing Applications (NCSA); and Christa Sommerer, who heads the Department for Interface Culture at the Institute for Media, University of Art and Design in Linz Austria. It is a long list, but also all are established practitioners in a volatile and emerging interdisciplinary field whose practice would have been squashed by undertaking a more traditional PhD programme. Back in the day, certainly in the USA, the MFA was the terminal degree for artists and designers, creating restrictions for those wishing to operate at the upper echelons of their universities. The Planetary Collegium provided a highly flexible and intense community which respected the individual practice of these artists. Some of these works and their influences were mapped in the Future History exhibition held at the end of 2019.[33]

This nomadic model at scale may well have a challenging ecological footprint, but the decentralised approach provided a credible alternative to the real estate infrastructure and the Zoomification of education. The networked approach also formed a significant social capital among the researchers and the audiences at the hosted events. The transcultural approach pollinated new ideas and collaborations, and the network of alumni continues to nurture the international field of practice-based research. It is easy to forget that a network is temporal and its connections can be sustained longer than the physical spaces that might once have hosted them.

Event/space

Composite Sessions were the core of the Planetary Collegium programme; each of the nine ten-day Composite Sessions over the first three years of a research student's registration (three Composite Sessions per year – normally April/July/November) consisted of collective research updates, individual tutorials, a critical co-ordination of researchers' feedback, and a short summary response to the group. These points of focused face-to-face networked activity were usually hosted by an international venue or linked to a conference or workshop, such as ISEA or Ascott's own series of conferences, Consciousness Reframed, which were established in 1997 and ran through to 2019 with 21 editions. These global Composite Sessions were hosted in Dublin, Valencia, Marseilles, Rio de Janeiro, São Paulo, Tucson, Paris, Turin, Los Angeles, Perth, Gifu, Zurich, Milan, Fortaleza, Dalla, Istanbul, Montreal, Munich, Johannesburg, Beijing, Shanghai, and Cairo, to name just a few.[34]

Synergetic with the network approach, these events become subsidiary nodes on the network in labs, studios, and conference spaces that formed a wider community of researchers, often hosting post-event residencies, collaborations, and commissions.

As well as establishing a nomadic networked culture which intimately supported and challenged individual practice-based methodologies, the Planetary Collegium was built on a unique transdisciplinary foundation. Back in 1992, I cultivated a covert group of artists to infiltrate the School of Computing with the explicit intention of establishing a framework for the Planetary Collegium to put down sustainable roots, which has thrived for 25 years. It was this culturally alien environment that proved so provocative, from the outset we could have drowned or evolved into amphibious creatures. To complement the nomadic qualities of the Planetary Collegium, i-DAT.org[35] was formed in 1998 in the School of Computing as a mechanism for growing a practice-based research culture. Operating as a collective of 'digital artists', it sponsored collaborations with other areas of the University, such as robotics, earth sciences, and engineering. As a space for interdisciplinary exchanges, it supported projects such as *Psalms* (below) and facilitated the activities of the Planetary Collegium whenever the spaceship landed back in Plymouth.

Although a challenge to many institutional managers, the relationship with other academics was based on a mutual need to access resources and skills to realise each other's ambitions. Most importantly, we developed the skill of speaking in tongues. It was strategically positioned itself as an instrument and lab to enable and cultivate these processes, including operating as an Arts Council England National Portfolio Organisation, a mechanism for Arts Council England to 'protect and develop our national arts and cultural ecology',[36] effectively bridging a gap between professional artists, their audiences, digital processes, and access to other disciplinary approaches and resources. This superimposition of a national funded arts organisation was an interesting experiment that facilitated rich interactions with the broader arts community in the UK and supported project development, collaborations, and residences for practicing artists, some of whom have since engaged with PhDs in practice-based research. It also allowed artists to negotiate access to the tools and instruments outlined next, which they would have traditionally been excluded from. I would like to think that it still effectively embodies the incoherence of the Art College. These qualities are of course not unique and can be seen in institutions at a global level, research groups inside the University, independent arts organisations outside, and those that attempt a superimposition of the two.

The international landscape of event/spaces is rich and volatile. i-DAT was able to reach across and outside of the institution, offering a permeable membrane for research collaborations, other event/ spaces to operate in an inverse manner, bringing artists and interdisciplinary researchers together within them, such as Ars Electronica[37] or the Tate Moderns Blavatnik Exchange space, which offers a 'collaborative interdisciplinary ecology' or, as Nicholas Serota described in an appropriately popularist way, 'a combination of the Open University, art school, TED talks, and Guardian debates, all wrapped into one'.[38] While others define a third space, a neutral space between the University and the independent arts organisation, such as the Pervasive Media Studios[39] emerging from a collaboration with the Watershed, University of Bristol, and University of the West of England, Bristol, UK. Or reaching from the University out, such as Creative Informatics[40] at the University of Edinburgh, which aims to be the creative industries and technology sector together through an engagement with data-driven design such as fintech and IoT. Increasingly, the realisation of the inadequacy of the University to engage with the fermenting culture outside of its ivory towers is forging collaborations designed to facilitate interdisciplinary innovations. One can also look to organisations such as ISEA[41] and Leonardo that cultivate these conversations at scale, also adopting a nomadic networked event/ space model at a global scale.

This practice-based approach engaged pragmatically with people, communities, and institutions through collaborative and participatory design methods. The research gives new insights into the social, cultural, and ecological possibilities of coupling a real-time sense (data) with physical objects and spaces. But, essentially, the fundamental ingredient in any practice-based research should be a sense of playfulness. Play itself is a transdisciplinary instrument for the manifestation of material and imaginary worlds. One could frame this synthesis of methods as a *Ludile* method: agile and ludic. It is ai method that embraces a transdisciplinary practice to incorporate collaborative and participatory processes between people, disciplines, objects and things, algorithms, and collective behaviours, something which resists the ossification of academic metricisation, offering an In vitro *umwelt* for the practice-based research creature.

Instruments/artefacts

'Science and technology multiply around us. To an increasing extent they dictate the languages in which we speak and think. Either we use those languages, or we remain mute'.[42]

Speaking in tongues is important to have a focus for discussions. Across the arts and sciences, the instrument plays a negotiable role, around which conversations can take place. Unfortunately, the Arts have, for too long, flustered around with a fascination for the hegemony of the eye and a preoccupation with lens-based technologies to inform or limit our world view. With imaging technologies that require no lens to see, we can now touch the atomic forces that bind matter or hear the cosmic microwave background radiation that frames our universe. In a post-ocular culture, there is a function for Art and Design and the practice-based research it embodies, to negotiate the fragility of the reality embraced by the other disciplines that form the University. Once wrestled from their hands, these instruments can become lingua franca for conversations that cultivate new knowledge. The following are two examples of instruments that are evolving through strong interdisciplinary entanglements; they both challenge the pigeonholes that disciplines create, whether the disrupting STEM dominance of planetariums or the conservation of an autonomous system. And, in this problem of classification, we can see the provocation they make as a challenge to practice. Neither fish nor fowl, they give the practice-based researcher a voice in an interdisciplinary conversation.

Fulldome: One such instrument, which is also happens to manifest as an event/space, is the *Fulldome*. *Fulldomes* are immersive dome-based projection environments, more commonly found in science centres as planetariums. The *Fulldome*, in its planetarium guise, has been dominated by a STEM agenda for decades, usually located in science centres and constrained to retell narratives around the public understanding of science. It has been liberated from this yolk by transdisciplinary teams of artists, VJs, coders, performers, producers, and curators, who have occupied the space to explore and rejuvenate the language of *Fulldome* by developing a range of new experiences and enabling technologies that are now deployed in creative and cultural contexts. This process has also enriched the planetarium domain, as they can now reach new audiences, and the modes of production for *Fulldome* content are enhanced and enriched by the rapid evolution of this immersive language. As a shared virtual reality, these experiences realise László Moholy-Nagy's desire for audio-visual coherence: 'seeing, feeling and thinking in relationship and not as a series of isolated phenomena. It instantaneously integrates and transmutes single elements into a coherent whole'.[43] The evolution of the *Fulldome* artistic language has been supported by significant developments in the event/space of *Fulldome* Festivals which are cultivating this explosion of creative content and international collaborations around production skills, technologies, methods, and research. This is creating opportunities for audience participation in navigating trans-scalar imaginary and (im)material environments within the projected space.

These spaces become a new interdisciplinary playground for practice-based research, providing a platform for performers, chorographers, game designers, coders, and modellers. It cultivates experimental approaches to capturing, synthesising, and re-visioning the world through the imaging and sonification methods, such as medical imaging technologies, non-lens-based microscopy, 3-D scanning, photogrammetry, and point cloud visualisation (Figure 1.3.1). Simultaneously interdisciplinary

Figure 1.3.1   Murmuration, 2015, volumetric *Fulldome* projection from MRI scan of the author.

interactions are enriching the visualisation and sonification of data for earth scientists, physicists, microscopists, and medics. The rapid shift in the democratisation of production technologies is radically changing production, licensing, and distribution models. This paradigm shift is opening a new transdisciplinary dialogue between creative practitioners with the skills to handle the production tools and the plethora of disciplines eager to immerse new audiences in their data. As a result, the *Fulldome* is increasingly a place where, as Vonnegut wrote, 'all the different kinds of truths fit together'.[44]

Significant nodes supporting practice-based researchers in *Fulldome* include Martin Kusch and Ruth Schnell in the Department of Digital Arts, Universität für angewandte Kunst Wien, who led the European Mobile Dome Lab project (www.emdl.eu) and the collaborating organisation in Montreal, the Société des Arts Technologiques,[45] whose vast expertise and experience in experimental immersive environments is driven by Monique Savoie and Luc Courchesne. Collaborations between planetariums and universities for festival activity form the basis of a strong international network, such as Micky Remann at the Bauhaus-Universitaet Weimar, who directs *Fulldome* Festival[46] at the Zeiss-Planetarium Jena as well as the Biennale *Fulldome* UK.[47] Each one of these, as well as the numerous *Fulldome* spaces around the globe, are actively facilitating practice-based research across traditional disciplines.

Psalms: The other example for transdisciplinary engagement is the problematic presented by *Psalms*,[48] the Autonomous Wheelchair developed through a collaboration with Guido Bugmann and myself for Donald Rodney's *Nine Night in Eldorado* at the South London Gallery, 1997 (Donald Rodney died in 1998). On this Bilton wrote:

> Our fear of automata is again harnessed in *Psalms*, as the empty wheelchair courses through its various trajectories on a sad and lonely journey of life, a journey to nowhere.

Figure 1.3.2 Psalms, 1998.
Source: Photographed by the author.

> Its movements repeat like an ever recurring memory, a memory of another life and another journey, that of Donald Rodney's father.[49]

Psalms (Figure 1.3.2) is a powerful and moving (emotionally and literally) artwork which exhibits peculiar behaviours as it performs. It is a complex assemblage of sonar sensors, neural network and microcontroller, all entangled into a powered wheelchair. The work has since been exhibited at the Atlantic Project 2018), Vivid Projects (2016), Wellcome Trust (2012), Beurs van Berlage, Amsterdam (2009), and Institute of International Visual Arts (INIVA) (2008).

At the time of writing, the Tate was going through an acquisition process for *Psalms* to include it in its permanent collection, as it has done previously with Donald's *Visceral Canker,*[50] also constructed in collaboration with the author. Both works are electronic, with *Visceral Canker* incorporating a Peristaltic Pump and control board, and both works are challenging the curatorial and conservation conventions of the Tate. As the electronics perish, critical decisions must be made about the presentation of the works: are parts replaced, which parts are the work, is it the aesthetic look of the work or the behaviour it exhibits, where does one begin and one end.

Equally, the conventions of authorship around the work are problematic; Bugmann and I are usually mentioned in the small print of its provenance, and yet Donald is mentioned in the robotics research paper produced around the work in *Stable Encoding of Robot Paths Using Normalised Radial Basis Networks: Application to an Autonomous Wheelchair.*[51] The work itself sits within a happy agreement with the Donald Rodney Estate and i-DAT (who maintain and chaperone the work). For the work to exist in the Tate collection, it must meet the 'gold standard' of museum conservation, but in doing so it may have to change. The potential for the current MS-DOS and CORTEX-PRO rewritten and rehoused in more contemporary and ungradable platform as well as a rewiring and general replacement with contemporary components, or not – these are the decisions to be made through a dialogue with the Tate and the Donald Rodney Estate.

The point of this example is to highlight where the research lies with this work. For Donald as an artist, the research was an artistic practice in envisioning and developing the artwork; for Guido Bugmann, the work was an opportunity to develop his research around autonomous robots and neural networks; for the conservators, the research sits within an engagement with digital and electronic artworks which have generally been excluded from their collection. For me, it lies with a complex set of memories of a friend, the development of a programme of research around the preservation of code and the hardware that enacts it, and a continued fascination with the behaviour exhibited by *Psalms* as it moves around the gallery space and interacts with audiences. If there was an element of the work that should be preserved at the cost of all else, it should be this behaviour, which Donald defined in a pencil sketch of a figure of eight. The artefact can be seen as an instrument for research, something that embodies multiple research practices and methodologies, all mutually complementary and the catalyst for new knowledge. It was formed through a collaborative interdisciplinary research process and now forms a platform for research for curation and conservation, as well as an historical artefact that makes a powerful provocation for issues around race and disability.

Fish nor fowl

Sitting in this newly hatched body is not necessarily a comfortable place to be. Without the certainties, convictions, and bias of tradition and the impenetrable walls of orthodox discipline for protection, the blasphemy of body parts that is the practice-based researcher will inevitably experience a certain dysmorphophobia. However, given the correct nurturing *umwelt* and the networks and instruments to perform with, the interdisciplinary conversations that can be ignited through the ability to speak in tongues will be provocative and catalytic. It is difficult to remember that it was not always this way, and is in fact in a constant state of flux, no matter how hard institutions try to constrain things. For, when art is a form of behaviour, keep misbehaving.

Notes

1 In biology, imago is the last stage of insect development during metamorphosis.
2 White (1941, p. xvii).
3 *Monty Python's Flying Circus* (1969).
4 William Shakespeare, *The Tempest*, Act I, Sc. II.
5 Eno (2014).
6 As transitionary as disciplinary is, the term should be seen as a verb; while cross and inter embrace the distinctions and boundaries, the trans – translation/transformation/transmogrification/transubstantiation – is an evolutionary process of disciplines and those that inhabit them.
7 Eno (2014).
8 Wells, H. G. (2005). *The Island of Doctor Moreau*. Penguin Classics. London: Penguin.
9 Ascott (1968).
10 Future History v1.0. (2019). i-DAT.org, The Levinsky Gallery, University of Plymouth, Friday 29 November to Saturday 11 January 2019. [Online]. https://i-dat.org/future-history/ (accessed August 5, 2020).
11 HM Treasury. *The Green Book*. (2018). [Online]. (Accessed August 10, 2020). www.gov.uk/government/publications/the-green-book-appraisal-and-evaluation-in-central-goverment.
12 Crossick and Kaszynska (2016).
13 www.theaudienceagency.org/.
14 www.impactandinsight.co.uk/.
15 The metricisation of academia was a slow, corrosive process that hid accounting behind a thin veneer of accountability to reveal the cost of everything and the value of nothing – a process that has overpowered the UK education system since the 1980s, with culture testing, measurement, and quantification. The frog is now fully dissected, its bits are spread out on the tray in an orderly fashion, and we still can't find that funny little ribbit sound.
16 www.ref.ac.uk/.
17 Wilsdon et al. (2015).

18 Manton (2017).
19 Extract taken from the film *Zoolander* (2001) by Ben Stiller, Los Angeles, Village Roadshow Pictures.
20 Father Ted. "Hell." Episode 1, Series 2. Directed by Declan Lowney. Written by Graham Linehan and Arthur Matthews. Performed by Dermot Morgan. Chanel 4, 8 March 1996b.
21 www.intellectbooks.com.
22 Glinkowski and Bamford (2009).
23 Snow (2012, p. 26).
24 https://leonardo.info/.
25 www.artistsinlabs.ch.
26 www.artsci.ucla.edu.
27 www.gluon.be.
28 https://i-dat.org/planetary-collegium/.
29 Shanken (2002).
30 Milroy (2001).
31 Kiendl (2013).
32 Ascott (1994).
33 *Future History v1.0.* (2019). i-DAT.org, The Levinsky Gallery, University of Plymouth, Friday 29 Novemberto Saturday 11 January 2019. [Online]. https://i-dat.org/future-history/ (accessed August 5, 2020).
34 www.i-dat.org/planetary-collegium/.
35 https://i-dat.org/.
36 Arts Council England. 2018. *How Arts Council England works with National Portfolio Organisations. Relationship Framework*, p. 5. www.artscouncil.org.uk/sites/default/files/download-file/NPO_2018-22_Relationship_Framework.pdf (accessed May 29, 2021).
37 https://ars.electronica.art/.
38 Wainwright, O. (2016). First Look: Inside the Switch House – Tate Modern's Power Pyramid. *Guardian.* www.theguardian.com/artanddesign/2016/may/23/first-look-inside-tate-moderns-power-pyramid n (accessed March 27, 2017).
39 www.watershed.co.uk/studio/.
40 https://creativeinformatics.org/.
41 www.isea-web.org/.
42 Ballard (1974, p. 4).
43 Moholy-Nagy (1946, p. 12).
44 Vonnegut, K. (1962). *The Sirens of Titan.* Victor Gollancz, p. 7.
45 www.sat.qc.ca.
46 www.fulldome-festival.de.
47 www.fulldome.org.uk.
48 Rodney, D. (1997). *Psalms.* https://i-dat.org/psalms/ (accessed August 13, 2020).
49 Bilton, J. (1997). Exhibition Catalogue for Nine Night in Eldorado at the South London Gallery.
50 Rodney (1990).
51 Bugmann et al. (1998).

Bibliography

Ascott, R. (1968) [1970]. *Behaviourables and Futurables.* London: Control, 5.
Ascott, R. (1994). *The Planetary Collegium: Electronic Art and Education in the Post-Biological Era.* The 5th International Symposium on Electronic Art Catalogue. University of Art and Design, UIAH, Helsinki, 180–184.
Ballard, J. G. (1974). *Le Livre de Poche.* Paris.
Bugmann, G., Koay, K. L., Barlow, N., Phillips, M., & Rodney, D. (1998). *Stable Encoding of Robot Trajectories Using Normalised Radial Basis Functions: Application to an Autonomous Wheelchair (83,209).* Proceedings of 29th International Symposium on Robotics, Birmingham, April 27–30.
Crossick, G., & Kaszynska, P. (2016). *Understanding the Value of Arts & Culture, the AHRC Cultural Value Project.* Swindon: Arts and Humanities Research Council.
Eno, B. (2014). Archive on 4. Art School, Smart School. *BBC Radio 4*, November 22. www.bbc.co.uk/programmes/b04pr1w2 (accessed August 5, 2020).
Glinkowski, P., & Bamford, A. (2009). *Insight and Exchange: An Evaluation of the Wellcome Trust's Sciart programme.* Wellcome Trust. https://wellcome.org/sites/default/files/wtx057228_0.pdf (accessed January 10, 2021).
Kiendl, A. (2013). *Q&A: Anthony Kiendl on Web-Art Pioneer Roy Ascott.* Plug In ICA, Winnipeg, Canada, July 11. https://canadianart.ca/interviews/roy-ascott/ (accessed August 11, 2020).

Manton, C. (2017). *Multimedia and Non-Text PhD Research Outputs*. Theses Research Project for the British Library EThOS web service. DOI:10.14324/111.9781911307679.23; https://ucldigitalpress.co.uk/BOOC/Article/1/65/ (accessed August 5, 2020).

Milroy, S. (2001). The Fine Art of Control. *The Globe and Mail, Toronto*, September 19. www.theglobeandmail.com/arts/the-fine-art-of-control/article4153010/ (accessed August 11, 2020).

Moholy-Nagy, L. (1946). *Vision in Motion*. Paul Theobald & Co.

Monty Python's Flying Circus. Episode 8. Full Frontal Nudity. (1969). December 7.

Rodney, D. (1990). *Visceral Canker*. Tate Collection, Presented by Tate Members 2009. www.tate.org.uk/art/artworks/rodney-visceral-canker-t12769 (accessed August 5, 2020).

Shanken, E. A. (2002). Cybernetics and Art: Cultural Convergence in the 1960s. In B. Clarke & L. Dalrymple (Eds.), *From Energy to Information: Representation in Science and Technology*. Stanford, CA: Stanford University Press, 255–278.

Snow, C. P. (2012). *The Two Cultures*. Canto Classics. Cambridge: Cambridge University Press.

White, E. (1941). *A Subtreasury of American Humor*. Edited by E. B. White and Katharine S. White. New York: Coward-McCann, Section: Preface, Quote, xvii.

Wilsdon, J., et al. (2015). The Metric Tide: Report of the Independent Review of the Role of Metrics. *Research Assessment and Management*. DOI:10.13140/RG.2.1.4929.1363.

1.4

THE STUDIO AND LIVING LABORATORY MODELS FOR PRACTICE-BASED RESEARCH

Linda Candy and Ernest Edmonds

Introduction

In this chapter, we describe organisational models of studio and public spaces for inter-disciplinary practice-based research in the field of art and technology. It charts how a 'studio as laboratory' was established in a university setting and how it was later extended to include a 'living laboratory' in a public museum. During that process, the methodology evolved from a focus on research *about* practice, carried out by external researchers, into research *through* practice by creative practitioners, most of whom were PhD candidates. In other words, research that was *led* by and *for* practice became primarily based *in* and *through* practice. From these initiatives, a strong community of practice, underpinned by a principled approach to research, arose that continues to this day and is reflected in several contributions by practitioner researchers in this handbook.

To understand how this got started and the lessons we learnt, a little background history is helpful. It all began with a small conference series that became an international success underwritten by the Association of Computing Machinery (ACM), and a series of artist-in-residences that found an unusual home in a computing research centre. The first Creativity and Cognition conference in 1993 was a forum for exchange between artists, technologists, scientists, engineers, historians and others. A community of practice was formed which provided the impetus for the establishment in 1996 of the Creativity and Cognition Research Studios (C&CRS), a joint venture between the School of Art and Design and the Department of Computer Science at Loughborough University. In 2003, C&CRS was re-established at the University of Technology Sydney as the Creativity and Cognition Studios (CCS).

The Creativity and Cognition Research Studios (C&CRS) was an experimental place where artists and technologists could work in collaboration, producing new forms of art and new knowledge through research. This was a form of 'practice-led research', used here to mean research that investigates practice with the primary aim of informing and improving that practice, as distinct from creating new artefacts as its main goal. 'Practice-led research', according to this definition, usually involves researchers who are not necessarily practitioners themselves, and their focus is on describing and improving the practice from an objective point of view rather than from the practitioner's perspective. In the C&CRS/COSTART case discussed later, we adopted a case study based on a three view-point perspective – artist, technologist, observer – to record data. The 'observer' researchers carried out the collation and analysis of the data.

 DOI: 10.4324/9780429324154-6

Although the studios were located at the LUTCHI (Loughborough University Computer Human Interaction) Research Centre in the Computer Studies Department, the network of associated people throughout the university gave access to many different kinds of expertise across disciplines as varied as ergonomic, engineering, social science and art. In this way inter-disciplinary work drew upon a broad range of disciplines, each contributing to the research.

Collaboration between artists and technologists at C&CRS was enhanced by a research grant awarded by the UK's leading science funding body, the Engineering and Physical Sciences Research Council (EPSRC).[1] The main goal of the COSTART (**CO**mputer Sy**ST**ems for Creative Work: An Investigation of **AR**t and **T**echnology Collaboration)[2] as the key project was to facilitate the co-evolution of digital art works and technological innovations and, in parallel, carry out research into how people from different disciplines collaborated. It gave rise to the concept of the 'studio as laboratory', which represented a new avenue of art and technology research at that time. Realising the concept as an actual place for practice and research to coexist was a process which involved considerable effort to bring to fruition, because it not only required significant human and financial resources, but also transgressed the boundaries of subject disciplines and demanded new ways of thinking.[3] However, the opportunity to work in this kind of environment proved to be very productive for the artists concerned, and the research yielded valuable insights that were published in several papers. In particular, working closely with practising artists taught the project organisers an important lesson which was to have repercussions in the future. They realised that that no matter how well the development of new works was supported through collaboration and technical expertise, the studio-based production was but the first step in a process that could only be completed in the public domain, for example in an exhibition space. This was particularly so for those artists who were developing interactive installations that were only fully realised when audiences were present. This often meant that observing the interaction process was needed before a work was finalised and having access to exhibition space was not only desirable, as it would be for most artists, but also vital to the completion of the artwork itself.

In 2003, C&CRS moved from Loughborough University to the University of Technology, Sydney, as the Creativity and Cognition Studios (CCS). In its new location, CCS continued the inter-disciplinary art and technology research initiated in C&CRS but with a more extensive PhD programme. At CCS, the programme of PhD research involved a carefully managed process of meetings, seminars, co-located working and the collective creation of material necessary for facilitating university research requirements. The first ten years of CCS are documented in a multi-authored book that describes and discusses the features and outcomes of a rich and varied set of research processes and artistic practices.[4] Not long after the inception of CCS, the Beta_Space[5] 'living laboratory' was created in collaboration with the Powerhouse Museum (PHM) (now the Museum for Applied Arts and Sciences (MAAS))[6] as a public exhibition space within the museum. The partnership between CCS and PHM provided a new model for combining art practice with research in public spaces where audience engagement with art installations under development could be studied.[7] The model not only enables artists to study audience engagement within a public space beyond the studio environment but, at the same time, provides the museum or gallery with a means of gathering information about visitor experiences. These initiatives and their outcomes, including the many lessons we learnt, are discussed in more depth later in this chapter.

In the next section, we take a brief look at some early precedents for embedding research into art and technology practice. The initiatives were a foretaste of the tendency for leading edge practitioners to find homes outside their normal disciplinary spaces where they could encounter people with unfamiliar ways of approaching research.

Some background to art technology practice with research

Respect for the idea of research in art, or even art as research, has a long history. In 1923, Kandinsky proposed very explicit art research tasks in his plan for work at RAKhN, the Russian Academy of Artistic Sciences.[8] The purpose of this research was to support the process of *making* artworks, as against *interpreting* finished artworks. In the 1960s, the collective GRAV-Groupe de Recherche d'Art Visuel saw research as a way of achieving a more 'objective' approach to art making. Also in the 1960s, an important inter-disciplinary initiative that combined practice and research was the Experiments in Art and Technology (EAT) group, founded by engineers Billy Klüver and Fred Waldhauer with artists Robert Rauschenberg and Robert Whitman. EAT brought artists together with engineers to explore possibilities in new art forms.[9] It also initiated inter-disciplinary events and projects involving artists and new technology, including: *9 Evenings: Theatre and Engineering* (1966); *Some More Beginnings* (1968), an international exhibition of art and technology, held at the Brooklyn Museum; these artist–engineer collaborations designed the Pepsi Pavilion at Expo 70 in Osaka, Japan.[10]

Later, still during the 1990s, industrially related research in art and technology was undertaken in Japan at the Advanced Telecommunications Research (ATR) laboratories[11] in the Art and Technology Project, during which artists were employed to work on collaborative projects. This work was particularly notable because it was grounded in a technological world into which artists entered as full collaborators and, in some cases, as project leaders. At ATR, new technologies were developed with, and by, artists at the same time as new art forms were evolved. The exposure to the public was successful in demonstrating an innovative research direction and that technological research could be undertaken in an art–technology context. The interactive sound and vision work *Iamascope*,[12] made at ATR by Canadian artist Sid Fels in collaboration with Kenji Mase, was installed in Beta_Space when it was established in 2004. It was included in the studies described later in this chapter.[13] This event was an indication of increasing international collaboration between practitioners researchers in creative art and technology.

Some centres also pioneered new programmes that explored the potential of such work that could be accredited as formal academic PhDs. A relatively small number of groups became centres of practice-based art doctoral research programmes. A notable example is the Centre for Advanced Inquiry in Interactive Arts (CAiiA) founded in 1994 at the University of Wales, which later became the Planetary Collegium,[14] with its hub at the University of Plymouth in the UK. Roy Ascott, leader of the group, and his colleagues facilitated many artist-led research programmes resulting in PhDs in digital art practice. Each group conducting this kind of research has evolved its own set of norms and no two are the same.

Some of these initiatives were a foretaste of the tendency for art practitioners, especially those who wished to experiment with new technologies, to find homes outside traditional spaces: for example, practice-based research groups studying art in universities were not always in art departments. Very often that meant artists found themselves working in more scientifically attuned environments alongside people with different approaches to research, and it is not unreasonable to suggest that this exposure to systematic forms of investigation may have helped to shape practitioners approaches to practice-based research.

The inter-disciplinary nature of much of the work often resulted in strong links with technology departments, and sometimes, as for example with C&CRS, the base was within technology. This also led to more collaboration between artists, technologists and others, a subject that is beyond the scope of this chapter but is nevertheless a key driver in some areas of practice-based research. To explore collaboration in art and technology further, and including contemporary cases, the reader is referred to *Explorations in Art and Technology* Part 4: Collaboration.[15]

The origins of the Creativity and Cognition Research Studios

The background to the development of one environment for bringing art and technology together began at the Creativity and Cognition conferences.[16] The first Creativity and Cognition (C&C) meeting was held in 1993 and the second in 1996. Both succeeded in bringing together people from the different fields of art and science, enabling experts from one field to mix with experts from other fields. As well as meeting at the conferences, artists and technologists also came together in artist-in-residence projects. By 1999, C&C had joined the ACM SIGCHI conference series, where it has continued ever since as a powerful forum for practice-based research in multiple forms of creative practice.[17]

The early C&C conferences were important drivers in the development of a strong inter-disciplinary creative community which, in turn, led to the formation of C&CRS. The Studios provided physical and technical resources and spaces that supported many kinds of art and technology projects. This was achieved by bringing researchers in the computing discipline of Human–Computer Interaction (HCI) together with established practitioners in the arts, from visual artists to musicians, animators and others. The idea was to provide a base for a number of interrelated activities consisting of networks of artists and artists groups, media centres and conferences as well as externally funded projects. C&CRS can be compared with the PAIR model of art and technology collaboration conducted at Xerox PARC.[18] These models have similarities and differences. PAIR was used to maintain and stimulate the parent organisation's culture as a fertile ground for new ideas and new forms of technological innovation as one programme among many others. By contrast, the ethos of C&CRS was to make the art practice the central focus of the research into innovative forms in digital technology and art practice. This was done by facilitating both research and practice through enabling artists to work in a technical and physical environment not normally available to them. The principal goal was to advance the creative practice in some way. The engineering and scientific collaborations were drawn from departments across the university. A striking feature was that this use of the full resources of a university proved quite easy to facilitate because when an engineer, for example, from another faculty was approached, co-operation was almost always generously offered.

The C&C conference series and C&CRS initiatives were influenced by the UK's Department of Trade and Industry (DTI) Technology Foresight Programme. The Programme's Access and Creativity Task Group met from 1997 to 1999 and, in its final report, recommended that collaborative art and technology research initiatives should be strongly encouraged in the UK.[19] As well as reporting to the DTI, members of the Group followed the recommendations within their own organisations. C&CRS was no exception, strengthening its collaborations with other teams, working with the Arts Council of England and enhancing the access to advanced technology by artists to encourage 'the development of new forms and new understandings rather than purely the generation of new works'.[20] This final point was specifically pointing to the value of research into technology-based art, i.e., research by artists at the same time as research by technologists.

C&CRS methodology: the studio as laboratory

At C&CRS, the primary aim was to facilitate the co-evolution of art and technology accompanied by studies of the processes and the outcomes. All of the research involved creative practice: making artworks. For creative practitioners in the interactive digital arts, the limitations of standard laboratory research led to a demand for different kinds of settings based upon models from art studio practice. The C&CR 'Studio as laboratory' provided opportunities for practitioners to carry out research that centred on their creative practice in an experimental space at the same time as generating and communicating new kinds of knowledge.[21] In order to achieve this, an environment for creative practice with digital technologies must be highly responsive, supporting iterative processes in which practitioner insights are fed back quickly into the development process. This co-evolutionary process is fundamental to practice-based research where the existing technology is used in a new way and from which research derives new answers; in turn, the use of new digital technology may lead to transformation of existing forms and practices in art. The continual iteration of the practice-based process in interactive digital arts is not unfamiliar in other technology fields such as interaction design and, similarly, it can lead to innovative outcomes.

In deciding what was needed in the technology-based environments for creative practice, C&CRS looked at the requirements for the physical and technological spaces that supported the making processes of art practice. When C&CRS was first set up, in the mid-1990s, the facilities and resources available imposed their own constraints. A significant aspect of the response strategy was to draw upon the full resources of the university, including the knowledge base of the personnel across many disciplines. Much equipment was available that had not been acquired for artistic purposes, such as virtual reality laboratories and computer-aided engineering design systems. These proved useful for some art technology projects. However, most of the C&CRS artists began and finished their work in the studios where the facilities existed for use primarily in art practice. It was an important strategic decision to ensure that artists were not placed in a position of begging or borrowing from scientists and technologists with other pressing needs for the facilities. The primary reason for this was to ensure that the responsiveness to changes required by artists in the environment could be undertaken without creating conflicts in existing hardware or software configurations. Indeed, the technology itself was often advanced in response to artist's requirements.

The methodological focus of C&CRS was residential case study work and surveys of existing practice. The studies included a survey of current digital art artists which revealed the existence of a very broad range of creative digital technology practices with diverse approaches.[22] From the initial studies, a number of questions were identified about future requirements of creative technology environments: for example, what are the opportunities for digital technology to support creative practice and what are the design features of a supportive creative art technology environment? In order to investigate the questions and pursue the ongoing work at C&CRS, further funding was acquired from the EPSRC and COSTART, (**CO**mputer SysS**T**ems for Creative Work: An Investigation of **AR**t and **T**echnology Collaboration) came into being.

The COSTART research was case study-based in a series of artist-in-residencies. Artists identified from the prior survey were invited to a workshop when the opportunities and requirements were discussed with prospective technology researchers. It was particularly important that artists were prepared to participate actively in the research, as it involved recording events as they happened, and being willing to discuss openly their emerging ideas and problems encountered throughout the time in residence. The people resources available to the artists included a designated contact person and specialist consultations with technical experts throughout the residency period and for follow-up work up to one year after. The initial technological requirements had been identified from the artists' individual project proposals.

The co-evolutionary process, in which practice-based research was used to develop the art practice and technology in parallel, led to an understanding of research needed for the provision of responsive environments for digital art practice. The findings arose from a co-evolutionary process, in which practice-based research was the method used to answer the kinds of questions posed previously. A practice-based research life cycle was devised that represented the overall methodology applied across the various art technology projects undertaken and can be found in Figure 1.4.1.[23]

The COSTART outcomes are described in a number of reports and papers; a summary may be found here.[24] The research findings and first-hand accounts of those explorations by the participating artists and technologists are documented in *Explorations in Art and Technology*, a book published in 2002[25] and updated in 2018 with many more international perspectives.[26]

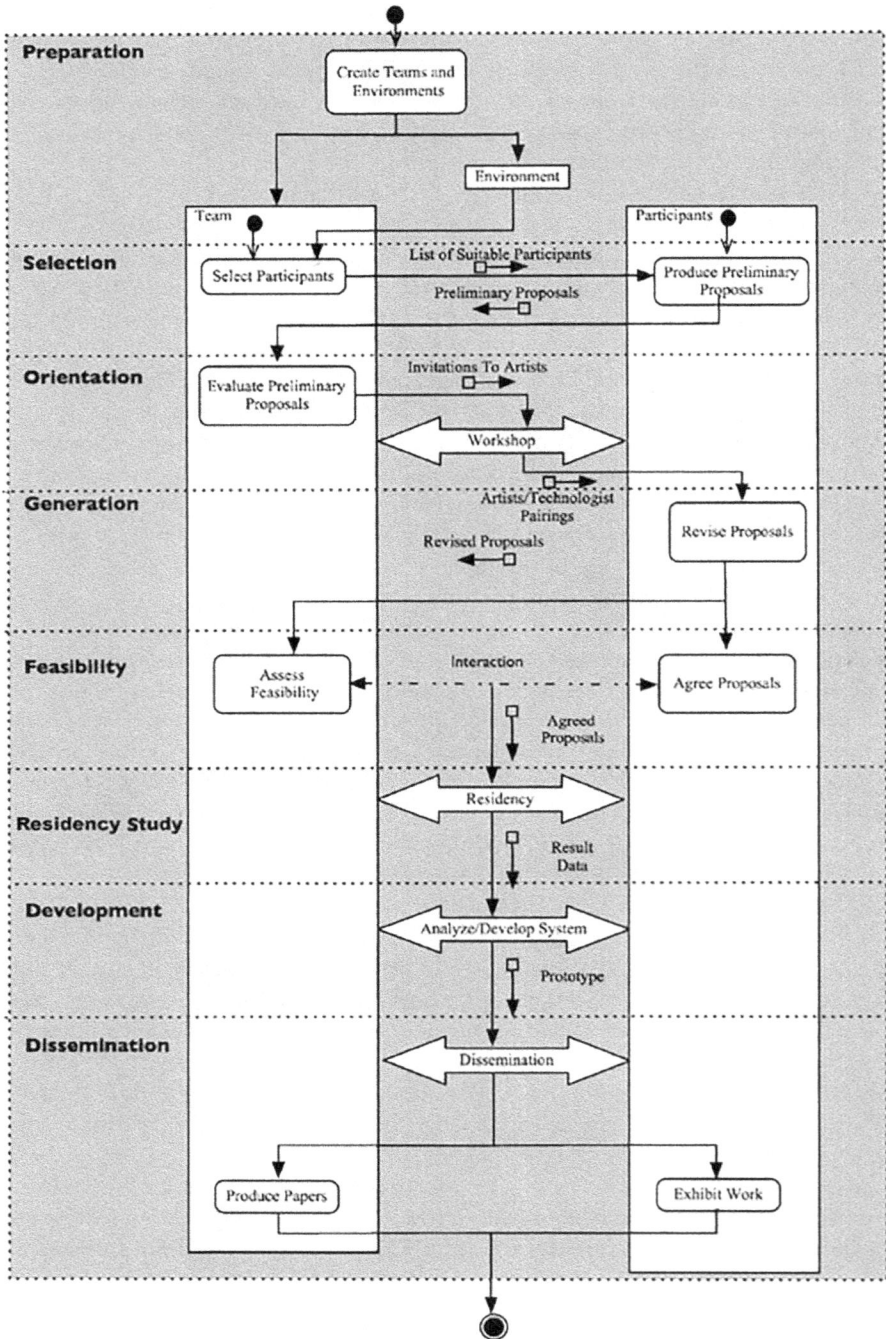

Figure 1.4.1 Practice-based research life cycle.

CCS and PhD research

In 2003, C&CRS was relocated as Creativity and Cognition Studios (CCS) at the University of Technology, Sydney. The work of C&CRS continued but with an enlarged PhD programme structured in a more formal manner. Candidates accepted into CCS followed a practice-based study structure that built on the existing PhD procedures at UTS, such as annual reviews and co-supervision as well as the standard university-wide research training programme. The CCS structure was based on a fundamental tenet that, from the start of the PhD, the research should include parallel creative practice and writing activities. The practice was central to the research and, while it was always closely discussed with supervisors, it had to be driven from within the individual's practice. The only extra advice that was always given was that each practitioner researcher kept detailed (informal) logs of what was done, what went well, what went wrong and, where possible, what thinking was behind the decisions.

For the writing component of the doctoral programme, more general advice was offered. A generic outline thesis was provided to each student when they began and, often, before that. This outline was not prescriptive but offered a first cut at a structure for what might be written. The actual structure, of course, evolved as the PhD progressed and was often only finalised in the final stages. A flexible writing plan of action was normally generated from the structure. For example, in the first year, the writing component would normally include a first draft of the state of the art, a refined description of the research problem or issue being addressed and a first cut at the methodological considerations. This approach is discussed in the lead chapter in Part IV of this book, evolved from the CCS structure document.

CCS PhD programme

In order to foster inter-disciplinary practice and research and the development of a strong community of practitioner researchers, having a clear and unambiguous approach to research is vital. At CCS, the programme of PhD research was established through a managed process that involved regular meetings and seminars for the presentation of ongoing work, made easier by co-locating the PhD candidates. The bringing together of students from very different backgrounds, ranging from computer science to English and fine art, formed a strong inter-disciplinary community. Through this, it was possible to engage the students in the creation of documents necessary for many aspects of the research, for example, documents to enable ethics clearance.

Regular meetings of the research group were an important part of the process. In the early period there was no prior agenda, and the choice of topics was open to research students and academic staff equally. These meetings were significant support to the community. Certain formal matters were covered, however, and the community engagement was strongly encouraged by giving each member equal status in such discussions. The two issues were ethics requirements for research studies and the selection of works to be installed in Beta_Space, the public space for practice-based research, discussed later. As time progressed, meetings expanded to include reading clubs that centre around the discussion of a research paper. Often, this is a published document from outside the group, but sometimes it is a draft paper by a CCS member to be submitted later to a journal or conference.

Ethics for practice-based research studies

Ethics approval is important when making any kind of study involving people. All universities have formal procedures for obtaining ethics approval. Often, these procedures seem distant and may, at times, seem a burden. CCS made the process more supportive and easier to implement. Each request for approval was, and still is, considered in a regular CCS meeting, although during the COVID-19 years this inevitably meant substituting an email process. A researcher provides written evidence of their proposed study and then it is discussed at a meeting and either supported or sent back for revision. This places a strong emphasis on the research methods and their justification in the context of the ethical issues. The discussion, therefore, becomes a positive supportive step, Fortunately, the University Ethics Committee, having attended CCS meetings, was willing to offer limited powers to grant approval. In practice, CCS can formally give ethics approval, retrospectively reporting to the Ethics Committee. There have to be certain restrictions; for example, approval to work with children can only be given by the University Committee. A CCS discussion will, even then, still take place.

Documents that supported practice-based research applications for ethical clearance were designed by the CCS researchers themselves with advice from the University staff tasked with this area of research compliance. The CCS Standard Operating Procedures Document for staff and post-graduate students of the Creativity and Cognition Studios outlines the practical aspects of applying ethics to research practice. It is used in conjunction with the *CCS Code of Ethics*, which details the legal codes and principles, including privacy and confidentiality, informed consent; approaches to the design, conduct and reporting of research, professional conduct; and further sources of information about the responsible conduct of research. For details, see Part IV, "The Practice-based PhD', in this handbook.

Beta_Space: a living laboratory for practice-based research

Art that depends for its full realisation on interaction with audiences requires an organisational model that brings together studio-based and public space research. Practitioners can begin making proto-types of their new works in a studio environment and then test them to a point where they might be ready for exposure to public scrutiny and interaction. The studio setting in a university, for example, can perform the role of a laboratory, because it is equipped with specialist tools and facilities. In a public exhibition space, both the works themselves and the accompanying research processes may need to be adapted to very different conditions. In 2004, CCS formed a collaboration with the Powerhouse Museum Sydney to address this issue, from which Beta_Space was created as a 'living laboratory' for practice-based research (Figure 1.4.2).[27]

Beta_Space proved to be a new and powerful organisational model that enabled practitioners to explore artistic concepts and prototype art systems and evaluate experiences with audiences. From the outset, the audience was involved in this process, changing the relationship of the artist to the audience and the relationship of the audience to the artwork. The participation of the museum team of exhibition curators, technicians, organisers and exhibition specialists was vital to the success of the research and the participant audience studies. Looking back, the senior curator responsible for the initiative, Matthew Connell, noted:

> It was a prototyping space for artists, it was a prototyping space for museum engagement, but it was not just about artists; there was something for us (museum workers) too, some-thing new for us to learn about ourselves too. Museums are significantly disrupted by new technologies and Beta_Space showed us a new way to engage this change by inviting peo-ple into the process rather than telling them what to think about it.

In this way, the museum played a vital role as a space for creating new works and new knowledge where the public was able to engage with the art and the research process. Beta_Space continued in the dedicated area until 2009, when a very large project, the Mars Yard, took over the space. It is

Figure 1.4.2   Entrance to Beta_Space, Spheres of Infl uence poster by Andrew Johnston exhibited September 2006.

Source: Photo by Julien Phalip 2007.

worth noting that an artwork that appeared in Beta_Space set a precedent for the introduction of a material never before permitted in the Museum; the material was sand:

> Jen Seevinck's work with the sand interface (+/- *now*, 2008) allowed us to lay the foundation work for the Mars Yard. Before that, there was no sand allowed in for archival reasons, but after Jen's work was a success with no fallout, in came the Mars Yard with a lot of sand.[28]

Work continued elsewhere in the museum until 2011, when limitations on funding led to closure.

The Beta_Space practice-based research contribution

As soon as a version of an artwork produced in the CCS was considered to be operational, it was made available for public exhibition in Beta_Space. During an exhibition, a research process was put in place that monitored audience engagement with the work (see Figure 1.4.1).

Prior to installing works in Beta_Space, criteria were applied that all aimed to meet,[29] for example, that a work was sufficiently well developed to be able to stand up to a significant amount of audience interaction. Sometimes, especially in the early stages, the sheer novelty of the art meant that risks were taken before the prototype was ready for public use, and inevitably there were problems:

> there was a real act of watching her work fail with our audience, even though it was beautiful and subtle and about not being able to sleep, so relatable. There were no cues (in the installation design), no pause for the audience for her constructed narrative to play out . . .

Figure 1.4.3 'Drifting' by Julia Burns.

. . . What that showed me is that prototyping is an essential part of creating museum experiences, artworks and any sort of design. . . . It showed me that museum lore (the rule of thumb) that we can't slow down people, that things have to take place in a set amount of time, isn't necessarily true either; that the way we design our exhibitions are open to new design solutions.[30]

Drifting is an interactive installation that explores human relationships in the transitional spaces between sleep and consciousness. It combines real world and virtual, screen-based technological worlds to create immersive and emotive experiences for participants and audiences. It was installed in Beta_Space for several weeks, during which time the artist observed people's experiences of the work and was able to identify unresolved aspects of its design and performance.

The practitioner researchers used data collection methods in Beta_Space, such as direct observation, audio and video recordings, including 'video-cued recall'.[31] As an example, the study of *Iamascope* analysed the interactive experience in light of four categories of response, control, contemplation and belonging, from which insights into the interrelationships between the four categories were identified as well as an understanding of the emotional states of the participants.[32] Evaluation studies of artworks installed in Beta_Space are described in Turnbull et al. (2011).[33]

Contributions to analysis frameworks for the evaluation and comparison of different audience experiences were developed by Zafer Bilda, a researcher who examined Beta_Space installations and the results of the individual studies of these works. Each artwork was characterised in terms of the feedback to the audience, the use of gestures and movements called 'bodily interactions' and goals and sense making referred to as 'cognitive indicators'. The study revealed that participants created unique situations out of their individual interaction experience and that triggered increased bodily and cognitive interactions or cognitive activity.[34]

As an example of more general results from this kind of analysis, an in-depth protocol analysis across a number of audience experiences helped us discover patterns we would not have been able to see simply through direct observation of individual events. The aim of protocol analysis was to analyse the period from when a participant enters the exhibition space to finishing his or her interaction with the artwork. The analysis helped us see and better understand participants' behavioural patterns, emotions and thinking process. Over the course of the research, a unified scheme was developed for studying audience experiences across different artworks: this embodied cognition framework was used to identify patterns of interaction occurring between the participant and art installation.[35] The research accumulated a body of knowledge which advanced our understanding of the nature of interaction between people and technology systems and in particular produced the Creative Engagement Model (CEM).[36]

CCS practitioner researchers discovered through placing works in the Beta_Space that there were different ways to approach and facilitate experimenting for audience engagement (see Figure 1.4.4). Moreover, people within the museum began to see what was possible when risks were taken:

there was something about the maverick nature of that particular project that was central to its success[37].

there was a freedom to not having things set out, to be able to try things out and see what happened[38].

The Ian Gwilt exhibition (*save_as*, 2007), was our first encounter where something from the virtual world turned up in the real world. There was a sort of moment of realising

Figure 1.4.4   Save_as by Ian Gwilt. Photo by Ian Gwilt. Mundane Traces at Beta_Space, August 2007.

that the boundaries between the virtual and the real were coming unstuck – a difference in the relationship between the two (upon reflection). It foreshadowed the post-digital.[39]

What Beta_Space and CCS together offered was a model for practice-based research that enabled professional practitioners to carry out their research in a manner that not only harmonised with their normal practice but provided extra dimensions to learning, experimenting and networking:

> I think back on Beta_space in a reflective research way, about process and foundational knowledge . . . as a curator, and a museum staff member, it gave me so much more than just foundational knowledge . . . apart from providing me with a learning space, a platform to experiment with and through as an emerging art-tech curator, a site for networking within the museum industry in a performative way that was appreciated, and where I learned about practice-led and -based research in an applied manner.[40]

For details of the lessons learnt, see Connell and Turnbull (2021).[41]

Recommendations for fostering practice-based research

Looking back on the history of bringing C&CRS, CCS and Beta_Space into existence, it is sometimes easy to forget the many pitfalls and potholes on the journey. Getting there was often a hard road to take, not least because it was a struggle to obtain resources, but mostly because at the beginning, there were few models to draw on. Having a vision for facilitating places where people who called themselves artists were equally comfortable with those who called themselves engineers was only a first step, and articulating a clear sense of direction, backed by a track record of exemplary work, only came with time. Additionally, as anyone who has been involved with breaking through boundaries in the past knows, you cannot expect acceptance, let alone support, from those who are comfortable with existing organisational structures: in universities in particular, the way disciplines are funded sometimes makes it difficult to achieve buy-in for inter-disciplinary ventures. Nevertheless, it was possible, and with resources from external funding bodies much can be done.

In establishing centres for practice-based research, it is important to put in place clear strategies that support practice-based PhD programmes, including ethical procedures that accord with practitioner researcher needs. This goes beyond simply incorporating practice-based research into general purpose research training. Every initiative will have its own starting points and drivers conditioned by the specific situations of practice and fuelled by the demands of disciplines and working practices. For that reason, in order to foster vibrant practice-based research communities, it is important to develop a set of principles, values and positions in respect of ethical procedures that accord with the context; these will all inevitably be influenced by geography, culture and philosophy. Where a centre for practice-based research has the PhD as its main focus, there are documented lessons from studies of PhD experience carried out at CCS.

What follows is a selection of recommendations drawn from the experience of creating spaces for practice-based research, in particular within a PhD programme. A detailed account may be found in Candy and Edmonds (2018).

Develop a research culture

Practice-based PhDs in the creative arts are often undertaken by mature students, and therefore, consideration needs to be given to the kind of culture and social environment that is suitable for this kind of research. Supporting doctoral research requires not only more than library and laboratory resources but also the availability of an appropriate social context. The existence of a collegial environment around a research group is a vital feature of a successful research culture. This can be achieved only by enabling everyday exchanges, as well as regular meetings and research seminars, where students can test out developing ideas and experience constructive criticism from their peers. A genuine collegial culture works best where like-minded people can share interests on common ground.

The practice-based model of PhD research is characterised by a high level of personal engagement. At the heart of the practice-based PhD, practitioners aim to explore and enhance their practice, as distinct from becoming an expert in a research topic. For that reason, other models of PhDs, typically with a culture of peer review and generally agreed standards of evidence-based knowledge, do not necessarily provide an appropriate vehicle for practitioner researchers. Having said that, some important features of traditional PhD research, such as establishing research questions, generating results from studies and publishing in peer-reviewed journals, are equally valid activities of practice-based research. However, according to a 2008 study of UK practice-based PhDs by Clements and Scrivener, there is 'a lack of conformity to the discursive formation of academic research, suggesting that either artistic research is fundamentally different from other research domains or that the practice-based research community has not come to terms with the generic research discourse',[42] indicating there is no consistent model of PhD management across all universities.

Practitioner research supervision

A successful practice-based PhD programme depends upon the availability of supervisors with appropriate skills and expertise. An important question to consider is whether or not being a practitioner should be a requirement for supervisors of practice-based doctoral research. For practice-based supervision, academic qualities need to be combined with practitioner experience and the ability to engage with the particular characteristics of practice-based goals and methodologies. Supervisors must have a high degree of expertise and experience of the candidate's particular area of research as well as the ability to communicate, be open-minded and empathise with the student.

Being able to undertake a practice-based PhD is, as with other kinds of PhDs, dependent upon having studentships available with accompanying organisational support. The funding is very important but, in choosing between institutions that offer research studentships and a wealth of resources, a decisive factor is the quality of the supervision. Students are often prepared to relocate across long distances where the right supervisor, combined with student funding, is available.

PhD research and the artefact

Observing artefacts in an exhibition space as part of the research process is often part of the practice-based PhD. Showing works in public supports the development of research thinking as well as the progression of new works. It provides realistic preparation for an actual installation and an opportunity for reflection in the light of audience experience. It also enables the public space to be used as a research laboratory, providing a facility for more extensive evaluation studies as was shown in the Beta_Space model described previously.

The role of artefacts in a PhD submission and the way in which it supports a research contribution to knowledge are important parts of the practice-based discourse that is addressed in chapters by Scrivener and Edmonds.[43] The materials, systems and other documentation included in practice-based PhD submissions do not always enter the archival records in the same way as the complete written theses, of which copies are lodged in the university library to be available for borrowing in the public domain. This means that the archiving of the records of artefacts, installations and performances as valuable outcomes from practice-based research is often incomplete. This implies that the regulations for submission must take account of the central role of the artefact in the research in order that it can be included in public archives.

Regulatory codes of practice: including the artefact

For practice-based PhDs, a critical difference from standard PhD models is the existence of regulations that validate the contribution that artefacts can make to the research process and its outcomes. For practitioners, the creation of artefacts, installations and performances is central to the research aims. The rules and regulations that govern the place of artefacts in PhD research differ between institutions and from country to country. Good models of rules for governing the submission of works and written theses are needed if high standards are to be established and maintained. University regulations, such as those of De Montfort University,[44] not only allow the submission of artefacts but also prescribe the inclusion of a written element and the criteria to be met in assessing the originality and sustainability of the contribution of the work in a coherent field of research. This combination of written and artefact elements is essential to the of the practice-based PhD.

Strategies for running practice-based PhD programmes

Running a practice-based PhD programme in the style of CCS can be summarised by a number of strategies that were employed.

Encourage documented creative practice from the outset

Everything that happens in practice, included discarded avenues of exploration, can be important for the research, and so the common practice of keeping a diary or log is a valuable source of data. Students are encouraged to fully engage in this practice and to retain all of the notes that they make. The form in which the information is recorded varies considerably, depending on the individual practitioner's preferences, but the material is often made more valuable if creative reflective practice has been considered.

Provide students with a first draft thesis structure from the outset

In any PhD, it is good practice to start writing from the beginning. Completing the research and only then 'writing it up' tends to lead to many difficulties. In the case of practice-based work, the practice itself can be much more attractive an activity than the writing, and so the discipline of writing all the time is particularly helpful, especially as a first cut at the final thesis structure helps to identify what must be done in writing about the practice and what is learnt.

Use that structure to contribute to project planning

It is helpful to have writing goals as well as practice goals. For example, a draft of the 'state-of-the-art' or literature review chapter is good to have by the end of the first year. The thesis outline can be used as a document from which a target plan can be drawn.

Build the community with regular group meetings

Studying for a PhD can be a lonely activity, and that is a particular problem if some of the methods being used are unfamiliar or the work is inter-disciplinary. A community of students can support one another in ways that add considerably to what supervisors can offer. They all face similar problems but come at them with very different levels and areas of expertise and experience.

Share and discuss ethics proposals across the group

Beyond the informal exchanges, regular meetings can be used to discuss ethics proposals. In this way, what is sometimes seen as a burden can be a positive supportive step in the research process. Having peer review among PhD students, at the same time as advice from supervisors, benefits all.

Conclusions

We have described the establishment of the 'studio as laboratory' and 'living laboratory' through the history C&CRS and CCS Beta_Space. These organisational models represent innovative ways of fostering inter-disciplinary practice-based research. Many lessons were learnt that were an essential part of developing strategies and methods for achieving high quality outcomes. Strategies for running practice-based PhD programmes in this context were described, including support for inter-disciplinary community building, thesis writing and dealing with ethics approval.

Notes

1 EPSRC: https://epsrc.ukri.org.
2 COSTART: Summary of the Project and Outcomes February 2003. lindacandy.com/COSTART/pdfFiles/COSTARToverview.pdf.
3 Edmonds et al. (2005).
4 Candy and Edmonds (2011).
5 The name Beta_Space comes from the software development practice of releasing new applications and products to the public before they are completed in order to gather feedback and improve their quality.
6 https://maas.museum/powerhouse-museum/.
7 Connell and Turnbull Tillman (2021, pp. 79–93).
8 Kandinsky (1923).
9 ICC (2003) E.A.T. – The Story of Experiments in Art and Technology. NTT Intercommunication Center, Tokyo; Billy Klüver, E.A.T. Bibliography 1965–1980, New York: E.A.T., 1980, 87 p. 2.
10 https://monoskop.org/Experiments_in_Art_and_Technology.
11 Advanced Telecommunications Research International ATR:MIC www.atr.jp/about/atr_e.html.
12 Fels, S. & Mase, K. (1999). Iamascope: A Graphical Musical Instrument . . . *Computers & Graphics,* 23, 277–286.
13 Iamascope in Beta_Space: http://research.it.uts.edu.au/creative/betaspace/iamascope.html.
14 See www.plymouth.ac.uk/research/planetary-collegium; https://en.wikipedia.org/wiki/Planetary_Collegium.
15 In Candy et al. (2018, pp. 289–385).
16 Creativity and Cognition Conferences: ACM Digital Library: https://dl.acm.org/conference/c-n-c.
17 Creativity and Cognition Conferences: https://dl.acm.org/conference/c-n-c.
18 Harris (1999).
19 Edmonds (1999).
20 Ibid., p. 4.
21 Edmonds et al. (2005).
22 Candy and Edmonds (2000).
23 Edmonds et al. (2005, p. 461).
24 COSTART: Summary of the Project and Outcomes February 2003: lindacandy.com/COSTART/pdfFiles/COSTARToverview.pdf.
25 Candy and Edmonds (2002).
26 Candy et al. (2018).
27 Muller and Edmonds (2006).
28 Matthew Connell in Connell and Turnbull Tillman (2021).

29 Costello et al. (2005).

30 Matthew Connell in Connell and Turnbull Tillman (2021).

31 Video-cued recall or retrospective reporting is a method for collecting verbal data commonly used for investigating human cognitive processes. Because reports are made after the experience, this method is regarded as having less impact on cognitive processes than concurrent think-aloud methods. Reporting retrospectively, however, presents the risk that the participant will forget details and that their recall will be interpretively filtered. The video cued-recall method helps to avoid these pitfalls by using video to help the participant recall the detail of their experience and avoid selective interpretation. See Costello et al. (2005).

32 Muller et al. (2006).

33 Turnbull et al. (2011).

34 Bilda et al. (2006).

35 Bilda et al. (2007).

36 Bilda et al. (2008).

37 Matthew Connell in Connell and Turnbull Tillman (2021).

38 Deborah Turnbull in Connell and Turnbull Tillman (2021).

39 Matthew Connell in Connell and Turnbull Tillman (2021).

40 Deborah Turnbull in Connell and Turnbull Tillman (2021).

41 'When Matthew and I published our chapter in *Interactions* together (Libri, 2011), I could see that we had learned a lot in terms of museum curatorship, events production, networking, programming, and what the pieces were that made all of those things possible in both independent and institutional practice. This was supported within the museum in larger programs such as the annual Art/Science networking events as part of Science Week. As an independent curator, I was then able to bring this knowledge forward into my research platform where I played with it for 8 years before being invited to do a PhD on my curatorial process. This resulted in the first criteria for curating interactive art, and it all grew from my first practice: curating Beta_space'. Deborah Turnbull in Connell and Turnbull Tillman (2021).

42 Clements and Scrivener, S. A. R. (2008). *The Discourses of Practice-Based Arts Research and How Contribution Is Made*. Presented at Research into Practice, November 2008, RSA, London.

43 Scrivener (2022); Edmonds (2022). In Vear, C. (ed.) *The Routledge International Handbook of Practice-Based Research*. Routledge: London and New York.

44 Research Degree and Higher Doctorate Regulations: De Montfort University 2019 p35www.dmu.ac.uk/documents/dmu-students/academic-support-office/policies/dmu-research-degree-and-higher-doctorate-regulations-2019-final.pdf.

Exhibition, Performance, Creative Writing or Similar Work

19.2 A student may undertake a programme of research in which the student's exhibition, performance, creative writing or other similar work, forms as a point of origin or reference, a significant part of the intellectual enquiry. Such work may be in any field but must have been undertaken as part of the registered research programme. In such cases, the presentation or submission of work relating to exhibition, performance or other creative writing or similar work must be supported by documentation in the form of a thesis which sets the work in its relevant theoretical, historical, critical and/or design context . . .

Bibliography

Bilda, Z., Candy, L., & Edmonds, E. A. (2007). An Embodied Cognition Framework for Interactive Experience. *CoDesign: International Journal of Co-Creation in Design and the Arts*, 3(2), 123–137,

Bilda, Z., Costello, B., & Amitani, S (2006). Collaborative Analysis Framework for Evaluating Interactive Art Experience. *CoDesign: International Journal of Co-Creation in Design and the Arts*, 2(4), 225–238.

Bilda, Z., Edmonds, E. A., & Candy, L. (2008). Designing for Creative Engagement. *Design Studies*, 29(6), 525–540, November.

Candy, L., & Edmonds, E. A. (2000). Creativity Enhancement with Emerging Technologies. *Communications of the ACM*, 43(8), August. https://doi.org/10.1145/345124.345144.

Candy, L., & Edmonds, E. A. (Eds). (2002). *Explorations in Art and Technology*. London: Springer-Verlag, 1st ed.

Candy, L., & Edmonds, E. A. (2011). *Interacting: Art, Research and the Creative Practitioner*. Faringdon: Libri Publications Ltd.

Candy, L., & Edmonds, E. A. (2018). Practice-Based Research in Practice: Regulations and Recommendations. *Leonardo*, 51(1), February. https://direct.mit.edu/leon/article/51/1/63/46472/Practice-Based-Research-in-the-Creative-Arts.

Candy, L., Edmonds, E. A., & Poltronieri, F. (Eds.). (2018). *Explorations in Art and Technology*. Springer Cultural Computing Series. London: Springer-Verlag, 2nd ed.

Connell, M., & Turnbull Tillman, D. (2021). *Stand Outs for Beta_Space Email Communication*, forthcoming.

Costello, B., Muller, L., Amitani, S., & Edmonds, E. A. (2005). *Understanding the Situated Experience of Interactive Art: Iamascope in Beta_Space*. Proceedings of Interactive Entertainment, CCS Press, Sydney, 49–56.

Edmonds, E. A. (1999). *Final Report: Access and Creativity, Technology Foresight Creative Media*. London: Department of Industry.

Edmonds, E. A., Muller, L., & Connell, M. (2006). On Creative Engagement. *Visual Communication*, 5(3), 307–322.

Edmonds, E. A., Weakley, A. J., Candy, L., Fell, M. J., Knott, R. P., & Pauletto, S. (2005). The Studio as Laboratory: Combining Creative Practice and Digital Technology Research. *International Journal of Human-Computer Studies*, 63(4–5), 452–481.

Harris, C. (1999). *Art and Innovation: The Xerox PARC Artist-in-Residence Program*. Cambridge, MA: MIT Press.

Kandinsky, V. (1923) [1988]. English Translation Published as Plan for the Physicopsychological Department of the Russian Academy of Artistic Sciences. In J. E. Bowlt (Ed. & Trans.), *Russian Art and the Avant Garde Theory and Criticism 1902–1934*. London: Thames and Hudson, revised ed., 196–198.

Muller, L., & Edmonds, E. A. (2006). *Living Laboratories: Making and Curating Interactive Art*. SIGGRAPH 2006 Electronic Art and Animation Catalog, ACM Press, New York, 160–163.

Muller, L., Edmonds, E., & Connell, M. (2006). Living Laboratories for Interactive Art. *CoDesign: International Journal of Co-Creation in Design and the Arts*, 2(4), 195–207.

Turnbull, D., Connell, C., & Edmonds, E. (2011). Prototyping Places: Curating Prototype Interactive Artworks in a Museum Context. In *Proceedings of Re-Thinking Technologies in Museum 2011: Emerging Experience*. May, Limerick.

1.5

PRACTICE-BASED RESEARCH AT SENSILAB

Jon McCormack, Alon Ilsar, Tom Chandler, Mike Yeates,
Elliott Wilson, Camilo Cruz Gambardella, Nina Rajcic,
Maria Teresa Llano, and Sojung Bahng

Introduction

In this chapter, we examine practice-based research through the lens of a creative technologies research laboratory, SensiLab,[1] in which (at the time of writing) the authors are all active members. Our group is a multidisciplinary collective of researchers with expertise that includes media arts, design, performance, virtual world-making, computational creativity, artificial intelligence and machine learning. Established in 2015 by Professor Jon McCormack, the lab includes specialised working spaces and studios (Figures 1.5.1, 1.5.2 and 1.5.5) designed to facilitate hybrid research methodologies through active practice. Our research projects cover many different disciplines and application areas, from virtual heritage reconstructions using virtual reality (VR) technologies to specialist medical wearable devices, new performative musical instruments and creative artificial intelligence systems. Despite this diversity of research areas and applications, all our projects have their foundation in practice-based research.

We consider practice-based research[2] as a mode of enquiry that seeks to generate new knowledge through practice, i.e. by making, doing, building, experimenting, creating and experiencing, so it may often involve the use of *knowing-how* types of knowledge, gained through physical exploration and direct action. This mode of research has its origins with creative practitioners, such as designers, artists and architects, who were attempting to frame their activities as a form of research in academic settings. It has become an established research program at many universities, particularly in the UK and Australia. Creative practitioners intuitively understood that the generation of new knowledge was embodied in their practice and emerged in the studio as that practice developed. Here, practice was concerned not only in terms of final artefacts and outcomes, but also through the process by which the researchers achieved them.

While practice-based and practice-led research[3] have existed for many years in the creative arts – in areas such as design, fine art, architecture, music and performance – at SensiLab our mode of practice-based research also considers research practices outside these disciplines, including areas such as creative technologies, visualisation, artificial intelligence, human–computer interaction, interactive media and games, simulation, cultural heritage and information systems. With this expanded domain and enlarged scope of what research through practice is, our aim is to bring a successful research methodology in the creative arts to a broader group of researchers: those who can understand their research as practice, even if it doesn't fit into a traditional creative arts stereotype.

Figure 1.5.1   SensiLab is a contemporary response to conducting effective practice-based research in the internet age.

Source: Photography by Tom Roe.

For many researchers from discipline areas outside the creative arts, the idea of research through practice may seem unfamiliar or foreign, even if they have been implicitly undertaking it without thinking of it directly as practice-based research. So, a challenge for us is to demonstrate the effectiveness and value of practice-based research for these areas and to encourage thinking through practice explicitly to give recognition and bring value to this way of working in research.

So, here we will address the unique aspects of practice-based research in general and then look specifically at how they can be applied in a variety of disciplines, including those where practice is not traditionally associated with research. We primarily do this through examples of our own projects, since these are the ones we are most familiar with. We also introduce factors that we think are most important to facilitate and nurture a rewarding and effective research culture built around practice-based research. Again, we use our lab as the main example for expediency.

SensiLab is driven by the creative engagement of people with technology, so our examples will necessarily draw from our personal interests in this area. However, we hope this chapter will be generally useful, even inspiring, for anyone curious about how to undertake research through practice and what advantages it might bring to their own field.

Communicating and contributing to knowledge

A first step in understanding any mode of practice-based research is to examine the categories of knowledge that are most suited to research through practice. There are many types of knowledge

93

that are best understood and communicated non-linguistically (i.e. not through language); riding a bicycle is a simple example.[4] You can read all you want about how to ride a bicycle, but that will not give you the ability to ride one – you must learn *how* to ride through the act of *riding*. This kind of knowledge forms part of what is often referred to as 'knowing-how' – encompassing practical, embodied, implicit and tacit knowledge – as opposed to 'knowing-that', which includes theoretical, propositional, explicit and focal knowledge.[5]

At SensiLab, a lot of what we do involves *making*, that is, building physical or virtual artefacts and prototypes that in some way embody the ideas we are thinking about, or that explore the physical, functional or material aspects of our research. Making involves multiple types of cognition. As making generally involves the body, embodied cognition often plays a significant role in developing this kind of physical knowledge and applying it effectively. Even in the age of digital tools and virtual making, physicality and materiality play important roles in the design of artefacts, something that is often underestimated in technology-driven research.

The experiential and perceptual nature of designed artefacts and prototypes also forms an essential part of the communication of knowledge and ideas through practice. This involves phenomenological experiences and bodily interactions that are often multi-sensory, including kinaesthetics, proprioception, touch, sound, light, smell and taste.[6] They draw on the individual nuances and subjective nature of phenomenological experience that cannot be directly communicated linguistically or fully via secondary media, such as photography, moving image media or audio recording. Hence, it is the human *experience* of the artefact that embodies critical knowledge or ideas, and this direct experience is an important way of communicating and sharing this knowledge.

Scientific research emphasises the objective and rational. It strives for the repeatability of experiments and communication of knowledge (*knowledge-that*) through language, with this language often being formal (e.g. mathematics). While many areas of scientific enquiry involve making and building (for example, apparatus to support specific experiments or hypothesis testing), the results of this making and building are typically considered only as necessary tools – a means to an end – rather than the literal embodiment of research knowledge.

While different, research through practice is not in conflict with this, but gives voice to the subjective, individual nature of experience and the value of that experience in understanding the world.[7] Additionally, this mode of research through practice is not a binary nor mutually exclusive way of gaining and communicating knowledge. Knowledge gained through experience *can* be communicated, albeit imperfectly, linguistically, or at least via instruction on how to create or replicate the experience itself (again, imperfectly). Secondary media and documentation, such as photography, video, audio, etc. can also assist in communicating the knowledge embodied in practice.

Research based in sensory experience makes use of the researcher's 'phenomenological awareness': a sensory acuity and awareness of being in the world. A skill that can be learnt, for example, through critical listening, focused drawing, dance, performance or structured making. Such a skill differentiates the professional from the casual. Just as becoming a virtuoso violinist requires many thousands of hours of practice to build mastery of the instrument, so too the practice-based researcher must learn to enhance their phenomenological awareness as necessitated by their mode of research.

This awareness can be understood through a simple example: the substantive differences between taking a photograph of something and making a drawing of that thing. As German artist (and founder of the Fluxus moment) Joseph Beuys observed, 'drawing is a special kind of thought'. The photograph is instantaneous; the technology captures the light reflected or transmitted from the object and the image appears as a representation of that object in that moment. Making a drawing, on the other hand, requires time, and mental and physical effort. One must study the object carefully and mentally work out how to translate what is seen by the eye to movements of the hand. Details are not only 'seen', but also cognitively processed into representational and relational entities conveyed through movements of a pencil onto paper.

Now, some readers might have difficulty with this idea, particularly if they consider drawing as only something for 'creative people', suggesting creativity is something you either have or do not. The reality is vastly different; creativity is something you work towards, not a facility you just have.[8] Indeed, the goal in understanding through drawing has little to do with the perceived creativity, aesthetics or accuracy of the drawing to reality, although these naturally improve with practice and render the act more rewarding. It is about learning to see with a view to comprehending what is seen in a phenomenological way.

Tacit and implicit knowledge

Knowledge can be broadly classified into three types: propositional knowledge (facts), procedural knowledge (skills), acquaintance knowledge (objects). As we introduced in the previous section, tacit knowledge is, by definition, implicit knowing that cannot readily be communicated linguistically. Examples include riding a bicycle, playing a musical instrument, driving a car, learning a language or recognising a face. As tacit knowledge often involves kinaesthetic or body-based learning, it requires a person to directly experience or practice the task to become proficient.

Practice-based research programs typically exclude this kind of knowledge due to the difficulty of representing and communicating it, even through exhibition or direct experience. It should also be noted that knowledge gained through making is not necessarily directly communicated through the artefact being made. So, the direct experience of an artefact may not communicate the knowledge acquired by the maker to the observer.

So, in simple terms, implicit knowledge resists codification. One reason for this is that gaining tacit knowledge requires training and practice, often with a person who has that knowledge to assist on a one-on-one basis (learning to play a musical instrument, for example). It concerns a specific 'knowing subject' (knowing-how), which makes it practically impossible to disseminate through electronic or printed media, for example.

Tacit knowledge raises an important issue regarding the knowledge contribution of practice-based research: organising, communicating and archiving of non-linguistic knowledge. Universities typically archive and classify the written PhD research thesis or exegesis and make this publicly available through their libraries. For practice-based research, the physical, sensory and exhibition components of the research are normally displayed temporarily for examination, but they are not retained or archived after the examination process is complete. While institutional structures such as museums, archives and galleries exist for the preservation and dissemination of these media, they are typically confined to specific areas or media, such as fine art, cinema or design, and are not necessarily focused on collections of research.

In summary, while acknowledging tacit knowledge is valuable for practice-based research, and may be acquired during the research process, we have found that it is challenging for this to be articulated beyond the researcher due to difficulties with communication and preservation of this type of knowledge. Of course, the researcher may go out into the world and teach this knowledge to others directly (through making, for example, the traditional method of knowledge transfer for craft-based practices), but as this knowledge cannot be recorded, it relies on direct human-to-human transfer, which is often a risky proposition.[9] It is our experience that the role of the practice-based researcher is to find ways to use their writing and made artefacts in combination as a vehicle through which the material artefact can find a discursive form.

Situated cognition

Ask not what's inside your head, but what your head's inside of.[10]

Situated cognition proposes that all knowledge is situated in activity, bound to social, cultural and physical contexts, and knowledge is inseparable from doing. Subjects and objects are dynamically, agentially and iteratively co-articulated in *intra-action*.

According to physicist and posthumanist scholar Karan Barad,

> We do not obtain knowledge by standing outside of the world; we know because 'we' are of the world. We are part of the world in its differential becoming. The separation of epistemology from ontology is a reverberation of a metaphysics that assumes an inherent difference between human and nonhuman, subject and object, mind and body, matter and discourse. Onto-epistemology – the study of practices of knowing in being – is probably a better way to think about the kind of understandings that are needed to come to terms with how specific intra-actions matter.[11]

More recent perspectives of situated cognition have focused on, and draw from, the concept of identity formation as people negotiate meaning through interactions within their communities of practice. From our own practice-based perspective, we consider situated cognition an important landmark in how we frame and theorise about research practice. We won't go into great detail here, but the interested reader is referred to the vast array of literature on this topic.

Comparing traditional and practice-based PhDs

A common question asked by our potential students at SensiLab is: what are the differences between a 'traditional' and a practice-based PhD? The most obvious difference is in what is assessed to evaluate the candidate's contributions to research. In a traditional PhD, the candidate's research is presented as a written thesis; in the case of our university (Monash), typically around 80,000 words. The written thesis is the sole 'output' used to evaluate the candidate's contribution to knowledge in the discipline or field of their work.[12] Assessment for a practice-based PhD includes an artefact, or series of artefacts, that are presented for examination and that engage one or more of the senses. The artefacts are also accompanied by a written exegesis, typically 30,000 to 35,000 words in length. Examiners receive the exegesis six weeks before attending the exhibition of research artefacts. The contribution to knowledge is determined through both the written exegesis and the artefacts produced. Examiners consider all the materials provided for examination equally in determining the knowledge contribution of the research.

The practice-based exegesis differs from the traditional thesis in that its primary purpose is to contextualise and support the practice presented. It would typically include the research questions or a research statement, a field survey or literature review that situates the candidate's work in relation to prior knowledge, theoretic or practical knowledge that has informed the production of artefacts and the studio practice. It may also include documentation of the production process or procedures developed as part of the research. As with a traditional thesis, supplemental material, such as video or photographic documentation, may also be included.

Beyond assessment, there are differences in the approach and methodology typically used in traditional and practice-based research. Many of these differences will be explored in this chapter. Most obviously, practice-based research at SensiLab must include a studio practice, i.e. the development of research undertaken through making and the iterations of evaluation and modification used to arrive at a completed artefact. As the artefacts engage one or more of the senses, they bring into play the sensory and phenomenological aspects of experience already alluded to in this chapter.

Practice-based research in creative technologies

While practice-based research has a clear and strong relationship with the creative arts,[13] it also has a particular relationship with certain kinds of technology-based research. Practice-based research methodologies have their similarities to those of *design science*: a set of methodologies often used in technological research that focuses on the building of new realities rather than explanation of existing ones.[14] These similarities include employing iterative processes and creating and evaluating artefacts. However, our practice-based research attempts to make a shift from a goal-driven approach to a value-driven one, where our researchers use self-reflection to step back from the goal and suspend their judgement during the process. This often leads to more rigorous questioning of the goals set out and the broader implications of the use of specific technology. Indeed, technology may not always be the best way to solve a given problem or reach a certain goal. Moreover, a focus on solving current problems using technology can often limit broader conversations about the kind of technological futures we really want. Here, design methodologies such as *speculative design* allow us to explore the cultural, social and technological implications of probable futures before they become realities.[15]

There are established practice-based researchers whose practice revolves around utilising technology to not only drive the creation of new art, but also explore our relationship with that technology and challenge the way we use it. Within SensiLab, there are artistic projects that raise questions – through creative technology – about our connection to history, our bodies and our emotions (some are detailed further in Part IV). However, our practice-based research is not limited exclusively to creative outputs. It can bring a more fluid creative process to more traditional 'problem-solving' tasks. Our lab strives to also work on projects that do not have clear 'creative' outputs. Technology from these projects may sometimes end up being used in creative projects, but using practice-based methods in the projects themselves also helps to shift goals and often change or challenge the research questions being asked.

There is often an insatiable urge within our practice-based research community to collaborate with researchers in other disciplines. Artists often work in collaboration with technologists, and so develop their skills in both self-reflection and group reflection. While collaborating with engineers and scientists, new forms of communicating knowledge and new ways of challenging our notions of creativity often emerge. Furthermore, we have witnessed how collaborators not only think together; they make, do and practice together, not merely thinking their way through a problem but 'practis-ing' their way towards a resolution. This shared practice builds into 'communities of practice' where groups of people across diverse disciplines – who share a common concern or passion – interact regularly and learn from each other. SensiLab uses internal work-in-progress presentations and read-ing/discussion clubs to continually discuss and realign the group's focus. To complement this internal reflexive process, we regularly invite external practice-based researchers to present public forums on the application of creative technologies, expanding our community of practice beyond the lab.

Designing a research culture

A focus on technology means that technological research comes before wrting [sic].[16]
    That new knowledge is expected to have two characteristics: it is shared and it can be verified or challenged.[17]

We now discuss our own research culture and the ideas and values it is built around. Saying you are 'doing practice-based research' is fine, but what about the mechanisms to support and nurture it in a working research environment? In this section, we will look at how we have built our own research culture and the infrastructure, methodologies and community that support it. These are specific to SensiLab, but we have observed similar ways of working and thinking about practice-based research at a number of other institutions around the world.

Here we summarise our strategy for this, which can be divided into three core aspects:

1   The physical workplace design and the infrastructure it contains;
2   Articulating our goals and values in an open and clear way;
3   Ongoing critique and reflection on our processes and outcomes.

As will be explained next, this strategy includes both tangible and conceptual elements, which are designed to be complimentary and work together. We will now present each of these topics in more detail.

Workplace design

Our lab is dedicated to exploring the undiscovered creative opportunities of technology, and our mission is to understand the innovative creative applications of that technology through practice. We work across many different disciplines, approaches and materials, and approach each inquiry with curiosity and rigour in order to learn through creating. It is important that we acknowledge process as well as outcome and value the experiential and the sensory. Hence, we have tried to design a creative space where new connections – including personal, conceptual and practical – can be easily made. Our lab's design is built around the premise that new thinking emerges from diversity, genuine collaboration and an openness to the unexpected.

Our working space design takes the specialised requirements of practice-based research as fundamental. This includes specialised atelier areas ('Makerspaces', Figure 1.5.5) with communal access to digital fabrication (laser cutters, CNC machines, 3D printers), electronics hacking and physical making and building. In contrast to more public-oriented consumer makerspaces, we also include specialised equipment and facilities, such as high precision stereo microscopes, robotic SMT pick and place machines, advanced electronics diagnostic equipment, industrial robotic arms, multi-material 3D printers, programmable embroidery and sewing machines. Together, these facilities support an 'anything is possible' approach to making and building, at least at certain scales.

The physical design mixes open-plan working environments with more private spaces such as reading nooks and small studios, providing a balance between open collaboration and the need for focused concentration or privacy (Figure 1.5.1). A number of studios provide specialist facilities, including a sound studio for experimentation with spatialised sound (Figure 1.5.2), an imaging studio and a motion capture/immersive VR studio which also hosts performances and talks. It is important to stress the democratic and open nature of these spaces: all facilities and areas are accessible and available (with appropriate training). Nothing is 'locked away' and visibility is an important reminder that the resources you might need are available. There is no space or authority hierarchy; PhD students share the same work area and spaces as research professors, encouraging implicit learning through observation and interaction between experienced researchers, early career academics and researchers in training. Casual chats about inspiring artworks and impromptu conversations about works-in-progress are commonplace.

Goals and values

As part of our research culture, we have developed a series of four key values that we believe should underpin all our research:

Figure 1.5.2   SensiLab's sound studio was designed to allow researchers to experiment, create and produce using immersive sound.

Source: Photo by Jon McCormack.

Practice

We learn through making, work to a creative end and value process as much as outcome. We are open to the unexpected.

Diversity

We believe that new thinking requires a diversity of skills, experience and perspectives.

Collaboration

We understand that progress comes from genuine collaboration, and actively share our knowledge to educate, inspire and influence change.

Integrity

We reflect on our position, respect all voices and conduct our work with integrity.

While each of these is succinct and relatively broad, these four 'compass points' serve to help us navigate how we undertake our research. Together, these values guide the research process, causing us to reflect equally on the process and the outcomes; using, where possible, ecological validation[18] over laboratory-based 'user-testing'; adopting iterative design strategies over trial-and-error approaches;

considering voices not present in any design or system; seeking to undertake research that is ethical and positive; always seeking cross-pollination of ideas and artefacts between different projects.

Critique and reflection

Intertwined with these values and the methodology they support is the issue of failure. This is particularly important where technology is involved, as there is often an unevenly weighted imperative to deliver 'successful' results above any other concern. As part of our critical review process, we acknowledge and reflect on failure as a normal part of the research process: it is only by understanding *why* something didn't work that we can learn how to improve both process and outcomes. This is particularly important for new researchers, who may be intimidated by the (apparent) successes of their more experienced colleagues (or, more likely, the modern academic system) and view failure as a weakness.

There are other, more procedural aspects of our research methodology that are worth highlighting. The first is developing detailed documentation of the making process, systematically recording each iteration to develop a 'research archaeology' of a final artefact or development. For this we make extensive use of visual documentation (images, videos) alongside the digital aspects of development (CAD plans and drawings, 3D models, source code, etc.) and often sound recordings or spoken word data. We typically store these using revision control systems, such as GitHub,[19] which allow both sharing (public and private) and revision history. Such histories allow reflection on the development process, as well as assisting in communicating any practice-led aspects of the research outcomes.

Another aspect of our methodology is that of the *critical review*. Borrowed from the 'art crit' sessions common in art schools for decades, research work is regularly presented to the group during development for constructive criticism, feedback, questioning and review. In addition to the obvious benefit of peer review and critique, such sessions also assist in keeping all researchers up to date with current projects and allow the fluid sharing of ideas and breakthroughs (however small or incremental). Such sessions are augmented with online postings on internal messaging systems, keeping those unable to attend involved and active in ongoing research conversations.

In summary, researchers using our space should have the expectation that 'anything is possible' – that learning something is a matter of trying things out and seeing what works and what does not. There will always be someone next to you who can share their knowledge and expertise, so you are never alone along your research journey. The open sharing of ideas and experiences helps to minimise repeating mistakes or wheel-reinventions that are common with more unstructured approaches to creative experimentation and play. Doing practice-based research is not a panacea for doing good research – it still requires scholarship, dedication, effort and persistence – but our environment and methodology naturally support that culturally, technically and socially.

Example projects

Having now explained the basic design, culture and methodological issues, we turn to practical examples to further illustrate how practice contributes to research in our lab. So, in this section we briefly describe some example projects that illustrate both the interdisciplinary nature of our research practice and hint at the scope of possibilities for practice-based research beyond the traditional creative arts.

Hardware development at SensiLab

SensiLab has developed many bespoke electronic devices across a number of different domains. This includes wearable sensors, interactive art installations, medical devices and customised robotic arms. Development of these devices was initially driven by their necessity for a specific task or to solve a defined problem. As discussed, an important aim of our practice-based research culture is to foster a diverse 'maker' community of practice. Through available expertise, shared hardware and software libraries, and investment in infrastructure and equipment, creating a custom hardware project within SensiLab – regardless of final target domain – is very easy.

Some examples of the benefit of this community of practice can be seen in three example hardware projects that we will briefly introduce here: the *Melanopic Eye*, the *DrawBots* and the hardware development of the *AirSticks* (Figures 1.5.3, 1.5.4 and 1.5.6 respectively).

The Melanopic Eye is a small, wearable sensor used to measure light quality, which integrates into an app designed to help you get better sleep. The DrawBot is a robotics platform equipped with sensors and a pen-holding mechanism that can be used to explore creative ideas in a physical way. The AirSticks are a new kind of gestural performative instrument for musicians.

Even though the end goals of these projects are diverse, targeting completely different applications and research disciplines, they evolve with and feed from each other; major milestones and developments in one project enhance and refine others. Examples of this can be seen in projects like the Melanopic Eye Sensor (a wearable sensor developed to enable light quality logging; Figure 1.5.3), where a breakthrough in Bluetooth development led to significant performance increases in the next iteration of the AirSticks project. Lessons learnt through numerous iterations of the DrawBot circuit board design (Figure 1.5.4, left) saved us from making the same mistakes in a further design iteration of the Melanopic Eye.

This synergy between projects and the researchers involved with them is made possible through the 'Makerspace' (Figure 1.5.5), an atelier space within the lab that is accessible (and visible) to all researchers working in the lab. The SensiLab Makerspace is the home of our community of practice. The unrestrained collaboration and sense of 'anything being possible' is embodied in a space that is specifically designed for making and building. Such a space is central to how we innovate as a working laboratory. It is simultaneously the artist's studio, the engineer's workshop, it's never neat and tidy, there's always something happening, something being *made*.

AirSticks

The AirSticks project is constantly evolving and ongoing. It centres around the creation of a gestural digital musical instrument for sonic exploration. Originally designed by a professional percussionist for personal creative use to facilitate the improvisation of transparent and expressive real-time live electronic music, the project has recently shifted towards focusing on using the instrument to facilitate experiences of 'sound play' for people living with disability. The co-designer and primary performer on the *AirSticks* has brought his own creative practice, as a drummer and electronic producer, to the project, collaborating in hundreds of performances with musicians, dancers and visual artists since the beginning of the project in 2013. Some of these collaborations are one-offs and others are ongoing as the creative practice of the collaborators influences the re-iterations of the instrument.

Figure 1.5.3   The *Melanopic Eye* light sensor wearable developed at SensiLab in collaboration with researchers from Monash University's Design Health Collab.

Figure 1.5.4 SensiLab's Drawing Robots were designed from scratch after experimentation with existing robot platforms failed to achieve the desired combination of creative potential and technical functionality. The image on the left shows the current multi-layer circuit board design, and on the right, the current working prototype.

Figure 1.5.5 Example Makerspaces used to support our research. Left: temporary SensiLab makerspace set up in Shenzen, China, for research staff and students during the development of the Drawing Robots project. Right: Part of SensiLab's permanent Makerspace, which includes digital fabrication technologies, electronic components and communal work areas. Photography by Elliott Wilson (left) and Tom Roe (right).

Though user testing of the instrument has been implemented, inviting drummers, producers and non-musicians to explore the sonic possibilities of the instrument, most of the verification and validation of the instrument has been done within the music making ecosystem. Professional performances, recording sessions and TV appearances all require the instrument to function predictably and reliably. This focus on 'ecological validation' drives the project with deadlines and real-world outputs that lead to further knowledge gained and further knowledge shared. The shift to making the AirSticks a more accessible instrument also involves this form of validation, inviting one of the early adopters of the AirSticks, a boy living with leukadystrophy, to perform with the instrument firstly for his family in an interactive house concert, and later in a publicly attended music festival.

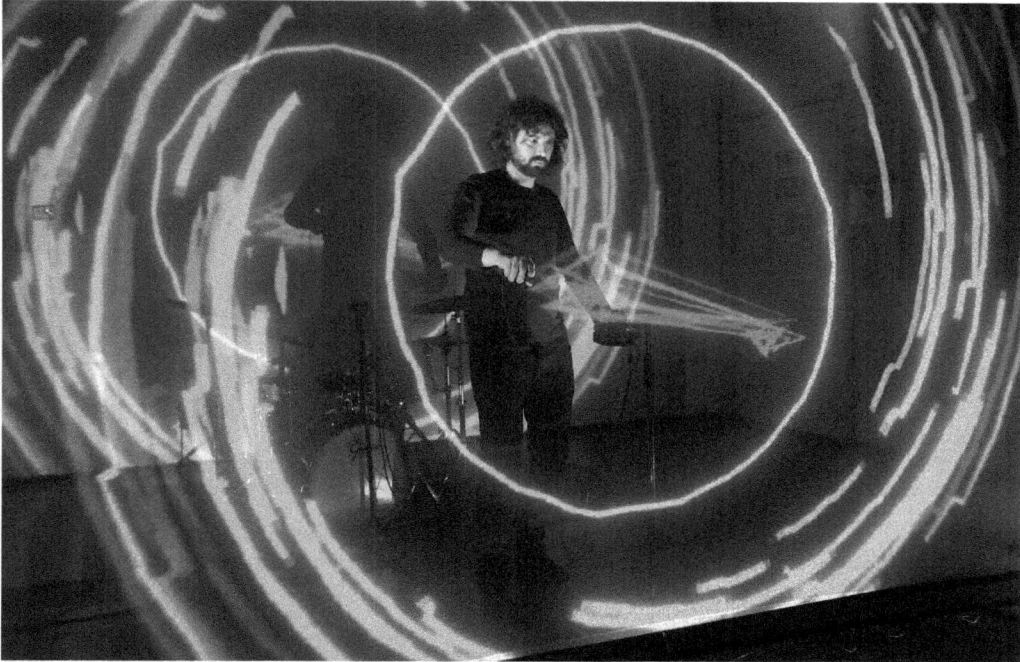

Figure 1.5.6   Alon Ilsar performing live with an early version of the AirSticks, which relied on an existing game controller. Subsequent developments at SensiLab have resulted in new, wireless hardware that is purpose built for music performance and interactive expression.

Virtual Angkor

The *Virtual Angkor* project (Figure 1.5.7) is predicated on a broad and deceptively simple premise; using the available evidence, what does it take to virtually model a medieval city?

The result has been the creation of an expansive virtual environment that owes everything to the practice-based approach; we learn through making and value process more than outcome. The project's co-creation with historians, archaeologists, 3D animators, ethnomusicologists, historians and cultural heritage technologists is a broad-spectrum collaboration of complementary domains of expertise and diverse skills, experience and perspectives.

The process of modelling an historical South East Asian metropolis in virtual reality is ongoing and has no definitive end. It embraces technology but redirects it backwards to dwell in a specific place and time. The venture is by nature iterative; the cumulative result is the body of work. While new discoveries, such as archaeological excavations, LiDAR surveys and translations of epigraphy continue to become available, so too do new technologies. The opportunistic adoption of AI in this project, for example, seeks to augment rather than redirect its overall aims. It may be a matter of path-finding algorithms for animated agents traversing the city, or how to train deep learning networks to extrapolate the vector-mapped boundaries of rice fields where the physical remains have been lost to history. Such techniques prompt questions that are themselves integral to the process. How does one weigh up the primary and secondary sources to arrive at an immersive historical recreation? Can the creation of an immersive world 'embody' historical evidence?

Advances in 3D graphics processing and 3D capture technologies also allow us to add new elements to the ever-evolving model of the city. Much of this is a dialogue between the tangible and

Figure 1.5.7 Still image from the Visualising Angkor project, which uses virtual reality and game engine technology to create an interactive recreation of the medieval Cambodian metropolis.

intangible, between the material present and the past. For example, physical artefacts – such as bronze palanquin fittings from museum collections – can be captured with a structured light scanner and then 'imported' into reconstructed scenes. Similarly, there is the opportunity to 'export' purely digital reconstructions, most of them made entirely within 3D modelling software, into tangible forms as 3D printed models. More recently, Virtual Reality hardware has enabled us to translate reconstructions of specific locations in the city into VR experiences. Yet none of these adventures would be feasible without the continuity of the project, especially the research into evidence-based 3D models that we have crafted years previously. So far, this organic process of development has yielded an increasingly rich digital environment which is perhaps best described as an exercise in historical 'world building'. We learn by making and, rather than seeking to bring forth a known end, follow an incremental approach that layers new methods of visualisation over a defined urban space.

Conclusion

Today, the generation of new research is greater than ever before in human history. The sheer amount of new information makes it difficult or impossible to know all the research available on a specific topic, even for experts. Simultaneously, many of the old divisions and canons of the academy are dissolving as formalised research is undertaken largely as team efforts whose members span multiple disciplines. Universities are no longer the exclusive innovators of knowledge, as global technology companies, for example, employ more PhD researchers than many universities. Rising superpower nations, such as China, are expected in the next decade to graduate more students annually than the entire population of the United States. With more people involved in research and producing more new knowledge than ever before, how do individuals and groups innovate within this ever-expanding global research ecosystem? One is tempted to recall the catchphrase of iconic designer Dieter Rams: *less, but better.*

We have highlighted many of what we believe are the unique aspects of practice-based research: its emphasis on making and leaning through practice, valuing implicit knowledge sharing, inclusion and diversity, building communities of practice, valuing process as much as outcome. Our results demonstrate the validity of this approach in a multi-disciplinary, collaborative creative technologies research environment. While in the past a number of scholars have raised issues regarding collaborations across disciplines, in our experience there is an increasing appetite for undertaking projects that cannot be progressed by a single discipline alone. As the projects we have briefly introduced demonstrate, practice-based research can accommodate medical researchers, technologists and engineers who work alongside designers, developers, luthiers and artists to deliver both practical and creative outcomes.

Practice-based research continues to grow in popularity beyond its origins in the creative and performing arts. As we have outlined in this chapter, research through practice can also benefit researchers in the technological disciplines. We hope the ideas introduced in this chapter will inspire other researchers – even those who do not consider their research as practice – to consider this expanding field of practice-based research and what it has to offer.

Acknowledgements

Jon McCormack was supported by a Future Fellowship Grant from the Australian Research Council, FT170100033.

Notes

1 SensiLab is a technology-driven, design-focused research lab based at Monash University in Melbourne, Australia. https://sensilab.monash.edu/.
2 Also known as *research-creation* in some countries.
3 Candy defines the difference between *practice-based* and *practice-led* research as follows: for practice-based research, the contributions to knowledge are through the artefacts developed, whereas in practice-led research the contributions concern new knowledge about practice itself (Candy 2011).
4 See Pears (1972) and Nelson (2006).
5 Borgdorff (2012).
6 McCormack et al. (2018).
7 And additionally, that subjective experience, by its nature, may be subject to the nuances of culture, gender, environment, affect and so on, hence there is no singular, universal individual experience.
8 Still and d'Inverno (2016).
9 An interesting example is that of 'Damascus Steel', used for sword-making and characterised by a specific mottled pattern. Around the time of the Crusades, the swords were coveted for their superior sharpness and strength over western swords. However, after around 1850, the method by which the swords could be made was lost for almost 150 years. It was only recently that metallurgists rediscovered the forging technique that allows Damascus Steel blades to be made (Verhoeven 2001).
10 Mace (1977, p. 43).
11 Barad (2003).
12 This is not entirely accurate; candidates may include supplemental material, including online links to source code, video examples, software, data, sound files, scores – essentially *any* kind of electronic media. However, as the name suggests, supplemental material only supplements the written thesis. It is not mandatory, nor does it replace the written thesis as the main articulation of the knowledge contribution.
13 Sullivan (2009).
14 Pello (2018).
15 Dunne and Raby (2013).
16 From Koskinen (2011, p. 26). Yes, that is a typo!
17 Candy (2006).
18 Understanding the ecological validity (Brunswik 1956) of any system or technique developed is an important aspect of our research verification and validation process. Ecological validity requires the assessment of artefacts or processes in the typical environments and contexts under which they are actually developed and used, as opposed to laboratory or artificially constructed settings. It is considered an important methodology for validating research in the creative and performing arts and naturally extends to practice-based research in creative technologies.
19 https://github.com.

Bibliography

Barad, K. (2003). Posthumanist Performativity: Toward an Understanding of How Matter Comes to Matter. *Signs: Journal of Women Culture & Society*, 28, 801–831.

Borgdorff, H. (2012). *The Conflict of the Faculties: Perspectives on Artistic Research and Academia*. Leiden, The Netherlands: Leiden University Press.

Brunswik, E. (1956). *Perception and the Representative Design of Psychological Experiments*. Berkley and Los Angeles, CA: University of California Press, 2nd ed.

Candy, L. (2006). *Practice Based Research: A Guide*. CCS Report: 2006-V1.0. Sydney: Creativity & Cognition Studios, University of Technology, November.

Candy, L. (2011). Research and Creative Practice. In L. Candy and E. A. Edmonds (Eds.), *Interacting: Art, Research and the Creative Practitioner*. Farindon: Libri Publishing, 36.

Dunne, A., & Raby, F. (2013). *Speculative Everything: Design, Fiction, and Social Dreaming*. Cambridge, MA and London: MIT Press.

Koskinen, I., Zimmerman, J., Binder, T., Redström, J., & Wensveen, S. (2011). *Design Research Through Practice (From the Lab, Field and Showroom)*. Waltham, MA: Morgan Kaufmann.

Mace, W. (1977). James J. Gibson's Strategy for Perceiving: Ask Not What's Inside Your Head, but What's Your Head Inside of. In R. Shaw & J. Bransford (Eds.), *Perceiving, Acting, and Knowing: Towards an Ecological Psychology*. Hillsdale, NJ: Erlbaum.

McCormack, J. et al. (2018). Multisensory Immersive Analytics. In K. Marriott et al. (Eds.), *Immersive Analytics, vol. 11190 of Lecture Notes in Computer Science*. Cham: Springer, 57–94. https://doi.org/10.1007/978-3-030-01388-2_3.

Nelson, R. (2006). Practice-as-Research and the Problem of Knowledge. *Performance Research*, 11, 105–116. DOI:10.1080/ 1352816070136356.

Pears, D. F. (1972). *What Is Knowledge? (Essays in Philosophy)*. London: Allen and Unwin.

Pello, R. (2018). Design Science Research – a Short Summary. *Medium*. https://medium.com/@pello/design-science-research-a-summary-bb538a40f669 (accessed May 18, 2021).

Still, A., & d'Inverno, M. (2016). A History of Creativity for Future AI Research. In F. Pachet, A. Cardoso, V. Corruble, & F. Ghedini (Eds.), *Proceedings of the Seventh International Conference on Computational Creativity*, 147–154. http://www.computationalcreativity.net/iccc2016/wp-content/uploads/2016/08/Proceedings_ICCC16.pdf.

Sullivan, G. (2009). Making Space: The Purpose and Place of Practice-Led Research. In H. Smith & R. T. Dean (Eds.), *Practice-led Research, Research-led Practice in the Creative Arts*. Edinburgh: Edinburgh University Press, 41–65.

Verhoeven, J. D. (2001). The Mystery of Damascus Blades. *Scientific American*, 281, 74–79.

1.6

WORKING THE *SPACE*

Augmenting training for practice-based research

Becky Shaw

Introduction: defining *space*

The practice-based researcher is located between the professional world of practice and the academic model of research. These historical cultural contexts act like magnets, generating powerful currents that influence all aspects of activity in their radius: how knowledge is understood, behaviour, collaboration, dissemination, voice, use of time, use of resources etc. Given, this, practice-based researchers are positioned within both of these powerful worldviews and have to build their *space* on the intersection. This *space* is needed so they can feel situated and so their work can be legible and transformational in its chosen location. While not all practice-based research happens within universities, undertaking research training as a PhD candidate often begins this process, and all candidates are tasked with developing the contexts and world that their research sits in. However, practice-based researchers have a particularly complex job as they are not just trying to find their place in the scholarly world but are also weaving this together with worlds of practice.

When talking about the task of building a group context between practice and academia, spatial metaphors come readily to mind. The researcher must locate themselves in a particular place in the intersecting parts of a Venn diagram. One or more of the circles of the Venn diagram will be in practice-based research where the terrains may have some blurry edges in the first place. These circles often then overlap with other fields of knowledge and disciplines. In the written *space* of the thesis, the researcher must build a new landscape of existing academic knowledge and practice knowledge with the circles of the Venn diagram and understand their position in it. This process involves the construction of the research knowledge *space* by the individual. Most of the policy and governance about supporting PhD researchers understands this endeavour as individual, supported by supervision and by some university training. The process of weaving a world between practice and research is a complex and creative act of construction. Not only is the context of study being built, but as researchers imagine themselves contributing to this, they are also bringing to life a new identity for themselves in this constructed landscape. While policy and governance understand the PhD as solo, perhaps it is useful to think about a shared endeavour for practice-based PhD candidates, where the navigation of practice and academia is collectively experienced and undertaken (although always with specific nuances). Given this, it seems important to explore how group learning could be used for building the *space* of practice-based research.

As *space* will be used frequently in this chapter, some definition is needed. I am first and foremost talking about a kind of social, as opposed to physical, space. This *space* is to be understood as a kind of

 DOI: 10.4324/9780429324154-8

group identity and belonging for those involved, but not one that has simple edges marking insiders and outsiders. This kind of *space* is not a fixed entity that individuals join but is being made all the time. There are extensive theories about the nature of *space* and even though it is exploring physical space, Doreen Massey's three-part definition[1] is particularly useful that:

- *space* is a product of social relationships and is formed through interaction
- *space* is a sphere of plurality: it contains lots of different, concurrent threads
- *space* is always being constructed – it is not finished.

While I am trying to describe the way a cultural social *space* can be built for practice-based research, it is still in the physical space of the university. In James Corazzo's[2] systematic review of learning space literature, social non-material spaces are not separable from the physical or virtual spaces where they form. The use of the term *space*, then, usefully acknowledges that there is a material dimension and also emphasises an active making where individuals and groups form together.

In the following section, I explore what is nationally expected in the UK as a quality doctoral training offer and question how it might, or might not, generate useful conditions to support the creative act of building the *space* that practice-based research requires. Following this, I offer some examples of ways to recognise, explore and create these *space* building processes with practice-based researchers and to think constructively about what can be done within training policy, or in the relationship between training and individual PhD candidates. By doing this, I hope to encourage experiment and even play with forms of *space*-making.

Practice-based research is a pan-disciplinary concept, and the conception of the term also differs between doctoral programmes internationally and nationally. Given the additional complexity of encompassing all these differences, I focus particularly on my experiences of working with practice-based doctoral researchers on an art, creative media and design research programme within an inter-disciplinary institute in the UK post-92 university, Sheffield Hallam (SHU).

The culture of art and design practice-based PhD candidates

Art and design doctoral research PhD candidates at SHU usually belong to a professional world before belonging to an academic research world, or if they have not been a professional, they have travelled through an undergraduate or master's education that constructs the habits and worldview of that discipline. For PhD candidates of vocational subjects (e.g., art, design, architecture, education, nursing, medicine, law etc.), this educational context is already entangled with practice, as teaching is orientated to 'the profession'.

When artists and designers (and film-makers etc.) begin a PhD programme they are often not fully prepared for the changes that are required, believing it will be business as usual but with a 'bit more writing'. They are not prepared for the upheaval doctoral research produces as it bends them to think, act and speak like researchers. In our experience at SHU, the processes of research training and doctoral management procedures generates critical moments where the researcher is torn between a long-cultivated way of being an artist/designer and the new, differently alien languages of knowledge, foresight, method, ethics and bureaucratic accountability, as well as the return of the classroom where doctoral training is often delivered. The principles and discourses of knowledge and methodology are vitally important to all researchers, and there is a tendency for art and design practitioners to think that 'they don't apply to them' for a while. Principles are not alien to the critical and informed art or design practitioner, but in research training, they come in linguistic and cultural forms that are at odds with many researchers (not just those in art and design).

The work to accommodate and translate this learning is often done by supervisors and is described well in John Hockey's[3] paper. Drawing from interviews of PhD candidates and supervisors, Hockey describes the five categories where candidates' disjuncture between the worldview of being a practitioner and a new PhD researcher are played out. These sections include problems with:

- regulations
- documentation of research evidence
- analysis
- academic writing
- balancing academic and creative work.[4]

Hockey describes how supervisors gradually encourage PhD candidates that an engagement with university research protocols is worthwhile and within their existing skills, making use of and stressing 'continuing well-established biographical strengths'.[5] Hockey also records how supervisors encourage candidates to see engagement with research practices as 'risk-taking' – a quality already valued (and indeed, valorised) in art and design practice. These key factors show a movement from resistance to adoption or adaptation and art and design researchers finding ways to merge and develop a new identity as an artist–researcher. Hockey's research focuses on the supervisor–candidate relationship to do this work, at a time when there was little requirement to deliver cohort-level research training. It is interesting to consider, then, how a training offer might also contribute to this adaption.

It seems that the deeply held professional identities of art and design practitioners generates a complicated and often difficult navigation of context, or *spaces* (introduced earlier). The practice-based researcher must navigate the terrain and boundaries between these different *spaces*: of university doctoral education and conventions of academic performance *and* the values and performances of art and design worlds. The act of working this *space* – the physical, intellectual and emotional moulding of a context where they can function and where their work is legible – has enormous impact on individual research journeys. It is easy to see the difficulties that arise from forming this space for the individual, and it is right to be attentive to the extra stress this causes. However, the production of this *space*, both for individual PhD candidates and also for institutions, could be seen as a dynamic, plastic, creative, critical and shared process. Before we think further about what might be developed, it's interesting to reflect on what is offered as cohort training and what is required by policy.

Generic training policy and relationship with practice-based research

The policies that structure the processes of doctoral education focus primarily on the individual PhD candidate. This is not surprising as, unlike a course, the doctorate is a singular programme of study, described, contextualised and carried out by the individual. Most UK universities work to meet the standards expressed by UKRI (UK Research and Innovation – the connecting organisation for all the separate research councils). Higher education (HE) organisations must meet these standards when providing doctoral provision in partnership with UKRIs. The guidance on the type of environment and support universities must provide includes 'excellent standards of supervision, management and monitoring', career advice that includes a range of sectors and not just academia' and 'in-depth advanced research training'.[6] The UKRI expects universities to operate a degree of flexibility and responsiveness specific to disciplines – such as practice-based research. The same policy usefully indicates that PhD candidates 'should, wherever possible, benefit from the advantages of being developed as part of a broader peer group (e.g., through cohort approaches and Graduate schools)'[7] but doesn't (for brevity, I assume) go into any detail about what purpose group work might serve.

Cohort delivery is often described as doctoral training. This always seems an uncomfortable word, producing images of shouting coaches and athletes competing against the clock. It also suggests there is a finite skill that has to be learned to become a researcher. Training differentiates the learning process from other standard models of education, perhaps to emphasise the non-taught nature of doctoral study, and also a professional transition to being a researcher. On one hand, this situates the process in more corporate models and implies there is a simple acquisition of skills to become a researcher, and on the other, it opens up the possibility of hybrid worlds between academia and professional contexts that could be useful and appropriate for practice-based researchers.

Research training: generic, practice-based, formal and informal

Most UK universities offer a generic training package for PhD candidates. 'Generic' research training includes skills that are deemed important for all researchers, regardless of discipline. Often a generic package includes: understanding of the principles of research design, bibliographic and computing processes, ethical and data management procedures, communication and engagement skills and also career development frameworks. However, all packages differ somewhat and are likely to be more extensive when national funding is in place (from AHRC or ESRC etc). The current Researcher Development Framework[8] specifies a detailed platform of skills and knowledges that individual researchers should be supported to develop during their doctoral study.[9] Many institutions take part in this framework, providing a way for researchers to map their own development, with supervisor input and some generic delivered training.

Art and design researchers seem to struggle to engage with the generic training on offer in institutions. This is certainly the case at SHU, but similar patterns are informally noted at other higher education institutions. While response will always be nuanced according to different programmes on offer, the 'one size fits all' way to be a researcher can seem a gulf away from the candidate's skills and usual disciplinary context and way of being. While candidates do adapt, as Hockey records, this is a slow process, and it is worth asking whether this generic training could attend better and quicker to the process of *space* building at work as part of its intellectual content.

As well as the difficulties faced by individuals in building a new *space* for themselves, it's important to question what type of researcher identity is being communicated by universities in the first place, and how well this might relate to practice-based research. In my experience, the notion of generic is inflected by the authority of certain subjects and their assumed defining of research training. In the case of many UK post-92 universities, a version of the generic may tends towards an engineering and biosciences discourse.

One of the problems with generic training is that it conveys a way of being a researcher that doesn't quite fit with the way of being a practice-based (and other) researcher. Like other forms of teaching, generic research training creates a form of 'disciplining'. James Corazzo[10] describes 'disciplining' as all the unspoken ways a discipline is communicated, such as via speech, behaviour, frames of reference etc. If generic training conveys a way of being a researcher that is too distant from any of the modes of research recognised by the practice-based researcher, then it becomes difficult to creatively re-form their identity and *space* with – or in relation to – this. The particular inflection of university generic skills training reflects how a university as a whole understands, represents and values research, so if PhD candidates feel alienated from this, it can make them feel alienated from the university as a whole, sending them retreating to the safety of disciplinary comfort, instead of doing the work to build a new *space* to inhabit productively.

Susan Carter reports[11] on an academic suspicion of generic training as 'a bolt-on process suspected by some of inherent flimsiness'.[12] However, the problem is not necessarily the idea of generic training per se, as there is value in striving to teach research skills or issues that all might have in common. In a paper on the generic provision of thesis writing skills for doctoral candidates, literature specialist Carter notes that the generic strangely means both a specific genre as well as the general or universal: almost opposite concepts.

Rather than arguing against the generic, then, perhaps the generic needs more critical and practical exploration, or could be understood as a goal to work towards or explore. Doctoral education is in UK terms *Level 8*: the highest educational qualification. The Researcher Development Framework accordingly sets out a sophisticated set of expectations of knowledge and critical thinking. Despite this, generic training is often delivered as a set of skills, without the same expectations of critical engagement and questioning that we would expect from undergraduate level education. Surely at *Level 8* we should be abandoning an expectation of education to give PhD candidates skills or certitude, and instead offer challenging questioning of the nature of research, knowledge, truth and process. If this level of questioning was an expectation of doctoral training, it would, by its framing, enable practice-based PhD candidates to reflect on how their worlds of practice and research inform their worldview.

Susan Carter describes the generic as a term that is 'promiscuous, slippery and generative',[13] suggesting that we could be seeing it as a site for imaginative and challenging conversations. Carter draws on the work of others, including Sharon Parry, and notes, warily, that doctoral candidates often vocalise their experience as a game – one or many meaningful social settings with rule/social codes that they have to learn to navigate. Navigating the building of a researcher identity and *space*, between profession (art and design) and university (generic conventions of being a researcher) is a key preoccupation of practice-based researchers; Carter feels that one thing that generic delivery could do better is to 'explicate the rules of the game'[14] rather than pretend that this particular world of academia is a universal given. I would go further than this and argue that exploring the possibility of the generic creatively and paying attention to how we form different research *spaces* would help all PhD researchers understand where they stand in relation to academia and practice.

In the SHU community of 30–40 art, film and design PhD researchers, the navigation of the *space* of researcher and practitioner is a key preoccupation. Rather than accepting this as a necessary evil of working in a post-92 university, we decided to explore the problem creatively and as part of our own doctoral training programme. There is a growing interest in ways that distinct practice-based training programmes can address the gap between the policies of doctoral training and the experiences of practice-based research PhD candidates: drawing together ways of being a practitioner, with recognisable ways of researcher performance (as in the excellent model at De Montfort University developed by Craig Vear, Sophy Smith and Ernest Edmonds). The benefits of making a formal programme are still being explored; however, one important aspect is that a formal programme makes this area of research visible, distinct, shareable, open to improvement and formally recognised by the university. It also performs a task of enculturation in which university needs and practitioner worlds are welded into a new form.

At SHU, we sought to develop training practices to inculcate an understanding of discipline-specific 'researcherly' behaviour, but without explicitly adopting the language of training or, if we did, to explore it critically. This approach was undoubtedly shaped by my experience as an artist–researcher exploring institutions and also as an artist–curator. I have been influenced throughout my art and education journey by artists like Andrea Fraser,[15] who was in turn influenced by Pierre Bourdieu. As in *Homo Academicus*,[16] where Bourdieu applies sociological method to sociology, it seemed appropriate to explore the conditions we found ourselves in using our own artistic methods, applying them to our understanding of what becoming a researcher might mean for the art and design practitioner.

By exploring how we train practice-based researchers, in relation to generic research training, we thought we might support practice-based researchers to build a *space* between practice and academia, and also potentially improve research training by using some of the strengths of practice-based researchers. Sensitivity to context is often abundant in art and design researchers, and we thought this could be employed critically to explore and challenge doctoral training and simplistic or out of date constructions of researcher identity. Instead, then, of feeling alienated or outside of the generic researcher, the critical, material thinking of practice-based researchers might build a relationship and *space* with it.

Recognising the need to think critically about training for practice-based researchers has led us to undertake a range of practical and material projects that approach conventional aspects of research training with a deconstructive and constructive eye. In the previous four years, we have playfully remodelled research training, the conference, the academic poster, definitions of data, method and research impact. In the following section, I explore some of the projects that tackle the plastic process of *space* building. This involves processes to build practice-based researcher identity: as a *space* or position between the generic and the specific and where practice meets research. These projects are not easy to order effectively as they overlap in time and one affects the other. However, for ease I organise them as:

1 making *space* through training
2 PhD researchers make *space*
3 making physical *space*.

Making space *through the time, structure and form of 'training'*

In this section, I discuss two projects, undertaken as doctoral training, that sought to construct researcher identities and *space* in relation to two different constructions of generic research presentation: the conference and the academic poster.

Universities often use conference models to train researchers in conference presentation, or to generate a showcase of PhD work. These conferences usually include PhD candidates of many disciplines and adopt a one-shape-fits-all unspoken expectation of standards of research communication and identity. In these contexts, it can be especially hard for practice-based research PhD candidates to communicate the landscape of their inquiry and their developing new researcher identity.

Our art and design doctoral candidates are (at time of writing) situated within an interdisciplinary research institute including computing, media and communication studies, and we are tasked with offering training for this wider community, filling in the gaps untouched by university-level training, as well as specialist art and design training. My focus in writing here is on art and design PhD researchers, but the purpose of this particular training initiative was for the wider interdisciplinary community. However, this interdisciplinary scope offered an excellent opportunity for practice-based researchers to explore how the identities and landscapes they were building differed or were similar to the researcher identities of different disciplines. Susan Carter notes a similar value in interdisciplinary dialogue and how 'discussion at the borderlands illuminates discipline-specific practice'.[17]

In response to our concerns about the operation of the generic (and as a productive challenge to the institution), we wanted to find the grounds for a critical research training that could be interdisciplinary without alienating specific research cultures. Instead of seeking a theme around content of research, it was decided to focus on the principle of research that is common in all disciplines: method. The simple notion of method was chosen over the more philosophical methodology to emphasise that research involves practical acts, informed by philosophical positions. Core alignment to methods is also a central part of different researcher identities; for example, it is the use of practice that makes practice-based research, and the use of ethnographic methods that makes anthropology research etc.

To use the framework of the conference to support individual learning and collective *space*-building, as opposed to dissemination, we looked closely at every expectation of conference production. We avoided a triumphing of successful outcomes and instead sought to make the critical thinking and creativity embedded in research methods tangible and explicit. Instead of focusing on research content, we shifted emphasis to the *practice* of research: how researchers *build* and relate structures of method. Every researcher, including practice-based, must be able to understand what methods are available and what will work effectively to answer their research inquiry. They must also be able to articulate this decision-making journey, make a case for their chosen direction and reflect on whether it was the best path. Method offered a clear training opportunity for researchers to articulate their process while also challenging and enriching it by exposure to methods from other researchers. As well as supporting all researchers, this frame would ensure that practice-based researchers understood all aspects of their process as part of their practice-based research: the structure of time, the ordering of materials, attention to participants and the processes of making, thinking, review and analysis.

The conference was built by using a PhD candidate steering group of diverse disciplines so they could devise the event in a way that would appeal to their disciplinary communities, and the difficult process of speaking across disciplines would be embedded in the development of the event from the outset. Some members of the steering group were uncomfortable with the need to talk about method, as this type of reflection was outside of the conventions of their discipline, where method is seen as a means to an end and just 'how things are done'. Members were also concerned that the focus on *how* rather than *what* might undermine the expected, desirable academic status of a conference, or generate a situation where they might feel ashamed or uncomfortable. We encouraged them to understand that communication of problem solving, working with problems or honestly reflecting publicly on flaws is part of a researcher's integrity, regardless of discipline. We avoided making our focus on language itself so we didn't seek to form any agreed cross-disciplinary terms; instead, we recognised the value of an open debate about different understandings.

The first abstracts submitted to the conference call focused on reporting findings, or only described method briefly. It took several weeks of supported intervention from staff for research PhD candidates to get inside their method and to understand their research practice as a process rich with philosophical and practical decisions and implications. Interestingly, the practice-based researchers tended to be better at discussing method, perhaps because they were practiced in the defence of their process as research, and the 'how' is an integral part of any artistic language. However, this entwinement of doing and analysing is also what makes articulating practice-based research in an interdisciplinary frame difficult.

We wanted to make the construction of research visible, so that no aspect of the building of a research project could be taken for granted as 'just what you do in my discipline', or 'just what my supervisor told me to do'. To do this, we established key processual themes that went across disciplines. These processes included *Analysing Narrative, Constructing Data, Making Models, Questioning Method, Exploring Where Real and Virtual Collide, Observing and Ordering* and *Translation*. These categories were intended to draw attention to and intensify focus on what makes up the research process. The presenters would have to 'see' and articulate their disciplines while also recognising the key shared ingredients of researcher process.

During the event itself, there were moments when the interdisciplinary clash caused caution and uncertainty. After the first panel on *Constructing Data*, there was a silence and then an audible intake of breath as the audience was tasked with beginning to unpick the relationships between a project that used numerical data collection methods and a project where an artist researcher understood their artwork as data. Fruitful dissonance was generated powerfully in the session on *Making Models*. Here a computing research project exploring the creation of a digitally constructed online classroom environment was bombarded with questions and input about the aesthetic visual and spatial language of the classroom. The online classroom hung in a simple visualised digital environment of a horizon line where sea meets sky: a token *anyplace*. This generated considerable challenge and debate as communications and creative practice candidates debated what this *anyplace* meant for education. All the candidates could not rely on daily disciplinary doing of their researcher identity and world, but had to pay attention to every sharp edge and blurry corner of it.

The apparently simple focus on method illuminated research process (and research training) as a site of creativity, innovation and discovery, and also built epistemological awareness, drawing attention to what is being researched and how knowledge is being produced. Art and design researchers contributed greatly to these conversations as they recognised the practical activities in the work of others and the creativity, fluidity and problem finding involved in other disciplines. They also contributed significant insight into the non-neutral effect of visual communication in presentation that might have remained overlooked by others.

Method sought to challenge the tendency for generic doctoral training to present simplistic generic formula of how to speak, perform and get on, and an additional tendency to present research careers as a strategic, linear manoeuvre for visibility and success. This focus also generated an ability to appraise the quality of the research of others and to understand the formulation of different researcher identities and the construction of different *spaces* that can be occupied. After 'Method', candidates spoke frequently about their excitement about getting to know each other, recognition of similar outlooks in unexpected communities, and also about the way that exploring work criticality builds both a shared interdisciplinary learning community and a new sense of the *space* of the practice-based researcher community. The oscillation between the different communities and the assumptions of generic research identity and communication felt like an active 'working the space' where positions were seen, felt and articulated.

Case study: the poster show: title, formatted in sentence case
(the title of the project)

The academic poster is a familiar part of the academic conference in many disciplines (including in design and film theory but rarely in visual arts). The poster is used as an additional presentation format alongside verbal performance of papers, keynotes and workshops etc. The posters themselves usually adhere to particular conventions of form, including common fonts, language, spatial layout, digital software, the use of diagrams and visual representations. There are many online examples of 'how to do' academic posters suggesting a fairly narrow set of generic criteria for success. The conventions differ across disciplines but are, by and large, framed by an idea of scientific language and delivery. Like any other form of visual communication, the academic poster is subject to changing fashion and style, but its parameters remain fairly constant.

The poster performs an act of compression and completion, reducing years of work and potentially huge geographic space into a sheet of paper. The poster acts as a 'point of visibility'[18] for processes, spaces and contexts otherwise unavailable to us. It is usual to find a whole series of affiliation logos on posters, including from universities, fieldwork sites, labs, sponsors etc. As well as registering investment, these are badges of honour, belonging and endorsement.

In art, design and film, the poster is an historic and common part of professional practice. The poster can be the form of an artwork or can refer to another place and time, inviting us to go and see the 'real thing'. The art and design poster extends/compresses place and time in the same way as an academic poster does, and it must also convey the atmosphere of the event it refers to in every bit of its form – its paper quality, its font, language, spatial arrangement and location.

Academic posters are required to be clear and to carry a large amount of information about the process and progress of research, in an accessible form. While all PhD researchers seek clarity and precision, language is never solely about reportage after the event. For example, ideas (and researcher identities)[19] are built and discovered through writing. This is even more complex for creative practice researchers, as the visual language of the poster can be both the mode of communication and the medium of the research. The art and design researcher may also be looking for forms to communicate many different things at the same time, using creative mediums and methods to compress ideas, not pull them out in sequential points. Being able to communicate clearly is a vital part of research training, but it is also an important arena for thinking critically and sensitively, to challenge partial, exclusive definitions of universal clarity and norms.

Instead of simplifying or ignoring the relationship between worlds of academia and professional practice, we decided to explore it head-on by taking on the challenge of the academic poster. To the eyes of professional designers, artists and film-makers, the academic poster is often clunky, gauche, stuffy, uncommunicative and just plain ugly, with the demands of 'clear information' overwhelming all other concerns. Working with the poster, then, offered a context to materially construct, explore, experiment with, attack and enjoy a *space* where professional practice and research collide, offering frames for the communication of practice-based research.

During one afternoon, art and design researchers worked together to explore existing academic posters and to begin new ones for their own research. This process involved cutting, collaging, sampling, enlarging, copying etc. examples of existing posters and then playfully building in the researcher's own material. The resulting posters (as seen in Figure 1.6.1) adopt different levels of adherence to academic poster standards, such as the expectancy of a high volume of text. The fourth poster from the left combines a full-size image with little explanatory text, the sixth image from the left shows text fragmented across the image and the seventh image from the left contains no text at all. The researchers also experimented with art and design standards of a successful poster (already complicated because they may well be seeking to challenge these conventions, too). The process of discussion and practical workshopping that produced the posters was extremely rich, as we reflected on how to use the medium of a poster to convey practice-based researcher identity and how to use the place of the exhibition to convey the *space* of practice-based research.

Figure 1.6.1   Image by Sarah 'Smizz' Smith, including artwork by Jonathan Michaels, Jo Ray, Rachel Smith, Sarah 'Smizz' Smith Diana Taylor and Julie Walters.

In this example, and the method conference, training situations offer different forms to produce the *spaces* of practice-based research. In doing this, the practice-based research PhD candidates conceived themselves within new shared discourses, and also understood themselves as on borders, valleys, cliff-edges or within territories. The academic poster offered a physical, plastic canvas where the landscape of the PhD candidate's own research could be viscerally moulded with composition, surface and information. Every material decision the candidates made when organising different parts of the poster could be considered as an expression of the terrain: the relationships between practice knowledge and academic knowledge being materialised as sharp cuts, overlaps, a blur, a gradation, a fold, a void etc. The group of posters also articulated a *space* of a particular research culture, one that defined itself by its active and critical exploration of its own culture and that of the university.

PhD researchers make space

Alongside experiments that explore generic concepts of research and researcher training, we developed contexts where candidates might lead on delivering training. The most constant of these is a fortnightly seminar that actively follows the conventional ingredients of generic research degree training: *research structure, methodology and method, the literature review, analysis, ethical consideration, intellectual property* and *the submission*. For each session, one or more PhD candidates is invited to reflect on one of these through the lens of their research. This could be a simple recounting of *how* aspects are done, but more usually this involved experimental ways to learn and reflect.

One notable example included a researcher who explored the wrapping and unwrapping, storage and appraisal of differing sculptural objects as a way to engage deeply with ethical concerns. Through handling the different surfaces, weights and textures of the objects, she invited us to think about the ethics of moulding, shaping or reconfiguring material that belongs to another, the way material is formed through the research process itself and the inter-relation of sharp, intractable or fragile material with its environment. Another researcher used her own Deleuzian, rhizomatic study to challenge expectations of the literature review. These examples show how fundamental research issues can be re-considered, using practice.

We have also been lucky to have energetic and ambitious candidates who have used some of the sense of possibility grown from the close seminar relationships (and funding from the university) to foster additional projects and question academic life. A candidate[20] led an extraordinary project (outside of the seminars and for a wider audience) that invited people to reflect honestly on failure and shame. In a later project, four art and design researchers worked together to explore the experiences and value of artist researchers who have dual identities as healthcare professionals. In one summer, to address the gap of staff-led provision, a group of PhD candidates sought university funding to develop their own entirely run candidate programme, using a deliberately DIY-maker aesthetic. In addition to projects that sat between community building and training, PhD candidates have delivered conferences on visibility and orchestrated significant exhibitions questioning what the purpose of the exhibition is for practice-based art and design researchers.

These examples all depend on committed, energetic PhD candidates, who recognise that building the wider researcher *space* in turn contributes to their own research experience and learning. Craig Batty and Marsha Berry[21] note that despite the tradition of the lone scholar, PhDs are completed in teams and communities where 'generative ideas emerge from joint thinking'.[22] Without candidates who understand this, none of these activities happen. For a short period, SHU offered funding for PhD candidate-led training, as an addition to staff-led. There seems to be a lot more scope for candidates to do this and to deliver effective learning that supports the development of a specific practice-based researcher space. These models sit in odd contrast to the policy defined cohort training, yet they take seriously the importance of peer learning as a vital part of researcher *space* creation.

Making (physical) space for practice-based research

There is growing interest in the intellectual space of a research culture, as explored by Craig Batty and Marsha Berry (2016). They offer an analysis of the playful space of the creative practice research degree, for example, deliberately conflating the cultural space of the research degree, the space of the inquiry of the PhD candidate and, occasionally, the physical space where research is carried out, as well as the 'constellations' and 'connections'[23] that it involves to both inhabit and construct these 'spaces'. Their emphasis is more on the *space* of research co-created with candidate and supervisors rather than via cohort training. However, relevant to the exploration of disciplines, and training, they reflect on the creative potential of the creative-practice researcher's occupation of dual worlds, noting that they are trained to 'use the playful space of the academy to develop and contribute new and innovative knowledge'.[24] They note that the *space* of the practice-based doctoral project can be 'cluttered and chaotic', a refreshing and critical counter to the desire for training to make things 'clear'. They also note that the space created in a training environment is 'only partially charted, where candidates and supervisors can find previously uninhabited spaces'. This sense of training as an adventure and an inquiry rather than an agreed delivered set of information is one that can be developed much further.

Following on from a conversation about the wider *space* of research culture, it is important to consider how physical learning space is used to support the formation of art and design practice-based researchers. Traditionally, doctoral researchers are situated with their supervisors, either spatially or organisationally. In this relationship, the PhD candidate is reliant on the supervisors for all their knowledge of the doctoral journey, perhaps with some opportunity to meet others via research training etc. If candidates aren't situated in offices with supervisors, they might be in shared occupancy offices. In a scientific model, doctoral candidates might be working in groups in labs. Where and what, then, is the physical space of the practice-based researcher?

Many aspects of undergraduate and taught post-graduate (e.g., master's) design teaching (for example) mimic the behaviours and values of the professional design world. James Corazzo collates a large body of research on (among other things) the value of the studio as producing educational 'disciplining' in design PhD candidates.[25] The physical space of the studio has an influence that is both tangible and intangible, contributing to forming the intellectual and cultural educational *space* of a course. The physical space teaches how it feels to be a designer and a design candidate, constructing how they and staff speak, perform, move, place materials, argue, relate to each other and understand quality (among many other things). While Corazzo's use of the term 'disciplining' refers to the effect of actual space (which will be returned to shortly), it can also be used to describe the way art and design candidates must become encultured[26] into a new identity of a practice-based researcher. Corazzo proposes that it is not just the curriculum that teaches PhD candidate to be designers, but also the other sensorial ingredients and spatial practices: for example, interpersonal interaction, speech, space habitation at different times of day etc. These ingredients are also at work when practitioners begin to be practice-based researchers. However, this leads us to consider what type of learning space the university might need to provide to achieve the 'disciplining' of the practice-based researcher.

Given the university's commitment to providing laboratories, it was considered essential that they provide some form of studio space for artist and designer researchers. A dynamic Head of Programme made a case for hiring space in a professional artists community. This decision sought to find a spatial form to fuse an identity between professional art and academia. However, we recognised that the standard professional 'solo' studio was not quite the right form for a practice-based researcher's space. Research candidates are not full-time practitioners; rather, the doctorate involves periods of practice, writing, reading, talking and presenting etc. mixed across a day or sometimes separated through planned weeks or months. Separate studios would also mitigate against building a collective identity. These needs led us to consider the occupancy of contemporary design practices in Scandinavia (following a research trip to Copenhagen to visit some co-operative design for health studios). These provide shared, flexible space but without distinct allocated individual work areas. This model allows PhD researchers to continue to be together while doing different things, including writing, reading and making work, and which makes a spatial identity for practice-based research which is different to that of workshops, desk-based classrooms and undergraduate studios. If we had wanted to find this space within the university, we would have needed to describe it as either classroom, studio or lab – not something in between. Obviously, some of the needs we describe are already commonly addressed in art and design teaching studio provision; however, practice-based doctoral candidates tend to be smaller numbers and with less consistent and conventional studio use. It is difficult to make spatial claims for conventional studios for practice-based researchers as their time also consists of other processes beyond making art – and design – work. However, it is the ways that practice-based researchers combine different modes of doing that also produces their particular space, as described by Batty and Berry:

> Whatever the form and whatever the methodology, the space of creative practice research encourages a critical engagement with doing, making, re-doing and remaking. It creates a place in which practice can be incubated alongside ideas, calling into question the past, present and future of that practice.[27]

The studio we have at SHU is in a professional artists studio complex and does not always work well. Provision away from the university makes it harder to manage and cold is sometimes an issue (in the art industry, frequent use of old concrete buildings makes this a common problem). Sometimes the distance from other university spaces dissuades occupation. However, when the physical-space does work well, it adds an extraordinary layer to building the highly specific identity and *space* of practice-based research. The physical-space makes possible, and actively constructs, both the identity of individual art and design researchers and an art and design research community. The community then makes new uses and understanding of the space possible. In this space, art and design researchers share knowledge and understanding that generate some independence from supervisors. The space is populated at odd times of night or in specific blocks of time; the PhD candidates write there as much as make, they discuss, they have parties, invite others, trial work, drink tea, take part in research methods training and lead their own seminars and exhibitions, sometimes with staff.

Conventionally, research space support is compartmentalised between the lab and the desk. As Hockey articulates, the bringing together of research and practitioner processes and identities seems to be the overarching problem/dynamic for practice-based researchers. Our research space refuses the separation of researcher and artist/designer identities and brings both together. Doctoral education and art practice are full of histories, experiences, metaphors and myths of isolation and separation. Labs are communal, desk spaces are separate, researchers work in libraries and offices and solo artists use cavernous New York loft studios. However, running parallel to well-trodden myths of seclusion and isolation, artists and designers have other histories of working collectively and practices that are made in communities, not studios at all. Artists and designers have always sought each other out to build new communities, use spaces to create visibility, work with different disciplines, lever funds, and to communicate externally. Historic versions of this include models of artist-led organisation and space, including examples like Casco projects in Utrecht,[28] Sheffield's digital Access Space[29] and Static in Liverpool.[30] These collective spaces overlap with histories of experimental and collective education like Black Mountain College.[31] These type of art spaces *make* artists: by making space and constructing a collective identity, they give artists permission to '*be*' in different ways: they are artists' engines. We wanted to harness this same process of using space to build practice-based researcher identity. One PhD researcher described it as follows:

> Our PhD studio space gives me a home to my constant context: the low hum that powers each day – which is the people who give my research process, art practice and life texture and joy and depth. The space gives me a place, and the gift of just being a part of a World.[32]

In addition to thinking about how artist's organisations influence how we might create researcher space, there is also a tradition of artists using their artistic processes to explore education, such as Annette Krauss' 2008 *Hidden Curriculum*[33] with school children. The exploration of training and spatial production of researcher space here might be considered to draw on the same tradition of using artistic methods to explore education. Nadine Kulin draws together some of these examples, including both art projects that take the forms of education and artists who experiment in education or propose to 'do it better'[34] than institutions as part of the 'pedagogical turn' in contemporary art. Kulin draws from the work of Lambert, who proposes '(dis)organising a course at the juncture of art and pedagogy may permit the generation of alternative ways of knowing as well as the critical inter-rogation of norms and sites within the University'. Kulin also draws on the work of L. E. Bailey's 'syllabus that seeks to unsettle "education-as-usual" in universities'. Sometimes generating a critical position can be a form of posturing or virtue signalling, an easy position of being the critic without commitment, a position we have sought to avoid here by exploring how space for practice-based research can built with, and in relationship to, more universal forms of doctoral training and learning space-making within the university.

Summary and recommendations

Policy and governance offer no specific requirements for practice-based research; however, there *is* recognition of the value of cohort training. The practice-based researcher (including many practices, such as nursing, law and education, as well as art and design) has a particularly interesting job at hand, to make a *space* that draws together worlds of practice and research. Understanding the significance of this process can provide inspiration to experiment with forms of researcher training.

In the previous text, I have explored a number of strategies for inculcating a *space* for practice-led art and design research: this includes finding ways to explore researcher training with practice, enlisting PhD candidates to produce training and considering how physical space can build practice-based researcher identity. These projects are grown in response to the particular context posed by the post-92 university and respond to the specific energies and sensibilities of a group of PhD candidates at a particular time. This specificity perhaps makes this body of researcher-development at odds with the need to agree on wider strategies that can be applied in different institutions and different disciplines; and at odds possibly even with the needs of a handbook. However, there are principles here that can be developed by others and that might contribute to practice-based researcher development. Given my own artistic identity and a pursuit of a conjoined identity that is also researcher and academic (perhaps no less difficult and evolving than the researchers) instead of principles, I offer a mani-festo for the shaping of practice-based researcher development that I hope can be of use to various practice-based researcher communities:

- See practice-based research *space* as an act of construction, not a given.
- See practice-based research *space* as always being built, never finished.
- Understand that all training is developed in context of disciplines and individuals – although this doesn't mean it can't be applied or adapted for others.

- See the construction of training as an experiment, where different approaches can be trialled.
- Approach every training opportunity critically: this is doctoral level education.
- Support PhD candidates to be critical of, and take ownership of, their own programme.
- Allow the professional methods and cultures of the practice-based field to inform how training can happen.
- Play with the spectrum between formal and informal training provision and the different opportunities and visibilities they offer.
- See the contribution of creative practitioners to training as offering a useful and productive critical perspective on the construction of research by the wider university/institution.
- Explore opportunities for interdisciplinary discourse as a way to understand and build disciplinary and interdisciplinary *space*s and identities.
- See training provision as expanding and joyous instead of procedural and repetitive.

Acknowledgements

Thanks and acknowledgements are due to the extraordinary community of art and design PhD candidates at SHU, especially Sarah 'Smizz' Smith, Debbie Michaels, Anton Hecht and Emma Bolland, who were instrumental in delivering some of the projects discussed here. Huge thanks also to Head of Research Degrees Kathy Doherty, who initiated 'Method' and builds space all the time for PhD candidates and for staff.

Notes

1 Massey (2005).
2 Corazzo (2019).
3 Hockey (2007).
4 Ibid., pp. 159–168.
5 Ibid.
6 (www.ukri.org/wp-content/uploads/2020/10/UKRI-211020-StatementOfExpectationsPostGradTraining-Sep2016v2.pdf).
7 Ibid.
8 www.vitae.ac.uk/researchers-professional-development/about-the-vitae-researcher-development-framework/the-vitae-researcher-development-statement.
9 Editor's note: As a contextual reference, you can read Pearl John's chapter 'The Impact of Public Engagement with Research on A Holographic Practice-Based Study' in Part I of this handbook as an exemplar of how the Vitae framework can be implemented in a PhD.
10 Corazzo (2019).
11 Quoting from Wingate (2006).
12 Carter (2011, p. 727).
13 Ibid.
14 Ibid.
15 Fraser (2007).
16 Bourdieu (1990).
17 Carter (2011).
18 Harrington (2016).
19 Carter (2011).
20 Sarah 'Smizz' Smith (2018).
21 Batty and Berry (2016).
22 Batty and Berry (2016, p. 186).
23 Ibid., p. 182.
24 Ibid., p. 192.
25 Corazzo (2019).
26 Batty and Berry (2016, p. 190).
27 Ibid., p. 185.

28 Choi and Wieder (2011).
29 https://access-space.org/.
30 www.statictrading.com/about/.
31 Harris (2002).
32 Smith (2018).
33 www.theshowroom.org/projects/annette-krauss-hidden-curriculum.
34 Kulin (2012).

Bibliography

Batty, C., & Berry, M. (2016). Constellations and Connection: The Playful Space of the Creative Practice Research Degree. *Journal of Media Practice*, 16(3), 181–194. www.tandfonline.com/doi/full/10.1080/14682 753.2015.1116753.

Bourdieu, P. (1990). *Homo Academicus*. Cambridge: Polity Press.

Carter, S. (2011). Doctorate as Genre: Supporting Thesis Writing Across Campus. *Higher Education Research and Development*, 30(6), 725–736. DOI:10.1080/07294360.2011.554388; http://hdl.handle.net/2292/9461.

Choi, B., & Wieder, A. (2011). *Casco Issues XII- Generous Structures*. Berlin: Sternberg Press.

Corazzo, J. (2019). Materialising the Studio: A Systematic Review of the Role of the Material Space of the Studio. *Art, Design and Architecture Education, the Design Journal*, 22(sup 1), 1249–1265. DOI:10.1080/1460 6925.2019.1594953.

Fraser, A. (2007). *Museum Highlights: The Writings of Andrea Fraser*. Cambridge: MIT Press.

Harrington, J. (2016). *Fantasies of Making*. SHU. https://extra.shu.ac.uk/transmission/papers/HARRING TON%20Jerome.pdf.

Harris, M. E. (2002). *The Arts at Black Mountain College*. Cambridge: MIT Press.

Hockey, J. (2007). United Kingdom Art and Design Practice-Based PhDs: Evidence from PhD Candidates and Their Supervisors. *Studies in Art Education*, 48(2), 155–171, January 1.

Kulin, N. M. (2012). (De)Fending Art Education Through the Pedagogical Turn. *The Journal of Social Theory in Art Education*, 32, 42–55.

Massey, D. (2005). *For Space*. London: Sage.

Researcher Development Framework. www.vitae.ac.uk/researchers-professional-development/about-the-vitae-researcher-development-framework/the-vitae-researcher-development-statement.

Smith, S. (2018). *C3RI Institute Research Blog*. https://blogs.shu.ac.uk/c3riimpact/smizz-blog-s1artspace/?doing_wp_cron=1526997206.7613630294799804687500.

UKRI Statement of Expectations for Postgraduate Research Training. www.ukri.org/wp-content/uploads/2020/10/UKRI-211020-StatementOfExpectationsPostGradTraining-Sep2016v2.pdf.

Wingate, U. (2006). Doing Away with 'Study Skills'. *Teaching in Higher Education*, 11(4), 457–469.

1.7

UNDERSTANDING DOCTORAL COMMUNITIES IN PRACTICE-BASED RESEARCH

Sian Vaughan

Introduction

The growth in practice-based research at doctoral level means that the thesis (as an argument and contribution) can now be articulated through a range of physical and even virtual forms. This complicates traditional assumptions of a written text as the thesis and of doctoral education as being designed to support the production of text. So, how can institutions develop doctoral provision that supports practice-based research? Investigating provision and community for doctoral researchers also provides a lens for understanding the needs of practice-based research in academia more broadly.

This chapter draws on the findings from semi-structured interviews with centre directors and coordinators of doctoral education in a number of institutions worldwide which focused on initiatives and developed models to support building community in the particular institutional settings for practice-based research.[1] The institutions selected represent a diversity of approaches to doctoral education in the broad fields of art, design and performance, although the initiatives developed to build community demonstrate that disciplinary boundaries are productively porous in practice-based research. The institutions are: Edith Cowan University (Australia); Goldsmiths, University of London (United Kingdom); Queensland University of Technology (Australia); the University at Buffalo (USA); the University of Plymouth (UK); and my own, Birmingham City University (UK), where I have personally been involved in trying to build community among doctoral researchers in a Faculty of Arts, Design and Media with many practice-based researchers. Given a growing emphasis on cohort-based approaches to doctoral education, particularly in the UK, the chapter also includes an exemplar of how a multi-institution Doctoral Training Partnership (DTP) has integrated support for practice-based research.

My focus is on lived-experience of trying to support doctoral researchers engaging in practice-based research. The quotations from the interviews included reveal human and individual perspectives rather than emphasising formal institutional policies or theoretical interpretations. The honesty with which interviewees responded also highlights some of the challenges in building community in practice-based doctoral research, with open discussion about what have been perceived as less successful interventions and thorny issues that have not been resolved. While the chapter focuses on institutional perspectives on doctoral education, this only emphasises that community for those engaged in practice-based research is an imperative that is not only an institutional responsibility. It is an imperative that we all have multiple roles to play in building and supporting – as supervisors, as PhD leads, as academics and researchers, as students and peers.

DOI: 10.4324/9780429324154-9

The role of community in doctoral education and practice-based research

A doctorate entails a contribution to knowledge, and by default this implies a community. A contribution to knowledge is a contribution to discourse and a community engaging in that discourse. Doctoral researchers are learning how to join and belong to those discourse communities as they are learning how to undertake research. As Mantai has stated, doctoral researchers need 'a growing sense of belonging to a scholarly, academic, or research community. This sense of community is a well-established requirement for the doctoral student's scholarly development'.[2]

Practice-based doctoral researchers have specific needs alongside those of other doctoral researchers due to the position and nature of practice-based research itself. Frick has described doctoral education as a process of doctoral becoming that is 'an ontological, epistemological, methodological and axiological concern'.[3] Practice-based doctoral researchers are grappling with the epistemology of practice as a form of knowledge and of the inter-relationship of engagements with theory, practice and text as methods in a practice-based methodological approach. They are learning how to frame their creative and research interests in relation to the values and ethics of an academic discipline, frequently challenging the axiology of one or more disciplines through their research. Often, however, the most challenging aspect for practice-based doctoral researchers is the ontology of becoming a researcher in the academy alongside maintaining an established identity as a creative or professional practitioner.[4] For practice-based doctoral researchers, understanding the relevance of doctoral processes and the requirements of academic research can be particularly challenging and a source of stress precisely because of the ontological challenge. Collinson's research found that practice-based doctoral researchers

> initially perceived the combination of bureaucracy and research protocols as fundamentally detrimental to their creative activity and consequently to their creative selves. . . . The intuitive, emotional, spontaneous and 'open' self was confronted by institutional processes and academic demands, which seemed intent on limiting, managing and packaging creativity into tight timescales and pre-defined forms.[5]

In many ways, for practice-based doctoral researchers, resolving this perceived tension with doctoral processes in relation to their own research is a moment of conceptual threshold crossing in their doctoral becoming.[6] As Newbury has highlighted in arguing for research training specific to doctoral researchers in the creative arts and design, it is imperative that 'the development of research skills takes place as part of an active research culture'.[7] As part of such an active research culture, practice-based doctoral researchers can be encouraged to share experiences, collectively negotiating ontological challenges alongside developing their understandings of the epistemological, methodological and axiological positioning of practice-based research. To put it more simply, feeling part of a community can provide reassurance as well as opportunities to test out ideas and learn from others.

Community as legitimising

Building a sense of community can have particular benefits for practice-based researchers within a doctoral education context. Opportunities to share experiences and to 'think aloud' about the articulation of practice-based research can be beneficial in that they increase the confidence of individual doctoral researchers as well as increasing the visibility of practice-based research in the institution. This is important because, despite several decades of precedent, practice-based doctoral researchers can still face anxiety about the legitimacy of their research and perceive university structures for doctoral education as disadvantaging and othering, which can be interpreted as prejudicial, in turn creating further anxiety.

At Goldsmiths University in London, concerns around support for practice-based doctoral researchers and a wish to influence the institutional ethos of practice-based research led to the creation of a forum in which these concerns could be explicitly explored. *The Goldsmiths Forum for Practice-Based Postgraduate Research* (2009–17) operated across the arts, humanities, engineering and physical sciences, encompassing studio arts and computational programmes as well as the social sciences. A key aim was to build community to support the heterodox nature of practice-based research at Goldsmiths and to balance this with its communicability as an ethical ethos that could be owned and promoted by the university. As founder Professor Janis Jeffries remembers:

> Initial discussions and observations suggested that whilst practice research was thriving amongst the PGR community at Goldsmiths and being supported in many different ways by individual supervisors, there was not a coherent strategy for its wider understanding within Departments, its support (financial and in terms of resources), and its strong promotion. This left some researchers feeling vulnerable and staff feeling confused about how to articulate practice research in a cohesive sense. The fact that there was a reluctance to articulate the relation between practice and theory on a postgraduate research degree signalled a deeper concern with the balance between individual inventiveness (on the part of staff and students) and the sharing of this across the institution and outside academia. It may also have signalled a concern about homogenisation of values and individual approaches.[8]

As well as having established the first practice-based PhD across Arts and Computational Technology in 2007 for the Department of Computing at Goldsmiths, Jefferies was also part of the team with the Graduate School who offered training and induction workshops to doctoral researchers, and she was aware of the lack of explicit provision for, and acknowledgement of, practice-based researchers.

The Goldsmiths Forum for Practice-Based Postgraduate Research was created with an explicit cross-disciplinary and interdisciplinary focus. Doctoral researchers from Art, Computing, Sociology, Music, English, Comparative Literature, and Design attended Forum events, and Jeffries was assisted in the later years in running the Forum by Dr Katrina Jungnickel from Sociology. Jeffries recalls the key themes that motivated discussions within the Forum:

> What does it mean to do a practice research PhD?
> What is the status of the value of practice within research?
> Wherein lies the speculative? Wherein lies reflectivity?
> What are some of the models and methods and forms of public engagement and forms of judgement across discipline boundaries?
> What does a Goldsmiths practice research PhD look like?

This agenda clearly reveals concerns around the identity and status of practice-based research, which persisted throughout the life of the *Forum*. In the last year of the *Forum*'s activities in 2016/17, for example, events led by doctoral researchers, staff and invited speakers focused on: life writing and narrative emotions; the ethics of collaboration in computing and design; rigor in practice research; the possibilities of presenting, exhibiting, installing practice-based work; and the challenges of the documentation of live performance. Events varied between the more traditional seminar format of presentation followed by discussion to more interactive sessions. Two more informal *Unblocking the Blocks* workshops were also run by the *Forum* in 2017, as explicit opportunities for doctoral researchers to be open about the challenges they were currently facing and to ask for peer support. Such sharing encompassed critical, creative and practical issues relating to practice-based research. One former participant, Dr William Goodin, was explicit about the benefits of these sessions:

> It was in one of these sessions that I had the breakthrough that helped me create what I feel were the best artistic works I used in my PhD and contributed to an entire chapter of my final thesis. One of the presenters at an *Unblocking the Blocks* event had an image in their PowerPoint that started a chain of events that helped me to find the key elements that I felt were missing in my practice up to that point. In addition, after speaking to this presenter about my research, they suggested a book that was well outside of my current cannon of texts. It was this book that ended up being critical to tying up many of the principal aspects of my research in to a tighter and more solid logical flow of ideas and concepts.[9]

Goodin demonstrates how the sharing of individual problems can have resonance for other practice-based research projects and how such cross-disciplinary discussions can be productive as resources and approaches are shared across disciplinary boundaries. Such sharing of experiences and approaches to the articulation of practice-based research can assist in an ongoing individual and collective renegotiation of progression and monitoring processes, developing understanding of the presentation and publication of practice-based research. The benefits are obvious in terms of the increase in an individual's confidence, but also in avoiding unnecessary 'redesigns of the wheel' as success and failures in approach are shared and reflected upon.

Other institutions have taken a similar approach to Goldsmiths in addressing perceived gaps in provision for practice-based doctoral education by creating discussion fora. At Edith Cowan University in Australia, contemporary artist Lyndall Adams, design strategist Chris Kueh, performance maker Renee Newman-Storen and environmental writer John Ryan collaborated to create *This Is Not a Seminar* in 2012[10] as a multidisciplinary forum 'to assist postgraduate research students in connecting their creative practices to methodological, theoretical and conceptual approaches while fostering an atmosphere of rapport across creative disciplines'.[11] While the Graduate Research School and library provide generic training and support, including seminars and workshops on topics such as writing, qualitative and quantitative research methods and data-management, Adams and her colleagues had recognised a lack of provision specifically for practice-based doctoral researchers. This was despite the fact that in the School of Communication and Arts, and the Western Australian Academy of Performing Arts (WAAPA), two of the three schools in the Faculty of Education and Arts, the majority of doctoral research students were engaged in creative research projects. The key initial questions that *This Is Not a Seminar* set out to explore were:

> What is research and what are the limits to what might be considered research?
> How can we develop creative research skills across a range of disciplines?
> How can practice-led research students benefit from a transdisciplinary and dialogic learning environment?
> What are the problems that practice-led creative researchers often experience, and are these issues related to feelings of isolation and inadequacy in the academy?[12]

There are clear similarities here with the concerns around articulation, legitimacy and practicalities that had informed the creation of Goldsmith's *Forum* a few years earlier. It is significant that in both institutions, the response was to create community-building events and spaces for conversations to explore these issues, rather than training to provide answers and solutions.

Reflecting the disciplinary diversity of its founders, *This Is Not a Seminar* was established as a weekly forum with a distinct focus on cross-disciplinary conversations:

> In naming the forum *This is Not a Seminar*, we set out to foster an environment of egalitarianism, dialogue, exchange and questioning between facilitators, guest practitioners and participants, rather than a traditional learning structure of 'students working with a professor. . . . As a 'breeding ground' (the botanical connotation of 'seminar' and indeed our preferred one), TINAS conversations were often unscripted and rhizomatic, leading to unforeseen realisations about the nature of creative research through a synergy of ideas.[13]

The weekly forum included a mix of different types of events. *In Conversation* sessions were open conversations held between academics and doctoral researchers from across creative disciplinary fields. The team soon realised that these were more successful if there was more than one panellist and if different disciplines were represented, as this broadened appeal and increased attendance, as well as providing fruitful cross-disciplinary fertilisation and reflection. The *This Is Not Theory* series was a set of critical reading exercises, each based on a few paragraphs of theoretically dense text: 'These sessions were fun, engaging and, best of all, noisy debates about meaning. Everyone dug in to unpack the dense material – to get to the heart of challenging concepts'.[14] *This Is Not Rocket Science* workshops had a more practical focus on training specific to practice-based research; topics have included reflective journals, public speaking, copyright in creative practice, ethics and even how to wrangle Microsoft Word to include film as well as images.

Adams and her colleagues have identified numerous benefits to *This Is Not a Seminar* which have repaid their investment of time and resources in the initiative:

> [it] has enabled a heightened rapport and a greater sense of community amongst researchers across creative disciplines; a broader acknowledgement of the range of work that constitutes practice-led and practice-based research; confidence in the development of documentation, communication and methodological skills; an appreciation for the modes through which creative practices can be theorised and contextualised in academic terms; and a stronger representation of practice-led and practice-based researchers in academic environments.[15]

The benefits are individual and institutional. As with the *Forum* at Goldsmiths, confidence both in articulation of practice-based research and its legitimisation via the institutional value accrued through the profile and visibility of the community have benefited academic staff as well as doctoral researchers.

While the *This Is Not a Seminar* initiative continues and evolves at Edith Cowan University, the *Goldsmiths Forum for Practice-Based Postgraduate Research* effectively came to an end in 2017, prompted in part by the retirement of now Emeritus Professor Jefferies. As former participant Goodin perceptively states:

> Initiatives and communities like this can only be sustained through the dedication of the presiding faculty members and with the support of the University. It takes actively engaged faculty members that believe in the process and stay engaged with the community at large to keep initiatives like these alive. Communities like these have to constantly evolve as the crop of students, their projects, and their individual needs change.

While arguably the practice-based research environment at Goldsmiths had evolved during the nine years of the *Forum*'s existence and the needs were different in 2017, the central role played by Jeffries is evident, and this points to questions of sustainability of community for practice-based doctoral researchers. It is a difficult balance to achieve, enabling and responding to a fluid group of doctoral researchers whose needs may change from year to year, while ensuring that community initiatives become embedded and sustained by the institution and participants. Currently there is more of a disciplinary focus to doctoral community events at Goldsmiths, for example the PhD Art programme running regular *Flashpoints* where doctoral researchers share their research and its challenges. This points almost paradoxically to the success of the *Forum*, in that the benefits of the activities have been recognised and replicated at disciplinary level, removing the gaps that the *Forum* filled in provision. However, arguably the potential of cross-disciplinary exchanges that the *Forum* embodied has diminished.

Opportunities to share experiences and to 'think aloud' are beneficial for practice-based doctoral researchers and can increase an individual's confidence and understanding of their own research project. Both at Goldsmiths and Edith Cowan University, facilitating discussions via seminars and workshops with cross-disciplinary and interdisciplinary communities had particular benefits in raising the visibility of practice-based research in the institution as well as addressing specific methodological challenges and training needs. There are challenges, however, in balancing the benefits of disciplinary communities based on shared language and understanding, with the benefits of outsider views and reconceptualisation via the effort of translation that come from interacting with researchers from other disciplines.

Spaces for community building across disciplines

Creating spaces for interdisciplinary and cross-disciplinary communities to mix and form can be challenging as well as fruitful. Described on its website as 'an Open Research Lab for playful experimentation with creative technology'[16] i-DAT at the University of Plymouth on England's South Coast is an evolution of the Institute of Digital Art and Technology established over 20 years ago. i-DAT hosts doctoral researchers from a diverse range of disciplines which can include digital design, communication design, sonic arts, digital art, interactive media, curating and architecture. Through relationships with other research institutes and groups at Plymouth and internationally, the broader pool of doctoral researchers with which i-DAT's academics engage includes those in the medical humanities, robotics, education and performance. Across this rich mix and key to i-DAT's research activity is the ethos of, and spaces for, openness and play.

As well as more traditional seminars in which doctoral students can present their work-in-progress and prepare for conference presentations, 'workshops are a fundamental weapon in i-DAT's creative arsenal'.[17] This spirit of openness and playfulness is evident in the organisation of these workshops, in which the doctoral students are joined by academics and post-doctoral researchers from not only the design disciplines but also the University of Plymouth's Sustainable Earth Institute and Impact Lab, and industry contacts through the South West Creative Technology Network.[18] Mike Phillips, Professor of Interdisciplinary Arts and Director of Research at i-DAT, explains what he sees as the importance of a playful, interdisciplinary mix for the workshops:

> They feed off each other and other projects feed in. . . . It is proactive thing to do the workshop methods, especially when you're doing a very practice-based work-shopping method. It breaks down so many kinds of boundaries between disciplines. Just the making of things is really crucial to what we do, and it does get quite disruptive and it breaks people's isolation. You know often peoples' PhD is to disappear into a cupboard for a time and things like that. The workshop method really challenges people in a very, I think, productive way because it's not like you're forcing people to do things, it's just that to engage you have to do something.[19]

The workshops i-DAT organises and hosts are designed to be hands-on and often include elements of training and exploration with particular types of software, coding or technology that are framed as experimental and playful. For example, a workshop hosted by i-DAT as part of the University of Plymouth's Design Research series in November 2019, which was on data to support applications to a funding call from the South West Creative Technology Network, was called the *DATA TA-DA!* and its schedule included a half-hour slot for what was described slightly tongue-in cheek as 'pizza & more fiddling'.[20] As well as the particular project, technology or strategic aim for an individual workshop, Phillips sees the interactive, practical nature of the workshops as serving a strong community-building focus:

> I take a lot from knitting groups actually where there is that tinkering process where you're doing something and talking at the same time. You need the seminar and more intensive things as well, but this is a great way of starting those things off and so it's quite agile as a way of engaging people from different backgrounds as well certainly the interdisciplinary thing which is quite important to us. Work-shopping with mathematicians, also stats people, it just sort of brings them together and also if they don't know how to make things which often other disciplines don't (like the humanities) they are scared of making things or breaking things. And, actually, it starts interesting dialogues between two individuals that can then enrich the symposia-type kind of model.

Not all of the doctoral researchers connected with i-DAT are practice-based researchers, and as Phillips suggests, the practical workshops can encourage skills exchange and respect across the different disciplines. The workshops are often led or facilitated by practice-based doctoral researchers, as Phillips believes that 'it gives them the chance to focus and to explain their research in a very pragmatic way to an audience who are keen to learn'. His analogy with knitting groups emphasises the discursive nature and sociality of the workshops, in which the doing facilitates discussions of different disciplinary and theoretical perspectives.

As well as the playful and open ethos, provisioning the environment has been a central feature of the research culture of i-DAT. Phillips reflected on the pitfalls of i-DAT's previous physical spaces, where the offices, workshop and lab spaces were too far away from each other, even in different buildings, and where it was difficult to build a sense of a cohesive community. i-DAT now has collocated office, lab and studio spaced, and he revealed:

> We've actually sort of been liberated by the space that we just moved into, we have a new office-lab space and conversations are happening in that space much more kind of in a self-generated thing where you might run a workshop and the remains of the workshop are there, you know the bits, and then other things take place so we have little pop-up things happening as well so it's becoming a more dynamic space.

Each full-time doctoral student has an allocated workspace, and there are hot-desks shared by part-time students and visitors. Phillips did offer a note of caution based on a previous experience:

> Well I think that probably the most destructive thing that can happen is when you mix PhD cultures in space. When you have very theory-based students who want to spend a lot of time reading and that sort of quiet stuff . . . that then became almost like fistfights in the office space, the shared space. Literally because someone was trying to make an Arduino squeak and somebody was trying to read something. And then you know people tried to create house rules, as soon as you started to do that it became, not an enforcement, but something that you were aware of all the time. That started to become unproductive I think.

It is important to Phillips that the research spaces and the research students are self-governing in their use of them. As he pointed out, there are plenty of quiet study spaces provided across the university:

> But there are actually very few kinds of cohabited making-spaces where you can exchange ideas. So that's I think the priority. I mean we do have quiet times there and occasionally an email has to be sent, you know which is usually when there's maybe a mass of students getting rowdy. But it's minor stuff but it's always done in a very open way as well.

For Phillips as Director, is it important that i-DAT maintains an open and playful ethos as a creative community, and that the environment reflects and enables the experimental, fluid interdisciplinary practice-based research that is i-DAT's research agenda. Doctoral students are entangled with academics, artists and industry contacts as an integral part of this community through shared spaces, events, seminars and workshops.

At Queensland University of Technology (QUT) in Australia, both the use of space and encouraging disciplines to mix have proved challenging, in ways that resonate with the experience at i-DAT. Established in 2016, the QUT *Design Lab* is the research centre in the School of Design in the Creative Industries Faculty. The *Design Lab* encourages transdisciplinary collaborations, and the website claims: 'our approach to research removes disciplinary silos and brings together all higher degree research students'.[21] The research environment into which doctoral researchers are embedded includes regular workshops, seminars and more experimental exploratory events described as rumbles and mudpits.

In contrast, QUT's *Creative Lab*, while being established for a similar period of time, is a more dispersed amalgamation of disciplinary communities, including art, theatre, dance, music and creative writing. Professor Gavin Sade, currently Associate Dean (Teaching and Learning) in the Creative Industries Faculty, has an informed perspective on both research labs. As an artist and designer specialising in interaction design and electronic arts, he has supervised doctoral researchers in both the *Design Lab* and the *Creative Lab*, as well as being Interim Director of the *Creative Lab* in 2018. Sade[22] noted that while *Creative Lab* also runs seminars for doctoral researchers and workshops, there was less sense of a cohesive community. He admitted that when he became Interim Director it was a challenge that he tripped up on. One of the first things he did was to bring all the PhD students in Creative Lab together to speak about research community, but 'many of them hadn't even seen each other, didn't even know who each other was, some of them didn't even know that they belonged to a research group called the *Creative Lab*'. He concluded that in this context his 'sort of elbow-forcing' approach to seeing community through the lens of an organisational structure was not appropriate. The *Creative Lab* contains a higher proportion of the practice-based researchers, yet Sade had to recognise that the *Creative Lab* was 'still a little bit earlier in that evolution'. While within the disciplines there were established and supportive doctoral research community events, it would take longer to build community across the *Creative Lab*.

The issues of both size of community and space for community building have been key at QUT. For example, in 2017 there were 175 higher degree researchers across the Master of Philosophy, Doctor of Philosophy and Doctor of Creative Industries programmes attached to *Creative Lab*.[23] Sade reflected on the issue of lab size and the optimal size for a research community:

> They're almost too large to be singular communities, there's a few different sub-communities in there, clustered around things, so some of those thematically organised groups have been very successful, especially where they have a culture of almost self-generating activity, so some of the really successful groups have had the students running a lot of the community style activity.

Sade gave as an example the *Design Lab's Rumbles*. He explained that the *Rumbles* were started with the directors and supervisors programming them but that over time, doctoral researchers in the group took on organising them.

While acknowledging that the provision of physical space is a vexed institutional issue that he personally tries to keep out of, he identified the benefits of colocation for doctoral researchers. He also identified the key role played by academic staff in investing in building community and gave the example of an urban informatics group within QUT's *Design Lab*:

> They had a director and associate director who were just dedicated to the lab, that were co-located, every Friday they ran things, they asked their students to be present, in the space which they fought very hard to keep, so they put a lot of effort into bringing their HDR students together, they worked with them together, they ran reading groups, they started this, and once that had happened consistently and regularly over a period of years it became a habit.

This example highlights the importance of staff engaging to create the impetus for community building and to boost activity until it becomes embedded when there is the potential for community to be self-sustaining. It is also clear that the issue of space, and the colocation of academic staff and doctoral researchers, was important in enabling the building of community. It is also interesting to consider the disciplinary differences that may affect the provision of space for doctoral research communities. Sade pointed out that practice-based research is less prevalent within QUT's *Design Lab*, which has more of a focus on written outputs, journal articles and monographs even where creative practice is part of the research. In *Creative Lab*, practice-based research is much more common. Undoubtedly providing space enables community:

> The urban informatics group that was very successful had an open plan office space in the same area of the building where the director and associate director and all the other researchers were, so they were part of that culture. Physically they were located there, so we do, we try and embed them into the research culture, but when it comes to it – the space conversation gets in the way a bit.

Resourcing practice-based research with physical space can be more difficult institutionally, as more than just desk space is required.

The provision of space for doctoral researchers can help to create a sense of community and belonging, and for practice-based researchers these spaces need to enable both quiet study and the making, doing and sharing of practice. As Phillips found with i-DAT in Plymouth, when space for collaboration and creative practice can be provided, it can be transformative and is an important enabler of community. The experiences at i-DAT and QUT demonstrate also the importance of academic staff being present and visible within the spaces, yet without imposing regulation or hierarchies with the spaces.

Doctoral researcher-led and inclusive communities

While community can be initiated and supported by academic staff, where the doctoral researchers become involved in running community events and initiatives, communities can become more sustainable and be perceived as more successful. In my own context, within a Faculty of Arts Design and Media at Birmingham City University in the UK, I have been involved in community-building initiatives that are the co-creation of staff and doctoral researchers.

In 2014, I was involved in setting up a peer-mentoring scheme for doctoral researchers in Art and Design. Traditionally, doctoral students have looked to supervisors for guidance in their academic career, and more recently, this has been supplemented in most institutions by a separate career development programme focusing on the development of transferable and employability skills. However, we recognised that this provision was not explicitly supporting the complexity of multiple, potentially conflicting identities of doctoral researchers in Art and Design in which academic, industry and practitioner roles are often entwined. This complexity is one that can be troublesome to navigate and comes in addition to the isolation and trepidations that are axiomatic of the doctoral experience regardless of discipline. Our peer-mentoring scheme was therefore designed to provide both psychosocial support in addition to the supervisory team, and also to the enhance the skills of mentees and mentors.[24]

From its outset, the mentoring scheme included group social gatherings as well as individual partnerships. This emphasised doctoral research as a collective experience as well as individual endeavour, enabling informal spaces to discuss experience and identity. Feedback from our participants in the pilot year was overwhelmingly positive about the benefits they perceived from such community events. Each individual mentoring partnership was also initiated with the gift of a voucher for a local independent coffee shop:

> We felt this would be an important part of the scheme: it could provide a neutral venue for the meetings and a means of easing the initial conversations. On a practical level, it would facilitate meetings between partners based at different campuses, as the coffee shop chosen was fairly equidistant from all sites.[25]

Somewhat unsurprisingly, feedback from participants was also positive about the vouchers. Beyond the obvious gratitude for free coffee and cake, the participating doctoral researchers stated that it both made them feel valued by the institution and encouraged them to value the mentoring process itself. The impact of the psychosocial support encountered through peer-mentoring seemed to emerge on two interrelated levels: through an enhanced sense of wellbeing and also increased confidence for both mentors and mentees. The mentoring scheme was explicitly designed and led as a staff–student partnership between myself, another academic and a doctoral researcher in the final stages of submission and viva. That the peer-mentoring scheme was jointly led by a staff–student partnership in a visibly non-hierarchical manner was crucial in enabling discussion and facilitating engagement around issues of professional and creative identity, and in creating the sense of a broader research community in which doctoral researchers were legitimate participants. The success of the scheme in its pilot year, as evidenced in participant engagement and feedback, also signified for us the obvious benefits of facilitating sociality and community through the provision of catering.

An institutional restructure in 2014 brought art and design doctoral researchers into closer contact with those in music and performing arts, media and cultural studies, and English. It gave us the opportunity to expand and embed our doctoral community across more disciplines. The resulting *PGR Studio* was the co-creation by academic staff and doctoral researchers of a postgraduate research community as 'a creative, collaborative and practice-based space of doctoral training'.[26] With all its activities designed and run in collaboration with doctoral researchers, *PGR Studio* aims to support all forms of research activity, not just practice-based research. We were fortunate in establishing *PGR Studio* to be able to employ our former student partner, now with a newly minted doctorate from her own Fine Art practice-based research, and we have employed some of our current doctoral researchers to help devise and facilitate the running of *PGR Studio*'s programme of events each year.

While the studio can be an artists' studio, a design studio, a recording studio or a rehearsal studio, it can also be a writing studio for those undertaking more traditional forms of research in the arts and humanities. This is important for a Faculty in which there are doctoral researchers working in English Literature, Art and Design History, Cultural Studies and Musicology as well as practice-based researchers in Fine Art, Jewellery, Composition, Performance, Fashion Design, Architecture and Creative Writing. As Coordinator Dr Jacqueline Taylor explains:

> The PhD as incorporating creative or artistic practice is not set up as separate to the 'traditional' PhD. Rather, all research is approached as part of a spectrum in which there are different nuances of practice to avoid setting up a binary between research involving creative practice and that which does not and risk 'othering' practice against more traditional research.[27]

The emphasis is on enabling a supportive, social and creative community for all doctoral researchers, in which research is research regardless of its methodological or conceptual approach and where creative encompasses intellectual and conceptual leaps as much as it does professional artistic practice. This ambition has proved tricky to enact and requires a fluid and flexible approach to doctoral provision.

PGR Studio still runs the peer-mentoring scheme with the associated social group gatherings; mentees have become mentors to newer doctoral researchers, demonstrating both their perception of benefit and their willingness to perpetuate the community. Other regulars in the *PGR Studio* calendar are an annual spring research festival and doctoral researcher-run summer conference. The annual conference explicitly encourages performances, participatory workshops and exhibitions rather than more traditional papers: 'The conference rethinks the conventional conference format and provides a vital platform for students to experiment intellectually as well as in the dissemination and form of the research itself'.[28] In fact, this focus on the experimental and playful presentations has led to grumblings from other doctoral researchers who feel that practice-based research is being over-privileged and who resent not being able to contribute more traditional conference papers in this annual doctoral research showcase. Arguably there are opportunities for doctoral researchers to present more traditional papers in the more discipline-specific seminar events run by research clusters elsewhere in the faculty; however, the resentment suggests that concerns about legitimacy and hierarchies in research can affect those doing more traditional forms of research as much as practice-based doctoral researchers.

PGR Studio does facilitate more traditional training and professional development events, including for example workshops on the use of specific referencing software, *Viva Survival* workshops, writing retreats and *Demystifying Progression Assessment* workshops. The practice-based ethos is never far away; for example, an *Introduction to Academic Conferences* workshop was run as a mini-conference, complete with awkward coffee breaks, a fly-in-fly-out out keynote speaker and staff role-playing the 'this is more of a comment than a question' audience contribution, all of which are then collectively discussed and critiqued. Provisioning the environment to be supportive and collegial is an important aspect of these workshops and events, and catering often is a critical element. A careers event on routes out of doctoral study is run as a panel discussion and networking event as *Careers and Cocktails* with a tailored menu of cocktails and mocktails whose names prompt discussion of potential career routes.

Significant features of the community-building activities include a lively blog on the website and a small *Researcher Development Funding Awards Scheme*. The funding scheme supports doctoral researchers to organise and facilitate development initiatives that benefit the doctoral environment and community. In 2017, practice-based art doctoral researchers successfully bid to hold a midweek meditation and mindfulness workshop which included a group outing to the Kadampa Meditation

Centre in central Birmingham, and then a group trip to a vegan café. As the organisers commented in a blog post:

> Aside from the obvious benefits of improving well-being, stress management and problem-solving skills, sessions such as these give a focused reason for students to come together and share the experience of the activity, allowing for new academic and social connections to be formed, reaching across faculties and year groups.[29]

This is reminiscent of both Professor Sade's reflections on the challenges of establishing community within *Creative Lab* at QUT, and Professor Phillips' analogy with knitting groups. Community cannot be forced and imposed. You need to give people a reason to come together and then, as connections are made, networks form and community can grow. *PGR Studio* shows that there can be clear benefits in giving doctoral researchers the agency to develop their own community and to find their own reasons to come together.

Similarly, Midlands4Cities (M4C) actively promotes community among all doctoral researchers in the arts and humanities, and an inclusive approach to practice-based research in a structure which enables the agency of doctoral researchers to identify and build the communities that they want to be part of. M4C is a consortium of eight universities across four cities in the midlands of England which form an Arts and Humanities Research Council-funded Doctoral Training Partnership (DTP).[30] As the expanded successor to an earlier, Midlands3Cities (M3C) DTP, the combined cohort that M4C supports is several hundred students at any one time, and the M4C has purposively encouraged the sense of an M4C cohort identity that bridges the institutions.

As well as the centrally supported events such as an annual *Research Festival*, M4C's Cohort Development Fund (CDF) was initiated to support training, development and research activities designed by award holders. The CDF explicitly promotes doctoral community, as each CDF application has to come from award holders in at least two of the consortium universities, and applicants are encouraged to include doctoral researchers outside the DTP wherever possible. In the first year of the CDF's availability, one of the funded events that was proposed was about practice-based research. In June 2016, Nottingham Trent University hosted *Critical Creativity*, a one-day symposium which brought together writers, scholars and postgraduate students to explore the relationship between critical writing and creativity. As the relation between creative-practice and academic writing or exegesis has been a frequent concern in the discourses of practice-based research, it is perhaps not surprising that the practice-based students proposed this topic for one of the early community events. The symposium included papers by academics from within and beyond the M3C consortium, as well as a practical workshop on understanding and developing creative-critical writing techniques. A wine reception and a showcase of readings in the evening encouraged networking and the sense of a practice-based research social community.

Cracking the Established Order (CtEO) was a two-day interdisciplinary conference held in June 2019 at De Montfort University and is another CDF-funded event that demonstrates M4C's support for practice-based research. However, as the title of the event reveals, it also demonstrates the ongoing concern of practice-based doctoral researchers with the positioning and legitimacy of their research in the academy, seeing themselves and their research as other. The symposium aimed to draw out the provocative potential of practice-based research and focused on the different approaches to the methodology of practice-based research. Purposefully avoiding the conventional academic conference format, *CtEO* consisted of a variety of workshops, performances, discussion groups, screenings and installations. *CtEO* was successful as an event in terms of attendance and engagement and led to a special issue of the *International Journal of Creative Media Research*. It has also generated longer term benefits for the practice-based research community within M4C, as an organising committee of doctoral researchers has been established to take *CtEO* forward as an annual event. This suggests that the practice-based researchers within M4C see the benefits for themselves as a community in coming together. It is something that they want to keep doing.

An important element of community for practice-based doctoral researchers is the psycho-social support that can be mobilised. The challenge for universities and those tasked with coordinating doctoral education is to find mechanisms that encourage supportive and collegial environments while allowing for agency. Both PGR Studio and the M4C consortium have mobilised relatively small funding opportunities to enable doctoral researchers to define and create their own community events.

Community as a professional skill for practitioners

Not all PhD programmes have the same structure. While in the UK the emphasis is on individual study and research supported through supervision, in the USA and Canada, PhD study involves assessed coursework, followed by comprehensive examinations and then a research project, with full-time doctorates taking on average four to six years. The coursework element of structured training means that the North American doctoral experience has a stronger cohort element, at least in the initial stages. In practice-based doctorates, this sense of cohort and community can serve multiple functions – pedagogic, social and professional. North American academic Professor Sarah Bay-Cheng explicitly links it to the skills and aptitudes needed for a career, whether in the arts, in academia or combining the two: 'I think the sense of cohort is really critical . . . how do you work collaboratively? How do you understand professional networks?'[31]

Motivated to close the gap between professional practice and scholarship, and working with a group of colleagues in the University at Buffalo, in 2010 Bay-Cheng founded an innovative PhD in Theatre and Performance for doctoral researchers in dance, performance studies and theatre practice. Keen to embed a community ethos as a professional skill, the coursework element coalesced around three credit-bearing seminars and a non-credit bearing weekly studio to encourage and support practice, all of which reinforce the sense of a doctoral researcher being part of a community. In the *Performance Research* seminar, the emphasis was on the mechanics of research as an introduction to postgraduate and doctoral study and to the field. So, as well as methods training in archival research, 'it was also a place to from the very beginning talk through ideas about how theatre practice and historical research might get it together and what a practice-based research model is'. The *Performance Scholarship* seminars were focused on dissemination, on the publishing process and on understanding the different audiences for scholarship. The *Pro-Seminar* had an explicit employability agenda and aimed to prepare doctoral researchers for the job market with sessions on applications and CVs, job talks and workshops on organising a research agenda. For Bay-Cheng, it was crucial the doctoral researchers had a collaborative space in which to explore and prepare for future careers:

> My goal was to create a programme that was rigorous, sustainable, and equitable. You know, it should be about you and the quality of your work, but your success should not depend on whether or not you get a supervisor who is super involved, and, because there are lots of great supervisors who are super knowledgeable, right, but they haven't been on the job market in 20, 30 years. Things are changing all the time. And so, the goal in that was that you wanted a space to be really attentive to what was happening in that particular moment.

Practice-based research was embedded throughout both the coursework and research stages of the PhD through the studio component, and this balance between professional skills and space for creative experimentation and failure is key to Bay-Cheng's approach to doctoral education. It is a tricky balance to achieve, however, particularly when the doctoral community brings together people from different specialisms and backgrounds, including experienced creative practitioners for whom failure may be conceptually and reputationally difficult:

> I think one of my tendencies was to go as quickly as possible to the fun stuff that I really liked, and I think one of the things that I frequently underestimated, was a) how people understood where they were coming from, and b) how they understood where everybody else was coming from, and so one of the things that we had to kind of retool was language, and making expectations really explicit, particularly around studio. . . . In a group dynamic, I think it's really important to – in the same way that you would working in a theatre company – there has to be some early investment in building the community and in building the group.

It is significant that Bay-Cheng draws on her professional arts background and identifies the commonalities between building community in academia and collaborating in theatre practice. As well as supporting doctoral researchers to make contributions to knowledge, doctoral education in practice-based research has to prepare researchers for collaboration and cooperation in their future careers, whether inside or outside universities, or the more common portfolio model of combining being an academic with a continuing professional creative practice.

Professor Sade from QUT in Australia echoed this view of community as a professional network and being part of a community as a professional skill for academics. He gave the example of the urban informatics research group in the QUT *Design Lab* who have established 'a community of alumni around the world that know each other, when they're anywhere they visit'. His view is that PhD students 'need to have a network, and build a network, if they are going to be successful afterwards.'[32] For practice-based doctoral researchers, these professional networks are likely to be both within and beyond academia, in communities that are disciplinary, cross-disciplinary and interdisciplinary.

There are inevitably professional research skills as researchers that practice-based doctoral researchers need to acquire and develop, whether about particular methods, ethical practices or the various forms that academic writing and dissemination can take. Supporting skills development through structured training can be a collaborative activity and engender community. Being a productive member of any community is a skilled activity, involving communication, interpersonal and professional skills – skills which need to be developed and supported as part of research activity. These are also skills that are valued outside of higher education and research careers. It is important, however, to recognise the current debates around doctoral graduates' employability and preparation for routes out of study may not have such resonance in creative subjects where the boundaries between academic and professional employment are not so absolute. Practice-based researchers need to be part of multiple communities as a professional activity.

Concluding thoughts: building community as rebuilding the academy?

My intention here in reflecting on different experiences and initiatives in doctoral education was to offer another lens to understanding practice-based research more broadly and the conditions under which it thrives. There are clearly significant benefits to building and supporting community among practice-based doctoral researchers. Community can provide the sense of belonging supportive of doctoral becoming, enabling productive conversations that aid the navigation of methodological, epistemological and ontological threshold crossing. It is important to create community-building events and spaces for conversations to explore issues, rather than training to provide answers and solutions. Community can enable safe spaces to vent as well as to share experiences. Feeling part of a community can provide reassurance as well as opportunities to test out ideas and learn from others.

It is also evident that building and supporting community for and among practice-based doctoral researchers is not without its challenges. The interviewees have honestly reflected upon occasions and initiatives which have floundered. Mixing disciplines and integrating practice-based researchers with other researchers can be challenging as cultures, languages and understanding may not be shared and may need to be negotiated. Community and individual anxieties around the legitimacy of practice-based research surfaced in all the interviews that I conducted, and in my own lived experience at my own institution. Community activities and conversations within communities can be a significant way to raise the visibility of practice-based research within institutions and to build confidence in its legitimacy. In my view, however, we need to leave such epistemological and methodological questions open, rather than presenting restrictive solutions. Echoing Søren Kjørup,[33] we need to plead for plurality in how our practice-based doctoral communities recognise and value research, focusing on questions of quality, significance and appropriateness that build confidence rather than increasing anxieties or suggesting hierarchies.

Encouraging community membership as a cooperative endeavour in practice-based doctoral education benefits the individual doctoral researchers but also has potential to benefit supervisors and other academic staff and to challenge institutions. Many of the benefits described and observed in the interviews would be as valuable for academic researchers as they are for doctoral researchers. All academics benefit if they can be part of truly collegial, creative and intellectually challenging research communities alongside doctoral researchers. Academia may increasingly be structured as competitive and hierarchical; however, those tendencies do not have to be reinforced, and instilling a cooperative and collaborative mentality in the doctoral researchers as our peers can benefit creative interdisciplinarity and the future of the academy itself.

Notes

1 While this chapter focuses on critical interrogation of different lived experiences of supporting practice-based doctoral communities, in the PhD section of this volume I provide some more pragmatic ideas for initiatives and considerations that can be applied in your context; please see 'Community-Building for Practice-Based Doctoral Researchers: Mapping Key Dimensions for Creating Flexible Frameworks'.
2 Mantai (2019, p. 368).
3 Frick (2011, p. 127).
4 Collinson (2005), Hockey (2008), Wisker and Robinson (2014).
5 Collinson (2005, p. 718).
6 Wisker and Robinson (2009).
7 Newbury (2010, p. 377).
8 This and other quotations are taken from the author's email correspondence with Emeritus Professor Janis Jeffries, December 2019–January 2020.
9 This and other quotations are taken from the author's email correspondence with Dr William Goodin, January–February 2020.
10 For a fuller discussion and reflection on *This Is Not a Seminar*, its pedagogic and philosophical underpinning and their positions as facilitators, practice-led researchers and ethnographer-participants, see: Lyndall Adams, Christopher Kueh, Renee Newman-Storen, & John Ryan (2015) 'This is Not an Article: A Reflection on Creative Research Dialogues' (This is Not a Seminar), *Educational Philosophy and Theory*, 47(12), 1330–1347.
11 Adams et al. (2015).
12 Ibid., p. 1336.
13 Ibid., p. 1334.
14 Ibid., p. 1341.
15 Ibid., p. 1344.
16 i-DAT 2019a
17 i-DAT 2019b

18 The South West Creative Technology Network is a £6.5 million project to expand the use of creative technologies across the south west of England, led by the University of the West of England and funded through Research England's Connecting Capabilities Fund. The University of Plymouth is a partner alongside Bath Spa University, Falmouth University and industry partners Watershed in Bristol and Kaleider in Exeter.

19 This and other quotations are taken from the author's interview with Professor Mike Phillips, 20 December 2019.

20 Design Research website (2019).

21 QUT Design Lab (2019).

22 This and other quotations are taken from the author's interview with Professor Gavin Sade, 16 January 2020.

23 QUT Creative Lab (2017, p. 12).

24 Boultwood et al. (2015).

25 Ibid., p. 16.

26 PGR Studio (2019).

27 Taylor (2019, p. 205).

28 Ibid., p. 213.

29 Bailey and Walden (2017).

30 The Midlands4Cities Arts and Humanities Research Council funded Doctoral Training Partnership (DTP) comprises: the University of Birmingham and Birmingham City University; Coventry University and the University of Warwick (both in Coventry); De Montford University and the University of Leicester (both in Leicester); and Nottingham Trent University and the University of Nottingham. Midlands4Cities is an expansion of the earlier Midlands3Cities (M3C) Doctoral Training Partnership (2014–2019) which had not included the two universities in Coventry. M3C-funded 439 doctoral awards between 2014 and 2018, and M4C intends to make approximately 460 awards for PhD study between 2019 and 2024.

31 This and other quotations are taken from the author's interview with Professor Sarah Bay-Cheng, 27 January 2020.

32 As with the earlier quotations from Professor Sade, this quotation is taken from the author's interview with Professor Sade on 16 January 2020.

33 Kjørup (2010).

Bibliography

Adams, L., Kueh, C., Newman-Storen, R., & Ryan, J. (2015). This Is Not an Article: A Reflection on Creative Research Dialogues (This is Not a Seminar). *Educational Philosophy and Theory*, 47(12), 1330–1347.

Bailey, S., & Walden, S. (2017). *Midweek Meditation and Mindfulness*, October 11. http://pgr-studio.co.uk/30454-2/?doing_wp_cron=1579110891.2238659858703613281250 (accessed February 6, 2020).

Boultwood, A., Taylor, J., & Vaughan, S. (2015). The Importance of Coffee: Peer Mentoring to Support PGRs and ECRs in Art & Design. *Vitae Occasional Papers Volume 2: Research Careers and Cultures*, 15–20.

Collinson, J. A. (2005). Artistry and Analysis: Student Experiences of UK Practice-Based Doctorates in Art and Design. *International Journal of Qualitative Studies in Education*, 18(6), 713–728.

Design Research Website. (2019). *DATA TA-DA!* https://designresearch.info/data-ta-da/ (accessed January 17, 2020).

Frick, L. (2011). Facilitating Creativity in Doctoral Education: A Resource for Supervisors. In V. Kumar & A. Lee (Eds.), *Doctoral Education in International Context: Connecting Local, Regional and Global Perspectives*. Serdang: Universiti Putra Malaysia Press, 123–137.

Hockey, J. (2008). Practice Based Research Degrees in Art & Design: Identity and Adaption. In R. Hickman (Ed.), *Research in Art and Design Education: Issues and Exemplars*. Bristol: Intellect, 109–120.

i-DAT (2019a). *About.* https://i-dat.org/info/ (accessed January 17, 2020).

i-DAT (2019b). *Workshops.* https://i-dat.org/environment/ (accessed January 17, 2020).

Kjørup, S. (2010). Pleading for Plurality: Artistic and other kinds of research. In M. Biggs & H. Karlsson (Eds.), *The Routledge Companion to Research in the Arts*. Abingdon: Routledge, 24–43.

Mantai, L. (2019). A Source of Sanity: The Role of Social Support for Doctoral Candidates' Belonging and Becoming. *International Journal of Doctoral Studies*, 14, 367–382.

Newbury, D. (2010). Research Training in the Creative Arts and Design. In M. Biggs & H. Karlsson (Eds.), *The Routledge Companion to Research in the Arts*. Abingdon: Routledge, 368–387.

PGR Studio. (2019). *An Introduction to PGR Studio.* http://pgr-studio.co.uk/ (accessed February 20, 2020).

QUT Creative Lab. (2017). *Creative Lab Annual Report.* https://research.qut.edu.au/creativelab/wp-content/uploads/sites/29/2018/11/23477_CI-Creative-Lab-Report-October-2018-Web-F2.pdf.

QUT Design Lab. (2019). *Study with Us*. https://research.qut.edu.au/designlab/studywithus/ (accessed February 20, 2020).

Taylor, J. (2019). Discourses of Dissonance: Enabling Sites of Praxis and Practice Amongst Arts and Design Doctoral Study. In M. Breeze, Y. Taylor, & C. Costa (Eds.), *Time and Space in the Neoliberal University: Futures and Fractures in Higher Education*. Cham: Palgrave Macmillan, 191–220.

Wisker, G., & Robinson, G. (2009). Encouraging Postgraduate Students of Literature and Art to Cross Conceptual Thresholds. *Innovations in Education and Teaching International*, 46(3), 317–330.

Wisker, G., & Robinson, G. (2014). Experiences of the Creative Doctorate: Minstrels and White Lines. *Critical Studies in Teaching and Learning*, 2(2), 49–67.

1.8

RESEARCH DOCTORATES IN THE ARTS

A perspective from Goldsmiths

Janis Jefferies

Introduction

In many countries, including the UK, independent art schools provided discipline-specific courses that mostly drew on the atelier traditions of the academy or on Bauhaus-inspired formalism. The debate that ensued about whether an artist was 'made' or could be 'taught' owes a great deal to arguments about what constitutes skill, whether it is part of an art education, training or a networking opportunity to enter the professional (and at that time) singular version of the art world, its institutions and processes of commodification. To my knowledge, the word research was never used within a studio context but referred to only that within which was described as complementary studies developing out of art history with a broader cultural remit.

The UK's now defunct Council for National Academic Awards (CNAA)[1] ratified my own degree in Fine Art at Maidstone College of Art in 1974. The following year, it was possible to have the degree award at undergraduate level if one's creative work was clearly presented in relation to the argument of a written thesis set in the relevant theoretical, historical or critical context. A key change that is particularly relevant to the discussion in this chapter is that this was the first time when doctoral programmes were permitted under CNAA rules, and significantly, it was possible to include artefacts i.e. outcomes from creative art, music, performance and other disciplines, in the submission for the PhD. The restructuring of art and design education along university models was instigated by the first of several Coldstream reports.[2] This heralded a new era of intellectually ambitious, critical self-reflectivity and cultural renewal.

Having had many years' experience working in higher education in the creative arts, I would argue that, nevertheless, an unresolved tension emerged between what was constituted as academic and scholarly research and the artist theorist who frequently raided across ideas, disciplines and practices to invigorate their research practices from within. As in the 1980s, the image of the artist-theorist as practitioner and researcher is taken as the focal point of how artists receive their professional induction rather than that of the arts educator. The practitioner theorist came to the fore during the 1980s. I was certainly included in that category, employed as both a visiting artist and a theorist across studio and critical theory. It was an era in which the scripto-visual/text and image production dominated debates within the studio and the academy.[3] For example, Victor Burgin in particular was courted by the academy, exploiting connections between practice and academia at the same time as occupying an uneasy position within artistic production.[4] It is possible that the institutional reception of his work helped to establish a template for artistic practice and the dominance

     DOI: 10.4324/9780429324154-10

of the text in the University.[5] Increasingly, visual arts as practice-based research and creative research has been positioned as needing to be grounded in practices from art itself, particularly if the research enquiry is studio based.

This concept is foregrounded in much of the available literature, from Graeme Sullivan's seminal 2005 study *Art Practice as Research: Inquiry in the Visual Arts* (substantially updated and revised in 2010) to Mick Wilson's 2008 highly provocative paper, 'Conflicted Faculties: Knowledge Conflict and the University', National College of Art and Design, Dublin, from the international conference Arts Research: The State of Play. James Elkins, speaking from a US perspective, argues that no one knew how to supervise these degrees.[6] Such publications have engendered a fervent debate around practice-based research in the UK, Northern Europe and Australia from the 1990s and through to the present day. Art and art-based research, it is argued, verges on uncertainties: it is 'from this sense of knowing and unknowing, and how we deal with it, that visual arts practice can be described as a form of research'.[7] Sullivan has a great deal to say on the subject of practice-based research, so I will return to his analysis later in this chapter.

Since the UK polytechnics operating under CNAA rules were absorbed into the university system in 1992, the general regulations covering research and research performance now apply equally to those institutions, including those formerly separate art colleges. These organizational changes have had profound and lasting effects. In the visual arts in particular, what we now call practice-based research must satisfy both the university rules and performance measures as well as the non-academic structures of art production. The scholarly value of art-based practices has been recognized in the academic accreditation of practice-based PhDs in the UK for almost 50 years. The argument presented in the artwork, whether it is music, design, performance, creative writing, must be accompanied by a written component – the thesis, the length varying according to discipline and university.

Sullivan suggests that rather than adopting research methods from other disciplines, practitioners need to insist on 'their own, yet different complementary paths'[8] and that judging matters of equivalence between practice and text 'according to rules that can only be changed by those who make them'[9] becomes a lop-sided affair: there is, as Sullivan again notes, 'an inherent folly in assuming practices from different field can be validly compared if criteria are drawn from the disciplines of authority'.[10] In this account, the text accompanying the artwork needs to be guided by the art practices that have given rise to it rather than to comply with other disciplines' conventions of academic writing. The task, then, is to not to explain the individual artwork nor reduce its meanings but rather open up a multitude of potential links to be made beyond the stated and verbalized intentions of the artist.

Defining practice-oriented research

Carole Gray sets out two enabling definitions for what she termed practice-led research:

> Firstly, research which is initiated in practice, where questions, problems, challenges are identified and formed by the needs of the practice and practitioners; and secondly, that the research strategy is carried out through practice, using predominantly methodologies and specific methods familiar to us as practitioners.[11]

In the context of practice, knowledge, tacit or otherwise, can be about how something is done within a professional and cultural framework and within dynamic systems of complexity and emergence. This view is consolidated by Delday and Gray,[12] who venture that doctoral candidates might 'make their own pedagogic experience within the doctoral framework'.[13] Estelle Barrett has commented on the unique knowledge that creative practice can afford: 'creative practice allows us to access such codes via the aesthetic image and to make visible knowledge that everyday use of language and discourse hide'.[14]

One of the other ways to explore these possibilities is through *creative research*, a term widely used in Canada and Australia. The Creativity and Cognition Studios (CCS) at the University of Technology, Sydney, has identified how the terms practice-based and practice-led research can be distinguished. Within the context of CCS's definitions, practice-based research is an 'original investigation undertaken in order to gain new knowledge, partly by means of practice and the outcomes of that practice'. Creative outcomes can include artefacts such as images, music, designs, models, digital media, performances and exhibitions. Consequently, new fields of understanding have been opened up. Practitioners find methodologies that can embrace the processual, elucidating practical outputs as new fields of knowledge. Out of some of the definitions from CCS, the Goldsmiths Digital Studios (GDS) was established in 2005. It too was dedicated to multi-disciplinary research and practice across arts, design, technologies and cultural studies.

There is confusion and conflict around terms such artist, artist as researcher, practitioner, practice-led, practice-based, as can be gleaned from the available literature. In the introduction, many terms were cited: practice-based, practice-led and creative practice. In 2021, a new term is introduced, *practice research* (author emphasis). In PRAG's recently commissioned report, authored by Goldsmiths alumni, Ozden Sahin and James Bulley (2021). Their key questions were: what is practice research? Why this report now? In what follows, the term practice research is used to include the various competing names. Practice research aims to unify the range of terminology describing the relationship between practice and research. It highlights the breadth of practices that exist in which the researcher is intrinsic in leading to new knowledge and understanding. It consolidates a range of conflicting terms, including practice-based research, practice-led research and practice as research.

A Goldsmiths roundtable

This section discusses the outcomes of a roundtable discussion held at Goldsmiths, University of London, on 25 September 2019 in person, online and by email. The aim was to exchange interdisciplinary experiences of the value of practice research. The participants were researchers across a range of practices based in different departments at Goldsmiths: the disciplines were arts and technology, music, design, drama and performance. The material in this chapter is drawn from a transcript of the roundtable and from the researchers' writing that was subsequently sent to the author.

The objective of the roundtable exchange was to discuss principles and values in an interdisciplinary context in terms of the differing research that, for example, involved participants outside of academia or focused on individual practices.[15] The group considered why they had undertaking practice research in the first place. Each was aware that developing their careers within the university sector now required a PhD. Chatzichristodoulou noted that at her institution, there were many senior researchers for whom a PhD is not considered viable at the latter stages of their career. She also pointed out that professional doctorates are available in many universities and have their own value.

The roundtable participants synthesized their descriptive data to produce their research narratives – also known as *restorying*. Many of their research narratives have common elements; for example, how their practices combine with methods are drawn from many sources, selected as appropriate. While practice was at the heart of their methodologies, other methods were drawn from co-design, autoethnography, live methods, diaries, journals, blogs and reflective commentary. Methods were eclectic and hybrid, or what Norman Denzin and Yvonna Lincoln have called 'a "bricolage", a dense, reflexive collage-like creation'.[16]

Methods

This section provides examples of the discussion about research into methods used in art practice from which new insights and understandings have been derived.

As Bulley suggested, 'you analyze your own method and communicate what happened during the process of making work'.[17] For him, practice is characterized by what theorist Donald Schön has called 'unique events', dealt with strategically by the practitioner. According to Schön, practice is the work of real-time cultural extension and transformation.[18] The ways of knowing that emerge in practice are shared in practice research through research narratives. For Stacey Pitsillides, stories of people, places and things were essential to retaining the life of co-designed, user-centred practices that constructed the knowledge she gained from listening to her participants as collaborators. By contrast, Goodin invited his participants to interact and evaluate his augmented vintage toys through online and written questionnaires.

The discussion further revolved around how different methods combine to respond to multi-faceted, multi-layered and complex questions, particularly in interdisciplinary research projects, such as the ones the researchers had taken on. For Psarras, the ontological turn in social sciences was a significant breakthrough; Lury and Wakeford's pioneering anthology, *Inventive Methods: The Happening of the Social*,[19] was one of the most helpful. It includes a range of genres, including autoethnography and different writing styles. They were deployed to explore how methods can be inventive and cut across disciplinary boundaries. Together with Back and Puwar's influential *A Manifesto for Live Methods: Provocations and Capacities*,[20] new forms of writing, telling stories with self and others enabled the researchers to reach successful completion.

Such anthologies underpin many of the interdisciplinary PhDs that are fairly common at Goldsmiths. In my definition, *inter* means 'between' or 'among'. In interdisciplinary research, each individual researcher or collaborative contributor talks from his or her expertise, so there is a conversation between and among disciplines, as those involved in this chapter pursued. Each researcher had ideas planned in advance, but their research questions evolved in the process of making and creative interactions. Their experiences aligned with practice research. Practice itself, whether individually (Oicherman, Bulley, Goodin) or collaboratively (Pitsillides, Chatzichristodoulou), was foregrounded in their research narrative, becoming the significant method of the research project. Bulley emphasized the point in discussion that the practice research field is unique in its focus on sharing often-unrecognized ways of knowing that emerge in the process of the research narrative.

Research narratives

The following research narratives also act as case studies. They operate within the widely contested nature of the practice research field. Alternative models that function as examples rather than paradigms are acknowledged. Readers will find extracted discussions of concrete examples of doctoral work and artistic/design/computational practices that have an explicit engagement with ideas of practice research, knowledge and enquiry. The goal was not, then, to establish a single fixed model intended to work for all practice research but rather to map their and my experiences as positions worth attending to as a provocation for further dialogue. There is no critical analysis or value judgement offered as the parameters of knowledge production, and are set according to each individual voice of experience.

Interdisciplinary PhDs: Katya Oicherman and Stacey Pitsillides

In terms of design, two different methodological models were pursued; one individual, devised by Katya Oicherman, and the other a collaborative process involving two different partners, a university and a hospice, devised by Stacey Pitsillides.

Katya Oicherman spoke of her *Binding Autobiographies: Torah Binders Revisited* (2014) as a multi-faceted research project. There were several interlinked stories. One was the biography of an object, an 1836 Torah binder from the Bavarian town of Furth, exploring its journey from everyday to ritual object and to museum artifact. Torah binders are textiles that carry text – they document the birth and circumcision of Jewish boys, functioning as embroidered certificates of birth. In light of this, the relationship of text and textile was central to her research. Yet aesthetic material exploration was also a significant part, alongside traditional historical research, emphasizing a performative reading of a Torah binder and related rituals. This performative reading opened up another story in the research – a creative response in which Oicherman designed and embroidered a binder referencing simultaneously the original object and her family history. Her research became a written interchange of the object's biography, Oicherman's own autobiography and family history and the 'biography' of the art project, conveying her own making process and reflections as an artist responding to a museum object. In that last context, her art practice was her research methodology, which in the final thesis was explained, contextualized and presented alongside the more traditional historical and anthropological chapters. The position of cloth, shifting between autobiography, art practice and history, was explained using the writings of Jacques Derrida on material, creative act and memory, notably the concept of *subjectile*. Oicherman kept a diary documenting the entire process, the iterations between the historical and the creative research, which formed part of her thesis chapters.

Post-PhD, Oicherman was appointed Head of Department of Textile Design in Shenkar College of Engineering, Design and Art in Israel. The thesis proved a very useful starting point to develop undergraduate courses in textile history, art and design for students of creative disciplines who often find the traditional art-historical teaching less appealing. The model and experience of the thesis allowed Oicherman to channel the aesthetic and material curiosity of the students into creative research that accepts and values the responsibility of historical research and finds outcome formats in creative projects that are meaningful for artists and designers.

For Oicherman, writing in research produces many sites of interdisciplinary exchange following what can be perceived as a spatial pattern of enquiry. In order to find a place from which to reflect upon new modes of enquiry, new ways of knowing and being, we have to draw on the patterns that emerge out of 'situated knowledges'.[21] The interrelations between where the 'I' autobiographically might be sited and located operate in the slippery exchange between practices of writing and the writing of practices. Interstitial relationships between patterns of thinking become a compelling device to ask: where does autobiography end and theory begin? Where does writing the self blur the boundaries of research as an academic convention and point towards the possible and the experimental?[22] Readers can explore her blog[23] which opens up the question of her practice as a locus for autobiographical writing, which was a principle, and value for her PhD research.

Oicherman posed this in her thesis, her textile work and the video she made which narrated her experiences of both. These were questions raised in supervisions and in discussing 'Autoethnography', as posited by researcher Carolyn Ellis in her 2004 book, *The Ethnographic I: A Methodological Novel About Autoethnography*. Individual experience became a reflective method that grew in complexity within many practice research submissions that I supervised during my tenure at Goldsmiths and in the international examinations I was asked to lead on.

The insertion of writing the self into storytelling is as significant in Katya Oicherman's journey as in Stacey Pitsillides' approach in her thesis, *Digital Death: The Materiality of the Co-Crafted Legacies* (2016). Pitsillides developed new modes of practice using co-design and situated design to produce knowledge through the forming of artifacts with live methods.[24] The use of co-design as a specific form of design research was key to the practice as it navigated the risks involved in working with partners: the Hospice of St Francis in Berkhamsted and in people's individual homes. Stacey worked with those bereaved at the same times as handling a two-year ethics debate between Goldsmiths, research governance and the Hospice.[25] The research took shape as each process solidified. It had to build in elements of collaboration, helping to craft research questions as an integrated part of the whole. This became evidenced in a final exhibition and series of short films that document the process of collaboration in both making and curating the artifacts on display (see Figure 1.8.1).

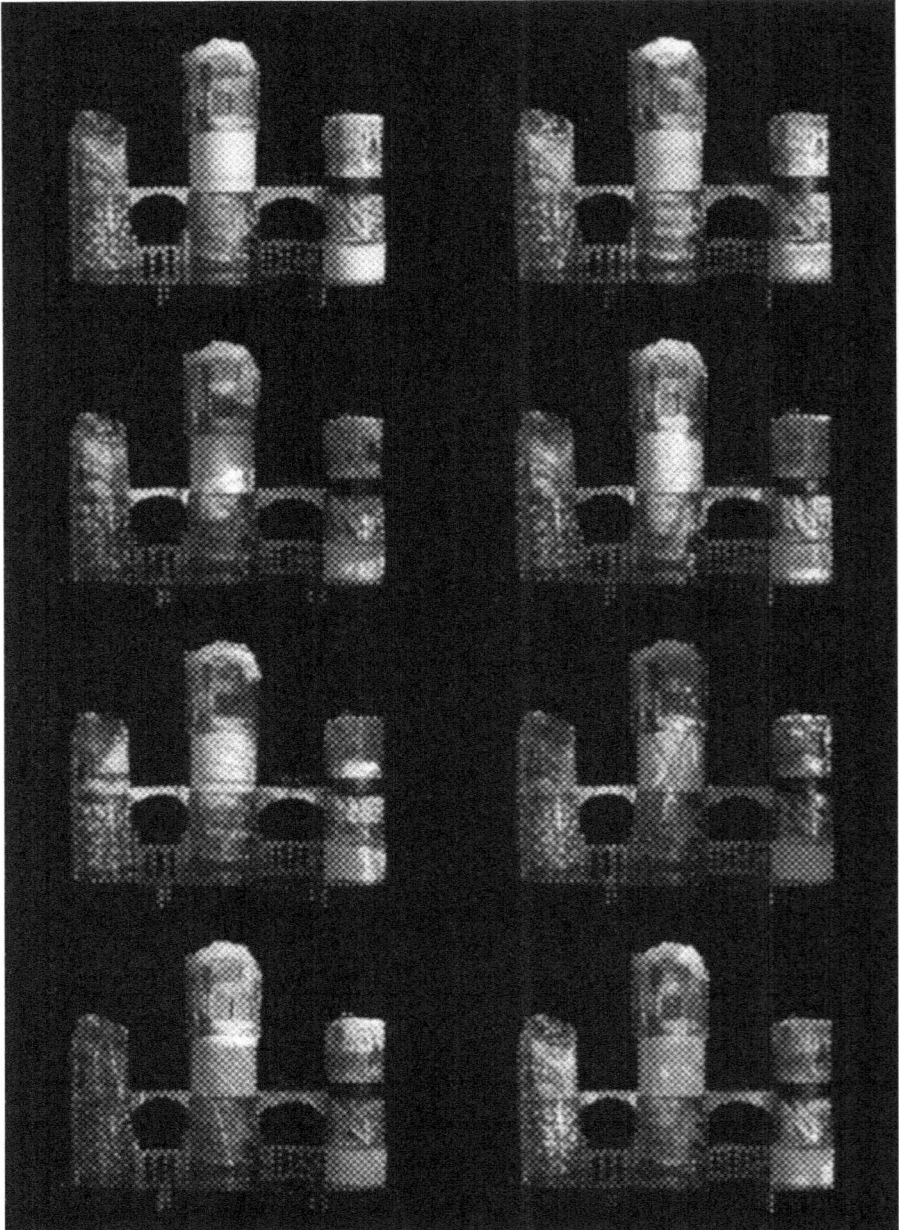

Figure 1.8.1 Above and Below. Detail of installation from Material Legacies Exhibition. Stephen Laurence Gallery, London. 28 February–24 March 2017.

Source: Photo credit to Tadej Vindiš.

Contextually, the research explores the role of digital legacies and the material agency of the dead through the objects and data left behind. It supports this investigation through three paired collaborations between a designer, an art therapist and three bereaved participants that through the process become collaborators and co-designers – exhibiting their work publicly to give people the chance to experientially encounter those they have lost. As Pitsillides' research began to recruit participants, the role of the design and artistic methods emerged. Interviews were used to support participants' reflection on the process, and this was then woven back into the making and curating of the exhibition. This responds to the need for a greater degree of individualization, where collaboration provides a more holistic view of the experience of systems and their usage.[26] Research through collaborating uses the qualities of crafting, where a research process is approached as raw material with a range of unfixed potentials,[27] and that outcomes are not pre-packaged into specific design outputs or limited by the researcher's own proposed outcomes, as evidenced in her website, www.digitaldeath.eu. The use of an exhibition as a mode of knowledge generation opens up the research without diminishing the individual approach to collaborations and relationships (see Figure 1.8.2).[28]

This is key to the value of the project in that the research does not aim to create or test a service. It is an exploration of how services could be grown or adapted in light of current developments within therapeutic and creative practices by flipping the approach to enquiry and focusing on live action rather than documentation. Core to this aspect of transdisciplinary research, practice research often exhibits what? (working with communities inside and outside of disciplinary boundaries and institutions). As James Bulley pointed out in an email exchange to the author on 8 December 2020, for him, transdisciplinarity is an extension of interdisciplinarity that transforms ways of knowing across and beyond disciplinary frameworks. In his post-doctoral research, he draws upon *The Handbook*

Figure 1.8.2   Installation shots of Material Legacies Exhibition. Stephen Laurence Gallery, London. 28 February–2 March 2017.

Source: Photo credit to Tadej Vindiš.

of Interdisciplinarity on the issues of transdisciplinary, which 'strives to grasp the relevant complexity of a problem, taking into account the diversity of both every day and academic perceptions of problems, linking abstract and case-specific knowledge, and developing descriptive, normative, and transformative knowledge for the common interest'.[29]

Practice researchers learn from people rather than about them[30] and invite them to be collaborators and co-creators of the research. This shifted the priorities of Stacey Pitsillides' research for the Hospice and involved quite complicated ethical procedures, which, in the end, took on a whole chapter in the final written part of the thesis. These issues speak to the role of partnerships in practice research PhDs, as an increasingly regulated university context has a preoccupation with 'ethical approval' and 'risk assessment' that increases anxiety among researchers who are not responsible solely for ideas but are also responsible to the ideas. While ethics forms were filled, completed and signed off by the institutions and organizations involved, research ethics is more than a tick-box exercise. It is worth stating that research integrity – the core of which is honesty, rigor, transparency, open communication, care and respect – applies to all those engaged with research but is particularly important to those developing a research project such as Pitsillides. They have a responsibility to consider how the work they undertake impacts on those they work with – participants and collaborators, the wider research community and society. Jefferies was chair of Goldsmiths, Concordat to Support Research Integrity (2013–2018), and in accordance with the UK Research Integrity Office (UKRI) *Policy and Guidelines on Governance and Good Research Conduct*, we encouraged all researchers to maintain integrity in all aspects of their research.

Arts and computational technology: Jack in the Box

In terms of a hyper-connected world that can afford new possibilities to re-image observation and the generation of alternative means of practice research data, the intersection of interactive digital art, collaboration and co-authorship further elaborates ethical debates and the challenges of collaboration. Although from a different perspective, there are connections to Pitsillides' methods, outlined earlier. The question of how traditional audiences can be extended and retained depends as much on context as on platform, as Chatzichristodoulou knew very well from her experience, gained from many years as a professional performer and curator.

William S. Goodin IV's Arts and Computational Technology PhD thesis *Reconceptualising Krapow: Creating Interactive 'Happenings' Using EEG Technology* (2018) explores concepts and strategies used to create moments of interaction that are themselves the objects of artistic production (see Figure 1.8.3).

Research methods and ethical concerns are centred on the creation of artefacts as integrated components of interactive computer systems utilizing EEG technology that collects the brainwave data of participants, which is then rearranged or reinterpreted. This reforming of brainwave data through the system allows the participant to have an active role in the creation of the art, which is the moment of interaction. It also performs as a reconceptualization of the precepts governing interaction established by Alan Kaprow's physical performance in the 1960s. The research utilized the theories for facilitating and enhancing the nature of the interaction: art as experience, play theory, affect and magical thinking. Throughout the creative and research process, the PhD project was documented through a website, blog and/or online journal system, allowing for self-reflection and feedback from peers that was incorporated in the evolution of the research. Transmission formed a balance between the input from the viewer and the artist's control over what was presented to view. The challenge was to make a participant aware of his or her contribution while still creating a sophisticated work of art.

Bill Psarras completed his PhD in Arts and Computational Technology in 2015. In his contribution to this discussion, Psarras sent me some of his thoughts via email on 1 December 2019 and again on 12 December 2020.

Figure 1.8.3 Jack in the box (2017), a participant-based interactive artwork created by Dr William S. Goodin IV.
Source: W. Goodin.

His research explored through interdisciplinary methods the physical and metaphorical ideas at the core of flaneur; a concept of cultural significance which reverberated across 20th- and 21st-century aspects of performance art, site-specific interventions and lately locative media practices (see Figure 1.8.4). Bringing together art practice and critical reflection was achieved through selected methodological devices, including conceptual walking performances often of site-specific character, audio-visual media and GPS data, selected cognitive metaphors and finally online blogging reflection. He also used a diversity of methods to reveal these performative combinations. Inspired by wider interdisciplinary schemes of 'mobile methods',[31] inventive ambulatory actions[32] 'practice-led research', his research formed a continuous dialogue between art practice and critical writing. Blogging as a parallel diary ran concurrently with writing methods. In the online space, a blog functioned as a platform of reflection, which facilitated a repetitive 'thinking by writing'.[33] This was exemplified through a blog section entitled *Footnotes*. The latter formed an ongoing juxtaposition and, as Psarras

147

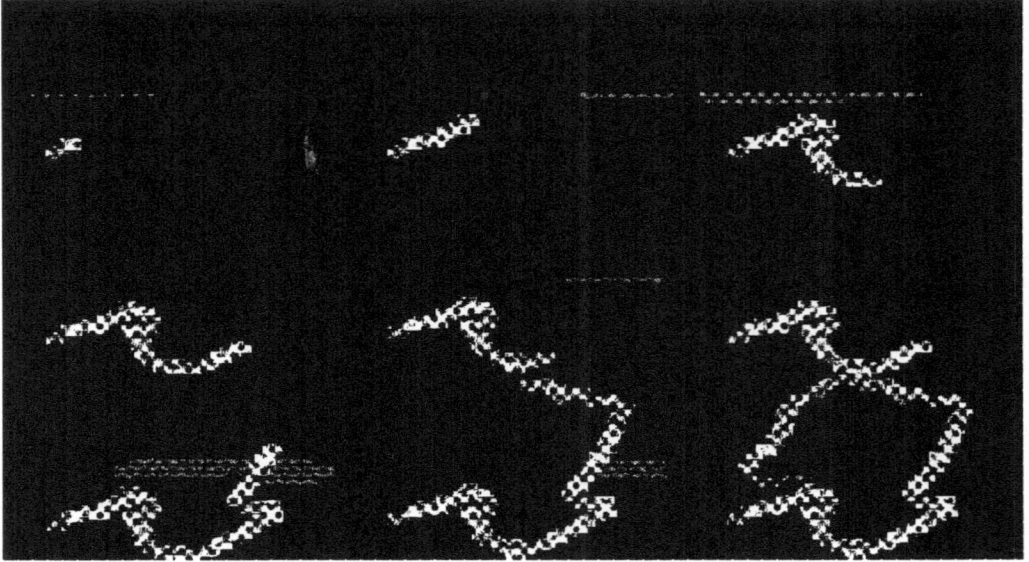

Figure 1.8.4 Weaving footsteps and confessions: emotive circle (2012), an artwork part of the PhD.
Source: B Psarras.

writes, may identify a potential resemblance between walking and blogging: as important to the thinking/writing dimension in online space was a sensing/walking relationship in the physical environment. Both revealed the capacity to produce new knowledge in art practice and/or wider practice research. For him, such an inventive method focuses on the potential of objects (whether they technological, organic, material as performative co-producers): creative apparatuses that drive, map, extend and often augment body gestures, intentions, auto-ethnographical narratives and knowledges between artist, audiences and site (see Figure 1.8.5).

Arts and computational technology: performance and the value of collaboration

Maria Chatzichristodoulou's thesis *Cybertheatres: Emergent Networked Performance Practices* (2010) also took up issues of knowledges and experiences between audiences, artist and site. With the support of two awards by AHRC ICT Methods Network and LCACE (London Centre for Arts and Cultural Exchange), as well as financial and in-kind support by many departments at Goldsmiths (including Computing, Theatre and Performance, Media and Communications, Music, Visual Cultures, Cultural Studies and the Graduate School), Chatzichristodoulou, in collaboration with Rachel Zerihan, developed the three-day practice research festival *Intimacy: Across Digital and Visceral Performance* (2007). External partners included Trinity Laban, The Albany and Home London – all cultural institutions in South East London. *Intimacy* featured a digital and live art program consisting of six workshops, four seminars, 36 performances/happenings, a marathon of 'show and tell' presentations and screenings, and a day-long symposium. It provided a platform for intimate interactions between audiences and performance artists working with live bodies and digital media in both visceral and digital performance encounters, as well as for the discussion of (sub-) cultural practices concerned with displaying intuitive and intimate relationships between artist and other. The program was designed to address a diverse set of responses to the notion of 'being intimate' in contemporary

Figure 1.8.5 Territorial Poetics (2019, performance for camera and drone).
Source: B. Psarras.

performance,[34] in one-to-one and micro-audience performances. Practice-led examinations and seminar discussions explored the diverse environments that play host to digital and visceral art works, raising issues around bodies of data and flesh; presence as aura and representation; desire as embodied condition and disembodied fantasy; the human and posthuman self. *Intimacy* also endeavoured to explore technologies that can enhance 'closeness': networking technologies such as the Internet, wireless networks, telecommunications and Web 2.0; sensor technologies; virtual reality and other digital multi-user environments, examining the synergies and different strategies between performance that takes place through visceral encounters and work that is mediated through the use of digital and networking technologies. The research also examined, through a practice-led investigation, the curation of live intimate encounters, whether those unfold face-to-face or through a physical distance – online, in virtual worlds or, simply, through the telephone. The final chapter of Chatzichristodoulou's thesis reflects on her practice as a producer and curator of media and digital arts through the practice of creating *Intimacy: Across Digital and Visceral Performance*.

Music: material thinking and the value of resonant practice

James Bulley's *Sounding Materiality: Explorations in Resonant Practice* (2018) was an account of specific arts research projects that operated at the intra-face[35] of theory and practice. Bulley discovered techniques of 'live composition' and 'locative sound' that forged a closer relationship with materiality in sound practice. Initially, an overarching methodology of 'resonant practice' inspired by Donald Schön's 'reflective practitioner'[36] was proposed, and Bulley employed it to activate 'material thinking'[37] in his practice. Bulley embraces a plurality of methods in his work, a particular facet of practice research, and the thesis details interdisciplinary methods that reference Norman Denzin and Yvonna Lincoln's ideas of 'bricolage' cited earlier (see Figure 1.8.6).

Methods included ecological surveying, atmospheric condition modelling, orchestral arrangement, sculpture and ambisonics. At the roundtable, Bulley described how specific sound art works

Figure 1.8.6   Living Symphonies – Crowd, Fineshade Woods Installation (2014).
Source: Jones/Bulley.

in the thesis were driven and defined by real-time natural processes, and that these works often went through multiple iterations at different installation sites. The practice was then captured in cross-media documentation, acting as a proxy for the highly site-specific works, and this documentation was then imbricated in a cross-media research narrative combining sound, text, video and illustration. From his research inquiry, Bulley established two methods of practice that have become core to his practice: 'live composition', a framework that uses parameterization to drive real-time sound composition based on live source data, and 'locative sound', a framework that explores the spatial and conceptual relations between sound, source and material. His thesis acted as a critique of his own subjectivity as an experienced practitioner and researcher, providing a learning curve that allowed him to find a personal voice to the research narrative within, which was then further expressed in an accompanying website that brought sound, video, images and words together for further context (see Figure 1.8.7).

Experimentation and transmission

Over the last decade, experimentation in research in general and practice research at Goldsmiths specifically has primarily focused on methods as evidenced in the research narratives cited. Just to give a few examples of experimentation from the dynamic turn in the social sciences: messy,[38] mobile,[39] live,[40] inventive,[41] digital,[42] sensory,[43] creative[44] and interdisciplinary methods. Yet, as Kat Jungnickel has asserted in her edited collection, *Critical Tactics for Making and Communicating Research*,[45] there are many opportunities to intervene in the ways standard and familiar knowledge is pursued and 'outputs' work to tidy up messy and multiple realities. The collection questions the idea that clear and direct finished practices and arguments require reduced complexity. Complexity is not reduced

Figure 1.8.7   Living Symphonies – Crowd, Fineshade Woods Installation (2014).
Source: Imogen Lloyd.

in the practices articulated by Bulley, Chatzichristodoulou, Goodin, Oicherman, Psarras and Pitsillides. Every discipline, whether music, theatre and performance or design, has established knowledge frameworks and systems opening up explorations and experimentations in sound, cybertheatre, arts, design and computational technology. They had to constantly recognize the value of their own work and make their stories of experience and transparency. As they discovered, no one single instrument or tactic can tell an entire story, only pieces of it. Becker writes about how 'experimenters and innovators don't do things as they are usually done', and that 'their solutions to standard problems tell us a lot and open our eyes to possibilities more conventional practice doesn't see'.[46] This conforms with my own experiences, supervising nearly 30 PhD students over 15 years, and is the value and the driving principle of all practice research at Goldsmiths over the last 20 years.

Conclusion

My own experiences, supervising nearly 30 PhD students over 15 years, echoes Becker's. I regret that the practice research forum, which ran wonderfully well at Goldsmiths for several years, did not continue in the spirit of cross-departmental engagement as one might have hoped. Personnel changed and institutional funding was not freely available. Academics have been hard pressed to find time for their own research, let alone give of their time voluntarily, which is how the practice forum was organized. As the researchers told me at the roundtable discussion, adequate financial support and access to resources are needed for bridging professional practice to practice research that is sadly not formally ratified by institutional pressure.

However, the collected essays in *Transmission: Critical Tactics for Making and Communicating Research*[47] reveal a tactical combination of making practice, theory, methods and data that gives shape to research and communicates. The anthology emphasizes how it is possible to value, share and entangle others in it. Publication is used here in the broad sense, to include exhibition, performance and recordings. Frequently, the practices that form the core component of research are presented for examination in a public and professional context, such as a public exhibition or performance. It is a dynamic method of communication research to many people. The challenge now is how to capture and maintain practice research, ensuring its openness, visibility and sustainability for future practitioners and researchers to access.

As with those who have addressed their practice research so eloquently in this chapter, experimentation meets dissemination, experiences meets values – a fusing of the usually separate focus on subject, method, values and principles. We make things to understand what we are doing and write, perform, play, speak enact and position them in a range of different ways across the arts, humanities and sciences.

So, if you were to start on your practice research project now, how would you begin? What values would you aspire to? This would be my top eight tips:

- Trust your practice and experience in whatever practice(s) you own
- Blog, log, write short entries on your practice and get peers to read and comment
- Tailor your research practice project and methods with creative and critical reflection
- Find your community of likeminded innovators
- Collaborate where you can and across disciplines
- Write, write, rewrite and learn to love words
- Speak with confidence and from different subjective viewpoints
- Find your own voice but be kind to your reader and viewer

Acknowledgements

Sincere thanks go to my former PhD students who participated so generously in the roundtable discussion, for sharing their experiences, values and ideas in person, by Skype, email text contributions, amendments and corrections. What joyous conversations we had and how wonderful they have continued. With particular thanks to William Goodin IV for his editorial assistance. Special thanks to Lesley Hewings, Head of Graduate School (until 2017) for her knowledge, support and editing during our time at Goldsmiths and since.

Notes

1 The Council for National Academic Awards (CNAA) was the largest single degree-awarding body in the United Kingdom. There were over 140 institutions offering first degrees and postgraduate level courses approved by the CNAA, including polytechnics, institutions of higher education, Scottish central institutions, colleges of art and various other colleges throughout the United Kingdom. CNAA awards are comparable to those of universities, and these are recognized by professional associations and employers. The CNAA was abolished by the 1992 UK Further and Higher Education Act.

2 In the 1960s, William Coldstream and his committee made a report which changed the landscape of art education in the UK forever. Did he foresee that the acceleration of technology and what was happening in art history and theory demanded that art schools would never become respectable unless they were able to have awarding degree powers? During the 1970s, critical and contextual studies were established on the newly formed honours programs that were to provide a research base for the new universities. For a full account of this debate, see Jon Thompson (2005), 'Art Education: From Coldstream to QAA', *Blackwell Journal, Critical Quarterly*, Vol 47, Issue 1–2, pp. 215–255.

3 One strand of art theory, sometimes referred to as 'scripto-visual' practice, is close to the semiotically informed critical practice outlined in Elizabeth Chaplin.

4 Victor Burgin's practice is inseparable from his theoretical writings, which are steeped in the ideas of many 20th-century poetical, psychoanalytical and linguistic theorists. During the 1970 and 1980s, his work was based on the juxtaposition of text and image names as the 'scripto-visual'.

5 Dr Maeve Connolly, 'Art Practice, Peer-Review and the Audience for Academic Research', Position Papers on Practice-Based Research, National College of Art and Design, Dublin, Ireland, 22nd April 2005.

6 Elkins (2009, Introduction: p. xii).

7 Sullivan (2005, pp. 1–15).

8 Ibid., p. 34.

9 Ibid., p. 89.

10 Ibid., p. 89.

11 Gray (1996, p. 3).

12 Delday and Gray (2011, p. 45).

13 Ibid., p. 288.

14 Barrett (2013, p. 115).

15 Oicherman used creative writing elements and her diaries while Psarras relied on journals and blogs to offer multiple strands of evidence in his work. Both were deeply engaged in their investigating of own subjectivities, in order to deliver an insightful analysis. Psarras' journal gives valuable insights to this process and is available at https://emotiveterrains.wordpress.com. Chatzichristodoulou and Pitsillides mobilized a range of collaborative partners and participants, while Goodin repurposed his practice to explore the value of co-authorship through online networks. For Pitsillides in particular, by 'following practice research . . . [and constructing it] through the collaboration, it solidified my belief that it's important for research today to take risks and to do things that push the boundaries'

16 Denzin and Lincoln (2010, p. 4).

17 Email to author from Dr James Bulley. 8 December 2020.

18 Schön (1987, pp. 16–17).

19 Lury and Wakeford (2012).

20 Back and Nirmal (2012).

21 Harraway (1988).

22 Jefferies (2012).

23 http://narrativembodiment.blogspot.com/2011/02/katya-oicherman-torah-binders-as-locus.html.

24 Back and Puwar (2012).

25 Janis Jefferies and Stacey Pitsillides presented their unpublished paper on 'Creativity and Ethics' and how it interconnected at a cross-Goldsmiths Research and Integrity event at Goldsmiths on 22 March 2017.

26 Bradwell and Marr (2008).

27 Alfondy (2007).

28 Jungnickel 2010).

29 Pohl et al. (2017, p. 32).

30 Ingold (2013).

31 Sheller and Urey (2006).

32 E.g. Pope (2000), Ingold (2004), Pinder (2005), Back and Nirmal (2012).

33 Nardi et al. (2004, p. 225).

34 All *Intimacy* footage is available from the British Library and the Live Development Agency. See www.writing.gold.ac.uk/intimacy/ (accessed 1 February 2020).

35 Barad (2000).

36 Schön (1987).

37 Carter (2004).

38 Law (2004).

39 Büscher et al. (2011).

40 Back and Nirmal (2012).

41 Lury and Wakeford (2012).

42 Orton-Johnson and Prior (2013).

43 Pink (2015).

44 Kara (2015).

45 Junhnickel (2020). Fourteen authors and researchers rethink tactics for inventing and disseminating research, examining the use of such unconventional forms as poetry, performance, catalogues, interactive machines, costume and digital platforms. Jefferies' chapter is called 'Performing, Provoking'.

46 Becker (2007, p. 7).

47 Junhnickel (2020).

Bibliography

Alfondy, S. (2007). *Neo-Craft: Modernity and the Crafts*. Halifax: The Press of the Nova Scotia College of Art and Design.

Back, L., & Nirmal, P. (Eds.). (2012). *A Manifesto for Live Methods; Provocations and Capacities*. Oxford: Blackwell.

Barad, K. (2000). Agential Realism. In *Encyclopedia of Feminist Theories*. New York: Taylor & Francis, 15–16 (*The Sociological Review Foundation*, 60(1_suppl), 6–17. Article first published online: June 1, 2012; Issue published: June 1, 2012).

Barrett, E. (2013). Materiality, Affect and the Aesthetic Image. In B. Bolt & E. Barrett (Eds.), *Carnal Knowledge: Towards a 'New Materialism' Through the Arts*. London and New York: I.B. Tauris.

Becker, H. (2007). *Telling About Society*. Chicago: University of Chicago Press.

Bradwell, P., & Marr, S. (2008). *Making the Most of Collaboration: An International Survey of Public Service Co-Design*. London: Demos.

Bulley, J. (2018). *Sounding Materiality: Explorations in Resonant Practice*. PhD Thesis.

Büscher, M., Urry, J., & Witchger, K. (Eds.). (2011). *Mobile Methods*. London: Routledge.

Carter, P. (2004). *Material Thinking: Collaborative Realisation and the Art of Self Becoming*. Carlton, Victoria: Melbourne University Publishing.

Chatzichristodoulou, M. (2010). *Cybertheatres: Emergent Networked Performance Practices*. PhD Thesis.

Delday, H., & Gray, C. (2011). A Pedagogy of Poiesis: Possible Futures for 'Artistic' Practice-Led Doctoral Research. *Art as Research*, 9, 45.

Denzin, N. K., & Lincoln, Y. S. (2010). Introduction: Entering the Field of Qualitative Research. *Strategies of Qualitative Inquiry*, 1–20.

Elkins, J. (2009). *Artists with PhDs: On the New Doctoral Degree in Studio Art*. Washington, DC: New Academic Publishing.

Ellis, C. (2004). *The Ethnographic I: A Methodological Novel About Autoethnography*. Oxford: AltaMira Press.

Goodwin, W. S. (2018). *Reconceptualising Krapow: Creating Interactive 'Happenings' Using EEG Technology*. PhD Thesis.

Gray, C. (1996). *Inquiry Through Practice: Developing Appropriate Research Strategies*. Keynote speech in Proceedings of No Guru, No Method? Discussions on Art and Design Research, University of Art and Design, UIAH, Helsinki, Finland, 82–95.

Harraway, D. (1988). Situated Knowledges: The Science Question in Feminism and the Privilege of Partial Perspective. *Feminist Studies*, 14(3), 575–599, Autumn.

Ingold, T. (2004). Culture on the Ground: The World Perceived Through the Feet. *Journal of Material Culture*, 9, 315–340.

Ingold, T. (2013). *Making: Anthropology, Archaeology, Art and Architecture*. London: Routledge.

Jefferies, J. (2012). Pattern, Patterning. *Invented Methods: The Happening of the Social*, 125–135.

Jungnickel, K. (2010). Exhibiting Ethnographic Knowledge: Making Sociology About Makers of Technology. *Street Signs*, 32–35.

Junhnickel, K. (2020). *Transmissions: Critical Tactics for Making & Communicating Research*. Cambridge, MA: MIT Press.

Kara, H. (2015). *Creative Research Methods*. Bristol: Polity Press.

Law, J. (2004). *After Method: Mess in Social Science Research*. London: Routledge.

Lury, C., & Wakeford, N. (Eds.). (2012). *Inventive Methods: The Happening of the Social*. London: Routledge.

Nardi, B. A., Schiano, D. J., & Gumbrecht, M. (2004). *Blogging as Social Activity, or, Would you Let 900 Million People Read Your Diary?* Proceedings of the 2004 ACM Conference, Chicago, IL; ACM, New York, 222–231, November 6–10.

Orton-Johnson, K., & Prior, N. (2013). *Digital Sociology: Critical Perspectives*. London: Palgrave Macmillan.

Pinder, D. (2005). Arts of Urban Exploration. *Cultural Geographies*, 12, 383–411.

Pink, S. (2015). *Doing Sensory Ethnography*. London: Sage, 2nd ed.

Pitsillides, S. (2016). *Digital Death: The Materiality of the Co-Crafted Legacies*. PhD Thesis.

Pohl, C., Truffer, B., & Hirsch-Hadorn, G. (2017). Addressing Wicked Problems Through Transdisciplinary Research. In R. Frodeman, J. Thompson Klein, & R. C. S. Pacheco. (Eds.), *The Oxford Handbook of Interdisciplinarity*. Oxford Handbooks. Oxford: Oxford University Press, 2nd ed.

Pope, S. (2000). *London Walking: A Handbook for Survival*. London: Ellipsis.

Psarras, B. (2021). Performing the Poetic and the Technological: Transmedia Artistic Approaches on Walking, Objects and Interaction into Place. In A. Maragiannis & E. Papadaki (Eds.), *Digital Technologies in Interdisciplinary Creative Landscapes*. LIBRI Publishing, forthcoming.

Schön, D. A. (1987). *Educating the Reflective Practitioner*. San Francisco: Jossey-Bass.

Sheller, M., & Urry, J. (2006). The New Mobilities Paradigm. *Environment and Planning A*, 38(2), 207–226.

Sullivan, G. (2005). *Art Practice as Research: Inquiry into the Visual Arts*. London: Sage Publications.

Wilson, M. (2008). *Conflicted Faculties: Knowledge Conflict and the University Held at the National College of Art and Design*. Dublin for the International Conference, Arts Research: The State of Play, Dublin.

1.9

THE PHD IN VISUAL ARTS PRACTICE IN THE USA

Beyond Elkins' *Artists With PhDs*

Bruce Mackh

Introduction

James Elkins published *Artists with PhDs: On the New Doctoral Degree in Studio Art*[1] as a resource to help artists, educators, academic administrators, and students to compare and evaluate emerging doctoral programs in visual arts practice. A second edition followed in 2014, revising and adding to the first volume. Re-examining both books today yields a much different picture of doctoral study in visual art than once predicted. In the US, the PhD has not supplanted the MFA as the primary qualification for faculty positions in studio art, nor has the degree altered prevailing views about the relationship between research and arts practice. Nevertheless, we can learn from Elkins and his contributors, further informed by lived realities, to piece together a more accurate view of the US PhD in visual arts practice and, by extension, doctoral degrees in other practice-based fields.

Before we enter into this discussion, perhaps we should pause for a bit of clarification. Many varieties of doctoral study are associated with the visual arts, which may or may not include an element of practice or art-making. Much like the contemporaneous terms "interdisciplinary," "multidisciplinary," and "transdisciplinary," which tend to be used as though they were synonymous, there is little differentiation between types of doctoral study involving art, especially when linked to words like "studio," "practice," "visual," or "fine." Although Elkins prefers the term "studio art"[2] for this discussion, I will use the phrase *visual arts practice* to describe a range of doctoral programs, including those involving direct participation in arts practice, those focused on the scholarly study *of* or *about* arts practice, and those including aspects of the two, thus covering many different configurations. The discussion will also encompass doctoral degrees that combine the practice or study of the visual arts with other scholarly or creative fields.

A personal statement

Like Dr. Elkins,[3] I admit I am not writing from a neutral position. I entered academia somewhat later in life, following a career in business. I earned my PhD in 2011 from Texas Tech University's Fine Arts Doctoral Program, immediately following an MFA from Tulane University in New Orleans in 2008 and a BFA from the School of the Art Institute of Chicago in 2006. The trend towards doctorates for visual artists caught my attention while I was an undergraduate, and by the time I embarked upon my graduate work at Tulane, I decided to pursue a PhD because I thought it would improve my chances of finding a tenure-track teaching position. I also felt an increasing desire to prove that artists can and should be scholars or philosophers, not just makers of art.

DOI: 10.4324/9780429324154-11

My doctoral research evolved partially in response to the opposition I encountered during my MFA that my photographs were "too documentary" and, therefore, "not art." The research methodologies courses I undertook at TTU and the instruction I received in the scholarship of art bore no resemblance whatsoever to my MFA studies. Although conceptually linked through the ethos of social documentary photography, my dissertation *The Documentary Aesthetic*[4] was vastly different than the photographs I produced for my MFA show, *No Direction Home: Images from New Orleans*.[5] As someone who has earned both degrees, I can definitively state that the MFA is not at all the same as a PhD, despite longstanding belief in their equivalency. Or, as Timothy Emlyn Jones explains in the second edition of Elkins' *Artists with PhDs*:

> some MFA programs have no requirement for students to engage in theoretical work, or at least to produce evidence of any reading in which they might have engaged. The gulf between the two degrees is deep.[6]

It would be fair to accuse me of being an opinionated overachiever, and I freely admit my pursuit of a PhD was an attempt to justify and enhance my art practice through scholarly research as well as an overt effort to become qualified to teach in a fine arts doctoral program after graduation. At the time I was pursuing my MFA, it appeared that many such programs were about to emerge, all of which would need qualified faculty members. History has proven otherwise, of course. In fact, my doctorate has had the opposite effect: instead of enhancing my qualification for university teaching positions in art, hiring committees comprised of MFA faculty view me with suspicion at best and with outright antagonism or sheer terror at worst.

Those fortunate few of us who find work in an arts department after earning an MFA and PhD often face opposition from colleagues. In each faculty position I've held, I've encountered bafflement or disdain that my ongoing research takes place through writing and publishing rather than creating and exhibiting new artworks. This experience is hardly unusual. Judith Mottram relates the story of a recent PhD graduate who told her colleagues in visual art that she was working on a research project connecting aesthetics and neuroscience. They responded, "What? Why aren't you doing anything in the studio?"[7] Mottram herself received similar treatment when interviewing for a faculty job shortly after earning her doctorate.[8] These attitudes demonstrate a basic unfamiliarity with the value of a PhD in visual arts practice. Mottram writes:

> What both of us had to offer was knowledge of our particular field of contemporary art that was of greater breadth and depth than could have been achieved through study at Masters level. We also had the capacity to undertake further research in subjects both close to and at a remove from our core interests, potentially enabling us to prepare teaching materials for a range of undergraduate or more advanced courses.[9]

My first job after earning my PhD was as Director of the Mellon Research Project (2012–2015), a study of arts integration at research universities in the US and UK, during which I visited 46 institutions of higher learning and interviewed more than 950 individuals from university presidents to students. Two quotes I collected from interview subjects during a 2013 site visit to a top research university[10] dramatically illustrate the problem that faculty members from departments outside of the arts simply do not understand art or artists and cannot conceive of creative practice as research. In the first, I asked the provost to address the value of the arts. He replied:

> Look, the arts are important; no one will ever tell you they aren't. But, the arts have to do more to support the institution's mission on the whole, far better than they have, or

they will continue to be mere recipients of goodwill in higher education, rather than valued partners.[11]

During the same site visit, I read the provost's quote to the president of the university's Faculty Senate, who responded:

> This is absolutely accurate! In fact, let me take this one step further. The arts continually fail to communicate their importance, demonstrate their contributions, and sustain their message. They think they do, but they don't. Instead, the arts rely too heavily on the hope that any cultural significance and contribution they make will be received and perceived by their audience and instantly appreciated.[12]

These statements starkly illustrate the fundamental misalignment between the practice-based scholarship in the arts and that of other academic fields. Across virtually every area of inquiry, knowledge-building commonly occurs through research communicated in a textual-verbal form, yet artistic products and performances intentionally keep the research components of the creative process out of sight. Consequently, most US academics and administrators outside of the arts do not perceive the value or importance of the arts as a legitimate area of scholarship, nor do they grasp the connection between the arts and the mission of higher educational institutions – producing or discovering new knowledge that improves the quality of human life on earth. Elkins asks:

> How is it possible to make sense of the claim that a single object – the visual art on display in the exhibition – is at once a visual object, the knowledge that is produced by that object, and the research by which the visual object produced that knowledge?[13]

After extensive study of the arts in higher education, I must say: it can't. No matter how excellent or innovative a work of art may be, it cannot comprehensively communicate the artist's knowledge to an external audience. Steven Scrivener expresses a similar viewpoint in his paper, *The Art Object Does Not Embody a Form of Knowledge.*[14]

> If we are to talk of something as communicating knowledge, or perhaps more literally engendering knowing, then we should expect to experience this knowing when reading it. In short, we should be able to say, "I know this or that as a result of viewing this artwork." . . . Artworks cannot be read, at least to the level where they are usually assumed to function, i.e., to endow deep insight into emotions, human nature and relationships, and our place in the World, etc. If this is the case, then individually we cannot "know" anything of deep significance through viewing an artwork.

Likewise, Linda Candy upholds the need for artworks produced during practice-based research to be accompanied by "a parallel means of communication" that conveys the new knowledge represented by the artwork that helps to shape the audience's view of the work and facilitate comprehension of the knowledge the artwork represents.[15] With regard to doctoral studies featuring practice-based research, Candy emphasizes that the artwork produced during the research process should be accompanied by a "substantial contextualization of the creative work" that allows viewers to judge the work's original contribution and serve as a means of evaluating whether the PhD candidate has met general scholarly requirements, demonstrating that the student has acquired doctoral-level analytical abilities and mastered requisite disciplinary knowledge, conveyed through a textual-verbal form accessible to and understandable by scholarly peers.[16]

If, as suggested by Scrivener and Candy, an artifact cannot adequately stand alone in testament to the research process by which it came into being, then practice-based researchers should consider how the artifacts they produce through their research could be perceived by others. Chapter 5 of *Artists with PhDs* provides a good illustration of the communication problems that can arise when presenting an artwork as the embodiment of knowledge. Victor Burgin writes of a 1996 conversation between Bernard Stiegler and Jacques Derrida, who tells of teaching a course in California in which two students, with Derrida's permission, responded to an assignment by submitting videocassettes instead of a written paper. Derrida was intrigued by this possibility but finally rejected their efforts, explaining to the students (emphasis added):

> If your film had been accompanied – or articulated with – a discourse refined according to the norms that matter to me, then I would have been more receptive, but this was not the case; what you are proposing to me is coming *in the place of* discourse but does not adequately *replace* it.[17]

I find this to be true in my artistic practice: I am a social documentary photographer, and no matter how self-explanatory my images may seem, there is inevitably more to them than meets the eye. For example, I produced a photo of a woman lying in the doorway of a business in the French Quarter of New Orleans. The viewer's immediate assumption is that she is drunk and has passed out on the sidewalk (a common sight in the French Quarter); however, only if I *tell* the viewer that the woman in the photo was the victim of a mugging and that I photographed her as I accompanied two police officers on their nightly patrol does the image achieve its full meaning. My 2008 MFA show featured 68 photographs of New Orleans during the second and third years post-Katrina, examining the social implications of politics, poverty, decay, and attempts at rebuilding amid this singular cultural and historical setting. These images have been the springboard for many discussions, yet despite the truism that "a picture is worth a thousand words," even 68 pictures are not equivalent to a 60,000-word dissertation.

The US PhD in visual arts practice: a brief history

Interest in doctoral programs in visual art spiked in the mid- to late 2000s, driven by the expansion of doctoral programs in the UK and giving rise to fears that the PhD was poised to displace the MFA as the primary qualification for faculty positions in studio art. However, doctoral programs in Japan, the UK, and the US had existed for decades, including the Fine Arts doctoral program at Texas Tech University (established in 1972) and the Interdisciplinary Arts doctoral program at Ohio University (established in 1968).

The second edition of *Artists with PhDs*[18] includes an extensive list of worldwide doctoral programs in operation or in development in 2014,[19] revising Elkins' initial assessment: "In the first edition of this book I had predicted, based on the previous decade's growth that there would be 127 programs in North America by 2012."[20] The revised list mentions just six US programs and rumors of a handful of other programs in development.[21] This tally has grown, but certainly not at the rates once expected. As of the 2020–21 academic year, at least 13 US doctoral programs that include some aspect of the visual arts exist, but as Elkins accurately asserts, identification is problematic because there is no comprehensive database or clearinghouse for such programs.[22]

Curiously, *Artists with PhDs* devotes scant attention to existing US doctoral programs. Elkins' editorial choices focus on examples from international programs, mainly those in the UK. Only one chapter of either edition is written by a US author about a US program (George Smith, founder of IDSVA), with additional suggestions for US institutions developing doctoral programs in visual arts practice provided by Timothy Emlyn Jones, one of Elkins' UK contributors. Chapters authored by

Elkins include discussion of matters he believes to be of concern to future US doctoral programs, but his choice to overlook those in continuous operation for half a century at Ohio University and Texas Tech University is conspicuous.

Table 1.9.1 lists doctoral programs available in the US as of the 2020–21 academic year, but they reflect neither the rapid proliferation of the degree in other areas of the world nor their substance. Furthermore, these programs differ widely in their emphases and configurations. None except Transart Institute exclusively focus on studio art practice, but since Transart is not accredited by a US institution, it might not qualify as a US program despite its mailing address in New York City. IDSVA features research about arts practice but disallows the inclusion of a creative work. All of the US programs demonstrate varying forms of interdisciplinarity: combining the study or practice of visual art with music, theatre, literature, aesthetic philosophy, etc.; integrating arts practice with art history, theory, or visual studies; or focusing on media arts, electronic arts, or the intersection

Table 1.9.1 US doctoral programs involving visual arts practice.

Program (year founded)	Emphasis	Website URL
Arizona State University (not stated)	Design, Environment, and the Arts (History, Theory, and Criticism)	https://art.asu.edu/deeree-programs/design-environment-and-art-history-theory-and-criticism-phd
IDSVA (2007)	Art Theory and Philosophy	https://www.idsva.edu/
MIT (not stated)	Media Arts & Sciences	https://gradadmissions.mit.edu/programs/mas
Ohio University (late 1960s)	Interdisciplinary Arts + Art History or Philosophy of Art	https://www.ohio.edu/fine-arts/interdisciplinary-arts/academics
Rensselaer Polytechnic (2007)	Electronic Arts	https://hass.rpi.edu/arts/electronic-arts-0
Texas Tech University (1972)	Interdisciplinary study in Music, Theatre, Visual Art, and Aesthetic Philosophy	http://www.depts.ttu.edu/fadP/
Transart Institute (2004)	Creative Research	https://www.transartinstitute.org/phd
University of California at San Diego (2002, concentration in art practice added in 2009)	Art History, Theory, and Criticism with a Concentration in Art Practice	https://ucsd.edu/catalog/curric/VIS-gr.html
University of California at Santa Barbara (1999)	Media Art and Technology	https://www.mat.ucsb.edu/
University of Texas at Dallas (not stated)	Interdisciplinary Visual and Performing Arts	https://www.utdallas.edu/academics/fact-sheets/atec/phd- atec/
University of Washington (2001)	DXArts (Digital Arts & Experimental Media)	https://dxarts.washington.edu/dxarts-phd
USC (University of Southern California) (2013)	Media Arts + Practice	https://map.usc.edu/phd/
VCU (Virginia Commonwealth University) (2006)	MATX (Interdisciplinary Media, Art, and Text)	https://matx.vcu.edu/?utm source=grad&utm_campaign =vc uartsweb&utm term=matx

Note: Other US programs may also exist.

of visual art and technology. All hold high expectations for scholarship consistent with other academic fields; require extensive coursework in theory, philosophy, and research methodologies; and maintain rigorous written dissertation requirements, reflecting Jones' observation that, "There is a distinctive American way of dealing with art research and research degrees in art."[23]

Moving forward

For many years, the only way for US artists to earn a doctorate was in art history, art education, or visual studies. Practice-based study was limited to the MFA, long upheld as the terminal degree in studio art, equivalent to a PhD in other fields. Despite such claims of parity, PhD-level faculty in other university departments have historically looked down on colleagues in the arts as being their intellectual inferiors due to their lack of this credential and the pervasive notion that artists are makers, not thinkers. Indeed, disharmony between the artists and art historians in university art departments dates back at least to the 1950s, as seen in Lester Walker's 1955 *College Art Journal* article, 'The Studio Artist and the Art Historian',[24] and is also mentioned in George Smith's chapter in the first edition of *Artists with PhDs*.[25] Furthermore, because the vast majority of current studio art faculty hold MFA degrees, they mount significant opposition to the PhD in their adamant insistence that doctoral study for artists is unnecessary.[26] The scarcity of these doctoral programs up to the late 2000s rendered such objections rather pointless, though: because only a handful of artists earned PhDs, the degree posed little threat to the dominance of the MFA.

Artists with PhDs raises questions about just what a doctorate involving visual arts practice should entail, yet doctoral study in most academic fields other than art has been fairly standardized since the early 20th century, characterized by requirements for:

1   advanced graduate-level coursework.
2   substantial independent research.
3   a written dissertation reporting the research process and its results, which must demonstrably contribute new knowledge to a field of inquiry (60,000 words, more or less, depending on departmental or institutional requirements).
4   an oral examination or defense of the dissertation.

The second and third points are the site of much discussion in both editions of *Artists with PhDs*, examining the meaning and usage of the terms research and new knowledge, considering how these may or may not apply to the process of artistic creation as research, the embodiment of new knowledge in the resulting work of art, and whether doctoral research should culminate in written text, works of art, or a combination of the two.

The overwhelming majority of Elkins' and his contributors' theoretical and philosophical models apply to degree programs outside the US. Particularly in the UK, academic programs are shaped by institutional requirements such as the former Research Assessment Exercise (RAE) and its successor, the Research Excellence Framework (REF), which have no parallels in the US. Under the RAE and REF, university funding is partially determined by the number of advanced research students in each department,[27] thereby fueling the proliferation of the PhD in UK universities. Literature justifying these research methodologies and upholding the idea that the creative work they produce also embodies new knowledge has a strategic benefit for visual arts departments.

The UK's wide acceptance of doctoral study in the visual arts offers a variety of relevant models. For example, Goldsmiths-University of London offers three pathways to a PhD in Art.[28] *Pathway 1: Thesis by Practice* requires a substantial body of creative works, curatorial practice, or art writing, along with documentation and a written component of 20,000–40,000 words. *Pathway 2: Thesis by Practice and Written Dissertation* includes a body of creative works, curatorial practice, or art writing and a 40,000–80,000-word dissertation, presented as an integrated whole. *Pathway 3* leads to a dissertation of 40,000–80,000 words alone. As another example, the Glasgow School of Art offers PhD programs in each of its five schools: Fine Art, Design, Innovation, Architecture, or Simulation and Visualization.[29] GSA's doctoral students can engage in practice-led projects combining a written thesis with objects, artworks, exhibitions, performances, or other artifacts representing their research, or they may choose a more traditional path and produce only a written thesis. Models such as these offer a valuable resource to US artists and scholars contemplating a PhD in visual arts practice or other practice-based fields. However, we must remain aware that significant differences between the philosophical, financial, and administrative models of UK and international doctoral programs and those of US higher education can complicate comparisons.

Subsequent discussion builds on Elkins' personal contributions to the second edition of *Artists with PhDs*, particularly Chapter 13: "Fourteen Reasons to Mistrust the PhD"; Chapter 15: "Positive Ideas for PhD Programs" and "Envoi." Chapters by Timothy Emlyn Jones, Mick Wilson, Judith Mottram, and George Smith also raise important points, especially regarding the relationship between the MFA and PhD. These discussions shed light on practice-based research and practice-based doctoral programs.

Makers and thinkers

International doctoral programs in which students have the option of producing a work of art in place of a written dissertation bear some similarities to the American MFA, which requires a final exhibition of the student's artwork, usually accompanied by a brief written thesis (sometimes little more than an elaborate artist's statement). Like the MFA, students in international PhD programs might spend the bulk of their time in seminars and in the studio. Justification of the American MFA as the terminal degree equivalent to a PhD in other fields rests at least in part on these presumed similarities. However, Jones suggests US art schools should

> question the idea of a "terminal degree," which is unknown outside the US. Employment should normally go to the candidate best qualified overall for a position by a number of criteria, not just to the person with the smartest certificate. Most of the world is familiar with the idea of a terminal illness, but not a terminal degree. It is worth asking whether it contributes any vitality to education.[30]

The title MFA itself is also open to examination. *Master*, an essential component of Master of Fine Arts, has its roots deep in the guild system of the Middle Ages, indicating an individual's completion of a masterwork demonstrating the person's skills and competencies, which conferred the title of *master* upon the craftsman after review by other masters. Medieval university students attained the status of *master* when they had achieved a certain level of competence in an area of study sufficient to teach it, just as an MFA confirms an artist's skills and competencies, bestowing the qualification to teach their artistic practice to others. For centuries, the terms master, doctor, and professor were synonymous, denoting a person of scholarly accomplishment who had earned the qualification to teach.[31] This changed by the early 20th century when the doctorate came to represent a higher level of study than a master's degree. According to Henry Bent, in the March 1959 edition of the *Journal of Higher Education*, a person who earns a doctorate is "able to do original work in at least one great field of study, and to supervise and criticize the work of others" and to "understand a subject as fully as its development may permit, to learn more about it through their own research, and to teach it with enthusiasm and effectiveness to others."[32] Or as Jones explains, "Conventionally, a Masters represents new perceptions of the current state of knowledge in the subject while a Doctorate represents new knowledge or significant contributions to understanding in the subject: not the same at all."[33]

Although the MFA is generally accepted in the US as the qualification to teach in a studio art program, just as most other departments require a PhD, significant differences exist between the two. Both degrees typically require 60 credit hours of coursework, but the MFA specifies that a minimum of 65% of total credit hours must be in studio courses, with 15% of credits earned in "academic studies concerned with visual media."[3435] Conversely, doctoral students typically spend about 80% of their studies in academic coursework, leaving the remainder for the development of the dissertation.[36] These differences continue into graduates' professional careers: as university professors, art faculty usually meet their contractual obligation for research through their exhibition record and evidence of their continuing art practice, whereas faculty in other departments are expected to engage in formal research, including peer-reviewed publication.

Inequalities between the MFA and PhD are felt most keenly within university art departments themselves, where schisms between studio artists (MFAs) and art historians (PhDs) often become acute. George Smith explains that professors of art are usually restricted to the studio, teaching the "labor" of art but not questions of philosophy or theory in contemporary art. Art students "learn to make art from artists, but they learn about what art means from scholars who generally do not make art,"[37] reflecting the dramatic divide between theory and practice. Non-art majors learn about art from courses in art history, usually in fulfillment of a liberal arts requirement, absorbing the point of view of art historians, not artists. This perpetuates the view of art as manual labor instead of an intellectual activity because "the art historian's essential story . . . tells American students that the artist works and the historian thinks."[38] Smith founded the Institute for Doctoral Studies in the Visual Arts (IDSVA) in opposition to this entrenched belief, seeking to promote the idea of the artist-scholar or artist-philosopher. IDSVA students do *not* create art as part of their studies. The program is designed to be a "mix of theory and practice, education and experience"[39] representing a conscious departure from the creative production typical of either faculty responsibilities or MFA requirements.

Opposition to the PhD

At its best, doctoral-level study of visual arts practice provides artists with opportunities to delve more deeply into their chosen field, to learn the history, theory, and philosophy behind the making of art, and to understand art in a broader context than the inside of their studios. Although the PhD may someday become the preferred qualification for teaching art at the university level in the US, as it has in other academic fields, the comparative rarity of available doctoral programs in the US makes it quite unlikely the PhD will supersede the MFA anytime soon.

Nevertheless, current art faculty still exhibit a great deal of nervousness around this issue. Jones explains:

> The case against PhDs in studio art in America has rested more on the anxiety of academics worrying that they might need to go back to art school to regain their credentials as teachers than on any academic grounds.[40]

Smith speaks of related suspicions:

> Art history faculty don't want to lose PhD status over MFA faculty. MFA faculty don't want freshly minted studio PhDs coming in as new hires and impressing art students with their doctoral airs. Nor do MFA job candidates want more years of study and more debt just so they can qualify for studio teaching jobs that are few and far between.[41]

Some of these concerns may be warranted. Adding PhD-qualified faculty to studio art departments could put the teaching of the "intellectual" aspects of art – history, theory, criticism, and philosophy – into the hands of artist-scholars and artist-philosophers instead of remaining under the control of "thinkers" outside the studio. Nevertheless, even if the PhD were to proliferate as predicted in the mid-2000s, universities would never require tenured studio art faculty to earn a PhD to keep their jobs. Instead, departments would bring in PhD-qualified faculty gradually, filling positions as they are vacated through retirement or voluntary departures. The "double threat" posed by artists with PhDs who would be able to teach both studio art and art history, theory, and criticism[42] has not materialized, either. Hiring committees continue to favor traditionally qualified candidates, maintaining familiar differentiation between the studio and the classroom.

Of more concern, Elkins' contention that further education might adversely affect an artist's practice deserves serious examination, revealing an uncomfortable prejudice lurking behind the surface of his book – that too much thinking can damage the artist's ability to make art. For example:

> I would say it is generally supposed that knowledge of art history is in itself not a bad thing: but for a working artist, it may also be that too much art historical knowledge might hamper or even ruin ongoing art projects.[43]

And:

> If your art is, say, Neoexpressionist, then an advanced degree may actually harm your practice by making you aware of historical and critical reasons to doubt your own interests. (I have sometimes advised artists who do expressionist work to drop out of school even before the MFA).[44]

In my opinion, the assumption that elevating artists to the status of scholars will be detrimental to their art sounds like an abhorrent variety of paternalism, reminiscent of similar biases against educating women or granting women the right to vote either because they were believed to be incapable of such lofty pursuits or because it could undermine their ability to perform their domestic duties. It also implies artists are unable to form coherent thoughts about their creative practice, working only from intuition, not intellect.

Although some artists might not be capable of PhD-level study, the same could be said of *all* academic disciplines: not all teachers, scientists, engineers, mathematicians, or sociologists choose to accept the challenge of doctoral study, either. Few people hold PhDs because it is one of the most challenging goals anyone can achieve, underlying the high esteem society bestows upon those who have earned a doctorate. When studio art faculty ask, "How can you expect art students to write 50,000-word dissertations when my students can barely write a short master's thesis?"[45] the answer is that it *is* unreasonable to expect most of them to do so, but not because they are artists. Only 2% of the US population over the age of 25 has earned a doctorate, placing this goal out of reach for 98% of the population in general.[46]

The prejudice that artists are not intellectuals underlies one of the most troublesome aspects of the MFA. As Smith explains,

> today in the US, the artist gets pretty much the same MFA that was offered at Yale in the early 1970s. Which is to say that even in the wake of postmodernism's deconstruction of formalism and aesthetic autonomy, the American MFA curriculum focuses almost entirely on form and technique.[47]

NASAD's guidelines for the MFA testify to programs' heavy emphasis on making art. Traditionally oriented programs infrequently offer opportunities for scholarly pursuits as is typical of other departments of the liberal arts and sciences (although exceptions do exist). Admission requirements of many US art programs exempt artists from the same levels of academic achievement expected in other departments. Instead, aspiring artists gain entry to MFA programs on the merits of their portfolios, much as the academic shortcomings of talented athletes are excused in favor of their prowess on the playing field.

The MFA serves as evidence of an artist's mastery of his or her artistic medium and fitness to teach the labor of the studio, but it offers no guarantee that the holder of this degree is a highly qualified scholar at the same level as a PhD, prepared to teach aspects of art such as theory, criticism, or philosophy. Elkins confirms this, stating:

> MFA [programs], despite their many virtues, simply do not produce graduates who really know art theory. I say this after twenty years teaching at the School of the Art Institute of Chicago: in all that time, I have seen no more than a couple of dozen students who were educated at the level of rigor that is expected of philosophy or political science students in major universities. MFA students are routinely given degrees even though they have only a sketchy, somewhat bewildered sense of such things as deconstruction, semiotics, or psychoanalysis.[48]

Moreover, because MFA faculty are expected to continue making art and submitting it to exhibitions as the research component of their professional duties, many have little incentive to delve into the scholarly aspects of art. Instructors teaching the newest generation of artists may have learned everything they know about art history, theory, and criticism when they were graduate students themselves, sometimes 30 years ago or more, thus remaining willfully unaware that their knowledge base may no longer be relevant. This reality underlies the tendency of many schools of art to uphold Enlightenment-era aesthetics in a postmodern world, such as perpetuating the worn-out prejudice that creative media such as ceramics, textile art, or photography should be excluded from identification as "fine art," even though this notion definitively changed in contemporary aesthetic philosophy long ago.[49] We would also do well to question the word *Fine* in the title *MFA*. Does this archaic term make sense in light of myriad developments in aesthetic philosophy in the contemporary era?

Master–apprentice models exist in many fields. For example, medical students progress from interns to residents to physicians, or students aspiring to careers in education complete observation hours, engage in supervised student teaching, and finally earn certification as qualified teachers. Developing disciplinary skills and knowledge, practicing the discipline under the supervision of a qualified expert, and earning a qualification to teach that practice to others exists across history and in many areas of human life, just as it does with the MFA. Doctoral-level study involving practice-based research extends this model beyond perpetuating a given discipline to engaging in practice to answer a directed research question which cannot be explored through other methods,[50] thereby extending the boundaries of knowledge in, through, with, or about that practice.

Questions and opportunities

As Jones and Elkins each suggest, doctoral programs in visual arts practice present an opportunity to re-examine art across higher education, but its underlying premises must also be brought to light and examined. For example:

- The idea of a terminal degree deserves a great deal of scrutiny: is it *ever* acceptable to stop learning, particularly for a professional educator?
- Can "too much learning" really be harmful to an artist's practice, or does harm originate in the outdated, sketchy (or absent) instruction in art theory, aesthetic philosophy, and art criticism typical of most MFA programs in the US?

- Is the role of "thinker" the sole provenance of art historians and other PhD-holders, or can a "maker" of art be a "thinker" as well?
- Are artists' voices so often ignored in discussions of art history, theory, philosophy, and criticism because of the "inferior" academic status of those holding MFAs, or does this prejudice go deeper, rooted in a longstanding bias against artists in general?

Despite the possibility that a PhD will someday become the preferred credential for university teaching positions in studio art, the majority of institutions have yet to acknowledge this development. For instance, the Professional Practices Committee of the College Art Association (CAA) published the following statement in the fall of 2008, shortly before the first edition of *Artists with PhDs* went to press:

> At this time, few institutions in the United States offer a PhD degree in studio art, and it does not appear to be a trend that will continue or grow, or that the PhD will replace the MFA. To develop a standard for a degree that has not been adequately vetted or assessed, and is considered atypical for the studio-arts profession, is premature and may lead to confusion, rather than offer guidance to CAA members, their institutions, and other professional arts organizations.[51]

To offer a personal perspective, I was a member of the CAA committee charged with re-examining this policy in 2014, just before the second edition of *Artists with PhDs* was published. As might be imagined, the statement's new wording engendered much debate among committee members, especially with regard to the word "the" in the first sentence of the new statement (emphasis added): "The College Art Association affirms that the Master of Fine Arts is *the* terminal degree in studio art practice." The revised statement, adopted by the CAA Board of Directors in 2015, goes beyond the 2008 version quoted in the preceding paragraph, finally acknowledging the existence of doctoral degrees but falling short of outright endorsement:

> At the same time the Association recognizes the existence of Doctor of Philosophy (PhD), Doctor of Fine Arts (DFA), Doctor of Visual Arts (DVA), Doctor of Studio Art (DA), and other doctoral degrees that incorporate art and/or design practice; in the United States these programs emphasize formal research and are often offered in combination with other disciplines. Doctoral programs in the visual arts may take varied forms dependent on each institution's requirements, reflecting specific academic opportunities and research instruction. CAA recognizes the unique prospects such programs offer for research-intensive study in the visual arts and design, and affirms that offering such opportunities is not only within the purview of individual institutions but has the potential to add to the diversity of research in higher education.

The relative scarcity of US doctoral programs and opposition to the PhD by MFA-qualified faculty each lead to an essential question that remains unanswered: why should anyone struggle and toil to earn a PhD when an MFA is still "good enough" to get a university teaching job? Some people will respond to this challenge with the same answer as "Why climb Everest?" Because it's *there*: certain individuals seek to reach the pinnacle of whatever achievement is set before them, and the PhD in visual arts practice represents that highest level. Others will take a more pragmatic view, hoping that earning a PhD will maximize their opportunity to secure a tenure-track faculty position in an extremely competitive marketplace already over-crowded with MFAs. Still others will respond to the PhD's promise of increased status, meriting the respect of colleagues elsewhere in the university: the community of scholars. It might also be viewed as a passport to positions in academic administration such as a deanship, since few such roles are filled by individuals with an MFA alone.

However, the intrinsic value of a research degree needs to be balanced against its instrumental value, and this needs to be acknowledged and, where possible, transparent. Any college degree requires a daunting investment of time, effort, and expense, made all the more significant when no certification is required for professional achievement. Art schools would do well to examine the real value of the degrees they offer, analyze the progression of learning from bachelor's through doctorate, and ensure they provide study of equivalent quality, value, and rigor to that of other academic fields.

In today's challenging world, students pursue college degrees because they want to qualify for careers that will lead to a better life.[52] However, the purpose and value of graduate study in the visual arts may not mirror other academic areas. A PhD in aerospace engineering could lead to employment with NASA or SpaceX as well as qualifying the possessor for a university teaching position, but we cannot say the same of either the MFA or the PhD in visual arts practice. Neither can we prove that earning any visual art degree from bachelor's to doctorate will enhance the possessor's likelihood of art world success. I suggest it is time to acknowledge that the instrumental value of an MFA or PhD in the US is to qualify graduates for tenure-track teaching positions in university art departments. Certainly, each degree is valuable for its own sake, but this intrinsic merit is undiminished by its concurrent value as a credential for a career in academia. Embracing this truth may be an essential step towards establishing robust doctoral programs in studio arts practice.

The dissertation vs the practice as sharable *new knowledge*

PhD programs culminating in an object, artifact, or performance instead of a written dissertation remain unlikely to find acceptance in the US. Academic pursuits must be of *like kind and quality* as those of other disciplines if we hope to be recognized as scholars in addition to receiving appreciation for our creative output. However, both editions of *Artists with PhDs* discuss doctoral programs that accept a work of art as a legitimate demonstration of research and embodiment of new knowledge. Jones offers this suggestion to those working to establish such programs in the US:

> The idea of art as a process of inquiry is the keystone of art research. Process supposes an aesthetic of method as against an aesthetic of style, a concept that has yet to be fully worked through, but one that places educational creativity at the center of aesthetic creativity in a way pioneered by Joseph Beuys. Look too to the place of John Dewey, David A. Kolb, and Donald Schön in American educational heritage: their precedents for learning through activity are the foundations of the studio art doctorate. . . . Art research, therefore, already has a strong provenance in US culture even if it is not yet widely celebrated.[53]

As in other contexts worldwide, a Doctor of Fine Art (DFA) might serve as an alternative designation if the research output takes the form of an art exhibition alone; however, this opens the degree to the same legitimate criticisms as its international cousins. Critics echo Elkins' complaint that such programs

> are a kind of prolonged MFA, with students just sitting in their studios another two or three years, producing more of the same art, writing about themselves, navel-gazing, trying to achieve a pinnacle of self-awareness that may or may not make their work more interesting.[54]

Traditional MFA programs focusing almost entirely on time spent in the studio deprive their graduates of the very skills and competencies necessary to the professorate, perpetuating the second-class status of studio arts faculty. The absence of instruction in research methodologies and a focus on the studio over and above active engagement in artistic theory, philosophy, and criticism underlie many of the problems in US art departments today. It's not that artists are incapable of engaging with these topics: many of them were never expected to learn them in the first place. Their training prepares them for the research of the studio that undergirds artistic production, not for the types of scholarly activity common to the rest of the university.

As an example, many of the studio art faculty with whom I've worked received little to no instruction in data analysis or academic writing during their graduate studies. These deficits carry over into other areas of faculty responsibility, such as service on program assessment committees or understanding institutional research data. When I've worked with visual art faculty in these areas, their lack of experience with formal research becomes all too apparent, especially on institution-level committees involving colleagues outside of the arts. We deny the instrumental reality of the MFA as a qualification for the professorate to our detriment: continuing emphasis on art-making over all else does not adequately prepare future faculty members with the academic skills and competencies routinely imparted through doctoral study and implicitly expected of faculty in other departments.

Many discussions in *Artists with PhDs* involve questions of how the creation of visual art relates to the genesis of new knowledge through research. I remain uncertain of the strategic value of these conversations. Perhaps terms like research and knowledge may be just as indefinable as the word "art" itself since they vary so widely in practice and application. We may never agree about definitions, but perhaps we can agree on operational principles for doctoral study in art and other practice-based fields.

Timothy Emlyn Jones and Mick Wilson each clarify the essential elements of doctoral scholarship. Jones explains:

> The PhD both provides a training in research methods and methodology that is achieved through a program of inquiry framed as a project, and it generates new knowledge or contributions to understanding through that program of inquiry. That is to say, learning of how to do research is as indispensable to a PhD as the new knowledge that is generated by means of it.[55]

In Chapter 17,[56] Mick Wilson explains how the Dublin Graduate School of Creative Arts and Media addressed this problem, relating the key questions doctoral students must answer through their independent research. These questions un-complicate and demystify the research process considerably, echoing baseline expectations consistent with Jones' suggestions for US PhD programs in Chapters 6 and 9. They are also useful when considering other practice-based research programs.

- What are you trying to find out?
- Why is it worth knowing?
- How do you go about finding it out?
- How will you know when you are finished finding out this "something"?

Wilson also writes that "research, while being a broad portfolio category, is not a completely elastic term."[57] Doctoral candidates "are expected to be able to articulate and defend an epistemic practice and orientation within a discursive exchange with a group of assessors."[58] In other words, doctoral students cannot present a work of art, art exhibition, or artifact generated by another type of project-based research as a *fait accompli*. Any item presented *as research* must be accompanied by a written document and verbal defense, articulating the research process and communicating the new knowledge produced through the work. Developing proficiency in effective academic communication about one's scholarship is a hallmark of doctoral study – a competency that graduate students and faculty alike across all creative and practice-based fields should develop.

After all, it makes little sense that any practitioners, artistic or otherwise, should expect the entire system of higher education to grant an exception to the norms of doctoral study in other scholarly fields. To paraphrase Jones, neither art nor any practice-based field is "so special" it should be exempt from established expectations for doctoral-level study – that is, if we want our PhDs to achieve parity with others.

The scholarship of art

Until quite recently, US students seeking opportunities for doctoral study in the visual arts had few options.

- They could leave their arts practice to study art history, theory, and criticism.
- They could pursue a PhD or EdD in visual studies or art education.
- Or they could enroll in one of a handful of interdisciplinary doctoral programs or combining art with a field such as media, technology, or other arts such as music, theatre, or dance (see Table 1.9.1).

Although these programs offer sound educational experiences and produce distinguished graduates, very few focus exclusively on practice-based research in visual art. Nevertheless, arts practice offers an incredibly rich and diverse field for scholarly study in its own right. Anthropologists, historians, paleontologists, and scientists have adopted art as a productive area of academic study, but each of these fields contributes to the understanding of art from an outsider's perspective. Their contributions to knowledge are valid and valuable, but the work of art becomes an inert specimen utilized as the site of an investigation into a related field. Where is the voice of the artist in these studies? For example, ancient artists such as those who painted the walls in the cave of El Castillo will forever go unnamed, but present-day artists could make meaningful scholarly contributions to the understanding of their long-ago predecessors' creations, bringing an alternative sensibility to this endeavor and, most importantly, an insider's perspective and sense of identity.

Scholarly research as conducted by creative practitioners necessarily differs from that of other fields, just as research in sociology differs from research in applied mathematics or quantum physics. If fewer pathways to doctoral-level achievement exist in arts practice than in other fields, it does not mean that these options are non-existent or that they are less worthy of attention than other areas of research inquiry. Furthermore, artists interested in pursuing scholarly research into their creative practice should not have to become anthropologists, art historians, or scientists for academia to see their work as equally valid or valuable compared to work conducted in those disciplines.

A seat at the table

Without a doubt, research in visual arts practice is a very small field of study in the US. The National Center for Science and Engineering Statistics and the National Science Foundation conduct annual surveys to collect data concerning research and development activity (R&D) at degree-granting institutions. Most of this activity occurs at a small percentage of institutions in US higher education: only about 1000 of the approximately 4400 postsecondary degree-granting institutions reported any R&D expenditures for 2018 (the most recent data available), and of these, the 115 institutions designated as having the highest research activity performed 75% of all academic R&D. Within this top tier, just 25 institutions account for nearly half of all academic R&D and one-third of the national total.[59]

In 2018, the NCSES reported a total of $79.3 billion in Higher Education R&D Expenditures. Of these, Science accounted for $62.2 billion, Engineering for $12.4 billion, and Non-S&E (all fields other than Science or Engineering) for $4.6 billion. R&D in Visual and Performing Arts accounts for $13.4 million[60] – just 3% of all Non-S&E and only 0.18% of total R&D in higher education. To put these figures into perspective, if all the R&D in higher education were worth $100, Science receives $78, Engineering $16, and all fields other than Science or Engineering receive $6, out of which R&D in the Visual and Performing Arts accounts for only 18 cents.

Furthermore, fully 70% of funding for Non-S&E R&D comes from non-federal sources, and of these, 67% is from institutional funds.[61] All told, research and development in the visual and performing arts represents an almost microscopic share of all R&D activity at US degree-granting institutions and is most likely to be funded primarily by institutions themselves rather than external sources. The same is true of other Non-S&E fields, however. The discipline of Communications occupies a comparably tiny share of R&D expenditures, as do Social Work and Law. Competition for institutional funding among Non-S&E fields is simply the way of things in academia. However, if patterns in the STEM disciplines are any indicator, practice-based researchers' successful attainment of research funding will raise the esteem they receive within their institutions, enhancing their ability to advance their scholarship and bringing attention to the merit and value of their research accomplishments. Therefore, a PhD is essential if practice-based researchers are to earn a seat at the table and a voice in the conversation as legitimate scholars deserving of respect equal to that of their peers.

Program structure and components

Taking all of this information into account, two basic principles governing a US doctoral program in practice-based research emerge:

1   It must involve the scholarly study of that practice.
2   It should include an option for direct engagement in that practice.

As discussed earlier in this chapter, very few such opportunities for doctoral study in visual arts practice exist in the US apart from interdisciplinary programs, but I believe we should approach interdisciplinarity with care. Combination degrees may adopt the rigors of the partner discipline and subordinate the arts practice element. As a parallel illustration, I visited Virginia Tech's Robotics and Mechanisms Program during the Mellon Research Project, speaking to researchers who created robots that could assist the elderly and infirm with household tasks and self-care. Their prototypes could perform all of the desired functions flawlessly until the researchers discovered an enormous drawback: the robots terrified their intended users. The team recruited artists to modify the robots' aesthetic appearance, making them less frightening. This art + science partnership seems like a promising collaboration, but it's not hard to see that art was an afterthought, not an equal partner from the beginning. I observed an amazing array of collaborations between art and science that made valuable contributions to knowledge, society, culture, healthcare, and more, improving the quality of human life on Earth. However, most of them could best be described as SCIENCE + art rather than a partnership between equals.

Interdisciplinary PhD programs run the same risk unless carefully planned and managed. Placing art in the service of another discipline, even if well-intentioned, misses an opportunity to legitimize arts practice as an equally valid site of scholarly inquiry. I'm grateful that the doctoral program

I completed at Texas Tech did not fall into this category. Its faculty were mindful of their graduates' potential to attain leadership positions in higher education such a future deanship of a college of visual and performing arts, so the required coursework in aesthetic philosophy, art, music, and theatre served to broaden students' understanding of the relationships between and among the arts while also nurturing students' acquisition of skills and competencies as researchers within a particular area of arts practice.

The University of California at San Diego serves as an excellent exemplar of what is possible within a mono-disciplinary focus on visual art in its PhD in art history, theory, and criticism with a concentration in art practice.[62] This program includes the elements typical of doctoral study in the US with additional components specific to artistic creation. These are:

- Completion of required coursework, including demonstrated knowledge of a foreign language.
- Qualifying examination consisting of two bibliographies, a practice-related bibliography, a dissertation prospectus that includes a practice-related component, a written examination, and a two-hour oral examination in the student's major field.
- Researching and writing a dissertation, along with producing a visual component decided by the student and their dissertation committee.
- Oral defense of the dissertation.
- Final dissertation published to ProQuest through the university library.

UCSD's program has two additional characteristics I believe are crucial to producing highly qualified artist-scholars. First, the program's faculty are well versed in the standards and expectations of doctoral study in visual art. Speaking from personal experience, this is not the case at most institutions: few artists either hold a PhD degree or possess experience in similar programs where they could have developed an understanding of how doctoral-level achievement in visual arts practice differs from study exclusively in art history, visual studies, or studio art. Second, UCSD is located in a major metropolitan area with a vibrant arts community and in close proximity to others. Not only are its faculty distinguished artists with national and international reputations who are qualified to evaluate doctoral candidates' creative works, but other accomplished artists live near the university who might be capable of serving as knowledgeable evaluators of a dissertation's practice-based components. No program's faculty could be prepared to evaluate every potential PhD candidate's work – they will inevitably encounter students whose interests supersede the faculty's collective knowledge and experience. In that case, the program must be capable of gaining access to experts who can serve as evaluators after receiving some directed preparation in the expectations for dissertation-caliber student work.

The issue of proximity may seem odd to readers in Europe, where metropolitan areas are often only a train ride away. However, in the US, universities in certain states or regions might be quite remote, requiring many hours of travel by car, train, or air. Some large, well-known universities lie hundreds of miles from major metropolitan areas, so their art students' access to world-class galleries, museums, or other cultural resources may be limited. Therefore, universities in or near cities like Chicago, Dallas, Denver, Los Angeles, Miami, New Orleans, New York, or San Francisco could mount a doctoral program more successfully than institutions without sufficient cultural assets nearby.

Coursework preparing students to conduct formal research is especially important since BFA and MFA programs do not typically address research methodologies, nor do they require students to become proficient in scholarly writing. Table 1.9.2 compares formal research and arts practice based on the observations I gathered during the Mellon Research Project. Bold type indicates areas of significant difference.

An ideal PhD in visual arts practice would merge the expectations in both columns, providing instruction in research methodologies, conducting a literature review, writing a dissertation, and

Table 1.9.2 Formal research and arts practice

Formal research	Arts practice
Preparation	
Identify a research problem or question.	**Identify the type and content of the artwork to be produced.**
Identify the methodological paradigms that will guide the study.	(no parallel in arts practice)
Conduct a formal literature review, searching for precedents and information to identify what is known about the topic and what has yet to be discovered; determine how the intended study might meet an otherwise unidentified informational need.	Search for similar works, precedents, and information that might inform the creative process or generate artistic inspiration.
State a goal or create a hypothesis.	Determine a general idea regarding the form of the intended artwork.
Plan the research project to address the goal or test the hypothesis.	Form a general plan regarding how to create, produce, perform, or accomplish the work of art.
Procedure	
Data collection (varies) • Experiment (physical and social sciences). • Ethnography, interviews, case studies (social sciences). • Examination of primary and secondary sources (historical research). • Other methods specific to a given discipline.	**Creative Practice** Creation of a work of visual art or design utilizing specific artistic media.
Conduct data analysis and interpretation according to methodologically specific procedures appropriate to the academic discipline in which the research is being conducted.	Conduct critique and self-assessment of the work in progress, adjusting and improving the work of art according to a cyclical heuristic process of action–reflection–evaluation–revision.
Determine whether the study has answered the research question, proved or disproved the hypothesis, or solved the problem. **State the study's original contribution to knowledge.**	Review and critique: determine whether the work is ready for exhibition or performance, making adjustments or changes as necessary (**typically no written component other than an artist's statement**).
Presentation	
• Publish in print [book, article in a professional journal, dissertation repository (ex: ProQuest)]. • Present verbally (dissertation defense, professional conference, or symposium; often accompanied by **written documentation**).	Present to the public: **exhibition, gallery show, website, or another public forum**.

explaining how the student's work makes an original contribution to knowledge. It would also provide an opportunity to exhibit the student's creative work consistent with the norms for the presentation of artworks produced in other university programs such as the MFA. Just as the student must mount an in-person defense of their research to demonstrate their attainment of expertise as a scholar, the exhibition of the physical work of art is equally crucial to demonstrate their expertise as a creative practitioner. Live presentation and a physically present artwork also mitigate the possibility of academic dishonesty or digital manipulation. Thereafter, images of the artwork should be included in the archived dissertation, serving as the permanent record of the student's scholarly and artistic achievement.

Furthermore, a US doctoral program in visual arts practice would do well to adhere to standards published by NASAD, the largest US accrediting body in art and design. Even if the program is not seeking NASAD accreditation, these standards provide a basis for consistency in what doctoral study should entail across institutions. These include:[63]

1   An emphasis on research or scholarship in some aspect of art or design.
2   The equivalent of at least three years of full-time graduate work.
3   Procedures specific to the institution.
4   Qualifying prerequisites:

    a   Intellectual awareness and curiosity sufficient to predict continued growth and contribution to the discipline.

    b   Significant professional-level accomplishment in one or more field(s) of study.

    c   Knowledge of analytical techniques sufficient to perform advanced research and produce scholarly work in one or more fields or specializations.

    d   Knowledge of the historical record of achievement associated with the major area of study.

    e   Knowledge of general bibliographic and information resources in art and/or design.

    f   Considerable depth of knowledge in some aspect of art and/or design.

    g   Writing, speaking, and visual skills to communicate clearly and effectively with members of the scholarly and research communities and the wider community.

    h   Research skills appropriate to the area of study as determined by the institution.

Item 4g is of particular interest. NASAD standards for the MFA require "writing and speaking skills to communicate clearly and effectively to the art and design communities, the public, and in formal or informal teaching situations,"[64] whereas doctoral standards expect communication with "members of the scholarly and research communities," underscoring the importance of a doctoral degree in preparing students for participation in the wider community of scholars beyond art or design. In addition, NASAD does not specify credit hour requirements in studio art for the doctorate as it does for the MFA, but neither does it prohibit engagement in creative practice as part of the student's program of study towards a doctorate. Therefore, a PhD in visual arts practice could adhere to NASAD standards and still allow for doctoral-level practice-based research.

7 key criteria for success

US institutions contemplating a doctoral program in a practice-based field such as visual arts might wish to consider the following series of questions by which they could assess their capacity to launch and sustain a PhD program featuring practice-based research.

1   **Academics.** Do we presently offer or do we have the capacity to develop programming and coursework aligned with standards for accreditation that provides sufficient depth, breadth, and rigor to impart expert-level status within a field of study and prepare students to conduct formal research within that field?

2   **Structure.** How would we create program components aligned with the expectations of the university's Graduate School?

3   **Faculty.** Do we employ or could we hire faculty who have attained expert-level status within the field(s) encompassed by the program who are also knowledgeable about the norms and

expectations for doctoral-level achievement and capable of assessing candidates' attainment of these standards? Do we employ or could we hire faculty who are proficient in communicating with audiences in fields outside our primary discipline(s) and who can teach effective cross-disciplinary communication to our students?

4   **Location.** Does our city or region offer access to disciplinary experts who would be available to serve as external evaluators of created products that may be outside the faculty's areas of expertise?

5   **Resources.** Will our university provide resources to launch and sustain the program?

   a   Full-time tenured or tenure-track faculty qualified to teach at the doctoral level, satisfying disciplinary accreditation standards.

   b   Funding for scholarships, graduate teaching positions, or other financial support for students.

   c   Facilities, equipment, and materials allowing students to create the practice-based components of their dissertation.

6   **Buy-in.** Do the current faculty, staff, and administrators of the college and the departments involved support the PhD program?

7   **Enrollment.** Can we be reasonably assured we will be able to recruit, retain, and graduate a sufficient number of students?

The program at UCSD meets these criteria, as does the Doctor of Design (DDes) program at Carnegie Mellon University. Carnegie Mellon's coursework begins with a thorough grounding in the research of design, for design, and by design and an understanding of how these intersect with design practice. As students advance through the program, they build their skills as designers, researchers, and scholars of design. Students have the option to create a dissertation alone or a combination dissertation and design project, making an original contribution to knowledge that advances the field of design.[65] CMU is located in Pittsburgh, a city of more than 300,000 people, so although Philadelphia, the largest city in the region, is more than 300 miles away, adequate cultural resources exist near the university. Furthermore, the program maintains a narrow focus on Transition Design and employs an accomplished faculty of internationally recognized experts in this field, reducing the necessity for external evaluators.

The seven questions align with the necessary and sufficient criteria under which a PhD program in many practice-based fields could become possible:

1   **Academics.** Students should receive specific training in research methodologies and scholarly writing in preparation for conducting doctoral-level research and communicating with external audiences. These skills are not just required for the degree – they're crucial to a career in the professoriate. (This is why I feel that a DFA in visual art practice is inadvisable in the US if it culminates only in one or more works of art without a written dissertation.)

2   **Structure.** The program's structure should align with other graduate programs offered by the university. The US Department of Education does not mandate the components of doctoral programs, but more than a century of history and tradition has created expectations that practice-based researchers in the US should not ignore if they wish to be taken seriously as scholars.

3   **Faculty.** Faculty should possess experience in doctoral programs, qualifying them to teach above the master's level and to evaluate doctoral students' scholarship and practice-based achievements.

4   **Location.** Physical access to cultural or disciplinary resources related to students' area(s) of study facilitates the program's capacity to bring field-based experts to campus to support and evaluate the products of its doctoral students' practice. Direct evaluation of the physical artifacts created through practice-based research is crucial.

5 **Resources.** The program must provide access to studio facilities, production laboratories, or other resources appropriate to practice-based researchers' areas of inquiry to support engagement in practice as part of the research process.

6 **Buy-in.** Faculty within the academic unit offering the program must support the program and its students. Covert or overt opposition creates conditions incompatible with success.

7 **Feasibility.** The program need not be large, but a certain level of enrollment is necessary to its feasibility, viability, and sustainability. Dependency on the university's largess would place the program in danger whenever budget cuts become necessary.

Since so few institutions could meet each of these criteria, it might be fortunate that the predicted proliferation of US PhD programs in visual arts practice failed to materialize. Political and economic upheavals, the relentless rush of technological development, and worldwide crises have changed higher education, and our dreams for doctoral programs have necessarily changed along with it.

Should a US PhD in visual arts practice exist? Adamantly, yes, it should, but perhaps not as broadly as once anticipated. Demand for such programs is still small. Few faculty are presently qualified to teach in or administer them, and other key criteria for success are not present nationwide. Unlike the CAA's assertion that "The MFA is *the* terminal degree in studio art practice,"[66] I believe academia can and should allow for scholarly and creative achievement beyond this level, opening wider possibilities for scholarship than have existed before. The MFA remains valuable because it continues to produce faculty members who are experts within a field of art or design practice, prepared to engage within that community and to transmit their expertise in visual arts practice to their students. The PhD, on the other hand, takes the value of the MFA a step further by preparing graduates to engage in the wider community of scholars, continuing their arts practice if they wish but also shaping larger conversations about the scholarship of art through ongoing involvement in formal research. The PhD will not replace the MFA. Rather, it lifts the limitations imposed by the MFA and opens a path for scholarly accomplishment beyond what was previously the terminus of achievement in visual arts practice.

Conclusion

I believe the PhD in visual arts practice will eventually earn broad acceptance in the US, following the lead of doctoral programs in the UK and other parts of the globe. However, we should also be aware that normalization of the degree bears a significant risk of engendering discord within art departments because it could impose yet another level of resentment and competition in already-contentious professional relationships. As enthusiastically as I support practice-based doctoral-level study, I also recognize that we have notable barriers to overcome before this dream can be realized for the visual arts. At every institution I have attended as a student, visited as a researcher, or worked for as a faculty member or administrator, I encountered palpable tension between the faculty in art history or visual studies, most of whom have PhDs, and the visual art faculty, the majority of whom hold MFAs. These resentments take root early in artists' education. For example, when I was a student in an undergraduate art history class at SAIC, the graduate teaching assistant arrogantly informed the captive audience of studio art majors, "*You* make the art; *we'll* tell you what it means." At Tulane, the art faculty summarily dismissed the idea of a doctorate in visual arts practice. And when I was at Texas Tech, where the Fine Arts Doctoral Program has been awarding PhDs for a half-century, the studio art faculty harbored such hostility towards the doctoral program that FADP students were barred from using the studio facilities and were begrudgingly permitted to serve as graduate instructors of studio courses only on the rarest of occasions.

These words of caution might apply anywhere a university is considering a new practice-based doctoral program, proceeding with great caution and measured wisdom until they are prepared to address the challenges posed by practice-based doctoral degrees. Academic administrators leading these efforts must be visionary and diplomatic leaders who inspire confidence, reassuring the practice-based faculty that their skills are still highly valued and their jobs remain secure. They must simultaneously convince the research-based faculty that a practice-based doctoral program will lead to greater respect for the department within the university, explaining that the research produced in the doctoral program will make significant contributions to the discipline at large. Perhaps most importantly, they must be capable of leading their dual faculties in becoming cohesive teams instead of warring tribes – no small task, to be sure.

Despite obstacles to increasing the acceptance and availability of practice-based doctorates, they open exciting opportunities to re-think and re-envision the entire system of higher education. Practice-based doctoral programs could allow us to build on what is good and remedy known flaws and weaknesses. A robust PhD will validate scholarship of, in, through, with, and about practice-based fields, helping to dispel disparities between thinkers and makers or doers. It will also allow practice-based researchers to stand on equal footing with peers in other disciplines – a goal that will be of benefit to everyone, without regard to their position or degree status.

Notes

1 James Elkins published *Artists with PhDs: On the New Doctoral Degree in Studio Art* (2009).
2 Elkins states that he prefers the expression "PhD in studio art" but also acknowledges that many such titles exist including "practice-based doctorate" and other variants (1st ed., p. xvi). His use of the term "PhD in studio art" is inclusive of doctoral programs in which the research output is only a creative work or exhibition of works, those that combine creative and written production, and those in which only a written document such as a dissertation is produced.
3 Elkins (2009, p. xi).
4 Mackh (2011).
5 Mackh (2008).
6 Elkins (2009, p. 170).
7 Mottram (2014, 2nd ed., pp. 40–41).
8 Ibid., pp. 40–41.
9 Ibid., p. 41.
10 According to IRB protocol, interview subjects involved in the Mellon Research Projected remain anonymous and are not identified by institution, only by position titles.
11 As previously mentioned, IRB protocol prohibits identifying interview subjects involved in the Mellon Research Projected by institution or name, referring to them only by position titles.
12 As with the previous quote, the institution and name of the interview subject are not identified in adherence to IRB protocols.
13 Elkins (2014, 2nd ed., p. 323).
14 Scrivener (2002).
15 Candy (2011, p. 54).
16 Ibid., p. 55.
17 Derrida quoted by Burgin (2014, 2nd ed., pp. 92–93).
18 Elkins (2014).
19 Ibid., Chapter 2.
20 Elkins (2014, 2nd ed., p. xvi).
21 Ibid., p. 30.
22 Ibid., p. 17.
23 Ibid., p. 172.
24 Walker (1955, p. 119).
25 Smith, in Elkins (2009, 1st ed., p. 91).
26 Ibid., pp. 90–91.
27 Ratcliffe (2014).
28 Goldsmiths, University of London, 2021.
29 Glasgow School of Art: Doctoral Degrees, 2021.
30 Elkins (2014, 2nd ed., p. 172).

31 Encyclopedia Britannica (1911).
32 Bent (1959, pp. 142, 143).
33 Elkins (2014, 2nd ed., p. 169).
34 NASAD Handbook (2020–21, p. 146).
35 NASAD, National Association of Schools of Art and Design, is the largest US accrediting organization for visual art and design.
36 Texas Tech University, Fine Arts Doctoral Program (2018).
37 Smith in Elkins (2009, 1st ed., p. 90).
38 Ibid., p. 91.
39 Ibid., p. 93.
40 Elkins (2014, 2nd ed., p. 170).
41 Ibid., pp. 131–132.
42 Ibid. (Elkins, "Reason 6" to "mistrust the PhD," 2nd ed., pp. 232–233).
43 Ibid., p. 312.
44 Ibid., p. 472.
45 Ibid., p. xii.
46 US Census Bureau (2020).
47 Elkins (2014, 2nd ed., p. 146).
48 Ibid., p. 473.
49 Carroll (2009).
50 Candy; see also Skains (2017).
51 CAA Statement on Terminal Degrees (2008).
52 Gallup (2018).
53 Elkins (2014, 2nd ed., p. 173).
54 Ibid., p. 469.
55 Ibid., p. 109.
56 Ibid., p. 343.
57 Wilson (2014, p. 346).
58 Ibid., p. 348.
59 National Science Foundation (2020).
60 Ibid.
61 Ibid.
62 University of California at San Diego (2020).
63 NASAD Handbook, (2020–21, pp. 147–148).
64 Ibid., p. 145.
65 Carnegie Mellon University School of Design (2020).
66 CAA (2015, emphasis added).

Bibliography

Bent, H. (1959). Professionalization of the PhD Degree. *The Journal of Higher Education*, 30(3), 140–145. www.jstor.org/stable/1978286 (accessed July 14, 2009).

Burgin, V. (2014). Thoughts on "Research" Degrees in Visual Arts Departments. In J. Elkins (Ed.), *Artists with PhDs: On the New Doctoral Degree in Studio Art*. Washington, DC: New Academia Publishing, 2nd ed., 92–93.

CAA. (2008, updated 2015). *Statement on Terminal Degrees*. https://www.collegeart.org/standards-and-guidelines/guidelines/terminal-degree-programs

Candy, L. (2011). Research and Creative Practice. In L. Candy & E. A. Edmonds (Eds.), *Interacting: Art, Research, and the Creative Practitioner*. Faringdon: Libri Publishing Ltd., 33–59.

Carnegie Mellon University School of Design. (2020). *PhD in Transition Design*. https://design.cmu.edu/content/phd.

Carroll, N. (2009). Les cul-de-sac of Enlightenment Aesthetics: A Metaphilosophy. *Metaphilosophy*, 40, 157–178.

College Art Association. (2015). *Statement on Terminal Degree Programs in the Visual Arts and Design*. www.collegeart.org/standards-and-guidelines/guidelines/terminal-degree-programs (accessed November 19, 2019).

Elkins, J. (Ed.). (2009). *Artists with PhDs: On the New Doctoral Degree in Studio Art*. Washington, DC: New Academia Publishing, 1st ed.

Elkins, J. (Ed.). (2014). *Artists with PhDs: On the New Doctoral Degree in Studio Art*. Washington, DC: New Academia Publishing, 2nd ed.

Encyclopedia Britannica. (1911). *Professor.* www.1911encyclopedia.org/Professor (accessed July 29, 2009).

Glasgow School of Art: Doctoral Degrees. (2021). www.gsa.ac.uk/study/doctoral-degrees/doctoral-study/ (accessed February 28, 2021).

Goldsmiths, University of London. MPhil/PhD Art. www.gold.ac.uk/pg/mphil-phd-art/ (accessed February 28, 2021).

Mackh, B. (2008). *No Direction Home: Images from New Orleans* [art exhibition]. Tulane: Tulane University MFA Exhibition, Newcomb Art Museum.

Mackh, B. (2011). *The Documentary Aesthetic.* Doctoral dissertation. Texas Tech University Electronic Theses and Dissertations. http://hdl.handle.net/2346/45222.

Mottram, J. (2014). Researching the PhD in Art and Design: What Is It, and Why Do a PhD in Art and Design In J. Elkins (Ed.), *Artists with PhDs: On the New Doctoral Degree in Studio Art.* Washington, DC: New Academia Publishing, 2nd ed.

NASAD Handbook. (2020–2021). Reston, VA: National Association of Schools of Art and Design.

National Science Foundation, National Science Board, National Center for Science and Engineering Statistics. (2020). *Table 9. Higher Education R&D Expenditures, by R&D Field: FYs 2009–18,* January. https://ncses.nsf.gov/pubs/nsb20202/academic-r-d-in-the-united-states.

Ratcliffe, R. (2014). REF 2014: Why Is It Such a Big Deal. *The Guardian,* December 17. www.theguardian.com/higher-education-network/2014/dec/17/ref-2014-why-is-it-such-a-big-deal (accessed November 19, 2019).

Scrivener, S. (2002). *The Art Object Does Not Embody a Form of Knowledge.* Working Papers in Art and Design 2. http://sitem.herts.ac.uk/artdes_research/papers/wpades/vol2/scrivenerfull.html; ISSN:1466-4917.

Skains, L. (2017). The Practitioner Model of Creative Cognition: A Potential Model for Creative Practice-Based Research, Part 1. *The Disrupted Journal of Medial Practice.* In Scalar: The Alliance for Visual Networking Culture, March 31. https://scalar.usc.edu/works/creative-practice-research/what-is-pbr (accessed February 28, 2021).

Strada-Gallup. (2018). *Why Higher Ed? Top Reasons US Consumers Choose Their Educational Pathways,* January. www.stradaeducation.org/report/why-higher-ed/#:~:text=Work%20outcomes%20are%20the%20main,four%2Dyear%20colleges%20and%20universities.

Texas Tech University, Fine Arts Doctoral Program. (2018). *Basic Requirements.* www.depts.ttu.edu/fadp/program/basicrequirements.php (accessed November 19, 2019).

University of California at San Diego General Catalog 2020–21. (2020), *Interim Update. Visual Arts PhD Program,* November 4. www.ucsd.edu/catalog/curric/VIS-gr.html.

US Census Bureau. (2020). *Educational Attainment in the United States: 2018.* www.census.gov/data/tables/2018/demo/education-attainment/cps-detailed-tables.html.

Walker, L. (1955). The Studio Artist and the Art Historian. *College Art Journal,* 15(2), 119–123. DOI:10.2307/772936

Wilson, M. (2014). Between Apparatus and Ethos: On Building a Research Pedagogy in the Arts. In J. Elkins (Ed.), *Artists with PhDs: On the New Doctoral Degree in Studio Art.* Washington, DC: New Academia Publishing, 2nd ed., 341–359.

1.10

THE RELATIONSHIP BETWEEN PRACTICE AND RESEARCH

Gavin Sade

Practice and research

For communities of researchers, of creative practitioners, a significant challenge as it remains today is how through practices we are able to make a material difference. In the face of a changing world, where there is a 'diminishing richness and vitality of life on Earth', Gablick asks 'whether art will rise to the occasion and make itself useful to all that is going on'.[2] Frayling described research in the creative arts and design, questioning stereotypes and outlining how artists and designers can contribute to new knowledge through their creative practices.[3] He describes research as addressing questions related to the aesthetic or perceptual aspects of creative work or theoretical perspectives on art and design. In combination, these two voices, from different communities and parts of the world, signal both a call for creative practitioners to engage in the urgent problems human society faces and a way to contribute through research.

Now, close to 30 years later, there is an increased sense of urgency to address the social, economic and environmental problems that face human society, brought to the fore by a global pandemic. Practice-based research has matured and become one of the avenues for practitioners to respond to such urgent problems and at the same time make original contributions to knowledge, whether visual and performing artists, designers, filmmakers, architects or musicians. However, creative practitioners do this differently than researchers in science and social sciences. They often employ creative methods that make the familiar unfamiliar and disrupt the status quo that allows us to explore alternate imaginaries and re-envisage our future. As a result, practice-based researchers within the creative fields can find themselves caught between the norms of academic research and their respective field. This presents unique challenges for the practice-based researcher, which are played out in the way each individual positions practice within the context of research.

Often artists feel compelled to rationalise subjective judgements in order to meet the demands of incompatible research models,[4] but to do this, as Cater argues, fails to understand the practices of research and, equally, the nature of artistic practice.[5] Where creative practice is transformed into a research method for collecting data in pursuit of an instrumentally posed question, it is no longer mobilised by artistic or creative intentions. On the other hand, as research is transformed into creative practice, there is a risk it will no longer secure claims to knowledge as expected within an academic context. These issues, identified by artists as they engaged in research within academic contexts, motivated communities of researchers to seek alternative frames of reference and to develop original approaches that navigate the movement between practice and research. There are

 DOI: 10.4324/9780429324154-12

different terms used for the resulting approaches to research, from practice-led research to artistic research, each developed by communities of researchers and practitioners within different disciplinary or intuitional settings. For clarity, the chapter will use the term practice-based research as an umbrella term to refer to all research forms that share this focus.

Over the last 30 years, practice-based researchers have established a place within the broader academy at a time where there has been an increasing recognition that solutions are not found within singular disciplinary fields, and that new modes of investigation are required to address the complex problems of our time. Artists and designers have turned to practice-based research as a preferred methodology due to the way it recognises the value of their specific methods of knowledge production and provides a framework for employing research methods from beyond the field of practice.

What makes practice-based research significant is how it places the practitioner at the centre of research. It recognises the practitioner's voice and provides a framework that supports claims to new knowledge resulting from practice and their creative outcomes. Thus, practice-based research positioning opens ways for practitioners to engage and apply their capabilities and working methods to address urgent problems of our time in a way that is recognised as contributing to the stock of human knowledge and the development of scholarly and academic fields. Conversely, research that employs practice has opened practice to new forms of interrogation within the context of academic scholarly inquiry and produced outcomes which transform practice itself.

However, as a new approach to research, relatively speaking, practice-based researchers face several challenges in establishing and maintaining this voice. In part, these challenges lie in the tensions between practice and research and how researchers navigate the movement between the established practices of research and the practices of their respective professional fields. Today the question is not about establishing practice-based research as a recognised field of research or arguing for its legitimacy. Practice-based approaches provide the practitioner researcher with the epistemological and ontological ground to support their expressions of new knowledge and claims to significance beyond the immediate concerns of practice.

This chapter sets out to consider approaches that have enabled practitioners to move from the immediate concerns framed around a specific practice, the studio, the creative outcomes, to develop studies which hold the potential for broader impact. To do so, the chapter begins with a brief introduction to the theory of practice and then discusses what it means to consider practice-based research as a new *paradigm*. Following this, the chapter considers the movement between paradigm and strategy, and the tension between the practice of academic research and those of the specific field of practice. The chapter concludes by discussing key questions as to how the specific approaches to research that practice-based researchers adopt are designed to support claims to new knowledge.

Why practice?

Practice is a diverse concept that has been the focus of philosophers and sociologists such as Giddens, Foucault, de Certeau and Bourdieu, who explore the relationships between human practices, self and societal structures. Practice has become more than an object of study, a strategy employed within research, where investigation is based in practice, employing the methods of practice, and research questions arise through the process of practice. For artists and designers, the significance is in a focus on practice that recognises practice as a critical context for asking questions and draws upon practitioners' embodied and imaginative abilities to find new problems to address. In this way, practice involves a questioning of the nature of practice through cycles of reflection and action.

The potential of practice-based research resides in an understanding of practice as a critical praxis, a hermeneutic interrogation of our practices and their relationship to the world. One that recognises the way our practices, those assumed as 'natural' or habitual, are implicated in the problems we are addressing. Fry argues that practice can be redirected to address the problem it is implicated in perpetuating through what he refers to as ontological designing.[6] For the practitioner, research provides one way of redirecting practice. This is because research directly engages in what can be described as ontology, asking questions about reality, existence and being. New knowledge resulting from research, in many instances, brings into question, challenges and changes our understanding of our lifeworlds, society or what it means to be human. This is how practice-based researchers can address the challenges of our time. However, to do so involves exploring the relationship between practice and research. Specifically, how research methodologies are developed to reveal the nature of our practices and how to present these in a way that recognises and holds up practice for critical analysis, revealing the economies of practice and producing alternative accounts of how practice can generate new knowledge and lead to meaningful change.

Schön makes an important observation:

> The problems of real-world practice do not present themselves to practitioners as well-formed structures. Indeed, they tend not to present themselves as problems at all but as messy, indeterminate situations.[7]

Through this discussion, Schön claims that a practitioner's ability does not depend on being able to describe what they know how to do, or to entertain in conscious thought the knowledge their actions reveal.[8] Practice-based researchers, especially those from the creative arts and design, draw upon Schön's concept of reflective practice to develop methods for reflection and to support the emergence of theory from practice. However, practice and reflection are both structured activities, shaped by habitual modes of thought and action.

Broadly, practice (both professional and research) can be considered as drawing on a silent reserve of activities and procedures that organise spaces and languages shaping both thought and action,[9] yet are often occluded by the narratives we construct about practice. Descriptions of practice are framed by research questions and often limited by our horizon of concern, for example, the phenomenology of the studio experience, the aesthetics of our creations or the immediate social relationships figured around practice and artefact. Artists and designers engaging in practice-based research over time have begun to move beyond simple descriptions of practice, which reinforce an assumed order, and instead critically engage with the broader economies of practice. In other words, practice-based researchers not only consider practice as methods for the conduct of research but also have developed ways to interrogate what Fry calls the structuring situations out of which practitioners, and their agency as active subjects, come into being. However, contradictions and problems associated with practice-based research often flow from the way practice is conceptualised within the context of research, and the resulting decisions made about what is and is not practice.

Practice itself can be a critical context for inquiry, and Sennett shows that when technical abilities reach higher levels, the focus of the practitioner shifts.[10] This shift, he suggests, is grounded in the care practitioners have for the qualities of the things they make, which leads to questioning and a desire to understand how practices generate social, political and ethical values. In this way, practice can be understood as inquiry led, where practitioners not only solve problems but also identify new problems and ask questions about the relationship between practice, its outcomes and the world. This inquiry-led nature of practice, of practitioners, could be considered as one of the motivations for a practitioner to engage in academic research. Where practice becomes a context for critical inquiry, it enables practitioner researchers to advance their practices and, in turn, to address the way practice and its outcomes shape and are shaped by ourselves and the world.

This relationship between practice and questioning, actions of the hand and thought, is central to any discussion about practice-based research. For research undertaken in or through practice, this is something that needs to be explicitly addressed. Not as a justification of the choice or mode of research, but to establish the philosophical position from which the study has been designed and conducted, and how this supports methodological decisions and underpins claims to knowledge. This provides a starting point for addressing the question of practice in the context of research methodology.

Practice-based methodology

While practice-based research is employed in a wider range of fields, including health, education and business, it has been in the creative arts where communities of researchers have taken up practice-based research and advanced it as a preferred methodology. Over the past decades, a growing body of literature has led to the development of a broader international community of scholarship and a sustained scholarly debate. In turn, this can be seen as developing a new shared language and methodologies for the conduct of practice-based research, and a recognition of the way practice and its outcomes produce new knowledge. Through the literature there is a thread, or common interest, in the philosophical implications raised by practice-based research – in respect to both the focus on practice and the turn to artistic methods.[11]

Artists can see practice-based research as a critical intervention where the resulting outcomes gain significance through the ways they rupture forms and significations circulating in the social field.[12] It is from this perspective that practice-based research is described as a new species of research,[13] one that has evolved from qualitative methods employed within the humanities and social sciences. Practice-based approaches have been positioned with respect to positivist, constructionist, critical and participatory paradigms of qualitative research.[14] Leavy describes arts-based research as one of the five approaches to research, aligned with theoretical schools of thought associated with phenomenology and embodiment.[15] As a new paradigm, forms of practice-based research are commonly described in contrast to positivist and scientific paradigms, which invert the dominant academic hierarchy of knowledge[16] and transform the nature of research through distinctive aesthetic, sensory, affective and material methodologies.[17]

Practice-based researchers have often found themselves between two world views or habitus: that of academic research and that of the specific practices within a field. This has been described as 'an ironic cleavage established between self-conscious "researchers" and the vast majority of creative practitioners'[18] and has led to concerns about the way academic research can distort practice. To address this tension, communities of researchers have drawn on creative practices to develop new methodological approaches and ways of understanding research. In doing so, they highlight the importance of understanding artistic modes of knowledge production and the resulting artistic expressions as more than simply objects of study, phenomena to be quantified or instruments to facilitate data collection. They argue that for

> research to deal adequately with human society, it needs to embrace those aspects of knowledge production that deal with human subjectivity and relationships, not as phenomena to be deduced and re-harnessed within human control, but open-endedly, as part of a process of creative construction and interpretation that is relative, specific to context and value-driven.[19]

Practice-based research: a new paradigm?

A paradigm can be considered as the way a researcher answers a series of philosophical questions. Questions such as, 'what constitutes the objects of reality and what is being (ontology)?' and 'what is knowledge and what does it mean to know something (epistemology)?' How a researcher answers these questions is important, as it establishes a 'starting point or givens that determines what inquiry is and how it is practiced'[20] and shapes methodology. Paradigm shifts are described as occurring as a result of recognising the limits to existing ways of knowing,[21] where new vocabularies, fresh assumptions and different explanations and interpretations replace and reconfigure the older ways of understanding the world.[22] Consequently, considering practice-based research as a paradigm shift that results from the limits of ways of knowing is very different to seeing it as emerging from an 'audit culture'[23] or as a result of the movement of art and design colleges into the university. These origins have been contributing factors to the development of the field and scholarly discourse. However, they do not account for the variety of different approaches to practice-based research, nor its continued application over the past decades across a range of disciplines.

Whether one considers practice-based research as a new paradigm or not, it is the capacity of its various forms to generate new possibilities for thought and action that has been critical in the development and uptake of it as a methodological approach. This has been most significant, and contested, in artistic or creative fields where practitioner researchers have developed approaches to research that have recognised the 'world-constituting and world-revealing power of art',[24] which Bennett (2013) argues is 'fundamental to any understanding of the connections between lifeworlds, disciplinary procedures and given problems'.[25] However, practice-based research has also incorporated features of other approaches to research, from scientific to research-orientated humanistic modes of investigation,[26] thus adopting and employing a mix of approaches and theoretical positions drawn from qualitative research more broadly. So, while in some communities specific instances of practice-based research can be seen through the lens of an 'artistic turn', not all view practice-based research as a paradigm shift from a Kuhnian perspective, which argues against relativism and is grounded in a belief in scientific progress.[27] From this perspective, it may be more useful to consider practice-based research as pluralistic or multi-paradigmatic.

The knowledge produced through research, whether by a scientist or artist, is highly situated within specific material, social and cultural contexts. Similarly, researchers are not neutral or purely objective observers but participate in a 'system of symbols and narratives that shape the culture' within their specific field of practice.[28] Haraway (1988) addressed the problems this presents by suggesting that researchers need to understand, and make visible, the contingency of the position from which they make claims to knowledge. This provides some guidance for reconciling the tensions and challenges that arise as practitioners engage in research. By recognising the situated nature of knowledge production and engaging in what might be thought of as a systematic interaction of differences, practice-based researchers are able to establish a position from which they can make claims. Studies of interdisciplinary research collaborations discern a similar interplay, which has been described as agonistic and orientated towards effecting ontological change.[29] Thus, to think about practice-based research as a new paradigm, in opposition to other paradigms, does not recognise its pluralistic origins and how it has influenced broader research practices. Importantly, it is such interaction of differences that has led practice-based research to 're-conceive both the object(s) of research and the relations between research subjects and objects'.[30] With this in mind, the next section will discuss methodological concerns and how practice-based researchers have addressed these.

Methodological considerations

Creative arts and design do not have a unique position with respect to the focus on practice within the research. Other fields from health, business and education engage in research about, through and for practice. In all of these areas, what distinguishes research from practice as it occurs outside the context of research is methodology. Methodology is more than just methods or a description of steps and stages. Methodology, as an 'ology', is the science and study of methods. Methodologies are continually critiqued by researchers and change over time. It is the 'contestation of methods'[31] that mobilises communities of researchers to interrogate and improve the practices of research. As a methodology, it is a mix of methods from respective professional practice and qualitative research with a linage in social science, drawing on methods developed in fields as diverse as anthropology, ethnography, psychology, media communications and technology studies. Various approaches to practice-based research draw on a mix of pragmatism, theories of experiential learning, social constructivism, phenomenology and aesthetics. Specific methods employed by practice-based researchers range from reflective techniques drawing on Schön to participatory and emancipatory forms of action research. There is no singular formula for practice-based research, and different communities and researchers have developed unique approaches in respect to specific contexts. Thus, any reading of practice-based research reveals a field which is continually reframing problems, generating new ideas and exploring different modes of representing knowledge and alternate ways of thinking and being.

Practice in itself does not constitute research, especially as it is defined within a university setting, and this is a question that all practice-based researchers inevitably address as they articulate a specific methodology. Denzin and Lincoln (2011) describe qualitative research as 'a situated activity that locates the researcher within the world', involving a collection of 'interpretive and material practices that make the world visible'.[32] This emphasis on making things visible is seen in the literature on practice-based research. For example, Bolt (2004) refers to practice-based research as 'materialising practices': as a dialectical relationship between artistic practice and research. Just as qualitative research is considered a series of interpretive practices, with communities of researchers developing shared methodologies that allow them to advance knowledge within a specific field, there is a wide range of approaches within practice-based research. Multiple disciplines from visual and performing arts to design and architecture employ practice within the context of research. Similarly, there are different models for PhD studies between universities and regions.[33] Across each of the variants, there is a shared focus on methodology, and especially the relationship between practice, research and creative artefact. This has led to the realisation that research insights often arise as a result of the tensions that occur at the intersection of research and practice, whereas McNamara (2012) suggests needs do not correlate.[34] Similarly, developments of practice-based approaches have in part been motivated by consideration of where the metaphors of practice as research, and research of practice, break down.

In the context of a specific research project or study, each research navigates the movement between practice and research and draws connection between the two which are designed to support their specific focus. This involves describing specific interpretive and material practices employed and how these are suitable for addressing the problem, supporting argumentation and claims to an original contribution to knowledge. This involves making a series of decisions about how the research will be conducted, often referred to as *research design*. Each decision needs to make sense in respect to the problem being addressed, and provide a foundation to support claims to knowledge. As such, methodology plays an essential role in shaping a study, determining what can be asked and how claims to knowledge can be made.

The use of the term design here is worth considering. Designers engaging in practice-based research commonly drawn on a body of design theory which interrogates what it means to 'design', and recognises that design is a subject-decentred practice. In the context of practice-based research, this is influential in the way it turns attention to the act of research design and the agency of research methods. Thus, drawing attention to the way research methods are consciously shaped and how this influences the specific *practice* within which research is based. Nelson (2009) describes research design as 'matching the means and intentions of research'[35] and shows that research design is just as critical for researchers in the creative arts as it is for researchers in other fields, even if their means and intentions are very different. In other words, methodology, questions and outcomes (whether research texts or creative works) are contingent upon each other. This is no different for a scientist within a laboratory setting or an artist working in the studio or social setting. All are engaged in an ongoing study and critique of methodology, both the material practice and the school of thought which guide their actions, and it is this feature of research which provides it with a means to redirect practice through an ontological re-designing.

While we have introduced what it means for practice-based research to be considered a paradigm, this remains a contested proposition. PhDs that focus on practice are not all framed within a practice-based research paradigm, but are often undertaken within existing interpretive paradigms or are multi-paradigmatic. Studies may equally draw upon other theoretical perspectives: for example, constructivist, feminism, critical theory, cultural studies, queer theory, post-colonialism etc. are all theoretical perspectives that might frame interpretation, guide thinking and writing and inform creative practice. Different theoretical or interpretive paradigms come with assumptions in respect to ontology and epistemology, as well as preferences for specific research methods, criteria for evaluation and the typical form statements take when making claims to knowledge. While practice-based research as a paradigm may have a grounding in other theoretical schools of thought, for example, aesthetics, phenomenology and embodiment, there are two aspects which in combination make it unique in the academy. The first is a focus on research in and through creative practice,[36] which by itself is not unique to practice-based research. The second feature, which in combination with practice makes the paradigm distinct, is the view that research outcomes can include outcomes from practice. Non-traditional outcomes such as artworks, performance or architecture, for example, are considered expressions of new knowledge in and of themselves, presented for examination alongside a written thesis or exegesis.

Research strategies

While many approaches that fall under the umbrella of practice-based research describe how the specific methods of practice are mobilised within the context of research, the range of disciplines that employ practice-based research and the variety of approaches suggest it would be useful to consider it as a *research strategy* or family of related strategies. From this perspective, practice-based research can be viewed as a bridge between a paradigm and the activities and methods of research. Similarly, action research, participatory design, case study, ethnography, grounded theory and ethnomethodology can all be considered research strategies. Different research strategies are employed by communities of researchers to provide them with ways of 'implement[ing] and anchor[ing] paradigms in specific empirical sites or specific methodological practices'.[37] In this way, research strategies bring with them a web of connections to different schools of thought, preceding literature and research undertakings. Thus, within a historical context, research strategies can be understood as being developed and refined by researchers to enable them to address problems of their time, providing new ways of grounding changing world views in the practices of research. For example, practice-based research undertaken within the creative arts has developed methods that Bennett describes as acknowledging artists' concerns for aesthetics as 'fundamental to any understanding of the connections between lifeworlds, disciplinary procedures and given problems'.[38] Practice-based research in other fields will draw upon different values, world views and schools of thought in a similar way, yet still focus on practice.

Describing practice-based research as a research strategy, or family of related strategies, distinguishes between the choice to conduct research in or through practice and actual practice. In other words, research strategies do not describe the specific methods of practice, data gathering or interpretation but rather outline the broader approaches that researchers employ as they move between interpretive paradigms and practice. That said, a specific research strategy may have a range of preferred methods that are consistently employed. These specific *methods*, such as observation, interviewing, textual analysis, and so forth, are the operational day-to-day research activities. In the context of practice-based research, methods have expanded to include the practice within a specific field or discipline. These activities result in *data* – or in the context of artistic research, creative outcomes – and are presented and analysed through the lens of an interpretive paradigm or stand as expressions of new knowledge themselves. Situating practice-based research as this bridge between paradigms and practical methods of research provides an adaptability that has been employed across a range of fields, from architecture to visual arts, each with different working methods. Equally, this enables researchers from across a broader range of disciplines to employ the strategy and contribute to its development. In each of these settings, researchers make the strategic decision to conduct research in and through practice. They develop ways to bridge their individual practice and practices in their field with ways of thinking about the world.

For many practice-based researchers the choice of practice is often made first, as they have come to research through their practices. However, it is the other decisions which shape research. In practice-based research, it is the choice of how to represent research outcomes, new knowledge, that has been most contested. Within the literature, there is commonly a distinction made between two variations of practice-based research, often described as practice-led and practice-based,[39] or, for Smith and Dean, practice-led research and research-led practice.[40] There are nuances, and in places contradiction, to the definitions; however, it is the way research outcomes are presented which is the salient difference. On one hand, through practice a researcher might produce qualitative and/or qualitative data that is evaluated to come to a set of conclusions, to support claims to new knowledge, and is expressed in traditional research publications. On the other, practice might be viewed as a site for the creative generation of new ideas that are represented in the form of a non-traditional research outcome, with accompanying exegetical publications. While very simplified, these are illustrative of two very different approaches; what is important is the realisation that there is no one right answer. Instead, it is the task of the practitioner researcher to articulate their methodology, one that is appropriate for addressing the specific question or problem at hand.

Some approaches to practice-based research, however, have problematised conceptions of research design and methodology. In some studies, the products of practice are more akin to research methods employed to generate 'data' that can be evaluated to support claims to knowledge. In other studies, the products of practice stand as expressions of new knowledge in and of themselves. This is an important distinction to understand and is at the core of concerns regarding the instrumentalisation of art within research and the misunderstanding of research by artists. Here Sullivan suggests that it is important to for a research methodology to describe how 'data' is 'translated into interpretive forms able to be communicated to others'.[41] Traditional publications are embedded within the practices of scholarly knowledge dissemination. Such academic writing is shaped by the way different disciplines engage in argumentation, critique and analysis in written form. To view non-traditional outcomes as 'interpretive forms' or 'assessable materials of research' is to recognise that research and the resulting new knowledge can be represented and communicated in a variety of ways. However, it is a mistake to think that these are directly analogous to traditional research outcomes. A non-traditional outcome is not merely illustrative of a theory or a research instrument that enables data collection and analysis methods. Instead, it represents thinking in another form and needs to stand as a defensible expression of new knowledge communicated in a shared language: a shared language other than the text of a traditional thesis, expressed through the creative form of a non-traditional outcome, such as music, dance, visual image, or architectural design.

Traditional and non-traditional outcomes differ not just in form, but also in how each can be understood as communicating knowledge. Non-traditional research outcomes do not follow the same logics and structures for presenting research as traditional publications. The most salient difference is that such epistemic objects are not closed texts. Instead, they are understood as holding 'capacity to unfold indefinitely' and are always in the 'process of being materially defined'.[42] Consequently, non-traditional outcomes do not necessarily communicate research outcomes didactically as might a traditional research publication. While this open nature might suggest that an outcome is somehow incomplete, it is problematic to consider this feature as an issue that needs to be resolved by employing other methods to empirically prove the efficacy of non-traditional outcomes.

Practice-based researchers commonly addresses this concern about the nature of non-traditional research outputs through the way they approach accompanying the exegesis, thesis or traditional research publication. Bakhshi et al. identified that, over time, the way practice-based researchers have engaged in writing has shifted from documentation of the research process towards approaches that involve more philosophical analysis and critical reflection.[43] This is movement from highly situated and contextual accounts of practitioners to writing that engages with contingencies of knowledge production. Here the important point is that writing provides practitioners with a way of thinking, and that this is not the same as the generative schemes thought, those of practice, that it claims to report upon.

To guide thinking about the relationship between the traditional and non-traditional research outcomes, practitioners have looked to beyond the academy. Artistic researchers, for example, have drawn upon 'art's own established rhetorical genres of textual practice',[44] recognising that many artists produce 'second manifestations of creative thinking' that go beyond 'idiosyncratic individual modes of production'.[45] Sullivan describes this movement between practice and writing as a way of 'making space' and 'central to theorising practice *as* research'.[46] Here writing becomes an essential part of practice-based research. It is where the different ways of thinking and world views of the practitioner and the researcher come together. This dialogue finds focus in writing that communicates a study's methodology, which necessarily engages in current debates about methodology, and in the process makes explicit the research design. In doing so, practice-based researchers are able to secure and share new understandings developed through practice, and their methodological innovations, in a way that makes a contribution to the ongoing development of their field of practice and the broader endeavours of research.

Concluding thoughts

What makes practice-based research significant is how it situates practice and practitioners at the centre of research by asking them to interrogate the ontological nature of their practices and resulting expressions, as both structured and structuring. The artistic turn in research exemplified in communities of practice-based researchers has seen creative practices and research not as antithetical, but as yet another approach to registering and responding to urgent problems or our time: an approach where the thinking that occurs through practice, and the resulting epistemic objects, are recognised as alternate ways of creating and sharing new knowledge. This differs from practices in other fields and disciplines, specifically in the way knowledge is represented in poetic, metaphoric and aesthetic figures. Such ambiguous forms of expression are an important strategy in estranging the habitual, enabling a critical reflection upon the habits of mind and body and the ways practice is structured by the habitus from which it arises. This approach to research suggests a new range of situated epistemological and ontological positions, and a necessary methodological disposition for those who are concerned about the ways both practice and expression present possibilities for thought.

The relationship between practice and research has been a focus for communities of practice-based researchers. Asking what it means to consider practice-based research as a new *paradigm* highlights the differences between previous approaches to research. As practice-based researchers developed new methodological approaches, they addressed several concerns which arose from the movement between practice and research – specifically, how practitioners position themselves and the resulting original works in the context of research. Irrespective of specific decisions, it is only through consciously designing a studies research methodology that it is possible to mobilise the differences between *practices* – the practices of research and those of a specific field of discipline – towards addressing a problem and securing claims to knowledge, perhaps even to 'ris[ing] to the occasion' and contributing to halting the 'diminishing richness and vitality of life on Earth', as Gablick asks.

Notes

1 Coessens et al. (2009).
2 Gablik (1993, p. 74).
3 Frayling (1993).
4 Jewesbury (2009).
5 Cater (2004, p. 10).
6 Fry (1999, 2009).
7 Schön (1987, p. 15).
8 Schön (1987, p. 22).
9 de Certeau (1984).
10 Sennet (2008).
11 Bolt and Barrett (2007), Nelson (2009), Coessens et al. (2009, p. 13), Borgdorf (2012a).
12 Guattari (1995, p. 130).
13 Bolt and Barrett (2007).
14 Gray and Malins (2004).
15 Leavy (2017).
16 Schön (1987).
17 Leavy (2017).
18 Butt (2017, p. 81).
19 Coessens et al. (2009, p. 176).
20 Guba (1990, p. 18).
21 Kuhn (1970).
22 Coessens et al. (2009, p. 13).
23 Cazeaux (2017).
24 Borgdorf (2012b).
25 Bennett (2013, p. 2).
26 Bakhshi et al. (2008, p. 17).
27 Kuhn (1970).
28 Wilson (2002, p. 9).
29 Born and Barry (2010).
30 Barry et al. (2008, p. 25).
31 Nelson (2009, p. 100).
32 Denzin and Lincoln (2011, p. 6).
33 Elkins (2013).
34 McNamara (2012).
35 Nelson (2009, p. 103).
36 Wilson and van Ruin (2013, p. 147).
37 Denzin and Lincoln (2011, p. 14).
38 Bennett (2013).
39 Candy and Edmonds (2018), Haseman (2007).
40 Smith and Dean (2009).
41 Sullivan (2009a).
42 Cetina (2005).
43 Bakhshi et al. (2008, p. 17).
44 Barker (2007).
45 Coessens et al. (2009, p. 23).
46 Sullivan (2009b).

Bibliography

Bakhshi, H., Schneider, P., & Walker, C. (2008). *Arts and Humanities Research and Innovation*. London: Arts and Humanities Research Council.

Barker, K. (2007). The UK Research Assessment Exercise: The Evolution of a National Research Evaluation System. *Research Evaluation*, 16(1), 3–12, March.

Barry, A., Born, G., & Weszkalnys, G. (2008). Logics of Interdisciplinarity. *Economy and Society*, 37(1), 20–49.

Bennett, J. (2013). What Is Experimental Art. *Studies in Material Thinking*, 8. www.materialthinking.org/papers/88 (accessed September 8, 2020).

Bolt, B. (2004). *Art Beyond Representation: The Performative Power of the Image*. New York: I.B Tauris & Co Ltd.

Bolt, B., & Barrett, E. (2007). *Practice as Research: Approaches to Creative Arts Enquiry*. New York: I.B Tauris & Co Ltd.

Borgdorf, H. (2012a). *The Conflict of the Faculties: Perspectives on Artistic Research and Academia*. Amsterdam: Leiden University Press.

Borgdorff, H. (2012b). The Production of Knowledge in Artistic Research. In M. Biggs & H. Karlsson (Eds.), *The Routledge Companion to Research in the Arts*. London: Routledge.

Borgdorf, H. (2013). Artistic Practices and Epistemic Things. In M. Schwab (Ed.), *Experimental Systems: Future Knowledge in Artistic Research*. Leuven: Leuven University Press.

Born, G., & Barry, A. (2010). Art-Science. *Journal of Cultural Economy*, 3(1), 103–119. doi:10.1080/1753035 1003617610.

Bourdieu, P. (1990). *The Logics of Practice*. Stanford: Stanford University Press.

Butt, D. (2017). *Artistic Research in the Future Academy*. Bristol: Intellect.

Candy, L., & Edmonds, E. (2018). Practice-Based Research in the Creative Arts: Foundations and Futures from the Front Line. *Leonardo*, 51, 63–69. doi:10.1162/LEON_a_01471.

Carter, P. (2004). *Material Thinking: The Theory and Practice of Creative Research*. Carlton: Melbourne University Publishing.

Cazeaux, C. (2017). *Art, Research, Philosophy*. Abingdon and Oxon: Routledge, an Imprint of the Taylor & Francis Group.

Certeau, M. (1984). *The Practice of Everyday Life*. Stanford, CA: University of California Press.

Cetina, K. (2005). Objectual Practice. In T. Schatzki, K. Cetina, & E. Von Savigny (Eds.), *The Practice Turn in Contemporary Theory*. New York: Routledge.

Coessens, K., Crispin, D., & Douglas, A. (2009). *The Artistic Turn: A Manifesto*. Leune: Leune University Press.

Denzin, N., & Lincoln, Y. (2011). *The Sage Handbook of Qualitative Research*. London: Sage, 4th ed.

Elkins, J. (2013). Six Cultures of the PhD. In M. Wilson & S. van Ruiten (Eds.), *SHARE, Handbook for Artistic Research*. Amsterdam: ELIA European League of Institutes of the Arts. www.sharenetwork.eu/resources/share-handbook.

Frayling, C. (1993). *Research in Art and Design*. Royal College of Art Research Papers Series 1.1. London: Royal College of Art.

Fry, T. (1999). *A New Design Philosophy: An Introduction to Defuturing*. Sydney: UNSW Press.

Fry, T. (2009). *Design Futuring: Sustainability, Ethics and New Practice*. Sydney: University of New South Wales Press.

Gablick, S. (1993). The Reenchantment of Art. In W. Kelly (Ed.), *Art and Humanist Ideals: Contemporary Perspectives*. Victoria, Australia: Macmillan Art Publishing.

Gray, C., & Malins, J. (2004). *Visualising Research: A Guide to the Research Process in Art and Design*. Aldershot, Hants and Burlington, VT: Ashgate.

Guattari, F. L. (1995). *Chaosmosis: An Ethico-Aesthetic Paradigm*. Bloomington: Indiana University Press.

Guba, E. G. (1990). *The Paradigm Dialog*. Newbury Park: Sage Publications.

Haraway, D. (1988). Situated Knowledges: The Science Question in Feminism and the Privilege of Partial Perspective. *Feminist Studies*, 14(3), 595–596.

Haseman, B. (2006). A Manifesto for Performative Research. *Media International Australia Incorporating Culture and Policy*, 118, 98–106, February.

Haseman, B. (2007). Rupture and Recognition: Identifying the Performative Research Paradigm. In B. Bolt & E. Barrett (Eds.), *Practice as Research: Approaches to Creative Arts Enquiry*. New York: I.B. Tauris Publisher, 147–157.

Jewesbury, D. (2009). Some Problems with 'Research' in U.K. Fine Art Institutions. *Art & Research: A Journal of Ideas, Contexts and Methods*, 2(2).

Kuhn, T. S. (1970). The Structure of Scientific Revolutions, International Encyclopedia of Unified Science. In *Foundations of the Unity of Science*. Chicago: University of Chicago Press, vol. 2, no. 2, 2nd ed.

Leavy, P. (2017). *Research Design: Quantitative, Qualitative, Mixed Methods, Arts-Based, and Community-Based Participatory Research Approaches*. New York: The Guildford Press.

McNamara, A. (2012). Six Rules for Practice-Led Research. *TEXT: Journal of Writing and Writing Courses*, 14. www.textjournal.com.au/speciss/issue14/McNamara.pdf (accessed January 1, 2020).

Nelson, R. H. (2009). *The Jealousy of Ideas: Research Methods in the Creative Arts*. Fitzroy: Ellikon.

Schön, D. A. (1987). *Educating the Reflective Practitioner: Toward a New Design for Teaching and Learning in the Professions*. Jossey-Bass Higher Education Series. San Francisco: Jossey-Bass.

Sennett, R. (2008). *The Craftsman*. New Haven: Yale University Press.

Smith, H., & Dean, R. (2009). *Practice-Led Research, Research-Led Practice in the Creative Arts*. Edinburgh: Edinburgh University Press.

Sullivan, G. (2009a). *Art Practice as Research: Inquiry in Visual Arts*. Thousand Oaks: Sage Publications.

Sullivan, G. (2009b). Making Space: The Purpose and Place of Practice-led research. In H. Smith & R. Dean (Eds.), *Practice-Led Research, Research-Led Practices in the Creative Arts*. Edinburgh: Edinburgh University Press.

Wilson, M., & van Ruiten, S. (Eds.). (2013). *Share, Handbook for Artistic Research*. Amsterdam: ELIA European League of Institutes of the Arts. www.sharenetwork.eu/resources/share-handbook.

Wilson, S. (2002). *Information Arts: Intersections of Art, Science, and Technology*. Cambridge: MIT Press.

PART II

Knowledge

2.1

KNOWLEDGE

Linda Candy, Ernest Edmonds, Craig Vear

What is knowledge?

A key question addressed in this section is, what is knowledge? In research, it is obviously important to ask a related question: how can we obtain new knowledge? This second question is also considered in Part III of the handbook dealing with 'Method', but there are some very special issues in practice-based research concerning how knowledge is acquired, and those issues cannot be entirely separated from the first question. For that reason, this chapter, and some of the other chapters in this section, will cross the boundary between these two issues: defining knowledge and acquiring it.

Research produces new knowledge. That is why it is done. If the research is public, formal or academic, then the new knowledge is made available to the world and archived for future generations to inspect, use or advance. Practice-based research is no exception to this, nor is the knowledge that it generates.

The word knowledge has many meanings depending on the field in question and the perspective with which you wish to view it, with only some of these relevant to this handbook. Knowledge can be private to the individual, or available to a discipline, community or field. Knowledge can advance the individual's practice into new areas and directions, or it can change the way a community of experts in a field conducts its practice. Knowledge can move the practice forward into new insights and areas of innovation or can bring small incremental nuance to the practice.

It is important to understand the difference between *knowledge* and *knowing*, as both are important within practice-based research. Simply put:

- *Knowledge* is externalised, validated and can be archived.
- *Knowing* is felt within and can be demonstrated by actions.

We summarise these in more detail below.

Explicit *knowledge* is expressible in some form. This is sometimes called propositional, codified, declarative or 'know-that'. In general, explicit knowledge describes a thing or a phenomenon. It generally uses language or some other descriptive tool to compare and relate this thing or phenomenon as a communicable construct. For example, I know that this page is made of paper (if you are reading this from a book) and can justify it by comparing other examples of paper with this one.

 DOI: 10.4324/9780429324154-14

The belief that this 'page is made of paper' is justified to be true because of the principle that our knowledge of what paper is contributes to the truth that this 'page is made of paper', additionally, it can also be justified by other's observations of the page.

Tacit *knowing* is by definition inexpressible, or is hidden inside your experiences and practice. Because of this, tacit knowing is really difficult to express as justified belief or in a shareable format as it defies the same comparative process that defines propositional knowledge. This is sometimes called 'know-how', i.e. I know *how* my judgement impacts my practice, or *how* I embody my instrument when performing in front of an audience. Knowing is often critically important in practice-based research and gets called by many terms, for example, insider, 'know-this', experiential, implicit, in-vivo. Because it is rooted in context, experience, culture, practice, values, experience, perception, consciousness and emotions (to name but a few), it can be regarded as elusive, intuitive, specific to an individual, hidden or personal. For the practice-based researcher, this is the place where their true and unique knowing of their practice is often based.

The various features and forms that knowledge and knowing take in practice-based research are now discussed.

On knowledge

Knowledge can be defined as something accepted as true through a 'right to be sure'[1] or, as has been proposed in the context of practice-based research, 'justified belief'.[2] In either case, the concept is that we need explicit reasons that make our belief at the very least 'plausible' and, ideally, using the legal phrase, 'true beyond reasonable doubt'. In research, we will normally be somewhere between those two extremes. For example, you could say "I know that I live in a particular country, the UK, and my belief in this knowledge is justified to be true by the many official documents I hold, and the evidence I can provide through social media, images, and affirmations through others who work at my university".

Although this example does not come from practice-based research, nor is it original in the world, it is an example of knowledge. Someone could feel justified in presenting as true knowledge, and attempt to validate beyond a single representation by providing corroborative objects, like certificates or personally addressed letters, so that others may equally trust this knowledge or reliably challenge it. Of course, there is always the slim chance that they have been tricked and that, like the main character in the film *The Truman Show*,[3] their world is, unknown to them, actually an elaborate TV set perhaps built in some other country. But we can reject this doubt, as there seems to be no evidence to support this as fact.

Research in its different manifestations is usually expected to lead to new knowledge that challenges existing theories and assumptions. Researchers everywhere seek to verify hypotheses or prove that existing theories are wrong. However, practice-based research has characteristics that do not necessarily conform to traditional norms about the nature of knowledge and how it is generated. For one thing, the practice that is so central to practice-based research may include making things, whether visual or sound objects or installations, exhibitions or performances. For practitioner researchers, although these works are at the centre of the research, there are also other kinds of outcomes which arise in parallel and are an essential part of any claims to have produced new knowledge.

In the context of practice-based research, we can both ask about the nature of the research knowledge that advances practice and the aspects of practice that offer new knowledge to research. Knowledge from practice includes all the normal categories that we see in research. In engineering, one such category is the definition of a new type of artefact, such as a solar-powered car. Outcomes of this kind are not uncommon in practice-based research, for example a musical instrument played by interacting with neural networks. Examples of knowledge arising from practice-based research are:

- Models, taxonomies, frameworks.
- Strategies, criteria for action/design.
- Exemplars, case studies.
- New artefact forms.

These outcomes from research in creative practice represent a wide variety of contributions to knowledge.[4] New artefacts can stand alone without explanation whether shown in public or not, but in practice-based research, they must be given context in written form if their contribution to new knowledge is to be shared successfully. Models and frameworks are equally likely outcomes. Andrew Johnston's own PhD research provides an interesting example. His practice-based research centred on the development of virtual musical instruments for live performance.[5] During performances, the virtual instruments responded to sounds produced on the acoustic instruments by simulating physical objects to create a dynamic virtual sculpture with which the musician interacted. The practice he underwent in making the instruments raised a number of questions that he was able to explore through research, for example: how did musicians interact with the virtual instrument? Did they approach it like a traditional instrument? To what degree did they seek to control it? Did it facilitate musical expression?

Johntson's research approach was to seek evidence about musicians' experiences with the instruments he created, by analysing observations and interviews. A study of expert musicians' experiences resulted in the development of a theory of musical interaction in which different 'modes of interaction' were identified, representing ways in which the musicians approached music making with the virtual instruments:

- **Instrumental:** the musician seeks to control the software.
- **Ornamental:** the musician allows the software to add ornamentation without controlling or responding to it.
- **Conversational:** the musician shares control with the software, sometimes steering the 'conversation' then relinquishing control and allowing the virtual instrument to take over.

This theoretical framework could be applied to designing and evaluating other kinds of interactive instruments.

Johntson's achievement was to establish an artistic, technical and methodological foundation that continued into his ongoing practice and research. The new virtual artefacts embodied theories with them, in this case the theory related to the nature of interaction and musical expression. The value of the practice-based research was to make the theory explicit, and therefore open to reflection, evaluation and validation. This was made possible by a framework that encapsulated practitioner understanding developed through research into a form of knowledge that could be applied by others.

In 'Theory as an Active Agent in Practice-Based Knowledge Development',[6] Linda Candy shows how practice-based knowledge emerges from research and introduces the problem of navigating between theoretical and practice knowledge. She introduces 'theory as active agent' and shows how practitioners evolve theories through practice-based research towards the generation of a new knowledge. This knowledge can take the form of frameworks that act as both instruments for thinking and making and examples of new methodological knowledge.

An example framework is presented by Craig Vear in 'Mapping Practitioner Knowledge: A Framework for Identifying New Knowledge through Practice-Based Research'.[7] The chapter addresses the difficulty of consolidating the nature of research with the nature of practice and presents a way of understanding practice-based research as a consolidation of *outside-looking-in* and *inside-looking-out*. The chapter then presents several examples of the framework in use. It will be of use to PhD students, first-time practice-based researchers and supervisors in developing an understanding of the specific nature of their research projects, and the development of a more systematic and targeted focus to investigating new knowledge in practice-based research.

The relationship of the artefact to the production of new knowledge is discussed in more detail in a later section of this chapter. We now consider the notion of tacit knowing in contrast to explicit knowledge.

On knowing

Knowing, defined as something felt within that can be demonstrated by actions, is often referred to by practitioners as a gut feeling or an intuition, something that is known but not consciously applied. For example, you can know how to balance a ball on your head, and this shows that you have possession of a certain knowing about yourself and the world and can put this into action. Similarly, you know how to judge your impression of a painting. This type of knowing is tacit and is discussed in more detail below. It is a key recourse in practice. But this knowing does not count as knowledge in a form that is accepted in research – although, as we will read below, this does not exclude knowing from being externalised and shared as knowledge – and this is one of the fundamental challenges for many practitioner researchers.

For the practice-based researcher, who is naturally inside the practice doing the research, tacit knowing can often be a rich area of study, and if externalised, validated, original and shared can present the field with significant new knowledge. However, explicit knowledge and tacit knowing often work together. Their interaction can be a vital process in the development of new knowledge for both the individual and for the wider community.

In 'Mapping the Nature of Knowledge in Creative and Practice-Based Research',[8] Kristina Niedderer discusses tacit knowing, which she calls 'tacit knowledge', both pointing out its importance and arguing that ways must be found of making it a significant element of practice-based research. Her discussion proposes a new framework to relate the different types of knowledge and knowing and presents different examples to highlight how important tacit knowing is in enabling the practitioner to develop high levels of expertise and connoisseurship. The conclusion is that there is no issue, in principle, with including the tacit in research, but there remains a tension between the use of tacit knowing and the research requirements of knowledge, i.e. justification, sharing and archiving, and that of precisely representing tacit knowing in such a format.

Garth Paine, in 'Un-knowing: A Strategy for Forging New Directions and Innovative Works through Experiential Materiality',[9] continues the subject of tacit knowing, particularly investigating the search for, and in, the unknown (what he calls *un-knowing*) that is shown to be a key part of creative practice. He concentrates on the private research that is part of practice, the results of which remain hidden except for their manifestation within creative artefacts, with examples being taken from music. He asks the reader to consider the 'complex coalescence' of different forms of knowing that go into skilled practices. His chapter is concerned with research methods within and for artistic creation. He uses the term *un-knowing* to describe a 'state of productive allowing', in which the practitioner searches for new directions that can lead to unexpected and stimulating outcomes. He makes a strong argument for the importance of this to creative practice and the contribution to the general development of a community's understanding of the possibilities in their practice. As such, he provides examples of how they can represent changing practices that have led to inspiring new genres, each addressing a different social milieu, giving voice to social, cultural and economic challenges.

From this example, we see that a common challenge for the practitioner researcher is that tacit knowing can be embedded in, embodied with and enacted through the individual doing the practice. It is, by its very nature, tacit, hidden and personal to the individual. Attempting to extract this as a form of knowledge is extremely difficult, if not impossible. Therefore, trying to externalise this tacit knowing and represent it as close to its true form as possible presents the practitioner researcher with challenges, but also an opportunity to support growth and rigour in the field.

To explain this, we can use the example of riding a bike. Our general knowledge of riding a bike is well documented, and any parent who wishes to train their child to ride a bike can reach for online videos and helpful lists to guide the factual (how a bike works), conceptual (gravity and balance) and procedural (the bio-mechanical process of pedalling and leaning) types of knowledge that make up explicit knowledge of riding a bike. But this knowledge alone is not enough to make your son or daughter ride a bike (unlike, say, setting up Wi-Fi router in your home). But as a parent who can ride a bike, you have implicit (tacit) knowing of how you do it. Furthermore, you can help your child to build their tacit knowing in many ways, depending on the difficulties that your child is encountering. In this process, you might reflect on your movements as you turn a corner and realise that you can explain an important aspect of what you do: as you turn the front wheel to point into the corner, you also lean your body into the bend while keeping the bike in something much closer to an upright position. Your tacit knowing is now available as explicit knowledge that you can pass on to your child. This is the process of externalising knowing into knowledge and is discussed in detail below. Incidentally, now that you have this explicit knowledge, you also realise that the action is very different when turning on skis: your body leans the other way.

This everyday example highlights that tacit knowing is multi-faceted and has value. In acknowledging it in this way, we can address the challenge of externalising and sharing tacit knowing through practice-based research and journey to a position of validation and sharing required of new knowledge in research (discussed next).

Knowledge from knowing

As we have discussed, in practice-based research not all knowing is easily encoded or encodable as knowledge. Knowing can be embedded in, embodied with, enacted through or extended by[10] practice through touch, feeling, know-how, intuition, etc. For example, a nurse might know that the emotional feeling of touching a patient can be beneficial to the patient. But what, in this scenario, would constitute knowledge that can advance the practice of other nurses? It might be that understanding more closely the emotional feelings behind the duration of touch, or intensity of the action of touch, can benefit others. Or, responding to how the patient moves and displays a reaction to the touch can influence the emotional encounter for the nurse. These examples are not always easily encoded as knowledge. That is partly because this kind of activity is not a function of analytical thought alone ("if I hold this touch for three seconds it will be more emotional"). The experience involves the sense of touch and is, in part, entangled with kinaesthetic knowing (relating to a person's awareness of the position and movement of the parts of the body) and, in part, somatic knowing (how the body perceives from within), alongside the affectual feeling of touch and the response of the patient.

The challenge, here, is to identify within such an entanglement of different types of knowing, what precisely would constitute knowledge that advances the field, is justifiable true belief and can be effectively communicated to others with the 'right to be sure' (discussed earlier). If it can only be communicated by demonstration, this can be problematic, as it would need the person who knows to be available, and that cannot be relied on over space and time. However, we increasingly see demonstrable know-how recorded in video form. This is a path that is important for practice-based research and a significant example of where an artefact, rather than text alone, is important. The issue of moving to knowledge from knowing is discussed by Linda Candy in a chapter in another section of this handbook.[11]

As we can see in the example of the learning to ride a bike, thinking about the tacit knowing involved can lead to it being explicitly communicated as knowledge. Externalising tacit knowing is often addressed through reflection in and on action.[12] Schön argued that reflection in action was a 'legitimate form of professional knowing'.[13] In his work, the concepts of reflection in and on action are seen to be a very important aspect of professional practice in which that practice is advanced. Linda Candy has shown how these concepts are also significant in creative practice and discusses how the resulting knowing can be externalised to become explicit new knowledge.[14]

New knowledge

The purpose of research, of any kind, is to generate new knowledge. We have defined knowledge, but to bring us closer to the purpose of this handbook, we need to discuss new knowledge and what that means and doesn't mean to the practice-based researcher.

There are principles that can be used to shape any determination of new knowledge in the context of practice-based research; it must be:

- **Original:** it is new in the world.
- **Validated:** there are identifiable reasons to believe that it is true.
- **Contextualised:** intended beneficiaries can understand its relevancy and reasons why it is new.
- **Shareable:** others can benefit from it and understand it within its context.
- **Retainable:** in archives in sustainable material of any form.

Central to this list of principles is an understanding that new knowledge has beneficiaries beyond the individual and that it is communicated in a clear and sustainable way so that they can benefit from the knowledge. For example, there are many things that a concert pianist knows to be true about technique and interpretation, and these can be demonstrated through recordings, performances, writings, teachings and reflective evaluation that can be shared with the field. But to determine whether this is new knowledge, in the research sense, the principles earlier must be considered:

1 **Original.** The knowledge shared by the researcher should advance or enhance the field. It is possible that it could transform the practice for others or contribute towards a more rounded understanding of the field. It should not repeat existing knowledge, either willingly or unwillingly; therefore, it is important that the practitioner researcher has a grasp of the field and its discourse so that they can identify the gap in the community's knowledge and add to this.

2 **Validated.** This knowledge needs to attempt to go beyond a single instance of itself. Once it has been identified, the practitioner researcher should attempt to repeat or challenge the phenomenon through repeated instantiations and, if appropriate, other practitioners. This will help contribute to the justified belief of the knowledge and, primarily, can reinforce the value of the truth that the practitioner places in the knowledge.

3 **Contextualised.** The knowledge needs to be presented with a relevant context so that it can be understood by others as advancing the field or enhancing the practice. This also adds justification to the truthfulness of the knowledge and highlights the belief that the practitioner researcher places on their new knowledge, by inviting challenge or corroboration from the field. By stating the context, the researcher points to what is not claimed (it is outside the context) as well as clarifying what is claimed.

4   **Shareable.** The knowledge needs to be shared in a form that clearly articulates, illustrates and expresses the precise and original nature of the new knowledge. This will involve some form of written text that exposes the knowledge embedded within the practice and a range of documentary media that exemplifies the knowledge in the domain it was generated. The value of truth here is expressed in the various ways that the knowledge can be documented. There is no one size fits all, and this in itself provides opportunities for innovation and challenges.

5   **Retainable.** Linked to the shareable nature of the new knowledge, this knowledge should be retainable in forms and materials that preserve it beyond its immediate lifespan. Additionally, some new knowledge takes a while to permeate through the field; as such, it is feasible that innovations and new knowledge generated today will only find a hold in a decade or, sometimes, much longer timescales.

All of these principles have to be applied. One example of an easy mistake to make in practice-based research is to concentrate only on, for example, the first two or three of them. The researcher might survey the field, contextualise it and show that the contribution is not already described by a previous practitioner/researcher. Then they might validate it by employing the new result in practice, for example making an innovative sculpture or composing a new piece of music. If the work stops there, we certainly could see that something interesting has been done and that it had been valuable in practice, but if it has not been shared and retained according to the last two principles, the contribution would not count as delivered research knowledge.

This principled approach is exemplified in Kerry Francksen's chapter, 'The Curious Nature of Negotiating Studio-Based Practice in PhD Research: Intimate Bodies and Technologies'[15] in Part V. She outlines her investigation, which comprised a largely qualitative and empirical study into the 'dancer's insights and appreciation of perceiving and generating movement from inside the embodied experience'.[16] Through this, she developed an 'effective framework for generating new forms of embodied knowledge within a generative and dynamic [digital] system'. Gathering evidence through repeated iterations of her performance work helped to build a strong case for support and also 'became instruments for uncovering a number of key insights concerning the dancers' perceptual and experiential responses'. Her PhD thesis was presented as a 'portfolio of works accompanied by a written document that provided a textual analysis of the significant insights gained throughout the research process'. Her written submission was an 'analysis and document of the overall process or methodology' and included

> images from the three works and was interspersed with dancer testimonials and significant moments of interaction . . . and video recordings of some of the on-going practical investigations and crucially they were asked to experience a version of the work live.

Francksen's research highlights the hallmarks of new knowledge: it had beneficiaries beyond the individual and was communicated in a clear and sustainable way so that others can benefit from the knowledge. Notably, her thesis did not contain full-length videos of her final performances, as she felt that documenting the process, not the product, was the clearest way of communicating the new knowledge. It also highlights the five principles as original, validated, contextualised, shareable and retainable.

The role of practice in the generation of new knowledge

Practice is, by definition, an integral part of practice-based research, and research is expected to contribute to the generation of new knowledge. Knowledge in practice-based research can be about a range of activities, objects and outcomes of practice. This can include audience experience and reception, strategies for designing new operational processes, taxonomies of practice-based skills or models of behaviour, to name but a few. Practice-based research outcomes can also include the work itself, such as the music composition, the engineer's bridge, the software code, the surgeon's operation and the teacher's supervision.

The practice (the outcomes, the actions and the process) is not only often the subject matter of the research, or the motivating inspiration for an investigation, but is also the site and system for the research. From a methodological perspective, the practice forms a pathway through which new knowledge, insights and understandings are generated, and also the world-of-concern in, with or through which such products come into being.

Fabrizio Augusto Poltronieri's chapter, 'Dreaming of Utopian Cities: Art, Technology, Creative AI, and New Knowledge',[17] in Part V of this handbook, discusses how scientific methods help him to 'see new insights, which are put into practice and foster further theoretical discussions'. He uses the theoretical ideas of the Czech-Brazilian philosopher Vilém Flusser (1920–1991) and the notion of 'apparatus' to transform 'the world through plays of symbolic permutations'. He continues to discuss how new methodologies can be developed that can generate new knowledge, which in turn leads to significant questions of meaning and belief. The fruitful combination of practice-based research, art and technology provides him with 'the ideal scenario for playful experimentation combined with scientific rigour'.

The relationship of the artefact to new knowledge

The role of the artefact in research is a contentious aspect of the practice-based research debate, especially where the artefact is implicated in the kind of knowledge that is generated. The practitioner researcher may use artefacts as the object of study or as experimental apparatus, and in many cases the actual creation of an artefact is critical to the research process. Creating artefacts is a core activity in many forms of practice-based research, particularly in the field of creative arts, and for many practitioners, they embody complex layers of knowledge of their field that some argue do not require explanation.

Smith and Dean discuss different forms of research and their relationship to the generation of knowledge. 'Knowledge' is seen as verbal or numerical in expression and something that can be generalized to other processes or events that are outside those that gave rise to it. It is also considered to be transferable (and for that, verbal expression is needed). However, visual or sonic forms 'transmit' knowledge (as Smith and Dean term it) in non-numerical or verbal forms and, they argue, this needs to be acknowledged if it is to be given sufficient weight in practice-based research outcomes.[18] This argument raises the question of the form that transmissible knowledge should take.

How we 'transmit' or, rather, share knowledge with others is an important issue. Some argue for conducting empirical studies, the results of which are readily expressed in linguistic or numerical forms by way of explanation. This 'evidence' can be understood unambiguously, it is argued, whereas an artefact cannot stand on its own without an explanation of context. In many ways, this is fundamental to the whole question of the role of the artefact in research and knowledge generation. If the import of a painting has to be explained in words, it assumes that the viewer does not have access to what is sometimes called the 'language of painting'. If enough people know this 'language of painting' to understand what the creator is claiming to be new, why is there a need for linguistic explanation as well? However, not everyone can 'read' paintings, and even those who can will have different subjective interpretations of what they see. The key point is to understand that the word 'language' is being used by analogy here. It is used to indicate that there are aspects of some paintings that can be compared to certain aspects of a language. It does not refer to an actual language with a defined syntax, grammar and lexicon. There is no dictionary that we can refer to in order to look up the meaning of the elements of a painting.

The question of ambiguity is important here. Explanations expressed in notations such as mathematics use a universal notation that is unambiguous to those who have learnt it. Likewise, musical scores and choreographic dance sequences have similar characteristics. Of course, a musical score only defines part of what is to be played and there is plenty of room for interpretation. Without an unambiguous 'language' for all artefacts, there is room for multiple responses and interpretations and, after all, why not? Ambiguity and different interpretations are at the heart of our appreciation of art 'artefacts'. Artefacts like these embody many things, but it is the experiences that they facilitate rather than what they embody that we are more easily able to share and appreciate. A given artefact will give rise to different experiences in different people in different contexts. In the arts, for example, this is all to the good but contrasts with the communication of knowledge.

The oft-expressed view that artefacts embody knowledge is challenged directly in the chapter 'The Art Object Does Not Embody a Form of Knowledge Revisited',[19] by Stephen Scrivener. He discusses the nature of knowledge in practice-based research, firstly through his 2002 paper of this title and secondly in this section of the handbook that revisits it after a substantial gap. His 'provocation' is based on examples from the visual arts. He makes it very clear that the communicating of knowledge should not be seen as the main function of an art object; indeed, arguably, it is not such a function at all. He points out that 'claims to new knowledge require both the knowledge claimed and its justification to be communicated' and that this will not be found in the artwork itself. The original paper makes the strong claim that art making should not be justified in terms of research outcomes. He does not argue against practice-based research, only that it cannot be the core purpose of art practice. In revisiting the paper, Scrivener largely confirms that his earlier provocation is still valid and that, whatever benefits may come from research, it is not the case that 'new knowledge is the means by which the field of art advances'.

An important issue in the debate about the artefact in practice-based research is its role in contributing to new knowledge in academic doctoral research. In some practice-based PhD research programmes where the creation of an artefact has been central to the outcomes, it is not possible to gain a complete understanding of the significance of the research without direct reference to the artefacts. In these cases, a textual description setting out the context, the process and insights reached is not enough in itself to convey the full import of what has been achieved. However, while artefacts are certainly centrally important in conveying meaningful experiences, it is questionable whether they can transmit meaning.

The precise expression of the new knowledge embedded within the practice is not provided by the artefact in isolation, even to those in-the-know. The artefact can be used to provide some form of justification about a process, or as an object of representation, but even here, it is debatable whether the precise nature of such a claim is clearly exposed in the artefact, or whether it adequately deals with a gap in knowledge. This problem therefore implies a need for a parallel means of communication that exposes the precise nature of the new knowledge and how this relates to the artefact. In effect, text coupled with other media documentation can help to frame the way that we view the artefact and comprehend that which is truly the new knowledge.

A further issue is that an artefact can be perceived in many ways, and depending on who is observing it, will inspire different thoughts about what is new. In fact, it is possible to see different levels of innovation in an artefact, some of which may not be intended by the original practitioner researcher. Therefore, observing the artefact alone at best allows the new knowledge to be diffused or confused with the viewers own subjective interpretation. This further reinforces the fact that, if the artefact is the primary outcome of a practice-based research project, then the precise nature of it needs clarification through text and other forms, so that it unequivocally communicates the originality and contribution to the field.

When considering practice as a complex situation that is not only the object of study, and the objective, but is also the site, the context, the method, the outcome and the subject of research, it can be difficult to pinpoint the precise details of new knowledge. Some kinds of new knowledge are derived from empirical studies of audiences and operational systems, while others are more speculative and exploratory, employing documented reflection. The outcomes from practice-based research in a range of fields represent a wide variety of contributions to culture and knowledge advancement. The artefacts (music compositions, steel bridges, code, artworks and interactive systems, surgical procedures, caring principles) stand for themselves as field-specific treatise, but in formal practice-based research are placed in context and elucidated through written theses and published papers.

In 'Research, Shared Knowledge and the Artefact',[20] Ernest Edmonds looks at the role of the artefact and its relationship to practitioner knowledge from the point of view of research rather than practice, which is exemplified by examples from art. A number of famous examples of art innovations are presented in order to illustrate what new knowledge can look like in practice-based research. The formal nature of knowledge is discussed and examples of knowing, as against knowledge, are presented. New knowledge arising from practice is discussed, elaborating on the discussion earlier in this chapter. The issue of finding ways to generate new knowledge from tacit knowing is emphasised as important in much practice-based research. The role of the artefact and ways in which artefacts can form part of the outcome of a practice-based research project are elaborated.

The artefacts generated within a practice-based research programme may well represent the core of the 'new knowledge' generated by the research, but the clarity with which that knowledge is communicated directly through the artefact is debateable. While the art, for example, in itself is not directly concerned with 'communication', research that involves an artefact may produce claims for new understandings that require some form of justification. If we accept that the artefact can, in some sense, represent new knowledge, the problem of sharing what that knowledge is implies a need for a parallel means of communication that illuminates the knowledge, in effect, a linguistic one. The textual element is vital in completing the contribution to knowledge that the artefact may, in its core, represent. A doctoral submission or a journal paper, for example, must also include a substantial contextualisation of the work. This critical appraisal or analysis not only clarifies the basis of the claim for the originality and location of the original work but also provides the basis for a judgement as to whether general scholarly requirements are met.

The nature of publishing practice-based research is shifting in response to the changing needs of researchers, academic contexts and global issues around carbon footprints. This is offering an opportunity for publishers of academic journals and research societies to reconsider how they support the dissemination of practice-based research artefacts. For example, MIT Press provides the opportunity for authors to publish supplementary material, such as sound or video files, associated with journal papers and books, including the art/science journal *Leonardo*. In her chapter, 'Publishing Practice Research: Reflections of an Editor',[21] Maria Chatzichristodoulou – as editor-in-chief of the *International Journal of Performance Arts and Digital Media* (IJPADM), which also facilitates the sharing of multi-media artefacts – highlights this shift and presents solutions that offer 'care and consideration' to the specific way in which individual practice-based research needs to be communicated.

Topics on knowledge in the chapters to follow

There are many different categories of knowledge discussed in the chapters that follow, each of which relates to a different aspect of practice or formalising the nature of knowing. Taken as a whole, they provide a broad picture of what knowledge is and how it is advanced in practice-based research as well as the variety of perspectives that exist within the practice-based research community on knowledge and knowing. Among the many diverse aspects of this topic, they explore issues such as:

- How practitioners develop theories and new knowledge through practice-based research.
- Frameworks as support for thinking and making.
- How methodological knowledge arises from practice-based research.
- The value of tacit knowing to expertise and connoisseurship.
- New directions in creative practice leading to unexpected and stimulating outcomes.
- Empirical frameworks for generating new forms of embodied knowledge.
- Philosophy, science and theory in new knowledge.
- The artefact and the generation of new knowledge.
- Advances in art practice are different to knowledge from research.

Notes

1 Ayer (1956).
2 Frayling (1993).
3 https://en.wikipedia.org/wiki/The_Truman_Show (accessed May 16, 2021).
4 Candy (2011).
5 Andrew Johnston 'Almost Tangible Musical Interfaces'. In Candy (2011, pp. 2011–224).
6 Candy (2022b).
7 Vear (2022).
8 Ibid.
9 Paine (2022).
10 This categorisation is explained in more detail in Vear (2022).
11 Ibid.
12 Schön (1983), Candy (2022a).
13 Schön (1983).
14 Candy (2022a).
15 Francksen (2022).
16 Ibid.
17 Poltronieri (2022)
18 Smith and Dean (2009).
19 Scrivener (2022).
20 Edmonds (2022).
21 Chatzichristodoulou (2022).

Bibliography

Ayer, A. J. (1956). *The Problem of Knowledge*. London: Palgrave Macmillan.

Candy, L. (2011). Research and Creative Practice. In L. Candy & E. Edmonds (Eds.), *Interacting: Art, Research and the Creative Practitioner*. Farringdon: Libri, 33–59.

Candy, L. (2022a). Reflective Practice Variants and the Creative Practitioner. In C. Vear (Ed.), *The Routledge International Handbook of Practice-Based Research*. London and New York: Routledge.

Candy, L. (2022b). Theory as an Active Agent in Practice Based Knowledge Development. In C. Vear (Ed.), *The Routledge International Handbook of Practice-Based Research*. London and New York: Routledge.

Chatzichristodoulou, M. (2022). Publishing Practice Research: Reflections of an Editor. In C. Vear (Ed.), *The Routledge International Handbook of Practice-Based Research*. London and New York: Routledge.

Edmonds, E. (2022). Research, Shared Knowledge and the Artefact. In C. Vear (Ed.), *The Routledge International Handbook of Practice-Based Research*. London and New York: Routledge.

Francksen, K. (2022). The Curious Nature of Negotiating Studio-Based Practice in PhD Research: 'Intimate Bodies and Technologies: A Concept for Live-Digital Dancing' Retrospective. In C. Vear (Ed.), *The Routledge International Handbook of Practice-Based Research*. London and New York: Routledge.

Frayling, C. (1993). Research in Art and Design. *Royal College of Arts Research Papers*, 1(1), 1–5.

Paine, G. (2022). Not Knowing/ Meta-Materiality. In C. Vear (Ed.), *The Routledge Handbook of Practice-Based Research*. London and New York: Routledge.

Poltronieri, F. A. (2022). Dreaming of Utopian Cities: Art, Technology, Creative AI, and New Knowledge. In C. Vear (Ed.), *The Routledge International Handbook of Practice-Based Research*. London and New York: Routledge.

Schön, D. (1983). *The Reflective Practitioner – How Professionals Think in Action*. New York: Basic Books.

Scrivener, S. (2022). The Art Object Does Not Embody a Form of Knowledge. In C. Vear (Ed.), *The Routledge International Handbook of Practice-Based Research*. London and New York: Routledge.

Smith, H., & Dean, R. (Eds.). (2009). Introduction. In *Practice-led Research, Research-led Practice in the Creative Arts*. Edinburgh: Edinburgh University Press, 1–38.

Vear, C. (2022). Mapping Practitioner Knowledge: A Framework for Identifying New Knowledge Through Practice-Based Research. In C. Vear (Ed.), *The Routledge International Handbook of Practice-Based Research*. London and New York: Routledge.

2.2

THEORY AS AN ACTIVE AGENT IN PRACTICE-BASED KNOWLEDGE DEVELOPMENT

Linda Candy

Introduction

This chapter is about the nature of theory as characterised in scholarship and research, in the generation of knowledge about the natural and man-made world. The main focus is how practitioners can use and evolve theories through practice-based research towards the generation of a new kind of theory: the role of this kind of theory is as *an active agent*. In this process, existing theories, or theoretical constructs, play a dynamic role in the practice-based research process that can lead to the emergence of new practitioner knowledge. In a situation of practice, these constructs typically take the form of a framework devised by the practitioner which operates as an instrument for thinking and, where applicable, making something new. Any given framework may draw on existing theories from multiple sources that are combined into a unified structure for use within a particular situation of practice. Crucially, these target gaps in knowledge challenge assumptions in the field. (See Vear's chapter on 'Mapping Practitioner Knowledge' in this section). On a metaphorical level, the constructs can act like a scaffold that accumulate practitioner knowledge (inductive); or they can act as a mirror with which to draw attention to aspects of practitioner knowledge (deductive).

When a practitioner researcher uses *theory as an active agent* in practice, this can lead to questions about that theorysuch as does it work? Is it appropriate for my practice? Answering those questions involves observation, reflection and judgement, a process which can reveal the practitioner's implicit thinking and, in so doing, turn what might remain tacit into explicit understandings: i.e. a form of practitioner knowledge. This chapter will give examples of how *theory as active agent* is used and evolved into new knowledge as an outcome of practice-based research. In PhD research, that knowledge can then be shared with others and thereby become part of the collective knowledge of a domain.

The chapter begins with a discussion of theory as used in different contexts and then considers how theory operates as an active agent in practitioner research. Traditionally, theory has been juxtaposed with 'practice', which is seen as making/creating/acting rather than thinking/reflecting/ speculating. In more recent times, there has been a move to bring theory and practice together and to consider how that might generate new kinds of knowledge. Expanding on these ideas, I describe a trajectory model of theory in practice that represents how practice-based knowledge emerges from research. The trajectory model describes the interplay between theory and practice, which can reveal hitherto implicit knowledge developed by practitioners in the pursuit of their work. This knowledge takes different forms from documented artworks and exhibitions to new methods and tools. In this chapter, I focus on examples of practitioner frameworks which operate as tools for thinking, making,

 DOI: 10.4324/9780429324154-15

appraising and evaluating. This process and its outcomes are illustrated with examples from practitioner researchers who continue to extend and share their knowledge through publications, events and exhibitions. The trajectory model was derived by observing and interviewing practice-based research practitioners and identifying common and differentiating features from the data. The trajectory model has also proved to be a valuable tool for practitioners themselves to adapt to different contexts and situations of practice. What began as a researcher's method for generalising about the practice-based research process has proven to be usable by the participant practitioners themselves. This is possibly because it is flexible enough to describe different research pathways at the same time as representing key activities and outcomes common to much practice-based research.

Nets to catch the world

> Theories are nets cast to catch what we call 'the world': to rationalize, to explain, and to master it. We endeavour to make the mesh ever finer and finer.[1]

This quotation by Karl Popper wonderfully encompasses the potential and power of theory to explain the natural and man-made world and enhance our understanding of it. However, in the English language, the word 'theory' is often used in everyday speech to mean a hunch or supposition, and you may even hear people dismiss certain information because it is 'only a theory', meaning it is unproven or mere opinion. By contrast, a more accurate characterisation of theory, one that is used in scholarship and research, is the serious and systematic consideration of facts, observations, data, experiments, artefacts and evidence in the creation of well-founded understanding of the natural world, human society and the literary, cultural and societal environments that we humans create and inhabit. The word 'hypothesis' splits the difference and is a less ambiguous term for this 'hunch' kind of theory yet does not have the well-founded characteristics of more formalised theory.

Theories can be generalisations formulated on the basis of individual cases. Inference from individual cases to a universal statement is termed 'induction', while 'deduction' is the opposite, that is, to infer individual cases from a generalisation. In order to speak of 'accurate' or 'inaccurate' theories, they must bear close scrutiny in light of reality, or risk remaining hypotheses. However, arriving at verified knowledge is a problem, and we cannot be certain whether a generalisation applies in all cases: for example, take the generalisation 'all swans are white'. For hundreds of years, people in those parts of the world known to Europeans assumed this was an absolute truth, and the expression 'black swan' was used as a metaphor for something that was impossible. The 'all swans are white' theory remained unchallenged until the Dutch explorer Willem de Vlamingh sailed into the Swan River on the western coast of Australia (then New Holland) in 1697. His claimed sighting was, however, insufficient to convince everyone, and in 1726, two birds were captured near Dirk Hartog Island and taken to Batavia (present day Jakarta) as proof of their existence. Thus, it was not enough to just assert the existence of a new theory of swans: material evidence was required for verification. Providing evidence can confirm a theory but that, in turn, may be challenged and refuted. This is a fundamental aspect of scientific method.

Scientific theory

In science, theory refers to a comprehensive explanation of natural world phenomena supported by a body of facts that have been repeatedly tested through observation and experiment. Such theories may be of a descriptive or predictive kind, but whatever their purpose, they are, nevertheless, always open to challenge. An important strategy in scientific method is to actively seek to challenge the strength of the theory, and the more the theory stands up to what Karl Popper referred to as 'falsification',[2] that is, the more attempts to prove it is false, the stronger the theory is seen to be. As an example, the theory of evolution by natural selection is considered to be as factually reliable and consistent an explanation of the evolution of life on earth as the atomic theory of matter or the germ theory of disease. Evolution by natural selection is one of the best substantiated theories in the history of science, supported by evidence from a wide variety of scientific disciplines, including geology, genetics and biology. Nevertheless, evolutionary theory continues to be challenged as scientists look for ways to test and extend, through ongoing research, Darwin's original theory published in 1859.[3] This research not only seeks to refute the core theory but also to identify the things he missed. In the location where Darwin witnessed the natural phenomena that inspired his original ideas, Rosemary and Peter Grant and colleagues carried out many years of longitudinal field studies of Galápagos finch populations, seeking to understand why there were so many variations of beak design.[4] Having visited the islands in 2016, where I witnessed the extraordinary variety of wildlife that continues to be a rich source of evidence about the natural world, I now understand much better how important the location was then and is now as a crucible for the kind of ongoing investigations that are the bedrock of scientific theory.

Theories, whether longstanding or new, can only ever be described as 'provisional' from a scientific point of view because there is always the possibility they may be over-turned one day. Thus, the centuries-old theory that all swans are white was 'falsified' by new evidence that black swans actually existed.

Critical theory

A very different type of theory from scientific theory is 'critical theory' involving the critiquing of society, culture and the arts, drawing on the social sciences and humanities and inspiring ideological discourses that represent many divergent perspectives. The origins of critical theory can be traced to Marx's economic and social critique and those who were inspired by him to develop the critique further e.g. the Frankfurt School.[5] Today, it is used as an umbrella term to describe any theory founded upon critique, and social scientists and philosophers from different disciplines have adapted its goals and tenets to a wide range of approaches: e.g. feminist theory, gender and queer theory and media studies.

The primary method of critical theory is argumentation drawing upon lines of thought from sources that support the viewpoint presented. This differs from the scientific method of experimentation designed to produce reliable and tested findings based upon empirical data. In both areas of theory, there are accepted norms and conventions that are agreed within the community of experts within the field. Some disciplinary boundaries are tightly maintained, while others are more open to cross-disciplinary fertilisation of ideas and methods and the new perspectives that such openness can bring.

Critical theory pervades many disciplines, including my own original area of study. As a student of English literature in the late 1960s, like many of my contemporaries, I was introduced to that giant of literary criticism, F. R. Leavis, whose emphasis on a close reading of literature as the embodiment of the moral and social values of western culture was the dominant theory of that time.[6] I learnt to give precise attention to my reading of the great works of literature and to analyse the poems, novels and plays in an almost forensic manner. It was both enjoyable and disconcerting to view great works as exemplars of grand theory, an experience that imposed a heavy weight when first trying to teach literature to secondary school students. By the 1980s, I found that this kind of critique on the nature and role of literature did not help me advance the way I approached my teaching curriculum, which I was trying to develop and adapt for new ways of working, principally digital ones. I turned, instead, to a different kind of theoretical discourse that addressed the role of developing new knowledge about educational practice through research. This experience alerted me to the need to raise questions about how best to inform practice and especially how to bring about improvement through change. It was the discovery of reflective practice and action research that alerted me to the kind of theory that was more appropriate for educational practice.

Pragmatist theory

Pragmatist theory of knowledge holds that knowing is an ongoing process of inquiry that relies on experience and evidence and that knowledge should be checked by the force of experience. This will be discussed here through three different key developments:

a   Reflection in action (Schön)

> The origin of the philosopher Donald Schön's ideas was based on pragmatism. He first introduced the term 'reflection in action', by which he meant the integration of thought and action within a given context, and proposed a relationship between this and the development of practitioner knowledge.[7] When a practitioner reflects in action, he becomes a researcher within that particular context and 'constructs a new theory of the unique case'.[8] Thinking and action are a form of dialogue through which the practitioner assesses his or her actions and, in doing so, learns how to develop better ways of addressing the problem faced. This is discussed in detail in 'The Creative Reflective Practitioner', Chapter 2.[9]

b   Experiential learning (Dewey)

> In the foundational work of John Dewey,[10] his exploration of thinking through experience provided the foundations of a new theory of experiential learning.[11] Dewey's theory of knowledge and the role of experience, interaction and reflection are fundamental to the work of Kurt Lewin, known for his development of 'field theory' and the need for collective decision making over autocratic approaches to solve social and industrial problems.[12]

c   Action research (Lewin)

> Lewin coined the term 'action research' to describe a process where research is: 'a spiral of steps, each of which is composed of a circle of planning, action, and fact-finding about the result of the action'.[13] This process relies on the researcher taking actions in order to understand how intervening in a given situation can lead to insights that may be used to make improvements in that situation. Here the researcher can be a participant in the subject of investigation rather than an impartial observer, as is the case in traditional forms of science where distance is considered necessary to achieve more objective results. Action research has been successfully used to underpin a stream of research dedicated to improving practice and has, in combination with Schön's concepts of reflective practice, become an invaluable feature of practitioner research.[14] Contemporary thinking on action research may be found in 'What Is Action Research and How Do We Do It?' by Smith (2017).[15]

Action research and reflective practice are critical to the development of practice-based research methodologies. New understandings gained from intervening, and then observing the effects, are context- or situation-dependent but, if undertaken systematically and documented carefully, they can contribute to a type of theory that is highly suited to implementing change in practice. I refer to this as *theory as active agent*. It is this kind of theory that is to be found in practice-based research. In the categories of theory discussed previously, theory is the underpinning of knowledge, but *theory as active agent is* most directly relevant to practice because it is concerned with 'making new' by practitioners, not simply critiquing what already exists. I propose that we can extend theory further by considering how theory works as an *active agent* in a situation of practice. The question posed is about the role and value of theory in thinking, making and evaluating by practitioners. A reciprocal question, how practice informs theory, is a necessary counterpart to the first.

Theory as an active agent in practice

In practice-oriented fields such as art, design and engineering, the term theory is often used to refer to non-practical aspects of the making of works. Theory as a wrap up concept for everything *not* to do with making an artefact or artwork is often seen as tangential to the core business of creative practice. Practitioners may consult scientific or critical theory and even adopt particular concepts in pursuing their practice, but such theory is seen mainly as a source that inspires ideas rather than something that might be open to modification or even the spark for new type of practitioner theory. An alternative perspective is to consider theory in practice in a more integral and reflexive sense where, not only does the theory inform practice, but the practice informs the theory as a dynamic process which can inform and shape new constructs and processes and, in turn, be subject to further development. In this way, we can characterise theory in practice as an *active agent* that practitioners use to guide the thinking and making of new works.

Schön identified patterns across different fields of practice whereby the problem is approached by the practitioner as a unique case that does not lend itself to standard solutions, responses or techniques. Features are identified through a type of discovery process from which an intervention strategy is devised, or 'designed', to use his word. The very first thing the practitioner does is to identify the problem: problem finding as distinct from problem solving.[16] Once the initial interpretation of the problem has been found, then the way forward involves reframing it and trying to fit the features of the situation to the new frame. Reframing and testing the features against the new frame is a kind of experimental process (Schön calls it 'move-testing'). The process relies on drawing upon practitioner understandings derived from years of experience in practice.[17] Often the process leads the practitioner to knowledge that is new to them. Moreover, that knowledge can also be new to the world and, furthermore, be shared with the world: it is new knowledge in the research sense. I now focus on the development of new knowledge through practitioner research.

Practitioner research and new knowledge

The approach I am taking in this chapter stems from my research into the way practitioners develop new knowledge in different situations of practice. Most practitioners, regardless of field, domain or discipline, do much more than draw upon their existing knowledge. In practice, when they encounter new questions or challenges, they first explore the nature of the specific situation, identify the problem and then consider various options they are aware of, or as discussed previously, they 'reframe' the problem. Having identified the key salient issues, they might seek further information through personal research before going on to devise a way forward. This process inevitably leads to an understanding of what works in a given context or situation, and their overall expertise is enhanced through these new insights.

For the most part, the practitioner knowledge that emerges from the practice-based process described previously remains implicit. It is clear that tapping into the implicit element is hard to achieve because it is buried deep in the practitioner's thinking and action. Yet this is the key to a better understanding of the nature of the knowledge that is derived from practice. If we can find a way to uncover the implicit and tacit elements of practitioner knowing, we can begin to understand how it might emerge from practice. A way forward is discussed in the following section on the trajectory model of practice and research. Before that, let's explore the issue of implicit knowledge a little more and consider two kinds of theory in practice.

Uncovering and articulating what practitioners are thinking as they act is not easy, especially where there are high levels of cognitive demand, not uncommon in many situations of practice. Asking them to explain how they solve problems as they act is not necessarily a reliable guide to what is really going on. Consider first the difference between what people say they do and what they actually do. Argyris, Putnam and McLain-Smith distinguish between two kinds of theories in practice: 'espoused theories' and 'theories-in-use'.[18] Espoused theories are those that (when asked) a person claims to follow. However, this theory is not necessarily what the practitioner actually does in practice. Although people often do different things to what they claim to do, there is, nevertheless, a theory that is consistent with what they do – a 'theory in use'. For example: a designer meets a client's brief by proposing a design solution based on certain materials, not because a deep analysis shows them to be best, but because years of experience have shown they will achieve the condition and style he prefers. Argyris argues that practitioner theories-in-use happen not just in any way but rather in a 'designed' way that is specific to the individual concerned. Theories-in-use are a kind of tacit 'cognitive map' by which practitioners design their actions. However, in each situation of practice, the 'theory in use' is not usually made explicit in such a way as to be understood and shared by other people. It remains private and, therefore, does not enter the collective knowledge of a given field of practice.

This raises the question as to whether or not communicating these theories to others is feasible. Argyris thinks the answer lies in reflection on action but with the caution that this also embodies implicit assumptions: 'theories in use can be made explicit by reflection in action but reflection itself is governed by theories in use'.[19] The implication is, if they wish to share what they know, practitioners may need first to reveal to themselves the 'theory in use' that governs their actions, and this means it has to be articulated, that is, made explicit in some way.

Theory-in-use as characterised in the discussion earlier relates to theory as active agent insofar as it is an explicit acknowledgement of the role of theory in practitioner actions. However, theory in practice and its role in developing practitioner knowledge has additional features that arise from the dynamic nature of the world of practice where the theory is an active agent that can be manifested in different ways. One such manifestation is in the creation by practitioners of 'frameworks' which operate as a method or tool for thinking and making. An initial framework may include collections of theories from multiple sources or disciplines that are assessed and compared before being combined into a unified structure ready for use. The source theories may have canonical, discipline-specific features that are selected according to the aims and intentions of the practitioner working within a situation of practice that they have defined.

But how are these frameworks developed in the first place? And how do they emerge from the implicit elements of practitioner knowledge? The approach I have adopted to understanding more about how this takes place is described in the following section on the trajectory model of theory in practice. This model shows how practice-based knowledge emerges from research by representing the interplay between theory and practice, which can reveal hitherto implicit knowledge used by practitioners in the pursuit of their work. Embedding research in the practitioner process and undertaking a principled form of research through practice is practice-based research as characterised in the lead chapter of this handbook.

The trajectory model of practice-based research

Reflection in practice can be of benefit to the individual practitioner at a personal level, but this process in itself is not one that automatically leads to sharing the new ideas that emerge from practice. To make tacit understandings explicit, the process and its outcomes have to be externalised. One way of achieving that is for theory in practice to be an integral part of a documented research process.

A crucial aspect of any model of practice-based research is the presence of theory building. Making theory *in* practice is very different to applying theory *to* practice. Sullivan, in his book on art practice as research, called it 'making in systems', by which he means moving 'beyond discipline boundaries and into areas of inquiry that interact and intersect and require new ways to conceptualize forms and structures'.[20] From studies of practitioner researchers, who might be said to be 'making in systems', it is now possible to better understand patterns of practice-based research to be found in formal research PhD programmes. Features of the way practitioner researchers develop frameworks have been represented as models in which the reflexive relationship between practice and theory follows a 'trajectory' or pathway, each one influenced by individual practitioner goals and intentions.

The core trajectory model of practice-based research has three elements: theory, practice and evaluation, representing activities undertaken by the practitioner as they draw on and extend existing theory, create and evaluate artefacts and develop frameworks for ongoing work (Figure 2.2.1). Trajectories of practice-based research work in different ways. Where the primary driver is theory, a framework is developed that draws on existing knowledge and is used to shape the evaluation process and the creation of works. A second type of trajectory is one where the practice drives the development of theory. In both cases, the process is cyclical, and there is often a tighter iterative sub-process in which the theory and practice develop together. The trajectories represent different kinds of relationships between theory, practice and evaluation as exemplified in the cases described in Edmonds and Candy (2010).[21]

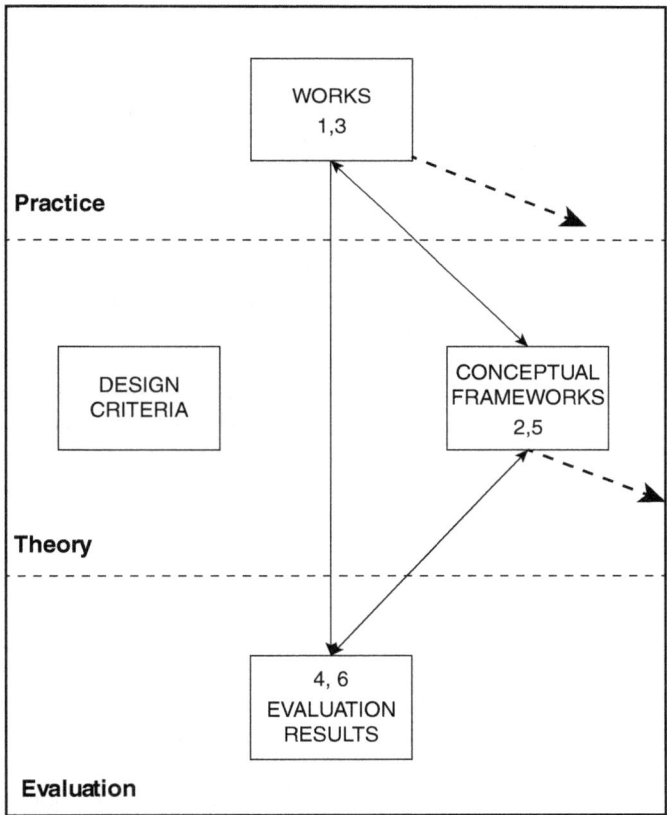

Figure 2.2.1   Trajectory model of practice-based research.

Theory in the context of practice-based research consists of examining, critiquing and applying areas of knowledge that are considered relevant to the individual's practice. This theory is an active agent that gives rise to the creation of 'frameworks' that include theories from canonical or discipline sources that are combined into a unified method for ongoing thinking and making.

Practice is a primary element in the trajectory, comprising activities such as making artworks, exhibitions, installations, musical compositions etc., which provide the impetus and central elements of the research.

Evaluation facilitates reflections on practice and a broader understanding of audience experience of artworks. It may involve direct observation, recording, analysing and documented self-reflection as part of a semi-formal approach to generating understandings that externalise reflections on personal practice.

The pathways traversed between these elements by practitioner researchers are guided by differing motivations, goals and methods. This means that the modelling must represent multiple degrees of variability in the individual's intentions and goals and the process they undergo in order to achieve them. The trajectory model enables practitioner researchers to examine their own processes in the light of the model and identify the differences and similarities. As more practice-based research takes place, this will lead to a further evolution of the model by practitioners adapting it to their own personal situation of practice. Two examples are described next.

Example 1. A practitioner model of theory and practice in research has been developed by Sean Clark, drawing on the original trajectory model described earlier. In his PhD thesis, Sean describes his research process as a series of cycles, the first of which was initiated by an analysis of his previous artworks undertaken over several years.[22] He calls it his 'Theory–Create–Exhibit–Reflect' model, which represents his practice of making, exhibiting and documenting new digital artworks or "art systems". These works were designed according to features identified from previous works and then evaluated – the "Reflect" part of the cycle. The second cycle was informed by the results of this evaluation, which revealed how critical a role the exhibition space played and how different digital artworks interacted with one another as well as with people. These insights were then investigated through observation and analysis from which evidence confirmed the initial insight and also provided a basis for further understanding about the nature of the interactions, extending Sean's existing knowledge. On this basis, a 'Framework for Connected Digital Artworks' was developed combining the properties and principles identified earlier. In the third cycle, through research, Sean had modified his practice, leading to the development of new digital technology for connecting digital works. His new knowledge arising from practice-based research consisted of the framework and exemplars of its use, as well as the innovative technology. The framework not only provides tools that other practitioners can use but also makes a direct contribution to a more general application of 'systems theory'.[23]

Example 2. Another practitioner researcher is Alice Bell, who adapted the original trajectory model to include the participant alongside the researcher as an integral element of her new participatory practice-based framework, hereon referred to as 'Part-PbR'. This addition had the effect of expanding the specific behavioural values required by a practice-based researcher when operating in a Part-PbR trajectory. This implied the need to differentiate between three researcher types: analytical-researcher (AR), facilitator-researcher (FR) and practitioner researcher (PR); and three operational styles that move outside (O), beside (B) and inside (I) the process of Part-PbR artefact generation. Bell's chapter demonstrates an application of her new Part-PbR framework to the project Transformational Encounters: Touch, Traction, Transform, hereon known as (TETTT).[24]

Practitioners are driven by differing motivations, goals and methods, and therefore models of practice-based research must be flexible to the extent that the model is more akin to a scaffold than a fixed representation of desired features. The trajectory model reveals what matters most in a practice-based research process where the goal is to develop new understandings that may be shared with others. It is a way of arriving at a language with which we can make connections between what appear on the face of it to be very different processes, but which have underlying common elements. The point of developing models like this is so that others can examine their own processes in the light of the model and identify the differences and similarities. More cases will inevitably lead to the further evolution of the model.

The models described earlier represent the interrelationship between research and practice in the process of developing practitioner theory, a form of new knowledge arising from practice. The models can be seen as an outcome that in itself is a theory of practice-based process that may be continually tested in practice, evaluated and modified in the light of experience. Practitioners such as those described in the following section develop personal frameworks as an integral part of the practice-based research process. The role of frameworks as examples of practitioner theory will now be discussed by examining different kinds of frameworks that emerge from the research.

Practitioner frameworks and new knowledge development

In practice-based research, practitioners develop frameworks that are used to guide practice and research in ongoing work. The practitioner framework is a theoretical structure used to inform the development of practice, and in turn, the practice informs the content and evolution of the framework and hence, the theory. A framework may consist of many different things according to the individual practitioner's goals and intentions. They take different forms and play different roles, whether they are expressed as criteria for informing and shaping design or classifications (taxonomies) for creating and evaluating new work. These frameworks embody practitioner theories in practice that may be used as tools or methods and act as new ways to improve practice. They can be subject to continuous evaluation and may change according to the situation of use whether used by the originator or other practitioners. Unlike individual works which are unique and embody the individual practitioner's personal knowledge, the frameworks are more likely to have general characteristics (see Figure 2.2.2).

Practitioner frameworks are defined by those who invent them and the purpose they serve, for example to inform the works under construction. The practitioners whose work is described next are working primarily within the field of interactive art systems, using forms of digital technology to create experiences for direct audience participation in the creation of visual and sound artworks. These practitioners were engaged in doctoral research that involves a cyclical process of putting theoretical knowledge into practice and revising theory as a result of the outcomes. Theory and practice are intertwined in the development of their practice.

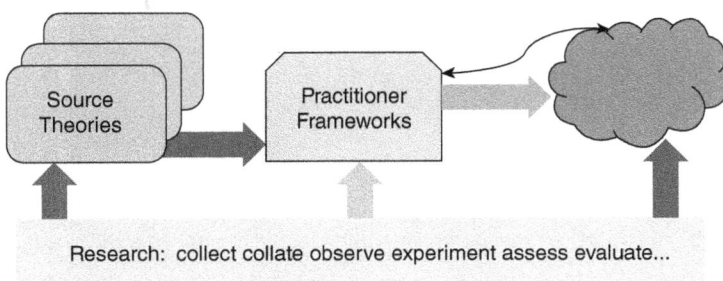

Source Theories

Practitioner Frameworks

Research: collect collate observe experiment assess evaluate...

Figure 2.2.2 A research process for practitioner framework development.

The following two case studies show how the practice-based research process can have variable trajectories in different situations of practice, sometimes driven principally by theory and in other cases by practice. Whichever is the case, the outcomes contribute to practitioner knowledge in a way that can be used by other practitioners. From these trajectories, practitioner frameworks have been developed that represent a type of theory on practice that is used and evaluated through ongoing practice.

Case study 1: a framework for emergent experience

Jennifer Seevinck explores how interactive artworks can stimulate the appearance of new shapes and forms or 'emergent' experiences that are not apparent in the original work. An analysis of her trajectory of practice-based research revealed that theory drove the practice for the most part. Her practice-based research trajectory is represented in Figure 2.2.3. The numbers and lettering correspond to the description below.

---

1   **Practice:** as Jen created **W**orks, she addressed questions as to whether or not they fulfil her expectations with regard to the audience or viewer.
2   **Theory:** based on literature of emergence, she derived a **F**ramework for describing the compositions and shapes observed in audience interaction based on a range of criteria from different sources.
3   **Evaluation:** using the Framework, she evaluated her existing **W**orks. These Works had been designed to stimulate emergent responses in audiences according to a working hypothesis.).
4   **Practice:** The **R**esults of the **E**valuation and the refined **F**ramework were used to inform and guide the making of the next **W**ork

---

The **F**ramework is a taxonomy of properties for describing the compositions and shapes observed in audience interaction. The qualities of emergence were structured according to origin (e.g. perceptual and physical) and intrinsic and extrinsic structures. Theory provided the ingredients of her Framework for evaluating of audience experience and both informs the art making process and provides a means of interpreting the results of observing audience response and behaviour through

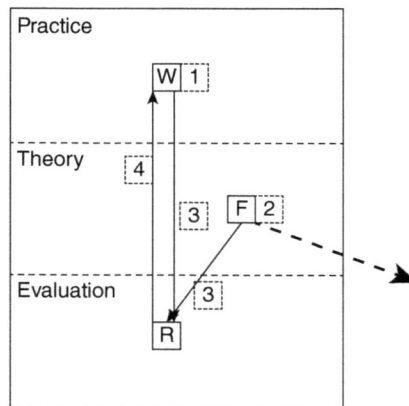

Figure 2.2.3   Trajectory model variant: theory drives practice.

Evaluation. Once fully tested and refined, the **F**ramework, which takes the form of a Taxonomy of Emergence in Interactive Art (TEIA), provided a contribution to ongoing studies.[25] It combines various positions on emergence and synthesises them into one view such that it enables comparison between them.[26]

Jennifer has disseminated the knowledge derived from her practice-based research beyond her personal artistic purpose into the field of interaction design in her book on emergence in interactive art.[27] In writing a book on this work, she has also given other practitioner researchers new tools for thinking and making: 'A practitioner seeking to instantiate the characteristics of emergence can look towards the models and approaches collected here to identify possible approaches or mechanisms'.[28]

Jennifer has continued to expand the range of works and evaluation studies after the PhD. As an integral part of her working method, she has put the framework through its paces by testing it out in different contexts for a variety of applications. Over time, she has adopted a more flexible approach, viewing the role of the TEIA framework as a tool that guides and supports her creative explorations rather than imposing strict definitions and constraints on the scope and implementation of her actions. In this way, it acts as a tool for reflective thinking and making that is open to further change.

Case study 2: a framework for play experience

Brigid Costello, an artist researcher, makes artworks that enable playful interactive experiences for audiences.[29] Her trajectory of practice and research moved through several stages: from making artworks, from which questions were formulated, to generating design strategies, which were tested with existing artefacts, to the creation of new works using the tested (and modified) strategies. The numbers and lettering in Figure 2.2.4 correspond to the following description:

---

1   **Practice:** Brigid created a number of interactive **W**orks that enabled her to explore audience experience using **C**riteria for design (strategies that shape her works so that they engendered or encouraged playful experiences).

2   **Theory:** from an exploration of theoretical literature about play, pleasure and related phenomena, she developed a **F**ramework of play based on 13 pleasure categories.

3   **Evaluation: W**orks created using the modified **C**riteria were studied, in which the **F**ramework was used to support the evaluation of observational data gathered from audience experience studies.

4   **Theory:** from the **R**esults of the audience studies, new understandings about the capability of interactive works for play experience were derived and the **F**ramework was refined.

5   **Theory:** a relationship between the refined **C**riteria and the final version of the **F**ramework was established. The 'play framework' of 13 pleasure categories provides a structure both for creation and evaluation of work.

From an exploration of literature about play and related phenomena, she developed a Pleasure Framework (originally called a 'play framework'), which she used to support the evaluation of data gathered from studies of audience experiences with her artworks. From this, she gained new insights into the capability of interactive works for play experience and was able then to refine the framework, which also provided a structure that guided the making as well as evaluation of her works.[30]

This example of theory in a practitioner framework offered a new method for creating works in terms of different qualities of audience experience (see Table 2.2.1). Although the Pleasure Framework had a crucial role in evaluating audience experience, it was intended primarily to be a way of framing the creative thinking in terms of the kind of experience the artist wanted the work to create.

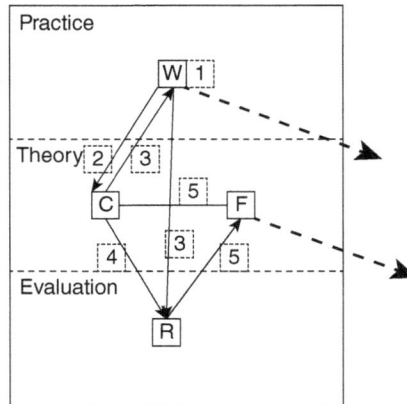

Figure 2.2.4 Trajectory model variant: practice drives theory.

Table 2.2.1 The pleasure framework revised.

Pleasure Framework Category	Description
Mastering and competing	The pleasure of playfully developing a skill, performing something skilfully and/or trying to achieve a defined goal. The pleasure of playfully competing against someone or something.
Making and building	The pleasure of playfully creating or building something and of expressing oneself creatively.
Imagining and acting out	The pleasure of playfully acting out something that mimics real life or is a creation of the imagination.
Exploring and discovering	The pleasure of playfully exploring something, making a discovery or working something out.
Leading, following and collaborating	The pleasure of feeling captivated or controlled by something or someone during play. The pleasure of playfully taking the lead, following a leader and/or collaborating with others.
Feeling and sensing	The pleasure of physically feeling one's own emotions, sensations or movements made during play. The pleasure of empathetically feeling the emotions, sensations or movements of someone or something else while at play.
Acting up and taking risks	The pleasure of playfully feeling scared, surprised, in danger or at risk. The pleasure of oneself, or someone else, playfully breaking rules, behaving mischievously or acting subversively.

It has been applied extensively and found to be very useful as an evaluation tool, particularly where facilitating dialogue between artist and audience was required. Through testing in use, modifications that gave more depth to the categories were identified. These developments are described in detail in Costello and Edmonds (2009).[31]

The ongoing evolution of the framework and its extension and dissemination has entered the collective knowledge of the field of interactive design and art in particular.[32] In her book *Rhythm, Play and Interaction Design*,[33] she has taken the new knowledge further, having been inspired to focus on rhythm and play and, at the same time, developing knowledge derived from practice by focusing on revealing practitioner thinking and working methods through their first-hand accounts. The original framework continues to be a live entity and is in use by others who are testing its applicability to their own situations of use. In an attempt to examine more specific playful experiences for the evaluation of interactive products, Lucero et al.[34] expanded the framework to include game experiences, emotions, play and reasons why people play, building on the original model and extending the number of categories in order to examine the wide range of experiences elicited by interactive products when they are used in a playful manner. The focus shifted from pleasures to experiences to indicate that not all such experiences are always pleasurable in the context of play. The categories are intended to identify the most playful features of different products.

Brigid Costello's framework continues to be used for evaluation of aspects of playful interaction.[35] She continues to advance the modelling of the practice-based research process. In her chapter in this book, she relates the trajectory model to the time-based rhythmic structures of practice-based research where practice and research interweave as if on a journey through a landscape.[36]

Practice-based research is making important contributions to the development of practitioner knowledge expressed as frameworks for thinking and making. In the examples described earlier, we can see how new theories developed through practice-based research and are refined and strengthened for use in practice, a process that leads to the continuing evolution of practitioner knowledge. In other publications, it is illustrated in the research of Zafer Bilda, Andrew Johnston and Lizzie Muller.[37][38]

Applying and adapting the trajectory model

This chapter has described how *theory as active agent* works in the development of new knowledge through practice-based research and the role of frameworks as examples of practitioner theory. It has described the trajectory model of practice-based research, representing the process in which practitioner frameworks are created and modified, followed by examples of frameworks that represent new forms of knowledge.

Practitioner researchers embarking on new projects, whether as part of a PhD or other reasons, can apply the modelling approach to scoping and defining how they go about their own practice-based research processes. A starting point is to ask questions, for example:

- How do I draw on and extend existing theories in other disciplines as well as my own?
- What working methods do I use and what happens when I reflect on the results of that use?
- Do I have a nascent framework that I can use in different situations of my practice?
- Does that lead to changes in the framework, and if so, is that new knowledge?
- What kind of new knowledge do I need for my ongoing work and how much does that augment my practice more generally?

In practice-based research, new knowledge across different domains is being generated that is made available to other practitioners to use and develop according to their situation of practice. The frameworks described are manifestations of practitioner theory embodied in tools and methods for thinking, creating and evaluating in practice. They represent an advance in understanding about the nature of practitioner theory and contribute to the collective knowledge of a given field or discipline.

Notes

1 Popper (1959, p. 59).
2 The theory of falsification was proposed by Karl Popper: for any theory to be considered scientific, it must be subject to tests that aim to disprove its validity. In general, science should attempt to disprove a theory, rather than support theoretical hypotheses; it is a way of demarcating science from non-science.

3 Darwin (1859).

4 Grant, Peter R., & Grant, B. Rosemary (July 14, 2006). Peter and Rosemary Grant and colleagues studied Galápagos finch populations every year since 1976 and have provided important demonstrations of the operation of natural selection. The Grants found changes from one generation to the next in the beak shapes of the medium ground finches on the Galápagos island of Daphne Major.

5 The Frankfurt School was a group of scholars known for developing critical theory and popularising the dialectical method of learning. It is associated with the work of Max Horkheimer, Theodor W. Adorno, Erich Fromm, and Herbert Marcuse: www.thoughtco.com/frankfurt-school-3026079.

6 Leavis (1974).

7 Schön (1983).

8 'Reflection in action': Schön defined reflection-*in*-action as:

> "When someone reflects-in-action, he becomes a researcher in the practice context. He is not dependent on the categories of established theory and technique but constructs a new theory of the unique case. His inquiry is not limited to a deliberation about means which depends on a prior agreement about ends. He does not keep means and ends separate but defines them interactively as he frames a problematic situation. He does not separate thinking from doing, ratiocinating his way to a decision which he must later convert to action. Because his experimenting is a kind of action, implementation is built into his inquiry.

9 Candy (2020).

10 John Dewey (1859–1952) was one of American pragmatism's early founders, along with Charles Sanders Peirce and William James, and arguably the most prominent American intellectual for the first half of the twentieth century. Dewey developed systematic views in ethics, epistemology, logic, metaphysics, aesthetics, and philosophy of religion: https://plato.stanford.edu/entries/dewey/ (accessed November 1, 2020).

11 Dewey (1933).

12 Adelman (1993).

13 Lewin (1948, p. 206).

14 The relationship of Dewey and Schön is neglected, argue Colucci and Colombo (2017), adopting a historical and theoretical perspective. Their relationship is discussed with particular reference to a common theoretical basis and conception of activity. Dewey and Lewin can be seen to contribute to contemporary psychology – action research in particular – by furthering its "emancipatory social relevance."

15 Smith (2017). Action research: Smith explores the development of different traditions of action research. Lewin's action research spiral "is not a 'method' or a 'procedure' for research but a series of commitments to observe and problematise through practice a series of principles for conducting social enquiry. . . . While Lewin does talk about action research as a method, he is stressing a contrast between this form of interpretative practice and more traditional empirical-analytic research": https://infed.org/mobi/action-research/.

16 The relationship between problem solving and problem findings has been explored in creativity research. Getzels and Csikszentmihalyi studied the way problems are envisaged, posed, formulated and created. Whether working in art, business or clinical investigation, great problem finders can look at the same phenomena that others have observed, identify gaps in understanding and come up with new perspectives that lead to a creative solution. Getzels, J. W., & Csikszentmihalyi, M. (1976). *The Creative Vision: A Longitudinal Study of Problem Finding in Art*. New York: Wiley.

17 Schön (1983 chapter 5).

18 Argyris et al. (1985).

19 Ibid., pp. 82–83.

20 Sullivan (2010).

21 Edmonds and Candy (2010).

22 Clark, S. (2018).

23 Bertalanffy (1969). Systems theory is the study of systems and there are many domain-specific versions. A system is a set of interrelated and interdependent elements which can be natural or human-made. Every system is bounded by space and time, influenced by its environment, defined by its structure and purpose and expressed through its functioning. See https://en.wikipedia.org/wiki/Systems_theory.

24 Bell (2022).

25 Seevinck and Edmonds (2008).

26 In the book Seevinck (2017, pp. 25–26), the classification of eight types of emergence are described in full.

27 Seevinck (2017).

28 Ibid., p. 21.

29 Costello (2009).

30 Costello and Edmonds (2007).
31 Costello and Edmonds (2009).
32 Seevinck's papers on the framework have been cited by others within and without the originating field.
33 Costello (2018).
34 Lucero et al. (2013).
35 Duarte et al. (2020).
36 Costello (2022).
37 Bilda et al. (2007), Johnston (2009), Muller (2008).
38 See details in Candy and Edmonds (2011).

Bibliography

Adelman, C. (1993). Kurt Lewin and the Origins of Action Research. *Educational Action Research*, 1(1), 7–24. https://doi.org/10.1080/0965079930010102 (accessed November 2, 2020).

Argyris, C., Putnam, R., & McLain Smith, D. (1985). *Action Science*. San Francisco: Jossey-Bass (In chapter 3 Theories of Action, 82–83).

Bell, A. (2021). *The Angel of Art Sees The Future Even As She Flies Backward: Enabling Deep Relational Encounter Through Participatory Practice-Based Research*. PhD thesis. London: De Montfort University.

Bell, A. (2022). A New Framework for Enabling Deep Relational Encounter Through Participatory Practice-Based Research. In C. Vear (Ed.), *The Routledge International Handbook of Practice-Based Research*. London and New York: Routledge.

Bertalanffy, von L. (1969). *General System Theory: Foundations, Development, Applications*. New York: George Braziller.

Bilda, Z., Candy, L., & Edmonds, E. A. (2007). An Embodied Cognition Framework for Interactive Experience. *Co-Design: International Journal of Co-Creation in Design and the Arts*, 3(2), 123–137.

Candy, L. (2020). *The Creative Reflective Practitioner*. London and New York: Routledge.

Candy, L., & Edmonds, E. A. (2011). *Interacting: Art, Research and the Creative Practitioner*. Faringdon: Libri Publications Ltd.

Clark, S. (2018). From Connected Digital Art to Cybernetic Ecologies, PhD Thesis De Montfort University.

Colucci, F. P., & Colombo, M. (2017). Dewey and Lewin: A Neglected Relationship and Its Current Relevance to Psychology. *Sage Journals Theory and Psychology: Research Article*, November 10. https://doi.org/10.1177/0959354317740229 (accessed November 2, 2020).

Costello, B. M. (2009). *Gestural Interfaces That Stimulate Creative Play*. PhD Thesis. Sydney: Creativity and Cognition Studios, University of Technology.

Costello, B. M. (2018). *Rhythm, Play and Interaction Design*. Springer Series on Cultural Computing, Springer Nature. Cham: Springer. http://doi.org/10.1007/978-3-319-67850-4.

Costello, B. M. (2022). Finding the Groove: The Rhythms of Practice-Based Research. In C. Vear (Ed.), *The Routledge International Handbook of Practice-Based Research*. London and New York: Routledge.

Costello, B. M., & Edmonds, E. A. (2007). A Study in Play, Pleasure and Interaction Design. In *Proceedings of the 2007 Conference on Designing Pleasurable Products and Interfaces, DPPI'07*. New York: Association for Computing Machinery, 76–91. Presented at 2007 Conference on Designing Pleasurable Products and Interfaces, DPPI'07, Helsinki, Finland, August 22–25.

Costello, B. M., & Edmonds, E. A. (2009). A Tool for Characterizing the Experience of Play. In *Proceedings of the Sixth Australasian Conference on Interactive Entertainment*. New York: ACM. Presented at Australasian Conference on Interactive Entertainment, Sydney.

Darwin, C. (1859). *The Origin of Species: By Means of Natural Selection, or the Preservation of Favoured Races in the Struggle for Life*. Cambridge Library Collection – Darwin, Evolution and Genetics, 2009. Cambridge: Cambridge University Press.

Dewey, J. (1933). *How We Think: A Restatement of the Relation of Reflective Thinking to the Educative Process*. Boston: D. C. Heath, revised ed.

Duarte, E. P., Méndez Mendoza, Y. L., & Baranauskas, M. C. C. (2020). *InsTime: A Case Study on the Co-design of Interactive Installations on Deep Time*. Proceedings DIS'20, ACM, Eindhoven, Netherlands, July 6–10.

Edmonds, E. A., & Candy, L. (2010). Relating Theory, Practice and Evaluation in Practitioner Research. *Leonardo Journal*, 43(5), 470–476.

Getzels, J. W., & Csikszentmihalyi, M. (1976). *The Creative Vision: A Longitudinal Study of Problem Finding in Art*. New York: Wiley.

Grant, P. R., & Grant, B. R. (2006). Evolution of Character Displacement in Darwin's Finches. *Science*, 313(5784), 224–226, July 14.

Johnston, A. (2009). *Interfaces for Musical Expression Based on Simulated Physical Models*. PhD thesis. Sydney: Creativity and Cognition Studios, University of Technology.

Leavis, F. R. (1974). *The Great Tradition*. London: Pelican Books, Published by Penguin, new impression ed.

Lewin, K. (1948). *Field Theory in Social Science; Selected Theoretical Papers*. Edited by D. Cartwright. New York: Harper & Row.

Lucero, A., Jussi Holopainen, J., Ollila, E., Riku Suomela, R., & Karapanos, E. (2013). *The Playful Experiences (PLEX) Framework as a Guide for Expert Evaluation, Designing Pleasurable Products and Interfaces'13*. Newcastle upon Tyne: ACM, September 3–5.

Muller, E. (2008). *The Experience of Interactive Art: A Curatorial Study*. PhD thesis. Sydney: Creativity and Cognition Studios, University of Technology.

Popper, K. (1959). *The Logic of Scientific Discovery*. New York: Basic Books.

Schön, D. (1983). *The Reflective Practitioner*. New York: Basic Books.

Seevinck, J. (2017). *Emergence in Interactive Art*. Springer Series on Cultural Computing. Cham: Springer International, Springer Nature.

Seevinck, J., & Edmonds, E. A. (2008). Emergence and the Art System "Plus Minus Now". *Design Studies Special Issue on Interaction Design*, 29(6), 541–555.

Smith, M. K. (2017). What Is Action Research and How Do We Do It? *The Encyclopedia of Pedagogy and Informal Education*. https://infed.org/mobi/action-research/ (accessed October 28, 2020).

Sullivan, G. (2010). *Art Practice as Research: Inquiry in the Visual Arts*. Thousand Oaks, CA: Sage Publications, 2nd ed.

2.3

MAPPING PRACTITIONER KNOWLEDGE

A framework for identifying new knowledge through practice-based research

Craig Vear

Part 1: problem definition: the dilemma of consolidating practice and research

Through my conversations with practitioners curious about expanding their practice into research, I have recognised that a core concept under discussion is their perceived difference between the nature of knowledge in practice and that of knowledge through research. On the one hand, their practice is generally understood as a whole world-of-concern that they invoke to create, explore, or be curious within, whereas their perceptions of research are that it is generally clinical, systematic, scientific and at odds with practice. This dichotomy can seem like an unsurmountable challenge, and consolidating them near impossible. But, as I will argue through this chapter, the two processes can be closer that we first realise.

Through my own experiences of supervising PhDs and conducting practice-based research over the past 30 years, I have at times experienced a lack of clarity about the precise nature of what I am investigating. Upon reflection, I realise that I struggled through some of my PhD (2008–11), blindly following my practice and intuition, secretly wishing for someone to point to a framework that would help me understand the precise nature of my research question, and a strategy for unpacking my practice to reveal the *where* and *what* of new knowledge. This was not the fault of the supervisors (far from it; they enriched me in many ways), but more a reflection on how I understood, and needed scaffolding for, a type of practice-based research process that was more specific and targeted. In a sense, bringing the systematic nature of research into the nature of my practice. This wasn't helped by the authoritative text books at that time, especially in my field (cybernetics and music performance), which generally avoided such a specific approach to identifying new knowledge in practice and instead offered broad, poetic or scientific terms and definitions. I feel that I could have been more productive if I had a clear focus at the start of my research. Also, I would have benefited from a clearer, more systematic approach to the research investigation, ultimately presenting more valuable research findings.

The problem for me lies in the way knowledge is defined and discussed, with terms such as *tacit* and *propositional, meta-cognitive* or *procedural* not really meaning much from my experiences as an experimental practitioner and as a relative late-comer to academia. Furthermore, as a practitioner, I don't really see a significant difference between what I do when I explore an idea through practice and investigate a question in practice-based research. The only significant change is focus and intent:

 DOI: 10.4324/9780429324154-16

but, at its core, I am still entering my world-of-concern (e.g. composing music) and, for example, manipulating materials (e.g. sound recordings), or narrative ideas (e.g. compositional structures), or technical procedures (e.g. adopting machine learning) to produce new understandings and insights. I am still thinking in-action and reflecting on perceived success against parameters (e.g. the aim of the investigation or the exploration of a creative idea). Obviously, the process that follows the practice is different and more involved, but it has an equally significant role in informing my practice or the next stages of the process. And yet, I don't feel that these are significant enough to warrant considering my practice-based research as something other than investigatory or curiosity-driven structured practice.

I concluded from this that my thinking about the difference between practice and research needed to change; and this is the guidance I now give to others. To this end, I developed a mapping framework, described here from the practitioner's perspective, with the aim of addressing this shift and, at the same time, providing a simple framework with which to navigate the practice domain in search of new knowledge.

Introduction: three modes of knowledge

It has been suggested that human knowledge expresses itself in three modes.[1] These are basically described as:

1   Explicit, semantic, or propositional knowledge (sometimes called 'know-that') which, at its basic level, describes things and phenomena using proven belief, with clear and unequivocal explanations.
2   Implicit, intuitive, or tacit knowing (sometimes called 'know-how', 'knowing in action', or 'know this'), which is really difficult to describe because it is inside our practice and actions.
3   Visual, pictorial, or episodic memory, which is the mental representations of events and experiences of a person's life.

To a practitioner entering into research for the first time, the characteristics of these first two modes seem to align to the perceptions of research and practice. And in a way they do: explicit/propositional knowledge feels scientific and observational in its nature, and implicit/tacit can describe the world of implicit knowing of one's unique practice which is difficult to categorise and describe. The challenge taken by this chapter is to consolidate these without losing that which is core to the practitioner's world of exploration.

I wish to discuss these two knowledges from a perspective argued by the French philosopher Henri Bergson (1859–1941). This has many parallels with how I, as a practitioner researcher-then-academic understand the relationship between research and practice, as I explain next. Furthermore, I will discuss how to use his philosophy as the basis for consolidating these two modes of knowledge so that the practitioner can maximise their practice-based research.

Bergson proposes that these two types of knowledge be understood as:

'Analysis'; 'intuition'[2]

Analysis is knowledge that explains things and phenomena from a distant perspective. It clearly aligns to the notion of propositional knowledge but places the viewer's perspective centre stage. *Analysis* is constructed by using different points of view and is expressed using symbols and codes (e.g. written words), concepts (e.g. theories and frameworks) and representation (e.g. images and media). This type of knowledge moves around an object, practice, or phenomena, describing or comparing it with an outside-looking-in perspective. The substance of this knowledge depends on the point of view that we position ourselves in relation to this thing, and on the symbols, concepts and representations that we use to express ourselves from such a position.

For Bergson, this first kind of knowledge is therefore relative, insofar as it uses something else to relate to the thing in question in order to bring clarity of understanding, and is told from an outside perspective looking into the thing. Bergson states that *analysis* is 'the operation which reduces the object to elements already known',[3] and even if we were to use tens of thousands of symbols and descriptors from an equal number of different perspectives, we would never form a unique understanding of what this thing is from the inside perspective.

This presents some problems for the practitioner researcher, whose perspective, and knowledge, from inside their practice is unique to them, and only them. When I think of my practice (music and creative AI) and the knowledge I have of doing practice, it is impossible for me to use outside analysis to get to the truth of my inside knowledge, because I would always be using comparative and relative symbols, concepts and representations of what it is from an outside perspective. Even if I was to use an infinite number of symbols to represent it, I would only achieve a complex reduction of it through elements already known. In short, this knowledge will always be relative: a translation into symbols from different perspectives outside the thing.

> We will call this perspective and type of knowledge *outside → in* or **in-vitro** (Latin: within the glass), as conceptually it resembles the type of knowledge generation that is developed by analysing from outside (in this case the practitioner researcher), by looking into the test tube (the practice).

Intuition, by contrast, is the knowledge gained from being one with a thing through direct lived experience. Bergson describes this as a method of feeling one's way intellectually into the inner heart of a thing to locate what is unique and inexpressible in it.[4] This has many similarities to the definitions of tacit knowing and similarly is very difficult to express. Examples of intuition can be viewed in jazz musicians, gardeners, cyclists and caricature artists, to name but a few. These practitioners generally just 'get on with it', and although they may be aware of their intuitive skills and creative processes, they will find it difficult to articulate precisely what is going on, and more than likely will gloss over it as a tacit (implicit) part of their creativity. Of interest to us here is the contrast in perspective: if *analysis* is knowledge from the outside, then *intuition* is implicit knowledge of entering into the subject: from the inside.

> We will call this perspective and type of knowledge *inside → out* or **in-vivo** (Latin: within the living), as it is developed from the unique knowing from being inside (the practice). The challenge is how to discuss it without losing this inside perspective or reducing it to *in-vitro* observations

Challenge 1: consolidation

Our unique knowledge of our individual practice is *in-vivo*. Everything we know about it in the moment and the memories we have about prior experiences are from the *inside → out* perspective, and that contributes to a whole-knowing of what our unique practice and knowledge are. Crucially, this perspective places our sense-of-self at the centre of this knowledge. We cannot extricate this sense-of-self from our practice. This presents two problems:

- First, if we wished to discuss or share *in-vivo* knowledge of our practice, then we would need to find the most appropriate symbols, concepts and representations to enable another to understand that knowledge from the inside. If we describe it through *analysis* or *in-vitro* perspective, then we do not specify this unique knowledge as it reduces our inside knowledge by comparing it to something else, rendering it no longer unique, but general.

- Second, how can we share this knowledge given that one's sense-of-self is implicated in it?

For the latter problem, Bergson proposes a solution: if we were to describe the *in-vivo* knowledge in such a way that another person could, through some form of 'intellectual sympathy',[5] project themselves into the practice and identify themselves with its uniqueness, then this would invite them into our *in-vivo* knowledge in such a way that they could implicate their sense-of-self and prepare them for understanding the unique knowledge.

For the first problem, and also as a strategy for the latter, Bergson proposes that we utilise the potential of *analysis* to describe the complexity and uniqueness of *in-vivo* knowledge. While it is true that no image, symbol or concept can combine to create such unique knowing, it is possible to use these as tools, not to describe directly, but to:

Zero-in on a point of in-vivo knowledge that gets the beneficiary closer to this inside → out knowledge

This is not to say that *in-vitro* knowledge in isolation is not an appropriate means with which to share practice-based research. On the contrary, *in-vitro* analysis is an ideal way to share empirical results and generalisations of *outside → in* knowledge using sense perceptions, observations and rationalised explanations, including written and verbal explanations, photographed images, media examples, frameworks and working models. But if the new knowledge from practice-based research is unique to the *in-vivo* perspective, then sharing its richness and uniqueness from *inside → out* is critical. Although the analytical tools to achieve this may be similar to those used for *in-vitro* analysis, the next challenge is to identify the *what*, *where* and *for whom* of this new *in-vivo* knowledge in order to isolate and identify the precise tools of analysis to accomplish such a process.

Challenge 2: the *what, where* and *for whom* of new knowledge – a 4D visual metaphor

The core premise of this mapping framework is that *in-vivo* knowledge is not one thing but an entangled meshwork of many different strands of knowing types in constant flux and interaction.[6] By extension, that new *in-vivo* knowledge generated through practice-based research is not an amalgam of this entire meshwork, but is rather identifiable in a single or a handful of knowing types and is fluid and malleable over time and iterations of practice. And that crucially, sharing this new knowledge benefits the field by advancing or enhancing the field or its practice; and this demands precision.

To draw these elements together, I developed a visual metaphor (see Figure 2.3.1). The many different strands of coloured lines (here in grayscale), represent the different threads of *in-vivo* knowledge types that are present in practitioner's knowledge. This includes knowing of technical practice, process, personal belief, implicit, professional, situational, complex, conventional, trained, cognitive, codified, public, experiential, perceptual and sensual knowing, to name but a few, and at a basic level they can be categorised into three core domains: psychomotor (doing), affective (feeling) and cognitive (thinking/ reasoning) (discussed further later). They also intersect at nodal points, representing the points in our practice when new insights might emerge. If this image were animated, we would see these lines in constant motion as we engage in our practice, with these nodes firing at different times and in different combinations as new insights and instinctive ideas emerge.

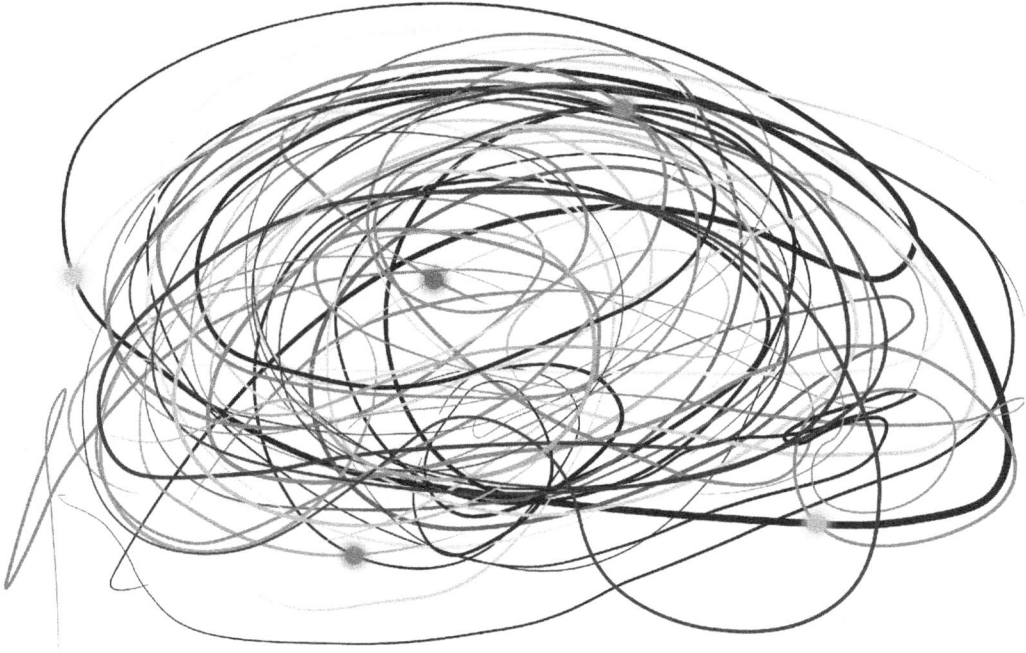

Figure 2.3.1 Visual metaphor for practitioner knowledge. 'Busy Mind' (2021).

Source: Serena Menezes Poltronieri.

Identifying and targeting new knowledge

The intention with this visual metaphor is that it acts like a gateway for the novice practitioner researcher to start to strategise about what the new *in-vivo* knowledge is in their practice. To illustrate this, I wish to describe how I understand practice-based research through this visual metaphor (Figure 2.3.1). To set this up, I will use the image of travelling to Barcelona as a metaphor for me travelling into a single instantiation of my practice. Through such a trip, I would have unique knowledge of my experience of Barcelona, only through this single trip, and from my perspective. But there would be millions of other people here in Barcelona, each of whom would be creating their own unique *in-vivo* experience. Therefore, this knowledge would be uniquely mine, and for that unique duration.

I could use the tools of *analysis* to describe the *in-vivo* perspective of my trip to Barcelona by presenting a range of videos, images, senses, tastes, sounds, feelings, perceptions, stories, histories, conversations, emotions, affectual encounters, flirts, exchanges etc. These combined would give the beneficiary an opportunity to get a sense of what it was like for me during the duration of my visit. The challenge here is to present enough materials and symbols to adequately invite the beneficiary into this experience with enough vitality so they can identify themselves within its uniqueness and prepare them for a similar *in-vivo* experience (inside looking out). This would need to stretch to the entirety of human experiences (metaphorically the totality of the intricate mesh of lines in Figure 2.3.1 for the duration of the visit), and perhaps wouldn't contribute to an enhancement or advancement of their understanding of Barcelona.

Now let's imagine that I wished to target some new understanding of Barcelona from my experience. I have identified that a new restaurant has opened and I wish to contribute to a traveller's guide to

Barcelona and to write a review of a new restaurant. This is metaphorically new knowledge, effectively shared. I could describe this meal using the tools of *analysis* from an *in-vitro* perspective, but if I wanted to share the uniqueness of my experience of that meal, I would use a combination of several different types of knowledge-threads as a tool to bring the reader closer to meal invoking a sense of its *in-vivo* experience. This might be taste, smell, ambience. I wouldn't need to use the entirety of my human experience of my whole trip to Barcelona to achieve either of these types of knowledge (*in-vitro* and *in-vivo*); instead I would strategically target those knowledge-threads and insights (the lines and nodes in the visual metaphor) that most effectively evoked the uniqueness of my experience for this new knowledge.

The key take-away here is that the practitioner researcher needs to be able to target specific aspects of the practice as their contribution to new knowledge for the betterment of others. Not everything that I do in my practice can be considered new knowledge, so I would need to identify what insights I could use as a contribution, and how best to share this through *in-vitro* and *in-vivo* tools of analysis. This is one of the key arguments, for me, for why the artefacts of practice cannot speak for themselves as new knowledge. Simply put, it does not in itself evoke the precise conditions for 'intellectual sympathy' that specifically evoke the unique *in-vivo* experience that has been defined as the new knowledge as there is so much other knowledge surrounding it. Hence the need for some sort of exposition or other analytical tools that points to and illuminates the uniqueness of the researchers intended new knowledge.

Part 2: mapping practitioner knowledge

In this section of the chapter, I outline the mapping framework. It does rely on a familiarity with the visual metaphor concept and the targeting new knowledge part. Additionally, the authorial voice has shifted its perspective so that it speaks directly to the reader as a practice-based researcher.

This mapping framework is designed to be universal, and not discipline dependent. Like the underlying concepts discussed earlier, it places your sense-of-self at the centre of this exercise, and so is meant to be personal and subjective to you. Also like each instantiation of practice being a unique experience through time, this mapping is not an anchor around the practitioner's ankle, but should be an agile and flexible method.

The mapping process consists of five layers:

1   Mode
2   Knowledge-type
3   Perspective
4   Preference
5   Verification

These are defined as:

Mode

This layer has two states and asks you to consider whether you are inducing new knowledge or deducing new knowledge through practice.

- *Inducing* means that you have identified a gap in the knowledge terrain of the field and can use your practice to provoke or generate new knowledge or insights. An example of this is: there is a weakness in a technical area of the practice in your field and you can use your skills to enhance or advance this type of knowledge.
- *Deducing* means that you have identified or reasoned a unique way of conducting your practice which can enhance or advance the field. An example of this is: there is a new or innovative

way of you conducting a technical area of practice in your field, and you study these in such a way that through sharing them with the field you can enhance or advance this type of knowledge.

These two modes have very different feelings to them when conducting your research. The inducing mode feels more open-ended and exploratory, especially at the start, whereas deducing is more concerned with testing and confirming theories and hypothesis. It might be that you shift modes through the process of practice-based research or run them in parallel.

Knowledge-type

This layer identifies the specific type of knowledge-threads that are best associated with the new knowledge. For example, technical skills, procedural narratives, material conversations, collaborative agents, affective feelings, cognitive constructs, kinaesthetic sensing, distributed patterns, engrained belief etc. From this identification process, it is possible to build the symbols, concepts and representations that describe the *in-vitro* and/or *in-vivo* perspective specific to each thread, thereby presenting enough relevant and targeted symbols, concepts and representations that can describe the knowledge from an outside perspective (*outside → in*); and/or adequately invite the field into this experience with enough vitality so they can identify themselves within its uniqueness and prepare them for a similar *in-vivo* experience (*inside → out*).

Through editing this handbook, I became aware of some of the generalisations and specificities of how the contributing authors describe and define new knowledge in practice-based research. From this range, and also in consultation with colleagues, I ended up with a list of commonly understood types of knowledge that are generally the focus, or result, of practice-based research. I stress that this list is not exhaustive, or finite, and really only covers the bare essentials to get the beginner or novice practitioner researcher thinking in more detail about practitioner-knowledge. These are roughly categorised into three types:[7]

Psychomotor knowledge (e.g. knowing through doing)

- **Technical skills:** the abilities and knowledge needed to perform specific tasks associated with the technical practice in your field. For example, the technical skill of making a violin sound in a musical manner.
- **Operational knowledge:** the abilities and skills needed to perform general tasks associated with the working practice in your field. For example, the ability to sequence melodic notes on a violin in a musical way.
- **Distributed knowledge:** in a multi-agent system, this is the knowledge that is shared across the multiple-agents. For example, the violinist working in a string quartet which is operating as a single agent in the flow of music-making, or how the physical presence of a musician's instrument informs and instruct their individual practice with that instrument.
- **Embodied knowledge:** the knowledge within the body about how to act, how to touch, how to know what to do, which is seemingly independent of the reasoning mind. For example, how the violinist uses certain ways of moving their body with certain techniques to stress musical inflection, without the mind controlling the process.

Affective knowledge (e.g. knowing through feeling)

- **Material affect:** the use of the things in your practice. For example, how these materials and their technical properties influence or shape surgical practice, or how the feel of the experiential qualities of surgical tools influences technical skills, and the ways the surgeon processes these materials using the tools of her field.

- **Spatial affect:** the relationship between the practitioner and the environment of their practice. For example, the relationship between a dancer and the physical performance space and the solutions and knowledges of this ecosystem.
- **Temporal affect:** the relationship between the practitioner and the durational dimension of their practice. For example, how the dancer organises and makes sense of their movement trajectory through time and the feeling associated with organisational strategies.
- **Belief knowledge:** the knowledge associated with knowing who we are and what our beliefs are. For example, the knowledge associated with the gut-feeling that a creative-coder has when she organises an algorithm in a specific way.
- **Aesthetic knowledge:** a practitioner's ways of understanding the sound, smell, taste, feel, look of the inherent elements of their practice. For example, how an engineer feels when they interpret mechanical components in technical drawings.
- **Kinaesthetic knowledge:** sometimes called tactile knowledge. Binds cognitive, perceptual and affective knowing through the acts of doing and experiencing. For example, how a dancer is able to approach an improvisation in an open and exploratory way having learnt core skills and techniques through years of training.[8]

Cognitive knowledge (e.g. knowing through thinking)

- **Engrained training (discipline theory):** the abilities and skills developed, and reinforced, through formal training. For example, the way a designer will develop a product using specific knowledge that has been agreed on and assumed by their communities' operational methods (e.g. higher education, work-place or geography).
- **Logical constructs:** the theories and concepts that guide the deductive reasoning about our practice. For example, the way a guitarist conceives of and justifies the way he composes pop songs, i.e. his formulae for success.
- **Implicit technique:** the embedded ways-of-doing our individual practice that affect what we do, and how we do it. For example, the habitual or unconscious decisions that an individual artist will make with a given brush or material.
- **Perceptual inference:** the immediate inference of meaning by the practitioner. For example, the way a software developer will demarcate and prioritise one property of their code over another in a given situation, and this is not acquired from prior assumption.
- **Narrative devices:** the structural concepts that a practitioner uses to organise their practice. For example, how an engineer plans the design of bridge, or a creative writer constructs a story.
- **Ludic and tactical systems:** the methods of play applied by the practitioner through their practice. For example, how an improvising musician improvises with other musicians and the mental processes that govern the decisions they make in-action.

Perspective

This layer highlights the different relationships each of the new knowledge types has with practice. It is adapted from cognitive science studies and the phenomenology of creativity.[9] In this work, they discuss creativity from the perspective of 4 Es: embodied, embedded, enacted and extended. They recognise that cognition is not solely in the realm of some thinking mind in which an understanding of cognition is only dealing with information-approaches to knowledge. In contrast, the 4 Es approach supports an approach to cognition as 'distributed across the entire body of a living system and its surrounding environment, and as continuous with the fundamental adaptive biological processes required for survival and flourishing'.[10]

Practice-based research will naturally generate knowledge and cognition through the phenomenological involvement of the practitioner researcher. New knowledge, therefore, can be distributed across different domains of practice. For example, the knowledge of doing a specific technical task can be understood from the perspective of physically doing the task (embedded), or applying this task in an applied situation (embodied), or evaluated as a model of thinking-through-doing (enacted), or in an ecological relationship between the practitioner and the environment of their practice (extended). Each of these dimensions is fluid and will naturally be reflected across each as they overlap. But each of these in isolation can specify some form of new knowledge in the field and help the practitioner-research zero in on a point of *in-vivo* knowledge that gets the field closer to this *inside* → *out* knowledge.

The difference in these perspectives can be understood as follows:

1 **Embedded *in* practice:** the specific skills and technical procedures of directly doing the practice.
2 **Embodied *with* practice:** carrying out or applying a procedure through executing, or implementing.
3 **Enacted *through* practice:** put into practice ideas and new models of thinking/approaching the practice, equally new concepts, new embodiments or ontologies of doing or behaving.
4 **Extended *by* practice:** the shifts in the practitioner and their ways-of-doing afforded by the relationships and possibilities generated between the practitioner and their world of practice.

To put these perspectives into practice, here is a hypothetical example. Elisabeth's practice is theatre making, and she wishes to explore the role that AI can play in generating narratives in collaboration with school children. Within this scenario, there can be lots of new knowledge e.g. advancing technical skills in the making of AI-theatre, developing theoretical frameworks for organising narratives in AI-theatre or affectual understanding of the child's relationships between AI narratives and audience reception, to name but a few. But we can go deeper and zero in on the perspective of how Elisabeth's knowledge relates to the practice. In this sense, is the precise new knowledge that Elisabeth seeks:

1 *In* her acts of doing as she generates new technical skills, for example the technical requirements and process of training neural networks and embedding AI into real-time narrative generation.
2 *With* her acts of creation as she extends her technical skills or understands more deeply the role of creator, for example working with AI in real-time narrative generation unveils new feelings or new possibilities about practice in her field.
3 *Through* her acts of making as she generates conceptual or philosophical understandings of this type of practice, for example the shifts in the ways of thinking creatively with AI and real-time narrative generation in digital theatre.
4 *By* her world of practice as these new technical agents and creative possibilities reach out to her and suggest, stimulate and shift her through 'suggestions or clues as to how to use the properties'[11] which are independent of her 'experience, knowledge, culture or ability to perceive'.[12]

It is possible that, depending on the size and scale of the research project, that two or all of these dimensions are the focus of the research investigation. But it is important to note that they should be considered discreet objectives so as to maximise the potential of new knowledge and the shareability of such knowledge to the field.

Preference

This layer is person-specific and relates to the unique ways each individual processes knowledge. It relates to the psychology of the individual and is a factor that is generally overlooked when discussing research. This adopts Kölb's cognitive processing model,[13] which maps an individual's cognitive processing preference as a construct of four key attributes:

- Activist (doing) e.g. problem solver.
- Theorist (thinking) e.g. new or improved models.
- Pragmatist (experiencing) e.g. new concepts and experimental.
- Reflector e.g. deep studies and observations.

It is important to stress that an individual's cognitive processing preference will be a combination of these four basic types. This is generally determined through a questionnaire,[14] the results of which are mapped onto a visual matrix highlighting the personal preference of an individual. For example, mine is (out of a possible 100% for each preference) 70% pragmatist (experiencing), 60% activist (doing), 55% theorist (thinking) and 40% reflector. This means that I naturally, or have bias, towards seeking and finding new knowledge as new concepts and an experimental approach, with a strong element of problem solving in order to present theory. This seems to align to the contents of my last book, *The Digital Score*,[15] which presents experimental practice, a typology of digital score types and a taxonomy of experience.

Verifying practice-based knowledge

New knowledge that emerges through practice-based research should be verified. The practice-based researcher needs to go beyond a single instantiation of an insight in order to reveal the precise nature of the new knowledge and to strengthen their belief in its truthfulness. This requires a process of verification through which the insight is challenged. This can be done, for example, through conducting multiple instances of the activity that generated the new knowledge, and could involve other practitioners. This mapping framework and the implementation of the various tools of *analysis* can help with this verification process, ensuring that the precise nature of the new knowledge is articulated through different lenses, and that the field can be assured that it has been challenged through the investigation process.

Two examples in practice

Before I move onto the exercise of mapping real-world practice-based research PhDs, here are two examples of this mapping framework in action. They are based loosely on real-world research, although have been simplified for the purposes of this exercise.

Example 1

Sara is a flutist and a practitioner researcher with many years of professional experience in contemporary music performance and commissioning new works. She has identified in her playing a new technical approach to her instrument that has yet to be documented in the field. Her research aim is to articulate this new technique through documented examples and thorough instructions. Additionally, she would like to present a critical commentary about how to enhance existing practices to best accommodate this new technique, so that others may adopt this technique and composers may start to apply it in their compositions. This is a small-scale project with modest goals, but she wishes to use this as a foundation for a larger study of the effect of this technique upon compositional process.

Using the layers listed above, her investigation through practice-based research and the *in-vivo* and *in-vitro* knowledge she would need to develop are:

1   **Mode:** her mode is *inducing* as she has identified a property of her practice that has hitherto not been presented to the field and intends to build a theoretical structure to share with others.
2   **Knowledge-type:** the basic type of knowledge would be *technical skill* as she is seeking to advance technical skills of an existing system beyond the known usage.
3   **Perspective:** Sara's aim is to present a taxonomy of new skills for her instrument and a reflection on how this can transform practice for others. The knowledge perspective is two-fold: *embedded in practice* for the technical skills, and *embodied with practice* for her critical reflection on the practitioner.
4   **Preference:** Sara has a preference towards *pragmatist*, with a secondary preference for *reflector*. Her preferred way is to experiment with a new concept of technique and to push this inquiry to breaking point.
5   **Verification:** the symbols, concepts and representations she would need to use as tools to best describe the *in-vitro* and *in-vivo* experience could include:

   a   *Embedded in practice* for the technical skills

   - Audio and video recordings of her technique in action (*in-vitro*)
   - Video recordings of her describing how to form the new shapes on her instrument (*in-vivo*)
   - Annotated images articulating the motor movement of her hand and lips (*in-vitro*)
   - Documented descriptions of her musical focus while working with the technique (*in-vivo*)
   - Instructional video or documented how-to's for other to follow (*in-vivo*)
   - Case studies of her training this technique to another musician (*in-vivo*)
   - Interviews and conversations with other instrumentalists on the translation from her playing to theirs (*in-vivo*)
   - Website containing a multi-media resource supporting this new technique (*in-vitro*)

   b   *Embodied with practice* for her critical reflection on the practitioner

   - Audio and video recordings of her applying this technique through an improvisation or an open score (*in-vitro*)
   - Video recordings of her describing the technique to a composer (*in-vivo*)
   - Critical commentary of the potential of how this technique can be best applied by composers (*in-vivo*)
   - Annotated video recording of her subjective thoughts as an over-layer to a recorded performance (*in-vivo*)
   - Auto-ethnographic/poetics of the practice
   - Interviews and discussion between her and the composer (*in-vivo*)
   - Website expanding the multi-media resource from above to include this example application as a case study (*in-vitro*)

Evaluation of example 1

Although similar analytical tools are applied to both dimensions, the purpose of them changes the subject of the study and the potential for the beneficiary to place themselves into the *in-vivo* experience. In this example, it might have been feasible to extend the *knowledge-type* layer to include kinaesthetic involvement as a knowledge-trait, in which case the symbol, concepts and representations would need to be augmented to capture both an inside and outside perspective. But what I hope is clear is that identifying the *what*, *where* and *for whom* parameters of this research project and mapping these using this framework allowed Sara to zero in on a point of *in-vivo* knowledge that gets the field closer to her *inside → out* knowledge in both dimensions as discreet contributions to knowledge for the field, rather than smudging them together as one objective in the investigation.

Example 2

William is a creative AI researcher, digital performance maker and PhD candidate. He is investigating how the design of director AI used in computer games design can be transformed into an active agent with its own needs and trajectories. His proposition is to conceptualise the director AI using embodiment theory from digital performance in order to assign it behavioural attributes beyond the current state of the art. He wishes to articulate this as a framework that defines behavioural traits developed through in-the-loop performance and affective immersion, so that gaming design teams can enhance their implementation of director AI into next stage games development. This is a PhD project and a collaboration between industry and academia.

Using the layers listed earlier, his investigation through practice-based research and the *in-vivo* and *in-vitro* knowledge he would need to develop are:

1   **Mode:** his mode is *deducing* as he has identified, reasoned and deduced a unique way of conducting the practice of game design which can enhance or advance the field through a framework.
2   **Knowledge-type:** the basic type of knowledge would be both *operational knowledge* as he is seeking to advance the abilities and skills needed to perform general tasks associated with the working practice in your field; and *ludic and tactical systems* as he is also seeking to design a framework that can control the behaviour of the agents in the practice.
3   **Perspective:** William's aim is to present a framework to guide future games designers with a new model of thinking about their field and a new conceptual way of achieving this. The knowledge perspective is: *enacted through practice* for the new conceptual way of doing the practice. This will be achieved using in-the-loop frameworks by placing himself into the role of director AI using digital performance techniques to gain the *inside → out* perspective of *in-vivo* knowledge.
4   **Preference:** William has a preference towards *reflector*, with a secondary preference for *theorist*. His preferred way to conduct the research investigation is through deep practice-based studies and iterative design processes in order to develop new models of practice or that guide the practice of others.

The *what*, *where* and *for whom* are therefore:

- **What:** a framework for guiding new ways of doing and conceptualising practices.
- **Where:** enacted through practice (evaluating).
- **For whom:** professional games developers.

5   Verification: The symbols, concepts and representations he would need to use as tools to best describe the *in-vitro* and *in-vivo* experience could include:

1   *Enacted through practice* for the new conceptual way of doing the practice.

- Audio and video recordings of his in-the-loop experiments of operating like a Director AI (in this case using Dungeons and Dragons as a platform) (*in-vitro*)
- Annotated description overlaid onto the video recording from the perspective of the director AI (*in-vivo*)

- Schematic/branch-tree flow chart of basic director AI needs and wants (*in-vivo*)
- written blogs describing his ongoing findings and framework development (*in-vitro*)/ (*in-vivo*)
- Comments from industry and stakeholders through the blog and social media (*in-vitro*)
- Written framework and theoretical foundations (*in-vivo*)
- Test and implementation of framework in real-world game-development (*in-vitro*)
- Annotated dialogues and discussions with industrial partners through development process (*in-vivo*)
- Recorded evidence of the test implementations (*in-vitro*)
- Analysed and evaluated documented consultations with a wider group of industrial stakeholders on the value of his framework (*in-vitro*)

Evaluation of example 2

The knowledge generated by William's *in-vivo* experimentation of in-the-loop play as if he was a director AI has a profound influence on his framework. This unique knowledge from the inside enabled him to get a sense of what it was like to be a director AI and to analyse from the inside the needs and wants, behaviours and decision-making processes that contributed to a rounder construction of the framework. This, in turn, was realised by the game developer who implemented it as a beta version of a game, from which both perspectives were used to evaluate its effectiveness and innovative qualities. William was able to assess it from the *in-vivo* perspective of knowing what it was like to be a director AI, and the game developer equally was able to assess it from the inside perspective of the maker of the game.

Part 3: case studies

The following case studies are real-world mappings of practice-based research PhDs from different fields, using this framework.

Case study 1: Bruce M. Mackh (USA)[16]

Bruce M. Mackh (USA) is head of the Department of Art and Design at Valdosta State University, USA. His PhD, *The Documentary Aesthetic*, was submitted at Texas Tech University in 2011. His research identified new theories regarding identification of objects as works of art within the context of art programs in higher education. The beneficiaries were students whose creative practice of visual art may not conform to faculty expectations of what "is art" or "is not art". Through his retrospective mapping of his PhD, he identified the following:

1 **Mode:**

   a   *Inducing:* I identified a gap between definitions of art as used by faculty when evaluating student artworks and objects accepted as works of art in the wider art world, specifically as this gap relates to perceptions that social documentary photographs can or cannot also be identified as works of art

2 **Knowledge-type:**

   a   *Aesthetic knowledge:* practitioner understanding of the elements of practice in creating photographic imagery

   b   *Cognitive knowledge:* logical constructs in prevailing theories of art and aesthetic philosophies governing identification of objects as works of art

3 **Perspective:**

    a   *Enacted:* I conducted ethnographic and auto-ethnographic research through interviews of practicing social documentary photographers reflecting on the relationship between their practice and their identification of their photographs as works of art

4 **Preference:**

    a   *Theorist:* presenting an improved model for identifying objects as works of art
    b   *Activist:* identifying the causes underlying the gap between faculty knowledge of art theory an art theory as it exists outside higher education

5 **Verification:**

- Formal research into aesthetic theory and art history, seeking to formulate a comprehensive definition of the word "art" and, by extension, parameters for identifying an object as art (*in-vitro*)
- Ethnographic interviews with practicing social documentary photographers inquiring about their creative process and their perspective on how they define their photographs as art (*in-vitro* and *in-vivo*)
- Auto-ethnographic of my own creative process and why I identify my photographs as art (*in-vivo*)
- Examination and synthesis of data gathered across the research into art theory an art history, the data resulting from the interviews, using "inverse fractal concept analysis" to generate insights into how prevailing norms in faculty members' education, training and professional expectations may fail to impart sufficient knowledge of contemporaneous art theory, which then affects faculty attitudes towards students' creative works (*in-vivo*)

Case study 2: Fania Raczinski (UK)

Fania Raczinski (UK) is a creative generalist/coder, transdisciplinary. Her PhD (submitted in 2018) investigated search algorithms and computer creativity through a poetry-generating website. The new knowledge generated was a framework for interpreting creative artefacts; this was embodied in her practice. The intended beneficiaries are artists, coders and poets. Through her retrospective mapping of her PhD, she identified the following:

1 **Mode:**

    a   *Inducing:* e.g. read a lot of OULIPO literature, got inspired to do poetry search results, combined with my web dev skills I ran experiments, then added the poetry as a feature retrospectively when I knew it worked
    b   *Deducing:* e.g. algorithm design – come up with idea in shower first, dictate into phone, write it out on paper next, then pseudo code, then code snippet, then prototype, then live site, etc.

2 **Knowledge-type:** the main ones would be *technical skill* (programming), *aesthetic knowledge* (web design and thesis design) and *logical constructs* (evaluation framework)
3 **Perspective:** *embodied with practice*
4 **Preference:** very strong preference for *reflector* type (synthesis of information) and *theorist* (analysis of philosophical computer zombies, evaluation framework), although I certainly had periods that felt more pragmatist (code, art, experiments)

5   **Verification:**

- Public website with main artistic artefact (poetic search engine) (*in-vivo* and *in-vitro*)
- Source code and documentation publicly available for artefact (*in-vivo* and *in-vitro*)
- Documentation of development practice (e.g. code snippets, flow charts, screenshots and mathematical equations) (*in-vivo* and *in-vitro*)
- Thorough analysis of artefact (e.g. benchmarking, lots of numbers and tables) (*in-vivo* and *in-vitro*)
- Inclusion of artefacts output (poems) in thesis (*in-vitro*)
- Source code version control logs (*in-vivo* and *in-vitro*)
- Theoretical foundations for creativity evaluation in computers (*in-vitro*)
- Philosophical issues with interpretation of creativity in computers (*in-vivo*)
- Evaluation framework for artefacts of computer creativity (*in-vivo*)
- Impact case studies, where I discussed two real-world applications of how my work was used (*in-vivo* and *in-vitro*)
- Discussion of ongoing/iterative nature of work, aspirations, limits, bias (*in-vivo*)

Case study 3: Kerry Francksen (UK)[17]

Kerry is an independent dance-artist researcher. Her PhD,[18] *Intimate Bodies and Technologies*, was submitted in 2016 through the Institute of Creative Technologies at De Montfort University. Her new knowledge was a methodological approach and framework for new ways of perceiving and experiencing live and digital materials. This was enacted through the practice of experiencing and perceiving (continuous evaluation) for the benefit of dance makers, choreographers, students and digital-dance practitioners. Through her retrospective mapping of her PhD, she identified the following:

1   **Mode:**

a   *Inducing:* my research observed a problematic relationship in live and digital dance performance and explored and developed an evolving framework for movement invention. The aim of the research was to identify and then implement a methodological approach, and to model new ways for experiencing and devising digital choreographies. This offered the field a new conceptual method for achieving synchronicity between live and digital materials.

2   **Knowledge-type:**

a   *Psychomotor knowledge (specifically embodied knowledge):* the dancer's perception and understanding of the elements of moving in digital environments
b   *Affective knowledge (specifically spatial effect, temporal effect and kinaesthetic knowledge):* the dancer's experiences of the complex relations between live and digital materials in practice, resulting in a methodological approach
c   *Cognitive knowledge (specifically ludic and tactical systems):* implementing technological and improvisational methods to provoke the dancer's thought processes and performative decision making in-action

3   **Perspective:**

a   *Enacted through practice:* my research established new models and systems for approaching live and digital materials in practice, which gave rise to new behaviours and ways of thinking about live-digital choreographies b   *Extended by practice:* the advances made by the dancers in terms of establishing new ways-of-doing was critical in shifting the world of practice for dance in media–rich environments

4   **Preference:**

a   *Reflector* and *activist:* my desire was to deeply understand the processes for experiencing and perceiving digital systems, and this drove the research imperatives. I was also constantly problem-solving throughout and this helped to inform both the theory and critical review processes too.

5.   **Verification:**

a   *Enacted through practice* for the new methodological way of doing the practice

- Continuous development of digital systems and choreographic scores to inspire new experiences for the dancer (*in-vitro*)
- Video recordings and testimonials from the dancers of their experiences of moving within the digital systems (*in-vivo*)
- Development of performance frameworks and structures for initiating the practice (*in-vivo*)
- Test and implementation of methodology via performance events (*in-vivo*)
- Audience surveys and documentation from WIP sharing's (*in-vitro*)
- Ongoing analysis of practice and context for building appropriate theoretical frameworks (*in-vitro* and *in-vivo*)
- Visual illustrations (annotated) of the system patches and descriptions of the technological set-up (*in-vitro*)

b   *Embodied with practice* for the critical reflection of the dancers' perceptual and embodied responses to the process

- Video recordings of the dancers' responses to each system development, and their subsequent movement decisions (*in-vitro*)
- Verbal reports from the dancers describing their reactions, and analysis of the appropriate technical and performative behaviours being developed (*in-vitro* and *in-vivo*)
- Critical observations of the dancers' embodied reactions to the changing environments and the impact this has on the development of technological systems (*in-vivo*)
- Annotated recordings of the dancers' subjective thoughts and ongoing experiences of the mediated environment (*in-vivo*)
- Auto-ethnographic reflection and analysis of the practical discoveries in context with peers and contemporaries within the field (*in-vitro* and *in-vivo*)

Case study 4: Andrea Bolzoni (Italy)

Andrea is a PhD candidate in his first year of study. He is a professional musician and researcher in music-AI seeking to find an embodied AI system for music improvisation. His aim is to investigate novel techniques to analyse and model human creativity and expand it in the form of an AI creative system, by exploring creative practice in musical improvisation through the interaction between a human musician and the digital system. The intended beneficiaries are professional musicians and music students of all ages. Through his mapping of his PhD he identified the following:

1 **Mode:** *inducing*
2 **Knowledge-type:**

   a *Aesthetic knowledge:* to be able to build an improvising system that has an identifiable and coherent aesthetic
   b *Ludical and tactical systems:* to define effective strategies to interact through improvisation

3 **Perspective:**

   a *Embodied with practice:* exploring improvised music creativity through interaction with the system
   b *Enacted through practice:* evaluation of the impact of the interaction with a creative AI improvising music agent in expanding human creativity, and production of pieces of music

4 **Preference:** I have a preference towards *pragmatist*, with a second preference for *activist*. My approach is to experiment to find new effective ways to interact with the AI improvising system and to find strategies to build upon the best ones
5 **Verification:**

   a *Embodied in practice*

   - Audio-video recordings of other musicians testing the AI improviser (*in-vivo*)
   - Interviews of the other musicians that tested with the AI improviser (*in-vivo*)
   - Documentation of the information collected in the interviews (*in-vitro*)

   b *Enacted through practice*

   - Audio-video recordings of music students exploring their creativity interacting with the AI improviser (*in-vivo*)
   - Interviews of the students that played with the AI improviser and/or their teacher (*in-vivo*)
   - Documentation of the information collected in the interviews (*in-vitro*)
   - Audio-video recordings of pieces of music produced by musicians playing together with the AI improviser (*in-vivo*)

Case study 5: Fabrizio Augusto Poltronieri (Brazil/UK/Germany)

Fabrizio's PhD investigated the role of chance in computational art, through a PhD programme in semiotics. His research questions revolved around the search for answers to how computers can be approached not only for their technical aspects but for the creative contributions provided by contact with them. From this perspective, the philosophical concept of chance plays a central role, and it was this philosophical basis that provided the initial ideas for my practical explorations. His new knowledge was a practical philosophical body of work to approach the role of chance in computer art; this was enacted through his practice. The intended beneficiaries were digital artists and creative computer practitioners. Through his retrospective mapping of his PhD, he identified the following:

1 **Mode:**

   a *Inducing:* I identified a gap in the theoretical and technical understanding of what computers are and how they can be used creatively

2 **Knowledge-type:**

a *Ludic and tactical systems:* I see computer systems as being, primarily, apparatus dedicated to playful play; to write code is to tell a narrative

3 **Perspective:**

a *Enacted through practice:* the way I approach problems is by trying to create new paradigms that can lead to new models of thinking and new concepts, taking as a starting point my knowledge accumulated over years of research and practice

4 **Preference:**

a *Pragmatist:* I do believe that trying to put theories into practice is the best way to act on the world, changing it through the logical, sensitive and symbolic tools we possess. Pragmatic thinking takes into account action on the world and its future consequences.

5 **Verification:**

- Annotations about my practice in pocket journals (*in-vivo*)
- Reading classical and contemporary philosophical texts on the concept of chance (*in-vitro*)
- Reading academic and technical texts regarding the application of chance in computer systems (*in-vitro*)
- Writing computational programs challenging theoretical concepts, exploring my vision of how to tackle the question of chance in a computational art practice (*in-vivo*)

Conclusion

This mapping framework is useful to those embarking on a project, such as PhD candidates, or reflecting on a research process. Equally, it can be a useful method through a research process to zoom into the precise nature of new knowledge that is emerging through the process of their research. Overall, it is a framework that hopefully leads the practitioner researcher towards precision and focus, and can guide a systematic and principled research investigation for those whose wish to do so.

At the heart of this framework is a conceptual understanding of the difference between two modes of knowledge associated with practice-based research: explicit/propositional and implicit/tacit. I discuss these in terms associated with Bergson's philosophy and draw a conceptual link between these two types based on perspective. This, I have experienced, has been useful in consolidating practitioner-knowledge with research, and in zooming in to the precise nature of new knowledge and identifying the suitable apparatus for drawing this out from inside so that it benefits the field.

The final part of this chapter presented a series of case studies from a range of practitioners associated with either my research institute or this handbook. The over-arching response was generally positive, with all practitioners stating that its use was effective and insightful. Interestingly, the retrospective mapping exercise of the four who had completed their PhDs unearthed new perspectives about their research. Andrea Bolzoni, who has only just embarked on a PhD study, did confess that the process was demanding as she needed to 'think hard through it'. But she did state that it is a framework that

> helps to clarify how to structure a research path through practice and, even if it is a little hard to pull out from the research project all the single components you are asked to fill it up, it will make life much easier, and research more effective, in the future.[19]

Through retrospective mapping, Francksen reflected that 'I really wish I'd had sight of this during my studies. It's super helpful as a scaffold, and going through the mapping process brought about some insights I could have had much earlier'.[20] Going further, she discussed that while the 'fundamental building blocks for the research were centred on embodiment, and thus the practice was pivotal for the knowledge discovered, considering how the process emerged through the lens of *enacted through practice* and by considering *extended by practice* usefully identifies the nuances of the evolving methodology', and ultimately led her to new ways of understanding and relating her PhD research.

In analysing the responses from these case studies, it is clear that the framework with its five layers can zoom in to precise types of new knowledge. Additionally, it doesn't seem to have been a limiting factor in the richness of the practice nor the range of outcome and documentation afforded to the practitioners. When I reflect on my own PhD journey, I cannot say with any conviction that my thesis was as precise as these: it wasn't bad, or sub-standard, but I wished I was able to bring as much precision through the process and in the exposition text. It is for this reason, and for those who equally strive for such precision and systematic approach, that this framework has been created.

Notes

1 Pöppel and Bao (2011).
2 Bergson's use of the word 'intuition' is not used in the same as everyday use of the word i.e. instinctive knowing without conscious reasoning.
3 Bergson (1949).
4 Ibid.
5 Bergson's Doctrine of Intuition Author(s): C. A. Bennett (1916).
6 Thanks to Tim Ingold's book *Lines* (2007) for inspiring this mental image presented here.
7 This division is based on Bloom's taxonomy of learning types.
8 Because this knowledge is felt through doing, it has been categorised in this domain, but is closely related to the psychomotor and cognition.
9 See Menary (2010a, 2010b), Rowlands (2010).
10 van der Schyff et al. (2018).
11 J. Gibson (1979).
12 Norman (1988).
13 Kolb (1984).
14 For example, www.emtrain.eu/learning-styles/.
15 Vear (2019).
16 Mackh (2022).
17 Francksen (2022).
18 www.pbrcookbook.com/intimate-bodies-and-technologies.
19 Personal email correspondence 12 March 2021.
20 Personal email correspondence 25 March 2021.

Bibliography

Bennett, C. A. (1916). Bergson's Doctrine of Intuition. *The Philosophical Review*, 25(1), 45–58, January. Duke University Press on behalf of Philosophical Review Stable. www.jstor.org/stable/2178562.
Bergson, H. (1949). *An Introduction to Metaphysics*. Translated by T. E. Hulme. Introduced by Thomas A. Goudge. New York: Liberal Arts Press.
Francksen, K. (2022). The Curious Nature of Negotiating Studio-Based Practice in PhD Research: 'Intimate Bodies and Technologies: A Concept for Live-Digital Dancing' Retrospective. In C. Vear (Ed.), *The Routledge International Handbook of Practice-Based Research*. London and New York: Routledge.
Gibson, J. (1979). *The Ecological Approach to Visual Perception*. Boston: Houghton Mifflin.
Ingold, T. (2007). *Lines: A Brief History*. London and New York: Routledge.
Kolb, D. A. (1984). *Experiential Learning: Experience as the Source of Learning and Development*. Englewood Cliffs, NJ: Prentice-Hall, vol. 1.

Mackh, B. M. (2022). The PhD in Visual Arts Practice in the USA: Beyond Elkins' Artists with PhDs. In C. Vear (Ed.), *The Routledge International Handbook of Practice-Based Research*. London and New York: Routledge.

Menary, R. A. (2010a). Introduction to the Special Issue on 4E Cognition. *Phenomenology and the Cognitive Sciences*, 9, 459–463.

Menary, R. A. (Ed.). (2010b). *The Extended Mind*. Cambridge, MA: MIT Press.

Norman, D. A. (1988). *The Psychology of Everyday Things*. New York: Basic Books.

Pöppel, E., & Bao, Y. (2011). Three Modes of Knowledge as Basis for Intercultural Cognition and Communication: A Theoretical Perspective. In S. Han & E. Pöppel (Eds.), *Culture and Neural Frames of Cognition and Communication: On Thinking*. Berlin, Heidelberg: Springer. https://doi.org/10.1007/978-3-642-15423-2_14.

Rowlands, M. (2010). *The New Science of the Mind: From Extended Mind to Embodied Phenomenology*. Cambridge, MA: MIT Press.

van der Schyff, D., Schiavio, A., Walton, A., Velardo, V., & Chemero, A. (2018). Musical Creativity and the Embodied Mind: Exploring the Possibilities of 4E Cognition and Dynamical Systems Theory. *Music & Science*, 1, 1–18. doi:10.1177/2059204318792319.

Vear, C. (2019). *The Digital Score*. New York: Routledge.

2.4

MAPPING THE NATURE OF KNOWLEDGE IN CREATIVE AND PRACTICE-BASED RESEARCH

Kristina Niedderer

Introducing the relationship of knowledge and practice in research

Knowledge plays a vital role in our lives, in that it reflects how we understand the world around us and determines how we act upon it. In this sense, knowledge is of particular importance for practitioners because they draw upon it to shape our world. While knowledge creation has traditionally been assumed by research, the use of practice within practice-based research has pointed to knowledge creation *through practice*.

Researchers in the creative and practice-based disciplines (PBDs) introduced the use of creative practice in research in the 1990s to be able to draw on and include practical knowledge. This has caused debate about what is formally accepted as knowledge in research[1] and raised the question about the nature, role, and format of knowledge in both research and practice, their compatibility, and integration for the PBDs. In the UK and many other countries, research regulations require a contribution to knowledge and understanding but remain silent about the nature of knowledge in the context of their specifications while implicitly prioritising propositional knowledge.[2] This has led to a number of problems concerning the role and format of knowledge in creative and practice-based research (PBR). For example, the language-based mode of propositional knowledge leads to its implicit prioritisation, which seems to exclude certain kinds or formats of knowledge associated with practice, which are often called practical, experiential, personal, or tacit knowledge and which evade verbal articulation. Polanyi puts the importance that practitioner researchers assign to practical knowledge succinctly into words:

> Rules of art can be useful, but they do not determine the practice of an art; they are maxims, which can serve as a guide to an art only if they can be integrated into the practical knowledge of the art. They cannot replace this knowledge.[3]

With 'rules of art', Polanyi refers here to subject-specific knowledge expressed in form of models and theories. He indicates that, while useful, there is another kind of practical knowledge, often called tacit knowledge, that is necessary to complement theoretical knowledge to make it applicable in real life. However, what exactly 'practical knowledge' is and how it can be included in research has remained elusive. This lack of understanding and recognition has created challenges for PBR and, in turn, with the applicability of research findings in practice. To address these challenges, this chapter

 DOI: 10.4324/9780429324154-17

examines current concepts and understandings of knowledge and their implications for PBR. It explores the nature of knowledge, referring to its philosophical foundations to reconsider and clarify the role and format of knowledge in relation to research and practice. The analysis aims to offer a deeper understanding of the foundations of knowledge and research, and of ways of including tacit knowledge within research through the use of practice to help researchers from the PBDs in defining their knowledge position and conceptual approach to research. The focus on the foundations of knowledge and research entails that this chapter cannot offer a detailed discussion of practice-based methods within its scope. Instead, these are further developed in following chapters, e.g. by Ernest Edmonds.

While this investigation has evolved from a national problem in the UK, the problem has also proven to be one of international significance. This is attested by international discussions (e.g. PhD-Design discussion list)[4] and conferences (e.g. Research into Practice,[5] EKSIG,[6] NordFO) as well as journal and book publications[7] concerned with this problem since the early 2000s, including this present volume.

The problem is discussed here, on a level that aims to be specific enough to be helpful but sufficiently generic to maintain its international relevance as well as its currency with the broader PBD community, including art, architecture, craft, design, education, engineering, music, nursing, and other PBDs. In explanation and justification of such a generic understanding, which can accommodate subject-specific individualities, Starszakowna has argued that:

> the concept of knowledge in art and design is, or should be, no different from the concept of knowledge in other disciplines. It is the constant search for, and ultimately the acquisition and dissemination of, a body of knowledge within particular areas or parameters which signifies a specific discipline. While the particular form that this knowledge might take will therefore vary, both between disciplines and within specialist areas within disciplines, such acquisition of knowledge is universal.[8]

A final aspect that might need clarification is the distinction between research and practice, which is used in this chapter, because one may occur in the context of the other. For example, a practitioner might also work in the academy and pursue research to inform their practice. Therefore, as distinguished by Niedderer,[9] the term 'research' is used to denote the systematic inquiry to the end of gaining new knowledge for one's field of study, and a 'researcher' is a person who pursues such research. 'Practice' is used to refer to professional practice, or to processes usually used in professional practice, including any creative practices, to produce professional work for any purpose other than the (deliberate) acquisition of new knowledge. 'Practitioner' accordingly refers to anyone who pursues professional practice. Practice-based research, then, refers to the pursuit of research, which uses practice, creative or otherwise, within and for the purposes of research, i.e. for the specific purpose of finding out new knowledge for one's field of study.

The two dimensions of tacit knowledge

This section examines two examples from the PBDs in order to draw out more clearly the nature of knowledge in research, the prioritisation of propositional knowledge and related problems, and how practice knowledge is different. Before looking at these examples, it is important to clarify the meaning of propositional knowledge, which is most commonly defined as 'justified true belief'. Grayling explains that

> this definition looks plausible because, at the very least, it seems that to know something one must believe it, that the belief must be true, and that one's reason for believing it must

be satisfactory in the light of some criteria – for one could not be said to know something if one's reasons for believing it were arbitrary or haphazard. So each of the three parts of the definition appears to express a necessary condition for knowledge, and the claim is that, taken together, they are sufficient.[10]

Despite continued criticism, the definition of knowledge as 'justified true belief' has remained the prevailing definition, and Niedderer[11] has shown that this understanding of propositional knowledge is implicit in the definition of research. This is because of requirements for research such as:

- the textual/written presentation of an intellectual position in the form of a proposition or thesis ('true belief')
- the logic of verification and defence of this intellectual position through argument and evidence (justification)
- the generalisability/transferability of research
- explicit and unambiguous communication

which are all criteria of, or indicators for, propositional knowledge.

The following two examples show that this understanding causes difficulties at different stages, and that it is at those stages that practical knowledge is missing. One of the two examples is from design/engineering, the other from art and music. The examples have been drawn from existing literature which is concerned with the problem of knowledge in relation to practice. They have been chosen because they offer discussion of two important generic knowledge areas of PBDs, one of which is related to procedural knowledge and expertise using the example of technical development; the other is related to experiential knowledge and connoisseurship, using the example of aesthetic evaluation and judgment.

Example 1. In the 1960s, a Canadian research laboratory successfully developed and built a so-called TEA-laser.[12] British attempts to replicate the laser on the basis of written information, or a third-person informant, however, failed as long as informants who had participated in building the original laser were not included personally in the replication-project.[13] Collins' (1985) study of the replication attempts further showed that an extended period of contact was required between the expert and the learner to transfer the practical knowledge, and that the learner could not tell whether they had acquired the relevant knowledge or skill until they tried it.

This example suggests that such practical knowledge is largely tacit, that such tacit knowledge is developed and plays an important part within research (here: the development of the laser), and that it evades the conventional textual communication of research. Polanyi describes this as follows:

> An art which cannot be specified in detail cannot be transmitted by prescription, since no prescription for it exists. It can be passed on only by example from master to apprentice.[14]

This kind of tacit knowledge is usually associated with practical knowledge and skill, and with vocational training. In many disciplines, it is regarded as distinct and excluded from academic research – or possibly overlooked and ignored – because it withstands articulation and argumentation and thus wider dissemination.[15] Only in the last 15 years or so has the importance of tacit knowledge been recognised and explored in PBR, often in educational and practice contexts (e.g. Wood, Rust, and Horne 2009; Warner, Seitamaa-Hakkarainen, and Hakkarainen 2021). This research has shown that the inclusion of tacit knowledge is often essential for success in terms of the communication of any insights, whether in research or practice. It is therefore associated with expertise, which has been defined as 'an intuitive grasp of the situation and a non-analytic and non-deliberative sense of the appropriate response to be made'.[16]

Example 2. In addition to successful communication, tacit knowledge seems essential also for successful inquiry itself, which the second example of knowledge from experience demonstrates. Reflecting on the 'signature touch' that distinguishes one pianist from another, Polanyi makes the observation that, technically, it is difficult to account for the difference in touch, even though we can clearly hear it.[17] This makes it difficult to describe or measure it sufficiently, e.g. for the purposes of teaching or evaluation.

This becomes clearer in the context of a more recent study. In her PhD *Delineating Disease: A System for Investigating Fibrodysplasia Ossificans Progressiva*,[18] Lucy Lyons used drawing as a tool to investigate and communicate the tissue changes associated with the disease. Working with pathologists, she drew bodies, using her aesthetic sense as an artist to interpret what she saw and to represent it through drawing. Through doing so, she was able to uncover changes not obvious in photographs of the same subject.

In both these cases (the pianist and the drawing), judgement relies on perceptual appreciation, also called connoisseurship, in which personal judgement is applied to sensory experience. Connoisseurship in the context of this investigation is referring to an ability for very fine (qualitative) discrimination that is (usually) beyond scientific measurement and that is acquired through extensive training.[19]

In these two examples, we have seen that tacit knowledge is an important requirement for creating new experiences, abilities, and knowledge and achieving best results, not just in practice but also in research. It plays a significant role in research in the process of generating and evaluating research and its results as well as of communicating and applying research outcomes and insights.

The following section examines what exactly we mean by practical or tacit knowledge, why and how it has this important role, and how it relates to propositional knowledge, drawing on philosophical sources in order to give the discussion a sound grounding.

Different types of knowledge and knowledge systems

Previously, we have discussed two different kinds of knowledge, tacit and propositional, but these two terms are not usually paired. Rather, tacit knowledge tends to be paired with explicit knowledge.[20] Propositional knowledge is variously paired with non-propositional knowledge, including experiential knowledge (also: knowledge by acquaintance) and procedural knowledge.[21] While the explicit–tacit knowledge-pair has been formed to denote and distinguish knowledge by the characteristic of communication, propositional and non-propositional knowledge pairs provide distinctions concerning their nature. However, the relationship between propositional and non-propositional knowledge seems not as clear-cut as that of explicit and tacit knowledge, because there are a number of different kinds of knowledge clustered under non-propositional knowledge (illustrated in Figure 2.4.1 below).

In the creative and practice-led disciplines, a variety of further terms are being used, such as practical or practice knowledge, skills knowledge, process knowledge, personal knowledge, implicit knowledge, professional knowledge, situational knowledge, control knowledge, complex knowledge, conventional knowledge, cognitive knowledge, codified knowledge, public knowledge.[22] Most of these terms have been created to describe different aspects or purposes of knowledge. While some of these terms offer important distinctions for their field, to discuss all of them in detail is beyond the scope of this chapter. This research therefore focuses on the distinction between propositional and non-propositional, explicit and tacit knowledge, which are the most significant pairs from a conceptual point of view, to examine their meaning and relationship in more detail. In addition, I will point out and differentiate synonymous terms where relevant.

Since knowledge is essentially a philosophical concept, in the following, the investigation is looking at philosophical approaches to knowledge before re-introducing them into the problematic of PBDs. Although there are a number of different types of knowledge discussed in philosophy, there seems to be some consent about what the key terms are: propositional knowledge, experiential knowledge, and procedural knowledge (e.g. Hospers 1990, e.g. Williams 2001, Grayling 2003).

Grayling explains that the definition of knowledge as justified true belief 'is intended to be an analysis of knowledge in the propositional sense'[23] rather than of knowledge that one might gain by being acquainted with something or someone, or that enables someone to do something (skill).

While there has been much debate about this definition of knowledge in the attempt to defeat or improve it, it has remained the central definition. An extensive and plausible defence is provided by Williams (2001), whose approach can be seen as a mediation between the two opposing positions of Foundationalism and Coherentism. While Foundationalism relies on foundational beliefs based on empiricism for the justification of knowledge ('prior grounding requirement'), which creates problems with accounting for any internal reality or the reality of other minds,[24] Coherentism relies on an intrinsically coherent system of beliefs that in turn has difficulties with accounting for our knowledge of (external) reality.[25]

In mediation of these two positions, Williams has proposed a third approach, which he calls 'Contextualism' and which assumes that one can rely on one's experience of external reality until there are reasons to challenge it (default and challenge requirement).[26] Context dependent, this allows assuming certain beliefs as foundational without the requirement of foundational atomism. It also avoids the circularity of Coherentism in that assumed foundational beliefs may be opened to scrutiny if the context changes. Williams argues that this approach is permissible because of the normativity of knowledge, which is not some a priori given but itself a human construct. The Contextualist approach seems to describe the way in which research operates in that it takes certain beliefs as foundational, on which it then tries to construct a coherent argument.[27] The following discussion therefore adopts Williams' contextualist approach to knowledge.

Relating propositional, experiential, and procedural knowledge

While propositional knowledge has been at the centre of epistemological discussion, experiential and procedural knowledge have been underrepresented. Philosophers have looked at these concepts separately,[28] but their satisfactory integration within epistemology so far appears outstanding. This section therefore examines the intrinsic characteristics of these concepts and their relationship.

Propositional knowledge, which is also associated with 'knowing-that', is usually expressed in the form of statements that can be verified or falsified and that allow in credibly believing that something is one way or another. In contrast, procedural knowledge refers to knowing 'how to do something in the sense of an ability or skill'.[29] Sternberg[30] associates procedural knowledge with tacit knowledge because its essence is difficult to put into words, as we have seen in Example 1. Drawing on Anderson (1976), Reber characterises procedural knowledge further by distinguishing it from declarative knowledge, a term used synonymously with propositional knowledge, pointing to its explicit character:

> Anderson's key distinction is that between declarative knowledge, which is knowledge that we are aware of and can articulate, and procedural knowledge, which is knowledge that guides action and decision making but typically lies outside of the scope of consciousness.[31]

While Reber makes a generic distinction between procedural and declarative knowledge as tacit and explicit respectively, I argue that this is not the complete picture because some parts of procedural knowledge can be made explicit. For example, in Example 1, it is possible to have explicit instructions of how to build the TEA-laser. This explicit part of procedural knowledge is called propositional or conceptual content.[32] The other part, which is tacit, is accordingly called non-conceptual or non-propositional content. It is less well understood because it persistently evades articulation and lies beyond the norms of declarative knowledge. The concept of non-propositional content is also more commonly associated with experience or perception. This brings us to the third of the three recognised categories: knowledge by acquaintance.

Knowledge by acquaintance is more often talked about as experiential, perceptual, or sensual knowledge. The term sensual knowledge is used to connote the unmediated reception of external reality through the senses. Perceptual knowledge is used to connote the reception of external stimuli mediated through human faculties.[33] Experiential knowledge is used by Williams[34] to connote the entirety of both, and therefore is the term and notion used here. However, the notion of experiential knowledge is not uncontentious, either. Because of its phenomenal nature, experiential knowledge is sometimes disregarded in terms of having any status as knowledge:

> Having a headache isn't knowledge though you certainly experience (are acquainted with) the headache; but knowing *that* you have a headache is. Seeing some colours in your visual field isn't knowledge; but forming concepts from your sensations and recognising that it's an animal stalking in the underbush, is. You couldn't have knowledge without acquaintance, but acquaintance alone is not yet knowledge.[35]

This problem leads back to, and is solved by, the idea of propositional and non-propositional content, because like procedural knowledge, experiential knowledge can be associated with displaying propositional and non-propositional content. For example, in Example 2 about piano playing, one may be able to experience the quality of a certain sound. One may also be able to recognise what one's experience means (non-propositional content) and thus to name it and to describe it (propositional content). However, one may not be able to justify one's experiential knowledge other than through pointing back to one's experience, which means that it is not necessarily possible for others to follow one's judgement. Also, one may not be able to describe one's experiential knowledge adequately with regard to replication. These reasons may be seen to validate and distinguish it from propositional knowledge, especially in the context of research and its requirements.

This indicates that the part of experiential knowledge that allows us to make sense of our experience seems elusive to articulation. Equally elusive to articulation is the part of procedural knowledge that allows us to act upon it, as we have seen. Thereby, the latter seems based strongly on the former. Further, both experiential and procedural knowledge become graspable through their propositional content, while propositional knowledge seems to receive its meaning from the experiential content that tacitly underlies it.[36] For example, written language – as a prevalent means of communication and storing knowledge – is constituted by arbitrary, socially agreed signs, which mean nothing until one knows and understands that these signs express certain concepts and what they mean, including the various layers of associations and connotations.[37]

Lyons' study (2009) expands Neuweg's point in that it elucidates the process of meaning making of experience through interpretation and conceptualisation as knowledge through language. It does so through its explicit two-stage process, firstly, the experience/reading and interpretation/representation of patterns through (the language of) drawing by the artist and, secondly, the interpretation of these patterns with regard to insights about the disease by and through the pathologists. This highlights the importance of experiential knowledge, in particular the (training in) fine discrimination required for identifying and eliciting such new knowledge.

Based on these considerations about the three kinds of knowledge and their relationship, I would like to propose that experiential knowledge can be understood as the basis for the other two kinds of knowledge (see Figure 2.4.1); that:

- procedural knowledge can be understood as experiential knowledge-in-action
- propositional knowledge can be understood as the norms or principles by which to understand experiential knowledge in that propositional knowledge helps to grasp and formalise experiential knowledge by conceptualising it in form of language

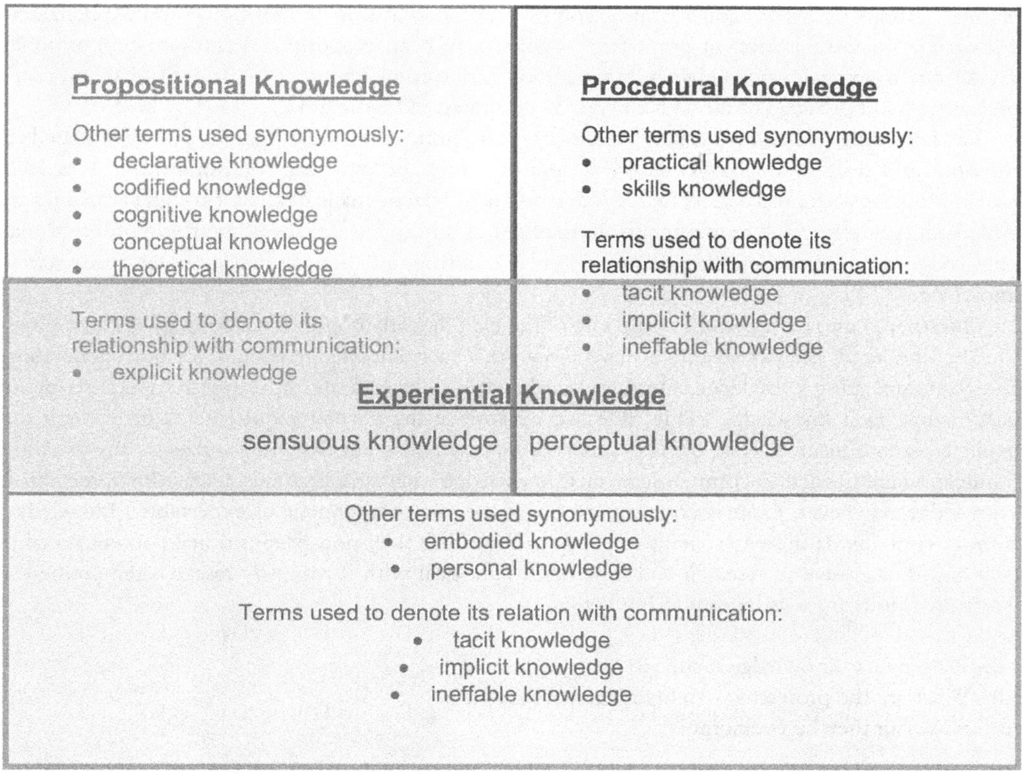

Figure 2.4.1 Relating the three types of knowledge: propositional, procedural, and experiential knowledge.

Furthermore, this would seem to indicate that non-propositional content (experience, procedural action) can exist without propositional content, but propositional content cannot exist without non-propositional content, because even Descartes' 'Ergo cogito, ergo sum'[38] is based on the experience of his self.

Positioning tacit knowledge within research

Having discussed the nature of, and a relational model for, the three different kinds of knowledge, the next step is to examine the format of these different kinds of knowledge in order to be able to consider the benefit of this inquiry for the PBDs in terms of including tacit knowledge within research.

The previous discussion has shown that propositional knowledge is usually associated with explicit knowledge, while non-propositional (experiential/procedural) knowledge is usually associated with tacit knowledge. It has also shown that there is a tacit component (non-propositional content) to propositional knowledge, which allows it for us to become meaningful, and there is an explicit component (propositional content) to non-propositional knowledge, which allows grasping and communicating the experiential content in form of concept(s). This indicates that the notion of explicit and tacit knowledge cannot simply be associated with propositional and non-propositional knowledge respectively, but that these concepts overlay one another orthogonally, and that the concept of explicit and tacit knowledge rather pertains to the notions of propositional and non-propositional *content.* Assuming an orthogonal relationship of propositional/non-propositional and explicit/tacit knowledge links the notion of propositional content of both propositional and non-propositional knowledge to explicit knowledge and that of non-propositional content of propositional and non-propositional knowledge to tacit knowledge as illustrated in Figure 2.4.2.

The preceding discussion has indicated that current notions of research are intrinsically related to the notion of propositional knowledge because of matters of logic and communication. Research has no problem with propositional knowledge because it can be made explicit through verbal means, which adheres to research requirements. In the light of this discussion, the reference to propositional knowledge must be understood to pertain to the notions of propositional content and explicit knowledge.

This, then, raises the question of the role of the tacit or non-propositional content of knowledge. On the one hand, the two examples have shown that tacit knowledge is vital, for both generating and communicating knowledge. On the other hand, it lacks recognition in research because, by its very nature, tacit knowledge evades research because of the current requirements of research for explicit communication. The obvious answer, then, would seem that non-verbal means of communication are needed to communicate tacit knowledge such as in Example 1 regarding procedural knowledge. However, explicit communication of the tacit component of experiential knowledge is more complex. It therefore seems important to establish why non-propositional content remains tacit and thus elusive to research and how one might deal with it to satisfy research requirements. Three questions are worth considering here:

i) Why is tacit knowledge tacit?
ii) What are the problems with being tacit for research?
iii) How can they be overcome?

The following reflections seek to answer these questions:

i) There are several sources that consider why non-propositional content is tacit. Most prominently, Polanyi explains it with the concept of focal and subsidiary awareness.[39] A common example is driving a car, where one needs to be aware of the road and the way one is going (focal awareness) while operating the car without being conscious all the time of single actions with the pedals, gearstick etc. (subsidiary awareness). This kind of split awareness has the great benefit that we are able to

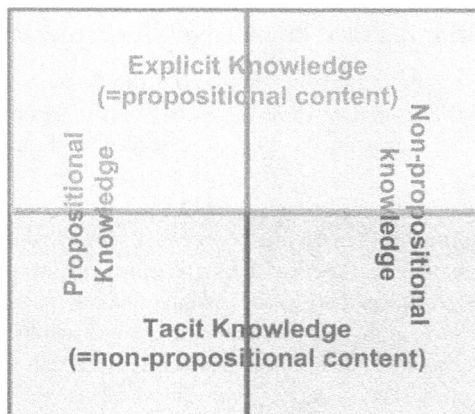

Figure 2.4.2   Knowledge to explicit and tacit knowledge.

act, because if we had to be aware of all stimuli and subsidiary actions all of the time, we would not be able to act at all.

ii) If tacit knowledge has this great benefit, the next consideration has to be why there are problems with research and what they are. The main argument is that research requires the conscious scrutiny of knowledge for the purpose of verification. While propositional content is open to this scrutiny, because it can be made explicit by verbal means, tacit knowledge seems to evade it. This has raised the question of whether tacit knowledge can be regarded as knowledge at all. If we follow Williams, who argues that we can speak of beliefs as knowledge if they can be verified, we may conclude that tacit knowledge should be regarded as knowledge if we can show that it can be verified.[40]

I would argue that tacit knowledge can be verified – and this is obvious in the two examples earlier – and not just that but that it is essential for meaning making and the creation of knowledge as a whole. Expanding on Example 1 on procedural knowledge (knowledge-in-action), in the most basic sense, every action constitutes a judgement over what is right (to do) in every given moment, and thus the knowledge is tacitly verified within and through action and its result. This can be explained further if following Williams' assumption that the two content states of knowledge are inseparable, and that therefore even where one speaks of tacit knowledge, propositional content is involved albeit has not been made explicit.[41] One can therefore assume experiential and procedural knowledge to adhere to notions of normativity and judgement – even when tacit – and that such judgement can be made explicit 'posthumously' through reflection, analysis, and explanation even though the underpinning experience cannot.

This analysis suggests that there are no intrinsic problems with the understanding and inclusion of tacit or non-propositional knowledge(s) in research. Indeed, such knowledge seems to be an intrinsic and essential part. The significance of this holistic understanding of knowledge for PBD research is expressed in Winch's writing (1958) as summarised by Smeyers:

> Winch's position implies that the discussion has to start from a particular social intercourse or 'practice'. It follows that the empirical observational methods (and statistical techniques) cannot possibly be the only yardstick. Instead, the human situatedness of the phenomena being researched requires that all our observations, arguments, and considerations must be based in our practices. Normative and value-laden elements have to play a crucial role throughout educational research and not just in the first or final stages.[42]

Despite this positive assessment, some practical problems with the integration and communication of tacit knowledge in research remain, because of the requirement for explicit analysis, explanation, and justification, which continue to be required by university regulations[43] as well as regulations for national research funding in the UK even today.[44]

iii)   Next, we need to look at how to overcome the problems regarding the integration of tacit knowledge into research and its communication as part of the research findings. Concerning the inclusion of tacit knowledge in research, practitioners from the PBDs have taken to using practice as part of their research in order to be able to draw on the tacit knowledge inherent in their practice.[45] While the discussion has shown that in principle there is no problem with using practice, the lack of clarity about knowledge has led to a lack of clarity about how to use practice.[46] This lack is inherent in both research regulations and research practice. Although research regulations formally allow practice into the research process,[47] they do not specify its purpose or role within research. This lack of specification, and of understanding knowledge, has led to an at times casual use of practice within research, which in turn has caused problems with the recognition of the use of practice as a valid and necessary means and method for making tacit knowledge available to the research process.

These considerations suggest that the explicit acknowledgement through research regulations and research requirements regarding the importance and intrinsic role of tacit knowledge for the research process as well as the provision of clear guidelines of how to do so would be of benefit to research in the PBDs. In the absence of such guidelines, one way to overcome this problem is through appropriately framing the use of practice within research (practice-based inquiry) by explicitly stating the role and purpose of practice as well as its processes etc. In other words, by drawing the related propositional content out explicitly and offering examples of the non-propositional content, both creative methods and processes and any resulting artefacts can be integrated successfully. A testament to this is the many successful practice-based PhD studies completed over the last 25 years. Good examples include Whiteley's research on prosthetic limbs using drawing,[48] Lyon's research into drawing to create new insights for pathologists,[49] Warpas' inquiry into designing for social dream spaces,[50] or Nyangiro's inquiry into food as a material for art practice.[51] Significantly, each of these studies includes examples of practice, in the form of drawings, video footage of an installation, or materials (spices in Nyangiro's case) to include and communicate the experiential/tacit component within the research.

This leads to the second problem, which concerns the communication and sharing of tacit knowledge. The above examples of PhD studies demonstrate that creative and non-verbal methods can be used and integrated with research publications, framed by explicit accounts of the propositional content, to communicate experiential and propositional knowledge quite comfortably. However, conveying the tacit content, which is necessary for the successful application of research findings in practice (cf. Example 1), has proven more difficult. Neuweg sums this problem up in pragmatic terms, declaring that 'although tacit knowledge is not teachable, it is coachable'.[52] This points to an intrinsic problem of research with tacit (embodied) knowledge, in that current research requirements do not acknowledge this vital component, which has consequences for the application of research in practice as well as for research education. Awareness of this issue on the part of supervisors and researchers involved in knowledge exchange will be a first step. The deliberate and explicit use of non-verbal means of communication to facilitate coaching where needed may be the second.

That this question about tacit knowledge is being asked may signal a positive development: a more critical engagement with prevailing research paradigms as well as an associated shift in research practice with which research requirements still have to catch up. Indeed, a gradual shift can be observed in a number of ways: there has been a rising number in literature over the last two decades concerned with knowledge and the use of practice in research in the creative disciplines[53] as well as other practice-led disciplines.[54] There has also been a rising number of PhDs in the PBDs, esp. in Europe and Australia.[55] There is even an emerging recognition of the applied nature of research, manifest through the introduction of impact planning and impact case studies by UK funding bodies.[56] Nevertheless, an explicit acknowledgement and guidance relating to the conduct, methodology and passing on of tacit research knowledge is still not available. Therefore, there is no unified understanding and approach, and the use of practice and the conveyance of tacit knowledge are still very much dependent on local practices of individual supervisors and doctoral training. This makes the present handbook important in paving the way towards such guidelines.

Conclusion and future research

This research has investigated the meaning, role, and format of knowledge in research and practice, with particular reference to research in the PBDs. The discussion has explored problems with the recognition of tacit knowledge within research, which have arisen because of the implicit prioritisation of propositional or explicit knowledge. Using various examples to analyse the nature of these problems has established that tacit knowledge plays an essential role in our ability to obtain highest achievements in practice as well as in research, often expressed as expertise and connoisseurship, and that therefore the deliberate inclusion of tacit knowledge within research is important and necessary.

In order to understand better how to include tacit knowledge within research, this chapter has examined the meaning and relationship of prevalent concepts of knowledge in philosophy. The discussion has proposed an orthogonal relationship between propositional and non-propositional knowledge and explicit and tacit knowledge in alignment with propositional and non-propositional content. These insights have enabled revisiting the role and format of knowledge in research, in particular of tacit knowledge, with regard to its inclusion and communication.

The conclusion is that there is no fundamental problem with including non-propositional knowledge in research, e.g. by means of using creative or professional practice, because non-propositional knowledge, too, has propositional content, which can be made explicit and by means of which it can be acknowledged. This also allows for communicating the propositional content of non-propositional knowledge, e.g. for the purposes of writing the thesis.

However, there are problems with conveying outcomes of research that are relying heavily or solely on non-propositional content, i.e. tacit knowledge, with regards to requirements of justification and practical application. With regard to justification, as mentioned earlier, while the propositional content part of non-propositional knowledge can be made explicit, the tacit part cannot, and the acceptance of it as satisfactory evidence within research may rely on pointing at, and sharing of, a common understanding and interpretation of, the tacit content. This nevertheless leaves room for contention, and there has been ample evidence of this in the creative disciplines.

With regard to practical application of any knowledge, the discussion concluded that the tacit, non-propositional component of knowledge is as important as its explicit counterpart. Furthermore, this tacit component evades articulation and therefore requires non-verbal means of communication to convey it. This may be pointing at, and/or a shared understanding of, the phenomena seen or experienced e.g. through empathy.[57] Although by now there are a growing number of doctoral studies and other research which are explicitly or implicitly concerned with this aspect,[58] it has yet to be formally acknowledged in research regulations and guidance.

In conclusion, from a methodological point of view, the suggestion therefore is that it would be desirable for future research to review and analyse existing studies from PBDs with regard to different methods and approaches for verbal/textual and non-verbal communication (e.g. description/narrative, examples, models, prototypes, case studies, [video] demonstration, coaching etc.) according to the four categories of knowledge content:

- propositional content of propositional knowledge
- propositional content of non-propositional knowledge
- non-propositional content of propositional knowledge
- non-propositional content of non-propositional knowledge

Such a mapping could lead to better understanding of current methods and approaches, their applications, and benefits for research in the creative and practice-based disciplines.

With regard to research policy (regulations and requirements), it would be important to acknowledge the existence and importance of non-propositional content/tacit knowledge, how it can be included under current requirements, and how research results can be communicated inclusive of its tacit component to facilitate consensus and best application in practice.

Acknowledgement

The original version of this chapter was presented at the Nordic conference in 2007 in Stockholm and published in the *Design Research Quarterly Journal of the Design Research Society* (DRS) in the same year, and the original paper is available in the DRS archive (designresearchsociety.org). I have revisited this paper after 13 years to take stock of the developments and changes in thinking in the field regarding our understanding of knowledge in the PBDs. Clearly, the use of creative practice in research is much more accepted, although the dichotomy of knowledge explored in this chapter remains: tacit knowledge is everywhere – it underpins all research whether this is conducted in a Foundational, Coherentist, or Contextualist framework, but too often is not acknowledged. Interdisciplinary research, and the drive towards it, may yet help to advance our thinking, since explanation(s) tend to be required where they cross over, improving integrity through acknowledging and justifying one's approach. In this vein, I have rewritten large parts of this paper to update it and reflect the developments that have occurred in research since its original inception.

Notes

1 Niedderer (2007).
2 Ibid.
3 Polanyi (1958, p. 50).
4 www.jiscmail.ac.uk/cgi-bin/webadmin?A0=PHD-Design.
5 www. herts.ac.uk/artdes1/research/res2prac/confhome.html.
6 https://eksig.org.
7 E.g. Durling et al. (2002), Koskinen (2011), Leavy (2020), Sullivan (2010).
8 Starszakowna (2002, Abstract).
9 Niedderer (2008).
10 Grayling (2003, p. 37).
11 Niedderer (2007).
12 A TEA laser is a gas **laser** energised by a high voltage electrical discharge in a gas mixture generally at or above atmospheric pressure.
13 Collins (1985), Neuweg (2002, p. 42).
14 Polanyi (1958, p. 53).
15 Herbig et al. (2001).
16 Berliner (1994, p. 110; cf. also to the understanding of expertise in the 5-stage model of Dreyfus and Dreyfus 1988).
17 Polanyi (1958, p. 50).
18 Lyons (2009).
19 Polanyi (1958, p. 54), Beeston and Higgs (2001, p. 110).
20 Neuweg (2002).
21 Williams (2001, p. 98), Grayling (2003).
22 Polanyi (1958), Reber (1989), Higgs and Titchen (1995), Nonaka and Takeuchi (1995), Refsum (2002), Eraut (2003), Abidi et al. (2005), Miles et al. (2005).
23 Grayling (2003, p. 39).
24 Williams (2001, p. 81ff).
25 Ibid., p. 117ff.
26 Ibid., pp. 159–172.
27 Niedderer (2007).
28 E.g. BonJour (2001); Gunther (2003); Maund (2003); Crane (2005).
29 Grayling (2003, p. 38).
30 Sternberg (1999).
31 Reber (1989, p. 16).
32 Williams (2001, p. 140); Gunther (2003).
33 Maund (2003, pp. 58–59).
34 Williams (2001, pp. 69–80).
35 Hospers (1990, p. 19).
36 Neuweg (2002).
37 Neuweg (2002, p. 45).
38 Latin: 'I think, therefore I am' (Descartes 1644).
39 Polanyi (1958, p. 55).
40 Williams (2001, p. 175).
41 Williams' (2001, p. 100).
42 Smeyers (2006, p. 479).
43 Niedderer (2007).
44 E.g. in the UK: AHRC (2006), RAE (2005), REF (2021a, 2021b).
45 Durling and Niedderer (2007), Niedderer (2013), Kokko et al. (2020).

46 E.g. Biggs (2002, 2004), Durling et al. (2002).
47 E.g. AHRC (2006), REF (2021a, 2021b).
48 Whiteley (2000).
49 Lyons (2009).
50 Warpas (2013).
51 Nyangiro (2015).
52 Neuweg (2002, p. 45).
53 E.g. Durling et al. (2002), Koskinen (2011), Leavy (2020), Sullivan (2010).
54 E.g. Higgs and Titchen (2001), Neuweg (2004), Nonaka et al. (2006).
55 E.g. Niedderer and Townsend (2014), Kokko et al. (2020).
56 E.g. AHRC (2013), REF (2014).
57 Shusterman (2011).
58 E.g. Eriksson et al. (2019), Wood et al. (2009), Warner et al. (2021).

Bibliography

Abidi, S. S. R., Cheah, Y. N. & Curran, J. (2005). A Knowledge Creation Info-Structure to Acquire and Crystallize the Tacit Knowledge of Health-Care Experts. *IEEE Transactions on Information Technology in Biomedicine*, 9(2).

AHRC. (2006). *Details of the Research Grants Scheme*. www.ahrc.ac.uk (accessed July 2006).

AHRC. (2013). *Research Funding Guide*. www.ahrc.ac.uk (accessed August 2013).

Anderson, J. R. (1976). *Language, Memory, and Thought*. Hillsdale, NJ: Erlbaum.

Beeston, S., & Higgs, J. (2001). Professional Practice: Artistry and Connoisseurship. In J. Higgs and A. Titchen (Eds.), *Practice Knowledge & Expertise in the Health Professions*. Oxford, MA: Butterworth, Heinemann, 108–120.

Berliner, D. (1994). Teacher Expertise. In B. Moon & A. S. Hayes (Eds.), *Teaching and Learning in the Secondary School*. London: Routledge, 107–113.

Biggs, M. A. R. (2002). The Rôle of the Artefact in Art and Design Research. *International Journal of Design Sciences and Technology*, 10(2), 19–24.

Biggs, M. A. R. (2004). Learning from Experience: Approaches to the Experiential Component of Practice-Based Research. In H. Karlsson (Ed.), *Forskning-Reflektion-Utveckling*. Stockholm: Swedish Research Council, 6–21.

BonJour, L. (2001). Epistemological Problems of Perception. *Stanford Encyclopedia of Philosophy*. http://plato.stanford.edu/ (accessed February 2006).

Collins, H. M. (1985). *Changing Order: Replication and Induction in Scientific Practice*. London: Sage.

Crane, T. (2005). The Problem of Perception. *Stanford Encyclopedia of Philosophy*. http://plato.stanford.edu/ (accessed January 2006).

Descartes, R. (1644). Ego Cogito Ergo Sum. *Principia Philosophiae: Apud Ludovicum Elzevirium*, 30–31.

Dreyfus, H. L., & Dreyfus, S. (1988). *Mind over Machine: The Power of Human Intuition and Expertise in the Era of the Computer*. New York: Free Press.

Durling, D., Friedman, K., & Gutherson, P. (2002). Editorial: Debating the Practice-Based PhD. *International Journal of Design Science and Technology*, 10(2), 7–18.

Durling, D., & Niedderer, K. (2007). The Benefits and Limits of Investigative Designing. In S. Poggenpohl (Ed.), *Proceedings of the IASDR International Conference 2007* (CD). Hong Kong: Hong Kong Polytechnic University. ISBN:988-99101-4-4.

Eraut, M. (2003). The Many Meanings of Theory and Practice. *Learning in Health and Social Care*, 2(2), 61–65.

Eriksson, L., Seiler, J., Jarefjäll, P., & Almevik, G. (2019). The Time-Space of Craftsmanship. *Craft Research*, 10(1), 17–39. DOI:10.1386/crre.10.1.17_1.

Grayling, A. C. (2003). Epistemology. In N. Bunnin & E. P. Tsui-James (Eds.), *The Blackwell Companion to Philosophy*. Oxford: Blackwell Publishing, 37–60.

Gunther, Y. H. (Ed.). (2003). *Essays on Nonconceptual Content*. Cambridge, MA and London: MIT Press.

Herbig, B., Büssing, A., & Ewert, T. (2001). The Role of Tacit Knowledge in the Work Context of Nursing. *Journal of Advanced Nursing*, 34(5), 687–695.

Higgs, J., & Titchen, A. (1995). The Nature, Generation and Verification of Knowledge. *Physiotherapy*, 81(9), 521–530.

Higgs, J., & Titchen, A. (Eds.). (2001). *Practice Knowledge & Expertise in the Health Professions*. Oxford, MA: Butterworth, Heinemann.

Hospers, J. (1990). *An Introduction to Philosophical Analysis*. London: Routledge.

Kokko, S., Almevik, G., Høgseth, H. B., & Seitamaa-Hakkarainen, P. (2020). Mapping the Methodologies of the Craft Sciences in Finland, Sweden and Norway. *Craft Research*, 11(2), 177–209.

Koskinen, I., Zimmerman, J., Binder, T., Redstrom, J., & Wensveen, S. (2011). *Design Research Through Practice*. Waltham, MA: Morgan Kaufmann.

Leavy, P. (2020). *Method Meets Art, Third Edition: Arts-Based Research Practice*. New York: Guildford Press, 3rd ed.

Lyons, L. (2009). *Delineating Disease: A System for Investigating Fibrodysplasia Ossificans Progressiva*. PhD thesis. Sheffield: Hallam University.

Maund, B. (2003). *Perception*. Chesham: Acumen.

Miles, M., Melton, D., Ridges, M., & Harrell, C. (2005). The Benefits of Experiential Learning in Manufacturing Education. *Journal of Engineering Technology*, 22(1), 24–29.

Neuweg, G. H. (2002). On Knowing and Learning: Lessons from Michael Polanyi and Gilbert Ryle. *Appraisal*, 4(1), 41–48.

Neuweg, G. H. (2004). *Könnerschaft und implizites Wissen*. Münster and New York: Waxmann.

Niedderer, K. (2007). A Discourse on the Meaning of Knowledge in Art and Design Research. In *7th International Conference of the European Academy of Design*. Izmir, Turkey: European Academy of Design, May 11–13. http://citeseerx.ist.psu.edu/viewdoc/summary?doi=10.1.1.535.3521 (accessed August 16, 2020).

Niedderer, K. (2008). Practice in the Process of Doctoral Research. In *Focussed – Current Design Research Projects and Methods*. Berne, Switzerland: Swiss Design Network, 199–212. https://swissdesignnetwork.ch/src/publication/focused-current-design-research-projects-and-methods-2008/SDN-Publication-2008_Focused.pdf (accessed August 16, 2020).

Niedderer, K. (2013). Explorative Materiality and Knowledge: The Role of Creative Exploration and Artefacts in Design Research. *Formakademisk*, 6(2), 1–20.

Niedderer, K., & Townsend, K. (2014). Designing Craft Research: Joining Emotion and Knowledge. *Design Journal*, 17(4), 624–648.

Nonaka, I., Krogh, G. V., & Voelpel, S. C. (2006). Organizational Knowledge Creation Theory: Evolutionary Paths and Future Advances. *Organization Studies*, 27(8), 1179–1208.

Nonaka, I., & Takeuchi, H. (1995). *The Knowledge-Creating Company*. Oxford: Oxford University Press.

Nyangiro, E. (2015). *The Materiality of Food: Investigating the Potential of Food as Material for Art Practice*. PhD. Wolverhampton: University of Wolverhampton.

Polanyi, M. (1958). *Personal Knowledge*. London: Routledge & Kegan Paul.

RAE. (2005). *RAE 2008: Guidance on Submissions*. www.rae.ac.uk/pubs/2005/03/ (accessed August 2007).

Reber, A. (1989). *Implicit Learning and Tacit Knowledge*. Oxford and New York: Oxford University Press.

REF. (2014). *Research Excellence Framework*. www.ref.ac.uk/2014/ (accessed August 16, 2020).

REF. (2021a). *Guidance to Submissions: Research Excellence Framework*. www.ref.ac.uk/media/1092/ref-2019_01-guidance-on-submissions.pdf (accessed August 16, 2020).

REF. (2021b). *Panel Criteria and Working Methods: Research Excellence Framework*. www.ref.ac.uk/media/1084/ref-2019_02-panel-criteria-and-working-methods.pdf (accessed August 16, 2020).

Refsum, G. (2002). Bête comme un peintre? *Working Papers in Art & Design*, 2. ISSN:1466-4917.

Shusterman, R. (2011). Somatic Style. *The Journal of Aesthetics and Art Criticism*, 96(2), 147–159.

Smeyers, P. (2006). 'What It Makes Sense to Say': Education, Philosophy, and Peter Winch on Social Science. *Journal of Philosophy of Education*, 40(4), 463–485.

Starszakowna, N. (2002). The Concept of Knowledge. *Working Papers in Art & Design*, 2. ISSN:1466-4917.

Sternberg, R. J. (1999). Epilogue – What Do We Know About Tacit Knowledge? Making the Tacit Become Explicit. In R. J. Sternberg & J. A. Hovarth (Eds.), *Tacit Knowledge in Professional Practice*. Mahwah, NJ: Lawrence Erlbaum Associates, 231–236.

Sullivan, G. (2010). *Art Practice as Research: Inquiry in Visual Arts*. London: Sage.

Warner, L., Seitamaa-Hakkarainen, P., & Hakkarainen, K. (2021). Quiltmakers' Meaning-Making in Aotearoa New Zealand: Social Interactions, Embodied Experiences and Material Mediation. *Craft Research*, 12(1).

Warpas, K. (2013). *Designing for Dream Spaces: Exploring Digitally Enhanced Space for Children's Engagement with Museum Objects*. PhD. Wolverhampton: University of Wolverhampton.

Whiteley, G. (2000). *An Articulated Skeletal Analogy of the Human Upper-Limb*. PhD thesis. Sheffield: Hallam University.

Williams, M. (2001). *Problems of Knowledge: A Critical Introduction to Epistemology*. Oxford: Oxford University Press.

Winch, P. (1958). *The Idea of a Social Science and Its Relation to Philosophy*. London: Routledge & K. Paul.

Wood, N., Rust, C., & Horne, G. (2009). A Tacit Understanding: The Designer's Role in Capturing and Passing on the Skilled Knowledge of Master Craftsmen. *International Journal of Design*, 3(3), 65–78.

2.5

UN-KNOWING

A strategy for forging new directions and innovative works through experiential materiality

Garth Paine

Introduction

The explorations of artists can be systematic, methodical and deeply considered, but they can equally be intuitive and utilise an intentional strategy of *un-knowing*, which I define as seeking new directions that can lead to unexpected and stimulating outcomes by distancing themselves from well-hewn paths of previous practice. These endeavours are, nevertheless, based within the context of developed understandings of the materials at hand with the evolutions in practice and philosophy being embodied in the artwork itself. Not every artistic process is revolutionary, or indeed even evolutionary, in its influence, but the arts, in this case music, is peppered with many exceptional examples that have changed the course of musical practice. I introduce how exploration, interrogation and invention in the arts can lead to certain aspects of practitioner knowing that are communicated through the individual's unique creative process and the artistic works that emerge from it.

Furthermore, that the process of *un-knowing*, of eschewing known techniques, workflows and approaches in favour of unbounded exploration, can be a valid investigatory method in creative practice-based research. The process of *un-knowing* is to allow the material of the work (sound, gesture . . .) to guide the artist without conscious intervention. It can appear to be largely spontaneous. One might have a much broader intention which is to make a new work and to expand the horizons of previous outcomes, or it might simply start as the exploration of a sound or feeling. This embodied sense of material can be described as a form of meta-materiality, the situated, embodied engagement with the material itself. This somatic engagement with the material is subtle and abstract but, for the experienced artist, can be a powerful avenue for research and development.

This chapter provides several historical examples and finally demonstrates the proposition through the author's own research and creation as researcher-in-residence at IRCAM (Paris) and ZKM (Karlsruhe), during which time he created the 50-minute immersive media work *Future Perfect*. The chapter does not seek to encompass all possible examples or to draw parallels outside of music, of which there are many, but space is short and so a focus on music is the most expedient and pragmatic vehicle for this discussion.

Knowing through *un-knowing*

The arts are an incredibly broad collection of practices, from classical to jazz to punk music, theatre, dance, film, ceramics and painting, to mention but a few. This diversity also implies that forms of

 DOI: 10.4324/9780429324154-18

knowledge in the arts may vary greatly. One might consider the embodied knowledge of the dancer and musician, the spatial awareness, whether in a physical space or a musical sound-space, whether the forces combined for a dance step or the air pressure, embouchure and finger positions combined for an overtone on an instrument. For both artists, different forms of knowledge co-exist and are necessary for high levels of performance. Consider the studied and practiced technical skills, the somatic qualities of the felt experience as a pathway to gauge the outcome(s) of enaction in real-time, the responsive possibilities of the performance medium and the associated prioritisation of intuition. These seem on the surface almost contradictory, that an artist would execute highly practiced skills intuitively, but various forms of knowledge are in fact enmeshed to form the foundational practice for both artists. Intensive training over time enables the high-level artist to enact them without conscious thought. The same may be said of research and development for these artists: that they may oscillate between these multitude sensibilities, somatically sensing the energy of the material while also intuiting a larger form and structure, while simultaneously developing technical solutions.

The combination of multiple channels of perception and explicitly learned skills, as outlined, represent a unique aspect of artistic practice, where years of technical discipline of the highest professional standard, a form of concrete logical knowledge, must be deeply integrated with real-time experience (abstract, felt knowledge) in order to create successful art works. These skills are also combined with a historical knowledge, providing context and interdependent references, illustrating the ecosystem of ideas and the practice-based influences that make up the current culture of enquiry, the melting pot of ideas and innovation that leads, looks forward.

This complex coalescence of different forms of knowledge is further complicated for many artists by the fact that artistic investigation can also often involve an intentional act of *un-knowing*. The distancing of the familiar, eschewing the well-honed techniques, the years of apprenticeship and development, the known anchors of safety or security, in order to generate an environment of multi-instabilities, to allow new paths to form, to allow the art making process and conceived aesthetic objects to impact on and shape a unique journey of discovery. This means that an artist may not clearly know or be able to define the outcome of a creative process until the work is finished. Even knowing when the work is finished is a complex decision, often un-known in advance. Guiding principles may be established at the beginning of the creative process, but the material of the work itself will direct the artist in any number of unexpected explorations, creating a final product that was unknown and unknowable at the outset.

In addition, artists regularly disrupt known methods of exploration and creation by changing the means of production or the environment in which the work is made, or by collaborating with other artists. Through these collaborations, new paths are hewn, new aesthetic outcomes emerge that can be tested against known expressive tools and somatic knowing for value and creative potential. If these explorations lead to new insights and contribute to an evolution of the field's practice, then they may represent a type of research investigation. This investigation may not have taken the form of a hypothesis, a specific intention or a controlled experiment (although the intention to freely explore artistic practice could be considered a form of experiment), and they may not be quantifiable through explicit/formal knowledge, but they can lead to discovery, invention and insight that is communicated through the artwork itself and becomes transferable, feeding into the community of inquiry associated with that art practice and others in the contemporary cultural milieu. Such investigations are often disruptive of the traditions of practice and may also be evolutionary, and in some cases such explorations inspire new genres of music or cross-pollinate other art forms, whether painting, dance, film, literature etc.

An example of this intentional intervention in known practices is Bob Dylan's move from acoustic band to electric on July 25, 1965, at the Newport Folk Festival. His style of folk music was epitomised as acoustic and intimate, aligned with the traditions of that genre. He had released *Like a Rolling Stone*[1] only weeks before the Newport performance on his album *Highway 61 Revisited*,[2] creating a sonic experience far removed from the accepted Dylan acoustic, folk sound. It was electric, loud and raucous. This change of creative resources (moving from acoustic to electric instruments) immersed Dylan in a heavier, denser, more raucous, guttural sonic world, facilitating intense power which he channelled into protest-driven sonic qualities and songs, songs that were attuned to the sentiment of the times. The change from acoustic to electric instruments radically changed his available potentials for music making and performance, producing a new set of instabilities that he formed into musical expression. He disrupted his fans' expectations with such force that he was met with anger and riots, booed off stage, rejected, but he persisted, knowing that the change produced a creative power not previously available to him, a power that was both timely and liberating. It continued to be a journey of discovery for the rest of his musical career. In addition, it is noted that during this first tour, the band was continuously exploring and seeking to understand the new musical language they had tapped into. They received a raucous and often riotous response to the new sound but played on, committed to its possibilities. It was, nevertheless, clear to many other musicians that Dylan's exploration exposed a wide range of new possibilities.

This is only one of a plethora of examples that mark the development of musical practice through the ages, the creation of fresh voices and expression aligned with both the social mores and challenges of the time. It points to a development, formed by a type of research that brought forth an evolved form of knowledge (performance technique, technology, musical language etc.) and practice, that was shared widely with the community of practitioners and in this case the public and inculcated fellow musicians, broadly changing the future of music making.

Dylan was not the only one; the history of music is marked by such investigation and development. Earlier examples of creative exploration leading to new knowledge, techniques and forms of expression can be identified in an increasingly broad focus on harmonic and timbral qualities in the music of Satie,[3] Liszt,[4] Scriabin[5] and Debussy.[6] They extended the accepted harmonic language of the time into rich timbral techniques, a kind of tone painting that laid the foundations for works such as Debussy's *Pelléas et Mélisande*,[7][8] which the composer Olivier Messiaen described as having 'extraordinary harmonic qualities and . . . transparent instrumental texture'[9] or Satie's *Gymnopédies*,[10] with its unresolved major seventh chords and open textures, hitherto unknown to the Western art music ear. Debussy's later works, *La cathédrale engloutie*[11] or *Jeux*,[12] although harmonically tame by today's standards, also ventured into uncharted territory for the time, producing a radical new expressive voice[13][14] and greatly influencing his contemporaries.

These works result from an exploration into the *un-known*, an iterative testing of boundaries of existing musical knowledge, intuitively seeking a new voice that more closely reflected broader social change. Debussy was looking for wider, more open harmonies, breaking down the orthodoxy of accepted musical language and form (form in music refers to the temporal structure of the work). To reach into this exploratory space, Debussy, I presume, had to let go of his formative training and his entrenched practices and venture into the unfamiliar, the unheard. To explore a somatic space, that is, to find what felt right, what fitted the evolving sensibilities of the time, to give rise to a new energy. He could not have known the end point at the outset. Debussy's *un-knowing* is documented in his statement that:

> We must agree that the beauty of a work of art will always remain a mystery . . . we can never be absolutely sure 'how it's made.' We must at all costs preserve this magic which is peculiar to music and to which music, by its nature, is of all the arts the most receptive.[15]

In decades prior to Debussy's work, Liszt (1811–1886), Scriabin (1872–1915) and Satie (1866–1925), among others, also moved from the strictures of traditional harmony and counterpoint to the experiential dimensions of timbre, from the functional to the somatic, exploring the materiality of sound, the surface texture, mass, density, viscosity, velocity and acceleration, the phenomenology of listening, of reception. The explorations I am outlining happen not so much in the known practices of the artist, but in a felt space, an abstract experiential, somatic space, where the artist is looking for a voice with new energy. While for some composers this might be an intentional, rigorous and systematic process, for many others it is not. Many artists constantly seek an evolving voice but reach those epiphenomena through unstructured exploration, intentionally avoiding known techniques, being receptive to an indefinite and often unresolved corporeality.

The musical works of Russian-born composer Igor Stravinsky (1882–1971) are an exemplary example of such influential development of harmonic structures. Stravinsky's use of polytonality (the use of more than one musical key, or tonal centre simultaneously) may not have eventuated if not for the explorations of Debussy, Bartok and Messiaen. Although Franz Liszt and then the American composer Charles Ives[16] (1874–1954) explored polytonality long before Stravinsky,[17] Stravinsky's *Le Sacre du printemps*[18] (*The Rite of Spring*) brought another shift, the elevation of rhythm, to front and centre as core musical material. No longer the servant of melody, or the subtle tricks of syncopation (placing the accent off the beat), the punchy rhythms of *The Rite of Spring* were a central, unruly and boisterous vehicle of expression, forming an often demented, intense energy hitherto unknown in Western art music. The twentieth-century composer Philip Glass comments that the legacy of this work was that 'the rhythmic structure of music became much more fluid and in a certain way spontaneous'.[19] This demonstrates that the work itself communicated aspects of Stravinsky's investigations through the language of the music score, which influenced a revolutionary change in music.

This elevation of rhythm to core musical material spawned an ongoing exploration of rhythm that laid the ground for composer John Cage[20] to de-emphasise pitch. Cage collected together unusual objects to form his percussion ensembles, for which he wrote several compositions for percussion instruments only. These explorations led to Cage thinking of a mixed percussive and pitch space, something he realised in his works for prepared piano. Prior to these works, percussion instruments were seen in Western art music as effects – added colour but never core instruments. The notion of a percussion ensemble was unfathomable. Cage's research can be experienced in works such as *First Construction (in Metal)* (1939),[21] and his use of divisions of duration, micro-macrocosmic rhythmic structures with palindromic rhythms and meter as in *Imaginary Landscape No. 1* (1939)[22] or the Sonatas and Interludes for prepared piano. These were radical works, challenging accepted definitions of music broadly by unburdening music from melody and harmony as primary characteristics, discarding key harmonic and structural practices accepted for centuries as essential to all Western art music. Again, Cage's musical works communicated the developments. Through careful exploration, the rejection or acceptance of experimental outcomes, Cage came to a new musical language that was taken up by others and broadly liberated musical practice in important ways during his life time and beyond.

These examples are not intended to show explicit and direct influence, but to illustrate how artistic research through exploration and investigation of creative materials can produce new forms, new vehicles for expression, pertinent to the concerns and general milieu of the field. It demonstrates how the communal psyche of artists is shifted by these explorations and how this provides a fertile ground on which artistic practice shifts and morphs. These explorations are not only individual – they are not lost after execution but feed an ongoing communal interest in the act of creation and both an individual and communal interrogation into the materials of production, in this case sound and music.

Meta-materiality as active agent

What these examples demonstrate is that evolutions of artistic practice can lead to an existential change in the perception of the material itself, the very nature of music or the inherent value of the omnipresent sonic environment. A reinterpretation of the meta-materiality of music or sound.

Music is, at its most fundamental, sound – 'organised sounds', in the words of Edgard Varèse. Varèse (1883–1965), like those artists discussed previously, also radically changed musical practice. He amplified rhythm and timbre as key musical parameters, viewing 'sound as living matter' and 'musical space as open rather than bounded'.[23] These new conceptual frames produced sound-masses that minimised the impact of individual pitches and led to the creation of new electronic sounds for works such as *Déserts* for wind, percussion and electronic tape (1950–1954) and *Poème Électronique* for electronic tape (1957–1958), among others. *Poème Électronique* was commissioned for the Philips Pavilion at the 1958 Brussels World's Fair and was one of the first musical works to be spatialised, with loudspeakers placed throughout the Le Corbusier-deigned Pavilion.

These artistic developments, although all musical, represent different forms of new insights and forms of tacit knowing, whether the raucous expression of the Dylan electric band or the rich harmonic language of Debussy, or the strident rhythmic intensity of Stravinsky and the sound masses of Varesé. As Debussy comments – it happens through a state of *un-knowing*: 'we can never be absolutely sure "how it's made".' This *un-knowing* points to non-conscious, non-rational, embodied and felt forms of knowledge through the materiality of the practice. Of course, documentation, critique and theorising followed the performance presentation of each of these works. The score as artefact was added to the music literature, although not in the case of Dylan and others like him, where the oral tradition predominates. But in all cases, it was the experience of the music that was the instigator of change. The new approach to musical material is held therein and communicated through its performance.

These epistemic phenomena do not arise independently, a fact that Steven Feld illustrates so well when discussing his term 'acoustemology', which he defines as 'one's sonic way of knowing and being in the world' (Feld & Brenneis, 2004). Feld goes on to say that

> Implicitly, [. . . sound] recordings ask what it means to live and feel as a person in this place. Voiced in a more contemporary way, . . . recordings signal that the concept of 'habitus' must include a history of listening.[24]

The use of the word 'habitus' implies embodied dispositions as a filter to perception. The process of high-quality sound recording was a new resource for creative practices in the twentieth century, one that gave rise to Music Concreté and evolved through Luc Ferrari[25] and R. Murray Schafer[26] into the foundations of Acoustic Ecology. Indeed, it was also critical to Varèse's *Poème Électronique*. But Feld's statement also draws out the perceptual frame. The state of *un-knowing* I have discussed involved an eschewing of known techniques and the artist's prior habits of production and re-situated the artist in a felt, embodied, somatic state of enquiring into the journey of creation and its relationship to the world around them, the ever-evolving communal psyche and their creative voice. That exploration seeks to be freed of the constraints of existing musical language, to find a new voice unbounded by history, a contemporary, experimental, ground-breaking and avant-garde utterance.

Epistemic cultures vary from one form of artistic practice to another. But what is clear is that each develops over time through exploration, interrogation, experimentation and investigation. That the resulting new insights and forms of tacit knowing are transmitted through the experimental forms of that discipline and shared in associated creative communities demonstrates that this knowledge is transferable and replicable. These insights are the lingua franca of that creative community; they are the product of a research process and are shared as common understandings, their evolution illustrated in the historical record of each discipline.

They may not be quantified or reported in the publications of the time, but they do represent important fractures and the expansion of key conceptual frameworks and emerging techniques that mark the evolution of artist practices. The subject-specific language of knowledge is generalisable and transferable, a fact well illustrated by the number of artists working in evolving musical genres and the inheritance acknowledged by younger generations of musicians to the iconic artists who proceeded them. Critically, this evolutionary knowledge may be arrived at through an intuitive, unconscious interplay, an interrogation that is not spurred on by intentionality, which from the outside may not seem rigorous, systematic or precise, but where the artist is engaged in a journey that is much more abstract, focused on an embodied sensation that is not predefined and may not be communicable directly to others except through the artwork itself. Other artists will recognise these new qualities and seek to work out how they might apply them in their own work, or perhaps, less specifically, be spurred on to also engage in open exploration, seeking that same felt, somatic sensation, through unstructured experimentation, interrogation, invention and a process of un-knowing.

In practice

In 2018–19, I was researcher-in-residence at IRCAM,[27,28] Centre Pompidou, Paris, and ZKM, Karlsruhe, Germany. I had proposed the creation of a large-scale musical and virtual-reality (VR) work, *Future Perfect*,[29] and although I had a general idea how the work could be realised and experienced, the technologies for its production and performance did not exist. I had no idea how to realise my vision at the time of proposal, but did have a notion of the energy of the work, what it would feel like to experience *Future Perfect*,[30] and I had outlined some of the key aesthetic and philosophical ideas the work would explore:

> *Future Perfect* is an immersive 3D audio visual performance and installation work that could also have an afterlife as a smart phone virtual-reality audio-visual, musical album.
>
> The project explores the seam between virtual-reality as a documentation format for environmental research and archiving nature, with the idea that 'nature' as we know it may, in the near future, only exist in virtual-reality archives, and the notion of the virtual, a hyper-real imaginative world contained by a technological mediation presenting that world to the individual as a private experience.

While the research drew on existing practices such as acousmatic music composition and performance, sound spatialisation, field recording, animation and virtual-reality development, the year of exploration and creation was one of un-knowing, of looking for potential solutions but not falling back on known techniques, of exploring sound recordings to understand and then expand their material properties, of developing technologies that accessed the audience's smartphones as a loudspeaker, expanding the spatial audio resolution to become personal to each member of the audience, moving them from passive listener to an active agent in the work and creating a performance system that allowed a dynamic execution of sonic materials in each concert. I had no concrete idea what the music would be or how I would create it, but I held an energetic sense of the work – a notion of how it might feel to be present within the work. *Future Perfect* was premiered at the inSonic Festival at ZKM in December 2018 (Figure 2.5.1).

As a composer, I often think of myself as an experience designer. This aspect is very critical to me as an artist in that when I'm setting out to develop new techniques, in fact to compose any new work, I am not laying out a concrete specification but am rather conjuring up an emotional and energetic state that I wish to experience and communicate. This energetic kernel of the work is what drives the conceptualisation and guides my explorations as I experiment and search for sound materials and compositional frameworks that represent the imagined dynamic energy state, that give it voice in the temporal domain.

The research process therefore is very abstract. Much of the production of the work takes the form of exploration of energy states, formed of musical and sound materials, constantly shifting in a dynamic morphology and forming and dissipating in a temporal frame that is also dynamic. It

Figure 2.5.1 Premier performance of *Future Perfect* at ZKM as part of the inSonic Festival 2018.

takes the form of investigating techniques to find outcomes that match the conceived, imagined, proposed experience and combine through the act of composition to create a work with many layers of simultaneous expression, reflecting those energetic states at different scales and intensities and in dynamic relationship.

The temporal form is informed by how the sound material unfolds. The layering forms an interplay between characteristics of the materials that when superimposed bring a new and unexpected quality, an energy that transforms itself. The creative materials guide the process. I allow them to unfold at their own pace and to develop relationships that were not preconceived, interactions and exchange between those materials that could not have been known in advance. In many ways, it is like hearing an author report that they did not expect a character to do X and that they did not expect certain relationships to develop and take the story line in new directions. Allowing the creative materials to dictate their evolution is of course not to give up everything, but to bring to bear one's highly developed skills and experience as a listener, observer and master craftsman. Like a sculptor who feels the direction of the wood grain and allows that to drive the form of the sculpture, there are myriad highly refined decisions being made as the materials are crafted.

These sensibilities feel the inter-relationships of the current material to all other matters in play; they understand the immersive, enveloping qualities of multi-channel sound spatialisation, the scale of the performance venue, the inter-relationship of different weights of material being presented over differing systems (large speakers in the speaker dome and the small speakers of the smartphone) and the ways in which the image world and the sound worlds co-create an experience rather than simply reflect one another.

The gathering of source material also reflects the not-knowing focus of this chapter. A vast array of field recordings was made in preparation for *Future Perfect*. At the time of recording, I did not know where the material would be used in the final work, but understood that the context I was recording in resonated with the energy of the work. That those sound environments could help contextualise the over-arching intention of bringing the audience to engage directly with and interrogate their place in the climate crisis, their inter-connectedness with the environment around them. Reflecting Feld's reflection of the uniqueness of 'one's sonic way of knowing and being in the world'.[31]

Future Perfect evolved into a 50-minute composition, because the musical materials dictated the rate at which they were revealed – something I discovered as the work evolved. The piece starts in the city and morphs to the nature sound recordings of national parks, from the human to the biosphere. The human returns through a James Joyce poem sung by the extraordinary baritone singer Gordon Hawkins,[32] who recorded the poem as a whispered monotone, with the material later subjected to frequency transformation through electronic processing.

Future Perfect is performed on a 59-loudspeaker 3D sound diffusion system, augmented by everyone's smartphone as a loudspeaker (see Figure 2.5.2 for initial system design), meaning there can be

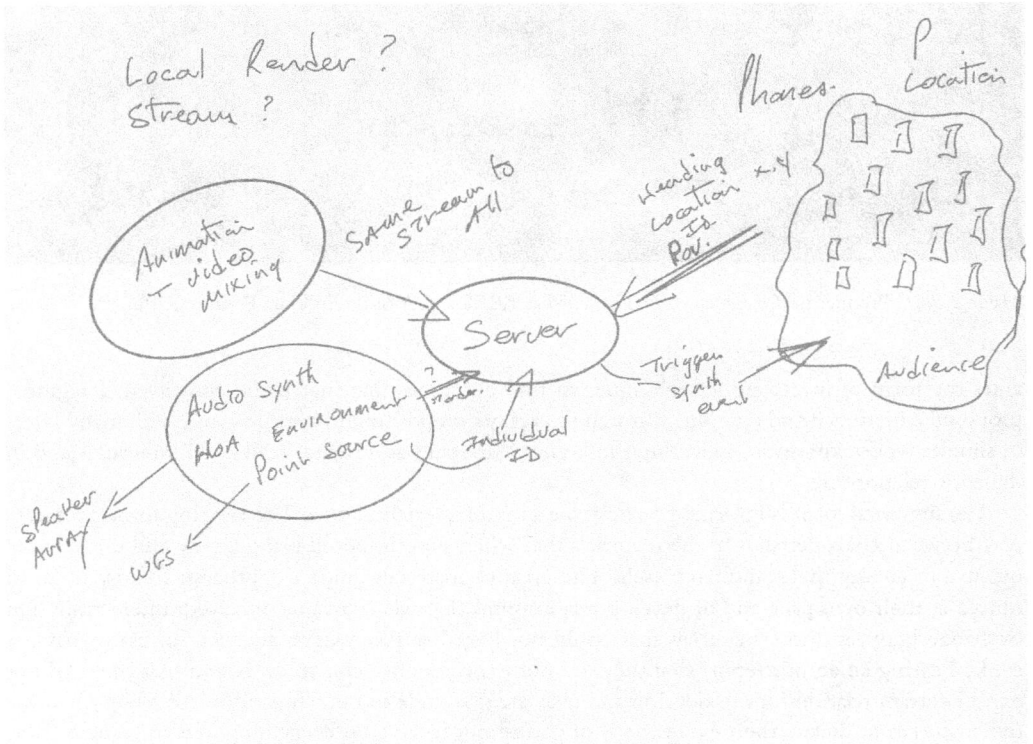

Figure 2.5.2 Initial system design for integration of high order ambisonics and audience smartphones.

hundreds of loudspeakers in the space, some personal and some part of the institution, some large and resounding, some small, individual and intimate in scale. The work is further supplemented by a 3D virtual-reality film. My research produced not only a large-scale immersive concert work but also a new technology to send sound to the audience's smartphones, to track the location of those phones in space and to allow the performer to dynamically draw spatial vectors that move the sounds through the audience (see Figures 2.5.3 and 2.5.4 for performance interfaces).[33] In addition, granular synthesis engines were developed so a granulation layer could sound on the phones to generate textures of varying densities and internal rhythms. The phones could run multiple musical layers synchronously and asynchronously, producing a very rich dispersed texture across the audience. These technologies will be shared as open source and have already filtered out into the broader community.

It is hard to describe the binding concept that drove me in producing *Future Perfect*. It certainly was not a hypothesis that I was testing, or even a proposition, but an embodied sense of the potential for audience immersion when several existing delivery techniques were combined and extended with new technologies (Figure 2.5.5). It was a remapping of the empirical, the experimental and the functional. This is a space I call the techno-somatic,[34] by which I mean a simulacrum of felt, somatic knowledge, the materiality of the sonic forces being mustered and the functional, technical potential of the delivery/performance frameworks being used. However, in order to get to this point, I had to put aside my pre-existing practices, to remain for long periods of time in a very uncomfortable space of not knowing, refusing to fall back on familiar techniques or forms of performance and presentation, and accept that the gathering and generation of material that addressed the somatic properties of the imagined experiential outcome would lead to creative discoveries.

Figure 2.5.3 *Future Perfect* performance setup with custom controllers for the web audio phone layers of the work.

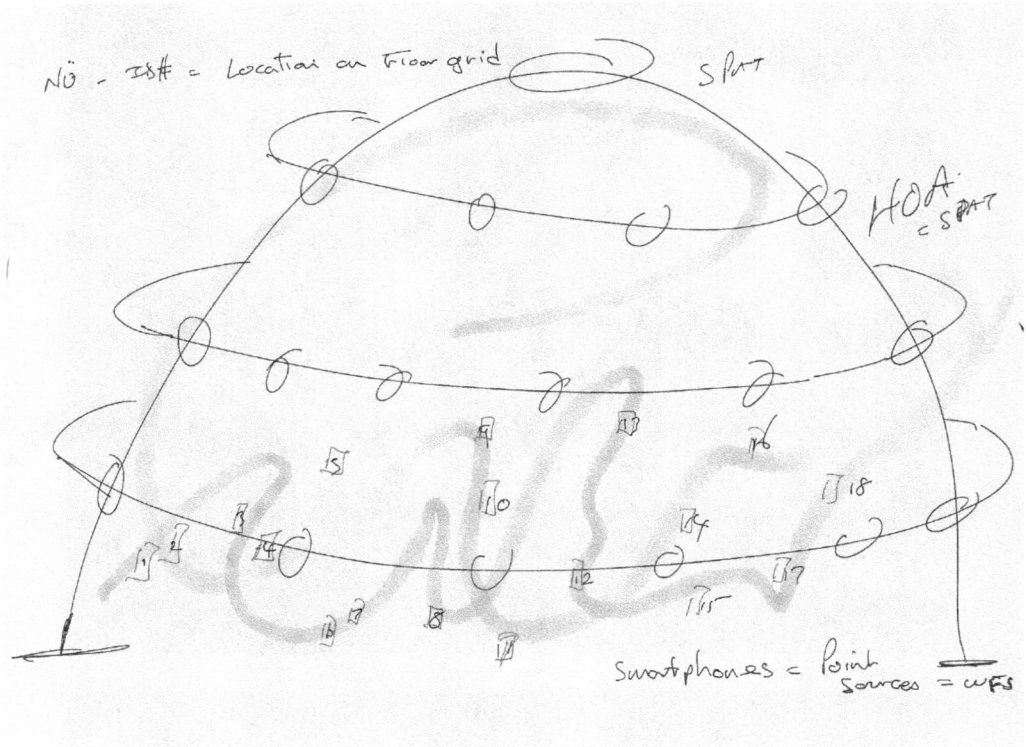

Figure 2.5.4 Early consideration of 3D spatial audio presentation of *Future Perfect*.

Conclusion

The creative intention to utilise a state of *un-knowing* as part of my artistic process, as outlined at the opening of this chapter, is not one of amateurism or lacking training or skill, but rather of allowing the materials at hand to determine their relationships and expressive form rather than applying, existing techniques and forms to the material without sensitivity. One might define this process as operating from within the material rather than through external imposition.

It is clear that this process of *un-knowing* is a strategy of investigation. If the artist is simply enacting a well-worn path to creation, then there is little room for discovery, let alone innovation. In this scenario, the techniques are well known, their outcome can be projected and the work therefore is largely known at the outset. It might be argued that research in the arts sets out to create the conditions for the unexpected. The process is not linear; it does not need to revolve around a test and evaluate method but rather seeks to open out to the myriad, complex interdependencies that exist in the materials at play. Such a method acknowledges from the outset that all phenomena are enmeshed in some way, that a work is by nature a set of multi-stabilities, or multi-instabilities. The work grows from unconstrained and unintentional exploration, seeking to create a quality of experience. The idea is simply a starting point, stabilised by years of technical acquisition, artistic experience and refined sensibilities to the inherent potential within the materials being gathered and generated. It is a dance between material, design, letting go and outcome, where the outcome is a somatic, felt energy, a multi-dimensional experience.

It is only with experience that I have developed the faith to trust that *un-knowing* is a state of productive allowing – that it opens up the creative process to innovation, to the unexpected, to new pathways and methods and by doing so makes much stronger work, work that is informed by the experiential dimensions of the material, from the functional to the somatic, exploring the surface texture, mass, density, viscosity, velocity and acceleration, the phenomenology of listening, of reception, in essence the energy of the work as a whole, be that with, or without a temporal vector.

Notes

1 See www.allmusic.com/song/like-a-rolling-stone-mt0007760873 for more information. (accessed January 4, 2021).
2 Dylan (1965).
3 For biographical information, see www.oxfordmusiconline.com/grovemusic/view/10.1093/gmo/9781561592630.001.0001/omo-9781561592630-e-0000040105 (accessed January 4, 2021).
4 For biographical information, see www.oxfordmusiconline.com/grovemusic/view/10.1093/gmo/9781561592630.001.0001/omo-9781561592630-e-0000048265?rskey=5euBWe&result=2 (accessed January 4, 2021).
5 For biographical information, see www.oxfordmusiconline.com/grovemusic/view/10.1093/gmo/9781561592630.001.0001/omo-9781561592630-e-8000900129?rskey=VEyoBm&result=4 (accessed January 4, 2021).
6 www.oxfordmusiconline.com/grovemusic/view/10.1093/gmo/9781561592630.001.0001/omo-9781561592630-e-0000007353?rskey=sCxS3V&result=1 (accessed January 4, 2021).
7 See www.opera-online.com/en/items/works/pelleas-et-melisande-debussy-maeterlinck-1902 for a synopsis of the opera (accessed January 4, 2021).
8 Debussy (1907) (begun 1893, staged 1902).
9 Samuel and Messiaen (1976).
10 Listen at www.naxos.com/catalogue/item.asp?item_code=CDS7820 (accessed January 4, 2021).
11 Listen at www.naxos.com/catalogue/item.asp?item_code=DE3006 (accessed January 4, 2021).
12 Listen at www.naxos.com/catalogue/item.asp?item_code=8.570759 (accessed January 4, 2021).
13 Many of the classical music scores mentioned here are available for viewing online at https://alexanderstreet.com/products/music-online-classical-scores-library (accessed January 4, 2021).
14 Further information on these composers and works is also available at www.oxfordmusiconline.com/grovemusic/ (accessed January 4, 2021).
15 Nichols (1992).
16 For biographical information, see www.oxfordmusiconline.com/grovemusic/view/10.1093/gmo/9781561592630.001.0001/omo-9781561592630-e-1002252967?rskey=8v8Yxg&result=2 (accessed January 4, 2021).
17 For biographical information, see www.oxfordmusiconline.com/grovemusic/view/10.1093/gmo/9781561592630.001.0001/omo-9781561592630-e-0000052818?rskey=Edn5QK&result=1 (accessed January 4, 2021).
18 Listen at www.naxos.com/catalogue/item.asp?item_code=8.112070 (accessed January 4, 2021).
19 Glass (1998).
20 For biographical information, see www.oxfordmusiconline.com/grovemusic/view/10.1093/gmo/9781561592630.001.0001/omo-9781561592630-e-1002223954?rskey=8OG5Ja&result=1 (accessed January 4, 2021).
21 Listen at www.naxos.com/catalogue/item.asp?item_code=8.574244 (accessed January 4, 2021).
22 Ibid.
23 Wen-Chung (1966).
24 Feld and Basso (1996).
25 For biographical information, see www.oxfordmusiconline.com/grovemusic/view/10.1093/gmo/9781561592630.001.0001/omo-9781561592630-e-0000009520?rskey=SxFnIu&result=1 (accessed January 4, 2021).
26 For biographical information, see www.oxfordmusiconline.com/grovemusic/view/10.1093/gmo/9781561592630.001.0001/omo-9781561592630-e-1002258098?rskey=B85I3a&result=1 (accessed January 4, 2021).
27 See www.ircam.fr/person/garth-paine/ (accessed January 4, 2021).

28 See www.ulysses-network.eu/profiles/individual/26512/ (accessed January 4, 2021).
29 For score and additional information, see www.activatedspace.com/Music/Future-Perfect/ (accessed January 4, 2021).
30 See documentary on development of *Future Perfect* at https://zkm.de/en/media/video/garth-paine-future-perfect (accessed January 4, 2021).
31 Feld and Brenneis (2004).
32 See http://gordon-hawkins-baritone.com (accessed May 24, 2021).
33 Paine (2015).
34 Ibid.

Bibliography

Debussy, C. (1907). *Pelléas et Mélisande*. Paris: Durand.

Dylan, B. (1965). *Highway 61 Revisited*. New York: Columbia University Press.

Feld, S., & Basso, K. H. (1996). *Senses of Place*. Santa Fe, NM and Seattle: School of American Research Press; Distributed by the University of Washington Press, 1st ed.

Feld, S., & Brenneis, D. (2004). Doing Anthropology in Sound. *American Ethnologist*, 31(4), 461–474. doi:10.1525/ae.2004.31.4.461.

Glass, P. (1998). The Classical Musician Igor Stravinsky. *Time-New York*, June 8.

Moreux, S. (1953). *Béla Bartók*. London: Harvill Press, 91–92.

Nichols, R. (1992). *Debussy Remembered*. Portland, OR: Amadeus Press.

Paine, G. (2015). Interaction as Material: The Techno-Somatic Dimension. *Organised Sound*, 20(1), 82–89.

Samuel, C., & Messiaen, O. (1976). *Conversations with Olivier Messiaen*. London: Stainer & Bell.

Suchoff, B. (2000). Background and Sources of Bartók's Concerto for Orchestra. *International Journal of Musicology*, 9, 339–361.

Wen-Chung, C. (1966). Open Rather Than Bounded. *Perspectives of New Music*, 5, 1–6.

2.6

APPRECIATIVE SYSTEMS IN DOING AND SUPERVISING CURATORIAL PRACTICE-BASED RESEARCH

Lizzie Muller

Introduction

The term *appreciative system* was coined and developed by the English systems scientist and polymath Geoffrey Vickers to describe a complex and powerful concept of how humans apprehend and make meaning from their world.[1] It was used by philosopher Donald Schön in his seminal work *The Reflective Practitioner*,[2] a foundational text in the epistemology of practice, to describe one aspect of how professional judgements are made. I begin this chapter by briefly examining Vickers' and Schön's use of the term and its wider implications for practice-based research. I argue that despite what Schön describes as its relative constancy, the expansion and adaptation of appreciative system, as both a personal and collective artefact, is crucial for practitioners and professions to respond ethically to a changing world.

Situating this idea within the growing field of curatorial practice-based research, I identify the different structural models emerging for curatorial research projects and the importance of "curatorial authorship" as a through-line in such projects. I analyse some of the factors that make it challenging for curators to explicitly recognise and discuss authorship, and the impact this can have on their research. Through the example of my supervisory work with curatorial researcher Dr Bec Dean, I show how working with the notion of appreciative system enabled Dean to acknowledge her individual curatorial authorship and articulate the unique contribution of her research to the field.

I conclude by arguing that an understanding of appreciative systems exceeds judgement of whether the results of practice are good; instead, it can be understood as the capacity to examine judgement itself, and how judgement has been challenged and enlarged through engagement with the world. Cultivating attunement to appreciative systems thus supports the doctoral candidate, the candidate–supervisor relationship and the supervisor's own reflective growth.

The work of appreciative systems

In 2009 I completed a PhD focused on my own experience of curating. I used the term "reflective curatorial practice" to describe my approach, which drew heavily from Donald Schön as interpreted and developed by artist, computer scientist and early pioneer of creative practice research Stephen Scrivener.[3] Scrivener had adapted Schön's work to use as a methodological guide for artists undertaking practice-based doctorates, and I further adapted Schön, via Scrivener, as a basis for my own methodology.[4]

 DOI: 10.4324/9780429324154-19

As leading Australian theorist of creative practice research Estelle Barrett has argued, the "generative strength that distinguishes practice as research from more traditional approaches" lies in its "subjective dimension" and its "emergent methodologies".[5] A large part of my own PhD research concerned developing methodological rigour in relation to curatorial authorship, which was at that point largely unchartered territory. This included the use of appreciative system as a tool for critical reflection on my subjective experience and cultivating insight into my own curatorial stance with support from my supervisors.[6]

My transition from student to supervisor has required me to revisit and expand my notion of reflective curatorial practice with each new candidate. In every project I have been fascinated to see how appreciative systems act as both a compass to guide and lens to interrogate curatorial agency. In returning to Schön now to develop this concept further, it is important to re-iterate that Schön defines an epistemology of practice, not a method for practice-based research. Those of us who have worked to establish such methods have elaborated on Schön's ideas, and often developed more out of them in terms of their methodological application than Schön put in. My use of appreciative system is such a case, though my brief engagement here with Vickers' original concept shows how the elaboration of an idea often requires an excavation of meanings that are already implicit and at work in its use.

Schön introduces the concept of appreciative system as one element of how a practitioner will judge the outcomes of her experiments-in-practice. He doesn't explain his use of this term, which could be taken simply to mean what a practitioner values. However, it is likely that Schön's usage also implies Vickers' more profound notion of the "code" by which an individual interprets or parses their social world, a kind of filter or rubric for determining what is "appreciable'" or can be apprehended in the world around us.[7] Some authors have likened Vickers' notion of appreciative system to a social version of German biologist Jakob von Uexkull's ecological/perceptual concept of "umwelt", which describes the unique lived-reality that different species may have within the same ecosystem, depending on their sensory adaptations.[8] The appreciative system therefore describes more than what a practitioner considers good: it defines what can be understood or even acknowledged as existent with that practitioner's social reality.

As an evaluative tool for experiments-in-practice, the appreciative system determines what the practitioner can recognise as pertinent to or part of the change in the situation engendered by her experiment. As Schön explains, these experiments "stimulate a situation's back-talk",[9] which leads to new discoveries. Following the American pragmatist philosopher John Dewey, Schön describes the "transactional" relationship between a practitioner and a situation, by which both the world and the practitioner are changed.[10] New knowledge is generated through the expansion of what Schön calls "repertoire": the range of actions available to a professional based on previous experience and enlarged through experiments-in-practice. For Schön, countering the flux of real-world engagement is the relative constancy of appreciative system. The following passage is perhaps the fullest description Schön offers of the work of appreciative system as a compass to navigate the dynamism of knowing-in-action, in which he briefly describes features of the appreciative systems of one of the practitioners he considers in his book, the architect Quist:

> Constancy of appreciative system is an essential condition for reflection-in-action. It is what makes possible the initial framing of the problematic situation, and it is also what permits the inquirer to reappreciate the situation in the light of its back-talk. Thus Quist's valuing of coherent design, nooks and soft back areas, artifices, and the softening of hard-edged forms makes it possible for him both to give his initial framing of the problem of the design of the school and to reframe that problem in the light of his discovery of the meaning of the gallery. . . . If, in the midst of such inquiries as these, there were a sudden

shift of appreciative system, inquiry would no longer have the character of a reflective con-versation. It would become a series of disconnected episodes.[11]

Quist's appreciative system is sketched briefly here as combining sensory, aesthetic, attitudinal and dispositional elements, which together constitute a through-line that turns a series of interactions with a situation into a "conversation". Constancy of appreciative system is an essential "condition" for a practitioner to judge the outcomes of their actions.

But the appreciative system can, and in fact must, also change, albeit at a slower rate. As a frame-work that determines what is thinkable and doable, the appreciative system can be understood as a kind of moral lens. In Schön's words, the appreciative system combined with other constant ele-ments (including an overarching theory and a stance of reflection-in-action) becomes an "ethic for inquiry".[12] Vickers describes appreciative systems as both individually and collectively held. Col-lectively, they are the basis for shared cultures, including professional communities of practice, such as that of curating. The capacity for a society or a professional community to change their collective appreciative systems is vital to adapt to a changing world. In fact, serious failures can arise when the appreciative system of a professional community renders pressing problems unappreciable.[13]

For Vickers, a collectively held appreciative system is both "sustained and changed by transmis-sion through the series of personal appreciative systems it develops and supports in succeeding generations'".[14] Individual practitioners must expand their appreciative systems in order for their community of practice to collectively adapt. This slow change is something that Schön both recog-nises as necessary and identifies as being beyond the scope of the work he has done in *The Reflective Practitioner*:

> But the constants – media, language, repertoire, appreciative system, overarching theory, and role frame – are also subject to change. They tend to change over periods of time longer than a single episode of practice, although particular events may trigger their change. And they are sometimes changed through the practitioner's reflection on the events of his prac-tice. The study of these sorts of reflection, crucial both to professional development and to the epistemology of practice, would require a more sustained longitudinal analysis than any I have attempted in the chapters of this book.[15]

As Schön identifies here, such changes occur across multiple episodes of practice, may be "triggered" by events and may result through a practitioner's reflection. In my experience, the kind of sustained and often iterative experimental work undertaken in a PhD, combined with a rigorous framework of reflection-on-action, provides an ideal ground for this slow expansion to occur and be recognised.

A decade has passed since I completed my curatorial practice-based PhD. Over that period, I have supervised several curatorial doctorates. In each case I consciously learned to tune into the prac-titioner researcher's appreciative system and closely attend to its expansion through their research. I have been fascinated to see what happens to doctoral candidates as their project and their frame-works for judgement enter that transactional relationship with the real world and have come to understand these expansions of judgement as closely related to the rather vexed notion of curatorial authorship.

Authorship in curatorial practice-based research

When Scrivener adapted Schön's work to creative practice-based research, he focused primarily on its use by artists. Applying his insights to curatorial practice in my own PhD required some transla-tion and adaptation. A key point that was necessary for me to articulate at the time was the question of curatorial authorship – i.e. the idea that a curator's work could be considered creatively authorial in the same way as an artist's.

This point – which is now much more established – was important in the mid-2000s as the curatorial practice-based PhD was at that time relatively rare. While the creative, artistic or studio-based PhD was rapidly gaining traction in the UK and Australia, there were few programmes offering curatorial PhDs. Beryl Graham, in Sunderland, UK, had completed one of the first in 1999, and by 2004 was beginning to build a cohort of practice-based curatorial students focused on media art curation. The Curatorial/Knowledge PhD Programme at Goldsmiths College London was founded in 2006. Primarily, however the curatorial PhD at that point was (and arguably continues to be) art-historical or theoretical. Establishing the authorial nature of curating enabled me to engage with and apply the burgeoning new thinking about the nature of creative practice-based knowledge being developed by many of the authors in this book, as well as key thinkers in Australia such as Estelle Barrett and Barbara Bolt.[16]

Since then, there has been a groundswell of curatorial PhD programmes and of curators undertaking practice PhDs internationally.[17] Across these contexts there is a wide range of structures for curatorial PhDs. Following curator and theorist Jens Hoffman's commitment to the exhibition as the primary unit of curatorial output, some supervisors and programmes emphasise the final exhibition as an examinable product of a curatorial PhD that embodies knowledge. My own focus on reflective practice, follows Schön in emphasising the interaction between the situation (construed broadly) and the practitioner. Exhibitions play a variable, sometimes non-existent part in that process. I have supervised or assessed PhDs encompassing all kinds of activities as legitimate parts of curatorial labour, including arts-in-health, archival interventions, collaborative systems and building infrastructure for creating art in precarious circumstances.

In this broad approach, the emphasis moves away from the success or failure of the public outcome of curatorial practice to the knowledge that emerges through doing. This is liberating for the practitioner as the potential for failure in outcome is crucial to risk taking and learning and should not be a barrier to producing and articulating knowledge. In fact, Schön's account of reflective practice shows us the opposite: that sometimes the most important insights are generated at moments of break down or failure.

Without a necessary public outcome to function as the locus and evidence of knowledge, the authorial nature of the practitioner's work becomes the through-line that provides structure and coherence to the overall research project. Schön's appreciative system is a productive way to understand and investigate that authorship, which I have used with PhD candidates to help them recognise their own creative actions and choices as a legitimate focus for reflection and enquiry.

There are a number of reasons why curators, including those pursuing practice-based doctorates, have difficulty in recognising and articulating their own authorial contribution. The first problem is the thorny issue of the artist–curator relationship and the power dynamics that are often at play in curatorial situations. Many curators describe themselves as "artist-led", in that they aim to follow, support and service as closely as possible the practices of artists. This stance often calls for an abnegation of curatorial agency and authorship in deference to the artists. This position is an understandable corrective to the fact that curators are in a position of considerable power in the artworld. For many curators, particularly those working in secure institutional positions, their regular wage and ability to influence artists' livelihoods and success through selecting them for collection and exhibition puts them in a somewhat uncomfortable position. Downplaying their own authorial agency can ease that discomfort and counteracts the powerful (if somewhat mythic) caricature of the international, superstar curator, more interested in his or her own vision than in serving artists. Disavowing curatorial authorship can be a stance of humility. By contrast, I would argue that the most ethical way to address this power dynamic is to make rigorously and explicitly clear the authorial decisions that underpin curatorial work and to be clear about the appreciative system that drives those decisions.

This rigour creates the possibility for adjusting or evolving that system when required – particularly if it reveals biases (such as the tendency not to include artists of particular backgrounds or genders, for example) that should be challenged.

A second problem is the way curating within institutional contexts can erase a sense of individual authorship. Where curators work frequently in large teams, or on exhibitions that are assigned to rather than devised by them, it can be difficult to recognise and unpick the way in which individual appreciative systems operate, influence outcomes and evolve over time. It is still common for exhibitions in large institutions – particularly projects like collection hangs – to not include specific curatorial credits. Curatorial agency is nevertheless very much at work in institutional settings. In the interaction between a particular curator's practice and the complex and multivalent nature of the institution, there is a rich seam of the particularly pragmatic form of creative work that is curating in action. Navigating the constraints of institutional contexts is a crucial part of that work. It can be empowering for curators to articulate their own appreciative system in such contexts, and in many ways essential to stay alive, and where necessary resistant, to institutional pressures.

Other curators, myself among them, tend to de-emphasise curatorial agency in deference to the will, desire or needs of the audience. In my own work I have championed the role of the audience in curatorial processes and argued for a form of curating that respects the reality of audience experiences. My research, like that of several PhD candidates I have worked with, has involved extensive qualitative data gathering with audiences as well as developing methods to involve audiences in curatorial processes. As a result, the will of the audience is often privileged over curatorial agency in the story of how decisions and actions are made, similarly to the way artist or institutional will is privileged in the situations described earlier. Note that this is only in the way the story is told. Curatorial agency is of course equally at work in all these contexts and understanding appreciative systems is an effective way to excavate and analyse its operations.

I have supervised curatorial candidates from all the contexts described earlier and worked with them to find ways to recognise, articulate and interrogate their own authorial agency in specific appropriate ways. This variety of contexts shows how responsive and idiosyncratic the structure of curatorial PhDs must be. An understanding of appreciative system creates a reference point against which the supervisor and student can together evaluate methodological options and determine the methods that will be appropriate to each study. In the following example of PhD student Bec Dean, I will show how appreciative system leads to certain methodological choices, and how those choices then – in the dynamic feedback loop of situational back-talk – create opportunities to challenge and enlarge a researcher's appreciative system.

Bec Dean: biomedical art and curatorial care

Bec Dean came to her PhD as an established and experienced curator who had worked in numerous institutional positions, including most recently as the co-director of *Performance Space* in Sydney, an innovative mid-scale contemporary arts organisation. Dean was trained as an artist, and she made body-based performance and video art influenced by feminism before becoming a curator in the late 1990s. Over several years in the lead up to her PhD, Dean began to identify an emerging field of art-practice that she called "biomedical art". For her, biomedical art engages fundamentally with not only science and medicine but also the lived experience of biomedical contexts, including the laboratory, clinic, hospital, sick room and death bed, as well as their wider social settings and networks. Describing and investigating this field – its historical antecedents as well as its expanding contemporary significance – was a primary contribution of Dean's research.

As a curatorial researcher, rather than an art historian, Dean came to this field in a practical and engaged way. Her well-established curatorial approach involved longitudinal relationships with artists, whose careers and creative development Dean had supported in numerous ways and through several different institutional roles. Her research questions arose from observing specific artists over long periods. Dean considers herself as "artist led", and her insights into biomedical art were, for her, a recognition of a common set of aesthetic and methodological concerns among artists that interested her.

Her doctoral research was founded on the creation of a touring exhibition and series of commissions – entitled *The Patient*[18] – which included many of these artists. *The Patient* concerned "the embodied experience of the artist as medical patient and the medical patient as living subject in contemporary art".[19] In this context, Dean's deep relationships with artists in the exhibition was amplified, as many were themselves chronically ill or living with pain. Dean's engagement with the artists often went beyond what might be thought of as a typical curator/artist relationship (if such a thing exists). For example, John Douglas, who had a new commission in the exhibition dealing with his experience of renal failure and kidney transplant, is a neighbour of Dean's, and over the period of the exhibition Dean was practically involved in his treatment and health care.

The Patient was an ambitious touring project that began at the UNSW Galleries in Sydney and then travelled to three regional Australian galleries. This was a courageous decision that required a subtle and sensitive approach to exhibiting challenging art works (many containing nudity and the body in extremis) to audiences that are sometimes assumed to be more conservative than those that frequent experimental urban galleries. In making this choice, Dean willingly took on an extended relationship of care for the works in the exhibition, the artists involved (particularly those creating new commissions) and the audiences who would experience the show. This willingness reveals something of the values that drive Dean's practice and underpin her appreciative system. She is "mission-focused" in seeking to bring the work of the artists she admires and believes in to wide audiences. She sees her role as one of extended, long-term, deeply engaged service to those artists.

Her close relationships to artists are therefore a fundamental influence on her own work. A key challenge we identified in her research was to unpick her own authorial agency from theirs. In her first PhD annual review (the milestone that in Australia confirms your candidature), Dean gave an attentive, subtle and detailed presentation on the practices of the artists she was working with. For Dean, this close attention to the labour of artists is the fundamental activity of her own curatorial labour. Through this close attention, she develops an intimate understanding of their motivations and their needs. Combining this with her years of experience in facilitating the creation and presentation of art, she finds highly expert, effective ways to support artists to make and display their work. However, this can leave very little room for articulating her own research contribution.

My challenge to her was that by the next year's annual progress review, the artists' projects should form only part of the overall explanation of her research. She needed to actively focus her attention on her own authorial agency and expand that part of her reflection and writing to equal her attention to the artists. A key part of this was in recognising a situation that Dean came to describe as "reciprocity with artists".[20] Drawing from Schön's idea of a transactional relationship with a situation, this formulation saw the relationship with artists as bi-directional and mutually impactful. Rather than seeing artistic authorship supported by non-authorial curatorial labour, this allowed her to see curatorial and artistic authorship in constant dialogue – with each affecting the other. Together we debated what data and data gathering methods were needed to investigate this reciprocity and decided on semi-structured interviews with the four artists who were creating new work for the exhibition. As Dean wrote in her thesis, "the interviews were not intended as evidence of curatorial efficacy, rather they add a discursive and enriching dimension to my reflections upon my own practice".[21] To put this another way, these interviews were not evaluative but rather an open-ended investigation of the relationships at the heart of Dean's research.

The concept of appreciative system was an invaluable tool in analysing these interviews as it allowed Dean to recognise and separate out the artists' systems from her own and identify moments when either or both of these systems had flexed or enlarged due to their interaction with each other.

As Dean observed, the interviews enabled her to "bring the artists' voices into my reflective process and into their case studies for [the] thesis".[22] Stylistically, she included frank, first-person descriptive accounts of working with the artists as pivotal sections in each chapter.

Through this close attention to the experiences of both the artist and the curator/researcher, Dean began to articulate her appreciative system in the context of "care". Traditionally, care (for buildings, people and things) has been part of the inherited understanding of the purpose of curating, which derives, etymologically, from the Latin verb "curarer": to care. The question of the contemporary role of care in curatorial practice is widely debated, with some authors seeking to move away from its entanglement with conservatism and or control and others seeking to expand the notion to a more fluid and relational interpretation. The notion of care in a broader societal sense was central to Dean's research into biomedical art. The dynamics and aesthetics of care in medical settings and situations featured strongly in most of the artists' works.

Through our supervisions, and by watching the kinds of literature, ideas and artworks that Dean was drawn to, I became increasingly familiar with her disposition, values and interests, and slowly began to form an impression of her appreciative system. As a curator and probably formerly as an artist, she is fascinated by experiences that reveal – in critical, compassionate and embodied ways – the complex interpersonal entanglements that care creates and sustains. Further than that, in her own work she values and seeks to create those relationships of mutual interdependence. In her practice – in *how* she does what she does – she is attentive to the ethics, struggles and rewards of these relationships. It is likely that she moved from being an artist to a curator because curating affords so many opportunities to care for (support, facilitate, enable) others to achieve their work.

This insight into Dean's appreciative system allowed us to begin to discuss the nature of care as a topic (within the exhibition), a guiding methodological principle for how she works and a key focal point for her contribution to curatorial knowledge. With this focus, we were able to actively and consciously articulate, challenge and expand her understanding of the operation of care in her practice. This became a through-line in her reflective process and in her thesis, which she described in her introduction:

> I view my curatorial practice as embodied principally around my care for others. I do not seek to present this way of curating as a "better" way of curating, but one that is particularly suited to my repertoire of practical skills, and to the kinds of artistic projects I am interested in. I make myself more adept to the service of others, by engaging with artists and their works throughout the entire life-cycle of their projects.[23]

She was able to identify different forms of care in operation in her work, each with its own definition and structure, including: longitudinal care, situational care, material care, conceptual care, receptive care and emotional care. Each of these gave her access to, and language for, a different aspect of her work. This approach brought rigour to experiences that were often very personal, allowing Dean to speak explicitly about the emotional labour of curating: an aspect of curatorial work that is rarely acknowledged.

Conclusion: expanding appreciative systems

Over the course of Dean's doctoral research, her appreciative system was placed in dialogue with three elements that are common to all curatorial projects: the *work* (in this case commissioning artists and creating a touring exhibition), the *situation* (including the broader cultural ecology of galleries, funding systems etc.) and *other actors* in the situation (including audiences, artists and professional colleagues). Through this dialogue, appreciative systems flex, resist and ripen. With attentive reflection, and with support from the external perspective of the supervisor, this process will allow students to apprehend a more fully developed sense of what they value and how they make judgements about the success of their practice.

The challenge for the supervisor is to help the doctoral researcher to grasp the partiality of their own judgements, and to understand that it is from that partiality that their unique contribution to knowledge arises. To take the example of Dean: it seems obvious to a highly empathetic and care-oriented practitioner that attending to the bodily needs of an artist should be part of her curatorial labour, and visible in the resulting exhibition. It was my challenge as a supervisor to continually remind her that this is not obvious to others, but instead that it arises from her own specific and personal perspective. If a curator fails to see this, they will likely miss what could be most valuable in their contribution to knowledge. They could fail to claim their own ground because they cannot recognise it as their own. The supervisor must continually challenge the researcher to see their own approach as partial and personal, and therefore unique and valuable.

Sometimes this challenge can cause conflict because it can be extremely difficult for a curator/ researcher to justify and explain something that to them seems obvious. The values, beliefs and preferences that underpin an appreciative system are deeply held and can seem incontrovertible to the practitioner themselves. Pushing them to articulate values can be interpreted as a challenge to their validity, or a "playing devil's advocate". In my experience, however, when a curator recognises and begins to work explicitly with her own appreciative system in this way, it is a powerful enabler for interrogating and explaining the wider value of her work. In many ways, the more idiosyncratic and individual this framework, the more impactful to the field it is.

Because of its many forms of disavowal, curatorial authorship is often heavily defended. It is challenging to bring it into discussion, let alone to recognise ways it may have expanded or changed. Paying attention to appreciative system is a means to reduce the resistance, offering a technical lens for researcher/practitioners to access and articulate their curatorial authorship. As the example of my work with Bec Dean shows, it is also a valuable shared concept for a PhD candidate and supervisor to employ together. As part of a supervisory "tool kit", it provides a discursive framework for bringing questions of authorship and judgement into view.

Understanding the meaning of appreciative system as developed by Vickers shifts the focus from a simple question of value judgement to a more profound sense of what questions or phenomena are "appreciable" within a researcher's practice. The slow change in system is experienced as an enlargement of what it is *possible to ask or know*.

Within this relational work, the supervisor's appreciative system is also in operation. Reflecting on my own experience with Bec Dean, I can recognise moments in our supervision processes where my own system shaped my perception of and reaction to her work. Encouraging a researcher to articulate their own system can counter the possibility that a supervisor will muffle their authorial agency. Further, the process of supporting a PhD candidate student to articulate their own appreciative system can, and often must, lead the supervisor to challenge and enlarge their own. Having supervised and assessed several curatorial doctorates, I increasingly recognise how it feels when my own appreciative system needs to expand in order to be able to appreciate (not just in the sense to value, but in that deeper sense of recognition or apprehension) the questions or phenomena with which a practitioner student is concerned. I have described this process as attunement. It is not that our two appreciative systems need to become congruent; rather, I need to expand my own system enough to be able to recognise, even while holding separately, that of the researcher I am supporting.

Working with appreciative systems enriches 1) the candidate's own reflective capacity, 2) the candidate/supervisor relationship and 3) the supervisor's own reflective growth and ability to supervise and support subsequent researchers. Beyond the doctorate, this has implications for the structural and temporal frames overlaid on research projects and careers. It calls for longer-term thinking in understanding the nature of knowledge produced through curating, an expanded notion of what that knowledge might consist of and how we articulate it and an expanded recognition of the interconnection between projects over time – both within a practitioner's long-term research trajectory and across an academic's supervisory career.

Notes

1 See for example Vickers (1965).
2 Schön (1983).
3 Muller (2009).
4 Muller (2011), Scrivener (2000), Scrivener (2002).
5 Barrett (2007, p. 135) (quotation page number from paperback edition 2010).
6 I was fortunate to have three supervisors, Ernest Edmonds, Ross Gibson and Toni Robertson, who each challenged and supported different aspects of my work. Though beyond the scope of this chapter, it is worth noting that, from my experience, the appreciative system can also be used in supervisory teams like this to create a shared focus for synthesising different inputs.
7 Although Schön does not attribute the term to Vickers, he does cite Vickers elsewhere – including unpublished memoranda – and was clearly very familiar with his work.
8 Novossiolovaa et al. (2019).
9 Schön (1983, p. 174).
10 Schön wrote his doctoral thesis on John Dewey and many of his ideas are indebted to Dewey. Contemporary readers may interpret the idea of "transaction" as a limited form of exchange between two parties that does not acknowledge the wider effects of interaction within a complex system – however, John Dewey's use of the word was more expansive than its current usage implies.
11 Ibid., p. 316). In the original quotation he also describes the appreciative system of the scientist Wilson, omitted here for brevity.
12 Ibid., p. 193.
13 Novossiolovaa et al. (2019).
14 Ibid., p. 55.
15 Schön (1983, p. 319).
16 Barrett and Bolt (2007).
17 Two examples of named programmes are the Zurich PhD in Practice in Curating (established in 2012) and the Monash Curatorial Practice PhD programme (launched in 2014). At UNSW Australia, numerous practice-based curatorial Higher Degree by Research projects are complete or underway in the School of Art and Design. In 2014 I convened a symposium, Knowing Through Showing: Conversations on Curatorial Knowledge, which featured nine UNSW curatorial projects (http://niea.unsw.edu.au/events/knowing-through-showing-conversations-curatorial-knowledge).
18 https://artdesign.unsw.edu.au/unsw-galleries/the-patient.
19 Ibid., unpaginated (web source).
20 Dean (2019, p. 10).
21 Ibid., p. 10.
22 Ibid.
23 Ibid., p. 14.

Bibliography

Barrett, E. (2007). Foucault's "What is an Author": Towards a Critical Discourse of Practice as research. In E. Barrett & B. Bolt (Eds.), *Practice as Research: Approaches to Creative Arts Enquiry*. London and New York: I.B. Tauris.

Barrett, E., & Bolt, B. (Eds.). (2007). *Practice as Research: Approaches to Creative Arts Enquiry*. London and New York: I.B. Tauris.

Dean, R. (2019). *The Patient: Biomedical Art and Curatorial Care*. PhD thesis. Sydney: UNSW.

Muller, L. (2009). *The Experience of Interactive Art: A Curatorial Study*. PhD thesis. Sydney: UTS.

Muller, L. (2011). Learning from Experience – A Reflective Curatorial Practice. In L. Candy & E. Edmonds (Eds.), *Interacting: Art, Research and the Creative Practitioner*. Berlin: Springer, 94–106.

Novossiolovaa, T., Whitmanb, J., & Dando, M. (2019). Altering an Appreciative System: Lessons from Incorporating Dual Use Concerns into the Responsible Science Education of Biotechnologists. *Futures*, 108, 53–60.

Vickers, G. (1965). *The Art of Judgement: A Study of Policy Making*. London: Chapman and Hall.

Schön, D. (1983). *The Reflective Practitioner How Professionals Think in Action*. New York: Basic Books.

Scrivener, S. (2000). Reflection in and on Action and Practice in Creative-Production Doctoral Projects in Art and Design. *Working Papers in Art and Design*, 1, unpaginated.

Scrivener, S. (2002). Characterising Creative-Production Doctoral Projects in Art and Design. *International Journal of Design Sciences and Technology*, 10(2), 25–44.

2.7

THE ART OBJECT DOES NOT EMBODY A FORM OF KNOWLEDGE REVISITED

Stephen Scrivener

Introduction

This chapter is concerned in particular with visual arts research (although I would like to think that the argument could be extended to art generally and certain kinds of design). Why, in the context of debates about visual arts research, has knowledge become such a hot topic? At least part of the answer can be found in the very idea of research, which is generally understood as an original investigation undertaken in order to gain knowledge and understanding. Given this definition, visual arts research must contribute to knowledge. However, the visual arts community places great significance on the art object and the art making process. Consequently, many visual artists wish to see a form of research in which art and art making are central: that is to say, the art making process is understood as a form of research and the art object as a form of knowledge. If one takes this position and accepts the common understanding of research, then one must be able to explain how visual art contributes to knowledge.

In this chapter, I start from the position that the proper goal of visual arts research is visual art. An alternative position is that the art making process yields knowledge that is independent of the actual art objects produced. However, this relegates the art object to that of a by-product of the knowledge acquisition process, and, in my view, places visual art making in the service of some other discipline. Notwithstanding the fact that valuable knowledge may be acquired in this way, from my standpoint it would be undesirable for this to become the dominant mode of arts research. Therefore, from my position the most interesting proposition to explore is the claim that the art object is a form of knowledge since it locates the art object as a central and fundamental component of the knowledge acquisition process.

Nevertheless, as you will see, in this chapter I argue against this proposition. I will not claim that the visual art object cannot communicate knowledge – it can. Instead, I will argue that this knowledge is typically of a superficial nature and cannot account for the deep insights that art is usually thought to endow into emotions, human nature and relationships, and our place in the World, etc. In short, I aim to demonstrate that visual art is not, nor has it ever been, primarily a form of knowledge communication; nor is it a servant of the knowledge acquisition enterprise.

I will claim that the objectives of visual art are different to the objectives of the sciences, etc., and that art may lose sight of its own objectives by adopting predefined notions of research. Indeed, art needs to ensure that what it chooses to call research contributes to its interests first and foremost. Hence, my aim is not to diminish the significance of visual arts research as compared to scientific

 DOI: 10.4324/9780429324154-20

research, for example. Quite the contrary, I hope to show that art research performs an equally important but complementary function to that of the knowledge acquisition research domains.

As noted, in the following I shall restrict my discussion to visual art and painting in particular, since I will only drawing on examples from this domain. Henceforth, I shall use the artworld, art and artwork to stand for the community of interest, the discipline and products of painting respectively.

Change and the context of change

Why has knowledge become a topic of so much discourse, particularly among artists and theoreticians in academe, and particularly in the UK? Indeed, outside of academe the issue would appear to raise little sense of urgency or heat. The 1990s saw radical change in the UK Higher Education system. During this period, the proportion of 18 year olds undertaking degree education more than doubled while the unit of funding per student decreased. In 1992, the binary divide was removed; polytechnics became universities and the artworld found itself firmly embedded as an "equal" player in the academic world of the university. As a consequence, the artworld became entitled to research funding, distributed via the Research Assessment Exercise (RAE). Not surprisingly, perhaps, the word "research" has become part of the artist's vocabulary and the artworld has committed wholeheartedly to the competition for research funds provided by HEFCE and the Arts and Humanities Research Board.

Since 1992, the UK Higher Education Funding Councils has recognized some activities of artists as research:[2]

> Research for the purposes of the RAE is to understood as an original investigation undertaken in order to gain knowledge and understanding. It includes . . . the invention of ideas, images, performances and artefacts including design, where these lead to new or substantially improved insights

Research, thus, is a knowledge derivation enterprise, and by definition, art is a part of that enterprise. Here, then, is the first reason for having an interest in knowledge. Research gains knowledge, and although images and artefacts are acceptable outcomes, it would appear that they are only relevant to research if their production leads to knowledge. In parallel with the changes noted earlier, there has been an enormous growth in the number of artists progressing on to PhD programmes, and the acquisition of new knowledge invariably figures as a central component in institutional regulations governing the PhD award. So, we have a second reason for being concerned about knowledge: to offer a PhD, a discipline needs to understand how its research contributes to knowledge, what kind of knowledge is produced and, where the creation of artefacts is central to the process, how these artefacts convey knowledge.

There is much anecdotal evidence of a steer toward research in academe, perhaps at the expense of art making (i.e., practice). If so, who is holding the harness? Is this the artworld driving forward in response to internal needs, or is the artworld being pulled along by largely external forces? There is a clear danger if the latter is the case: driven by external rather than internal imperatives, the artworld may lose sight of its own purposes. For example, let's assume for argument's sake that the removal of the binary divide has stimulated a desire in the artworld to be seen as equal to other academic disciplines. This being the case, undertaking research (perceived as the highest form of intellectual inquiry) and offering and obtaining doctorates (perceived as the highest form of academic qualification) might be seen as one way of achieving such equality. However, the artworld will need to show its research and its doctorate to be the same or equivalent to those of other disciplines. If research as defined in these other disciplines serves the purposes of those disciplines and if the purposes of those disciplines are different to that of the artworld, then unless great care is exercised the artworld might find itself drawn to an activity that fails to serve its purposes. This being the case, a shift to research in the arts could be extremely damaging as it has implications for academic competence, through its graduates the non-academic world and through them the general health of the arts.

It can be argued that this danger has been recognized and is reflected in the plethora of recent literature debating the nature and role of art and design research.[3] Artists make things; this is what they do and value. Consequently, some artists and designers would like to place making and the products of making at the centre of their research, i.e., practice-based research. This notion that art making is a form of research and that the art produced enables knowledge to be extracted has been and continues to be hotly debated. In fact, to believe in making as knowledge derivation does not require one to believe in the artefact as a conveyor of knowledge. For example, it is not inconsistent to argue for the former idea while making no claims for the latter. However, this would relegate the art produced to that of a by-product of the knowledge derivation process. If we start from the position that art is the proper goal of arts research, then knowledge derivation through art making would be research in some other discipline.

Taking art as the proper goal of arts research and gaining knowledge as the goal of research, we will need to consider whether art making can be seen as generating knowledge and how the art object might be understood as a knowledge transfer medium, or knowledge artefact. (By knowledge artefact, I mean an artefact designed with the intention of communicating knowledge. However, I do not mean to imply that knowledge is stored in these artefacts.)

Knowledge and the experience of knowing

According to Dancy,[4] the standard account defines knowledge as justified, true, belief: it hold that "a" knows that "p" if and only if

1 "p" is true,
2 "a" believes that "p",
3 "a"'s belief that "p" is justified.

For example, it's true that my partner's name is Deborah, I believe that her name is Deborah, and I'm justified in my belief on the basis of my having seen her birth and marriage certificates, heard her identify herself as Deborah, etc. In short, I know that my partner's name is Deborah. The tripartite account of knowledge defines propositional knowledge; knowledge THAT "p" is true. It does not define knowledge by acquaintance, as in "Stephen knows Walter", nor knowledge-how, e.g., how to drive a car, unless it can be reduced to knowledge-that [I'm driving, or have driven, a car].

As defined previously, knowing concerns the individual, e.g., "a". Philosophically, there are problems with the tripartite definition of knowledge and indeed with the whole epistemological enterprise of determining what it is for a belief to be justified.[5] Consequently, the sceptical conclusion that we do not know because we cannot know has not been conclusively countered. Insofar as it effects our personal and professional lives, the philosophical debate is somewhat peripheral, except at times such as these when there seems to be a pressing need for definitions of knowledge.

In our everyday lives, we have a sense of knowing, and this is distinct from, for example, merely thinking or feeling that something is the case. For example, sitting typing this paper at 4.30 pm, I know that there is a cup of tea on the table (I can see it). I think that I have the conference *information for contributors: full papers* document on my desk (I vaguely remember it being there earlier today, but I can't see it now). And I don't know where my partner is (I can't remember when she finishes work and so she could be at work, shopping or on her way home). I can distinguish many different mental states and "I know" seems appropriate to some mental states, whereas "I think" (unjustified belief), "I'm not sure" or "I don't know" seem appropriate responses to others. Asked to distinguish one state from the other, I might appeal intuitively to the tripartite definition of knowledge. I would claim to "know" if I hold a proposition to be true, believe it to be true and can provide a justification for believing it to be true, as for the cup on my table. With the *instructions to authors*, it was true they were on my desk (concealed by other documents), but I didn't believe they were on my desk, although I had some justification for belief: in short, I didn't trust my memory of past events.

Clearly, this sense of knowing is not knowing as defined previously, because in many cases I will not be able to distinguish, at the moment, true belief from false belief. For example, had I been asked at 3.00 pm today where my partner was, I would have said with confidence that she was at her place of work, but in fact she might have had to go to a meeting in another building. Nevertheless, when responding to the question, my experience would have been that of knowing. Generally speaking, this knowing might better be described as justified belief. Hence, knowing makes sense to me as a cognitive experience, and it is this sense of knowing that enables me to function in the world. By the same token, I would require any definition of knowledge or claim for a type of knowledge to be capable of being known in this sense. In other words, it must be possible to specify a situation consistent with a definition of knowledge which when confronted would engender knowing in me.

Some would argue that knowledge only exists in the minds of humans:[6] knowledge is not a stuff that resides in artefacts, such as books. Colloquially, we talk about books and other artefacts as repositories of knowledge, implying that such artefacts store knowledge that may be merely extracted. But this cannot be the case if we accept that only humans can have knowledge. Instead, it is argued that information rather than knowledge is stored in artefacts and humans derive knowledge by extracting it. For example, let's say on arrival at my local railway station I don't know the time of the next train to London. I walk to the timetable and find that the next train is in 30 minutes. I would now say that I know the time of the next train to London and would draw little distinction between being informed and knowing. The latter seems merely a restructuring of the former. In other cases, I might have to work harder to know. For example, arriving at Piccadilly Circus underground station I find myself confronted by two exits, one labelled Piccadilly North and the other Piccadilly South. I'm heading for Jermyn Street, which I know runs parallel to Piccadilly, but I don't know whether it runs to the North or South side. I return to the local area map in the station foyer, locate Piccadilly and then Jermyn Street, noting the latter's location relative to the former. Examination of the magnetic North indicator tells me that Jermyn Street runs to the South of Piccadilly. I now know that I should leave the station by the Piccadilly South exit. In either case, the knowledge is extracted by processing stored information. In this sense, we might claim that artefacts embody knowledge, which merely needs to be extracted. From this perspective, since art objects are artefacts, it is possible that they can embody knowledge in this way.

Ways of knowing

We can know in at least two ways. First, we can know through direct experience. I can know, for example, that my watch is on my wrist. Second, I can come to know through communication (knowing can be the result of a transfer of knowledge from one individual or agency to another individual). For example, by inspecting a train timetable I can know that there is a train to London at 9.00 am. In this instance, I have not acquired this knowledge via an exploratory, investigative interaction with the world, e.g., by testing a hypothesis. Rather, I have used something (let's say stored information) passed to me from another via the artefact to extract knowledge. (Henceforth, I shall use the word derive to signify a process of knowing through discovering and extract to signify a process of knowing by "reading" from artefacts.)

Clearly, since knowledge artefacts at the very least communicate information, this must involve representation, as information is about something that is other than the actual information. Young[7] defines a representation as follows:

> R is a representation of an object O if and only if R is intended by a subject S to stand for O and an audience A (where A is not identical to S) can recognize that R stands for O.

A knowledge artefact is a special kind of representation meeting the following additional conditions: it is intended by a subject to inform an audience and an audience can recognize that it is intended to inform. Consider, for example, the London Underground map. The map is an intentional representation of the London Underground system and I recognize what it stands for. Furthermore, the map is intended to inform and I recognize this intention. As far as I am concerned, this artefact meets the conditions of being a knowledge artefact, i.e., an artefact designed to engender knowing. On a recent visit to London, I used one of these maps in the foyer of the St. Pancras underground station to establish which line to take for Oxford Circus. The map informed me that both St. Pancras and Oxford Circus are on the Victoria line. Given my situation and goals, I utilized this information to extract the knowledge that this was the line to take to achieve my goal: I knew which route to take. I say "knew" because I experienced this cognition as a certain fact about the World: there were no ifs, buts and maybes.

By way of contrast, consider Figure 2.7.1, which illustrates a fragment of an artwork created by Simon Patterson called *The Great Bear*. Here I recognize the object as an artefact (an art object). However, the fact that it is an art object does not allow me to assume that it is intended to represent. Even if I take it to be a representation, I am not sure what it represents; I can't read it (i.e., its meaning is not obvious to me). In fact, I can read the picture to a certain extent. I recognize the names of famous historical celebrities (i.e., I take the words to represent these figures). I recognize lines and by association with different conceptions of lines, including underground lines, I can see that the location of celebrities on lines implies some kind of relation between them (i.e., I take the lines to represent some kind of relation). However, when I examine the picture further, I find it difficult to resolve what is represented (and hence what the elements of the picture mean). For example, the "circle" line seems to be concerned with philosophers, so perhaps each line represents a discipline or profession. Further examination reveals this to be a reasonable interpretation. However, what do the intersections between lines mean? The artist (art) line connects with the philosopher (philosophy) line. Does this mean that Raphael was also a philosopher (as distinct from the other artists depicted), that art and philosophy are related or both? Consequently, my experience of the picture is not one of being informed or of extracting knowledge; rather, it is one of possibilities and multiple potential meanings.

It might be argued that this is simply a matter of a lack of familiarity and that it is possible to learn to read *The Great Bear*. Although the meaning of the London Underground map appears clear to me now, there must have been a time when it wasn't, when I had to grapple with its meaning in much the same way as that described earlier in relation to *The Great Bear*. Of course, I simply don't know how I learnt to read the London Underground map. Maybe I already understood a lot about the world of the Underground, about stations, lines, etc. Maybe I discovered this world by using the Underground system and with this discovery the relation between the world and its representation. Whatever the case, it is highly likely that the map's meaning was not self-evident and that I had to learn how to relate representational components to the objects represented. Perhaps, but this assumes that Patterson intended the map to represent something specific and intended to inform us about something specific, and this does not appear to be the case. The artist has said of his work that "the idea of the viewer finishing the work is important . . . meaning is always shifting, anyway you can't

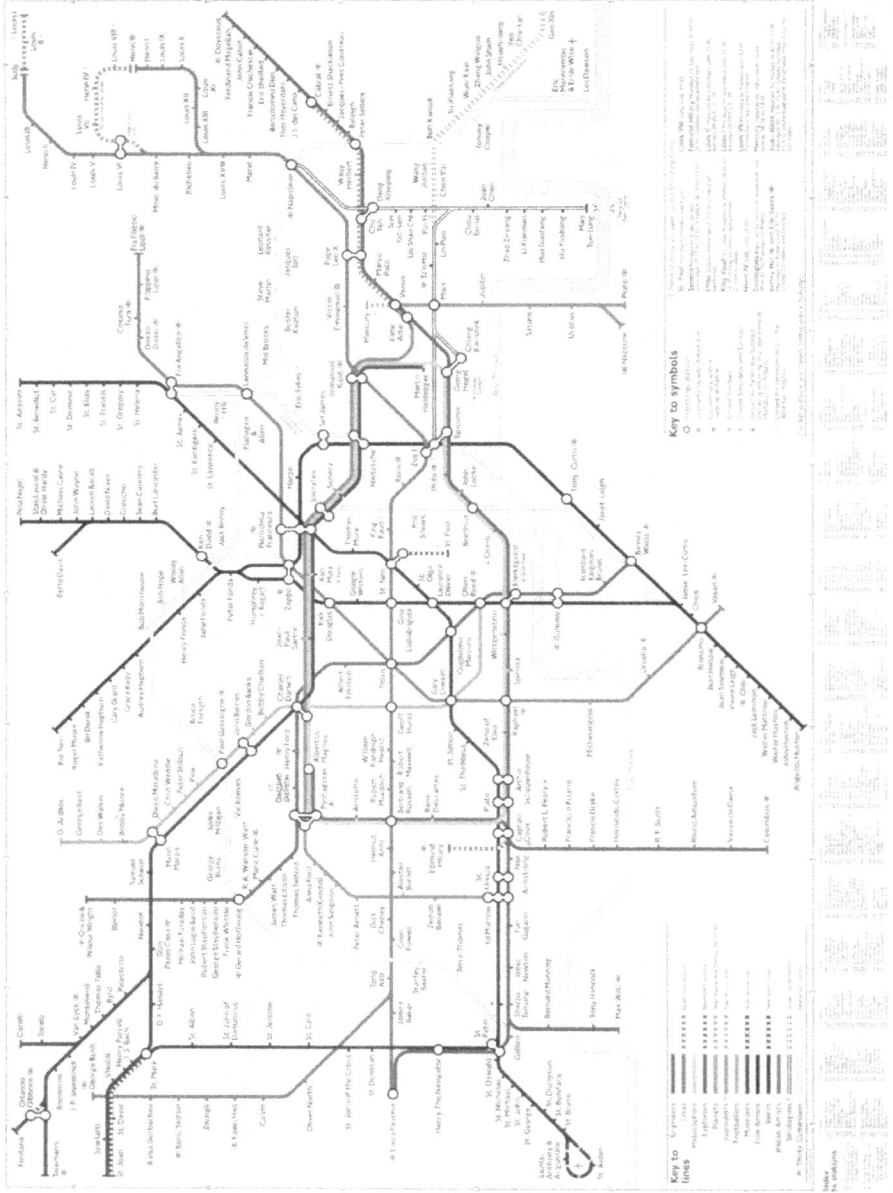

Figure 2.7.1 Simon Patterson, *The Great Bear*, 1992, print on paper, London.

Source: TfL from the London Transport Museum collection.

control the meaning of a work".[8] It would appear that he had no intention of conveying information: hence, the work is not a knowledge artefact according to my definition.

Again, we might want to argue about what Patterson means by the viewer finishing the work. Perhaps by this he just means the process of learning to read the meaning of the picture. In other words, that the work has specific meaning but the viewer has the task of discovering that meaning. If this is what he means, isn't his view about meaning and the work surprising? I would find it so and for this reason believe that Patterson means that the viewer completes the work by postulating meanings. That is to say, since the work's creator does not intend it to convey any specific meaning, if a viewer arrives at specific meaning, then it is he or she who has given it this meaning. Earlier, I implied that arriving at an understanding of the intended meaning of informational elements is a prerequisite for extracting knowledge from a knowledge artefact. I have argued that *The Great Bear* is not a knowledge artefact because it was not intended to convey specific information and hence knowledge. However, in each case I have acknowledged that the viewer has to assign specific meaning to pictorial elements. This being the case, and given that knowledge can be extracted from the London Underground map, is it not reasonable to argue that an individual who is able to assign specific meaning to the pictorial elements of *The Great Bear* can then proceed to extract knowledge from it, and hence know?

The fact of something being a representation does not imply that what is proposed is knowledge (i.e., true, justified belief). Clearly, fictions are represented. I would argue that to take a representation as knowledge, we have to recognize an intention to communicate knowledge: along with learning to read a representation, we need to establish that it is intended to communicate knowledge. On the basis of what we have learn about Patterson's interests, it can be argued that anyone who takes *The Great Bear* to be a knowledge artefact has made an error of judgement. Such a person may acquire what they take to be knowledge, but they would be mistaken in their belief that the artefact communicated true, justified belief.

From a personal perspective, my experience of *The Great Bear* is not one that I would describe as knowing. Rather, if asked I would say that I thought it was about this or that. In other words, my interpretation is a kind of hypothesis or possibility. I accept that we probably find ourselves in this position whenever confronted by an unfamiliar form of knowledge artefact (i.e., in the sense of an unfamiliar representation). But the purpose of the artefact is to communicate knowledge, not to engender possibilities, and any exploration involved on the part of the receiver in order to assign meaning is a means to this end. In the case of *The Great Bear*, we have seen that the artist makes no claims for communicating knowledge and instead seems to imply that the purpose of the work is to stimulate the viewer to construct possible meanings. Indeed, the artist makes no claims for any specific meanings and hence no claims for their truth or otherwise. In effect, the artist provides perspectives or ways of viewing the world, which may or may not be true.

To illustrate, let's say that I take lines to mean disciplines and intersections to mean individuals who contributed to the intersecting disciplines. Given this interpretation, I might infer that that Raphael was an artist–philosopher, that Titian was an artist–scientist–magnate and that Nietzsche was a philosopher–comedian. All of these inferences seem slightly odd to me, particularly that regarding Nietzsche. Nevertheless, given this perspective, I might read up on Nietzsche and discover that in his spare time he was a stand-up comic. In other words, I might discover that the perspective on Nietzsche that I registered as a consequence of assigning meaning to *The Great Bear* was true. However, as argued earlier, it would be wrong to take this as knowledge communicated to me via the representation. In effect, I would have assigned meanings that offered a perspective on Nietzsche and led to my obtaining knowledge new to me. In short, I'm suggesting that the function of *The Great Bear* is to offer hypotheses or possibilities rather than conclusions or certainties. Furthermore, I do not think this is peculiar to this work; rather, I would argue that it is characteristic of what is important about most, if not all, art objects. Additionally, I would argue that, generally speaking, art objects are not understood as knowledge artefacts.

Notwithstanding the above point, it is clearly possible to know something through pictures. Consider, for example, Millais's *Return of the Dove to the Ark* painted in 1851. Inspecting the picture, Figure 2.7.2, I can see that it depicts two young girls and a dove in a stable-like space. The girls stand side by side. The girl to the left of the picture leans back slightly, as does the bird, which she holds in one hand, while holding a flower in her other hand which rests on her friend's shoulder, who in turn leans forward in the act of kissing the dove. I would claim that I know (i.e., my experience is that of knowing) these things, *inter alia*, about the objects depicted and their spatial arrangement. Assuming that the Millais intended to convey knowledge, do you think this is the knowledge he intended to convey?

Clearly, there is more to the picture than this, as indicated by the title. I'm inclined to the view that the girl holding the bird is using her arm to protect it from the embrace of her friend, while recognizing her need to acknowledge the debt owed to the bird for bringing news of salvation. On the other hand, perhaps she is seeking to stay the embrace in order to pass the news on to the other occupants of the Ark. My point is that once I move beyond the physical objects represented to the meaning of their interaction and beyond that to the meanings implied by those meanings, then my experience is one of possibilities rather than certainties. I have ideas about what the picture is designed to communicate or engender in me; indeed, I might want to hold to several different interpretations of the work, but I wouldn't claim to know what it is about, or to know anything as a result of viewing it. Even if I arrived at a certain interpretation of the picture that engendered knowing, it seems to me that this knowledge is likely to be different to that intended by the artist. Once again, I would claim that this discussion reflects a general feature of my experience of art objects. Although we may be able to talk of knowledge being conveyed by art, this tends to be of a superficial nature that doesn't approach the deep insights that art is usually thought to endow into emotions, human nature and relationships and our place in the World, *inter alia*. In short, knowledge does not seem to be the right term for the cognitive states experienced when viewing Millais's work, Figure 2.7.2, which are better described as possibilities, or potentialities.

Shared knowing

The ability to convey knowledge means that we can share it and pass it from generation to generation. Hence, we talk about "bodies of knowledge" associated with different disciplines, such as physics, geography, etc. These "bodies of knowledge" can be refined and elaborated over time. This is only possible if individuals can derive knowledge consistently from a knowledge artefact, i.e., if there is a high degree of agreement between the knowledge that different people read from it. If the knowledge received by each person through inspection of a knowledge artefact is different to that of every other person, and if these differences cannot be resolved, then we can't talk about them having shared knowledge. This consistency of reading seems to be true of pictures such as the London Underground map. For example, I think that you will agree with me that three lines pass through Oxford Circus station: the Central, Bakerloo and Victoria lines. However, as implied earlier, consistency of interpretation is unlikely to be a characteristic of a group's experience of an art object. For example, take a few moments to write down what message you think is conveyed by Millais's *Return of the Dove to the Ark*. Now look at the bottom of the paper, where you will find what I think is the message of the painting: I suspect that our interpretations differ, probably quite markedly.[9]

Indeed, I would argue that, in general, artists do not strive to control the meaning that a work of art can have (cf., Patterson's comments earlier). Few artists, to my knowledge, make claims for their

Figure 2.7.2 John Everett Millais, *The Return of the Dove to the Ark*, 1851, oil on canvas, (WA1894.8).
Source: Ashmolean Museum, University of Oxford.

work as having particular meaning, and many appear to revel in the fact that a work can engender multiple and even inconsistent interpretations. The painter Eric Fischl has said of his work:

> Well, I'm not interested in narrative in the strict sense, as a kind of linear progression. I try to create a narrative whose elements have no secure, ascribed meanings so that an effect of greater pregnancy can be generated than in customary straightforward narrative.[10]

This being the case, it seems implausible to claim that the primary function of an art object is to communicate knowledge and of the art making process to create knowledge artefacts.

Justification

Darcy (1985) has noted that it is generally held that claims to knowledge demand justification. To what extent do we usually possess justification for our shared knowledge? For example, what grounds do any of us have for believing that the Oxford Circus underground station can be reached from St. Pancras by taking the Victoria line? I think we possess little real justification for believing this to be true, particularly when using the system for the first time. This, like a lot of knowledge, is taken on trust of authority. Through education we receive, via our teachers, what we accept as knowledge. Rarely does this knowledge come with justification. For example, I learnt at school that Henry VIII had six wives, but I didn't see the marriage certificates. Nevertheless, if asked whether I know how many wives Henry had, my answer would be "Yes, six". In such instances, it is not the justification that causes us to believe the truth of such statements; it is our trust in the communicator of the knowledge, i.e., the educational system. In effect, we take it on trust that someone in the system has justified the knowledge we receive. Thus, in the case of the London Underground map, we trust that the London Underground Authority has knowledge of the underground system and has established that the lines and stations, etc., are in reality as they are represented on the map. Hence, in many cases we get by with the statement component of knowledge without the justification of that statement.

However, sometimes we may feel that this trust cannot be taken for granted, leading us to question the status of a given piece of shared knowledge. To do this, we must be able to gain access to its justification preserved somewhere in an artefact. In most disciplines, "bodies of knowledge" are systematically organized such that it is possible to do just this: it is possible to trace through a "body of knowledge" to authenticate a component of it for oneself. I cannot see that art artefacts are organized in this way. We might wish to claim that the art literature provides the material that glues together the belief and justification embodied in art. However, the art literature is largely comprised of claims to knowledge of what artworks mean, etc., and justifications of those claims. In short, the art literature comprises a "body of knowledge" about art and artists produced by a separate knowledge acquisition discipline that takes these phenomena as matter for study. Therefore, I would argue, that there is little evidence to point to the existence of an organized, systematic "body of knowledge" in which art objects function as claims to knowledge and justification of those claims.

New knowledge

Earlier I suggested that there is a difference between knowledge extraction and derivation. The former belongs to the realm of learning and education and the latter research. Gaining knowledge that is new to the world is a process of knowledge derivation. Claims to new knowledge require justification since they are not already articulated within the knowledge system, and hence cannot be taken on trust. Most of us have direct experience of this need. For example, we are likely to take a sceptical position when a friend or colleague makes a claim to new knowledge – we would expect that person to justify that claim.

Research is concerned with new knowledge and justification has to come with the knowledge claimed. If an artwork is to communicate new knowledge, it must communicate both the knowledge claimed and the claim to the knowledge, i.e., its justification. It would appear that we have little difficulty in recognizing claims to new knowledge when represented and justified by linguistic statements (i.e., propositions). Here the justification of new knowledge is understood as an argument.

It follows that if an art object is to function as a means of conveying new knowledge, it must comprise both the new knowledge claimed and its justification. In what sense, then, can an artwork be understood as an argument? Young[11] claims that "the suggestion that paintings, sculptures, works of architecture and musical composition provide arguments is, however, frankly incredible".[12]

Young's conclusion follows from his belief that artworks (of all kinds) are essentially illustrative rather than semantic representations that cannot be used to make statements and therefore rational demonstrations. Personally, I find Young's argument against the propositional theory of art persuasive. In addition, the evidence of my own experience supports Young's conclusion. That is to say, I cannot personally ever recall viewing an artwork as an argument. More objectively, the notion does not appear to have featured prominently in the discourse on the role, nature and value of art. Therefore, I am drawn to the conclusion that the history of art cannot be understood as a process of building a "body of knowledge" through the acquisition of new knowledge because the contributions were not accompanied by justification.

Recapitulating

The question I have asked is whether art objects can be said to embody knowledge. First, I have suggested that, although the epistemological enterprise remains unresolved, we experience knowing as a distinct mental state where the thing that we know appears true, we believe it to be true and we can justify our belief. If an art object is to be accepted as a vehicle for engendering knowing, the least we should expect from engagement with a work of art is that it is accompanied by states of knowing.

I have suggested that there are two ways of knowing: we can come to know through active exploration of the World (i.e., by experience) or through messages conveyed to us by others (i.e., by communication). If art provides a way of knowing, then I take this knowing to arise by the latter means, i.e., communication. I have argued that if we are to talk of something as communicating knowledge, or perhaps more literally engendering knowing, then we should expect to experience this knowing when reading it. In short, we should be able to say, "I know this or that as a result of viewing this artwork". I have suggested that artworks cannot be read to the level where they are usually assumed to function, i.e., to endow deep insight into emotions, human nature and relationships and our place in the World, etc. If this is the case, then individually we cannot "know" anything of deep significance through viewing an artwork. Therefore, I am claiming that knowing is not the primary or significant cognitive state when viewing artworks. If an individual cannot read an artwork, then there is unlikely to be consistency of interpretation between individuals. Since this is a proposed prerequisite of shared knowing, it is unlikely that artworks function as a means of sharing knowledge. I have argued that a feature of a "body of shared knowledge" is that it is organized such that an item of knowledge can be authenticated by recovering its justification, i.e., the "body of knowledge" comprises both knowledge and justification. I have argued that, at least for the great mass of artworks, such justification doesn't exist. Finally, I have argued that claims to new knowledge require both the knowledge claimed and its justification to be communicated. Again, this does not appear to be a general characteristic of artworks.

Additionally, I have argued that the artworld (including artists, critics and historians, etc.) has not presented itself as if it were in the business of knowledge acquisition. Artists, the few that do write or talk in any great length about their work, tend not to make claims that a given work has specific meaning or that it is intended to communicate knowledge. Nor do they attempt to justify specificity of meaning or knowledge. Similarly, writers on art rarely make claims for artworks as knowledge or attempt to organize the items of knowledge conveyed via artworks into a coherent whole. If the artworld was primarily concerned with knowledge and knowledge acquisition, surely this would be clearly evident in its discourse. To insist that this is the proper function of the artworld is to accuse the community of longstanding and persistent incompetence. This being the case, I am drawn to the conclusion that it is implausible to claim that the primary function of an art object is to communicate knowledge and of the art making process to create knowledge artefacts.

Ways of seeing and ways of being

You will notice that I have not elected to argue that artworks cannot communicate knowledge, nor have I argued that they cannot communicate justification of that knowledge. If someone set their mind to it, I'd guess that they could communicate knowledge together with justification pictorially (it is another matter whether it would be regarded as art). Rather, I'm advocating that art does not appear to have been concerned with this and if so then it must have been concerned with something else. A wholesale shift toward knowledge and knowledge acquisition is likely to be at the expense of art's longstanding, but perhaps, implicit value.

What significant role does art perform, if it is not about acquiring and communicating knowledge? Earlier, I noted that it has often been understood as providing deep insights into emotion, human nature and relationships and our place in the World, etc. This view encompasses much of my appreciation of art; however, I do not need to dwell on its accuracy or otherwise for the purposes of my discussion. Instead, I wish to focus on cognitive status of these deep insights. Earlier, I suggested that, generally speaking, we experience these insights as possibilities rather than conclusions: as, "I think that" rather than "I know that". In this sense, artworks offer perspectives or ways of seeing. These perspectives may concern, for example, the way the World was, is or might be. So far, I have not discussed the ability of works of art to affect our perception, emotion and aesthetic sensibility. Because artworks have the potential to arouse such responses, we are able to associate sensation and feelings with how things were, are or might be. In this sense, artworks provide both ways of seeing and ways of being.

Through original investigation (i.e., research), we arrive at knowledge and understanding of the natural and artificial worlds, past and present. In contradistinction, art making brings into existence artefacts that have to be interpreted. Drawing on the natural and artificial worlds and imagination, the artist generates apprehensions (in the sense of objects that must be grasped by the senses and the intellect) which when grasped offer ways of seeing and being. Whereas original investigation is concerned with acquiring knowledge of what is or was the case, art making is concerned with providing ways of seeing and ways of being in relation to what is, was or might be.

The practical value of art

It may be argued that a fine line separates "reading" from interpretation and knowing from "seeing", that it to say conclusion from possibility. Nevertheless, these distinctions can be experienced and they have practical consequences. If I know something I am certain of the fit between this knowledge and the World, and consequently I can apply this knowledge and understanding unquestioningly. Returning to my earlier example, having acquired knowledge of the line to take from St. Pancras to Oxford Circus, I did not entertain it as a possibility that Oxford Circus was on another line: I took myself to Oxford Circus. Knowledge, then, can be taken as a given which may be applied to familiar situations without testing. Clearly, life would be very burdensome if every moment was one in which possibilities had to be tested prior to each move.

However, we can only be sure of the application of our knowledge in familiar situations. In novel situations, we may be forced to explore. In these circumstances we may draw on stored ways of seeing the World. Seeing the World in a particular way, we can do in the World in that way.[13] In short, I'm suggesting that art is one of those modes of experiencing that, rather than providing givens for dealing with situations, offers apprehensions that provide potential ways of seeing situations. Only experience will establish whether *can be* is, i.e., whether the apprehension fits the World. Furthermore, ways of seeing can be applied proactively. For example, an artwork might engender in me a way of seeing and feeling such that I register the latter as desirable. Consequently, I might choose to take this view on the World to see whether it has the perceived desirable consequences. From this perspective, art contributes to a mode of cognition that is crucial to our development and survival. The experience of artworks provides material for seeing and the experience of knowledge artefacts provides material for knowing. Each material contributes in its own way to our behaviour, the former dealing with the known world (in the sense of current beliefs) and the later the unfamiliar or unknown World (in the sense of a situation that confounds one's current knowledge). One is not a substitute for the other and each needs to be garnered.

Arts research

So, what does this all mean in terms of arts research? As noted earlier, HEFCE asserts that research is an original investigation undertaken in order to gain knowledge and understanding. Given the argument set out previously, can making art be described as an original investigation, and is this undertaken in order to gain knowledge and understanding? I have argued that art making is undertaken in order to create apprehensions (i.e., objects that must be grasped by the senses and the intellect) which when grasped offer ways of seeing the past, present and future, rather than knowledge of the way things were or are. Hence, in the context of making art I would define research as original creation undertaken in order to generate novel apprehension.

Why, then, should this definition be regarded as research, since it could be just as easily be said to describe "everyday" art making? In reply, I would argue that the researcher intends to generate novel apprehensions (by novel I mean culturally novel, not just novel to the creator or individual observers of an artefact) by undertaking original creation, and it is this that separates the researcher from the practitioner. Furthermore, the researcher would seek to comply with accepted ways of generating apprehensions and to meet discipline-determined norms of original creation.

In conclusion, I would propose that we should not attempt to justify the art object as a form of knowledge and should instead focus on defining the goals and norms of the activity that we choose to call arts research.

Epilogue – nineteen years on

In hindsight, I think that I wrote "The Art Object Does Not Embody a Form of Knowledge" as a provocation. By 2002, I was an experienced human–computer interaction researcher practiced in what might be called problem-solving research. In 1992, I transferred from a computer science to an art and design department. Some of the doctoral students I supervised there warmed to problem-solving research; others were chilled by it, primarily because this mode of research relegated their passionate commitment to art and design, their desire to contribute to it materially, their creative competences and the outcomes of their practice to supporting roles in both the acquisition and communication of new knowledge and understanding. This chill was felt across the whole art and design education sector, and not just in the UK; and already there was a lively debate around the subject of what has become known as practice-based research.

The paper was written in response to a call for papers for the second of a conference series called *Research into Practice* organized by Michael Biggs of the University of Hertfordshire. In the call and the subsequent editorial to Volume 2 of the e-journal *Working Papers in Art and Design*, Biggs[14] wrote:

> If we say there is no such thing as knowledge in art and design we may be putting all art and design researchers out of business, unless we also have an alternative description of what the field of art and design contains, and the relationship of research to how the field is advanced.

Previously, at the *Doctoral Education in Design: Foundations for the Future* conference, held at La Clusaz, France, between 8 and 12 July 2000, I had been forcibly struck by a comment made by one of the speakers. He observed that within five years his faculty staff would all hold doctorates but none would have professional design experience, owing to the requirement in his country that all university faculty appointees must hold a doctoral degree and the tendency for doctoral candidates to progress directly from master's to doctoral degree. My thought was this: if doing research in art and design means not doing art and design practice, then what is the future of art and design? Art, for example, is an ongoing venture, but its actors don't live forever, so to stay alive it must reproduce: nascent practitioners must enrol in the enterprise to be prepared by those who preceded them to carry the practice forward and, in due course, to act as docents to the next generation. What would the art and design life-cycles become, then, if their teachers were not themselves experienced practitioners?

I have never doubted that there is such thing as knowledge in art and design. What I questioned then (and still question now) is whether the acquisition and communication of knowledge is the way in which the fields advance. This is why early on in the paper I assert that "I start from the position that the proper goal of visual arts research is visual art"; implicitly, I'm suggesting that the proper goal of visual arts research is not new knowledge as we tend to think of it, i.e., new, believed to be true and justified propositions. Later in the paper, I claim that "art making is undertaken in order to create apprehensions (i.e., objects that must be grasped by the senses and the intellect)", meaning experiences that cannot be interpreted unequivocally by anyone and everyone as justified propositions. Later in the paper, I quality this claim as to what art does to define visual arts research as original creation undertaken in order to generate novel apprehension, meaning art that we have not experienced before: art that goes beyond what we know art to be.

The ninth impression of the paperback version of Russell's *Problems of Philosophy* contains the foreword to the German translation of the book in which he writes, "I should have chosen certain expressions differently if I had then taken account of relativity theory".[15] Likewise, if I were writing now, there are things that I would say differently. In particular, rather than relying on the philosophical definition of knowledge, I appeal to what knowing means to us, i.e., justified belief, given that we often are not in a position to affirm that a proposition is true. That is to say, we believe a proposition is true because we can provide reasons for believing it is true. At the time that I wrote the paper, I had not come across Peirce's definition of truth:

> The opinion which is fated to be ultimately agreed to by all who investigate, is what we man by the truth, and the object represented in this opinion is the real.[16]

Viewed in this way, the vast bulk of what we believe has not been brought to the end of inquiry, as ultimate agreement has not been reached and we are, therefore, not in a position to assert that a given proposition is true. In general, then, justified belief is hypothetical to one degree or another and, at best, merely established belief, i.e., belief that is accepted because it is supported by a wealth of experience and experientially affirmed deductions and inferences, rather than knowledge. As such, I am inclined to think that we would do better to talk about justified belief, rather than knowledge, because it allows us to contemplate that even a novel hypothesis can be viewed as a contribution to understanding: not because it can be deposited as a settled justified belief but because it stimulates further inquiry.

To conclude, what motivated me to write the paper in 2002 was an ambition to shift the focus of discourse from the problem of knowledge acquisition to how the field is advanced, which I took to be, first and foremost, through material rather than conceptual novelty. This is why I concluded the paper by saying, "I would propose that we should not attempt to justify the art object as a form of knowledge and should instead focus on defining the goals and norms of the activity that we choose to call arts research". So, the provocation had two prongs: the first sought to puncture the belief that art objects communicate new knowledge that we can hope to ever agree on; the second sought to puncture the belief that new knowledge is the means by which the field of art advances. I remain of the opinion that change in art, design and creative production is largely driven, in the first instance, by material, not conceptual novelty.

Notes

1 First published in www.herts.ac.uk/__data/assets/pdf_file/0008/12311/WPIAAD_vol2_scrivener.pdf.
2 HEFCE (1999, p. 5).
3 Cf., Strandman (1998); Buchanan et al. (1999); Korvenmaa (1999); Biggs (2000); Durling and Friedman (2000); Pizzocaro et al. (2000).
4 Dancy (1985, p. 23).
5 Ibid., p. 53.
6 cf., Freidman (2002).
7 Young (2001, p. 24).
8 Pirman (1997, p. 21).
9 I think Millais's *The Return of the Dove to the Ark* shows that we will be saved if we put our trust in God.
10 Kuspit (1987, p. 38).
11 Young (2001, p. 70).
12 Ibid., p. 71.
13 Schön (1983).
14 Biggs (2002, p. 1).
15 Russell's (1912/1989, p. 98).
16 The Essential Pierce (1992, p. 139).

Bibliography

Biggs, M. (2000). *Editorial: The Foundations of Practice-Based Research.* www.herts.ac.uk/__data/assets/pdf_file/0003/12279/WPIAAD_vol1_biggs.pdf (accessed January 30, 2021).
Biggs, M. (2002). *Editorial: The Concept of Knowledge in Art and Design.* www.herts.ac.uk/__data/assets/pdf_file/0005/12299/WPIAAD_vol2_biggs.pdf (accessed January 30, 2021).
Buchanan, R., Doordan, D., Justice, L., & Margolin, V. (Eds.). (1999). *Doctoral Education in Design 1998: Proceedings of the Ohio Conference.* Pittsburgh: Carnegie Mellon University Press.
Dancy, J. (1985). *Introduction to Contemporary Epistemology.* Oxford: Blackwell Publishers.
Durling, D., & Friedman, K. (Eds.). (2000). *Doctoral Education in Design: Foundations for the Future.* Staffordshire: Staffordshire University Press.
Houser, N., Kloesel, C., & The Peirce Edition Project. (1992, 1998). *The Essential Peirce.* Bloomington, IN: Indiana University Press, vol. 2.
Friedman, K. (2002). *Knowledge or Information.* Posting to the PHDDesign@ jiscmail.ac.uk discussion list, May 13.
HEFCE. (1999). *Research Assessment Exercise 2001: Assessment Panels' Criteria and Methods.* RAE 5/99. Bristol: HEFCE.
Korvenmaa, P. (Ed.). (1999). *Useful and Critical: The Position of Research in Design.* Helsinki: University of Art and Design Helsinki.
Kuspit, D. (1987). *Fischl.* New York: Vintage Books.
Pirman, A. (1997). Names are Worlds for Thing. *Sculpture*, 16(1).
Pizzocaro, S., Arrunda, A., & De Moreas, D. (Eds.). (2000). *Design Plus Research: Proceedings of the Politecnico di Milano Conference.* Milan: Politecnico di Milano, May 18–20.
Russell, B. (1912/1980). *The Problems of Philosophy.* Oxford: Oxford University Press.
Schön, D. A. (1983/1991). *The Reflective Practitioner: How Professionals Think in Action.* New York: Basic Books.
Strandman, P. (Ed.). (1998). *No Guru No Method: Discussion of Art and Design Research.* Helsinki: University of Art and Design Helsinki.
Young, J. O. (2001). *Art and Knowledge.* London: Routledge.

2.8

RESEARCH, SHARED KNOWLEDGE AND THE ARTEFACT

Ernest Edmonds

Introduction

This chapter discusses the nature of knowledge arising from practice-based research, where the production, modification or manipulation of an artefact is important. The word 'artefact' is used here very broadly. It can refer to a physical object, such as a piece of sculpture, but it can also be a novel, a musical performance, a computer program, a dance, a lecture – and so on. The chapter discusses the nature of such knowledge, the significance of artefacts and the way that research into and about making has developed, using the example domain of art and design. While the most significant aspect of practice-based research is in the way that it is conducted, through practice, the way that the results are presented can also be special. An artefact, or its documentation, is often included in the delivered results.

Both 'research' and 'knowledge' are words with multiple meanings, and it is important to be clear that, in the context of this handbook, only some of those meanings are relevant.

'Research' is used in the sense of an activity that delivers new knowledge that is expected to have key characteristics: it is shared, has been validated and can be challenged. The knowledge is new (in the world), can be shared with others over time (again, in the world) and has been validated in some way. The sharing is more than communicating with a few others. In research, sharing must be open and permanently available to unknown people. The knowledge must be available to archive in a sustainable material.

A concept or a form can be thought of as knowledge, providing that the preceding criteria are met. In science, the invention of the concept of gravity was new knowledge. In mathematics, the concept of zero was new knowledge. In practice-based research, the invention, for example, of a new art form can count as new knowledge. When such a new form is documented so as to explicitly meet the preceding criteria, it can be deliverable as a research outcome.

The words 'knowledge' and 'knowing' are used in many ways, only some of which are relevant to this chapter. Tacit knowing, for example, is by definition not expressed in a shareable form (tacit means silent or unspoken) and, while it might be critically important within practice, or even within a research investigation, it cannot be delivered as a research result. It cannot be shared in the way that is required in research. Tacit knowing, such as an expert's judgement, or knowledge embodied within an individual's skill can be shared through demonstration and training, for example, but not as research knowledge. However, such knowledge might have critical importance in other respects, as shown in two examples discussed in the chapter 'Mapping the Nature of Knowledge in Creative

DOI: 10.4324/9780429324154-21

and Practice-Based Research' by Kristina Niedderer in this volume.[1] It is important to recognize that 'we can know more than we can tell'.[2]

One form of knowing that is shared, but not in a way that meets the needs of research, is what we might term *songline knowing*, taking the name from indigenous Australian culture. This is a form of knowing, often mixed with myth, that is passed down generations through songs and stories. By its very nature, it lacks clarity but is rich, particularly because of the mix of myth and fact. The importance of songline knowing in life should not be under-rated, but it is not generated or shared through research. Clearly, research is not the key to everything that matters in life.

Kristina Niedderer's chapter also argues that we need to extend the scope of research in ways that incorporate a form of tacit knowing. In some way or another, this must depend upon finding ways of turning such knowing into explicit knowledge and so moving it out of the tacit category. In her chapter in this volume, 'Theory as an Active Agent in Practice-Based Knowledge Development',[3] Linda Candy shows a significant way in which this can be done in the context of practice-based research. She draws upon the work of Lewin, Schön, Argyris and others in which reflection in action and action research have formed the basis of the developments of both research methods and knowledge forms. Candy takes theories-in-use and extends that concept to theories-as-an-active-agent in practice. This leads to the notion of frameworks: collections of theories, criteria, taxonomies and strategies that can be used to inform practice. A framework is an item of knowledge that can be both used in practice and delivered as a shared research result.

In general terms, Frederick Crews put the point about how to approach knowledge nicely. He said that we should follow 'the ethic of respecting that which is known, acknowledging what is still unknown and acting as if one cared about the difference'.[4] A helpful book on writing about these topics is Francis Wheen's *How Mumbo-Jumbo Conquered the World*.[5] He provides a light-hearted, but careful, guide to spotting delusions. Chapter Four, called *The Demolition Merchants of Reality*, should be required reading for any researcher! This is a guide to convincingly sharing what has been discovered, and being convincing as is every researcher's desire. However, my chapter will focus on the special case of practice-based research, paying particular attention to the role of making or manipulating an artefact. Particular research methods may be used and different approaches to delivering the results can be required.

The problem of knowledge

To provide a background, I briefly discuss the problem that defining research knowledge presents us with, focusing on those aspects that are relevant to forming a full understanding of sharable research results. This section looks at the more general theory of knowledge without highlighting the special concerns of practice but provides a basis for the rest of the chapter.

Research is generally an ever-evolving process. Certainty is only possible if the knowledge is produced by deduction, where a logical argument is used to conclude the new knowledge from a set of premises. Even in this case, the certainty is only as strong as the certainty of the premises.

Deduction is used to show that B follows from A; then, if A is true, B must be also be true. Such an argument does not appeal to any facts about the world but relies entirely on its strength. This is not a matter of rhetoric; it has to be a matter of logic. In mathematics, for example, such research is common. The crucial issue is to show that the system of reasoning and the assumptions used do not contain or imply a contradiction. As an example, I can cite the first paper that I ever published.[6] The paper was about symbolic logic and I used a purely symbolic argument to show that my new knowledge was true. I did not refer to the world in any way and did not use a diagram to help. In fact, my training in logic at that time had taught me not to rely on looking at a diagram in case it led me to assume something that was not implied by the logic. The knowledge that I delivered was about as far from artefacts and the practice of making anything as one could be. Deduction may well be used within practice-based research, but in this context, it is not the central method for generating and validating the new knowledge.

In contrast to deduction, induction is the process of generating generally true statements of knowledge from a finite set of known facts. Induction generalises from sample facts. The general may apply to infinitely many cases, but the number of sample facts will be finite. Thus, the knowledge produced by induction can never be certainly true, as the next fact that is discovered may contradict the current generalisation.

A classic text on the subject is A. J. Ayer's *The Problem of Knowledge*.[7] He argued, pragmatically, that we needed to find a 'right to be sure' to support a belief in order to call it knowledge. It is a matter of debate how strong the backing needs to be for a believer to have the right to be sure that it is true. However, the setting of a standard requiring the impossibility of error should be resisted. His view was that one can have the 'right to be sure' even where error is possible. However, as well as being produced, research knowledge needs to be validated so that we can be sure that, as a community, we can reasonably agree to believe it.

A huge influence has been Karl Popper's view of a 'right to be sure' expressed in his description of scientific enquiry.[8] While he was concerned with science, his argument applies to research in a much wider context. His key point is that we cannot know any general truth about the world with absolute certainty. Philosophers, such as Bertrand Russell,[9] have pointed out that we can only observe a finite number of events and that, for all we know, the next observation will contradict any theory we have based on the earlier ones. Thus, Popper argued that the pursuit of new knowledge is based on attempting to falsify our current hypotheses or beliefs. The longer we go on failing to falsify them, the more we can claim the right to be sure. Once falsified, we need to find a new or modified theory. The main thing is to be open to, and even invite, criticism and attempts to disprove our theories. This is one reason explicitly sharing the knowledge is so important: to invite scrutiny.

Much of the understanding about having knowledge rests upon what we perceive in the world and what others report perceiving. The problem is that, while we have direct knowledge of what are often called 'sense data',[10] we only have indirect knowledge of the world through those data. A significant stance in this context is constructivism[11] (not to be confused with movements of the same name in art and in mathematics). In this view, in our pursuit of knowledge about the world, we construct rather than uncover it. One might argue that being sure is about being sure that what we construct is plausible rather than that it is certainly true.

Beyond knowing what is, this chapter is also concerned with what causes or influences what is. David Hume's work is classic here.[12] He explained the 'problem of induction'. Basically, we can never be sure that **X** causes **Y** because we cannot reliably induce the general case from specific instances (as earlier). However, he gave a valuable lesson in how to deal with such philosophical problems. He said that we had to rely on such induction in ordinary life or we would 'perish and go to ruin'. Some philosophical problems about knowledge may be intractable or devastating, but we must carry on anyway. We trust our knowledge of physics sufficiently to be willing to drive a car or fly in an aeroplane.

Research from practice is a special case but does not conflict with or fundamentally differ from research knowledge in general. Validating the new knowledge that practice-based research generates is not a matter of proving that it is certainly true; it is a matter of showing clearly that we have a right to be sure about it. This can be done in many ways, and all that is required of the researcher is that they state and justify whichever one they choose to use.

New knowledge through practice

I take art as my example of practice in this section. In the creative arts, a practitioner makes a step forward, from time to time, that changes our notion of an art form or of its place in society. Such steps could often be presented as new research knowledge but normally are only promoted as new concepts within art. A new art form can be new knowledge when the concept is delivered appropriately. While there is no need whatsoever to frame art innovations in research terms, looking at a few examples will help to illustrate what new knowledge in the arts can look like and hence will illuminate the meaning of new practice-based knowledge.

The first example is of such research carried out by a practicing artist through a PhD. None of the other three examples come from formal research, but each of them shows innovation that could have been explored and presented in research. Hence, they are important examples of practice that, beyond their undoubted fame, would also have been significant if conducted within a practice-based research project.

I will now describe an innovation in a recent practice-based research example and look at three historical innovations in the arts through a modern practice-based research lens. I include quotations from the artists and others that give a flavour of what, in a research context, was said about, or might have been added to, the artefacts in words.

Ian Gwilt researched the problem of 'augmenting the white cube' with augmented reality.[13] He looked at the addition of augmented reality to physical objects in a gallery context and at the implication for what was, in the early 2000s, a new mixed-media art form. The key artwork that formed part of this research consisted of a collection of his physical objects (in a 'white cube' gallery) and augmented reality software in a smartphone. When someone points the phone at a particular object, virtual objects are added to the physical one on the screen. The interactive artwork unites physical and virtual objects. He argued that this research allowed

> a much richer set of insights to be generated, beyond the capacity of an either exclusively text- or practice-based enquiry [. . . it] demonstrates how a combination of augmented reality content mapped to the location of a specific physical object can be used to engender cross-media experiences, narratives and layered readings[14]

Russian artist Kazimir Malevich is most famous for his first Black Square painting, from 1915. The painting consists of exactly what the title promises, a square painted solid black. This was seen as the ultimate step in abstract art. It was not as simple as it perhaps sounds. A reproduction hardly shows the reality of the work where, in 'the canvas's texture, created in rapid almost chaotic strokes, we can see the fingerprints of Malevich, who was in such a hurry that he outstripped his own brush'.[15] Borchardt-Hume describes the work thus:

> Rather than going through a lengthy process of gradually simplifying representational schemata . . . in the manner . . . of . . . Piet Mondrian, Malevich . . . invented at breath taking speed a new painterly language made up solely from shapes and colours. . . . No other work exposes the radical nature of this experiment more unequivocally than the Black Square.

And goes on to explain that when it was first exhibited, Malevich

> placed the Black Square high up on the wall across a corner . . . evoking the sacred place of the icon in a traditional Russian home . . . making a tabula rasa to ring in a new uncompromisingly modern age.[16]

Malevich's innovation was to make a painting that was purely abstract, not abstracted from nature, nor using forms that bore a resemblance to natural ones, nor making symbolic references. The Black Square was pure 'concrete' painting. Even its hanging in the traditional icon's position did not make it symbolic. It was a way of signifying the claimed importance of the innovation.

In the same year that Malevich produced his innovation, Marcel Duchamp began a series of works known as 'readymades'.[17] These were objects, already made, that were selected by the artist and exhibited as his work. The act of selection was the artist's creative contribution:

> The French avant-gardist signed his name to a bottle dryer and a fan and dated to 1917 his famous gesture of dispatching his sculpture *The Fountain* to a New York exhibition. The exhibition committee refused to recognise the recumbent urinal . . . as a work of art.[18]

Duchamp's 'readymades' introduced a new kind of art, one in which the artist made nothing. All they did was select something and assert that it was art, their art. The innovation was in the uncompromising conceptual nature of this new art from. It denied the necessity for any craft to be involved or for the artist's hand to play any part in the art making. This new form could not have been said to be validated after the initial rejection, but as years passed, 'readymades' and their descendants have become normal exhibits in art galleries around the world.

In 1952, the composer John Cage produced his piece 4'33". This was for piano, in three movements, lasting 4 minutes and 33 seconds. The instructions were that the pianist should open and close the lid at the start and end of each (precisely timed) movement. No notes were to be played on the piano. The timings were determined by chance procedures and the work was 'silent'. Cage said of the first performance, given by David Tudor at Woodstock:

> They missed the point. There is no such thing as silence. What they thought was silence, because they didn't know how to listen, was full of accidental sounds. You could hear the wind stirring outside during the first movement. During the second raindrops began pattering on the roof and during the third people themselves made all kinds of interesting sounds as they talked or walked out.[19]

In relation to 4'33", Cage acknowledged Robert Rauschenberg's White Paintings, saying 'To whom it may concern: The white paintings came first; my silent piece came later'.[20] The relationship between these paintings and Malevich's Black Squares is clear, but they might have been made independently. The year 1952 was during the cold war and Russian art was not prominent in the USA. However, Rauschenberg did study under Josef Albers that year at Black Mountain College.[21] It seems quite likely that Albers was well aware of Malevich. In any case, Cage's innovation related closely to Malevich's, albeit through Rauschenberg:

> As Rauschenberg postulated the dynamic of his paintings as their reaction to the changing light in the rooms in which they might be hung, and the conditions created by people coming and going in those rooms, so Cage conceived 4'33", as a piece of music constantly in flux, subject to the ambient sounds surrounding each performance.[22]

Rather like Duchamp and the 'readymades', Cage introduced a new musical form that did not require the composer's intervention in determining the sounds to be listened to. The composer's work was to encourage and facilitate listening rather than to provide the sounds that should be listened to. Also like Duchamp, it is not clear that this innovation was validated at its first performance, but subsequently it has become recognised as a significant work, not infrequently included in concerts.

While none of the last three examples are the result of practice-based research, each of them far exceeds the degree of innovation necessary within such research. A successful practice-based researcher does not have to be a Malevich or a Cage. Research often produces a less famous innovation than these. Gwilt's work, for example, has not been recognised to the same degree. At least not yet. However, as indicated earlier, he did provide the documentation, validation and justification that qualified it as a research result. It is worth noting that had 4'33" been reported as part of a research project, its relationship to the Black Squares would have been expected to have been noted, even though Cage may have had no knowledge of that connection. A research report of context is text that places the work in its international position, describing the state of the art. One of the research tasks is, then, to uncover previous work even if it had no influence on the practice involved.

Knowledge held within practice

In this section, examples of types of knowledge and knowing that cannot count as knowledge in research are identified in order to clarify what the chapter does **not** cover. Knowledge that is used within practice, not shared but held there, may or may not be new but cannot count as knowledge in our strict sense.

As mentioned earlier, knowledge comes in different forms, some of which are not of central interest in the context of research in general or practice-based research in particular. Examples of knowledge beyond the scope of this chapter are tacit knowing and private knowledge such as what I termed 'songline knowing'.

Some knowledge types cannot be shared and validated, such as:

Subjective knowing and private knowledge. Knowledge that is private to an individual is not shared and so is not delivered from research. For example, I have looked at many brands and types of acrylic paint and have found that a particular manufacturer makes the best type for use in the paintings that I am working on. I know that this choice is the correct one for me. As far as I can recall, I have not mentioned this to anyone apart from my supplier and I do not particularly intend to. Probably very many other people have made the same investigation and some will have come to the same conclusion for their own work. My knowledge is essentially private and sharing it would have little purpose unless I was teaching a follower in the old studio concept. I also make no claim that my knowledge is particularly new. I hope that my paintings make a new contribution to the world, but I make no such claim for my use of that particular acrylic paint. Hence, this knowledge cannot count as knowledge in research.

Tacit knowing. That we have achieved new implicit or tacit knowing is clearly not relevant because, by definition, it is not shared: it is not expressed. Tacit knowing becomes of interest in our context once it is made explicit and hence is no longer tacit. In her chapter,[23] Niedderer gives the example of the building of a TEA-laser where the reported research results proved insufficient when a second team tried to build one.[24] One of the first team's experts' tacit knowing had to be used and this required that expert to have an extended period of contact with the new team explicating this knowing. This demonstrates both that there was knowing that could not be delivered as research knowledge and that tacit knowing can be very important in practice; hence Niedderer's concern for it.

Candy's aforementioned chapter[25] shows how tacit knowing can be made explicit and describes, for example, Jennifer Seevinck's research into interactive art that stimulates the participant into seeing emergent forms.[26] Working from documented reflection on her own practice, adding knowledge from the literature and studies of audience experience, Seevinck generated a framework, including a taxonomy of emergence, as research knowledge that was delivered.

Songlines. Songline knowing can be incorporated into a song, a story or a dance.

> Songlines are maps of the land Aboriginal people live on. People sing as they walk about the country they are passing through and the stories and their relationship to it. They are connected to Dreaming Stories and to the stories told in dot

paintings. Many songlines are transmitted orally from generation to generation through songs, stories and dances.[27]

Such knowing exists in some form in all cultures. For example, it surely can be seen in the Bible, particularly in the Old Testament. There is no denying the importance of songline knowing. In Australia early in 2020, fires raged and it was clear to many that the Aboriginal peoples had known, through songlines, about fire and managing the land that European settlers had much to learn from. Its importance, however, does not in itself enable us to term it knowledge. Rather, it shows again that, as important as knowledge from research is, it does not include everything that we need or want.

The next section looks at research knowledge from various points of view in order to frame our concern for it within practice-based research and, in particular, in considering the place of the artefact.

New knowledge arising from practice

From the examples in section three, we see that, beyond knowing what is and knowing what causes what, we are interested in knowledge about action, about or arising from making. We are clearly able to find new knowledge about how to better make an artefact or how to make a new kind of artefact. Practical knowledge of this kind can still be shared, verified and criticised. 'Knowledge how' may not, however, provide the degree of explanation that 'knowledge that' does.

One form of research that is conducted within practice is action research, a process in which the researcher intervenes by taking actions within the world of concern. It is often used in education, medicine and management, for example.[28] The action researcher might, by changing what is being studied, generate new knowledge about how to do something, but leave it open to others to discover why it works. It is important to recognize that principled approaches to action research have been developed and practiced. Thus, this method is not problematic even though the researcher acts within the world being studied.

Phenomenology is a theoretic subject that is also relevant to much practice-based research. Phenomenologists[29] argue that 'knowing how' precedes 'knowing that'. From this point of view, action research should come before experimental research. Until the action research is complete, it could be argued, we do not know what to study experimentally. If we were starting from a clean sheet of zero knowledge, perhaps that would be true, but reality is more complex. All research builds upon what others have done before, so it never starts from a pure clean sheet.

Embodied cognition, a theory based on the idea that cognition depends on the entire body, is important for some practice-based research programmes. Discussing the work of Merleau-Ponty in the context of art and embodied cognition, Linda Candy explained that 'while perception is the origin of both the act of making art and its end product', there are 'important changes that occur in the translation and extension of perception into the physical process of art-making'.[30]

Phenomenology has a number of strands, but one important concept is that, to put it very briefly, the body is important in perception. A reliable account of perception must take action into account. From this perspective, action, cognition and perception must be considered together in any adequate description. If we accept this view, then research about human interaction with art works, for example, must try to capture information about all three aspects and unify them.

There is obviously more to practice than just thinking; however, reflection during practice is an important component.[31] Making the results of such reflection explicit often contributes to research, particularly for the creative practitioner.[32] Such a process in practice-based research draws attention to the place of artefacts in relation to knowledge and leads to certain questions, such as, to what extent can an artefact be said to contain or communicate knowledge? In the chapter 'The art Object

Does Not Embody a Form of Knowledge Revisited',[33] Scrivener makes a strong argument that an artefact does not embody knowledge. There is no way in general that an artefact can clearly communicate knowledge in the way that is required in research. The question is more about how the artefact, including the making process, can generate knowledge and to what extent we may need to see, touch or otherwise observe the artefact in order to fully understand that knowledge. As mentioned previously, in her chapter, Candy shows how 'frameworks' applied by and emerging from practice-based research can be examples of knowledge generated in this way.

A specific question in practice-based research is the presentation of the results. May it be that an artefact, other than a body of text, has to be included in the presentation? If the three historical examples given in section three had been part of research projects, surely the answers would have been 'Yes'? When many polytechnics were formed in the UK around 1969, they gave degrees under the auspices of the Council for National Academic Awards (CNAA), who allowed a research degree submission to include an artefact. For example, some fine art research students mounted exhibitions of their work for examiners to review, and this was much discussed in the early years.[34]

A review of the role of the artefact in art research stressed the individual differences between the artist researchers and advocated the use of clear frameworks within which they could work: respecting their differences but enabling comparisons and the sharing of knowledge.[35] As Candy put it, 'the problem of sharing that knowledge implies a need for a parallel means of communication, in effect, a linguistic one that can help to frame the way that we view the artefact and grasp the knowledge'.[36]

Bruce Archer and Chris Frayling were particularly influential in developing models of art and design research at the Royal College of Art, London. Archer's founding paper set the debate going in the 1970s,[37] in which he argued for the inclusion of design, alongside science and the humanities, in education (including research education). Archer went on to head the RCA's Department of Design Research, overseeing many PhDs in both design and art in which 'planning and making' formed a significant part. Frayling, Rector of the Royal College of Art from 1996 to 2009, has written widely on the nature of research in art and design, challenging many commonly held views.[38] Frayling follows Ayer's notion, discussed earlier, of 'the right to be sure', terming it 'justified true belief'. While the UK's growth in practice-based art and design PhDs was strong in the polytechnics, and then the new universities that they became, the RCA had a significant influence. Archer, in particular, advised many researchers nationally about the treatment of artefacts. Personally, as a supervisor, I was guided by Bruce Archer while developing methods for dealing with practice-based PhD studies.

One thing was to find ways of explaining the contribution to knowledge that is, in some sense, related to an artefact. What about it is new? How does that generalize beyond this specific case? How does the artefact illuminate the argument, the new knowledge? It is important to distinguish this last question from one about illustrating the argument. An illustration helps the reader understand the points being made but is not technically a necessary component in the argument. By 'illuminate the argument', we mean add to it in some way. Consider the following two very simple examples.

First, let us assume that we wish to argue that bisecting a rectangle with a straight line produces two rectangles. A geometric description can show this, but could be illustrated by Figure 2.8.1.

Second, let us assume that we wish to argue that if two straight lines are drawn from each side of a rectangle they can look as if they form a single line going behind the rectangle. This can be illuminated by Figure 2.8.2 but is difficult to demonstrate without showing that figure. It is needed in order to communicate the knowledge. In a similar way, it may be necessary to provide a sound file or a movie, for example, in order to communicate the results of practice-based research.

In this second case, although it is very simple, it might be helpful to aid the reader by saying 'imagine that the picture is in three dimensions'. In a more complex example, it might be very important to offer such help in knowing how the image should be viewed.

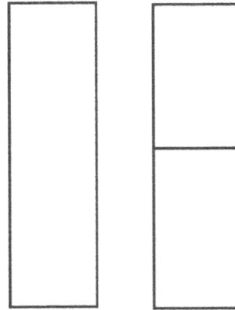

Figure 2.8.1 Split rectangle.

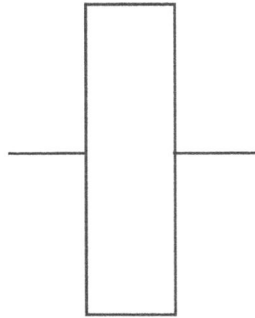

Figure 2.8.2 Rectangle and two lines.

Conclusion

Research generates new knowledge. That is why we do it, to expand our shared understandings and capabilities. Practice-based research does this through or within practice. It raises particular issues about the nature of the knowledge and about the ways in which it can be shared.

In sharing knowledge from practice-based research, there is often the additional issue of presenting artefacts or, at least, the documentation of artefacts. The accompanying text will have to describe the viewpoint that should be used in observing the artefact, helping the reader to understand questions such as those mentioned previously. What about it is new? How does that generalize beyond this specific case? How does the artefact illuminate the argument, the new knowledge? Answers to these questions enable innovation embodied in an artefact to be a component of the knowledge that a research project delivers.

Notes

1 Niedderer (2022).
2 Polanyi (1966, p. 4).
3 Candy (2022).
4 Crews (2006).
5 Wheen (2004).
6 Edmonds (1969).
7 Ayer (1956).

8 Popper (1959 – English translation).
9 Russell (1912).
10 Swartz (1965).
11 Bruner (1986).
12 Hume (1777).
13 Gwilt (2008) and Gwilt (2011).
14 Candy and Edmonds (2011, p. 262).
15 Shatskikh (2012, p. 45).
16 Borchardt-Hume (2014).
17 Cabanne (1997, p. 7).
18 Shatskikh (2012, p. 12).
19 Kostelanetz (1994, p. 65).
20 Cage (1966, p. 98).
21 Perl (2006).
22 Rich (1995, p. 142).
23 Niedderer (2022).
24 Collins (1985).
25 Candy (2022).
26 Seevinck (2017).
27 TIK (2020).
28 Argyris et al. (1985).
29 Lewis (1946).
30 Candy (2020, p. 97).
31 Schön (1983).
32 Candy (2020).
33 Scrivener (2022).
34 Jones (1980).
35 Candy and Edmonds (2010).
36 Candy (2020, p. 248).
37 Archer (2005).
38 Frayling (1993, pp. 1–5).

Bibliography

Archer, B. (2005). The Three Rs. In B. Archer, K. Baynes, & P. Roberts (Eds.), *A Framework for Design and Design Education. A Reader Containing Papers from 1970s and 80s*. Warwickshire: The Design and Technology Association.

Argyris, C., Putnam, R., & McLain Smith, D. (1985). *Action Science: Concepts, Methods, and Skills for Research and Intervention*. San Francisco: Jossey-Bass.

Ayer, A. J. (1956). *The Problem of Knowledge*. London: Macmillan Press.

Borchardt-Hume, A. (2014). An Icon for a Modern Age. In *Malevich*. London: Tate Publishing, 24–29.

Bruner, J. (1986). *Actual Minds, Possible Worlds*. Cambridge, MA: Harvard University Press.

Cabanne, P. (1997). *Duchamp & Co*. Paris: Terrail.

Cage, J. (1966). *Silence*. Cambridge, MA: MIT Press.

Candy, L. (2020). *The Creative Reflective Practitioner*. London: Routledge.

Candy, L. (2022). Theory as an Active Agent in Practice based Knowledge Development. In C. Vear (Ed.), *The Routledge International Handbook of Practice-Based Research*. London and New York: Routledge.

Candy, L., & Edmonds, E. A. (2010). The Role of the Artefact and Frameworks for Practice-Based Research. In M. Biggs & H. Larsson (Eds.), *The Routledge Companion to Research in the Arts*. London: Routledge, 120–137.

Candy, L., & Edmonds, E. A. (2011). *Interacting: Art, Research and the Creative Practitioner*. Farringdon, Oxfordshire: Libri Press, 267.

Collins, H. M. (1985). *Changing Order: Replication and Induction in Scientific Practice*. London: Sage.

Crews, F. (2006). *Follies of the Wise*. Emeryville, CA: Shoemaker and Hoard.

Edmonds, E. A. (1969). Independence of Rose's Axioms for m-Valued Implication. *Journal of Symbolic Logic*, 34, 283–284.

Frayling, C. (1993). Research in Art and Design. *Royal College of Arts Research Papers*, 1(1), 1–5.

Gwilt, I. (2008). *Mixed Reality Art*. PhD thesis. Sydney: University of New South Wales.

Gwilt, I. (2011). Augmenting the White Cube. In L. Candy & E. A. Edmonds (Eds.), *Interacting: Art, Research and the Creative Practitioner*. Farringdon, Oxfordshire: Libri Press, 257–267.

Hume, D. (1777) [1999]. *An Enquiry Concerning Human Understanding*. Edited by T. L. Beauchamp. Oxford: Oxford University Press.

Jones, T. (1980). A Discussion Paper on Research in the Visual Fine Arts Prepared for the Birmingham Polytechnic, England, in 1978. *Leonardo*, 13(2), 89–93.

Kostelanetz, R. (1994). *Conversations with Cage*. New York: Limelight Editions.

Lewis, C. I. (1946). *An Analysis of Knowledge and Valuation*. La Salle, IL: Open Court.

Niedderer, K. (2022). Mapping the Nature of Knowledge in Creative and Practice-Based Research. In C. Vear (Ed.), *The Routledge Handbook of Practicer-based Research*. London and New York: Routledge.

Perl, J. (2006). The Teacher as Magnetic Field: European Exiles and the American Avant-Garde. In *Starting at Zero: Black Mountain College 1933–57*. Cambridge: Arnolfini and Kettle's Yard, 75–82.

Polanyi, M. (1966). *The Tacit Dimension*. Chicago: University of Chicago Press, 4.

Popper, K. R. (1959) [2002]. *The Logic of Scientific Discovery*. London: Routledge (English translation).

Rich, A. (1995). *American Pioneers: Ives to Cage and Beyond*. London: Phaidon.

Russell, B. A. W. (1912). *The Problems of Philosophy*. London: Williams and Norgate.

Schön, D. (1983). *The Reflective Practitioner – How Professionals Think in Action*. New York: New York University Press.

Scrivener, S. A. R. (2022). The Art Object Does Not Embody a Form of Knowledge: Revisited. In C. Vear (Ed.), *The Routledge International Handbook of Practice-Based Research*. London and New York: Routledge.

Seevinck, J. (2017). *Emergence in Interactive Art*. London: Springer.

Shatskikh, A. (2012). *Black Square*. New Haven and London: Yale University Press.

Swartz, R. J. (Ed.). (1965). *Perceiving, Sensing and Knowing*. New York: Doubleday.

TIK. (2020). *Teaching Indigenous Knowledge: Songlines*. https://teachik.com/songlines/ (accessed December 15, 2020).

Wheen, F. (2004). *How Mumbo-Jumbo Conquered the World*. London: Harper.

PART III

Method

3.1

METHOD

Linda Candy, Ernest Edmonds, Craig Vear

Methodology and methods in practice-based research

As practice-based research is a relatively new mode of research, it means that there is no single, generally agreed methodology for practitioner researchers to draw on for designing their research investigations and choosing methods to suit their particular aims and objectives. For practice-based researchers, determining a methodology includes considering how the research methods apply alongside existing practice methods. As practitioners, they are already well equipped with well-tried methods used for their practice, and understanding what else might be needed for the research is often a challenge. This is a crucial issue and one many find hard to resolve because there are fewer precedents to call on.

This chapter discusses what we mean by methodology and methods in the context of practice-based research. Although the words 'method' and methodology' are often used interchangeably, they do not mean the same thing. At a basic level we define them as:

- **A methodology** is the study of methods, to use its literal meaning. The second meaning, that of a set or system of methods, is commonly used.
- **A method** is a process, a technique or a tool used in a research investigation.

In order to explore these terms in more depth, let's start by going back to the roots of the word methodology before discussing how to characterise it in the context of practice-based research.

The 'ology' in methodology means 'study or science of' from the Greek 'logia'. It is close to another Greek word 'logos', which means reason or idea. These etymological sources explain why words ending in 'ology' or 'ogical' have connotations of something based on reason and sound ideas, something thought out with precision and removed from emotion. Taken literally, at a basic level, the term 'methodology' refers to the *study of methods* but equally, it is used to refer to a *set* or *system of methods*.

Methodology as a *study* of methods

Methodology viewed as the *study of methods* involves examining the various options available to satisfy the aims of the research, and in the case of the practitioner researcher, their individual aims and intentions. At the start of a research project undertaken as part of a PhD or a funded research project,

 DOI: 10.4324/9780429324154-23

it is not unusual for the candidate or applicant to begin with a study of methods in order to establish what is appropriate for the particular purpose of the planned investigation. From that study, a methodology comprising the research strategy and the selection of procedures, tools and techniques to be used will emerge. Thus, methodology defined as the *study of methods* can lead to knowledge that informs about the rationale for the research design and can be the basis for answering the question: Why did you choose this approach?

The methodology, as a study of methods, will begin by exploring known research methods and the theories that underpin them, in order to choose the approach that best matches the practitioner researcher's objectives. A principled approach to the methodology is important. This is also an ongoing process and may need to be revisited as new insights and needs emerge. In addition, a properly defined methodology brings about clarity of the research process. These elements of research are usually highly influenced by the norms of the discipline within which the research is undertaken.

Methodology as a *system* of methods

In practice-based research, the role of the individual practitioner researcher is central to the shaping of the methodology as a *system of methods* for a specific investigation. When researchers talk about their methodology, they are usually referring to the rationale for the approach taken, the overall research design that outlines the way in which the research is carried out, the aims and objectives, the questions to be addressed, the procedures, the strategies that connect activities, timeline, tasks and expected outcomes and the methods chosen to achieve the goals. The research strategy is regularly evaluated and amended as the investigation progresses and new insights and priorities emerge.

In some modes of research, the methodology has already been determined as best-practice, and it is expected that the researcher will follow them. For example, in research areas, such as those associated with the natural and life sciences – chemistry, physics, biology, pharmacology, psychology etc. – experimental research design and the techniques needed to generate reliable results have been established through tried and tested processes that are well understood within the research communities. If the results of such research are challenged on the grounds of the misapplication of a method, the researchers concerned have to justify whether or not they have applied it correctly.

In practice-based research, however, the field is relatively new and operates under different norms. In particular, the role of the practitioner researcher is central to the shaping of the methodology (*system of methods*) for a specific investigation and is generally informed by the originating field of practice and bringing a range of existing methods into the research process. As such, practice-based research is an area where the methodology itself is under development and not easily captured in standardised procedures that can be taken off the shelf and applied. By implication, it is also a field of research where the methodology can be very innovative and, in some cases, make an original contribution to new knowledge.

Designing a practice-based research methodology

Designing a research methodology may begin by exploring the research methods used in a particular field and the theories or principles that underpin them, in order to choose the approach that best matches the practitioner researcher's objectives. A research design defines how the research is to be conducted, including when activities such as setting up investigatory tasks, the collection, analysis and interpretation of data, writing and reviewing are carried out. It will also consider the larger questions of ethics and data protection, documentation, archiving, timelines, milestones and evaluation of the research design and strategies and periodic reviewing points defined by academic rules. In any formal programme of research such as a PhD, this an on-going process and will need to be revisited as new insights, new goals and new needs emerge.

For practice-based research, a specific and focused research design draws together not only the needs and goals of an individual research investigation but also the defining character and principles of practice-based research. For illustration, by way of contrast, we can describe the basic principles of an ethnographic methodology in this way:

An *ethnography* methodology aims to explore the human social world and its culture, as well as shared beliefs and behaviours. Ethnographic research is undertaken in the field and attempts to capture and understand social action and the meaning of this action. The researcher is not based in a constructed setting or experiment but is part of the everyday natural situation within which those under investigation or those involved with the research project exist. For an ethnographic researcher, designing the strategy and choosing the methods of a specific investigation would need to keep these defining characteristics in mind.

In the case of a practice-based research methodology, there are some general principles within which the practitioner researcher designs their individual research programme:

Practice-based research methodologies incorporate research strategies and methods undertaken by means of practice. A basic principle is that the researcher is a practitioner who designs the research process, sets goals and determines the methods needed for producing outcomes of various forms including new artefacts, frameworks, taxonomies and theories. A crucial defining characteristic of practice-based research is that, not only is practice central to the research, but in certain cases, the knowledge that emerges can only be fully understood by access to the artefacts that are central to the research process in parallel with written texts that explain the context and results of the research. Insights and understandings arising from the thinking, making, reflecting and evaluating that take place may be viewed as new kinds of practitioner knowledge. All outcomes arising from practice-based research are relevant to the specific context but, at the same time, may offer knowledge that transcends the primary practice domain.

An example of a practitioner-defined methodology is that of Jennifer Seevinck. She defined her PhD practice-based research methodology in terms of integrating her creative practice making interactive art with research that evaluated the results of that process through iterative reflection-in-action. This also included theoretical work in areas that were directly related, such as interaction design and grounded theory.[1] Her research design incorporated practice, evaluation and theory, where although art practice was the springboard, the process was driven by theory. Having determined her overall design, she was then able to select appropriate methods that delivered her aims: this included reflective practice method and design and evaluation methods. The design and evaluation methods were drawn from iterative prototyping in software design combined with an ethnographic method for data collection and grounded theory for data analysis.[2] As this example illustrates, the choice of methods depends upon the practitioner's aims and approach. Seevinck's ongoing practice-based research is discussed in the section on reflective practices that appears later in the chapter 'Making Reflection-In-Action Happen: Methods for Perceptual Emergence'.[3]

Example research design models, strategies and frameworks

In the context of practice-based research, the research design is driven by practitioners' aims and the conditions and constraints of their specific field of practice, such as whether it is collaborative, self- or audience-focused, studio or public space located. This means that any model, strategy or framework that attempts to support the development process must be adaptable and flexible enough to be useful in a wide variety of situations. Examples from the chapters in this section follow.

'The Common Ground Model for Practice-based Research Design' chapter by Falk Hübner[4] offers a flexible method for enabling researchers to design and communicate their practice-based research strategies and methods needed for their particular situations of practice. The model is not a simple 'how to do it' approach but has a number of elements, including the preparatory stage of establishing research questions, points of departure and constraints, the choice of method or a set of methods and the structure of the research process, including how information flows. It is intended to take account of unforeseen events and therefore requires a high degree of flexibility in order that time and space for emergent thinking and making can happen. Falk proposes a *reframing* method as a means to explore research questions in the form of a network consisting of 'entities, activities, documentation, forms of reflection, learning, experiencing and knowing'. He stresses the importance of documentation of the research of processes as well as any artefacts or events, such as performances, workshops, exhibitions. The intention is to maximise the perspectives through which practice-based research design is considered and make explicit some of the underlying assumptions that would otherwise remain unspoken. In doing so, the model can act as a *framework* for supervision of practice-based PhD programmes.

In her chapter entitled 'Practice-Based Research in the Visual Arts: Exploring the Systems of Practice and the Practices of Research',[5] Judith Mottram explores the context for practice-based research in the visual arts in relation to cultural, social and economic frameworks. She points out that methods of practice-based research are no longer confined to the studio, an argument based on her wealth of experience in research, practice as a painter, research supervision and exhibition organisation among other things. Taking a long view from historical precedents and then bringing it up to the present day, Mottram raises issues about the nature of private studio-based art practice research that imply taking a broader view of method:

> In this way our notion of what a practice-based method is must move beyond a notion that what artists do is constrained to the manual production of art works. It may well include the writing of the exegesis, or the use of different information-gathering strategies such as surveys or interviews of questionnaires, or the setting-up of experiments to see 'what happens if . . .?'[6]

Mottram argues that, because today's generation of practitioners is no longer working exclusively with the traditional tools, materials and conventions associated with art, this raises questions about how practice-based research is taking place more broadly. As it happens, in certain fields of creative work, practice-based research methods have already moved beyond the studio practice focus at the same time as retaining the primacy of making artefacts as central to the process. Empirical study methods such as interview, questionnaire, survey and observation are well established in art and technology and new media art, for example.[7] Several chapters in this handbook mention the plurality of methods they draw on: this is discussed in the sections on empirical studies and reflective practice to follow.

The issues raised by Mottram indicate how the arrival of practice-based research in the traditional visual arts has raised broader considerations regarding method in any field of practice, particularly those that are exploring their practices through academically formalised practice-based research.

In her chapter, 'Finding the Groove: The Rhythms of Practice-Based Research',[8] Brigid Costello explores the rhythms of doing research and how this affects the practitioner researcher's effectiveness in the making, evaluating and reflecting process. She proposes that the value of identifying your own rhythms of time and effort is in the way it can support planning of a research timeline, giving a different perspective on what lies ahead and maintaining the kind of momentum that can be elusive in a research programme. By focusing on rhythm, she argues, the practitioner researcher is better placed to identify obstacles and how to resolve them. Advice on how to do this, such as developing strategies for varying the speed and energy of practice, using 'time-pacers' to match the rhythms and building in moments of reflective calm, are ways of using rhythm to analyse the progression of the work.[9]

The role and value of practice-based research to enhancing practice is an important motivating factor in the choice of methods for many practitioner researchers. In the chapter 'Crafting Temporality in Design: Introducing a Designer-Researcher Approach Through the Creation of Chronoscope',[10] Amy Yo Sue Chen and William Odom describe the role that time can play in the design of human–technology systems and a method for incorporating evaluation methods into the design process. They argue that first-person experiences of time in the design process are of great importance and can benefit practice-based research. The approach focuses on the experimental and novel outcomes of the design process arrived at through a reflective design practice using a reflective conversation with materials approach to the creation of a product, in this case, the creation of *Chronoscope*. Criteria (called 'lenses') are applied to the evaluation of the research design process: for example, it should be 'reproduceable', by which they mean giving a clear rationale concerning decisions taken and how the final insights are reached. This implies building on concepts that address the research questions directly and are applicable to future work. The approach is one which attempts to insert clear criteria for evaluation in terms of the rationale and traceability of the research design process. This is a method for ensuring that progress is monitored in a considered, transparent way. It has affinities with the design rationale method found in software system design.[11]

Methods for practice-based research

In practice-based research, the practice takes centre stage, but practice alone does not make it research. In a practice-based context, the practice involves turning concepts into artefacts, products or some other outcome, using methods that have been developed through the practice itself. It is important not to confuse the methods of practice with the methods of research, although in many practice-based research cases, a method may contribute to both. Practitioners coming to practice-based research are already well equipped with methods for practice, which will be specific to the field they inhabit, and those methods will continue into the research. Methods and techniques used in practice are designed for specific tasks and may be refined over time to suit the particular context. Research methods, tools and techniques are drawn from a broad range of fields such as social science, humanities, technology and scientific research and may be adapted to many different research purposes. Very often practitioner researchers customise such methods according to the needs of their specific situation of practice and the type of investigation they wish to undertake. For example, combining self-reflection with methods for empirical studies is not uncommon. For the practitioner researcher, looking to adopt pluralistic approaches from different disciplines often means learning new skills. Being principled when designing a research methodology applies equally to the selection of methods and a recognition of the effort required to master the skills needed for effective use.

Research methods are used to generate insights into the practice and potentially contribute new understandings that extend beyond the personal knowing of the practitioner, and may be shared, rather than private, knowledge. For example, a typical practitioner method in art or design or architecture is the making of models in wood in order to envisage the three-dimensional qualities of a concept. In the same practice context, adding a research method to that process would be to record observations of the tactile or visual qualities that the model evoked and compare it with models made of a different material. The collection of responses could be extended to other observers, or the model could be subjected to various stress tests in different physical locations. Through this process, new insights are generated, which are then evaluated by the practitioner researcher, who in turn folds them into their on-going study in order to generate new knowledge for the field.

In practice-based research, the identification of the appropriate methods is determined by the practitioner researcher working within the principles and strategies of their designed methodology. Judith Mottram describes methods as ways of developing our thinking and actions that enable new contributions to knowledge to be made:

> A method is a tool for investigation, a means of arranging our ideas in an orderly way, or a special form of procedure, so potentially requiring both mental and practical activity. Within the context of research, our methods are those processes of thought and practice that allow us to make contributions to the field, to move on or develop in new ways how we think, understand, or act. In both creative practice and in research, there is an element of generating new thinking or novel contributions that are recognised as such by appropriate audiences.[12]

Methods are procedures, tools, techniques and activities needed in order to carry out a specific aspect of research and to arrive at a specific type of outcome. They may be used to gather information needed by the practitioner to satisfy their goals in any research investigation: for example, this might be to construct a sound and visual installation or to design a new curriculum based on the preferences or learning targets of a particular group of people. They enable practitioners to satisfy their goals and may be tailored to the situation in hand. Sometimes this involves using methods for self-discovery through reflection; at other times, if there is a need to acquire information from or about others, this means using methods such as experiments, interviews, surveys etc.

Research methods for collecting and analysing data are an integral part of any research design. Where people are directly involved in the research process, the use of participative design method may be appropriate. In most research modes, the methods selected will depend on the type of data needed to answer research questions. Where the research questions have been set down and remain largely unchanged for the duration of the project, this is a fairly straightforward matter, but in practice-based research, having a clear set of fixed questions from the outset is rarely the case. One of the key distinguishing features of practice-based research is that research questions arise from the practice itself, usually but not always, in the making of an artefact. If this is the case, it means being aware of alternative possibilities, and for this, it may involve bringing new methods into the research process.

An example is the collective Cesar & Lois, who view the creation of spaces for raising challenging questions and experimenting with new methods for making and reflecting as a key characteristic of practice-based research. They engage with artists and practitioners from different disciplines and domains in a continuous quest for new knowledge. Their chapter, 'Thinking Together Through Practice and Research: Collaborations Across Living and Non-living Systems',[13] introduces their ideas, artworks and methods as a set of considerations for prospective practitioner researchers looking for ways of conducting interdisciplinary and collaborative practice. They propose a method for creative practice that is grounded in experimentation through making works combined with investigations in biology and technology. Their way of designing a research methodology involves finding the questions or issues, proposing challenging questions, developing prototype models or artworks and identifying ways to move ahead in new directions through reflections and the insights the thinking through making generates. They advocate a collaborative approach because different perspectives and expertise can provoke greater divergence that, in turn, can stimulate more creative thinking. This form of practice-based research aims to produce new knowledge through 'non-linear' artmaking and practice-research procedures which they see as accessible to a wide range of practitioners.

In his chapter 'Site: An Inventories Approach to Practice-Led Research',[14] Graeme Brooker, a design practitioner and educator, describes a method for practice-based research using 'site' as the location of the material to be handled. He outlines the elements of his method as an iterative and cyclical process that can be relevant to practitioners in any field of work that utilises existing materials. These are practices that prioritise the *thinking through making* approaches, characterised by Ingold,[15] and acknowledge the novel and less well travelled forms of research that practice-based research represents. The chapter aims to provide the reader with a broad knowledge of approaches to analysing existing material, as a provocation to reflection and its subsequent actions. The practitioner researcher is encouraged to develop 'sensibilities which reinforce enquiry, judgement, editing and decision making, as to the continuation of these distinct characteristics of the existing material'.[16] This is designed to enable better responses to the emergence of unexpected phenomena and offers alternatives to conventional design methods partly by releasing the practitioner from authorship and paying attention to the repurposing of what is already present.

Practice-based research has drawn on the methods of many other disciplines and domains, such as human–computer interaction, curatorial research, design science, educational and health research, anthropology and ethnography, all of which have provided a rich source of inspiration as well as practical methods to the practitioner researchers represented in this book. There are different pathways to be taken in practice-based research, a process which informs the methods employed. In many cases, practitioners combine and/or customise approaches from other disciplines or create new ones in response to the demands of the situation. Sometimes new understandings arise about how to deploy a method in a practice-based context, and this may constitute a novel outcome itself: for example, using video-cued recall in the interactive arts for audience research. Thus, in many cases, practitioner researchers combine existing methods, apply them in new ways or create new ones in response to the demands of the situation.

The subject of methods for evidence-based studies and reflective practice is discussed next.

Evidenced-based and reflective methods

There are two distinct pathways in the research process which inform the methods employed. These can be identified by asking these two questions:

1   Is the focus of the investigation on understanding the audience or user experience of the outcomes of the practice, for example, artworks or design products?
2   Is the focus of the investigation on understanding or advancing the individual's practice, which might include the making of an artefact?

The researcher will need to adopt evidence-based methods if the answer is 'Yes' to question 1, and reflective practice methods if 'Yes' to question 2. A combination of methods is needed if the answer is 'Yes' to both. These methods are discussed in the following section, with examples from contributing chapters.

Evidence-based research methods

We seek 'evidence' when we wish to provide more substantial grounds for believing something rather than simply relying on personal opinions or anecdotes. There are, of course, many degrees of evidence, and if we wish to base our actions upon reliable information, we need to carry out research into what others have found out or, if that is not available, do it ourselves. New evidence can be obtained through data collection methods such as direct observation, experiments, surveys or interviews. The aim is to produce knowledge based in actual experience, rather than in theory or pure reasoning. For some practitioners, the notion of *evidence-based* research connotes science, and to recover the concept, Linda Candy refers to it as *practice-based* evidence in relation to creative reflective practice.[17]

Whether it is called *evidence-based* or *practice-based*, in practice-based research, evidence is acquired by gathering data by means of practice over time. The practitioner sets a target or goal, decides on a set of actions, carries them out and then appraises the result of those actions. Many practitioner researchers have adopted evidence-based methods, sometimes in parallel with reflective methods.

Evidence-based research is used when the practitioner researcher is seeking to answer questions that require other viewpoints: for example, audience experience of artworks. Investigating audience experience requires a research process that draws upon actual events and situations as distinct from laboratory-based scenarios. Audio and video data are gathered so as to provide as accurate a picture of events as possible. The data analysis that follows must also be carried out in a manner that affords genuine insight into the nature of the picture that has been obtained.

There is value in practitioners observing their works in a real context and in progress, not only at the end point of completion. If a practitioner researcher seeks to move beyond personal reflections to include observations from other practitioners, experts, observers or audiences, for example, there is sufficient existing know-how about how to do this. For that purpose, methods can be identified for obtaining independent perspectives that offer the practitioner researcher an expanded view of what they are doing. Prior to devising a set of methods, practitioner researchers should identify the key elements of the research process, for example:

- Initial starting points or motivation for the project
- Theories and studies identified from relevant literature
- Time frame for the work or works to be created, performed, realised
- Environments and tools required to achieve the artefact, performance, outcome . . .
- The kinds of data to be collected
- Anticipated outcomes of the research process

Many studies involve observing closely and asking the right questions. To be able to do this successfully requires an understanding of appropriate research methods and how to apply them. Methods come in all varieties, but most fall into the qualitative area such as observation and interviews; others are qualitative, such as statistical tests. Some, like surveys, can produce both qualitative and quantitative forms from equivalent data depending on how the survey has been designed. Identifying well-tried methods is only a first step, however. Learning how to adapt and customise the methods to suit the particular context is a necessary second step.

Information gathering and analysis methods used by practitioner researchers are drawn from ethnographic research,[18] qualitative research[19] and grounded theory.[20] There is a very large set of standard methods for gathering and assessing information that can be useful to the practice-based process.[21]

The discussion that follows gives a brief summary of methods that have been applied in practice-based research in PhD programmes and other practice-related studies with reference to chapters in this book.

Methods for data collection

Methods for collecting and analysing data include recording self-reflections, observations of activities, tasks, behaviours, eliciting viewpoints etc. Primary data is collected through formal and informal procedures involving direct observation, verbatim reports, questionnaires and structured interviews. Written data consists of field diaries, transcripts from sound files and questionnaires, verbal protocols of discussions and interviews and visual data of still and moving snapshots, images and prototypes. These form a set of data showing the progression throughout the project, which can be compiled in a timeline. The data are compiled and structured in chronologically ordered records. This provides an overview of events indicating the activity, the length of time and whether a written, visual or audio record was taken.

Field diaries or personal logbooks may be collated into a single record base. Text formats enable methods such as lexical and protocol analyses to be applied. Different researchers carry out this work in order to arrive at independent viewpoints. A table can be compiled to keep account of the various types of data being gathered and where it could be found for the researchers to access. This provides a structured record of all the data relating to each stage of the project for periodic review and other purposes.

Obtaining multiple viewpoints can be very valuable to the thinking and making process of any outcome: artefact, installation, performance, for example. One such approach is a multi-view method that took a three-pronged perspective (in Figure 3.1.1, we have used *artist, curator, evaluator*, but other perspectives can be employed) on the development of prototype digital artworks.[22] Data was gathered from two main groups of participants during the creative process rather than at the very end when the final version of the artwork had been resolved.

The practitioner researcher's understanding can be greatly enlarged through the application of methods for facilitating dialogue between themselves and audiences. In this multiple viewpoint approach, the methods used were an experience workshop, semi-structured interviews and video-cued recall.

- The *experience workshop* method brought together experts from art, curating, human movement, science and interaction design. The method enabled the participants to consider their own experiences in relation to those of others in relation to the same artworks.
- *Semi-structured interviews* were undertaken with members of the general public in the art space where they could explore their experiences further, leading to richer experiential information. In this type of interview, the questions are framed in a way so as to provide structure but at the same time allow room for deviating from the script to follow lines of enquiry that seem interesting and relevant.
- *Video-cued recall* provided a combination of video data and verbal reporting, a method for closely investigating the detail of individual experiences. Video-cued recall involves replaying a video recording of a participant's actions and asking them to verbalise what they were thinking and doing at the time. The video-cued recall method helps participants recall details of their experience and avoid selective reporting.[23] This supports Lucy Suchman's[24] argument that verbal data obtained using video recordings can more accurately reflect lived experience than verbal data from interviews.

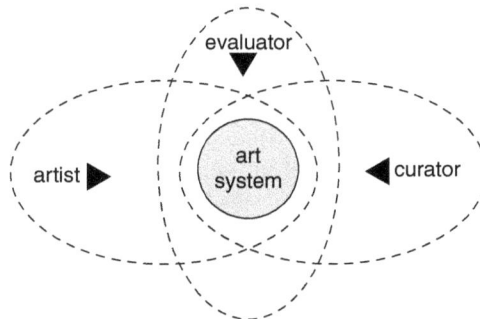

Figure 3.1.1 Three viewpoints on interactive art experience.

In her chapter, 'The Impact of Public Engagement With Research on A Holographic Practice-Based Study', Pearl John describes how she gathered multi-viewpoint data that encompassed her own personal reflections with audience responses to the works as well as expert and peer group opinions.[25] Her methods included surveys, direct observation documented in sketch books, field diaries and weblogs, and a method called the 'Silent Researcher Critique'.[26] She discovered that while audience experience of what they saw was both novel and emotional, only experts in art and holography were able to identify and comprehend the novel conceptual use of the technical aspects of her research. Her finding, that there is a distinction between expert and non-expert experience of some kinds of interactive artworks, tallies with the results of Zafer Bilda's study carried out at Beta_Space.[27]

John's practice-based research process demonstrates the value of combining different methods to obtain sources of data beyond the individual researcher's reflections. By engaging with public audiences, she was able to obtain feedback that supported the ongoing advances in her artistic work; moreover, the process led to a better understanding of research questions she needed to address and the methods best suited to answer them.

Deborah Turnbull Tillman's case study *ISEA2015: Disruption*, described in her chapter in Part V, 'Curating Interactive Art as a Practice-Based Researcher', also adopted a multiple viewpoint approach to data collections for her research.[28] The voices of the audience were documented through survey method, the curator's own voice through documented auto-ethnography and reflective practice and other curators' viewpoints working across art, science and technology with interactive, engaging or media-based artists and their work by interview analysed using conversation analysis.[29] Conversation analysis data usually takes the form of video or audio-recorded conversations, collected through semi-structured interviews from which detailed transcriptions are produced. This then forms the basis of an analysis using codes or criteria based on the researcher's questions to be addressed.

In her chapter, 'Please Touch!', Marloeke van der Vlugt describes the variety of methods she uses to answer her research questions through interdisciplinary artistic practice.[30] An important feature of her approach to practice-based research is the bringing together of social science and artistic research in making and exhibiting artworks, literature and theoretical background studies, observations, reflections, documentation and dissemination of the works through writing, filming and teaching. The project *Thresholds of Touch* deployed a multi-disciplinary, experimental method in a performance experiment that aimed to heighten participant awareness and document reflections on the experiences in notebooks. Research materials include participant notebooks, researcher field notes, and the video recordings of the participant experiences. The method documents informal practitioner insights and artistic strategies into experiential and material forms. For her practice-based research, she employs various methods throughout the process in an iterative manner, during which each step informs the next. This leads to a reframing of the methods in use and can lead to outcomes that are important to the contribution to knowledge.

Methods for data analysis

Analysing the data/evidence captured during these methods is a crucial activity that is often neglected in the rush to gather multiple forms of information. To this end, data may be analysed by hand or with digital systems that have been developed in the social sciences. The pioneering work of Lyn and Tom Richards[31] has given rise to mature and easy to use computer systems, such as NVIVO, that enable effective analysis of text, sound and video data.[32]

For comparative analysis based on larger data sets, it is worth developing a coding scheme based on 'events' derived from the data, such as the one developed by CCS researchers,[33] where audience experience protocols were coded by two researchers individually and then arbitrated into a final coded protocol. The coded protocols from the quantitative data can either be statistically analysed or used to produce charts and graphs which facilitate the identification of patterns of behaviour as well as the thoughts, feelings and perceptions of individuals during their art experiences.[34]

An example is the way Yun Zhang was able to carry out research into collaborative teamwork, building a public art installation using conversation analysis techniques with software support.[35] This research adopted a participative practice-based methodology in which the researcher was embedded in the team on a day-to-day basis, where she was able to observe closely how creative ideas were developed and implemented over a three-month period. The approach is one which can be readily adapted to any practice-based situation.[36]

These kinds of data analysis exercises should not be entered into without a serious attempt to grasp the basics of the analytical methods required. Practitioner researchers going down this route will need to acquire new skills. Where the questions to be addressed can only be satisfied by analysing significant amounts of data, putting effort into the preparation and learning can prove well worthwhile. Today's support systems are vastly easier to use than the precedents that existed in the 1980s and 1990s, although it is fair to say this has yet to become commonplace, even where the situation could be greatly facilitated by more digital support. Practitioner researchers do not need to limit their options to familiar paths and, indeed, as the pattern of practice-based PhDs indicates, they are more than willing to experiment with new methods.

The question inevitably arises, why would a practitioner choose to embark upon a lengthy process of gathering data, devising analysis frameworks and analysing examples across different cases? What is the value of evidence-based research conducted by practitioner researchers? Not only does it require time and effort to learn the skills, but it is a way of thinking that some fear could have unforeseen effects on the creative practice. Balancing the amount of effort needed is difficult to achieve. That said, depending on the particular practice-based research context, there may be good reasons to adopt a more structured approach, not least of which is the possibility of producing fresh insights into the processes of practice and its outcomes.

As an example, Brigid Costello created artworks that encouraged audiences to behave in playful ways. She also carried out studies of participant interaction with the works in order to explore whether or not she had achieved her aim of encouraging playfulness. She used observational techniques to study several of her own artworks, gathering data using video-cued recall and interviews and analysing the data using qualitative analysis methods and software analysis software in combination with mind-mapping software. However, the degree of effort required to carry out such studies raised questions in her mind about its value and in particular what effect was it having on her creative practice:

> I was concerned that the multiple opinions produced by a formal audience evaluation process might confuse or muddy the artistic aims of the artwork. . . . I was also concerned that the process of conducting and analysing audience evaluations might take precious time and focus away from my creative practice.[37]

In the end, she found the process 'surprisingly rewarding and creatively inspiring'.[38] She had also learnt new ways of designing interaction with her works that worked better with audiences.

As these examples indicate, there are different pathways to articulating a personal methodology in practice-based research. It is not enough to identify an approach and simply appropriate something: adapting and tailoring to meet one's own particular requirements is essential. Finding out what methods best match the aims of the situation and being able to draw on the methodological framework set out at the beginning of the programme provide a 'lodestone' that helps you keep sight of your aims at the same time as being a way of reflecting on the process. Such reflection is a vital part of practice-based research, and for some, it is the mainstay of their methodology, as discussed in the following section.

Reflective practice methods

Reflective practice involves a process of reflecting on one's actions and learning how to act differently as a result. Donald Schön's ideas on reflective practice have been influential because he asserted the value of practitioner knowing as having distinctive contributions to make to professional expertise.

The terms reflection-*in*-action and reflection-*on*-action are key to understanding what Schön meant by reflective practice and its relationship to tacit knowing in action.[39] Reflection-*in*-action is characterised as an intertwined and reflexive process of thinking about the actions being taken, or about to be taken, in a unique situation. Thinking and taking actions are a form of dialogue through which the practitioner assesses his or her actions and, in doing so, learns how to develop better ways of addressing the problems faced in professional practice.[40] Reflection-*on*-action, on the other hand, involves reflecting on how practice can change by evaluating a situation after it has happened in order to discover

> how our knowing-in-action may have contributed to an unexpected outcome. The reflection takes place after the event and draws on knowledge of previous events and their connection to an unexpected event; it includes working out what has to now be done to address this in the future.[41]

In her chapter, 'Reflective Practice Variants and the Creative Practitioner',[42] Linda Candy revisits Schön's original concept of reflective practice, which was derived from protocol studies of professional practitioners and goes on to describe the range of reflective practices found among creative practitioners. From interviews with creative practitioners, she has extended the range of reflective practices in the context of creative practice in music composition, visual and sculptural art, design and public art installations This reveals a more nuanced picture of how practitioners reflect as they explore ideas and experiment with materials in the creation of artworks. Variants, such as reflection for action and reflection in the making moment, she suggests, are likely to be found in any field where the design and construction of a new artefact or product is involved. The understandings that have emerged from this research have implications for applying reflective practice as a method in practice-based research.

Schön's concepts of reflective practice and its role in the development of expert professional knowledge are revisited in full in *The Creative Reflective Practitioner*.[43] Because reflective practice has contributed to the methodological foundations of practice-based research in a substantial way, it has become a familiar method to be found in practice-based PhD programmes.

Examples of reflective practice in practice-based research

In practice-based research, it is commonplace to adapt existing known methods, and Schön's concept of reflective practice has been readily embraced. However, this is only a first step towards developing a suitable method for practice-based research. In a certain sense, this requires the creation of new 'norms' out of old ones. One such new norm is the repurposing of the existing concept of reflective practice as a method for documenting insights and processes in practice-based research.

The impact of reflective practice on creative practitioners doing practice-based research is an important contribution to the methodological grounding of practice-based research. It has been a key pillar in the research of many practitioner researchers who created personal conceptual frameworks that informed the making and evaluation of the outcomes of their practice-based research.[44] Reflective practice, as a strategy for challenging existing practice and at the same time generating new understandings, is a pathway of choice being followed by many practitioner researchers. One of the most appealing aspects is that it validates their intuitive instincts within a framework of reflective enquiry. Moreover, it also provides an opportunity to document the process of reflecting in action as it takes place. Documentation can then be returned to later for further reflection. The introduction of structured documentation using diaries, weblogs and other recording methods are invaluable to rendering the process more transparent and, at the same time, making it available for scrutiny.

In practice-based research, reflective practice and making new works go hand in hand as the practitioner explores new ideas and reflects through the making process. This often leads to challenging questions that require different responses, which cannot always be answered by self-reflection alone. For example, in the chapter 'Reflection in Practice: Inter-disciplinary Arts Collaborations in Medical Settings',[45] Anna Ledgard, Sofie Layton and Giovanni Biglino describe the key principles of a reflective practice approach to arts engagement. A reflective ethos is at the heart of their creative collaborations:

> What is described is a process of fine-tuning, of calibration, of reflecting through another lens – a patient's, a collaborator's, an institutional partner's. . . . This is ultimately why projects of this nature cannot be fixed in their expected outputs from the onset, because such iterative processes will inform the framing, the development and the dissemination of the work.[46]

Sofie Layton's practitioner voice, 'Bearing Witness – the Artist Within the Medical Landscape: Reflections on a Participatory and Personal Research by Practice',[47] is a companion to Anne Ledgard and colleagues' account of their collaborative projects. As an artist, she has a deep interest in the medical world and how medical concepts are seen and absorbed by patients into their own stories. Her research is an investigation into the human body when sick. Her method is to bring patient experiences to light, conducted through workshop activities in the different spaces of the medical environment: the bedside, the hospital ward, the clinic. From the stories that arise through the workshops, she gathers inspiration for tangible artworks and the making process begins, a process that involves collaborations with people of different skills, such as sound designers and physical object fabricators. The end point of each project is a presentation of the artwork which becomes a reflection of the patient's journey through a particular illness and at the same time opens up the dialogue to a wider audience. This last part of her research process is a form of reflection at a distance, as described in the chapter on reflective practice variants.[48] Reflection-at-a-distance is a type of reflection-*on*-action that occurs when a degree of detachment from the making process is warranted. For the practitioner, taking an artwork out of the studio into a public space is a way of stimulating reflection by breaking with the familiar ground and transforming the practitioner's perspective; this can throw it into relief and reveal aspects not previously considered that feed into the research.[49]

In her chapter, 'Making Reflection-In-Action Happen: Methods for Perceptual Emergence',[50] Jennifer Seevinck describes methods from Schön's reflective practice repertoire such as framing, reframing and frame experiments, moves and situation talk-back, and relates them to emergence theory and features of the practice-based research process. She describes a cycle of emergent reflective practice, identifying reframing for emergence and emergent outcomes along with other methods from reflective practice, for example, situation talk-back. Situation talk-back takes place via feedback to the practitioner researcher, leading to a reframing of that situation. The resulting new frame is an emergent understanding, while the reframed situation can also produce new outcomes. For this to occur effectively, she argues, the practitioner researcher must adopt an exploratory, open-minded approach that is receptive to ambiguity and unanticipated possibilities. Cultivating an open-minded approach and sensitivity to potential multiple interpretations enables the practice-based researcher to first identify situation talk-back, then continue with subsequent cycles of reframing. Interpretation of the feedback can generate new understandings about emergent themes or patterns.

In Lizzie Muller's chapter, 'Curatorial Authorship and Appreciative Systems in Reflective Curatorial Practice', she refers to the concept of an 'appreciative system', first coined by Geoffrey Vickers 'to describe a complex and powerful concept of how humans apprehend and make meaning from their world'.[51] Donald Schön uses the term to describe the way professional practitioners develop their professional knowledge; it acts as a reference tool against which they can assess whether events are pertinent to the particular 'situation talk-back'. Muller has extended the concept of appreciative system from the relatively stable constant Schön envisaged, to one that is characterised by continual change as the practitioner advances their judgement through reflective practices. In her PhD research, she developed her own appreciative system through reflective practice that became a key contribution to new practitioner derived knowledge.[52] The influence of this concept from her PhD to her current academic supervisory work has been a thread that has been woven into the ongoing development of Muller's practice-based research methodology:

> I show how that an understanding of appreciative systems . . . can be understood as the capacity to examine judgement itself, and how judgement has been challenged and enlarged through engagement with the world. I show that cultivating attunement to appreciate system thus supports the student, the student-supervisor relationship, and the supervisor's own reflective growth.[53]

As an evaluative method, the appreciative system determines what the practitioner can recognize as pertinent to, or part of, the change in the situation engendered by actions in practice. Lizzie Muller argues that paying specific attention to appreciative systems focuses practice-based research and can support research supervision.

In his chapter, 'Co-evolving Research and Practice – _derivations and the Performer-Developer', Ben Carey acknowledges how Schön's concept of 'reflection-on-action' was vital to being able to make sense of the new ideas and insights that were emerging through his practice, undertaken as a central part of his doctoral research process. It was through sustained reflective practice that he learnt how to contextualise his work in the scientific and technological theory of human–machine symbiosis and musical interpretation. He learnt more about himself as a software developer as he grappled with the degree of agency that he built into his software code as he developed new forms of performance practice. The interrelationship of the practice and the theoretical research was facilitated by a reflective practice method which was faithful to his artistic practice, and which proved to be a contribution to a novel way of working. He writes:

> Having developed a new software artefact and an associated performance practice, I was now able to reflect upon the outcomes of my work in some depth, enabling me to connect these emergent ideas with literature in the field of science and technology studies, and musicology. . . . Alongside the developed software artefact, associated recordings and the novel methodological approach, these reflections became some of the core research contributions of my PhD.[54]

Carey describes how it 'provides opportunity to contextualise one's practice with broader theoretical concerns' and can also lead to the development of novel methodological approaches.[55]

As we can see from the novel ways in which the practitioner researchers referred to here have deployed reflective practice, this is a method that is proving highly adaptable and appropriate to very different situations of practice and research and leads to new and novel forms of knowledge. We have described the basics of methodology and methods in the context of practice-based research and differentiated between these terms. The methodology provides a principled framework that supports the design of the research inquiry. The methods are the means of data collection, the evaluation of results and the reflections on the processes and outcomes leading to new insights, artefacts and knowledge.

Topics on method in the chapters to follow

Distinctive features of practice-based research method are explained through the various forms of practice-based research discussed in the chapters that follow. These are:

- Proposed models for designing a practice-based research project
- Identifying questions through making, appraising and reflecting
- Frameworks articulating and sharing the practice-based process
- The role and value of practice-based research to practice
- Evidence-based and reflective practice methods with examples
- Methodologies as novel contributions to research

Notes

1 Seevinck (2011, pp. 242–256).
2 Edmonds and Candy (2010), Candy (2022).
3 Seevinck (2022).
4 Hübner (2022).
5 Mottram (2022).
6 Ibid.
7 Candy and Edmonds (2011), Candy et al. (2018).
8 Costello (2022).
9 Ibid.
10 Chen and Odom (2022).
11 Moran and Carroll (1996).
12 Mottram (2022).
13 Solomon and Baio (2022).
14 Brooker (2022).
15 Ingold (2013).
16 Ibid.
17 Candy (2020, pp. 244–247).
18 Crabtree (2003).
19 Richards (2009).
20 Glaser and Strauss (1967): Grounded theory is a method which enables the researcher to identify patterns in a set of data that can be studied across a range of data for comparative analysis: www.groundedtheoryonline. com/what-is-grounded-theory/.
21 See www.pbrcookbook.com/resources.
22 Edmonds et al. (2009).
23 Video-cued recall: Costello et al. (2005).
24 Suchman (1987).
25 Ibid.
26 John (2019).
27 Bilda (2011).
28 Turnbull Tillman (2022).
29 Sidell (2009).
30 Van der Vlugt (2022).
31 Richards (2002).

32 NVIVO is available here: www.qsrinternational.com/nvivo-qualitative-data-analysis-software/home.
33 Bilda (2005).
34 Bilda (2007).
35 Zhang and Candy (2007a, 2007b).
36 Zhang (2011).
37 Costello (2011) 'Many Voices, One Project'. Chapter 4.2, p. 182.
38 Ibid.
39 Schön (1983).
40 Schön (1991, pp. 68–69). Schön defined reflection-in-action in this way: When someone reflects-in-action, he becomes a researcher in the practice context. He is not dependent on the categories of established theory and technique but constructs a new theory of the unique case. His inquiry is not limited to a deliberation about means which depends on a prior agreement about ends. He does not keep means and ends separate but defines them interactively as he frames a problematic situation. He does not separate thinking from doing, ratiocinating his way to a decision which he must later convert to action. Because his experimenting is a kind of action, implementation is built into his inquiry.
41 Schön (1991, pp. 68–69).
42 Candy (2022b).
43 Candy (2020): Chapters 2 and 3.
44 Candy and Edmonds (2011) for chapters by Dave Burraston, Jen Seevinck, Andrew Johnston, Brigid Costello, Mike Leggett and Lizzie Muller.
45 Ledgard et al. (2022).
46 Ibid.
47 Layton (2022).
48 Candy (2022b).
49 Candy (2020, pp. 63–67).
50 Seevinck (2022).
51 Vickers (1968).
52 Muller (2008, 2011).
53 Muller (2022).
54 Carey (2022).
55 Ibid.

Bibliography

Bentley, T., Johnston, L., & von Baggo, K. (2003). Affect: Physiological Responses During Computer Use. In *Australasian Computer Human Interaction Conference, OZCHI 2003*. Brisbane, Australia: University of Queensland, CHISIG.

Bilda, Z., Candy, L., & Edmonds, E. A. (2007). An Embodied Cognition Framework for Interactive Experience. *CoDesign*, 3, 123–137.

Brooker, G. (2022). Site: An Inventories Approach to Practice-led Research. In C. Vear (Ed.), *The Routledge International Handbook of Practice-Based Research*. London and New York: Routledge.

Candy, L. (2011). Research and Creative Practice. In L. Candy & E. A. Edmonds (Eds.), *Interacting: Art, Research and the Creative Practitioner*. Faringdon: Libri Publishing Ltd.

Candy, L. (2020). *The Creative Reflective Practitioner*. London and New York: Routledge.

Candy, L. (2022a). Theory as an Active Agent in Practice Based Knowledge Development. In C. Vear (Ed.), *The Routledge International Handbook of Practice-Based Research*. London and New York: Routledge.

Candy, L. (2022b). Reflective Practice Variants and the Creative Practitioner. In C. Vear (Ed.), *The Routledge International Handbook of Practice-Based Research*. London and New York: Routledge.

Candy, L., & Edmonds, E. A. (2011). *Interacting: Art, Research and the Creative Practitioner*. Faringdon: Libri Publishing Ltd.

Candy, L., Edmonds, E. A., & Poltronieri, F. (Eds.). (2018). *Explorations in Art and Technology*. Springer Cultural Computing Series. London: Springer -Verlag, 2nd ed.

Chen, A., & Odon, W. (2022). Crafting Temporality in Design: Reflecting on and Extending the Creation of Chronoscope. In C. Vear (Ed.), *The Routledge International Handbook of Practice-Based Research*. London and New York: Routledge.

Costello, B. M. (2022). Finding the Groove: The Rhythms of Practice-Based Research. In C. Vear (Ed.), *The Routledge International Handbook of Practice-Based Research*. London and New York: Routledge.

Costello, B., Muller, L., Amitani, S., & Edmonds, E. A. (2005). Understanding the Experience of Interactive Art, Iamascope in Beta_Space. In Y. Pisan (Ed.), *Proceedings Interactive Entertainment*. Sydney: Creativity & Cognition Studios Press, 49–56.

Crabtree, A. (2003). *Designing Collaborative Systems: A Practical Guide to Ethnography*. Heidelberg: Springer.

Edmonds, E., Bilda, Z., & Muller, L. (2009). Artist, Evaluator and Curator: Three Viewpoints on Interactive Art, Evaluation and Audience Experience. *Digital Creativity*, 20(3), 141–151.

Edmonds, E. A., & Candy, L. (2010). Relating Theory, Practice and Evaluation in Practitioner Research. *Leonardo*, 43(5), 470–476.

Glaser, B. G., & Strauss, A. L. (1967). *The Discovery of Grounded Theory Strategies for Qualitative Research*. New York: Aldine Publishing Company.

Graham, B., & Cook, S. (2010). *Rethinking Curating: Art After New Media*. Cambridge, MA: MIT Press.

Hübner, F. (2022). The Common Ground Model for Practice-based Research Design. In C. Vear (Ed.), *The Routledge International Handbook of Practice-Based Research*. London and New York: Routledge.

Ingold, T. (2013). *Making*. London: Routledge.

John, P. (2019). The Silent Researcher Critique: A New Method for Obtaining a Critical Response to a Holographic Artwork. In A. Pepper (Ed.), *Holography: A Critical Debate Within Contemporary Visual Culture*. Basel: MDPI, 38–49 (A Reprint from Arts 2019, 8(3), 117).

John, P. (2022). The Impact of Public Engagement with Research on a Holographic Practice-Based Study. In C. Vear (Ed.), *The Routledge International Handbook of Practice-Based Research*. London and New York: Routledge.

Layton, S. (2022). Participatory and Personal Research by Practice. In: C. Vear (Ed.), *The Routledge International Handbook of Practice-Based Research*. Routledge: London and New York.

Ledgard, A., Biglino, G., & Layton, S. (2022). Reflection in Practice: Inter-Disciplinary Arts Collaborations in Medical Settings. In: C. Vear (Ed.) *The Routledge International Handbook of Practice-Based Research*. Routledge: London and New York.

Lewis, C., & Rieman, J. (1994). Task-Centered User Interface Design – A Practical Introduction. *Shareware*. http://hcibib.org/tcuid/.

Moran, T. P., & Carroll, J. M. (1996). *Design Rationale Concepts, Techniques and Use*. London and New York: Routledge.

Mottram, J. (2022). Practice-Based Research in the Visual Arts: Balancing the Systems of Practice and the Practices of Research. In C. Vear (Ed.), *The Routledge International Handbook of Practice-Based Research*. London and New York: Routledge.

Muller, E. (2008). *The Experience of Interactive Art: A Curatorial Study*. PhD Thesis. Creativity and Cognition Studios. Sydney: University of Technology.

Muller, L. (2011). Learning from Experience – A Reflective Curatorial Practice. In L. Candy & E. A. Edmonds (Eds.), *Interacting: Art, Research and the Creative Practitioner*. Faringdon: Libri Publishing, 94–106.

Muller, L. (2022). Curatorial Authorship and Appreciative Systems in Reflective Curatorial Practice. In C. Vear (Ed.), *The Routledge International Handbook of Practice-Based Research*. London and New York: Routledge.

Nelson, R. (2013). *Practice as Research in the Arts. Principles, Protocols, Pedagogies, Resistances*. London: Palgrave Macmillan.

Richards, T. (2002). An Intellectual History of NUD*IST and NVivo (2002). *International Journal of Social Research Methodology*, 5(3), 199–214.

Richards, L. (2009). *Handling Qualitative Data: A Practical Guide*. London: Sage Publications Inc.

Schön, D. A. (1983). *The Reflective Practitioner: How Professionals Think in Action*. New York: Basic Books (reprinted Aldershot, Hants: Ashgate Publishing Ltd., 1991, 2003).

Seevinck, J. (2011). The Concrete of NOW. In L. Candy & E. A. Edmonds (Eds.), *Interacting: Art, Research and the Creative Practitioner*. Faringdon: Libri Publishing Ltd., 242–256.

Seevinck, J. (2022). Making Reflection-In-Action Happen: Methods for Perceptual Emergence. In C. Vear (Ed.), *The Routledge International Handbook of Practice-Based Research*. London and New York: Routledge.

Sidell, J. (Ed.). (2009). *Conversation Analysis: Comparative Perspectives*. Cambridge: Cambridge University Press.

Solomon, L., & Baio, C. (2022). Thinking Together Through Practice and Research: Collaborations Across Living and Nonliving Systems. In C. Vear (Ed.), *The Routledge International Handbook of Practice-Based Research*. London and New York: Routledge.

Suchman, L. (1987). *Plans and Situated Actions: The Problem of Human-Machine Communication*. New York: Cambridge University Press.

Turnbull, Tillman D. (2022). Curating Interactive Art as a Practice-Based Researcher. In C. Vear (Ed.), *The Routledge International Handbook of Practice-Based Research*. London and New York: Routledge.

Vlugt, M. van der. (2022). Please Touch! In C. Vear (Ed.), *The Routledge International Handbook of Practice-Based Research*. London and New York: Routledge.

Zhang, Y. (2011). Investigating Collaboration in Art and Technology. In L. Candy & E. A. Edmonds (Eds.), *Interacting: Art, Research and the Creative Practitioner*. Faringdon: Libri Publishing Ltd., 122–135.

Zhang, Y., & Candy, L. (2007a). A Communicative Behaviour Analysis of Art-Technology Collaboration. In M. J. Smith & G. Salvendy (Eds.), *Proceedings of the 12th International Conference on Human-Computer Interaction*. Beijing, China: Springer, 212–221.

Zhang, Y., & Candy, L. (2007b). An In-Depth Case Study of Art- Technology Collaboration. In *Proceedings of Creativity and Cognition 2007*. New York: ACM Press, 53–62.

3.2

THE *COMMON GROUND* MODEL FOR PRACTICE-BASED RESEARCH DESIGN

Falk Hübner

Introduction

This chapter describes the *Common Ground* model for designing practice-based research.[2] The aim is to present a flexible framework in the design of practice-based research. It is equally useful in the designing of peer-feedback, supervision and teaching contexts, including a conceptual-philosophical grounding and contextualisation.[3]

The underlying hypothesis of this chapter is that the quality of research processes, outcomes and impact can be increased considerably through a thorough yet flexible approach to research design. This is not necessarily due to a "better" design (whatever this might mean), but because of a more explicit thought process that is brought into motion through a number of perspectives, lenses or categories. The proposed approach tries to accomplish two seemingly opposing aims:

- to provide some strictness, precision and clarity
- at the same time to be flexible enough to accommodate unforeseen events

A model such as this provides space for flexibility and emergence as essential parts of a research strategy (more about this later). The model does not propose a single notion of research design, but rather aims to offer a number of perspectives through which practitioner researchers can think about their desired approaches and methods. The model has a number of interconnected perspectives that afford a common ground between different types of practice-based research.

The model is intended for practitioner researchers in diverse disciplines – indeed, anyone who is thinking about research design or the methods needed. This includes experienced researchers for whom the model can present a way of questioning choices, thereby adding a reflective layer in addition to personal experience. These specific attributes aside, more importantly, the model's characteristics facilitate the notion of *inclusiveness*. Although the ideas presented here have their origin in the fields of artistic research, this model is not limited to a particular context – hence its name, *Common Ground*. Rather, it opens up towards other fields and disciplines and enables researchers of various kinds to communicate and disseminate their choices. It aims to create a cross-disciplinary shared vocabulary for designing and reflecting on research strategies and methodologies.[4]

 DOI: 10.4324/9780429324154-24

On the use of a model

The first elements of this model's design emerged through my teaching. I have been interested in teaching research strategy in a way that enables students to *play with* their ideas, rather than needing to work through a variety of traditional methods *first* (e.g., interviews, observation, literature review, etc.). From these initial design elements, the formation and visual representation of the model was developed through sketching, reading, teaching and dissemination activities.[5]

Formal models, specifically graphical ones (e.g., flowcharts), have a number of attributes, such as the abstraction and simplification of complex contexts or connections which may conceal important aspects of what they are intended to represent. By contrast, the *Common Ground* model avoids the simplification of complex relationships by presenting a framework to be filled with content and to unearth connections that have yet to be made. Crucially, this model is not directive and does not force the researcher into one particular way of usage; rather, it acts as a frame, a starting point, and supports the possibility of bending or, at times, even breaking it.

In the following sections, I will outline the various steps towards this model. After an introduction to the overall structure of the model, its various layers will be described in more detail.

Part 1: towards research design

The *Common Ground* model (illustrated in Figure 3.2.1) consists of six layers:

1   *Preparation* includes the aims and research questions, the points of departure and the conditions of various kinds.[6]

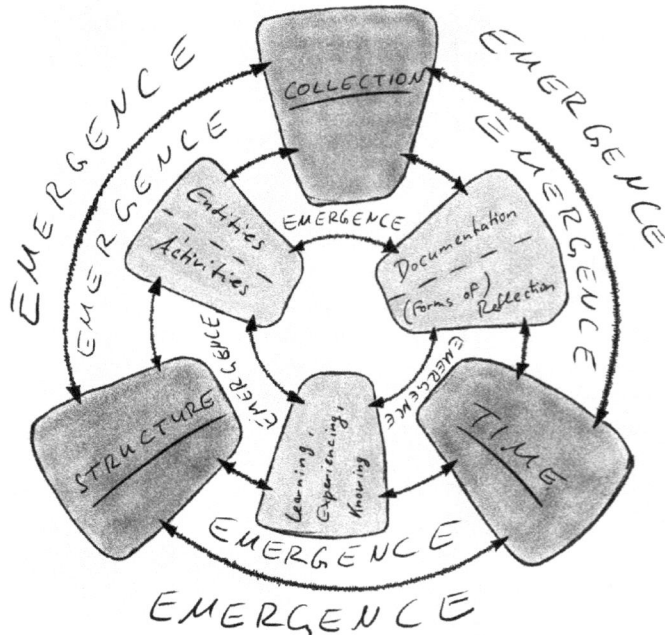

Figure 3.2.1   The *Common Ground* model. The three yellow planes represent the Crafting Methods framework, which will be elaborated on below.

2   *Collection* means, quite literally, a collection of methods: of different activities, non-hierarchical in principle and entirely based on the research subject and questions as framed in the preparation.

3   *Structure* aims towards some amount of ordering this collection, into what I call a "flow of data" – how information runs through a research process, which methods are carried out first before continuing to the next and the different kinds of structure that it can have, such as single threaded, parallel or feedback loops.

4   *Time* goes further than scheduling and planning and is instead motivated by content and the notion of *spending time* with something, how much time we want or are ready to give.

5   *Emergence* gives voice to the unexpected, to what "comes up" during the research process and within a given system; this layer is arguably the most complex one, and at the same time the most difficult (if not impossible) to actually design.

6   *Crafting methods* (the light grey planes in Figure 3.2.1) is the process of building methods "from scratch", from the very experience and reality of playing and making, radically oriented on research subject and question(s).

Preparation

Before making initial choices about methods or a particular strategy, a researcher needs to do pre-paratory work. First, foundational ideas and parameters need to be formed, developed, connected and related to each other.[7] The purpose and goals of the research need to be identified:

* Are the purpose and goals of the research in the areas of knowing, changing, learning, creating or designing?
* What are the points of departure in terms of both practice and theory?
* Which part of the researcher's practice is subject to the inquiry?
* What is already known and unknown about that matter?
* What kinds of (re)sources are available about it?

Developing and articulating research questions is an important part of preparation as well, as they offer the necessary direction for the researcher to move further towards outlining and collecting his methods of inquiry (see next section, "Collection") – even if subject to change during the research process. The research questions set the researcher into a mindset of not-knowing, despite being an expert in this practice and the sometimes-inviting notion of "knowing what happens". The researcher needs to be mindful and "think about strategies to make the familiar strange".[8]

As more formal aspects of preparation, the conditions, limits and possibilities for the research need to be clarified and articulated, such as duration, financial means, access to people, places and sources, necessary effort, contingency planning and ethical aspects.

Collection

Looking at it graphically, a collection of methods, in which each method is represented by a small letter, might look like Figure 3.2.2.

I define the term "method" as an *activity*, something that is a concrete technique carried out by the practitioner researcher in order to answer the research questions, or parts of them.

Collecting methods could, for example, begin with gathering ideas for how to work on one's research questions, such as "working with interventions" or "having conversations with a range of different professionals".[9] The nature of such a collecting activity entails aspects of brainstorming: items can be grouped and looked at, such as in the form of notes or other means, to come to new ideas. Collecting itself can be a creative and divergent activity.

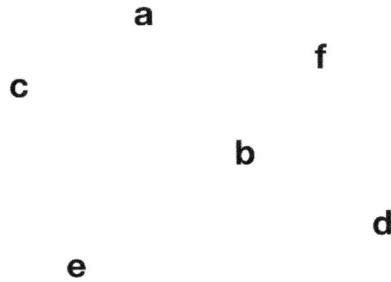

Figure 3.2.2 A collection of methods, represented by letters. Despite six methods being displayed here, research projects can obviously include less methods as well; quantity is not a criterium here.

It is essential here that the collection does not yet have a defined shape, structure or hierarchy. All methods, no matter how tentative, should be kept on the table. They can be either under or fully developed and are to be collected without thinking about the relations between the different elements.

Two aspects must not be forgotten while making this initial collection:

1   First, the close and articulated relation between each method to the research subject, area and its question(s). It needs to be clear for the researcher herself as well as the outsider why a certain method is part of this collection, and what its value and potential can be in the process of inquiry.
2   Second, the collection is viewed as non-hierarchical and in essence independent of traditional research paradigms. Literature research, for example, is by no means more important than an observation of a rehearsal or practical experiments.

Structure

This layer aims to process the collection of methods into a flow of data. For example:

*   What comes first, a practical experiment in a chosen location, or interviews with other practitioners who also work in a similar context?
*   Which steps come before or after other steps (i.e., which methods need to be carried out in sequence)?
*   Do the various activities flow straight into one another, one after the other?
*   Are there any parallel strands, two parallel methods that both need to happen before the next step, which synthesises both?
*   Are there any iterations or feedback loops involved?

Figure 3.2.3 illustrates several visual representations of research trajectories. These are meant to be an invitation, not only to think about the research design, but also to work with it, to use it through brainstorming or sketching or quick prototyping. Such visual representations can work particularly well in early stages of the design process, as they provide direct access and a certain transparency with regard to often complex and longer processes. As visual representations, diagrams can provide a more immediate entrance towards designing, particularly in modes of sketching or quick prototyping, rather than through text alone.[10] Therefore, the process of mentally visualising this flow can be helpful in designing a principal research investigation.

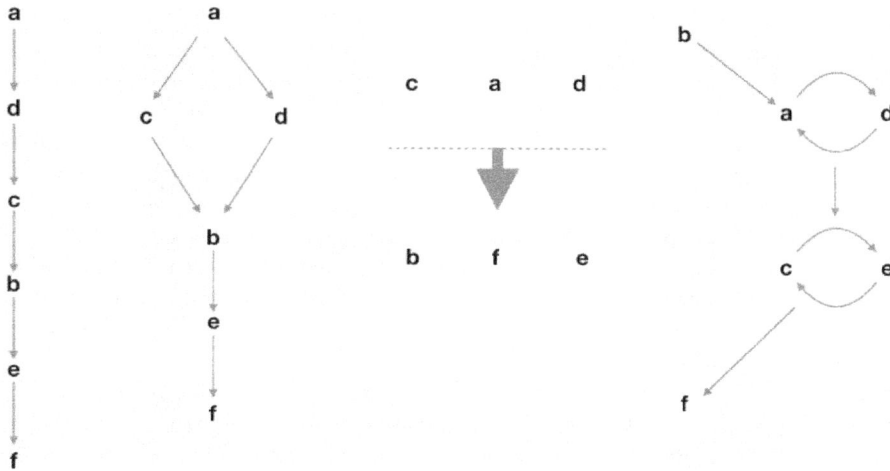

Figure 3.2.3   Different types of structures. The symbols are meant to be variables for methods or other research activities, for example, a could be – simplified – "study of book X", b "practical experiment", c "interview with Y" and d "observation of performance Z", and so on.

LAYOUT RESEARCH

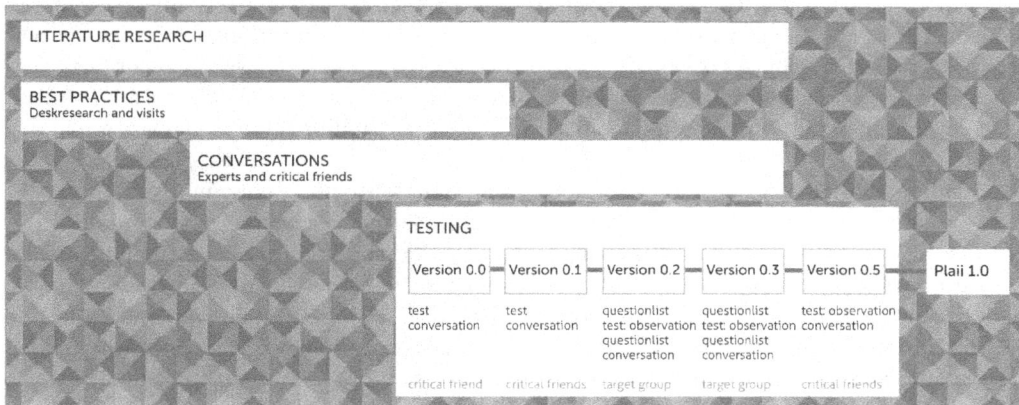

Figure 3.2.4   A more timeline-oriented, way of visually representing time.

Source: Geerling (2015, 11).

Time

In most research contexts, time plays a crucial role, determined by deadlines,[11] and should be acknowledged in the preparatory phase of a project. Time is often visualised by a certain kind of diagram (see Figure 3.2.4) that typically deals with time *planning*, scheduling – a timeline. Such a timeline is certainly important and can provide insight into how the different phases and activities relate to each other, in terms of time. This is what it adds to structure.

But, much more important, this visualisation signifies something else, namely that this layer is about *spending time* with something, or someone; about how much time we want to *give* to an activity, how much time we want or need to spend with a person or a group of people, in a particular space or surrounding.[12] Naturally, all this needs accounting for in the grand scheme and limits of the research investigation. This perspective relates to the discourse around the notion of *slowness*. The philosopher Paul Cilliers notes the importance of delay and iteration, "against the alignment of speed" and its accompanying notions, such as "efficiency, success, quality and importance".[13] In this sense, time as a category within the design model is not just a neutral/functional parameter but also includes a critical and ethical position. This resonates with writer Wendy Parkins, who argues that the actual question is probably not about fast versus slow, but rather about care as a central value,[14] in what she calls an "ethics of time". This understanding of time has relations to various movements concerned with notions of sustainability, responsibility for mankind and the environment.[15]

It might not be possible to avoid a tension between the notion of "quality of time" and a more pragmatic interpretation of the time layer.[16] The point here is that the category of time is what reveals this tension and makes it visible so that the researcher can work with this tension and develop a position towards it. Therefore, the category of time puts emphasis on the thinking about which methods, among all possible options in the collection, are the most important. In this way a true – and truly careful – relationship between the categories of collection and time can be established.

Making sense of emergence

In *Emergence*, Stephen Johnson describes emergence as a form of higher-level knowledge and behaviour, emerging from low or local level interaction in complex systems, based on "swarm logic, with no central office in command".[17] Its central elements are "tools of feedback, neighbour interaction, and pattern recognition". This resonates with Peter Cariani,[18] who remarks that "the full gamut of emergence encompasses new forms, new material structures, new organisations, new functions, new perspectives, and new aspects of being".[19] This includes new techniques or even (higher order) paradigms. Yet, it is crucial "that the higher-level behaviour is almost impossible to predict in advance. You never really know what lies on the other end of a phase transition until you press play and find out".[20]

Additional to these elements of higher order and unpredictability, Goldstein makes an explicit connection to creativity, concerning "how both emergence and creative processes are *creative*, that is, how they enable the coming into being of the radically novel".[21] To include the potential for emergence in research design means to give up some amount of control: emergent elements do not only appear unexpectedly, but also we have only limited (if any) capacity to truly understand it.[22] Where for some researchers the challenge might be to give space to emergent aspects despite their precise and thoughtful design, others might find it hard not to follow every new idea that comes up and actually find the dedication and focus to be faithful to one's design, to some degree. The main argument here is to find a balance and fruitful middle ground. A research design is always both precise and speculative to some degree, and it is this that creates the productive tension between design/planning and emergence.

Furthermore, this theoretical stance can provide a critical view on the difference between emergence that occurs and things that "just happen" rather randomly and are more of ad hoc decisions. The point I want to make here is that in rather unprepared research projects – or any projects, for that matter – all kinds of things can "just happen", simply because something always happens. But this is not emergence as discussed here. Arguing with Johnson, emergence happens in relation to and as a result of a complex network of decisions and activities, which are designed and carried out, initially as planned.

Emergence in research design needs a strong and solid (yet still flexible) design in order to occur, as otherwise there are no low-level interactions from which emergence can occur and literally do its work. Only then it can act, sometimes as a kind of counter-force to that which is already designed. However, there is a fine line between making use of emergence and just working ad hoc from one issue that comes up to the next. Still, I do not aim to imply that working ad hoc is negative in itself, but I do think that outcomes and impact can be seriously improved by paying attention and achieving a certain productive tension between planned and designed elements and emerging aspects.

Crafting methods

This layer is dealt with later in this chapter.

Part 2: the basic structure of the model

There are three core elements of the model: collection, structure and time[23] (see Figure 3.2.5). While it is possible to understand these elements as three subsequent steps of collecting, structuring and assigning time, it is important to see them as points of a flexible network. These points constantly shift and react to each other, rather than work in a hierarchical way. Philosopher Henk Oosterling has argued against notions of triadic societal structures and connections such as "top-down" or "bottom-up" and replaced them with the concept of networks.[24] By structuring or giving time to elements in a collection, to the methods that are envisioned to be used in a research project, the collection itself might change. Overall, this process works iteratively.

The process of collecting methods and structuring and assigning time to them implies describing what methods actually are, and for that purpose, the concept of crafting methods is in the centre of the design model. Before I return to the graphical form of the design model and assign this concept a place, I will elaborate on the elements on Crafting Methods in the following sections.

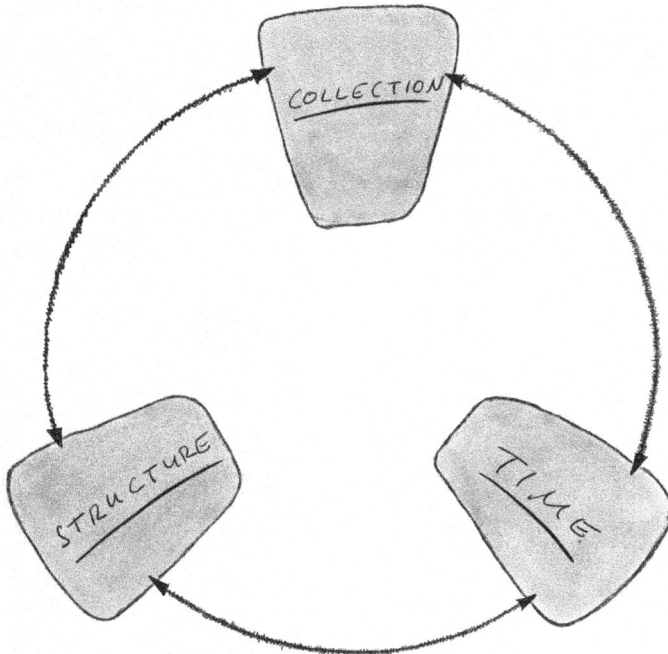

Figure 3.2.5   The basic structure of the *Common Ground* model.

Crafting Methods

The initial impetus for the concept of Crafting Methods[25] is a resistance against the notion of method as something predefined in terms of procedure, participants/actors and outcomes and taking shape by tradition. I regard methods as being devised from scratch, from the very experience and reality of playing and making, radically oriented on research subject and question(s).[26] A method is embedded with details of who or what is involved, in which kind of way and through which kind of activity, how this is documented, reflected upon and further processed towards an (intermediary) outcome.

As a point of departure, I define a method as a means to inquire into a research question and contribute to its exploration, potentially or partially answering it.[27] I suggest the re-framing of what a method is, in the form of a flexible network consisting of five elements (see Figure 3.2.6):

1   Entities
2   Activities
3   Documentation
4   Forms of reflection
5   Learning/experiencing/knowing[28]

1   Entities

The first element of the Crafting Methods framework concerns human and non-human *entities*. Entities describes the agents most central to crafting a method, who are indispensable for every activity of this part of the research. It is the entities, either human or non-human, that carry out a method and play an active or passive part in it, and is thus most closely connected to the element of Activities (see next section). The determining question is: *Who or what is involved, in which way, taking what kind of role or function?*

The researcher determines who and what are the important and necessary entities within a method (and within the entire research strategy), and what their function is. It is the researcher who assigns hierarchies, priorities and kinds of acting or non-acting to the various entities. In practice-based research, a crucial entity that is very often forgotten is the practitioner researcher herself. Other entities can, in no particular order, be and not restricted to these examples:

• authors of literature, or the texts themselves
• spaces or locations in which practice happen, such as classrooms, rehearsal spaces for musicians or theatre practitioners, elderly homes; however, it is important to be very

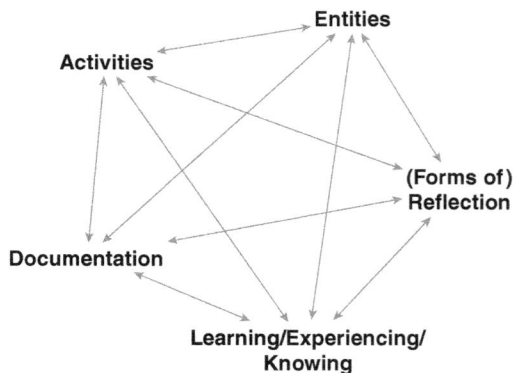

Figure 3.2.6   The network of Crafting Methods.

specific about spaces in order to maintain a precise idea how they function as elements of a research activity

- interviewees or other types of conversation partners
- musical instruments
- a virtual reality installation
- members/participants of group conversations or group activities
- crucial tools or other kinds of objects used by practitioners

Although my understanding of an entity is in its essence performative – an entity *does* something, is supposed to do something in the context of a method and in a feedback loop with other entities – that does not necessarily mean *at all times*, or in a completely equal and mutual relationship to eventually other entities present (or acting) within a certain method. In the case of a scenographer, there might be situations in which a space or theatrical objects suggest ideas, thus acting very actively. In other moments in which the scenographer directs the – at this point rather passive – space and its elements, she is doing so by design and on purpose, not just by chance or accidental intuition.

2 Activities

Activities are devised from a practical and performative point of view, literally in the sense of "what is to be done". For example, what is the researcher (or any other entity) going to do? How does she/it engage? In the case of traditional research methods, the name of a method often already frames the kind of activity, such as literature review, observation or interview. However, in the context of crafting a method from scratch, it might be more useful to describe and frame the kind of activity (even traditional/ conventional methods), to avoid the pitfall of just naming a method without being entirely conscious of its consequences. To clarify this, the process of traditional research methods can be defined by the guidelines attached to them; and these do not necessarily match with the kind of activity one wants to describe as part of the crafted methods. Overall, use this element to define the activity, and avoid merely reaching for the method name.

3 Documentation

Documentation is a key part of research and needs to be carefully considered. In general, in practice-based research this concerns the documentation of both processes and the artefacts or products, such as performances, workshops, session formats, teaching protocols and so on. In the context of this Crafting Methods process, I am referring to the documentation of the research process and insights, rather than documentation of the final products or artefacts of a research inquiry. Insights can naturally be generated in an activity itself – "in-action" so to speak – caused by or during a method or activity. But they can also be sparked by (the act of) documenting and reflecting itself. In this sense, there is a certain overlap between documentation and reflection possible, and at times, documentation might itself evolve into a kind of method, as Robin Nelson suggests: "modes of documentation constitute methods of capturing evidence".[29]

Documentation of process can have different functions, which essentially come down to evidencing the inquiry, and making it transparent, for oneself as researcher as well as for outsiders during or after a research project. Without documentation (as opposed to artefact), all is left to memory and to the researcher, participants or collaborators telling "what happened". This access is necessary in order to review, discuss and critique process, insights and arguments. It is especially relevant in an open format of designing research methodology, as transparency is an essential element to enable the reader to review and follow the research process and insights.

Documentation typically involves formats that are visual, aural and written. However, a number of different forms are possible:

- sketches
- scrapbooks, notebooks
- journals, diaries
- created materials, objects
- photographs, images
- video footage
- audio recordings
- scripts, scores

Some of these examples need to be produced as documentation; some (such as scripts or created objects) just need to be collected and stored, as they have been generated through the research process itself. Concerning documentation that needs to be produced, it is not always easy to define what exactly to document, as Nelson acknowledges:

> it is neither possible nor desirable to video every rehearsal of a performance production process. First, the presence of the camera can interfere with the process. Second, to record everything would be to end up with an amount of footage too massive to sift and edit in this context. So what to be documented and when?[30]

Practitioners usually have the best sense of their own practice based on experience, combined with "anticipating what kind of documentation might be useful to *evidence the research inquiry*".[31] Practitioners can and should generate ideas about how to document this practice, and to what end. However, as these moments are not always predictable, it is important for the researcher to stay alert and ready to document, in order to be able to capture unexpected insights, happenings, findings or "moments of discovery".[32] It can help to develop a sense of documenting-as-habit.[33]

Along with the questions of *how* and *what* to document, we should also consider the question of *who* will document? Sometimes it makes sense to let outsiders join specific activities, in order to document from their external perspective. For example, in a 2020 participatory performance and research workshop on *experiences of touch*, a collaboration between performance maker Marloeke van der Vlugt, sociologist Carey Jewitt and myself as composer,[34] we employed a variety of documentation modes (essentially all of the above-mentioned). However, we also implemented a dimension of perspectives, which van der Vlugt, Jewitt and myself documented through recording and writing as being the most direct insiders of our inquiry. Jewitt's team members documented by means of both images and video, as well as writing personal notes, themselves being professionals but not insiders to the research process. Additionally, the participants from the audience, both professionals and non-professional visitors, were also asked to keep notebooks while being in the performance workshop, to share ideas with each other (documented by Jewitt's team) and produce short imaginary accounts by themselves, either through writing, sketching/drawing or using provided video cameras. These varied perspectives generated a multi-faceted approach to documentation and provided we researchers with a rich and complex picture of what had happened in the performance workshop, which led to some surprising results.

4   Forms of reflection

Reflection in the Crafting Methods framework can essentially be understood as a form of processing collected data or experiences[35] towards the outcomes of a method, designated as learning, experiencing, knowing (see next point). *Reflection* can happen in many different situations: individual, collective, through conversation, writing, drawing, sketching, walking, meditating and so on. This element might be designed in itself, rather than just taken for granted in form and function. While reflection (and reflexivity) in itself is an area too extensive to cover here, I refer you to Linda Candy's chapter in Part III.[36]

5   Learning/Experiencing/Knowing

The final element, *learning, experiencing, knowing*, aims to indicate the outcome of a method: what is learned, what is taken further, possibly as input to the next method. What do we want to learn or get out of a certain activity? The three terms – *learning, experiencing, knowing* – indicate that this can vary greatly, according to where a certain method is located in a research strategy and what its function is. For example, a literature review can have a set of more thoroughly framed concepts as an outcome, whether a first visit at an institution or museum, say, can just provide a first (documented) impression of the place. However, this might lead to further insights as to what kinds of interventions might be necessary in order to provide a better infrastructure at this institution, for example, the method thus having a more inventory and experiential function.

Fitting in crafting methods

It is necessary to carry out yet one change of the visualisation of the Crafting Methods framework, in order to let it fit into the graphic representation of entire model. Through experience, especially with other researchers in conversations and in workshop contexts, it proved to be useful to create two pairs from four of the initial five elements of the framework, and thereby connect the levels of 1) entities and activities, and 2) documentation and reflection with each other, only separated by dotted lines. Figure 3.2.10 shows its graphic version, and how it is situated in the centre of the larger model.

Part 3: the *Common Ground* model in action

Figure 3.2.8 shows the entire model (already shown in Figure 3.2.1), in which the Crafting Methods framework is nested within the three main elements of Collection, Structure and Time; the multiple connections are represented by three "orbits". All elements surround an empty centre. Emergence does not have one designated space, colour or delineation around it, but all elements rather float in a "sea of emergence".[37] I developed this visualisation in order to take the many unforeseen possibilities and shapes of emergence into account and, most of all, despite the unknown direction, the certainty that in most cases some form of emergence will happen. This includes emergent developments or ideas that lead to potential "side tracks", as artist and researcher Annette Arlander calls them.[38]

Use of the model – three examples

There are two main perspectives from which the *Common Ground* model can be used:

• in the actual research design process
• in order to describe, analyse or reflect on a design, whether finished or in progress

As has become clear to me in a variety of cases, there is certainly not one (correct) way to use the *Common Ground* model, or even use it as a kind of recipe, or a method in and of itself. In instances in which the model is used, practitioner researchers have not adopted it in an entirely literal way,

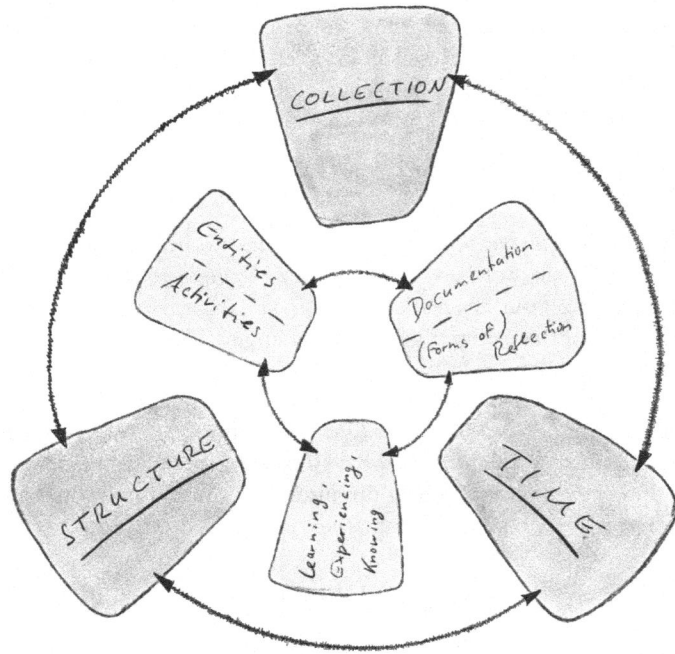

Figure 3.2.7   The graphical-organisational "edit" of the Crafting Methods framework makes it possible to slide the two levels of strategy (red) and method (yellow) into each other.

meaning to just take the model, fill in its different elements and "voilà", there we have a well-finished research design. This is not how the model should work. It is much more the case that researchers think about their research strategy *through the elements and perspectives the model offers*[39] and, from there, shape their design in forms that match their own thinking (rather than mine present through the model). For example, scenographer and PhD researcher Ariane Trümper was struck by the discussion on the element of time during a workshop, which brought her to create a "table of intensities" of her research, mapping the different phases and activities of her research design on the quantity and intensity regarding time (see Figure 3.2.9). This made it possible for her to think more deeply about the temporal relations between the otherwise parallel activities in her research design, and therefore organise them differently or in a more nuanced way.

Another example is how musician and PhD researcher Marco Fusi used the elements of the Crafting Methods framework in order to structure the research strategy of his doctoral research on the work of composer Giacinto Scelsi. It is interesting in his case that through work with the different elements of the model, Fusi has been able to discover a previously hidden two-phased structure of his research (see Figure 3.2.10).[40]

Another example of my own is when I analysed a performance experiment that I carried out with performance maker Marloeke van der Vlugt[41] and sociologist Carey Jewitt. Through this analysis, by means of looking through the lenses of the Crafting Methods framework, I was able to understand and focus on the diversity of techniques of documentation and reflection we utilised (see Figure 3.2.11). Additionally, I was able to provide an overview of the distinct perspectives of the different people who carried out both documentation and reflection; i.e., the three core researchers, the team of sociologists and the audience participants.

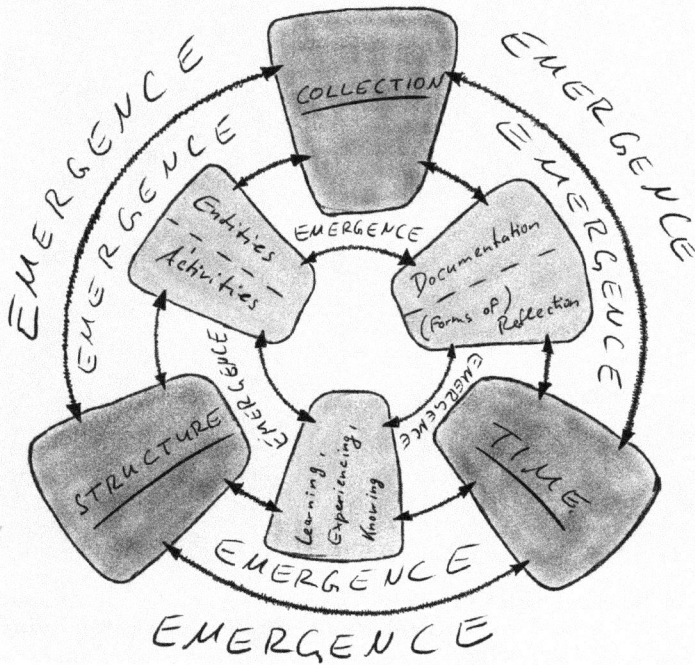

Figure 3.2.8 The complete *Common Ground* model, with the Crafting Methods framework in its interior, all floating in a "sea of emergence".

What the above-mentioned examples show (see Figure 3.2.11) is that the main point of this model is that is "works in action": it offers categories on the level of method and strategy, and connections between these categories, in order to *think through them* when engaging with research design, regardless of if this happens in the design phase or after a project has been concluded, or anywhere in between along the way of a research project. This means that the model can influence and guide the daily thinking of a researcher's practice while carrying out research (as happens in my personal research work, and in cases of colleagues I am close to).

Final thoughts

This chapter has defined the *Common Ground* model, which offers an approach to facilitating the research design process of the practitioner researcher in any given field or discipline. One crucial point addressed throughout the chapter is the fruitful tension between careful design on the one hand and space for emergence to happen and "do its work" on the other: in essence to control and let go and follow. Related to this is the question of "what to do" with emergence. Is it simply about acknowledging emergence, welcoming it and providing space and time for it? Or is it more about "managing emergence", as Haseman and Mafe (2010) suggest? It is the researcher herself who needs to answer this on a case-by-case basis, but dealing with it and designing with it are crucial.

This model is certainly not meant as some kind of unifying "how to" method. As for the purposes or applications, the model's openness results in various possibilities. First, it can support the process of designing research (thus for the researcher) to provide a number of perspectives through which a researcher can think about a research design. Second, it can help make the implicit components

Figure 3.2.9   Ariane Trümper's "table of intensities", in which she sketches the different amounts of time necessary to the parallel or integrated activities of reading, making work, learning, "hanging out" with space and conversations.

Source: Ariane Trümper.

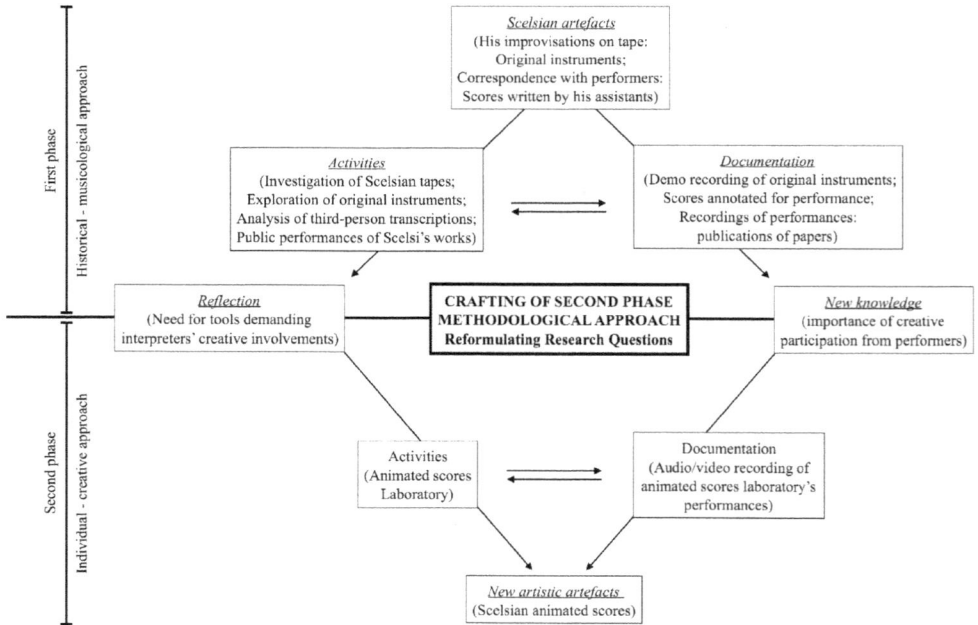

Figure 3.2.10   Schematic representation of the two phases in Marco Fusi's research process, using the terminology of the Crafting Methods framework.

Source: Marco Fusi.

participatory performance with various "touchy"
experiences manipulating sound through touch
conversation with participants
exhibition of various objects and videos related
to the perfomance and its creation process
reflective assignments

Activities

Entities

artists, researchers
theatre "black box" space
exhibition setting
circle of chairs
objects for performance, older
objects videos of process
videos of toher touch rituals

**(Forms of)
Reflection**

individual reflection through writing
(in various insider-outsider roles,
including the audience)
shared reflection via email and skype
discussion

Documentation

photos, video & audio
notes of researchers & participants
writing stories
sketches of future touch interactions
little ethnographic narratives

**Learning/Experiencing/
Knowing**

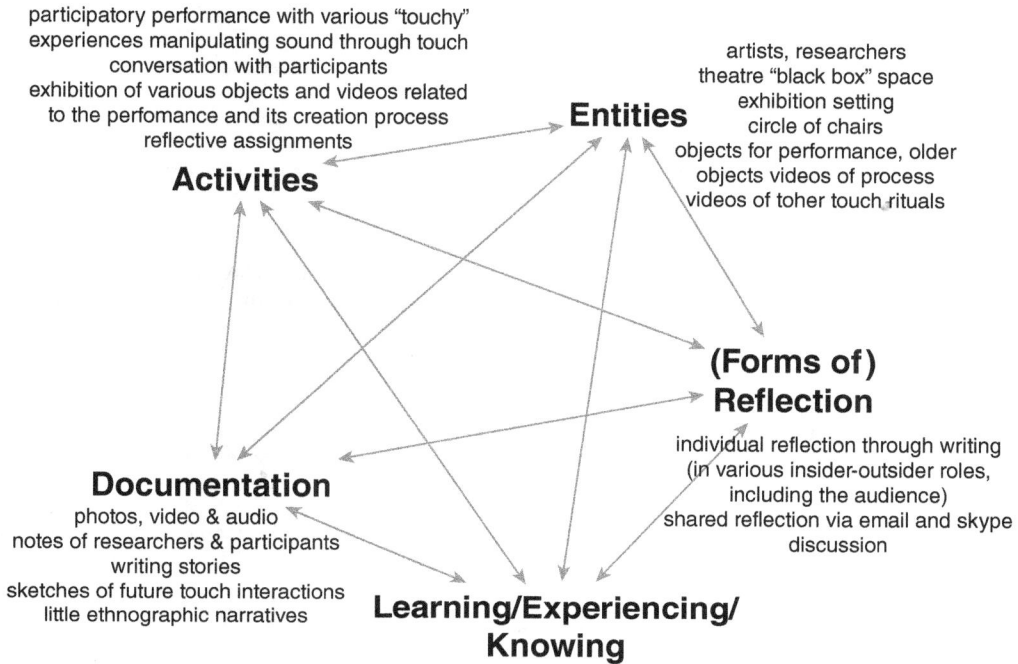

Figure 3.2.11   A look at the various entities, activities, ways of documenting and reflecting during the performance workshop *Thresholds of Touch* in Bloomsbury Studio Theatre London, through the Crafting Methods framework.

of research more explicit. Another way to see this is that the research process is being infused by the terminology introduced here, which makes certain ways of thinking and kinds of conversation possible that would otherwise unfold just accidentally or not at all – or, worse, be ignored.

At the same time, the model can support self- or peer-feedback, or the analysis and evaluation of a research design either finished or in process. It does so through its categories, perspectives or lenses, which, again, can make seemingly obvious parts of the research more explicit and thus available for feedback and evaluation as well as accessible for other researchers. As a third application, the model can work as a framework for supervision of student research projects.

It is not the intention that the model, or the approach to research design, will be actually recognisable as such by an outsider, in a final design. In fact, due to its flexible structure, the model can operate as a kind of vessel in which it encourages different ways to create a research design. In most, if not all, of the situations and applications I have seen this model being applied to, the model itself disappears after being used, providing space for what it is devised for: a convincing methodology that is well designed, yet leaves sufficient space for what emerges.

Notes

1 Borgdorff (2017).
2 This model is the conceptual core and outcome of a post-doctoral research project (2019–2021) on research methodology at HKU University of the Arts in Utrecht, generously funded by SIA, the Dutch governmental organisation for the support of practice-based research in the Netherlands.
3 In general, I identify two kinds, or strands, of literature on research methodology, directed at different target groups. On the one hand, literature largely conceptual and philosophical, targeted at professional artistic

researchers or doctoral students. In my opinion, this is very interesting and inspiring, but typically not very practical in the sense that (bachelor or master) students, or beginning researchers, can actually do a lot with it, in concrete terms of literally designing their research projects. On the other hand, literature that is very concrete, mostly practice-oriented in a hugely "how-to" fashion, mostly directed towards students and teachers/supervisors, about which kinds of methods can you use, how you conduct interviews, carry out observations, and these kinds of questions. Which is very concretely applicable, but at the same time lacks a certain layer of depth. Both kinds of literature have their advantages and disadvantages, but the actual problem that I am targeting is that there is only little connection, if any at all; although both literatures are directed at the same area of expertise, they do not overlap and actually seem to exclude each other.

4 I have experienced this build-up of a shared conversation, as well as fruitful exchange across disciplines myself, while running a methodology workshop of multiple sessions with teachers and researchers at the University of the Arts Utrecht. The participants came from various faculties – the music conservatoire, the school of theatre and school for art and economics – and one of the most useful experiences we had was the creation of an ongoing conversation that developed a shared repertoire of terms and perspectives; not in the sense of a limited set of approaches or techniques, but rather a way of thinking about research methods, strategies and methodologies as entangled spheres of research.

5 I have used teaching and supervising as deliberate *methods* to experiment with and develop my approach to research design and the Common Ground model. The students (or colleagues I have supervised) were well aware of this experimental approach and, in this sense, as ethnographer Raymond Madden refers to his exchange with a member of the Aboriginal community in the context of conducting an ethnographic research, were happy to "make our individual instrumentalities complementary" (Madden 2017, p. 64).

6 The diagram, as a visualisation of the Common Ground model, does not include the layer of preparation, as the actual model is dedicated to the process of designing research. Once the points of departure, conditions and aims of a research project are set, they are usually not changing to the same degree as the other elements of a design would.

7 In her chapter, "Navigating the Unknown", Hanna Slattne offers inspiring resonances and parallels between the dramaturgical process in performing arts and the research process in a practice-based research project. Her toolkit is hugely inspiring and insightful in this respect, in particular for researchers who are still developing their ideas early in the research process, even prior to the actual research design. See Slattne (2022).

8 Coffey (2018, p. 45).

9 See also the above-mentioned chapter by Hanna Slattne for the specific genealogy of ideas.

10 Gates (2018).

11 A nurse, for example, has her primary task of looking after patients in a hospital, of which research into a specific aspect of her practice is most likely just a part; the cases in which she might be able to spend her entire time on research could be relatively rare.

12 Figure 3.2.13, a sketch by scenographer and researcher Ariane Trümper, shows one possible way to visualise different intensities of time.

13 Cilliers (2006, p. 2).

14 See also Lizzie Muller's chapter, Curatorial Authorship and the Role of "Appreciative Systems" in Doing and Supervising Curatorial Practice PhDs', in which she positions care as a central value in the context of curatorial practice – "curatorial care". Her case study of the curator researcher Bec Dean is particularly insightful; Muller refers to Dean as beginning "to articulate her appreciative system [within her curatorial practice] in the context of 'care'".

15 While Cilliers refers mainly to the Italy-based slow food movement, there are other movements related to slowness, each with overlapping agendas, but also individual perspectives on the notion of slowness: the slow fashion movement accentuates sustainability, more respect and responsibility for man, environment and a changed consciousness towards the product and consumer behaviour (see http://slowfashionblog.de/slow-fashion/; accessed February 13, 2020). Slow science argues for a "non-real-time / offline, integrative and sustainable culture of thinking" and that science needs time without a constant pressure to publish at a too-high pace, for example (http://slow-science.org/slow-science-manifesto.pdf; accessed February 13, 2020).

16 This might be especially true in the case of students, who are always working according to a strict timeline, with a maximum of two years in the case of a master's, which is quite short to carry out a full research project. But the idea of spending time might also be understood not as some kind of intervention into working towards deadlines, but rather as an invitation, or provocation, to think in terms of achieving the greatest depth, or reflection, how to re-read, re-iterate, and so on.

17 Johnson (2001, p. 233).

18 Cariani (2008).

19 Cariani (2008, p. 2).
20 Johnson (2001, p. 233).
21 Goldstein (2005, p. 8).
22 See also Jen Seevinck's chapter (2022), in which she offers an in-depth view into one specific kind of emergence: "perceptual emergence". This kind of emergence happens *within* an artist or artist-researcher in situations of feedback and reflection, in moments when situations "talk back" and the artist-researcher interacts with a situation, changes this situation and/or is changed by this situation herself. Seevinck's and my view on emergence slightly differ in perspective: Her chapter offers strategies of how to work with aspects of emergence, mainly from the perspective or position of the researcher, whereas I look more broadly at emergence as a force that takes all kinds of agencies and interactions in a research project-as-complex system into account. I locate agency towards emergence in the researcher *to some degree*, whereas a space, object, animal or set of behaviours can just as certainly be at the inception of emergent aspects or behaviours.
23 For a better understanding of what is meant with these three terms, one might put an imaginary "of methods" behind these terms: a collection of methods, a structure of methods, the assigned time to the individual methods.
24 As for example in his 2013 lecture (in Dutch) at the symposium *Cultuur in Beeld. De kracht van cultuur* (*Culture in the Spotlight: The Power of Culture*). www.youtube.com/watch?v=vsTSasp8noE (accessed October 21, 2019). See also Oosterling (2013b).
25 The concept of crafting methods was inspired by artist and philosopher Erin Manning's book chapter "Against Method" (2015).
26 On the basis of this lies the issue of "why using methods at all", which was much more Erin Manning's initial thought, as she "was horrified that students were asking for methods" (Manning 2019 during audience conversation after my lecture "Against Method? Common Ground?" at CARPA6 in Helsinki).
27 I use "potentially" here as sometimes practice can be (part of) an answer to a research question, or one possibility in a spectrum of answers through practice. As such, it is my assessment that practice does not literally answer a question in the form of language and, in order to answer to criteria for being research dissemination, might be accompanied or complemented by writing.
28 For a general definition of what research methods are, see the leading chapter to Part III.
29 Nelson (2013, p. 99).
30 Ibid., p. 30.
31 Ibid., p. 31.
32 Ibid., p. 28.
33 Practically, this means to have a variety of tools and documentation devices at hand, which in today's times means that often a notebook, pen and mobile phone are sufficient. A phone nowadays easily functions as recording device of various kinds: video and photo camera, audio and voice recorder.
34 For more information on the project, see the website of the In-Touch research group at UCL: https://in-touch-digital.com/case-studies/threshold-touch-experiences/ (accessed February 13, 2020).
35 See, for example, the work of the HKU professorship art and professionalisation, where work forms, on the methodological basis of action research, were initially developed to provoke change in a huge variety of professional settings and have later become reframed as research methods. Particularly the form of (often collective) reflection is an essential part of this work. Another example is the reflection part during a workshop at the Symbiont conference on November 16, 2018 at Calgary University, where a group of participants (Claire French, Gretchen Schiller, Fredyl Hernandez, Maria Angelica Viceral and myself) carried out a workshop and chose to reflect with the entire group through writing and drawing with chalk on a board, entirely without talking, as a kind of meditative ending of an intense practice session.
36 Candy (2022).
37 I need to thank Judi Marshall for her feedback on the visualisation of the model, in which she coined this phrasing.
38 Arlander in an audience conversation during the CARPA6 (Colloquium on Artistic Research in Performing Arts) conference, Helsinki, August 30, 2019: "How can you still sort of do what you promise to do [in a research or grant proposal], but not abandon the side track?" Arlander's comment already suggests that this "side track" is not part of the original research design – and this is actually one of the crucial points of the process of doing practice-based research – or any research, for that matter. While generated by the original research project, a side track might become a parallel thread to this original trajectory, or an entirely new piece research on its own.
39 The words are not chosen accidentally here, as the notion of "thinking through" or "reading through" is inspired by Karen Barad's understanding of diffraction (see Barad 2007; Bozalek and Zembylas 2016;

Haraway 1997). The central idea of reading sources through each other, rather than letting one source dominate others, is central to the notion as I understand the use of the Common Ground model here.

40 See Fusi (2020). This resonates with what Jen Seevinck refers to as "situation talk-back" and reflection, in which the engagement with a situation or entity – the Common Ground model and its elements in this instance – results in feedback, reflection and finally a new emerging understanding of the situation and/or the researcher herself. In this case, the research design of the PhD project, including its overall logical and timely structure.

41 See Van der Vlugt (2022), in this handbook.

Bibliography

Barad, K. (2007). *Meeting the Universe Halfway. Quantum Physics and the Entanglement of Matter and Meaning.* Durham: Duke University Press.

Borgdorff, H. (2017). *Reasoning Through Art.* www.universiteitleiden.nl/binaries/content/assets/geesteswetenschappen/acpa/inaugural-lecture-henk-borgdorff.pdf (accessed October 6, 2019).

Bozalek, V., & Zembylas, M. (2016). Diffraction or Reflection? Sketching the Contours of Two Methodologies in Educational Research. *International Journal of Qualitative Studies in Education.* DOI:10.1080/09518398.2016.1201166.

Candy, L. (2022). Reflective Practice Variants and the Creative Practitioner. In C. Vear (Ed.), *The Routledge International Handbook of Practice-Based Research.* London and New York: Routledge.

Cariani, P. (2008). *Emergence and Creativity.* https://pdfs.semanticscholar.org/76b3/3ab7d44bd34b7c213997d953ed8feec38a8c.pdf?_ga=2.46681639.1089917717.1570250198-454848092.1570250198 (accessed October 5, 2019).

Cilliers, P. (2006). On the Importance of a Certain Slowness. *Emergence: Complexity and Organization,* 8(3). https://journal.emergentpublications.com/article/on-the-importance-of-a-certain-slowness/ (accessed November 1, 2019).

Coffey, A. (2018). *Doing Ethnography.* London: Sage.

Fusi, M. (2020). Customizing a Methodological Approach: Researching Giacinto Scelsi's Performance Practice Through Animated Scores. *Forum+,* 27(3). https://forum-online.be/nummers/herfst-2020/customizing-a-methodological-approach-researching-giacinto-scelsis-performance-practice-through-animated-scores (accessed November 17, 2020).

Gates, P. (2018). *The Importance of Diagrams, Graphics and other Visual Representations in STEM Teaching.* www.researchgate.net/publication/319086868_The_Importance_of_Diagrams_Graphics_and_Other_Visual_Representations_in_STEM_Teaching (accessed October 17, 2019).

Geerling, A. (2015). *Let's plaii. Een ontwerponderzoek naar interdisciplinair improviseren met kunststudenten* (*Let's plaii: A Design Research into Interdisciplinary Improvisation with Art Students*). Master thesis. Utrecht: HKU.

Goldstein, J. (2005). Emergence, Creativity and the Logic of Following and Negating. *The Innovation Journal: The Public Sector Innovation Journal,* 10(3), article 31.

Haraway, D. (1997). *Modest__Witness@Second__Millennium.FemaleMan©__Meets__OncoMouse™.* New York and London: Routledge.

Haseman, B., & Mafe, D. (2010). Acquiring Know-How: Research Training for Practice-led Researchers. In H. Smith & R. T. Dean (Eds.), *Practice-led Research, Research-led Practice in the Creative Arts.* Edinburgh: Edinburgh University Press, 211–228.

Johnson, S. (2001). *Emergence: The Connected Lives of Ants, Brains, Cities and Software.* London: Penguin.

Madden, R. (2017). *Being Ethnographic: A Guide to the Theory and Practice of Ethnography.* London: Sage.

Manning, E. (2015). Against Method. In P. Vannini (Ed.), *Non-Representational Methodologies.* New York: Routledge, 52–71.

Oosterling, H. (2013b). *ECO3. Doen denken* (*Doing Thinking*). Rotterdam: Jap Sam Books.

Seevinck, J. (2022). Making Reflection-In-Action Happen: Methods for Perceptual Emergence. In C. Vear (Ed.), *The Routledge International Handbook of Practice-Based Research.* London and New York: Routledge.

Slattne, H. (2022). Navigating the Unknown – a Dramaturgical Approach. In C. Vear (Ed.), *The Routledge International Handbook of Practice-Based Research.* London and New York: Routledge.

Van der Vlugt, M. (2022). Please Touch! In C. Vear (Ed.), *The Routledge International Handbook of Practice-Based Research.* London and New York: Routledge.

3.3

FINDING THE GROOVE

The rhythms of practice-based research

Brigid Mary Costello

Introduction

Artistic research is a convergence of materialities; sometimes a clash, other times a smooth flow, occasionally it is as if different rhythms play in counterpoint, pulling the researcher in different directions.[1]

At some point in the early stages of a practice-based research project, you will probably begin to plan your path by producing a timeline. This timeline will be written and rewritten multiple times across the span of a large project, with early versions usually (and necessarily) bearing little resemblance to later ones. As a student starting out on my PhD research, I found writing these timelines a bit terrifying, even though I recognised their value. There was just so much I didn't yet know about my project, and so I would produce very broad representations, like that in Figure 3.3.1. Now that I am a more experienced researcher with a better understanding of the methods of my practice, I am able to add more detail to the events on a research timeline. However, I have also come to realise that successful planning of a research process involves more than just identifying research events and their sequence.

As Kozel points out, a research process has a rhythm, one that fluctuates across the span of a project. It is the patterns of this rhythm that are also important. When planning a research process, you need to consider the rhythm of the practices of research events and the way those rhythms might intersect, combine, interfere, and play with each other. You need to consider that:

- Some of your research rhythms will be slow and some fast.
- Some will be predictable and some surprising.
- There will be moments of intensity and moments of calm.
- There will be times where you feel totally in sync with the flow of your research and times where you experience ruptures or a lack of synchrony.
- You will shift focus between the bigger picture and finer details, and you will transition between different research roles and tasks.

Together, all of these temporal patterns will form a rhythm that has a groove particular to your project: a rhythm that you can choose to compose and orchestrate. This rhythm may or may not be consciously discernible in whatever your research produces, but it will definitely influence its final

 DOI: 10.4324/9780429324154-25

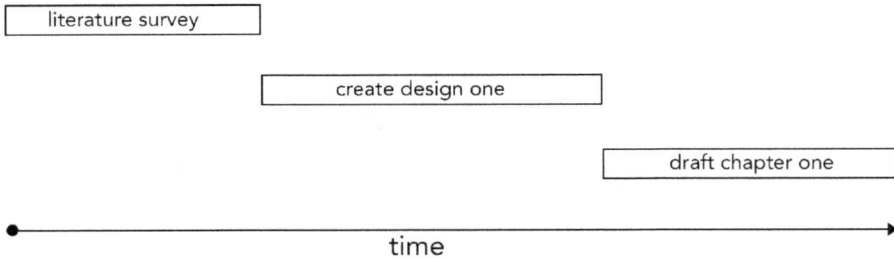

Figure 3.3.1   The type of research timeline I might have produced when I was a student.

Source: Image by Brigid M. Costello.

form and, perhaps just as importantly, will shape your experience as a researcher. This is why it is useful to pay attention to the rhythms of a practice-based research project.

As this chapter will argue, thinking about the rhythms of the research process will help you to plan when first starting out, find ways to keep going when in the middle of things, and make sense of where you end up. This will involve shifting your perspective away from the fixed outcomes of your research process to instead focus on the fluid processes of thinking, making, and talking that create these outcomes.[2] It is these fluid research processes and their resonances that form the patterns that will characterise your project's rhythms. Paying attention to the rhythms within your research project exposes the energies and speeds of the research process, allowing you to diagnose blockages, choose when to keep in step or break free, work with unpredictability, and transition between modes of practice. As you develop awareness and insight into research rhythms, you will also develop the ability to more consciously compose and orchestrate these rhythms to suit your particular context and your specific research practices. Because each research context has its own rhythm, this chapter will not focus on set patterns, templates, or defined strategies. Instead, the chapter is an exploration of some possible patterns and ways of thinking about the rhythms of a research process. As you understand these patterned possibilities, you will begin to develop an awareness of rhythm and become more sensitive to the particular rhythms within your own process and context. You will then be able to plan a research timeline that takes into account research events, their sequence, *and* the rhythms of their process.

What is rhythm and how is it experienced?

> Rhythm is timings, durations, intensities, speeds and accents . . . a rhythm tells you that something is changing or that we're somewhere else compared to where we were. That's something that you see, something that you perhaps hear, but also something you feel within the body.[3]

As you take a daily walk, you will sense the rhythm of your individual footsteps and also the overall pattern of the whole walk, including your lively steps at the start and possibly tired ones at the end. Rhythm occurs in:

- The moment-by-moment process of walking.
- Together, these patterns then form an overarching rhythm that encompasses the whole experience of your walk.

The word rhythm thus describes both the patterning of events as they occur in time and the overall shape of these events' unfolding: rhythm is both a process and a product.[4] Rhythm organises and give a felt quality to the temporal flow of our experience by creating mood, tension, intensity, and a feel or groove. A rush towards a soon-to-depart train and a sudden stop inside will have a different feel to a slow all-the-time-in-the-world saunter. Sitting alone in a room working uninterrupted on a project will have different feel to working in a room with several others whose actions interrupt you. We feel these rhythms, as Rhiannon Newton points out, not just through our eyes and ears but with our whole body. Our skin feels the whoosh of an arriving train. Our stomach feels the jolt of a heavy step upon a wooden floor; and these felt rhythms tell us something. Where one rhythm of a co-worker's tapping fingers will signal impatience, another will signal meditative thought. We are constantly sensing rhythms, feeling rhythms, and interpreting rhythms.

We must pay attention to a rhythm to perceive it, and this paying attention involves synchronising part of ourselves to the rhythms' events.[5] For instance, we might synchronise our hearing to the rhythm of a dripping tap, our eyes to the visual rhythm of a crowd of walking feet, or we might nod our head or clap our hands to help us 'hear' a rhythm. Becoming attuned to a rhythm synchronises our perception around its patterned cycles so that we will predict and anticipate the events that form each cycle. It is this that makes rhythmic perception able to be simultaneously voluntary and involuntary, something we open ourselves to and something that controls us. Our attention can be so bound to a rhythm that our feet will suddenly tap along to a beat, or so caught up with anticipating a rhythmic cycle that the drip of a tap will prevent us falling asleep.

In perceiving rhythm, we become alert to change and continuity, difference and repetition. We will spot a limp in a crowd of walking feet, hear the misfire of a car cylinder, feel the emotion behind the rhythm of a friend's fading smile, or identify an outlier in our research data. Although rhythm is expressive, what it expresses can be different from the events it contains. The same piece of music, for example, can be played rhythmically fast or slow and the melody of the note events will remain unchanged: it is the rhythm that expresses fastness or slowness. Similarly, the same steps of a research task can be performed fast or slow, with the speed of performance perhaps expressing the researcher's level of energy and interest in the task. Paying attention to rhythm develops in us an alertness to future events and a sensitivity to moments of change, to patterns of repetition, and to the meanings that the flow of rhythm expresses.

We are used to perceiving and reading meaning from rhythms in our daily life on many timescales, from the lengthening of days to the quickening of a heartbeat. Our ability to perceive and read meaning from these different rhythms is tied to the specific cultural context we live in. For example, the musical culture that we are born into has been shown to create perceptual templates of the 'rules' of that musical tradition, allowing us to easily recognise and perform its rhythms.[6] As a baby we learn our culture's rhythms of speech and facial expressions from our caregivers.[7] Later, we learn how to move our body in rhythms that match our culture's perceptions of gender.[8] Our ability to perform these rhythms in synchrony with others is what then marks us as belonging to a specific culture.[9] Rhythmic ability is also influenced by our position in cultural history and knowledge of the technologies of our age. This historical knowledge means we can distinguish between rhythms that to someone from another era would be indistinguishable.[10] Those of us born well after the invention of the motorcar can, for example, hear the difference between the rhythm of a motorbike engine and that of a car. This cultural specificity of our rhythmic knowledge means that we can easily recognise and perform rhythms from our own culture but may not recognise or be able to perform rhythms from another culture. This learnt rhythmic culture includes the rhythms of our research discipline. Thus, we may have trouble performing research rhythms from another discipline when we change fields or work on an interdisciplinary project.

Perceiving rhythms in research

The ability to perceive and perform rhythms is a skill that can be improved through practice. This practice is something you will need to consciously and reflectively do to be able to recognise and analyse the rhythms within your practice-based research process. To do this you will need to think about:

- Rhythm across different scales – minutes, hours, days, months, years – reflecting on the specific rhythmic traditions of your cultural, historical, social, and disciplinary context.
- Dynamics of its pacing. Is it fast or slow? And how does speed impact the rhythm's intensity and the pressure of your research process?
- Operating in sync with a rhythm and moments where this sync is interrupted; and to ask, can such rhythmic interruptions be useful?
- Unpredictable rhythms within research will also be useful, as will attending to moments of transition between the different modes of practice-based research. How do you develop the skills to improvise in the face of the inevitable unpredictability of the research process? And when and how should you transition between modes?
- The way that the dynamics, synchrony, unpredictability, and transitions within the rhythms of your practice-based research process weave together to create a sequence with its own groove. Is this sequence something you can or should try to compose and orchestrate?
- How does your research sequence intersect with other rhythms operating around you? For example, your process might intersect with other institutional rhythms, with the rhythms of your supervisor, with research colleagues, or with the rhythms of your family.

A rhythmic skill that you will need here is what Bluedorn and Standifer call the 'temporal imagination', a term describing the skill of being able to conceptualise and understand the ways your temporal rhythms intersect with the larger temporal rhythms you are acting within.[11] Developing these rhythmic skills is a process that the concepts within this chapter should help you to begin. In the following sections, we now take a deeper look at how rhythm might manifest within a practice-based research process.

The fast and the slow, the lively and the languid

Parts of your practice-based research process may require fast, intense energy in short bursts. Others may require less energy and operate across longer timespans. At times, you may need to perform tasks of different rhythmic characters simultaneously. Paying attention to such rhythmic dynamics involves thinking about both the rhythms you prefer to practice and the rhythms that your practice might demand. These two will most probably be entangled in such a way that it will often be difficult to tell which is driving the other. This entanglement is evident in fashion designer Emma's description of the rhythms of her process:

> So this idea comes to me, and the only thing that I want to do is rush over here and take a piece of cloth or paper or the paper-pattern, and do it over and over again.[12]

Energised by the intensity of a great idea, Emma prefers to move quickly, but it is the demands of her practice that seem to then drive the need for rhythmic repetition. The two rhythms, the researching body and the practice, are entangled together. This entanglement is one area where thinking about rhythm can be useful for diagnosing problems while researching. For example, the dynamics of a preferred way of working and those demanded by a practice can clash and be the reason why a research process may not progress effectively. Or perhaps outside rhythms are impacting dynamics by causing practice to occur in short bursts when longer durations are needed.

To counter these issues, experienced practitioners often have strategies for varying the speed and energy of their practice and thus modulating their practice rhythms. For example, in my own practice I might switch from a slow computing process to sketching on paper, to speed up the development of an idea. Faced with competing demands for my time, I might accept that I am not going to get a whole day for my practice but instead make a few hours stretch longer by blocking out all outside interruptions and scheduling the blocks regularly across a week so that the rhythms of the thread of practice are maintained. Paying attention to the speed and intensity of your research rhythms is a useful diagnostic tool and a way to develop strategies to solve any issues you uncover.

Working slowly during research is often associated with deep focus, attention to detail, and nuanced thinking. For instance, Bresler[13] talks about the value of slow rhythms for qualitative research processes that aim to develop insights through empathy. These empathetic insights may not occur if a researcher makes quick judgements or evaluations. Fast thinking might lead a researcher to jump to easy and predefined conclusions that do not necessarily reflect the data. Conversely, working slowly might reveal more nuance and develop greater empathetic awareness of what the data shows.[14] However, in other types of practice, speed can be useful. A management study of business teams observes that slow teams focus sequentially on single alternatives, whereas fast teams consider more alternatives and compare them. With more information to base their decision on, the fast teams were found to make more confident and effective decisions than the slow teams, who 'often panicked, making snap decisions in the face of looming deadlines'.[15]

The relationship between time pressure, narrowness of focus, and problem complexity was also a factor in another business study of teams.[16] The speed of time pressure was found to be useful when it caused teams to focus in on the most relevant information. However, if speed resulted in a smaller amount of information being considered less deeply, then it was detrimental. This detrimental outcome was connected to more complex problems, as solving these required teams to have deep knowledge of the intricacies of relevant information. What these examples indicate is that it is the context and type of problem that will determine whether a fast or slow, pressured or calm approach will achieve your practice goals. While we all know the unproductive panic of a looming deadline, we also know that such pressure can often be productive and spur us to make more progress or help overcome a lack of motivation. Similarly, while working slowly can be useful when generating concepts or finessing developed ones, working fast can generate interesting concepts through quick comparison of multiple alternatives and produce the energy required to bring a finessing process to an end (see Figure 3.3.2).

This discussion may make it sound like your job as a practice-based researcher is to find a single perfect match between the rhythmic dynamics of practitioner, practice, and problem. In reality, of course, these three contexts are themselves variable and dynamic, so a match that is perfect at one stage may not be at another. Like catching waves in the ocean, you will need to constantly adjust the dynamics of your rhythms as the waves of the practice-based research process roll in. Operating constantly at a fast pace can cause burn-out, while operating slowly for too long can cause boredom. Both will result in a lack of energy and engagement. By varying the dynamic pace of your research process, you will keep yourself focused and energised across the temporal span of a project. Also, as we saw in the discussion earlier, varying rhythmic dynamics is an important way to ensure that your research practice effectively navigates the changing contexts of your process.

Keeping step and breaking free

The rhythms of a practice-based research process can carry us along, motivating our actions like music does when it makes us dance. By keeping step with the rhythms of our process and synchronising our actions to its beat, we progress forwards through our research. This type of synchrony can operate across all scales of a practice-based research process. For instance, a project will usually have rhythms of internal and external deadlines that can operate across days, weeks, or years. The rhythmic cycles of these deadlines influence the dynamic pace of a project and act like a 'metronome' for

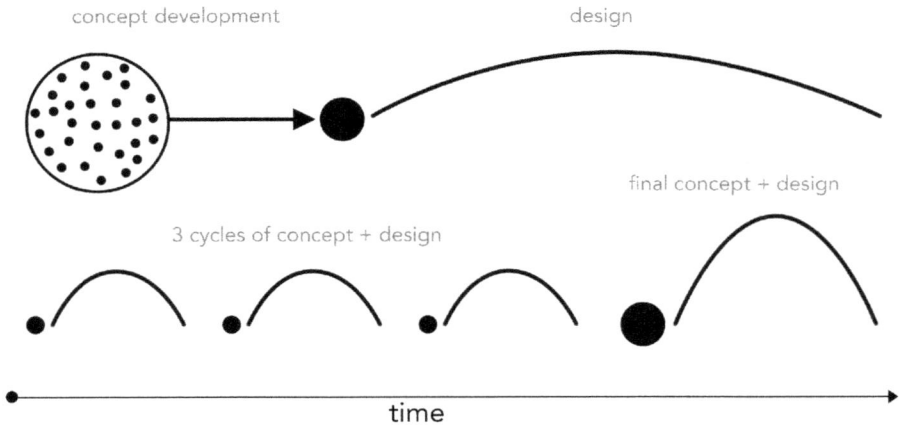

Figure 3.3.2 A slow process of concept development and design (top) compared to fast cycles of concept and design development that culminate in a final design cycle (bottom).

Source: Image by Brigid M. Costello.

the research activity.[17] Both a researcher and her or his research process can become entrained to the pace of such cycles, operating in synchrony with their demands. Projects can also consciously create such time pacers for themselves (e.g. Figure 3.3.3). These pacers form a consistent rhythm that then creates a flow with a momentum that drives the project forward. This type of consistent pacing can give researchers a feeling of control over the often-unpredictable rhythms of the research process.[18] Creating time pacers can be really useful for the beginnings of projects when directions and goals are not yet well defined. They are also valuable for times when motivation might be flagging, or you are feeling overwhelmed by the scale of what you need to do.

Another area where synchronisation plays an important role is in collaborative research projects where multiple researchers need to operate in sync with each other. A collaborative research project into psychotherapy describes this process using the metaphor of a dance, where

> clinic operations and research operations converged increasingly, creating an interactive process – a kind of research-practice tango – in which clinic staff and research staff learned to operate in ever closer synchrony with one another.[19]

This metaphor reveals the important role that sensitivity to rhythm can play in a collaborative research process. Dancing well with a partner requires the rhythmic sensitivity to predict your partner's rhythms and adapt your rhythms to theirs. If both parties simultaneously and skilfully perform this process of prediction and adaptation, then they will feel a satisfying sense of synchrony as they dance.[20] Another type of synchronous collaboration that occurs within practice-based research is that between a researcher and the materials of their practice. Here dance is also used as a metaphor by Ingold, who describes the relationship between practitioner and material as a 'dance of animacy': a dance where

> the mindful or attentive bodily movements of the practitioner, on the one hand, and the flows and resistances of the material, on the other, respond to one another in counterpoint.[21]

In Ingold's description, there is a similar synchronous rhythmic exchange of energy between researcher and material. There is also a similar need for a rhythmic sensitivity to predict and adapt to the movements of another. These various roles that synchrony can play in the rhythms of practice-based research are another place where potential issues can be diagnosed. When things are well synchronised, such rhythms might be less obvious, but when there is a break-down, a rhythmic perspective can help reveal what is needed to get the process dancing again.

A useful perspective to consider in relation to synchrony is the Greek distinction between the quantitative sequential time of Chronos and the qualitative time of Kairos.[22] To continue the dance metaphor, Chronos would relate to a focus on sequentially matching the beat of the music. Kairos, however, involves making a judgement about the rightness of the timing of your action, and that involves relating it to the whole context and its place within a time continuum of past and future. When choreographer Rhiannon Newton describes a dancer who has a good sense of rhythm, it is Kairos that plays a major role:

> You can watch certain dancers that just land the extension of the movement, or the shift of weight, in the centre of the beat. It means that often it's a quality of ease of going through the pathway and knowing that they just need to get there. As opposed to the dancer beside them that is going: 'I know the movement ends there, I know the end of the beat is coming, oh no, I was there early.' When you watch a dancer who has that feeling for timing, they have this sense of being incredibly grounded, but also floating.[23]

In her description, the dancer that doesn't hit the beat is focused on sequential quantitative time. The dancer who has good rhythm also has a sense of the quality of the whole timing context – where they have been, where they are, and where they are going.

To think about the qualitative rightness of timing leads to a consideration of the whole scale of synchrony within a research process. It leads to a consideration of the rhythms of research as they move across past, present, and future. It also leads to a consideration of whether there is a right time and a wrong time for synchrony. It asks you to think about whether a research process should play with timing and slide around the beat like a jazz musician, sometimes operating in sync and sometimes not. For instance, a research process might break out of the flow of synchrony and find this useful for developing ideas, finding inspiration or reinvigorating the rhythms of practice (e.g. Figure 3.3.3). Such moments of interruption and pause are often linked to reflection within creative

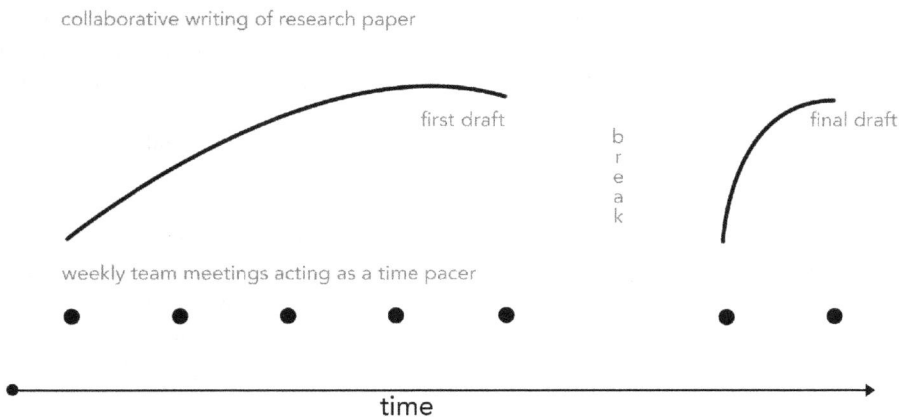

collaborative writing of research paper

first draft

break

final draft

weekly team meetings acting as a time pacer

time

Figure 3.3.3   A reflective break in the middle of a collaborative writing process with weekly meetings acting as a time pacer.

Source: Image by Brigid M. Costello.

practice and seen as opportunities to reveal and problematise tacit understandings.[24] They are also seen as opportunities to question and potentially change routines – a chance to step out of the micro performance of a process and reflect on macro rhythms.[25] Paying attention to synchrony within a practice-based research process involves thinking about when you might need to keep in step with synchronous flow and when and why you might need to break away from it.

Working with unpredictability

Rhythms of unpredictability are an inevitable feature of all research processes that seek to reveal, understand, and/or create new knowledge. Such rhythms can be even more prevalent within practice-based research, which has problems that are often 'messy', 'unclear', and involve 'conflicting agendas'.[26] Practitioners develop skills for producing 'knowledge in the face of uncertainty'.[27] They develop skills for dealing with the infinite openness of possible approaches and outcomes that their practice context presents them. Artist Anne Hamilton describes this as a conscious and 'rigorous' process of 'not knowing', which involves remaining receptive 'to all routes – circuitous or otherwise'. For her, this is a process of actively waiting and questioning.[28]

With a comparable focus on questioning, designers Bill Gaver and John Bowers speak of the value of processes of 'productive indiscipline' where flights of imagination meet eccentric sources of inspiration and experimental methods.[29] Productive indiscipline is also involved when improvising jazz musicians use 'aggressive emergence' to shift the direction of an improvisation, even if this sometimes is like throwing 'wrenches in the works'.[30]

Although it is not obvious, all of these practices for dealing with uncertainty are patterned with rhythms of unpredictability and predictability. Some kind of form is being produced by the practice, and this form emerges out of an interplay between predictable repetition and unpredictable novelty. If everything is unpredictable, that in itself becomes predictable: the unpredictable needs a contrasting field of predictability for it to be perceived as a wrench in the works. This contrast is described as involving a dialogue between instability and stability by composer and percussionist Bree van Reyk. As she puts it, her challenge while improvising is to keep her music 'open and free while knowing that it does need to eventually come into a structure that is repeatable and then can develop'.[31] Similarly, dancer Andrea Olsen describes improvisation as a process of 'holding uncertainty'.[32] The holding is a moment of stability against which the instability of uncertainty can operate. Stability and instability also operate together in my research practice. For instance, I might structure an interview with set questions but also ask impromptu questions. Or I might define a small set of parameters for a design project and then use them as an inspiration point for multiple experimental prototypes. Producing knowledge in the face of uncertainty involves sometimes opening to possibilities and other times closing them off.

The freedom of openness and the constraint of defined options can play off each other in productive ways during practice-based research. Paradoxically, freedom can emerge through constraint. We see this paradox in musical improvisation, where constraint is described as allowing musicians to access 'the freedom that is at the heart of improvisation'.[33] This paradox is also seen when practitioners constrain possibilities by developing a restricted palette to work with. The constraints of a restricted palette allow the 'what' of a practice context to be quickly investigated. A practitioner is then free to explore and improvise around the 'how' of practice, to experiment with the expressing, layering, transforming, and sequencing of that palette. This is not necessarily an easy process. As drummer Alon Ilsar explains, restricting his palette makes it at first 'harder' to improvise. After a while, however, he finds that 'there's actually a lot more to explore'. A restricted palette stops him from feeling 'at the mercy of possibilities'.[34]

By closing down the possibilities, the predictability of constraints gives a practitioner the freedom to then explore those constraints in unpredictable ways. Paying attention to rhythms of unpredictability within practice-based research thus involves thinking about when to open up the range of possibilities you are dealing with and when to close them down. It also involves developing skills and strategies for facing the unknown, the unexpected, and the uncertain. Here is where it can be useful to draw on improvisational methods from other practice traditions, like those described earlier.[35] Improvisation can develop skills that allow researchers to cope with unexpected situations within all kinds of research contexts.[36]

Moving between practice and theory

During practice-based research, a researcher will usually transition between many types of research activity. For example, they may move through phases of practising, data-gathering, theorising, reflecting, or writing. These transitions could be smooth or abrupt, but each will contain its own rhythms of arrival and departure. In many accounts of practice-based research, it is these moments of transition that can be the most difficult to navigate.[37]

One often difficult transition within practice-based research is the movement between processes of theoretical thinking and processes of practical doing. There are all kinds of metaphors for characterising this transition. In science, the transition is often described as having hard borders, with practice and theory each adding another brick of knowledge to a wall. However, for psychotherapy, Stiles argues that 'diffusion' is a better metaphor, because theory and practice can infuse and expand each other.[38] A similar sense of fluidity occurs in metaphors from design research. Here the theory/practice relationship is described as involving bi-directional flows of movement, and structural metaphors of extending, scaffolding, or blending the two knowledges are used.[39] Structural metaphors are also used in human–computer interaction (HCI) research, where many characterise theory and practice as being divided by a gap that needs to be bridged. However, others argue that the two should be seen as operating on a continuum.[40] For Beck and Ekbia, characterising the relationship as a gap leads to an overly narrow focus on bridging methods. They ask, how might we research differently if we instead see the relationship between theory and practice as a continuum with 'continuities and connections'? Building on this argument using HCI researcher and practitioner interviews, Colusso et al.[41] find that if there are gaps between practice and theory, they are multiple and of different types. The processes of translation and cooperation are therefore important, and this puts an emphasis on moments of transition. The notion of there being a gap between practice and theory is also rejected by performance researcher Kozel,[42] who emphasises that each has a materiality and motion:

> At first glance practice seems so heavy, and the theories so ephemeral. Yet in reality, ideas are felt, touched, lived, and breathed; practice is ephemeral, changeable, invisible, and disappearing.[43]

In each of these descriptions, we see rhythms. There is the staccato pattern of the brick and gap metaphors. Then there are the fluid and textured rhythms of metaphors involving diffusion, blending, continuums, and animated materiality. As Beck and Ekbia point out, the way a researcher conceptualises these transitional rhythms between theory and practice will impact the way they do research and potentially also the knowledge their research process produces.

An example of this potential impact can be seen in Wilson's account of his performance research process.[44] He began with a perception of his research process as being an 'excavation' where he was digging to find insights or truths.[45] This metaphor implied that there were pre-existing, defined truths to be excavated and caused him to transition constantly between performing the insider position of an artist and the outsider position of a researcher. For him, this created an unproductive distance between theory and practice. When he took the reflective position of an outsider, he felt 'alienated' from his work and consequently unable to think insightfully in either mode.

Brigid Mary Costello

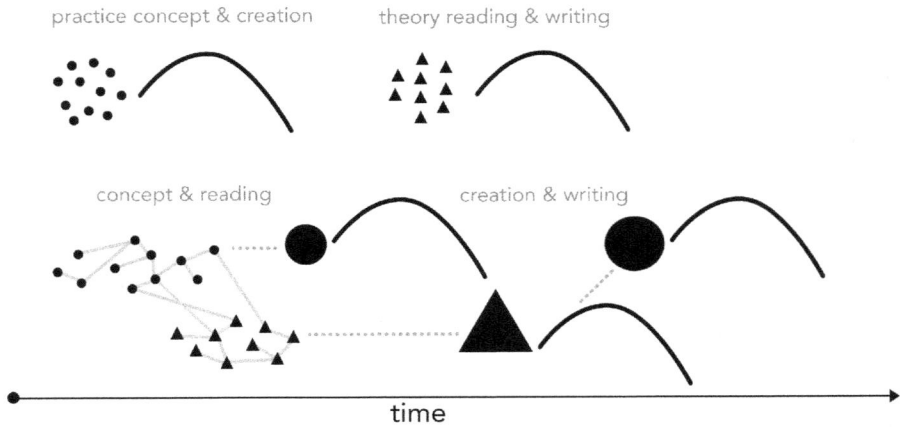

Figure 3.3.4 Theory and practice represented as separate research processes (top) or as intertwined processes that feed off and into each other (bottom).

Source: Image by Brigid M. Costello.

The solution to this impasse came when he realised that he needed to be both insider and outsider, 'not in alternation, but in one gesture'.[46] Wilson's new perception of his practice-based research process used the metaphor of 're-animation' to describe a process that used repetitive practice to produce less defined and more fluid truths than those implied by the excavation metaphor.[47] With this new metaphor and its new rhythmic relationship between practice and theory, he was able to successfully continue his research.

Thinking about the rhythmic quality of transitions within a practice-based research process is, then, another area where researchers can diagnose problems and find solutions to blockages in the flow of their research process. Your process of navigating and conceptualising these transitions will not necessarily be the same as that described by Wilson. However, like Wilson, you will find it useful to reflect on the way that you transition between practice and theory (e.g. Figure 3.3.4) and on which rhythms of transition are the most productive for your research process.

Weaving multiple rhythms together

Considering the rhythmic interweaving of dynamics, synchrony, unpredictability, and transitions across the long temporal spans of practice-based research reveals the rhythmic grooves their textures create. In a study of artistic research, these grooves are described as involving an 'interplay between practice, theory and evaluation', each patterned with rhythms of iteration and interaction and forming overarching 'trajectories'.[48,49] Trajectories also feature in anthropologist Ingold's description of knowledge development as 'wayfaring', a process that he says is like 'following trails through a landscape'.[50] In his metaphor, although the researcher's movement may form a continuous path, it is achieved by following, discovering, and rediscovering many trails, each one taking you a bit further along your research path.

A similar sense of movement that sometimes circles back on itself is found in artist and architect Beinart's characterisation of research. As she puts it:

the research process may not be a clear line of emergence but rather an eddy, a back and forth movement between clarity and murkiness.[51]

350

The murkiness within research is something that also features in designer Yuille's portrayal of the process as a journey into an abyss of the unknown.[52] The abyss is a 'messy, ill-defined and scary morass of ideas, theories, practices, fields, forces, pressures, emotions and opinions', but it is also a space that can be quiet and allow you to 'think differently'.[53] The rhythms in Yuille's research journey are cyclical, with the researcher moving from the known into the abyss of the unknown and back again.

Although each practice-based research process will be different and have its own rhythm, there are some common structural rhythmic elements revealed by these descriptions. Your research process will move along a path that in hindsight will form a continuous trajectory. However, in the present moments of the process, you will create patterns that sometimes move forwards and sometimes circle back. You will discover things, lose them, and find them anew – perhaps transformed. The predictability of repeated elements will lead you to focus more deeply on subtleties, while the unpredictability of the new trails you find will open your research to new possibilities. Together these rhythms will form a groove that describes the path of your research project.

Depending on your level of practice-based research experience, paying attention to this meta view of the overall rhythm of your research process may be something you find difficult.[54] However, it is a perspective that can be useful: for planning when first starting out, for helping to keep going when in the middle of things, or for making sense of where you have ended up. The metaphor of orchestration is useful here. Orchestration involves thinking of the rhythms of process at all layers and scales and the ways they combine together. In terms of a research process, it also involves considering how these might intersect with the rhythms within your institution, your family, your society, or your environment.

This type of complex rhythmic orchestration is something architects do when they design. For instance, architect Joe Agius describes how he might

> orchestrate a public space, be it a park, or a system of streets, or the rhythm of the street lighting, or the street trees . . . or the sequence of main road to laneway . . . or the positioning of towers within a terrain or a precinct.[55]

Joe's designs will not only interact with the rhythms of their environment but also work to generate new rhythms. For example, he might place a café at street level to enliven an area and form an ecology of rhythmic energies that 'feed off each other'. At times, Joe Agius' designs will orchestrate rhythmic energies so that they combine and grow. At times, they will focus on 'blocking or containing energies in order to shelter and support specific qualities of rhythm'.[56] This kind of architectural orchestration works across rhythms that are biological, material, aesthetic, environmental, institutional, social, and political. It is this complexity that makes architectural orchestration a useful metaphor to apply to the equally complex rhythms of the practice-based research process.

Orchestrating your research rhythms is not just about perceiving or analysing; more importantly, it is about composing the rhythms within your process. Doing this orchestration will involve thinking of rhythms at different scales and levels of nuance. It will require considering energies and the ways they might combine, grow, or fade. It will also involve thinking about strategies for dampening, amplifying, or nurturing your process rhythms. Key to all of this will be developing a sensitivity to Kairos, to the right time within your process that a rhythm needs to occur.

A timeline with rhythm

With a new awareness of the rhythms of practice-based research, we can return to our timeline example and re-imagine it (Figure 3.3.5). This re-imagined timeline now includes the regular beat of supervisor meetings and represents the speed and intensity of the three main activities. It

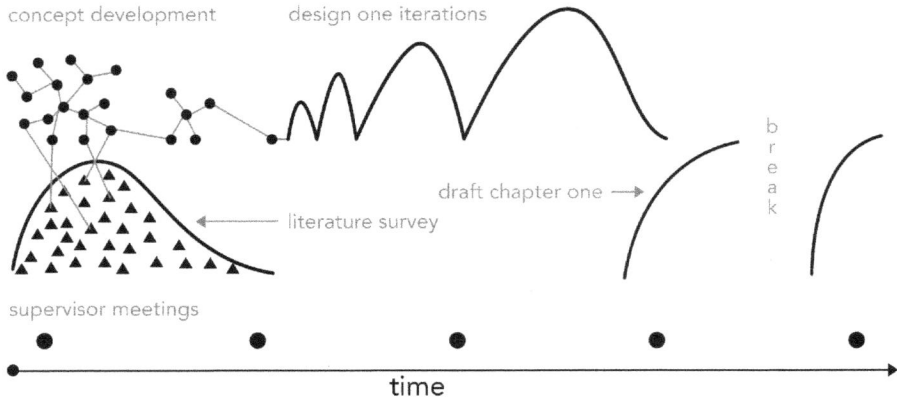

Figure 3.3.5 The timeline from Figure 3.3.1 re-imagined as a rhythmic process.

Source: Image by Brigid M. Costello.

shows the entanglement of theory and practice that occurs in the simultaneous processes of concept generation and literature survey. The concept phase is shown to move from the openness of many possibilities to the stability of a single choice. Similarly, the literature survey moves from an intense phase of quick reading of many papers to a slower deep reading of a select few. The iterative process of creating a design begins quickly and, as the design progresses, becomes slower and more intense. Lastly, the chapter writing is broken by a pause to allow for reflection and refreshment of perspective.[57]

Although this timeline is just a single simple example, it hopefully sparks ideas for ways you might take the insights from this chapter and now apply them to your own research process. Whether you produce a document that looks anything like that in Figure 3.3.5 is not important. What is important is that you pay attention to the nuance and detail of the rhythms of your own process. This will involve focusing on the patterns of your research trajectory and the layers of rhythms within it. It will require you to think about when to be quiet, when to inject energy, when to keep step, and when to break free. It will also ask you to think about how you might benefit from moving between the predictable and the unpredictable. Paying attention to these patterns will allow you to diagnose issues, develop solutions, and understand how you might orchestrate your process.

Conclusion

Each practice-based research process has its own groove, a groove composed of many interwoven rhythms, playing out across multiple scales. Developing an awareness of these research rhythms allows us as researchers to more effectively plan our process, diagnose and solve issues, and modulate momentum and energy.[58] As we have seen, in order to do this, we will find it valuable to:

- Develop and apply strategies for varying the speed and energy of our practice.
- Use time-pacers to help us keep in step with research rhythms.
- Schedule reflective moments of rhythmic calm.
- Exploit the improvisational dynamics of freedom and constraint.
- Conceptualise and apply productive metaphors for transitions between research roles.
- Take a meta view of the whole rhythm of our process and consider how we might orchestrate its patterns.

Within our research process, we can use rhythm as both a tool for expression and a tool for analysis. By paying attention to the rhythms of our research in this way, we will gain insight into the temporal dynamics of our research process and how its rhythms have played out, are playing out, and could and will play out. For us as practice-based researchers working on the edges of uncertainty, rhythm is a tool for effectively orchestrating the changing grooves of our process.

Notes

1 Kozel (2010, p. 204).
2 Bresler (2005, p. 169).
3 Rhiannon Newton, Choreographer, quoted in Costello (2018, p. 136).
4 Goodridge (1999, p. 44).
5 London (2004, p. 4).
6 Levitin (2006, p. 27).
7 Gibbs (2010).
8 Young (1980).
9 Turino (2008).
10 Emmerson (2007, p. 17).
11 Bluedorn and Standifer (2004, p. 161).
12 Gherardi and Perrotta (2014, p. 141).
13 Bresler (2005, p. 176).
14 Ibid.
15 Eisenhardt (2004, p. 273).
16 Karau and Kelly (2004, p. 187).
17 Zellmer-Bruhn et al. (2004, p. 137).
18 Eisenhardt (2004, p. 278).
19 Weisz and Addis (2006, p. 191).
20 Costello (2018, p. 87).
21 Ingold (2013, p. 101).
22 Thaut (2005, p. 16).
23 Costello (2018, p. 136).
24 Dix and Gongora (2011, p. 39).
25 Zellmer-Bruhn et al. (2004, p. 144).
26 Zimmerman et al. (2010, p. 310).
27 Kemmis (2005, p. 404).
28 Hamilton (2010, pp. 68–69).
29 Gaver and Bowers (2012, p. 42).
30 Dempsey (2010, pp. 125, 142).
31 Costello (2019, pp. 4, 7).
32 Olsen and McHose (2014, p. 67).
33 Berkowitz (2010, p. 180).
34 Costello (2018, p. 69).
35 Some relevant methods discussed in this book include Seevinck on emergence (Seevinck 2022) and Smith on risky play (Smith 2022).
36 Bresler (2005, pp. 174–175).
37 E.g. see Wilson (2019), Yuille (2017).
38 Stiles (2010).
39 Markussen (2017).
40 Beck and Ekbia (2018).
41 Colusso et al. (2019).
42 Kozel (2010).
43 Kozel (2010, p. 206).
44 Wilson (2019).
45 Ibid., p. 6.
46 Ibid., pp. 11–12.
47 Ibid., p. 6.
48 Edmonds and Candy (2010, p. 476).

49 For more on trajectories and the interweaving of theory and practice, see also Candy (2022a, 2022b).
50 Ingold (2011, p. 162).
51 Beinart (2014, p. 228).
52 Yuille (2017).
53 Ibid., p. 205.
54 Slattne (2022) in this book provides a perspective that may help develop this meta view.
55 Costello (2018, p. 54).
56 Ibid.
57 You might find inspiration for your own process diagrams in musical graphic scores and notation, e.g. the work of John Cage, Cornelius Cardew or Billy Martin.
58 These sensitivities about the process of research and the insights they reveal can, many suggest, also be a way of developing knowledge (see Bresler 2005; Frisk and Karlsson 2010; Veldt et al. 2017).

Bibliography

Beck, J., & Ekbia, H. R. (2018). The Theory-Practice Gap as Generative Metaphor. In *Proceedings of the 2018 CHI Conference on Human Factors in Computing Systems*. Montreal, Canada: Association for Computing Machinery, 11.

Beinart, K. (2014). Becoming and Disappearing: Between Art, Architecture and Research. *Arts and Humanities in Higher Education*, 13(3), 227–234.

Berkowitz, A. L. (2010). *The Improvising Mind: Cognition and Creativity in the Musical Moment*. Oxford: Oxford University Press.

Bluedorn, A. C., & Standifer, R. L. (2004). Groups, Boundary Spanning, and the Temporal Imagination. In R. L. Standifer & S. Blount (Eds.), *Time in Groups*. London: Emerald Group Publishing Limited, 159–182.

Bresler, L. (2005). What Musicianship Can Teach Educational Research. *Music Education Research*, 7(2), 169–183.

Candy, L. (2022a). Reflective Practice Variants and the Creative Practitioner. In C. Vear (Ed.), *The Routledge International Handbook of Practice-Based Research*. London and New York: Routledge.

Candy, L. (2022b). Theory as an Active Agent in Practice Based Knowledge Development. In C. Vear (Ed.), *The Routledge International Handbook of Practice-Based Research*. London and New York: Routledge.

Colusso, L., Jones, R., Munson, S. A., & Hsieh, G. (2019). A Translational Science Model for HCI. In *Proceedings of the 2019 CHI Conference on Human Factors in Computing Systems*. Glasgow, Scotland: Association for Computing Machinery, 13.

Costello, B. M. (2018). *Rhythm, Play and Interaction Design*. Cham, Switzerland: Springer.

Costello, B. M. (2019). Learning Through Improvisational Play: Design Strategies for Serious Games. *2019 IEEE 7th International Conference on Serious Games and Applications for Health (SeGAH)*, 1–7.

Dempsey, N. (2010). Digging a River Downstream: Producing Emergence in Music. In K. D. Norman (Ed.), *Studies in Symbolic Interaction*. London: Emerald Group Publishing Limited, 123–145.

Dix, A., & Gongora, L. (2011). Externalisation and Design. In *Proceedings of the Second Conference on Creativity and Innovation in Design*. Eindhoven, Netherlands: Association for Computing Machinery, 31–42.

Edmonds, E., & Candy, L. (2010). Relating Theory, Practice and Evaluation in Practitioner Research. *Leonardo*, 43(5), 470–476.

Eisenhardt, K. M. (2004). Five Issues Where Groups Meet Time. In R. L. Standifer & S. Blount (Eds.), *Time in Groups*. London: Emerald Group Publishing Limited, 267–283.

Emmerson, S. (2007). *Living Electronic Music*. Great Britain: Ashgate.

Frisk, H., & Karlsson, H. (2010). Time and Interaction: Research Through Non-Visual Arts and Media. In M. Biggs & H. Karlsson (Eds.), *The Routledge Companion to Research in the Arts*. New York: Routledge, 277–292.

Gaver, B., & Bowers, J. (2012). Annotated portfolios. *Interactions*, 19(4), 40–49.

Gherardi, S., & Perrotta, M. (2014). Between the Hand and the Head: How Things Get Done, and How in Doing the Ways of Doing Are Discovered. *Qualitative Research in Organizations and Management*, 9(2), 134–150.

Gibbs, A. (2010). Sympathy, Synchrony and Mimetic Communication. In M. Gregg and G. J. Seigworth (Eds.), *The Affect Theory Reader*. Durham: Duke University Press, 187–205.

Goodridge, J. (1999). *Rhythm and Timing of Movement and Performance*. London: Jessica Kingsley Publishers.

Hamilton, A. (2010). Making Not Knowing. In M. J. Jacobs & J. Baas (Eds.), *Learning Mind: Experience into Art*. Berkeley: University of California Press, 66–73.

Ingold, T. (2011). *Being Alive: Essays on Movement, Knowledge and Description*. London: Routledge.

Ingold, T. (2013). *Making: Anthropology, Archaeology, Art and Architecture*. New York: Routledge.

Karau, S. J., & Kelly, J. R. (2004). Time Pressure and Team Performance: An Attentional Focus Integration. In R. L. Standifer & S. Blount (Eds.), *Time in Groups*. London: Emerald Group Publishing Limited, 185–212.

Kemmis, S. (2005). Knowing Practice: Searching for Saliences. *Pedagogy, Culture & Society*, 13(3), 391–426.

Kozel, S. (2010). The Virtual and the Physical: A Phenomenological Approach to Performance Research. In M. Biggs & H. Karlsson (Eds.), *The Routledge Companion to Research in the Arts*. New York: Routledge, 204–222.

Levitin, D. J. (2006). *This Is Your Brain on Music: The Science of a Human Obsession*. New York: Dutton.

London, J. (2004). *Hearing in Time: Psychological Aspects of Musical Meter*. New York: Oxford University Press.

Markussen, T. (2017). Building Theory Through Design. In L. Vaughan (Ed.), *Practice Based Design Research*. New York: Bloomsbury Academic, 87–98.

Olsen, A., & McHose, C. (2014). *The Place of Dance: A Somatic Guide to Dancing and Dance-Making*. New York: Wesleyan University Press.

Seevinck, J. (2022). Making Reflection-In-Action Happen: Methods for Perceptual Emergence. In C. Vear (Ed.), *The Routledge International Handbook of Practice-Based Research*. London and New York: Routledge.

Slattne, H. (2022). Navigating the Unknown – a Dramaturgical Approach. In C. Vear (Ed.), *The Routledge International Handbook of Practice-Based Research*. London and New York: Routledge.

Smith, S. (2022). A Play Space for Practice Based PhD Research. In C. Vear (Ed.), *The Routledge International Handbook of Practice-Based Research*. London and New York: Routledge.

Stiles, W. B. (2010). Theory-Building Case Studies as Practice-Based Evidence. In M. Barkham, G. E. Hardy, & J. Mellor-Clark (Eds.), *Developing and Delivering Practice-Based Evidence*. New York: John Wiley & Sons, 91–108.

Thaut, M. H. (2005). *Rhythm, Music and the Brain*. New York: Routledge.

Turino, T. (2008). *Music as Social Life*. Chicago: University of Chicago Press.

Velt, R., Benford, S., & Reeves, S. (2017). A Survey of the Trajectories Conceptual Framework: Investigating Theory Use in HCI. In *Proceedings of the 2017 CHI Conference on Human Factors in Computing Systems*. Denver, CO: Association for Computing Machinery, 2091–2105.

Weisz, J. R., & Addis, M. E. (2006). The Research-Practice Tango and Other Choreographic Challenges: Using and Testing Evidence-Based Psychotherapies in Clinical Care Settings. In C. D. Goodheart, A. E. Kazdin, & R. J. Sternberg (Eds.), *Evidence-Based Psychotherapy: Where Practice and Research Meet*. Washington, DC: American Psychological Association, 179–206.

Wilson, J. A. (2019). Excavation and Re-Animation: Figuring Process in Practice-as-Research. *Studies in Theatre and Performance*, 39(1), 3–20.

Young, I. M. (1980). Throwing Like a Girl: A Phenomenology of Feminine Body Comportment Motility and Spatiality. *Human Studies*, 3(1), 137–156.

Yuille, J. (2017). Grokking the Swamp: Adventures into the Practical Abyss and Back Again. In L. Vaughan (Ed.), *Practice Based Design Research*. New York: Bloomsbury Academic.

Zellmer-Bruhn, M., Waller, M. J., & Ancona, D. (2004). The Effect of Temporal Entrainment on the Ability of Teams to Change Their Routines. In R. L. Standifer & S. Blount (Eds.), *Time in Groups*. London: Emerald Group Publishing Limited, 135–158.

Zimmerman, J., Stolterman, E., & Forlizzi, J. (2010). An Analysis and Critique of Research Through Design: Towards a Formalization of a Research Approach. In *Proceedings of the 8th ACM Conference on Designing Interactive Systems*. Aarhus, Denmark: Association for Computing Machinery, 310–319.

3.4

PRACTICE-BASED RESEARCH IN THE VISUAL ARTS

Exploring the systems of practice and the practices of research

Judith Mottram

Overview

In this chapter, creative practice is considered in relation to a process of research. After considering what might be meant by practice in the field of the creative visual arts, the role of novelty in the objects generated through practice is explored in relation to the gatekeepers in the field. A distinction is drawn between significant advances and incremental advances in a field, as described in theories of creativity. The context is objects or artefacts like paintings and sculptures which might be seen in galleries and might end up in museums of art. The focus is on these artefacts, drawing attention to artistic intentions and how innovation is recognised; we will return later to the thorny issue of the influence of the field on the reception and understanding of art. We are looking at the contemporary period, which is taken to mean artists working in the first 25 years of the twenty-first century and art produced since about 1950.[1] The players are artists and researchers, but we will also touch on the gatekeepers to recognition: the critics, curators, and others who influence what art reaches wider audiences. An analysis of themes and types of current art, to identify which artists might be referred to as contemporary, demonstrates how the gatekeeping function operates in respect of recognising achievement in the art market. This serves to remind us that the recognition of innovation is not necessarily tied to the intentions of the artist.

The chapter proceeds on the premise that those who might embark on practice-based research in the visual arts are already engaged in practice in the visual arts field, and that they anticipate that they might use some of what happens in their studio (or where- or however they make art) within their practice-based research. The core proposition is that if embarking on research in this space we should revisit our understanding of practices in the domain, and of how the system operates, in order to understand what tools practice-based research in the visual arts might call upon and to enable appreciation of its features and approaches. As the nature of practice changes, we may need to think about what evidence might be needed to investigate practice. In the visual arts, the particulars of practice could include the initial generation of an artwork, the practice of getting that artwork seen, of seeing and responding to that artwork, and the recording of some response in the world about that artwork. These may be carried out by a number of individuals and involve various institutions. If entering into the arena of research from practice within any field, consideration of the similarities and differences between normative professional practice and a research practice can be a useful platform for framing objectives and methods. It could be inferred that this implies no differentiation between the methods of practice and methods for research. The suggestion is, however, that there

DOI: 10.4324/9780429324154-26

may be differences in how one uses tools, regimes or actions in relation to different objectives. If the intent is to produce art objects for consideration within the social and cultural economy of the world of artistic exchange, the methods of practice might be employed to that objective. If, however, the objective is to intentionally engage in the exploration of a research question through practice-based research, some of those practices may be carried out for different purposes.

The articulation of research in creative subjects such as art and design made by Christopher Frayling in 1993[2] provides a useful backdrop to the current discussion. While Frayling had articulated how research might be carried out *for*, *through*, and *into* a discipline, the ensuing decades have seen increasing tendencies to conflate practice as research, blurring the understanding of what practice-based research might be. If creative practice in the visual arts is seen to constitute a research method (as is implicit in the notion of 'practice-based research in the visual arts'), there are some matters of how the field works that still need to be attended to. This chapter seeks to redress that omission, explore the implications for practice-based research, and hopefully enable future researchers to develop a clearer position on how to manage their intellectual and creative objectives.

Through discussing recognised innovation in the artistic practice of Frank Auerbach, the nature of that practice as a type of experimental method is proposed. Some tensions between recognition of the artistic oeuvre or 'signature' style and innovation are noted. This locates the discussion in relation to considering contributions to knowledge through research and to innovations in creative practice. What this discussion provides is a key question for practice-based research in the visual arts: is the intention to strive for paradigm-breaking innovation within practice?

More recent artistic practice that moves away from the material handling characteristic of modern and earlier artworks sets up a further conundrum. As the medium of some more recent artistic innovations may range way beyond the manipulation of 'stuff' to something more akin to event management, the tools for a practice-based research move well beyond the work that may occur in the studio. This prompts a question about what is happening as art is being made. Are we no longer dealing with the alchemy of the studio?

The conclusion suggests that there is scope to think of creative practice as analogous to an experimental method. A reconsideration of this term reflects the perspective of the author, as seeing her exploration in art practice through manipulation of colour and pattern as much a part of her personal research methodology as an interview transcript.

The perspective of this chapter may be useful for creative artists and scholars of art history and theory as well as other practitioners for whom untangling how things work might contribute to future practice, to research using the tools of that practice as method, and to an understanding of that practice.

Methods and systems in the visual arts and research

If creative practice in the visual arts is our starting point, we can think about a research methodology as a framework to undertake a practical 'thinking through' of questions in the specific domain of our practice. A method is a tool for investigation, a means of arranging our ideas in an orderly way or a special form of procedure, so potentially requiring both mental and practical activity. Within the context of practice-based research, our methods are those processes of thought and practice that allow us to make contributions to the field, to move on or develop in new ways how we think, understand or act. In both creative practice and in research, there is an element of generating new thinking or novel contributions that are recognised as such by appropriate audiences. The notion of new knowledge as the outcome of research is embodied in the definition of research and experimental development used by the Organisation for Economic Cooperation and Development (OECD): 'creative and systematic work undertaken in order to increase the stock of knowledge – including knowledge of humankind, culture and society'.[3]

As a foundation for thinking further about advances through research or practices in the field of the visual arts, Csikszentmihalyi's 'systems view of creativity'[4] gives us a system that accounts for the relationship between the individual creative person, the field of social institutions that provides the gatekeepers for the recognition of new contributions, and the domain to which field and actors relate. Thinking about method in this context, we can discriminate between the notion of method – our ordering of ideas or procedures – from overarching methodologies (often of a philosophical orientation)[5] that provide the theoretical framing of issues and questions of currency within the domain. For this discussion, methodology is used in the sense of the principles and specific uses of methods for research. While my core interest is in painting, the discussion also considers questions arising from other material practices emergent in the first 20 years of visual arts in the twenty-first century. In respect of practice-based research, the focus will be on the utility of the practices of art-making for research purposes.

Methods for research in practice-based research fields are situated within conceptual, theoretical, practical, or procedural systems. These may be at the scale of a domain, such as in archaeology, where objects or sites are scrutinised to make conjecture, through narrative and sometimes visual invention, about the past. They may be at the scale of a sub-domain, such as music composition, where the methods of enquiry employ different tools to other areas of musicological study such as computational sound generation. The 'system' of creative practice (in the visual arts) is seen here as essentially the overarching framework of the focus (visual creativity), the means (the generation of objects of attention), and the results of engagement in the domain (things being shared and seen). I take the system of practice here to include the range of actors – the artist and the audiences and other protagonists – who engage consciously or unconsciously, and closely or at a remove, within the domain and with its objects of attention or foci for thinking. For this chapter, we are looking at the system of the visual arts and the methods that might be relevant to research in that system, not the methodologies that might provide philosophical or theoretical framing devices.

Components of visual arts practice

When we are looking at methods for practice-based research, we might consider whether there are distinctions to be drawn between the methods of what I will provocatively for now label *everyday* artistic practice (by professional-orientated art-makers, for example), which generates novel outcomes to the individual, and artistic inquiry that realises something new and different for the field (whether intentionally or not), and whether either of these approaches can be used for research.

When mapping the terrain of artistic inquiry in the twenty-first century, to negotiate the breadth of thinking that ranges from Modernist determinism to the post-Duchampian assignment of artwork status, there is a shift with what the artist or researcher might be engaging with. For the purposes of this chapter, I will take art practice as making some accumulation of physical stuff, whether manipulated or not, and putting it in a place. The physical stuff could be human beings, or it could even be light or sound waves, or it could be pigmented paint arranged on a canvas (the notion of painting that might be most familiar beyond the field as an example of an artwork), but I'm assuming some intentionality on behalf of the artist for this stuff to be understood by some notional audience.

What is of interest is how this arrives through the agency of the artist, even if that agency is, in practice, a decision to allow chance or other systems to determine what occurs where in this artmaking activity. We can accommodate anything from action painting to flash mob opera in this scenario. I think we could probably extend the model to encompass team-intended artistic outcomes, whether consciously or unconsciously conceived. What is important when thinking about the potential for practice-based research methods in our field is how and why this happens and whether there are any differences depending on the aims and intentions of the creator.

Recognition of innovation

If we accept the notion that a principal characteristic of research is that it can make a recognised contribution to the field through new knowledge or new propositions or new questions, how might we view such outcomes realised through practice-based research in the visual arts? In the context of this chapter, we need to understand how innovation is recognised within the field of artistic practice, if our research is to be practice-based. The following five perspectives are suggested as a useful background to frame our thinking about practice-based research in the visual arts:

1   Novelty and ground-breaking innovation

The moment when new artistic works get recognised is covered by the model of creative innovation and change in creative fields expressed by Csikszentmihalyi.[6] In his model of creativity, derived from a study of Nobel prize-winners across disciplines, it was recognition of innovation by field gatekeepers that assured acknowledgement of contributions. Such contributions might be the merely novel, or those that were paradigm-breaking notions such as expressed by Kuhn.[7] The notion that revolutionary change to the systems of thinking or practice might occur was allied to the idea that change would be presaged by a 'proliferation of compelling articulations, the willingness to try anything, the expression of explicit discontent, the recourse to philosophy and to debate over fundamentals'.[8] In summary, there may well be a 'normal' or everyday practice in many fields, which is about not changing the way we think about a field but rather making an ongoing and useful contribution to the field through the supply of items that continue to be seen as having value (such as the drawings by Auerbach considered later). There may also be practice, often arising out of a confused period, which makes us think anew about a field. We may usefully think about the role 'normal'/personal artistic practice can play in research directed at making ground-breaking advances and how it can be a method with which to generate new knowledge.

2   Recognising changes within artistic practice

Recognising that there has been a ground-breaking change in creative practice is often associated with the attribution of a new name for a genre or category. These can arise through the ways in which artworks are discussed by critics, as exemplified by the term 'des Fauves' applied to Matisse and his fellows in 1905[9] or the term Young British Artists (yBA) first used by Simon Ford in *Art Monthly* in 1996[10] to describe Damien Hirst and his contemporaries. These labels are effectively a marker of recognition by gatekeepers of a new movement, genre, or class of works that carry some shared characteristics that mark them as distinct from what went before. In short, it labels them as recognisably new.

The recognition by gatekeepers such as critics, journalists, and the institutions of the art world has been explored by Marie Leduc[11] in a study that addressed the question of how the wider category of contemporary artwork might be recognised. She built upon the model proposed by Willi Bongard in the 1970s and 1980s through the Kunstkompass. In this model, the international art world's top artists are identified by the extent to which their work is included in collections and exhibitions or referred to in particular texts. Albeit that the Kunstkompass does have a bias towards German artists and a particular slice of the international art market that trades globally through a fairly specific social and commercial community, the analysis presented by Leduc does provide a context for another question about research using practice-based methods in respect of the sorts of practical work that might be undertaken.

3   Paradigm-breaking moments

Leduc's analysis of the changes in themes or modes of production reminded me of a paper published in *Art Monthly* in the 1980s[12] which looked at how models from science might be useful for understanding artworks. This was at a time when what we might now call professional magazines were one of the few means for a discourse within the field of visual art practice. Articles and letters responding to articles played out over periods of months, a very different speed of engagement to that afforded by the variants of these arenas in the early twenty-first century (e.g. web-based magazines such as *Abstract Critical*, or *AbCrit*, and *Turps Banana*). Richter's paper built on ideas from Thomas Kuhn,[13] considering whether new artworks were either within an existing aesthetic paradigm, as in providing exemplars or derivations of known idioms or ways of working, or whether they could be paradigm-breaking examples, which might first have an uncertain reception. The notion of whether or not a paradigm-breaking innovation was recognisable within the field was touched upon, but ultimately Richter concluded that both Clement Greenberg and Timothy Clark felt the 'post-modern' art of the 1980s did not fit their respective understandings of the purposes of art. As such, this provides an example of paradigm shifts not being recognised by the establishment within which they occur. So, if we seek big-C contributions to knowledge from our research, will we be able to recognise it?

4   Things that exist independently of the maker

The visibility of paintings or other sorts of artworks does rely upon gatekeepers as well as those who generate the things to be visible. As such, these players constitute part of the system of artworks as objects that have potential to become 'things that are products of the mind' but exist independently of their makers. In effect, artworks are emblematic of what Popper referred to as World 3 things.[14] Gatekeepers are pivotal in the identification of themes or discourses that serve to support the exchange of cultural capital among the different communities invested in contemporary art. As with magazines like *Art Monthly* and *Artscribe* in the 1980s, or *Abstract Critical* and *Turps Banana* in the 2000s, discussions of gatekeeping critics can provide one lens for understanding dominant modes of thinking. We can infer from their writings or even use quantitative measures to investigate the material they generate. While we may ask if this is reflective of the wider artistic community, a more pertinent question may be how artist-researchers operate themselves within such systems, what consciousness of and engagement with notions of innovation are within the field, and what thus the implications are for thinking about research methods within this system. Two matters come to mind: the argument that 'too much analysis causes paralysis', often cited as a reason why artists should not think too much about what they do; and that art's meaning is effectively more determined by the gatekeepers than by the intentions of the artist. This latter sets a clear challenge to any notion of research practices being replicable. It also gives an uncomfortable platform in respect of scope for mis-interpretation or misunderstanding of the outcomes of research. If the artwork is expected to embody the knowledge generated by the research, the fragility of the meaning determined by the originating artist feels problematic, and hence the need for clear contextual rationale to accompany the claim.

5   The continuum of knowledge creation activities

More recently, Biggs[15] has argued that knowledge creation activities are a continuum, from the truth claims of scientific method to socially or artistically created knowledge. Arguments for practice-based research methods which allow for the procedures of artistic practice to be seen as research methods mean we now have a generation of art practitioners who refer to their everyday applied artmaking as research. This chapter, however, takes the position that there may well be another paradigm shift in progress which moves beyond the acceptance of the potential of creative practice methods to add meaningfully to understanding of disciplinary-specific understanding. Given that art practices can now include almost any sort of activity, from having a conversation, taking a walk, or growing a culture of bacterial cellulose, to specifying the parameters for the construction of a three-dimensional installation of Perspex, wood, and steel or generating paintings by transcribing from time-lapse photography, we now have a tremendous range of practices which might be methods.

The following three sections will explore innovation and novelty in relation to specific examples of artistic practice and the challenges this generates for conceiving approaches to practice-based research. We will then return to these underpinning considerations in the conclusion, to test what we claim might be addressed through practice-based research and what might drive decisions about method.

Innovation and novelty in tension with seriality and attribution

The notion of practice-based research in the visual arts field carries the inference that the making of art would be a significant component of the method of such research. The crucial question is whether innovation arising from artistic practice can carry new knowledge such as is striven for as the necessary outcome of a research endeavour. The following discussion focuses on the visual characteristics of objects produced by an artist of reputation and finds some problems with the idea of novelty, expectations of innovation, and notions of verifiability. If we work with the model of the artist always being focused on producing revolutionary art, how can we account for the notion of oeuvre, or identifiable artistic characteristics in the outputs of individual artists? To look at identifiable characteristics and innovation, I will consider a catalogue of the work of Frank Auerbach.

In the opening words to a conversation between Frank Auerbach and Catherine Lampert, as reproduced in the catalogue to Auerbach's 1978 exhibition at the Hayward Gallery in London and Fruitmarket Gallery in Edinburgh, Auerbach said:[16]

> What I'm not hoping to do is to paint another picture because there are enough pictures in the world. I'm hoping to make a new thing for the world that remains in the mind like a new species of living thing.

These words have been reproduced several times, including by Alexander Moffat in his 1984 introduction to *The British Art Show* exhibition (Ikon Gallery Birmingham; Art Gallery of New South Wales, 1984) and in 2003 by Neil Mulholland, in *The Cultural Devolution: Art in Britain in the Late Twentieth Century*, giving some indication that these words are considered notable. The focus on the new thing being generated does appear to be a key part of the notion presented by Auerbach, and he talks elsewhere of the power of the pictures he considers of timeless quality in the same way.

It is useful to consider this notion of the new thing for the world when considering the claim for embodying new knowledge that might be conceived as part of the argument for practice-based research. The statement made could be read as a claim that each of his works were to be seen as individual new contributions to the world that could each be apprehended as a singular new species. Does the claim hold up to scrutiny? In the catalogue in which this quotation was first reproduced, we have a list of the 135 artworks presented. Of these works, 25 are titled *Head of EOW* (with various dates of execution from 1953 to 1973). There are six paintings titled *Head of J.Y.M.* (1970 to 1976) and two as *Portrait of J.Y.M. Seated* (both 1976). Compositionally, the works follow a very similar format, and there are clear similarities in some of the ways of rendering form. These do constitute something recognised as 'new' within the critical reception of his work, but the sense in which each individual work constitutes a new species or embodies any new knowledge is certainly less defendable. What we do have, though, is the sense of a practical problem of rendering or capturing that is carried through and supported by the written accounts of conversations with the artist. For each work, we get the sense that there may have been a question such as 'if I adjust the visual weight here, will that enable me to better capture X'? Later in the same conversation with Lampert,[17] we see Auerbach reflecting on the importance of understanding the context within which one might be attempting to contribute new perspectives:

> one's only hope of doing things that have a new presence, or at least a new accent, is to know what exists and to work one's way through it and to know that it is not necessary to do that thing.

So, in terms of the contribution to knowledge of how paintings might use form to convey specificity of form in space, we have some tangible elements that were exercised repetitively in the paintings of Auerbach. One could characterise the process as him repeating an experiment to see what result he might achieve, and the various accounts of his working process (in the recorded conversations with him and accounts by others) supports this proposition.

One way of thinking about experimentation through the repetition of motif is with reference to seriality, which was a significant approach particularly through the mid- and latter part of twentieth-century contemporary art but also present in earlier work such as Monet's haystack or waterlily paintings. Seriality has been manifested through both minimalist sculpture, such as Donald Judd, Sol LeWitt, and Dan Flavin, and through both conceptual and more expressionist work in painting. From the repeated drawings of the same motif by Matisse, through Jasper John's flag paintings, the visual persuasion of the order inherent in repetition of motif or format continues to stimulate artists and viewers. Hauser and Wirth's *Serialities* exhibition of 2017 in New York and Blain | Southern's *Concatation, Signature, Seriality, Painting* in London 2012 indicate the enduring interest in the use of seriality in contemporary and modern art. It is interesting for this discussion to note how Bochner opened his seminal paper, 'The Serial Attitude', with the claim that 'serial order is a method, not a style'.[18] He went on to distinguish the use of series as different versions of a basic theme, such as in the work of Willem de Kooning or Georgio Morandi, from that of 'serial attitude' as a 'concern with how order of a specific type is manifest'. The use of repetitive modules or logics had to meet three criteria to be considered properly serial. The modularity of Carl Andre and Andy Warhol or the logic basis for Jasper John's letter or number paintings were also seen as distinct.

The approach to the repetition of motif as discussed by Bochner does consider 'types of order as forms of thoughts',[19] and he provides an account of the practical underpinning permutations and sequences used in music, the isomorphic systems of linguistics, the row techniques of arithmetic progression, and projective geometry. While not necessarily apparent to the viewer, these systems of thinking employed in the generation of artworks give an indication of an approach to making where the artist might appear to be working within an experimental framework of 'what might happen if . . .?' In its most basic form, this could be a foundational premise for a practice-based research question.

Even though Bochner may have regarded Auerbach as a painter repeatedly using the same motif, both the notion of serial order as method and the small experiments on 'what might happen if . . .?' lead us back to the potential to ascribe a notion of experimentation to the procedure or system used in practice. What marks this thing or that as being a new thing in the field – the 'new accent', in Auerbach's words – is if a visual element is both observable and essentially measurable. So even though the claim to making a new contribution to the field through each work might be tenuous, we can see that each work is effectively addressing a question – 'what might happen if . . .?' – through the practice of making. And if these new insights were precisely communicated by some accompanying means and were validated, original, and contextualised, then we could reasonably call this practice-based research.

Before concluding the discussion of methods for contribution to knowledge through research, I want to consider further the changing nature of visual arts practice.

The changing landscape of what visual artists do: new materials and new practices for practice-based research

We noted earlier that the nature of what artists do seems to be changing, and this brings with it a need to check on our perceptions of what practice the practice-based researcher may be working with. Within the context of the visual arts, the practice-based researcher may be expected to justify any selection of methods from the available portfolio of the contemporary practitioner. Within contemporary visual arts practice, such methods now expand beyond the manipulation of stuff in the studio.

The picture of the several art worlds discernible in early twenty-first-century culture is complex. In the international art scene that is dominant in Western culture and now figures within both Middle Eastern and Far Eastern contexts, there is still a strong commercial art market where artistic novelty within recognisable parameters is what is expected of 'bank-able' artists. This is the market of the international art fairs and the major capital city museum show. The marketing machine of the commercial gallery needs a relatable and interesting story with intellectual credibility. Accounts of how seriously artworks are intertwined with contemporary thinking do help to reinforce the cultural capital with which works are imbued by connotation. When artistic practice has proved effective within this marketplace, the critic or curator or exhibition organiser writes or commissions the catalogue essay to place the artwork in the appropriate intellectual context to reinforce the role of the work within the cultural envelope.

We noted earlier the fact that the context for an artwork might be provided or explained by third-party gatekeepers, whether they are the gallery or the critic or the exhibition curator. Artists do also have a place in constructing such models of the context for their work, and the management of positioning a work to be seen against a distributed understanding of emergent constructs is as much a part of the method of the artist as is the specification of colour or the logistics of the photoshoot. This is arguably more about a strategic use of material other than the artwork itself and encompasses things like artistic affiliations, collectors, biography, and personality. We could frame this as a method of leveraging intellectual and social capital, perhaps, when considering artistic practices. But if we are considering practice-based research, we effectively need to expend our conceptions of methods to include this intellectual positioning, that colour specification, or that social event management as part of the arsenal of methods that may be employed in our practice-based research.

Moving into a new contemporary territory where the job of the artist is moving away from the producer of venerated objects does make things simpler when considering the role of research methods for practice-based research. New models of practice for the artist are becoming apparent as the forms and materials of art in the Western canon have shifted from that of the manipulation of stuff previously seen as the province of the 'artist', such as paint, clay, stone, etc., to the manipulation of any sort of stuff or things or space or people. The sorts of ideas about what artists do or stand for, however, are embodied within culture and society. Our dictionaries tell us that an artist is a 'person skilled in a practical art' (OED) where 'accomplished execution is informed by imagination', and that artists work 'with the dedication and attributes associated with an artist' (Concise Oxford) and have 'the qualities of imagination and taste required in art' (Chambers). While these definitions may have a reasonable degree of fit with the people self-identifying as artists, there are enough exceptions to the criteria to query the utility of the definition as providing a foundation for understanding what practices might inform a research method or research questions in the field.

Exceptions to the model of the artist as accomplished executioner in a practical art might currently include practitioners of relational aesthetics, or 'art and social cooperation'.[20] The skillset required to realise a piece within this arena might be closer to the project management skills for a live event. Digital technologies have also opened up the range of skills which might be required of artistic production, with coding and electronics becoming useful tools for enabling the construction of digital experiences or objects. The term chemist illustrates the difference between relatively stable and unstable disciplinary realities. There we have a definition that gives a clearer sense of fit when thinking about research method: 'A person who engages in the practice or study of chemistry . . .; a person who makes chemical investigations; an expert or specialist in chemistry' (notwithstanding the slight confusion in UK English of the chemist also being the outlet where drugs are dispensed). It is not such a clear position when thinking about the methods involved in the making of what might be encompassed by art.

Interestingly, a reflection on participation in an early participatory artwork,[21] Mark Dion's Chicago Urban Ecology Action Group,[22] comments on it having been 'archaic science':

> I always thought of success in terms of having participants be able to replicate Mark's system of archaic science as art practice, and with that comes a different understanding of knowledge production. . . . In the end it was about shifting perspective – about how we inhabit the world intellectually and ecologically.

We are now in a period where there is a slight dislocation between the understanding of what art is and does from the perspective of the players inside the field, from that of more general conceptions about art and artists in the wider population. The new open season on what artistic practice might entail brings with it a question about whether a notion of practice-based research continues to be necessary. As artistic practice can now seemingly refer to any activity intentionally undertaken as art, as long as one can convince some gatekeepers, is there any point in signalling any distinction as to the type of research being undertaken? Or, to put it another way, what is the key difference between artistic practice that seeks personal perceptions of innovation and artistic research that identifies questions, problems, or gaps in knowledge and uses a methodological approach to seek new knowledge for the field?

Conclusions

If we return to the five perspectives on recognising innovation or change in visual arts creative practice, we see that artists recognised as the leading practitioners in the field are no longer working exclusively with the materials and conventions popularly conceived as the tools of the artist. The focus for artists is changing with new generations, but also much previous material-based practice is enduring effectively in the marketplace. Along with a shift in materials and themes, we do have a strong continuum in practices. The small experiments of serial art practices and of the steps artists intentionally make to leverage their art-world position both have a place on the continuum of knowledge creation activities. If we accept that the 'what might happen if . . .?' in the studio setting is a question that prompts the realisation of something new to the field through its execution, we can legitimately incorporate the practical visual art-making activity as a research method. But we do this in recognition that the range of activities in visual art making also includes everything from project management to coding, thereby opening out the range of research questions that can be asked by artist-researchers.

In relation to some of the newer methods employed by artists, I reflect upon being interviewed by Samra Mayanja in December 2019 as part of her work. She was an artist-in-residence in an arts organisation, developing an artwork for exhibition. This artwork was to be an intervention taking place for a fixed duration in a building and was intended to give back 'something' to its occupants. The artist was forming the idea of what it might be from the discussions she had with various stakeholders. As such, the ideas and discussions were the medium through which the outcome might be realised. The method to be employed was inference and possibly emergence through those discussions and the thinking through of connections and repetitions of ideas, of both the artist and the stakeholders. So, if this is a new artistic practice, perhaps the portfolio of methods of practice might now extend to include the interview, the questionnaire, the survey, or the experiment? In short, the tools for enquiry get extended beyond the manipulation of stuff that has been the province of the artist for millennia. The implications of this for a conception of practice-based research are that the methods of research have opened up way beyond the physical art-making practices of the studio.

The implications for practice-based methods are that we need to think beyond the practices of the studio and think of the practice of the artist as encompassing other strategies and methods that could be put to use in a research context. In this way, our notion of what a practice-based method is must move beyond a notion that what artists do is constrained to the manual production of art works.

We have reached a place where we are seeing shifts in artistic practice that suggest an opening of the potential methods for practice-based research to include actions other than those of manual studio making, giving the prospect of replicability in method, albeit with unresolved implications for artistic intellectual property. We still have a slight impasse in terms of recognising artistic innovation and whether that is the sort of innovation that practice-based research aspires to – that would be an interesting study in itself. Similarly, while the notion of the artistic experiment appears foundational for both 'normal' and research-driven practice, we have not had space here to explore the bigger issue of the implications for artistic intellectual property.

There are two other foundational matters for a future research agenda. One is how to shape research questions in a context where the understanding or meaning of practical work by the creative visual artist is determined by a marketplace (whether a capitalist commodity trading market or a curatorial marketplace of ideas). The understanding or insights shared may well be other than what was intended by the artist-researcher.

Other questions emerging about what next for practice-based research in creative fields might include asking about the drivers to incorporate practice as a research method and whether the tension between knowingness and creative blocks is a truth or a fallacy. Detailed analysis of the visual components of artworks of any type could be done in new ways, using, for example, statistical methods of analysing two-dimensional visual arrays, such as box-counting methods.[23] If the practice is no longer bound by the idea of the artist manipulating stuff, what might be the data or evidence to investigate through the actions or encounters that occur as the art-making occurs? Digital media arts also flags another potential paradigm shift with the emergence of non-fungible tokens (NFTs) for born-digital artworks. Relating to many of the disciplines in this volume, incorporation of this new element into the landscape of practice-based research in our fields reminds us that, as researchers, we have to engage with the unknown.

At this interesting moment, we might also think about what we do together. This volume is sharing perspectives on what practice-based research might be or might not need to be. One of the lessons to be learnt of the explorations made within visual arts practice and research over the past 40 years is that the field is capable of much more than illustration or representation of innovation in other fields. Enquiry in the visual arts means that there is a developing capacity to reflect on or work through questions or understandings by visual means or analysis of visual material. This is more than complementary visualisation, rather an active part of cross-disciplinary collaboration. Future collaborators from outside the visual arts can help by asking us what we mean when we say the artist 'questions', 'interrogates', or 'explores' through their practice. We need to be asked for the evidence of our process or our claims or our findings, and together fields need to agree on evidential forms that have credibility across domains. We need also to be mindful of where the methods of artistic practices sit in relation to definitions of research by bodies like the OECD. They make a clear distinction in the Frascati manual between research 'for the arts, research on the arts and artistic expression',[24] echoing Christopher Frayling's notions of research for, into, or through art and design.[25] As such, the outcomes of practice can only be one element in our research methods.

The context of researchers and practitioners in the visual arts and the conventions of the social and economic worlds in which they live shape the focus of research questions and the means by which they might be explored. We see strong shared values and belief systems, inculcated through education and social norms. Expressivity and freedom from convention have often been associated with conceptions of the visual artist. Within the field, there is a fairly widespread understanding that the role of artists is to draw our attention to things, to make comment, to respond to the world we live in, while more commonly held views draw on romantic notions of genius and expressivity.

Notes

1 For the non-specialist, a good summary of what constitutes contemporary art at this point in time can be found in Wikipedia at https://en.wikipedia.org/wiki/Contemporary_art
2 Frayling (1993; as discussed in Mottram 2014).
3 OECD (2015, p. 44).
4 Csikszentmihalyi (1999, p. 315).
5 Savin-Baden and Howell Major (2013).
6 Csikszentmihalyi (1999).
7 Kuhn (1970).
8 Ibid., p. 91.
9 Chilvers (2004).
10 Marshall (2002).
11 Leduc (2019).
12 Richter (1982).
13 Kuhn (1970).
14 Popper (1972, p. 74).
15 Biggs and Karlsson (2011).
16 ACGB (1978).
17 Ibid., p. 22.
18 Bochner (1967, p. 28).
19 Ibid., p. 29.
20 Finkelpearl (2013).
21 Ibid., p. 89.
22 Dion (no date).
23 The box-counting method used by Richard Taylor in his analysis of the changing fractal dimensions of the work of Jackson Pollock.
24 OECD (2015, p. 64).
25 Frayling (1993).

Bibliography

Arts Council of Great Britain (ACGB). (1978). *Frank Auerbach*. London: ACGB.
Auerbach, F. (1978). Reinventing the World. In *The British Art Show 1984*. London: Ikon Gallery, 28 (originally quoted in 'Interview with Catherine Lampert'. *Frank Auerbach*: [catalogue of an exhibition held at the] Hayward Gallery, London, May 4–July 2, Fruit Market Gallery, Edinburgh, July 15–August 12. London: Arts Council of Great Britain).
Biggs, M., & Karlsson, H. (2011). *The Routledge Companion to Research in the Arts*. Abingdon: Routledge.
Blain | Southern. (2012). *Concatenation, Signature, Seriality, Painting*, September 7–October 5. www.blainsouthern.com/exhibitions/concatenation-signature-seriality-painting (accessed January 25, 2020).
Bochner, M. (1967). The Serial Attitude. *Artforum*, 6(4), 28–33, December. www.artforum.com/print/196710/the-serial-attitude-36677 (accessed January 25, 2020).
Chilvers, I. (Ed.). (2004). Fauvism. In *The Oxford Dictionary of Art*. Oxford: Oxford University Press, 3rd ed. www.oxfordreference.com/view/10.1093/acref/9780198604761.001.0001/acref-9780198604761-e-1238 (accessed March 29, 2021).

Csikszentmihalyi, M. (1999). Implications of a Systems Perspective. In R. J. Sternberg (Ed.), *Handbook of Creativity*. Cambridge: Cambridge University Press.

Dion, M. (n.d.). *The Chicago Urban Ecology Action Group*. https://distributedcreativity.typepad.com/on_collecting/files/Mark_Dion.pdf (accessed December 29, 2019).

Finkelpearl, T. (2013). . . . *What We Made: Conversations on Art and Social Cooperation*. Duke University Press. www.dukeupress.edu/What-We-Made/.

Frayling, C. (1993). Research in Art and Design. *Royal College of Art Research Papers*, 1(1).

Haseman, B., & Mafe, D. (2009). Acquiring Know-How: Research Training for Practice-led Researchers. In H. Smith & R. T. Dean (Eds.), *Practice-led Research, Research-led Practice in the Creative Arts*. Edinburgh: Edinburgh University Press.

Kuhn, T. (1970). *The Structure of Scientific Revolutions*. Chicago: University of Chicago Press, 2nd ed.

Leduc, M. (2019). Defining Contemporary Art: What the Kunstkompass Top 100 Lists Can Tell Us About Contemporary Art. *Journal of Visual Art Practice*, 18(3), 257–274. https://doi.org/10.1080/14702029.2019.1654204.

Marshall, R. (2002). An Interview with Simon Ford. *3am Magazine*. www.3ammagazine.com/litarchives/2002_may/interview_simon_ford.html (accessed March 29, 2021).

Mottram, J. (2014). Researching the PhD in Art and Design: What Is It and Why Do a PhD in Art and Design. In J. Elkins (Ed.), *Artists with PhDs: On the New Doctoral Degree in Studio Art*. Washington, DC: New Academia Publishing, 2nd ed.

OECD. (2015). Concepts and Definitions for Identifying R&D. In *Frascati Manual 2015: Guidelines for Collecting and Reporting Data on Research and Experimental Development*. Paris: OECD Publishing. https://doi.org/10.1787/9789264239012-4-en.

Popper, K. (1972). *Objective Knowledge: An Evolutionary Approach*. Oxford: Clarendon Press.

Richter, P. (1982). Modernism & After – 2. *Art Monthly (Archive: 1976–2005)*, 55, 5–8, April 1.

3.5

CRAFTING TEMPORALITY IN DESIGN

Introducing a designer-researcher approach through the creation of *Chronoscope*

Amy Yo Sue Chen and William Odom

Introduction and background

We now live in a world where digital technologies mediate many aspects of people's everyday lives, and this situation challenges design-led research as it struggles to adapt its methods to the speed of rapid technological change. The convergence of social, cloud, and mobile computing have made it easy for people to stay constantly connected and to create and share personal data at rates faster and scales larger than ever before. For example, social media services currently receive approximately 21,000 photo uploads per second and 657 billion photos annually.[1] These new technologies have enabled people to create vast digital archives that capture their personal history and life experiences, which can be valuable resources for connecting with others and reflecting on one's own life. Technological trends toward constant connectivity and the proliferation of personal data have opened many benefits. However, wide-ranging experiences of overload are emerging as people struggle to make sense of the masses of digital data they now create and receive. People are experiencing loss of control over the digital archives that capture their life experiences as they become oversaturated and fragmented.[2] Therefore, there is a clear need to develop new ways to support people in making sense of their ever-increasing archives.

Technologies are being designed without a clear sense of how people could develop longer-term experiences with them and the digital media archives that they produce. This threatens the ability for digital media and technology to be enduring, valued resources for meaningful activities, such as reminiscence, contemplation, and social connection.[3] These consequences are likely to become even more entangled and complex for future generations if current trends continue. As interaction design researchers, we cannot help but wonder: how can we advance design processes to enable digital artifacts to more appropriately support the meaningful activities of people's lives? How can technology be designed to become longer-term everyday resources that evolve over time? What kinds of interaction qualities and design decisions ought to be taken into account in design practice to engage with these issues?

In their original article on *slow technology*, Hallnäs and Redström[4] advocate creating technology that persists for long time periods in people's lives. At the heart of this design philosophy is the aim to encourage experiences of self-reflection as well as critical reflection on technology itself; and to investigate 'what it means to design a relationship with a computational thing that will last and develop over time'.[5] While slow technology offers promise for addressing digital overload, its conceptualization and translation into viable design strategies are still developing.

DOI: 10.4324/9780429324154-27

Research through Design (RtD) is an emerging interaction design research method that grounds theoretical investigations through the research-creation activity of design practice.[6] The highly polished and robust design artifacts produced through RtD each operate as exemplars of theoretical ideas and conceptual propositions. They also offer concrete ways to reveal new knowledge about how complex social issues, like digital overload and supporting longer-term human–technology relations, can be reframed and approached.

In parallel, recent work in the computing and interaction design communities has highlighted the need to design technologies that express alternative representations of personal data capable of enabling experiences that expand beyond 'an exclusive interest in performance, efficiency, and rational [self] analysis'.[7] Yet, the examples of demonstrating and unpacking how the slow technology design philosophy can be put into practice are sparse. Early cases such as Long Living Chair,[8] GoSlow,[9] Olly,[10] CrescendoMessage,[11] FamilyStories,[12] and Olo Radio[13] have begun to demonstrate how such rich and unique engagements with personal data can be supported through the creation of new design artifacts.

To bring together research across these two areas and to ground our own thinking in this space, we designed *Chronoscope* – a tangible photo viewer that embodies the lifetime of digital photos a person has accumulated over their lifetime. Inspired by prior research on designing for slowness,[14] key qualities of *Chronoscope*'s design include that it: takes time to understand; manifests change through time; and leverages different forms of time to prompt reflective experiences by manifesting their presence in everyday life. As Figure 3.5.1 shows, *Chronoscope* is a domestic technology that leverages temporal metadata embedded in digital photos as a resource to encourage more temporally diverse, rich, and open-ended experiences when re-visiting one's personal digital photo archive. Its scope-like form not only suggests rotation-based tangible interaction but also invites its users to view and contemplate the viewed phenomena in an intentional, inquisitive way.

Figure 3.5.1 Leveraging the metadata of each digital photo, *Chronoscope* is a tangible device that enables interactions through and across time in one's personal photo archive. The scope form and the monocular feature are designed to suggest its user a consciously focused viewing experience.

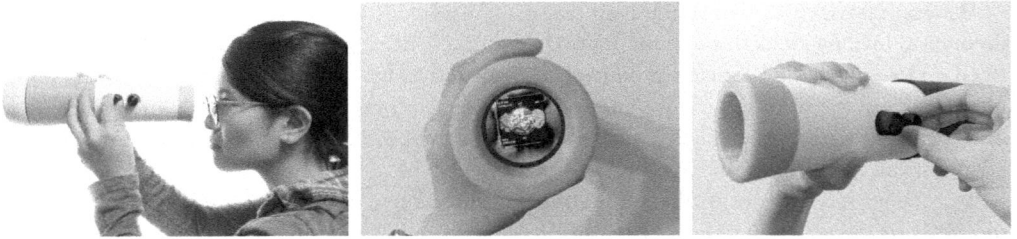

Figure 3.5.2 Left: The user manipulates a fully rotational black silicon surface (rotating clockwise moves 'forward' in time and rotating counter-clockwise goes deeper into the past). Middle: Peering into the turquoise eye piece through a magnified lens, the user views photos from his past. Right: The user manipulates a black metal knob that 'tunes' the granularity of photos that moved through in each rotation; the untouched knob toggles between timeframe modes.

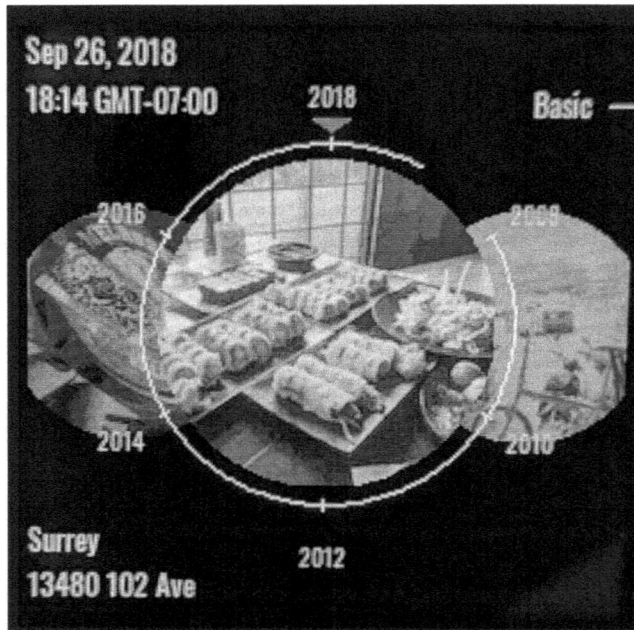

Figure 3.5.3 The *Chronoscope* UI visualizes the central photo's location in time and provides corresponding data around it.

When peering into *Chronoscope*, a single photo tied to the specific timestamp will be visible (see Figure 3.5.3). All the photos are reorganized in a chronological timeline. There are three rotational controls to navigate through the photo archive (see Figure 3.5.2). The primary rotating wheel controls the *viewing directions* in one timeline (clockwise to move forward and counter-clockwise to move backward). By rotating either direction, the user sees each photo in relation to a wide spectrum of other photos in the archive. The other two knobs on the side decide the *timeframe modes* and *viewing granularity*. For timeframe modes, users are allowed to switch between three timelines – *linear*, *date*, and *time*; each organizes photos based on different parts of the timestamp. As for viewing granularity, we use a rotational slider as a build-in support for 'tuning' the number of photos (from

1 to 300) that one could move through by each wheel rotation. More details could be found in our full paper, published in DIS'19.[15]

As a slow tech design practice, *Chronoscope* works slowly through offering alternative angles for people to curiously explore and reflect on their digital photo archives through and across time. Through designing in support for the presence of time and adding navigational controls for temporal exploration, several interaction qualities emerged during the design process and field testing of *Chronoscope*. Next, we draw on the design case of *Chronoscope* as a design example to introduce and reflect on our designer-research approach to practice-based research.

The designer-researcher approach

Over the past several years, we have adopted what we call a designer-researcher approach to making and reflecting on highly finished design artifacts as a form of practice-based research in and of itself. *Chronoscope* represents one of our latest projects in this line of work. The designer-researcher approach often involves a small but multi-disciplinary team that is reflexively focused on the experimental and novel outcomes of the design process that are critically and reflectively arrived at through design practice. To give a bigger picture, we introduce four main points that inspire and situate the position of our approach:

- research through design
- autobiographical design
- reflective conversation with materials
- research product

With these concepts in mind, we attend to key details in the creation of *Chronoscope* by oscillating between higher-level conceptual ideas and practice-based design decisions. In the following sections, we share with you a brief description of each point and how they specifically motivated the direction and development of our design process.

Research through design

We follow a research through design (RtD) method that directs interaction design researchers to engage 'wicked problems' through a real-world artifact to understand how the world could change from its current state to a preferred state.[16] Through the production of a design artifact, this approach creates the context for a collaborative interdisciplinary research environment that could involve (but is not limited to) designers, engineers, anthropologists, and computer scientists.

Most importantly, there are four lenses to evaluate a RtD approach from the original paper:[17]

- *process* that could be reproduced with rigor and rationale
- *invention* built on concepts that could potentially answer the research questions
- *relevance* of real-world knowledge and a preferred state to achieve via artifacts
- *extensibility* that aspires the research insights to be applied in the future

Due to the qualitative nature of our research questions and the increased complexity in designing digital technologies, we decided to adopt RtD in the *Chronoscope* project. The four lenses were carefully applied throughout the entire process, from our overarching research goal to every nitty-gritty design decision.

First, we make sure our research *process* is reproduceable. The term 'reproduceable' does not necessarily mean to produce the exact same results but to report rigorous details with clear rationale.

The gist is to clarify why the process proceeds with certain decisions in each step and how it arrives at the final insights. For example, we started the *Chronoscope* idea based on a rotational photo viewing control with an analogy of clock (e.g., clockwise to move forward in time). Therefore, we settled on its scope-like form because it suggests rotation-based tangible interaction and invites people to view their digital photos in an intentional, inquisitive way. With this decision, we moved on to polish its color and texture to ensure that every little design feature meets our research need. By reasoning details, the process can be examined, learned, and applied by other researchers.

Through another lens, *invention*, we conducted extensive literature reviews in relation to digital possessions, personal informatics, photographic practices, and memory studies to situate our subject in an academic standpoint. *Chronoscope* is deeply researched on the characteristics of data that have challenged people's original sense of owning physical objects,[18] such as being intangible and pervasive in its nature.[19] Without proper tools and structures, people have difficulty making sense of, organizing, and using photos.[20] Thus, we looked into how reflective and curious experience could be supported by exploring alternative ideas such as the 'circulation' concept between personal memories and memory objects.[21] Our design attitude for this exploration was also influenced by theoretical approaches such as ludic design,[22] reflective design,[23] and, especially, slow technology.[24] Building on these collective works, we invented *Chronoscope* to demonstrate its potential for advancing from the current state of the art.

As mentioned in the introduction, this project was initiated by both a real-world problem regarding accumulating digital photos and a goal of exploring innovative ways to build people's long-term intimacy with their photo archives. From this, *relevance* is achieved as *Chronoscope* was designed to fit in a domestic context and created as a vehicle to embody temporal relations among digital photos in its visual representation and tangible design. Without an actual artifact to interact with, it is exceedingly challenging for people to make sense of how big their digital photo archive is and how their photos could relate to each other in a much more macro view. Hence, a *preferred state* was explored through the *Chronoscope* design to enable a potentially more curious, reflective, and engaging experience with vast digital photo archives.

In terms of *extensibility*, the insights derived from *Chronoscope* are applicable to future interaction design research. Our inquiry led to *interaction qualities* that allow people to explore their digital possessions with ease and pleasure and grow more lasting and interactive relationships with them. The *temporal design features* could also inspire more application and development of any digital material that has a timestamp embedded in its file format. Overall, these insights were provided with an aim to open up new possibilities in future research.

In sum, our designer-research approach relies heavily on RtD as a research method that uses design artifacts as instruments to ask questions and test ideas. The four RtD lenses directly shaped the ways we connected *Chronoscope* to a real-world problem and communicated its value to the broader research design research community. Next, we unpack details of how our *process* was guided through first-person experiences shared among the design research team.

Autobiographical design

While our practice-led designer-researcher approach shares some similarities with first-person and autobiographical design (AD) approaches,[25] there are still some notable differences. Neustaeder and Sengers characterize autobiographical design as 'design research drawing on extensive, genuine usage by those creating or building a system'.[26] It focuses on the genuine needs of researchers, which embody their own experiences in the system design and concept exploration through a cycle of *building*, *learning*, *evaluating*, and *iterating* the design. While facing a challenge of reporting with objectivity, AD has benefits that include shortening the feedback cycle, anticipating and solving key issues before testing the design with research participants in the field, and uncovering the nuanced understandings of one's lived experience with the design artifact. AD is conducted by a single researcher, although others may also be present in the research (e.g., such as the researcher's family members).

In contrast with AD, our designer-researcher approach involves four special focuses:

- *originality* – initiating an idea that emerged from personal expertise and research trajectories
- *diversity* – forming a small but interdisciplinary team to ensure the quality of design artifacts
- *provocativeness* – pursuing both robustness and creativity reflexively through crafting and living with design artifacts
- *unity* – having a unified narrative voice to report our research outcomes

While AD emphasizes the genuine usage and exploration cycle in the long-term testing part, our approach values it and the technology creation part equally. Therefore, new ideas usually originate from design researchers' expertise in certain design materials and relate heavily to their own research trajectories. With a background in computer science and a special focus on digital photos, Amy Yo Sue Chen (co-author) joined Everyday Design Studio in 2017 as a PhD student supervised by William Odom (co-author), who co-directs the academic design studio and has investigated slow technology as a lens to design new kinds of experiences with photographs, music, and other digital media for several years. Motivated by the Photobox,[27] CrescendoMessage,[28] and Olo *Radio* prototype,[29] we (Chen and Odom) started the discussion of the *Chronoscope* project with an idea of having a handheld, telescope-like form as an alternative way to interact with digital photo collections. Without our research backgrounds, expertise, and positionality of accumulated works as key factors, we would not have initiated this difficult RtD project; however, because of it we felt positively that it would produce new knowledge that could only be explored through new design.

However, producing a highly resolved design artifact is not easy. In the *Chronoscope* project, we needed to involve other investigators and research assistants with expertise in electronics prototyping, form design, and digital fabrication to create the artifacts. By co-leading a small but interdisciplinary team, we were able to circulate thoughts and exchange *diverse* perspectives to overcome challenges. We see our designer-researcher approach as *reflexive* and with the goal of delivering first-hand insights through ongoing individual and collective design practice, group critiques, material explorations, and experiences of living with prototypes collectively in our studio as well as individually in our own homes. In addition to AD's core concept of trouble-shooting and iterating systems in their personal use and living with the design artifacts, our approach pursues unexpected but desired design features for longer-term lived experience that could not be imagined or foreseen without design researchers' keen observation and criticality. The sensitivity used to explore the design comes from both deep understanding of the design itself by the researchers and the intuition to critique and make changes as designers.

While embracing the benefits of having a wider range of expertise and opinions, a designer-researcher approach has a particular challenge of reporting research outcomes in a *unified* narrative voice across our design team. Hence, our approach requires one or two project leaders to constantly synthesize perspectives and distill insights from the various creative understandings among the team members. The attachment between researchers and the system design is formed through not only the individual practice-based creation process but also numerous collaborative discussions. We see this challenge as the most important part as it shapes how our first-person approach is interpreted and presented to the research communities.

Building on the first person-oriented nature of AD, the designer-researcher approach offers a related but different form of practice-based research. Although the designer-research approach requires extra work to unify voices across the design team, we find this most challenging part also to be the most rewarding. This reflexive process itself requires project leaders to reflect on their original goal, which can catalyze additional insights and research opportunities. Next, we talk about how *invention* comes into being in details.

Reflective conversation with materials

To begin our design research inquiry of *Chronoscope*, we explored potential design materials that pertain specifically to the temporal metadata in digital photos. We initiate our observation by virtue of trust in a reflective, practice-based process. We believe:

> the designer-researcher approach can contribute a highly insightful, first-hand, and reflexive view of practices of making design artifacts in relation to higher-level concepts framing key decisions in the design process and in light of attendant materials, tools, methods, and competencies.[30]

We see this approach as highly aligned with Schön and Bennet's characterization of design practice as *a reflective conversation with materials*.[31] A designer-researcher approach to practice-based research heavily emphasizes a designer's knowing-in-action that involves sensory and bodily understanding in relation to the team's evolving understanding of materials. This is a recurring process where the materials 'talk back' through a back-and-forth process that progressively leads to a refined understanding of the overall design and, ultimately, the final resolved design artifact. This approach gives prominence to first-hand insights that emerge through iteratively working with materials to ground conceptual ideas through the creation of new things: 'a process of moving from the universal, general and particular to the ultimate particular – the specific design'.[32]

In the *Chronoscope* project, the 'material' we worked with came in the form of digital photos and their attendant metadata, which is a series of information encoded into each digital photo at the moment the file is created. The digital photo metadata material includes information such as the camera specifications, geolocation, image size and compression style, and the timestamp capturing when the photo was created. We were most interested in 'speaking to' the timestamp data because it stands as a potential medium that provokes a tracing and reasoning experience of specific memories.

Timestamp is numeric and sequential in its nature, and the way to make use of this type of data requires a specified format (e.g., YYYY/MM/DD). From observing how various formats frame our cognitive understanding to systematically absorb and compare those photo content in a list view, we arrived at a linear 'timeline' design that visualizes clear *sequential events*.[33] We rapidly programmed an interactive processing[34] script and an iOS mobile application as prototypes in order to automatically capture, structure, and organize digital photos chronologically. By giving prominence to the presence of temporality hidden in every digital photo as an anchor point to make sense of the sorting order, we were enabled to actually see the possibly consequential connections between photos and to explore those connections through moving the photo collection backward or forward in time. This back-and-forth visual interaction provides a clear pathway for each specific photo memory to 'talk back' to us not just about their relative location in time but also their relations to other photo stories in a macro level.

However, the mere chronological timeline design felt somewhat limited and underwhelming. On one hand, the linear photo collection on the prototype has ends on both sides and therefore does not provide a sense of continuity to encourage exploration. On the other hand, those photo memories flow in an order of how we exactly experienced and remembered them. Crucially, the processing prototype enabled us to move quickly across a large number of photos in an archive and simultaneously saw timestamp information separated by time of day and date. The separation of information not only prompted us to stimulate recollection of past experiences peripheral to the central photo but also triggered our imagination of having non-chronological 'timeframe modes' that reorganize photos *across* the archive based on existing meaningful temporal patterns. Built upon prior works that have discussed how clock time[35] and digital time[36] have reframed the personal and collective rhythms of everyday life, we brainstormed in designing non-chronological ways of photo viewing that offer multilayer and cyclical perspectives (based on date or time of day, irrespective of year). Our interest in this notably wider spectrum of potential interactions is in part inspired by the concept of *ecphoria*,[37] which refers to the experience of recalling a fuzzy or entirely forgotten memory when prompted by sensory input – in our case, digital photos from one's past.

In a nutshell, the gist of having a reflective conversation with materials includes three parts:

- understanding the nature of design materials
- exploring various forms and expressions made of the materials
- observing the interaction dynamics with appropriate tools

We arrived at the final design features of *Chronoscope* that may not have been achieved without the support of powerful computing tools. These contributed to our dialogue with the materials and supported the potential of experience design into realms of reflection and curiosity through and across personal memories. Next, we discuss the benefits of shaping a design artifact into a research product.

Research product

In our design research studio, we polish our design artifacts to become *research products*[38] that fully support the final and actual deployment in the complex real-world setting to ask research questions. Research products are intended to be lived with over longer-time periods and achieve a high quality of fit in, and among, things in people's everyday environments. Thus, a key part of this process involves different design team members living with various prototype versions of the design artifacts we are making to fine-tune qualities of use (e.g., the pacing or rhythm of a slowly changing system), exploring living with different forms and materials, and field testing for technical robustness.

To put it concisely, there are two critical steps in the framework of *research product*:

- creating a highly finished research product that functions and lasts in an everyday context
- living with the research product over a long period of time

In order to inquire how people exactly experience *Chronoscope* and their relationship with it over time, a tangible version was created with full functionality for the everyday rather than laboratory/ studio context. We loaded programming code to a microcontroller board that controls electronic components such as a tiny display, three rotational actuators, and a light-weight battery. To store them in a small size form that best supports handholding, we meticulously designed the interior space of the scope and outsourced the 3D printing to a more professional service provider to achieve the highest resolution. This design process required numerous group discussions but also very personal moments of cultivating the sensitivity of feeling how tangible rotation manipulates the reaction of photos showing on the tiny display embedded in the scope. Through extensive trouble-shooting and fine-tuning in the iterations, we arrive at a robust and highly finished version of *Chronoscope*.

Before we deployed it to people's houses, our design team members lived with the product over a long period of time and made adjustments to counteract the frictions we encountered. The first friction was related to the sheer size of the photo archive. For instance, when a user has 20,000 photos and she aims to navigate to a specific time of her photo collection, it would take her about 2.77 hours to get there since each rotation moves through the photo archive by only one photo as a unit (and it takes about 0.5 second per rotation); this led to a sensation of being 'stuck in time'. This issue would very likely complicate people's ability to form long-term relations to *Chronoscope* and hinder our aim of making it a research product. This friction triggered our decision to include blending as a support for 'tuning' the number of photos (or *granularity*) that one moved through for each rotation. With this support, people could move through their photo archives in very slow, precise, and considered ways (i.e., one photo per rotation) to encounter a set of photos that triggered deep reflection or examination or, equally, quickly move across vast numbers of photos without an excessive number of rotations, while retaining a subtle awareness of what had been passed over (i.e., 100 photos per rotation). In order to solve the friction, the added 'tuning' feature opens up more freedom and flexibility for the user to move through photos from minutes in a day to years of one's life.

As the research product was designed to be a slow technology (see earlier), it is very likely that the artifact would be experienced differently as time passes by, and as researchers we need to negotiate this factor in order to make sense of micro-interaction. As a slow technology design takes time to understand and gradually evolved over time, *Chronoscope* introduced other challenges of the granularity control. If the upper threshold was too high, then a user could easily become 'lost in time' as they navigated a large number of photos in one turn (i.e., effectively flashing ahead into the future or back into the past without a clear point of reference). Through an iterative process, we determined that setting upper and lower boundaries of the granularity helped to mitigate these design issues in support of our higher-level goal of manifesting different forms of time in support of ongoing reflective experiences. However, such boundaries need to be dynamically alterable and able to evolve with the photo archive as it continues to grow over time. Our design team went through several rounds of prototyping with different types of interactions, forms, and levels of fidelity of the interface to arrive at thresholds that seem suitable to the modern average size of digital archives.

While the four points in our designer-researcher approach provide a foundation for making high-level decisions across all of our research projects, each project never quite follows the same pathway to form a unified narrative voice from synthesizing perspectives and experiences across the design team.[39] Our overarching goal in the *Chronoscope* project was to contribute concrete insights into unpacking how diverse temporal interactions might be designed with personal data through the form of an everyday artifact that is intended to be lived with over time. In this, we have emphasized where its key theoretical, methodological, and practical challenges emerged and our design moves to resolve them.

Discussion and conclusion

In this chapter, our piecemeal, practice-based designer-research approach enabled us to create a design artifact that extended core approaches to designing slow technologies by building in a high degree of end user control, while retaining design qualities that are closely tied to the original conceptual vision of slow technology. The case of *Chronoscope* illustrates that future practice-based approaches to crafting slowness and temporality into new artifacts may require added time for reflection and adaption that may be counter to the often frenetic, time-constrained norms of contemporary interaction design practice. This methodological insight is revealing about the need for design practice to evolve and is also reflective of the original vision of slow technology:

> As computers are increasingly woven into the fabric of everyday life, interaction design may have to change – from creating only fast and efficient tools to be used during a limited time in specific situations, to creating technology that surrounds us and therefore is a part of our activities for long periods of time.[40]

We have described and reflected on how slow artifacts can provide alternatives to how we might conceptualize supporting longer-term relations to our everyday technologies and personal data – and how such inquiries can be grounded in design practice. However, without our designer-researcher approach, the discoveries and insights would not have arisen.

Ultimately, we describe our designer-researcher approach and offer a reflection on how its four critical points – research through design, autobiographical design, reflective conversation with materials, and research product – could be exemplified through unpacking the process of crafting a design case. From these four points as keys to developing the practice-based insights, we would like to offer four takeaways of what practice-based research usually involves:

- shaping and manifesting the research questions and proposals into practices
- first-person perspectives both as an individual and a team
- deep reflection on the resources and people's perception of them
- adapting the practices to real-world contexts and negotiating them when friction arises

In conclusion, an additional goal here is to help readers to take a step toward understanding how our designer-researcher approach could potentially be applied and explored in their own fields. We see opportunities for bringing together philosophers and social scientists who explore differing theories of temporality with practitioners who are skilled in manipulating temporal media (e.g., musicians, composers, poets, new media artists, etc.) to collectively develop new ways of grounding theoretical concepts related to time through creative practice. Ultimately, we hope our research methods can support future practice-based initiatives that are aimed at investigating the complex and evolving subject of human relations with technology over time.

Acknowledgements

The Natural Sciences and Engineering Research Council of Canada (NSERC), Social Science and Humanities Research Council of Canada (SSHRC), and Canada Foundation for Innovation (CFI) supported this research. We also thank Ce Zhong and Henry Lin for their assistance in earlier parts of this project.

Notes

1 Meeker (2016).
2 Odom et al. (2014), Sellen and Whittaker (2010).
3 Whittaker et al. (2010).
4 Hallnäs and Redström (2001).
5 Mazé and Redström (2005, p. 11).
6 Frayling (1994), Gaver (2012), Redström (2017).
7 Elsden et al. (2016, p. 48).
8 Pschetz and Banks (2013).
9 Cheng et al. (2011).
10 Odom et al. (2018, 2019a).
11 Tsai et al. (2015).
12 Heshmat et al. (2020).
13 Odom and Duel (2018), Odom et al. (2020).
14 Hallnäs and Redström (n 4); Vallgarda et al. (2015), Odom et al. (2014).
15 Chen et al. (2019).
16 Zimmerman et al. (2007).
17 Ibid., pp. 499–500.
18 Belk (2013), Cushing (2011, 2013).
19 Odom et al. (2014, p. 2).
20 Kirk et al. (2006), Broekhuijsen et al. (2017).
21 Van House and Churchill (2008).

22 Gaver et al. (2004, 2011).
23 Sengers et al. (2005).
24 Hallnäs and Redström (2001).
25 Neustaedter and Sengers (2012), Rosner and Taylor (2011).
26 Neustaedter and Sengers (2012., p. 154).
27 Odom et al (2012, 2014).
28 Tsai et al. (2015), Chen (2015).
29 Odom and Duel (2018).
30 Odom et al. (2019b).
31 Schön and Bennett (1996).
32 Nelson and Stolterman (2012, p. 33).
33 Lundgren (2013).
34 Processing. [online] Available at: https://processing.org.
35 Martineau (2017), Lindley (2015).
36 Lindley (2015, p. 29), Rushkoff (2013).
37 Tulving (1985), Tulving et al. (1983).
38 Odom et al. (2016).
39 Odom et al. (2019b).
40 Hallnäs and Redström (2001).

Bibliography

Belk, R. (2013). Extended Self in a Digital World. *Journal of Consumer Research*, 40, 477.

Broekhuijsen, M., van den Hoven, E., & Markopoulos, P. (2017). From PhotoWork to PhotoUse: Exploring Personal Digital Photo Activities. *Behaviour & Information Technology*, 36, 754.

Chen, A. Y. S. (2015). *CrescendoMessage: Articulating Anticipation in Slow Messaging*. Taiwan: National Taiwan University of Science and Technology.

Chen, A. J. S. et al (2019). Chronoscope: Designing Temporally Diverse Interactions with Personal Digital Photo Collections. *Proceedings of the 2019 on Designing Interactive Systems Conference*. ACM. http://doi.acm.org/10.1145/3322276.3322301 (accessed June 28, 2021).

Cheng, J., et al. (2011). GoSlow: Designing for Slowness, Reflection and Solitude. *CHI'11 Extended Abstracts on Human Factors in Computing Systems*. ACM. http://doi.acm.org/10.1145/1979742.1979622 (accessed December 7, 2020).

Cushing, A. (2011). Self Extension and the Desire to Preserve Digital Possessions. *Proceedings of the American Society for Information Science and Technology*, 48(1).

Cushing, A. (2013). 'It's Stuff That Speaks to Me': Exploring the Characteristics of Digital Possessions. *Journal of the American Society for Information Science and Technology*, 64, 1723.

Elsden, C. et al. (2016). Fitter, Happier, More Productive: What to Ask of a Data-Driven Life. *Interactions*, 23, 45.

Frayling, C. (1994). Research in Art and Design. *Royal College of Art Research Papers*, 1(1). http://researchonline.rca.ac.uk/384/.

Gaver, W. (2012). What Should We Expect from Research Through Design? *Proceedings of the SIGCHI Conference on Human Factors in Computing Systems*. ACM. http://doi.acm.org/10.1145/2207676.2208538 (accessed March 31, 2021).

Gaver, W., et al. (2004). The Drift Table: Designing for Ludic Engagement. *CHI'04 Extended Abstracts on Human Factors in Computing Systems*. ACM. http://doi.acm.org/10.1145/985921.985947 (accessed April 24, 2021).

Gaver, W., et al. (2011). The Photostroller: Supporting Diverse Care Home Residents in Engaging with the World. *Proceedings of the SIGCHI Conference on Human Factors in Computing Systems*. ACM. http://doi.acm.org/10.1145/1978942.1979198 (accessed April 11, 2021).

Hallnäs, L., & Redström, J. (2001). Slow Technology – Designing for Reflection. *Personal Ubiquitous Computing*, 5, 201.

Heshmat, Y., et al. (2020). FamilyStories: Asynchronous Audio Storytelling Over Distance. *Conference Companion Publication of the 2020 on Computer Supported Cooperative Work and Social Computing*. ACM. http://doi.org/10.1145/3406865.3418563 (accessed November 16, 2020).

Kirk, D., et al. (2006). Understanding Photowork. *Proceedings of the SIGCHI Conference on Human Factors in Computing Systems*. ACM. http://doi.acm.org/10.1145/1124772.1124885 (accessed January 15, 2021).

Lindley, S. (2015). Making Time. *Proceedings of the 18th ACM Conference on Computer Supported Cooperative Work & Social Computing*. ACM. http://doi.acm.org/10.1145/2675133.2675157 (accessed April 24, 2021).

Lundgren, S. (2013). Toying with Time: Considering Temporal Themes in Interactive Artifacts. *Proceedings of the SIGCHI Conference on Human Factors in Computing Systems.* ACM. http://doi.acm.org/10.1145/2470654.2466217 (accessed November 27, 2020).

Martineau, J. (2017). Making Sense of the History of Clock-Time, Reflections on Glennie and Thrift's Shaping the Day. *Time & Society,* 26, 305.

Mazé, R., & Redström, J. (2005). Form and the Computational Object. *Digital Creativity,* 16(7).

Meeker, M. (2016). *Internet Trends Report (CODE Conference 2016, Rancho Palos Verdes, California, United States.* www.vox.com/2018/5/30/17385116/mary-meeker-slides-internet-trends-code-conference-2018 (accessed January 22, 2020).

Neustaedter, C., & Sengers, P. (2012). Autobiographical Design in HCI Research: Designing and Learning Through Use-It-Yourself. *Proceedings of the Designing Interactive Systems Conference.* ACM. http://doi.org/10.1145/2317956.2318034 (accessed January 22, 2021).

Odom, W., & Duel, T. (2018). On the Design of OLO Radio: Investigating Metadata as a Design Material. *Proceedings of the 2018 CHI Conference on Human Factors in Computing Systems.* ACM. http://doi.acm.org/10.1145/3173574.3173678 (accessed October 15, 2020).

Odom, W., Zimmerman, J., & Forlizzi, J. (2014). Placelessness, Spacelessness, and Formlessness: Experiential Qualities of Virtual Possessions. *Proceedings of the 2014 Conference on Designing Interactive Systems.* ACM. http://doi.acm.org/10.1145/2598510.2598577 (accessed February 24, 2021).

Odom, W., et al (2012). Photobox: On the Design of a Slow Technology. *Proceedings of the Designing Interactive Systems Conference.* ACM. http://doi.acm.org/10.1145/2317956.2318055 (accessed September 14, 2020).

Odom, W., et al. (2014). Designing for Slowness, Anticipation and Re-Visitation: A Long Term Field Study of the Photobox. *Proceedings of the SIGCHI Conference on Human Factors in Computing Systems.* ACM. http://doi.acm.org/10.1145/2556288.2557178 (accessed December 6, 2020).

Odom, W., et al. (2016). From Research Prototype to Research Product. *Proceedings of the 2016 CHI Conference on Human Factors in Computing Systems.* ACM. http://doi.acm.org/10.1145/2858036.2858447 (accessed March 3, 2021).

Odom, W., et al. (2018). Attending to Slowness and Temporality with Olly and Slow Game: A Design Inquiry into Supporting Longer-Term Relations with Everyday Computational Objects. *Proceedings of the 2018 CHI Conference on Human Factors in Computing Systems.* ACM. http://doi.acm.org/10.1145/3173574.3173651 (accessed May 24, 2021).

Odom, W., et al (2019a). Investigating Slowness as a Frame to Design Longer-Term Experiences with Personal Data: A Field Study of Olly. *Proceedings of the 2019 CHI Conference on Human Factors in Computing Systems.* ACM. http://doi.acm.org/10.1145/3290605.3300264 (accessed June 2, 2021).

Odom, W., et al. (2019b). *Reflective Knowledge Production Through a Designer-Researcher Approach.* Workshop Proceedings of First-Person Research in HCI, San Diego.

Odom, W., et al. (2020). Exploring the Reflective Potentialities of Personal Data with Different Temporal Modalities: A Field Study of Olo Radio. *Proceedings of the 2020 ACM Designing Interactive Systems Conference.* ACM. http://doi.org/10.1145/3357236.3395438 (accessed August 12, 2020).

Pschetz, L., & Banks, R. (2013). Long Living Chair. *CHI'13 Extended Abstracts on Human Factors in Computing Systems.* ACM. http://doi.org/10.1145/2468356.2479590 (accessed June 16, 2021).

Redström, J. (2017). *Making Design Theory.* Cambridge: MIT Press.

Rosner, D., & Taylor, A. (2011). Antiquarian Answers: Book Restoration as a Resource for Design. *Proceedings of the SIGCHI Conference on Human Factors in Computing Systems.* ACM. http://doi.org/10.1145/1978942.1979332 (accessed January 22, 2021).

Rushkoff, D. (2013). *Present Shock: When Everything Happens Now.* New York: Penguin.

Schön, D., & Bennett, J. (1996). Reflective Conversation with Materials. *Bringing Design to Software.* ACM. http://doi.org/10.1145/229868.230044 (accessed January 22, 2020).

Sellen, A. J., & Whittaker, S. (2010). Beyond Total Capture: A Constructive Critique of Lifelogging. *Communications of the ACM,* 53, 70.

Sengers, P., et al. (2005). Reflective Design. *Proceedings of the 4th Decennial Conference on Critical Computing: Between Sense and Sensibility.* ACM. http://doi.acm.org/10.1145/1094562.1094569 (accessed November 5, 2019).

Tsai, W. C. et al. (2015). CrescendoMessage: Interacting with Slow Messaging. *Proceedings of the 2015 International Association of Societies of Design Research Conference (IASDR'15),* 2078–2095.

Tulving, E. (1985). Memory and Consciousness. *Canadian Psychology/Psychologie Canadienne,* 86(1).

Tulving, E. et al. (1983). Ecphoric Processes in Episodic Memory. *Philosophical Transactions of the Royal Society of London B, Biological Sciences,* 302, 361.

Vallgarda, A., et al (2015). Temporal Form in Interaction Design. *International Journal of Design,* 9.

Van House, N., & Churchill, E. F. (2008). Technologies of Memory: Key Issues and Critical Perspectives. *Memory Studies*, 1, 295.

Whittaker, S., Bergman, O., & Clough, P. (2010). Easy on That Trigger Dad: A Study of Long Term Family Photo Retrieval. *Personal Ubiquitous Computing*, 14, 31.

Zimmerman, J., Forlizzi, J., & Evenson, S. (2007). Research Through Design as a Method for Interaction Design Research in HCI. *Proceedings of the SIGCHI Conference on Human Factors in Computing Systems*. ACM. http://doi.acm.org/10.1145/1240624.1240704 (accessed November 28, 2020).

3.6

THINKING TOGETHER THROUGH PRACTICE AND RESEARCH

Collaborations across living and non-living systems

Lucy HG Solomon and Cesar Baio (AKA Cesar & Lois)

Introduction

Formed in 2017 by Lucy HG Solomon (California) and Cesar Baio (São Paulo), our art collective *Cesar & Lois* probes societal relationships to nature by exploring the potential of new intersections between sociotechnical and biological systems.[2] We strive to think across disciplines and species, and in our artworks, we layer technological, social, and biological interactions. Our collective's work reflects on the societal structures that perpetuate inequities and threaten environments, and we strive to learn from living systems sourced in nature as well as from ancestral traditions that foster inter-species relationships. By inserting the logic of organisms like fungi and slime moulds into sociotechnical systems (e.g. slime moulds that tweet; fungi that grow on philosophy texts), we invite viewers to reflect on various methods of knowledge production. Through the unconventional layering of microbiological growth and human logic, our artworks suggest alternate logic models and embrace multidisciplinary research and practice. When we insert microbiological networks onto maps, texts, and communications systems, and when we as artists enter laboratory spaces, our artmaking pushes against disciplinary divisions and moves beyond the domain of art. We hope in the process to gain new insights, and in our particular practice, to learn from the wisdom of nature's growth algorithms.

We build *bhiobrid*[3] intelligent artworks by crossing different organisms, such as bacteria, fungi, and protists, with technological elements, including artificial intelligence, data mining, and social networks (see example in Figure 3.6.1). *Cesar & Lois* often interfaces and collaborates with philosophers, theoreticians, biologists, engineers, and students around the world. Such collaborations are important inputs as they can spur new thinking. Through interdisciplinary artmaking and research, we explore a range of scientific and theoretical concepts, which we classify as *broadening*.[4] Compared with strict methodologies which can constrain thinking to certain frameworks, this broadening in our artmaking, which you might think of as placing ideas in new contexts, can generate new insights. As part of this art making, there is a constant process of reflecting and writing during which we think about our own practices and work in relation to different contexts, and ultimately these texts feed back into our creative process and generate new works. In this way, more free-form research, crossing a range of methods without adherence to a single discipline, becomes a way of generating projects that relay knowledge about the world: a kind of knowledge that science or philosophy or art

 DOI: 10.4324/9780429324154-28

Figure 3.6.1   Cesar & Lois, 2019–2020, *Degenerative Cultures: Floresta Amazônica in Edital CoMciência – Ocupação em Arte, Ciência e Tecnologia*, MM Gerdau, Belo Horizonte, Brazil.

Source: Cesar & Lois.

may not be able to reach strictly on their own terms. What excites us about interdisciplinary collaboration is how the unorthodox mixing of methods can lead to the production of new knowledge. In the specific context of our works, our interdisciplinary and multispecies approach inserts the concept of thinking together into our art, a concept that originated in dialogue with people from many areas of research and operational collaborations among human and nonhuman intelligences. Through that dialogue and research across disciplines, we developed an artwork that also proposes new ways of thinking about networks. As this chapter intends to illustrate, this diversity of processes, reflective writing, and resulting artworks together form the operational system of our research and have the potential to lead to new knowledge, for the collaborators, for the fields of study, and for viewers.

Graeme Sullivan[5] asserts that the third domain of the visual arts research framework involves the repositioning of images and ideas from one context to another. When placed in a new context, familiar images and ideas can point to unforeseen insights. This makes it possible to challenge deeply embedded thinking and to trigger new ideas. Our collective *Cesar & Lois* crosses disciplines in order to learn from different areas and to move protocols, procedures, and concepts from one domain to another. In creating these short circuits between biology, sociology, computing, and art, we integrate discrete ideas, which often builds tension, and yet this can also produce new understandings. In our interdisciplinary process of artmaking, new possibilities and utopian ideas and images emerge. The manifestation of these in artworks can hopefully guide us as well as viewers towards new insights.

With this diversity of methods, we merge thinking and action in a single gesture, in a way that is close to what Donald Schön[6] calls *reflective practice*, a concept that Linda Candy explores in her chapter in this book.[7] Like Schön, we understand that creative reflective practice is essentially different from technical rationality, which Schön considers 'a process of problem solving'.[8] In building his critical theory of reflective practice, the author criticises the functional role of technique. As explained by Candy, awareness emerges through reflective processes, and this can occur before, in the moment or after the creative reflective practice. For *Cesar & Lois*, reflective process churns a circular feedback loop that explodes the separation between thinking and doing, or between theory and praxis. In alignment with theory of the embodied mind,[9] we understand the division between mind and body as one of the roots of the dualistic separation between praxis and theory. For us, reading, writing, interacting, and building are all part of our creative process. These activities inform our work and produce specific ways of knowing the world.

What follows is a discussion about interdisciplinary practice-based research as a vehicle for creating knowledge about the world, with examples drawn from our own art research and processes of artmaking. The text presents research grounded in experimental making that merges creative practice with investigations in biology and technology. This chapter is relevant to readers interested in contemporary art, media art, or biological art (bio art) as well as researchers from other areas with an interest in joining practice and making in their research, and for all practitioners working or collaborating across disciplines. The following text focuses on:

1) contemporary artmaking as a research process that extends beyond the domain of art
2) the artist/creator as a practitioner who collaborates across disciplines to shape the work in order to subvert/question the constraints of disciplinary knowledge
3) the artwork as a transient experiment grounded in research that gives shape to and extrapolates utopias
4) the process of making as the experimental ground where hypotheses are tested
5) the potential of practice-based research in shaping insights and building new ideas

These general takeaways go beyond our specific artworks and are intended to allow readers to reflect on their own creative (inquiring) practices as material methods of knowledge production.

The process of *Cesar & Lois*: research and practice beyond the domain of art

Our collective *Cesar & Lois* looks critically at layered aesthetic, political, social, and economic circumstances, drawing on knowledge from different fields to move through a conceptual process that includes creative experimentation and scientific protocols and practices. Our methodologies in research and practice are interdisciplinary and multispecies. In general, we engage in artmaking in order to think through challenging questions in open-ended ways; these include inquiries about the interrelationships between diverse life forms and consciousness, and the role of sociotechnical systems in perpetuating climate crises and accommodating global economic exploitation of resources and linked ecological extractivism.[10]

Examples of this eco-systemic approach and concepts can be found in our recent projects. Across our artworks, we strive to learn from biological systems. In the series *[ECO]nomic Revolution*,[11] we combine bio-data visualisation, happenings, objects, films, and performances that discuss what the egalitarian food distribution of the microorganism *Physarum polycephalum* (slime mould) can teach us about the economic disparity of abutting neighbourhoods. In the kinetic installation *Allochronic Cycles*, moving disks represent different life cycles embedded in plant life, environments, and the planet. The accelerated carbon output of specific groups of humans interfere with these life cycles, disturbing their synchronicity. This is reflected by physical disks embedded with plant specimens that move in and out of synchronisation. In the series *Degenerative Cultures*, books that document certain human compulsions to control nature (philosophical treatises and landscape design texts, for example) are used as substrates for the growth of microbiological organisms that literally consume these texts. A natural language processing algorithm (AI) linked to the microorganisms searches the Internet for texts that also subjugate nature to humans. Similar to how the physical book is consumed by microorganisms, the online texts are corrupted by the digital fungus linked to the living organisms. The living microorganisms and the AI tweet the progress of their destruction of the texts at the Twitter handle @HelloFungus, and users who mention @HelloFungus advance the AI's corruption of the online texts.

Many artists start their process of artmaking with a specific tool, a technology, or a creative procedure; as artists with a post-studio practice, we depart from a critical perspective, creating art as a way of examining and proposing *speculative apparatuses*.[12] For instance, we question how the ways in which we share information influence our actions, effectively restricting and driving our behaviour. What happens, then, when we alter those sharing modalities? As artists we look at and imagine new modes of technology and consider how those may change our behaviour. This examination and prototyping of a new method and approach requires:

1) identifying the issue
2) posing a question that challenges the existing state of things
3) experimenting with and positing an alternative through a prototype or model, such as an artwork
4) very possibly deviating from the proposal and moving in new directions

Throughout, we seek the method, skills, and potential collaborators that make the project viable. In the process of making, new ideas bubble up, transformations happen, and unforeseen possibilities are considered. In the end, the final result is a mix of the originally conceived idea and emergent concepts that are linked to material elements. Yet it would be a mistake to conceive of the above distillation of process as discrete and separate steps, as this process is in no way linear.

When applying theories of complexity to the act of artistic creation, Cecília Almeida Salles states that the process of creation is a complex system that follows a movement from chaos (disorder) to cosmos (order). For her, 'the artist's gaze transforms everything for the artist's interest, be it a broken sentence, a newspaper article, a colour or a fragment of a philosophical thought'.[13] The artist's thinking through making proves transformative for the ideas that the artist ponders and eventually materialises. This active ideation offers the possibility of thinking in altogether new and nonlinear ways, problematising as much as resolving the original driving issues. Salles describes the nonlinear pathways of creativity:

> From contact with different creative processes, one realizes that the production of a work is a complex mesh of purposes and searches: problems, hypotheses, tests, solutions, encounters and mismatches. Therefore, far from linearities, what is perceived is a network of tendencies that are interrelated.[14]

Such a process cannot be confused with a linear path, which arises from a problem or initial state and seeks to find a solution, as is often the case in the process of prototyping and developing technological products, for example. In a nonlinear approach, the problem itself is being redrawn as the project progresses, like a quantum particle that changes with observation.[15]

In the process of developing projects as *Cesar & Lois*, vibrations between modes of thinking and doing move the project's development along. The resulting artwork embodies a layering of experiment and concept, moulded in a spiral of practice and research. Conforming to Salles' model of nonlinear creativity, our iterative thinking and process of making entail enmeshed practice and theory, with output that embodies ongoing questions that incorporate philosophical research into technological and microbiological experimentation. Ideally, this results in artworks that pose the same questions that propel us.

Interdisciplinary collaboration: making through questioning

How might asking questions across disciplines instead of within one area of study challenge existing epistemological frameworks? What if your academic and creative inquiry were unfettered by disciplinary doctrines? What kind of new knowledge might be accessible? An interrogatory approach to thinking through difficult problems has epistemological foundations rooted in discrete disciplines such as philosophy and science. The methods for thinking through those problems are also defined by those same disciplines. The unrestricted interdisciplinary practitioner would be free to formulate questions that are beyond the bounds of specific disciplines. What an incentive to collaborate and think across disciplines!

Practice-based research is interrogatory in nature because the practitioner undergoes a hands-on – or material – exploration of questions. More than this, we point to the generative possibilities of an interrogatory approach that straddles different fields. Epistemological and methodological inheritances tend to lead us to analyse the world through discrete and independently organised areas: disciplines, which serve as boxes for frameworks, methods, and which, consequently, can generate preordained results and reinforce prevailing discourses. For us, to look at certain issues from the point of view of physics, for instance, brings a new set of references, information, and methods to probe and analyse. Collaboration and an interdisciplinary model give us access to questions with the power (and limits) of this discipline. The same happens when we examine issues from the points of view of philosophy, biology, the socio-political sciences, or economics. But would it be possible to look at the world and frame one's reality from perspectives that escape such disciplinary delineations? In fact, is it really possible to understand our existence (interrelated with others') in discrete concepts? And would it be possible to address big questions without using discipline-specific criteria and their respective knowledge sets?

Scholars who seek alternatives to the disciplinary thinking established in modernity – through interdisciplinarity,[16] the complexity of knowledge,[17] and the ecology of knowledges,[18] among others – point out successful incursions across disciplines. In the chapter of *Cannibal Metaphysics* aptly titled 'An Anti-Sociology of Multiplicities', Eduardo Viveiros de Castro's articulation of perspectivism[19] and multinaturalism[20] provides a foundation for us for thinking beyond disciplinary doctrine, lending a voice to the transformative potential of embracing the multiple (including in reference to culture, to nature, and to discipline). Viveiros de Castro pictures the benefits to the field of anthropology in, for instance, the adoption of the rules, premises, and questions of geometry:

> How such variety could be of service to anthropology is not very difficult to imagine, as everything ordinarily denounced in the discipline as scandalous contradiction suddenly becomes conceivable: how variations can be described or compared without presupposing an invariable ground, where the universals lie, and what then happens to the biological constitution of the species, symbolic laws, and the principles of political economy, not to speak of the famed 'external reality'.[21]

For us, to position ourselves outside of disciplines is to look at the world and formulate questions and hypotheses with multiple tools at one's disposal. Those disciplinary borders which were made distinct in modernity are softening from a surge in intrepid thinkers whose scholarship trespasses other areas[22] (see the artists trespassing in science laboratories in Figure 3.6.2). Working unbounded by a single discipline provides us with a substratum of overlapping knowledge to propel initial bursts of inquiry, respond to supervening questions, and test out possibilities.

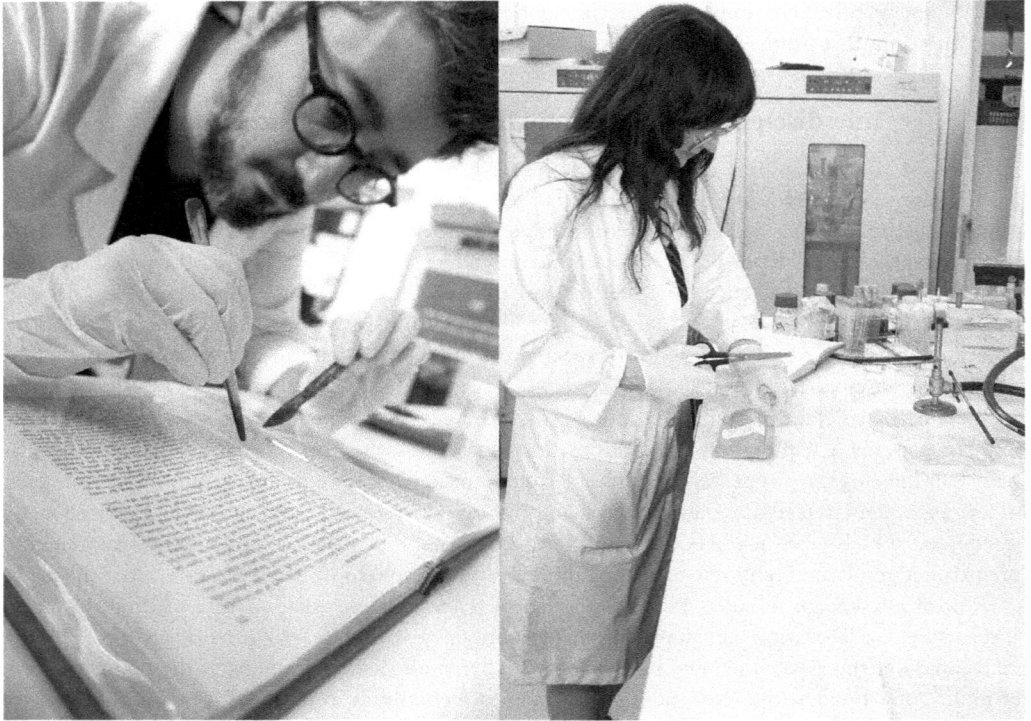

Figure 3.6.2   Cesar & Lois, 2019, Artists at work in the lab, Cesar Baio (left), Lucy HG Solomon (right).
Source: Cesar & Lois.

Through multidisciplinary inquiry and questioning from the vantage of conjoined disciplines, *Cesar & Lois* seeks to challenge the sociotechnical apparatuses of existing global information systems. In contemplating and constructing artworks that probe what it means to be human in the 21st century, we are continuously merging methods from different fields, sandwiching scientific procedures and artistic practices. Through experimentation that crosses the classical division between praxis and theory, we begin with questions that propel deep thinking and couple these with art practices in a nonlinear process of simultaneous conceptual and poetic construction. These questions tend to challenge modern disciplinary thinking[23] and are often broad in scope. Some of the questions that drive our practice-based research are: Given the importance of the human microbiome in human behaviour, what then does it mean to be human? How does time pass for the bacteria within our bodies, and how does time exist for other organisms, such as lichens? What would it mean for human agency and for machine thinking to integrate microbiological logic? How would an AI that is linked to microbiological systems base its decisions? Are there technological systems that take into account and support all of its constituents, without privileging one species and supporting only certain nodes of the network? How could we create new integrative *bhiobrid* systems, which is to say bio-digital hybrids (see Figure 3.6.3), to challenge networks that perpetuate human-centred, hierarchical decision-making? What decisions would such a system make?

The next section expands on the practice of collaboration as a methodology for thinking through such questions, identifying a collective interrogative approach by thinking across people, platforms, and even species. This approach involves multifaceted experimentation that is essential to this particular methodology of practice-based research.

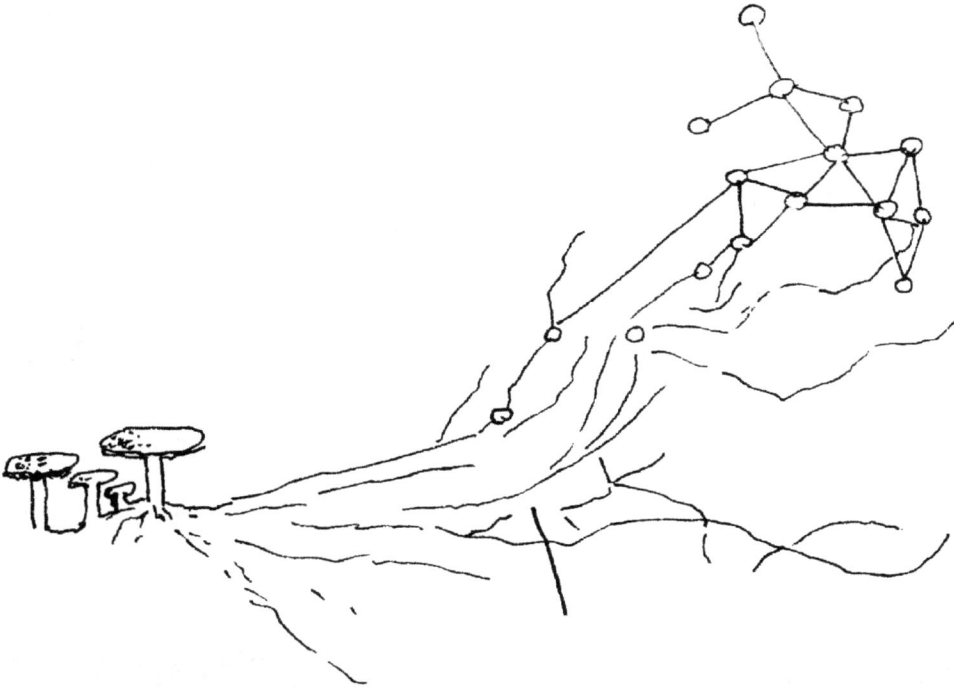

Figure 3.6.3 Cesar & Lois, 2018, Concept drawing of *bhiobrid system*.
Source: Cesar & Lois.

Art as transient experiment: transdisciplinary approaches

Our practice and research combine in an example of hybrid practice, which has parallels among the approaches of other artists whose work focuses on science and the environment, as well as those who engage in social practice. Many transdisciplinary makers address their own complex questions about the future through iterative processes, pushing beyond disciplinary boundaries with art that defies simple encapsulation as they explore social issues related to fundamental questions extending from the present moment. Contemporary artists are redefining borders through actions that provide forums for public participation, as with *Cog•nate Collective*'s reimagining of citizenship and land identity,[24] and Joel Tauber's *Border Ball*,[25] an art project that co-opts an American pastime of playing catch in order to foster cross-border community. Ecologically focused design, like that of Foreground Design Agency's *Posthuman Habitats* for embodying multispecies garments,[26] corrodes the boundaries between garment and environment. Artist Laura Beloff's wearable appendages shift the individual to a host, connector, nurturer.[27] Through the layering of human and nonhuman systems, artists Meredith Drum and Rachel Stevens' *Oyster City* layers living systems that act as marine filters and human mechanisms that produce waste, placing both into an interconnected cycle.[28] Cat Jones[29] questions ecology through a practice that involves rewilding the self and challenging preconceived notions of human perception by investigating *plant consciousness*.[30] With community-engaged projects and performative interventions (*Sensitive Territories*), Walmeri Ribeiro, along with a cohort of interdisciplinary artists, addresses the emergent societal and ecological concerns embedded in the waters off the coast of Rio de Janeiro.[31]

Together with many artists who are making art in a time of compounded ecological crises and global societal disparities, we question what kind of art we can produce in response to climate change, the possibility of mass extinction, and societal inequity marked by acute inequality in the distribution of resources across human populations involving the ecological destruction and skewed economics of global extractivism.

In our own practice as artists, we actively collect other ways of thinking. In a conversation with biologist Paul Cullen, we considered cell communication between and within yeast cells, pondering the ways that thinking is embodied or can exist as chemical exchanges and biological processes. We wrote a message with nonhuman entities in order to consider how human intention and microbiological logic converge and diverge (see Figure 3.6.4). Research consistently dovetails with our laboratory experimentation and conversations. For example, Francisco Varela broadens our understanding of self to include nonhuman entities when he describes the ecology of micro cellular exchanges in the body:

> In my view this identity is not, as traditionally stated, a demarcation of self as defence against the non-self of invading antigens. It is a self-referential, positive assertion of a coherent unity – a 'somatic ecology' – mediated through free immunoglobulins and cellular markers in a dynamical exchange.[32]

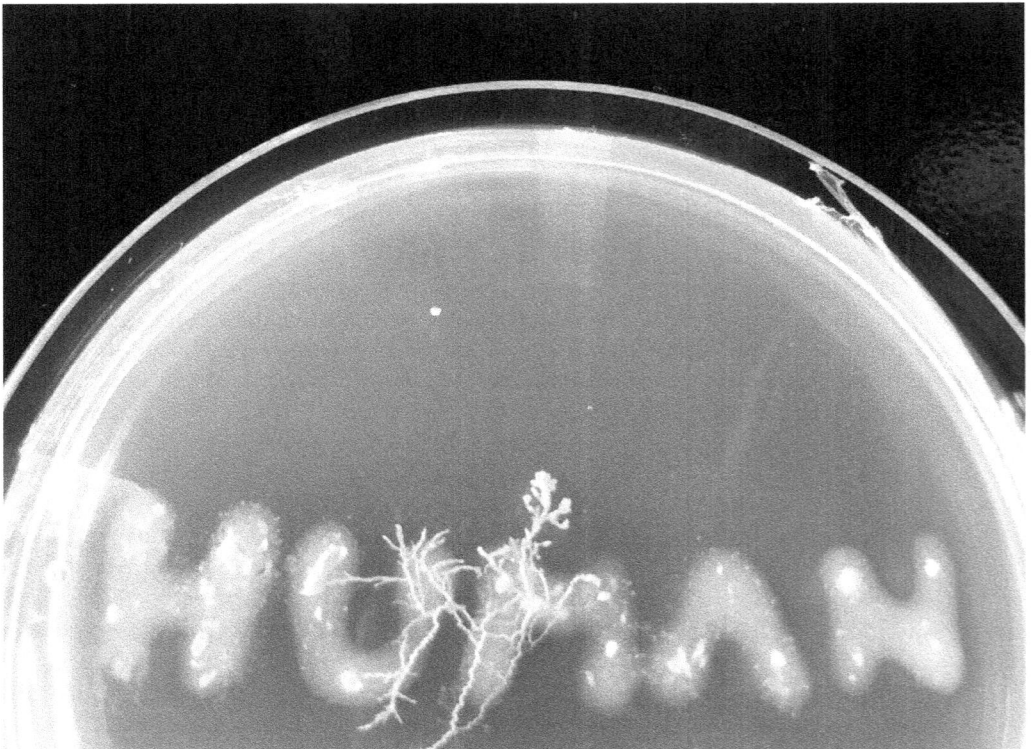

Figure 3.6.4   Cesar & Lois, 2020, *Physarum polycephalum* growth for *Collaborative Writing Workshop with Nonhuman Entities*, Coalesce Center for Biological Arts, University at Buffalo, NY.

Source: Cesar & Lois.

Our nonhuman cultures, the many beings that exist within our bodies, contribute information to their (and our) somatic ecosystem. The exchange of molecules in our cells – occurring within and in response to our external environments, and yet happening in our bodies' internal ecosystems, within which interdependent communities of species interact – embodies complex processing of information and cellular interactions. This focus on interspecies ecologies has parallels in anthropological and cultural understandings of the self as *more-than-human selves*.[33] Ancestral communities with multinaturalist understandings of human–nature relationships provide important models for how humans can and do think with natural systems.[34] Our initial foray into concepts of entanglement in anthropology, layered with our experimentation in the bio lab and constructive conversations with biologists, coalesced with additional research to cross concepts from neuroscience, philosophy, and biology. Our material experimentation, conversations, and research led us to arrive at an idea enmeshed in thinking from the many areas we had traversed: we considered that cellular cognition materialises a complexity of thought across species and environments. After this non-disciplinary-specific probing of our initial questions, we transposed our questions in artworks that ask viewers to *think together*. An example of these is *Thinking Like a Mushroom* by *Cesar & Lois* (see Figure 3.6.5), a participatory artwork with mushrooms growing within and through philosophy texts and a meditation that allows viewers to interact with and imagine themselves as spores.

By thinking through making and experimenting, we seek to develop new concepts. By taking on the challenges of thinking in new ways, creative practitioners can redefine the milestones that are sought by other fields, pushing against and trespassing conceived disciplinary boundaries. It is often the case that artists think and make before an idea becomes more mainstream and practical. xtine burrough's *Delocator*,[35] launched in 2005, is a crowd-sourced platform for finding local alternatives to corporate stores – a project that predated Google map's shopping features and which was at the forefront of the movement to 'shop local'. This project is an example of how the artist's transient experimentation can anticipate new futures. In creating discipline-defying projects and reflecting on, writing about, and discussing them, artists materialise their ideas and project new futures. We would argue that open-ended interdisciplinarity has the potential to push forward those futures. The next section identifies how interdisciplinarity has generated breakthroughs in our artistic practice.

Artmaking as hypothesis testing: imagining the future

In this section, we provide an example of how we design artworks to address pressing concerns and pose relevant questions around those. The artworks that we make are a response to a particular ecological and societal context, in which we challenge the logic systems that are perpetuated through and across our technologies. This reimagining of existing technologies answers calls for the development of ethical and equitable technology, in which potential societal outcomes are considered.[36] Through our speculative artworks, we propose a new shape for sociotechnical systems as networks that support ecosystemic decision-making and anticipate different outcomes from the logical decisions of current software.

An example of how we challenge the status quo through art is *Degenerative Cultures* (see Figure 3.6.6), which involves iterative experimentation across living and non-living systems and is manifested in a series of art installations with embedded microbiological and artificial intelligences. In this example of the artistic result of our speculations and imaginings, experiments are embedded in a series of living artworks. Each iteration of the artwork is responsive to local climate issues, and the artwork's artificial intelligence critiques the oft misguided human correctives to those. In each exhibition site, we work with local scientists and examine the contemporary climate concerns of that particular place. In this way, inquisitive interdisciplinary dialogue and research direct the artwork, and the outcome is dependent upon an array of inputs.

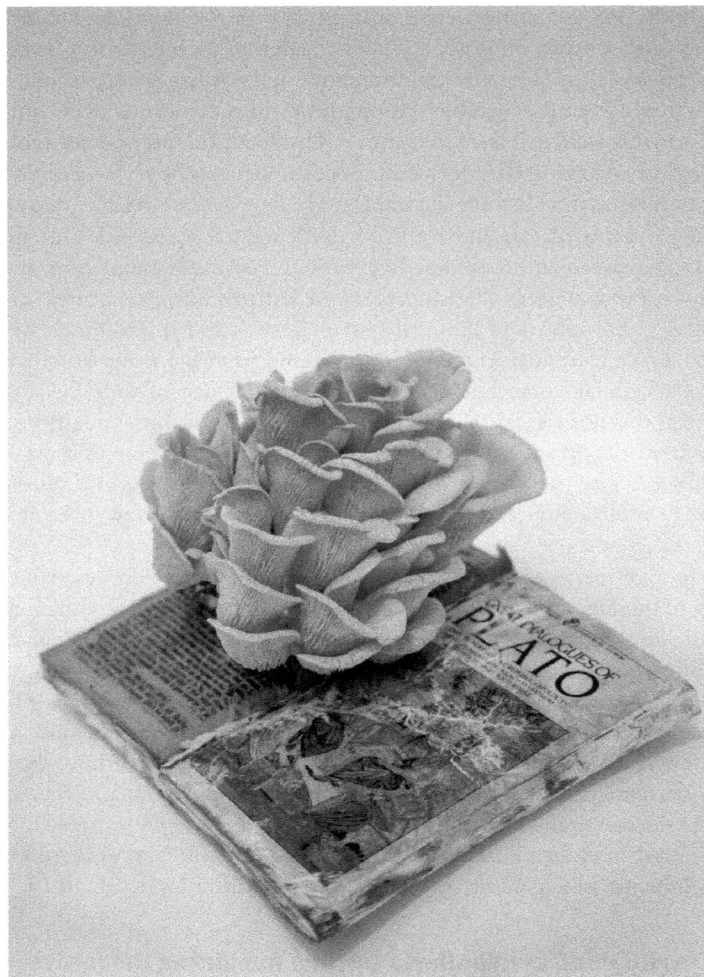

Figure 3.6.5 Cesar & Lois, 2019, *Thinking like a Mushroom*, fungal growth over philosophy text, mycelia layering the pages and mushrooms fruiting through the book in *Grey Matter* at A Ship in the Woods, Escondido, CA.

Source: Cesar & Lois.

By creating artworks in this mode of hypothesising with utopian goals, we suggest new futures: through *Degenerative Cultures*, we propose a new orientation for AI development. Many artists who work with AI ask if such an intelligence can create art (or even think) as humans do. Our work goes in another direction: we aim to create a non-anthropocentric AI that collaborates with nonhuman entities, challenging the conception of what *intelligence* is (this connects back to the originating question for our artistic inquiry, *what alternate ways of thinking can model new ways of being?*). The concept for a non-anthropocentric AI upends the idea that humans possess a superior intelligence that is the ideal model for a machine. The artwork *Degenerative Cultures* proposes a technological and biological relationship in which the AI's logic is oriented to a microbiological system. Enclosed in a glowing dome, the microorganism *Physarum polycephalum* grows over a text that asserts human authority over nature (see Figure 3.6.7); alongside the dome the AI destroys digital texts that assert this same human

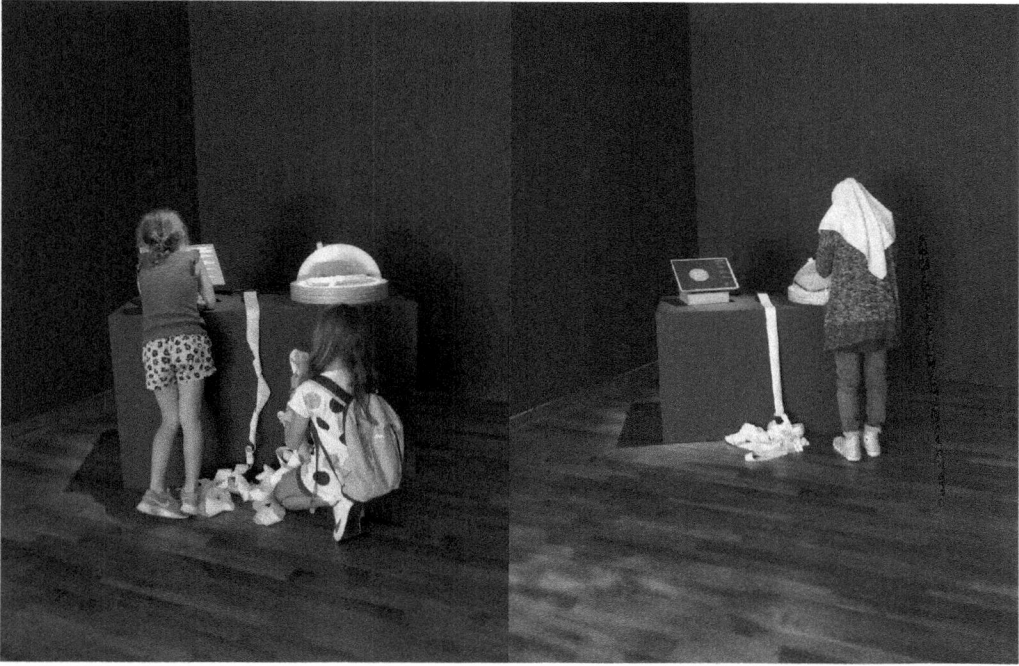

Figure 3.6.6 Cesar & Lois, 2019, *Degenerative Cultures: corrupting the codes of duality* in *Global Digital Art Prize Biennial*, NTU Institute of Science and Technology for Humanity, Singapore.

Source: Cesar & Lois

Figure 3.6.7 Cesar & Lois, 2018, *Degenerative Cultures*, detail of *Physarum polycephalum* growth over text on the classical human/nature divide, in *Uncommon Natures*, Lumen Prize Exhibition, UK.

Source: Cesar & Lois.

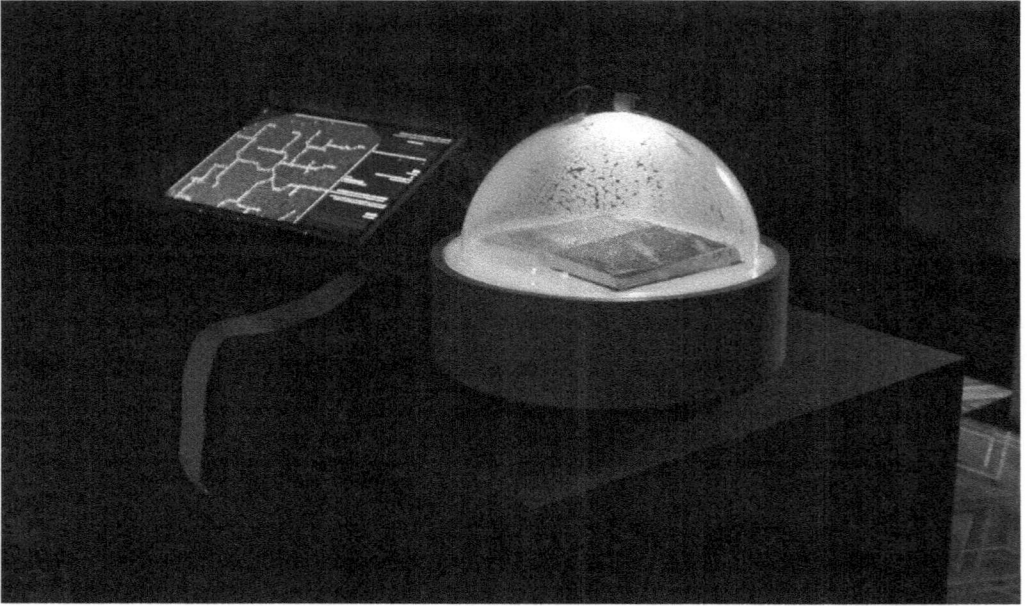

Figure 3.6.8   Cesar & Lois, 2019–2020, *Physarum polycephalum* grows over and tweets new versions of the text, while the digital fungus (AI) consumes online texts. *Degenerative Cultures: Floresta Amazônica* in *Edital CoMciência – Ocupação em Arte, Ciência e Tecnologia*, MM Gerdau, Belo Horizonte, Brazil.

Source: Cesar & Lois.

authority (see Figure 3.6.8). The artwork manifests the research leading up to it, out of which we arrive at the proposition that there are alternative logics to human thinking. It is in this way that our making and theorising twist and turn until each prods the other, concretising the art around new ideas. Playing with physical materials, coding, experimenting, and nurturing living organisms are all aspects of this kind of multifaceted making and hypothesising.

The materiality of the physical artwork both makes the technology work and suggests ideas for new ways of thinking about technology. In this way, the resulting artwork refocuses our original concerns stemming from ecological and societal considerations and offers the potential for disruption by radically different operating logics. The specific technology that we propose with *Degenerative Cultures* is a digital organism that incorporates AI and generative algorithms with a bio-digital interface. This interface connects the digital organism to the living organism. To materialise this, we developed a self-regulating cocoon capable of reading and writing to the digital system, with information passing back and forth between the growth environment and the technology. Ideally, the development and creation of this system is also a vehicle for thinking about the integration of technology and biology in new ways. Though contextualised as artmaking, our process included the development of new scientific protocols for growing microbiological systems on untraditional substrates, like books, and in non-scientific contexts with a variety of country-specific constraints (such as in an art space in Singapore). For the system to work, the art installation requires responsive lighting, a camera, and sensors. Alongside this digital technology, we designed scientific procedures which we shared with collaborating biologists and recorded in instruction manuals. The research and practice, though entangled, include the hypothesis of new forms of logic in conceptual terms, and also the development of the technology and processes needed for the artwork to function and the organisms to thrive.

In artworks like this, the creation of speculative apparatuses demands the development of new algorithms, electronics, procedures, methods, and interfaces. During this development, we become hybrids ourselves, as partial coders, biologists, interface designers, technicians, builders, etc. More than mere hands-on activities, these endeavours inform the conceptual part of the work while also producing new software, hardware, interfaces, and methodologies, which are shareable and applicable to fields beyond art. Along with the artworks and their catalogues, manuals, and the documentation of associated procedures and methods, the output of our research and practice is framed as emerging concepts that we relay in conferences and articles. Examples of these intellectual offshoots include the conceptual proposal for an interspecies remix, published in the *Media-N* (*The Journal of the New Media Caucus*);[37] and an argument for an ecosystemic artificial intelligence introduced in the *International Journal of Community Well-Being*.[38]

From the perspective of artists, we aim to generate subjective insights that can move others, including the viewers of our art and collaborating scientists, to new ways of understanding and imagining the world and its sociotechnical systems. By creating speculative technologies in the context of art, we focus on the creation of technology in order to probe at larger issues and to prompt new ideas, rather than creating tools in order to solve a problem. In this, we aspire to create the kind of open work that Umberto Eco describes:

> The open work assumes the task of giving us an image of discontinuity. It does not narrate it; it is it. It takes on a mediating role between the abstract categories of science and the living matter of our sensibility; it almost becomes a sort of transcendental scheme that allows us to comprehend new aspects of the world.[39]

The open artwork presents a whole world and yet is open-ended, allowing the viewer to engage with the piece. By opening up broad questions about AI's relationship to living systems and environments, we poke at the operating systems of future technologies and question their core principles. As artists who freely transgress disciplinary boundaries, we acknowledge that the assertions that our artworks make are speculative and, in practical ways, untested. Discipline-defying practice-based research can result in creative proposals that are multi-dimensional, built through a combination of conscientious questioning, intuitive development, experimentation, and making, yet not necessarily practical. Along the way, there are barriers and even failures. These, too, are essential to testing new possibilities and sparking new futures. The stages of development that may be construed as failure are in themselves generative.

Practice-based research: shaping insights, generating hypotheses, building ideas

Throughout this chapter, we present practice-based research that is based on collaboration and exchange across disciplines and presented as artworks. We detail some of the specific interactions that we have had while developing artworks, we reflect on the various theoretical inputs into our process, and we consider the contributions of scholars and practitioners from different fields. We think that this approach to research, which involves gathering and layering disparate thinking around a single (although nonlinear) line of inquiry, can be applied to practice-based research in many contexts.

Intellectual inputs that span disciplines and somehow converge around a shared spark do not materialise without effort. We consciously and methodically seek out those inputs that offer potential for growth. During this process, we continuously ask how we can push the project and our thinking further. This often requires posing unusual questions and asking others to join us in the sandbox of ideas, with no definitive promise of an outcome. But there is fun to be had – in our process, we speak with theorists and biologists, who prod us to consider the complexity of microbiological relationships within ecosystems. Such conversations across disciplines carry over within our art experiments, our written statements, and our iterative creative process. This practice-based research occurs within and outside of a collective, in collaboration with others and across conversations, continents, and disciplines.

Along our creative journey, we have encountered many artists whose work pushes us towards a post-anthropocentric relationship to our planet and co-inhabitants. This process of reorienting our thinking and ourselves through making results in a method of artistic production in which concept and context intertwine. The goal is to learn along the way and, if we are lucky, to create something novel that, in turn, reveals different ways of understanding and even imagining the world. In examining the deep and broad themes of decision-making across microbiological, human, and technological systems, we inevitably hurdle ourselves with neither enough information nor adequate knowledge at an audacious goal: conceiving of a new future. That self-propelled journey is shared by the conceptual artist, media artist, bio-artist, and experimenter across disciplines. This thinking and making is done hand-in-hand, and with multiple hands.

Similarly, the art collective as a structure for making and thinking together draws on plurality to make that difficult journey en masse: *Let's join together and arrive together* (see artists thinking with *Physarum polycephalum* in Figure 3.6.9). In this moment in time, when the trajectory of the growing network of the human species has overgrown the existing architecture for this sprawling global body and has effectively gentrified (through globalisation) so much of the planet, we collectively – our ecosystems and our intertwined selves – need emboldened practice-based research. The sociotechnical evolution of society, the development of a human network that supports rather than destroys eco-systemic relationships, depends upon it. Can practice-based research accomplish this, along with untold other aspirations of artists and thinkers the world over? We can at least try, and we can imagine: that is the definition of a speculative art practice.

In analysing our own practice, and in discussing an artwork, we intend to reveal how multidisciplinary inquiry has the potential to produce new knowledge and to move society in new directions. In our own practice, we develop original algorithms that do things that were not possible before, like the informational interface between *Physarum polycephalum*, AI algorithms, and Twitter in *Degenerative Cultures*. While in the context of technology this knowledge production may be viewed as an innovation, in the field of art what matters is not the technology itself but the artwork's capacity

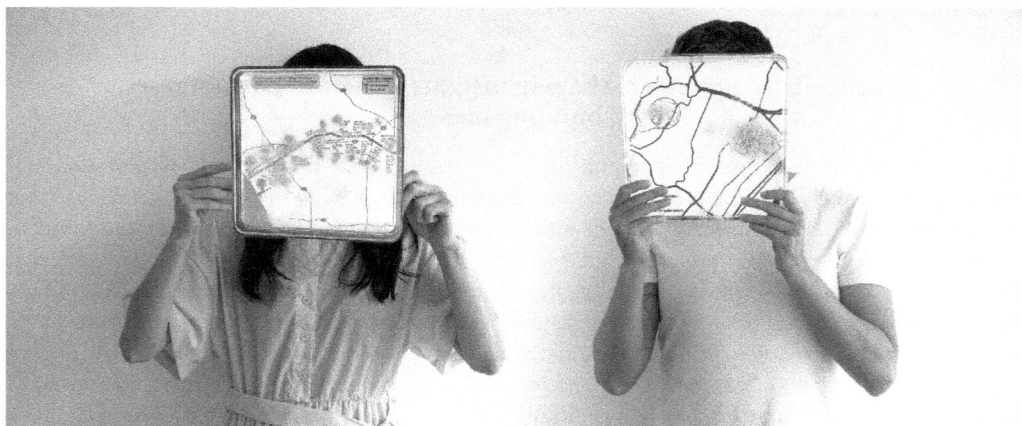

Figure 3.6.9 Cesar & Lois, 2018, Lucy HG Solomon and Cesar Baio pictured with microbio-logic.
Source: Cesar & Lois.

for generating insights, which in this instance give rise to new ways of understanding technology, human, and nonhuman lifeforms, and the relations between them. For us, when these insights stretch beyond our own artworks and into other disciplines, they gain the potential for generating new ways of understanding the world.

Our practical research involves obtaining knowledge through artmaking, with art–practice–research encompassing an inclusive set of procedures that moves from multidisciplinary inputs and collaborations to an experimental process intended to shape new ideas. These nonlinear procedures can be used by a wide range of practitioners, ranging from undergraduates to post-graduate practitioners and non-academic professionals and makers. Our reflection on our collective's practice reveals a particular multidisciplinary method for engaging with the world, with knowledge materialised through creative practice and the intuitive processes of experimentation. This form of art as activity is practiced and practiced and practiced, with knowledge accumulated from that activity. The multidisciplinary practice-research model, as a method for knowledge production, does not seek to take the place of a single discipline such as philosophy or science, nor can such inquiries be restricted by strict methodologies or contained within institutions. Rather, the practitioner of material research ideally engages in dialogue with other disciplines, within and outside of hallowed places for learning, in order to examine the world from a broad perspective, magnified and perplexed by an unfettered path to discovery.

Notes

1 With the term *networked thinking*, we refer to processing information across hybrid technological and biological systems, as when a technological node receives and processes a biological signal from a living organism.
2 See art projects by Cesar & Lois at cesarandlois.org.
3 We use the term *bhiobrid* to denote biological and digital hybrids, including integrated bio-digital systems and the combination of 'living organisms and digital agents' (Baio and HG Solomon 2018, p. 62). Cesar & Lois discuss their bhiobrid installation in *Generative Art* in Ravenna, Italy, in 2017 (HG Solomon and Baio 2018).
4 *Broadening* is a term also used by Epstein and Phan for an essential step for creativity that requires expanding one's knowledge base and skills (Epstein and Phan 2012).
5 Sullivan (2010).
6 Schön (1983).
7 Candy (2022).
8 Schön (1983, p. 39).
9 Varela et al. (1992).
10 *The New Extractivism* lays out the role of extractivism in advancing the inequities of global capitalism (Veltmeyer et al. 2014).
11 See more at: http://cesarandlois.org/economicrevolution/.
12 The term *speculative apparatuses* is related to artworks based on systemic worldviews that challenge current societal structures through a speculative approach. The use of *apparatuses* points to Vilém Flusser's concept (Baio 2018), discussed in the next section.
13 From the original: 'o olhar do artista transforma tudo para seu interesse, seja uma frase entrecortada, um artigo de jornal, uma cor ou um fragmento de um pensamento filosófico' (translated by the authors) (Salles, 1998, p.35).
14 From the original: 'No contato com diferentes percursos criativos, percebe-se que a produção de uma obra é uma trama complexa de propósitos e buscas: problemas, hipóteses, testagens, soluções, encontros e desencontros. Portanto, longe de linearidades, o que se percebe é uma rede de tendências que se inter-relacionam' (translated by the authors) (Salles 1998, p. 36).
15 In 1990, physicist Lawrence J. Landau made the case for examining quantum particles at the macroscale – a large space-time scale – to describe the particle trajectories.
16 E.g. Japiassu (1976).
17 E.g. Morin (2008).
18 E.g. Santos (2014).

19 In the context of anthropology (Eduardo Viveiros de Castro), *perspectivism* refers to the understanding that knowledge is shaped by cultural vantage points.
20 According to Viveiros de Castro, 'Multinaturalism does not suppose a Thing-in-Itself partially apprehended through categories of understanding proper to each species' (Viveiros de Castro 2014, p. 76).
21 (Viveiros de Castro 2014, p. 114).
22 Forman connects disciplinarity to the values of modernity: '*Disciplinarity* is an abstract noun referring to a cultural ideal, to a set of presuppositions about where the value of knowledge lies and what sorts of knowledge possess highest value, about the morally charged behavioural norms that producers and curators of knowledge must satisfy, and about the proper embodiments of knowledge in formal institutions. As a distinct cultural constellation disciplinarity began to take shape only toward the end of the eighteenth century' (Forman 2012, p. 59).
23 'Modern disciplinary thinking' refers to the division of disciplines codified during modernity (Forman 2012).
24 Cog•nate Collective et al. (2019, pp. 15–22).
25 Tauber (2019).
26 Foreground Design Agency (2018).
27 Laura Beloff (2012).
28 Drum and Haughwout (2018),
29 Cat Jones' project *Somatic Drifts* (2014) 'that grafts human and plant life into new collective experiences' (Goodman 2019, p. 138) is discussed in Andrew Goodman's text, *Black Magic: Fragility, Flux and the Rewilding of Art*. www.openhumanitiespress.org/books/titles/immediation/.
30 We use the term *plant consciousness* to refer to the interfacing between plant life with the environment and with other species, comprising an array of complex processes and responses; ecological artists who embrace the logic of nonhuman life forms assert nonhuman life consciousness.
31 *Sensitive Territories* is a Brazil-based multi-year project directed by Walmeri Ribeiro and includes many interdisciplinary collaborators, including Cesar Baio of *Cesar & Lois*.
32 Varela (1992, p. 9).
33 In this context, the concept *more-than-human selves* reframes the human being as a hub rather than an autonomous single (non-divisible) individual. See Eduardo Kohn's *beyond the human* anthropology that examines traditions of the Ávila Runa in Ecuador's Upper Amazon (Kohn 2013).
34 Kohn (2013), Viveiros de Castro (2014).
35 delocator.net, Xtine Burrough (2005–present).
36 Our argument for an ecosystemic AI insists that technological innovators anticipate and respond to the ecological and societal implications of their innovations (HG Solomon and Baio 2020).
37 HG Solomon and Baio (2021).
38 HG Solomon and Baio (2020).
39 Eco (1989, p. 90).

Bibliography

Baio, C. (2018). Speculative Apparatuses: Notes on an Artistic Project. *Springer Series on Cultural Computing*. doi.org/10.1007/978-1-4471-7367-0_38.
Baio, C., & HG Solomon, L. (2018). 'Culturas Degenerativas': Experimentações em torno de uma Rede Biohíbrida. *Revista Científica/FAP*, 19(2). http://periodicos.unespar.edu.br/index.php/revistacientifica/article/view/2433 (accessed January 23, 2021).
Beloff, L. (2012). *A Unit*.
Burrough, X. (2005). *Delocator*.
Candy, L. (2022). Reflective Practice Variants and the Creative Practitioner. In C. Vear (Ed.), *The Routledge Practice-based Research Handbook*. London and New York: Routledge.
Cog•nate Collective et al. (2019). Citizenship & Art in the Age of Mass Displacement: Visions from the Margin. In *Cog•nate Collective: Regionalia*. Santa Ana: GCAC, California State University Fullerton.
Debord, G. (1995). *The Society of the Spectacle*. New York: Zone Books, MIT, revised ed.
Drum, M., & Haughwout, M. (2018). *Mediated Natures – Speculative Futures and Justice Panel*. Proceedings: 24th International Symposium on Electronic Art, Durban, 430–435.
Epstein, R., & Phan, V. (2012). Which Competencies Are Most Important for Creative Expression? *Creativity Research Journal*, 24(4), 278–282.
Flusser, V. (2008). *O universo das imagens técnicas: Elogio da superfcialidade*. São Paulo: Annablume.
Foreground Design Agency. (2018). *Posthuman Habitats*. https://www.foreground-da.com/posthuman-habitats.

Forman, P. (2012). On the Historical Forms of Knowledge Production and Curation: Modernity Entailed Disciplinarity, Postmodernity Entails Antidisciplinarity. *Osiris*, 27(1), 56–97.

Goodman, A. (2019). Black Magic: Fragility, Flux and the Rewilding of Art. In E. Manning, A. Munster, & B. M. Stavning Thomsen (Eds.), *Immediations: Art, Media, Event*. London: Open Humanities Press, 134–160.

Haraway, D. (2016). *Staying with the Trouble: Making Kin in the Chthulucene*. Durham and London: Duke University Press.

HG Solomon, L., & Baio, C. (2020). An Argument for an Ecosystemic AI: Articulating Connections Across Prehuman and Posthuman Intelligences. *International Journal of Community Well-Being*, 3(4), 559–584. https://doi.org/10.1007/s42413-020-00092-5.

HG Solomon, L., & Baio, C. (2021). Case Study: Remixing Knowledge with Layered Intelligences. *Media-N | The Journal of the New Media Caucus*, 17(1), 29–43. https://doi.org/10.21900/j.median.v17i1.486.

HG Solomon, L., & Baio, C. (2018). Degenerative Cultures. *GASATHJ – Generative Art Science and Technology Hard Journal*. www.gasathj.com/tiki-read_article.php?articleId=101 (accessed January 23, 2021).

Huizinga, J. (2016). *Homo Ludens: A Study of the Play-Element in Culture*. London: Angelico Press.

Japiassu, H. (1976). *Interdisciplinaridade e Patologia do Saber*. Rio de Janeiro: Imago Editora.

Jones, C. (2014). *Somatic Drifts*. https://catjones.net/2016/08/11/somatic-drifts/.

Kohn, E. (2013). *How Forests Think: Toward an Anthropology Beyond the Human*. Berkeley: University of California Press, 1st ed.

Kopenawa, D., & Albert, B. (2013). *The Falling Sky: Words of a Yanomami Shaman*. Cambridge, MA: Belknap Press, An Imprint of Harvard University Press.

Maritain, J. (1944). *The Dream of Descartes*. New York: Philosophical Library.

Morin, E. (2008). *On Complexity*. Cresskill, NJ: Hampton Press.

Ribeiro, W. (2019). Em Tempos de Urgência: Territórios Sensíveis e a Poética de um Corpo em Performance. In W. Ribeiro & H. Briones (Eds.), *Arte: Novos Modos de Habitar/Viver*. São Paulo: Intermeios.

Salles, C. A. (1998). *Gestos Inacabados: Processos de Criação Artística*. São Paulo: FAPESP e AnnaBlume.

Santos, B. D. S. (2014). *Epistemologies of the South: Justice Against Epistemicide*. London and New York: Paradigm Publishers, 1st ed.

Schön, D. A. (1983). *The Reflective Practitioner: How Professionals Think in Action*. New York: Basic Books.

Sullivan, G. (2010). *Art Practice as Research: Inquiry in Visual Arts*. Thousand Oaks, CA: Sage Publications, 2nd ed.

Tauber, J. (2019). *Border Ball*. https://borderball.net/.

Tsing, A. L. (2015). *The Mushroom at the End of the World: On the Possibility of Life in Capitalist Ruins*. Princeton, NJ: Princeton University Press.

Varela, F. J. (1992). Autopoiesis and a Biology of Intentionality. In *Proceedings of the workshop 'Autopoiesis and Perception'*. Dublin: Dublin City University, 4–14.

Varela, F. J., Rosch, E., & Thompson, E. (1992). *The Embodied Mind: Cognitive Science and Human Experience*. Cambridge, MA: MIT Press, revised ed.

Veltmeyer, H., Petras, J., & Albuja, V. (2014). *The New Extractivism: A Post-Neoliberal Development Model or Imperialism of the Twenty-First Century?* London: Zed Books.

Viveiros de Castro, E. (2014). *Cannibal Metaphysics*. Edited by P. Skafish. Minneapolis, MN: Univocal Publishing.

3.7

SITE

An inventories approach to practice-led research

Graeme Brooker

Introduction

In this chapter, I draw on the importance of site as the location of the material to be *handled*[2] and then explore practice-led approaches, such as *precedent, spolia* and *superuse*, to explain how they engender the various processes of their reuse. Site is of importance to this work because it is the place where, along with intangible elements such as memories and atmosphere, provides what can be considered to be a unique sets of conditions. I use case studies to elucidate these ideas. This chapter title utilises the word inventories because practice-led research using these processes, is a generative approach, one that involves the indexing and cataloguing of the found. As we shall see, this produces renewed insights, arising from the scrutiny and reordering of the extant. The underlying theory concurs with Smith and Dean's descriptions of practice-led research[3] as methods that are characterised as iterative and cyclical, and which foreground contingencies and the diminution of a preconceived research framework.

The approaches described are relevant to practitioners in any field of work that utilises found or extant materials. These are practices that prioritise what may be called *thinking through making* approaches, characterised by Ingold, where theory may be applied subsequently to production.[4] It is also aimed at practitioners like myself, who arrived in academia after extensive practice experience. Practice-led research as a designer of interiors, and as an educator, in my view recognises that less-traditional forms of research can be overlooked in relation to more established theoretical or empirically based methods.

This chapter aims to equip its reader with a broad knowledge of approaches to working with existing material which offers an antidote to some of these traditions. Inventory-based approaches foreground analysis and the auditing of existing material, which prompts reflections and in turn suggests what forms of action can be enacted upon them. Therefore, the research answer may be considered to be already in existence. It is the formulation of the questions upon which actions are then based that form the primary motivation for the research. Therefore, inventorising asks the researcher to analyse and formulate responses to these features. Inventory-based approaches engage the person undertaking them in the development of particular sensibilities that reinforce enquiry, judgement, editing and decision making as to the continuation of these distinct characteristics of the existing material. I suggest that these qualities are of enormous benefit to research in climatically and social-justice challenged times.

DOI: 10.4324/9780429324154-29

The inventories-based approach to site

Inventarium . . . list of what is found . . . from Latin Inventus, past participle of invenire, find, to come upon.[5]

Normative linear-based processes of more established research methods, such as the devising or posing of a question, and then contextualising and answering it, can be considered to be less effective in inventory-based practice-led research. This is because utilising an *inventory* approach promotes contingency and mediation upon the material qualities of the found. It results in an aggregated combination of material, process, research and design. Inventories prioritise practice-led research that primarily takes place through the auditing and collating of extant on-site material, handling it, cataloguing it and then redeploying it back into the work. In essence, an *inventories-based* approach can engender ways of working that offer alternatives to conventional ways of researching.

The term *inventories* is borrowed from 15th-century practices of the recording of lists of goods, usually with their estimated values, which were often used to surmise the belongings and land of someone deceased. These were then used to assist in the distribution of the deceased's estate. Primarily, any inventorising approach prioritises auditing and valuing, and is also the method of collating, utilising and implementing the materials that have been collected, found or stored on a site.

Traditional research approaches in interiors, similar to much of the research that takes place in built environment subjects, such as architecture, are usually informed by social science and humanities methodologies. Engaging clients or communities in consultation, soliciting opinions through interviews or via responses to propositions and engaging in workshops with stakeholders are typical approaches. These are forms of investigations that usually incorporate the application of a pre-conceived framework of enquiry. Arguably, this situation has led to research in these areas being focused towards dominant discourses in the sciences and humanities: a situation that has arguably led to insularity, and potentially an over-reliance on already well-established processes. In *A Way with Words*, Jane Rendell supports this view when she suggests research in architecture 'has often been conducted in ways that is rather self-contained and which often follows accepted and long-standing methodologies'.[6]

Practice-led research in the built environment, and in particular in the discipline of interiors, differs from traditional modes of research. Practice-led research will often reverse normative ways of working. Rendell explains:

> Instead of posing research questions and then finding answers, in much design research the process operates through generative modes, producing works at the outset that can be reflected upon later.[7]

She goes on to say

> while a researcher in the humanities might first explore the context or background for a research question . . . in some cases design researchers will investigate ideas through the production of work first, and then later consider the larger field to ask who else is researching the same questions?[8]

Practice-led research can offer alternatives to normative processes. By this I mean the processes of generating a hypothesis and then elaborating a framework by which to explore and answer it. Instead of this, I will describe methods that explore the existing or found matter on a site, in order to then deduce and formulate approaches to their subsequent redesignation. In inventory-based processes, the condition of a site, and the material within it, is a situation that requires the practitioner, much like an archaeologist, to formulate new insights through contingencies. In other words, their exposure and subsequent indexing, cataloguing and potential reuse provide new insights and methods of further dissemination.

Site

Site is a critical matter in practice-led research. In numerous publications, I, and many others, have argued for the understanding of research and design, specifically in the interior, to be understood as concomitant with work on the reuse of existing buildings.[9] This is because, in essence, whether an architectured, designed, or decorated interior, all three essentially rely on being realised through the transformation of the *existing*. In other site-specific disciplines, such as installation art, archaeology, conservation architecture and geography, site is also prioritised as generative in practice. The processes of handling the material that is already on site is integral. Site is a complicated entity. It is one that contains all manner of both tangible and intangible elements. For example, Andrea Kahn and Carol Burns depict site as a series of *constructs*, describing them as relational, historiographical, socially, experimentally and materially structured places.[10] In their view, the site requires concentrated understanding, an interrogation that goes beyond the usual techniques of surveying. Instead, they advocate processes that enact profound quantitative and qualitative measuring of a place, in order to incorporate a wider range of characteristics and phenomena that can contribute to the understanding of a specific location.[11]

The exploration and analysis of these constructs form part of the processes of what Ewing describes as field/work.[12] Drawing on influences from anthropology and in particular the importance of being in *the field*, field/work draws on the ethnographic traditions of inhabiting within, and writing about, site-specific settings. Field/work coalesces around participant observation and often results in collections of information: for example, data on psychology, gender, sexuality, health, traditions, beliefs and so on. In all forms of site-related design, field/work is utilised as a way of exploring and making sense of a specific location. For instance, in interiors and architecture, these can be both pre-building processes, such as measuring through survey or setting out, and post-construction research, such as snagging, post-occupancy evaluations and so on. Essentially, field/work delimits and defines the area for research and practice.

Field/work is indispensable to practice-led processes in any discipline that prioritises site. Field/work can be characterised by the exploration and adaptation of obsolete matter: material that has lost its value, and resources that are considered waste. Sites for exploration in interior field/work are situations that have in common the proposition that an obsolete environment or element is not only a site of depredation but also a condition for mediation. It is the site of the enactment of research, and design processes will ensure that meaningful change through their reuse will take place. Chris Speed confirms this relationship: 'Field/Work and site reconfigures 'site' from the perspective of architectural design as socially, environmentally and economically contingent'.[13]

Because of the enduring relationship between the interior and site, the practice-led research processes of the interior ensure a close relationship with physical material, intangible matter and the cultures embodied within them. By this, I mean *material* as matter that is already on-site. Processes of research, in practice, therefore rely on the various ways of *handling* that material.

Handling

Because of the fundamental role of the site in practice-led research, the significance of sorting and ordering the site material emphasises the processes of *handling*. In *The Magic Is the Handling*, Barbara

Bolt describes how material thinking offers numerous ways of understanding relations of site and making, and its role in design research. Bolt uses Heidegger's ideas of *handling* from *Being and Time*,[14] where he described the understanding of the world not through theoretical contemplations, but instead through making sense of the materials, tools and ideas around their practice. Bolt suggested that:

> Words may allow us to articulate and communicate the realisations that happen through material thinking, but as a mode of thought, material thinking involves a particular responsiveness to or conjunction with the intelligence of materials and processes in practice. Material thinking is the logic of practice.[15]

This form of knowledge, which Bolt describes as *tacit*, implies that materials have an inherent intelligence. It is a condition formed from the combination of matter that creates an energy. It is what Jane Bennett refers to as *Vibrant Matter*: 'the enduring and persistent vitality that exists in all aspects of material'.[16] In practice-led site-based work, the extraction of *vitality*, in the form of both tangible and intangible narratives, and their sorting or cataloguing for subsequent reuse is essential. For instance, finding long-buried material, unearthed through excavation, explicitly demonstrates extraction processes in performative practices of handling, where expressive, vibrant and charged materials are extracted and then mediated upon in numerous and unusual ways.

One such practice of handling that clearly articulates the argument of this chapter is *mudlarking*. Mudlarking describes the processes of digging on the banks of the Thames and extracting the usually well-preserved artefacts from the anaerobic river-bed mud. For centuries the site of the city's landfill, the extreme tidal range of the river acts as a powerful force that exposes the discarded matter on its foreshores, dropped or dumped into the water throughout time. *Mudlarker* is the name given to people who, since the 18th century, have picked through the waste, selling it, or in this century, handing finds into the Museum of London for verification.[17] This type of practice-led work relies on how extant material, in this instance the wastes of the city and its inhabitants, is extracted, understood and then inventoried. Mark Dion used this approach in his work *Tate Thames Dig* which included mudlarking on the river in a three-phase approach: collecting from the river in specific places, during specific tidal times; establishing a field-centre for their cleaning, and taxonomising; and finally housing the collections in a series of cabinet *wunderkammers*. The cabinets, illustrated in Figure 3.7.1, were devised to be interactive. In contrast to a museum setting, viewers were invited to handle and reorganise the material themselves to make sense of it. Williams states with regards to the making of the cabinets that:

> There is no labelling, no chronology, and no interpretive text other than a reference to the sites where the material was gathered; in this the *Tate Thames Dig* Wunderkammer encapsulates the processes of their formation.[18]

Other performative practice-led materialist approaches can be seen in *fossicking*, a metaphor for rummaging around, or even *gleaning*. Preston describes practice-led fossicking as embodying 'a sense of un-doing, re-sorting, and making new ground'.[19] *Gleaning* incorporates the anticipation of things to be unearthed that provoke thoughts through unexpected associations. Articulated by Carless as 'both a spatial and material hacking but also an attitude',[20] like mudlarking and fossicking, gleaning prioritises the gathering of the unexpected and the insights associated with their extraction and sorting.

Whether *mudlarked*, *fossicked* or *gleaned* from a site, the material extracted forms the *exegesis*. This is a material, an element, a space or device which, rather than just explaining or contextualising the practice, acts as interpreter: the translator of the material, particularly when it is sorted or ordered.

Figure 3.7.1 Mark Dion Thames Tate Dig.

Source: Mark Dion Thames Tate Dig – Tate Images.

The exegesis instrumentalises knowledge through its extraction and subsequent narrative meanings. Bolt states:

> The task of the exegesis is not just to explain or contextualise practice, but rather is to produce movement in thought itself . . . such movement cannot be gained through contemplative knowledge alone, but takes the form of concrete understandings which arise in our dealings with ideas, tools, and materials of practice.[21]

The exegesis, the knowledge extracted from the site, is the material derived from the analysis and interpretation of the place. It is the translator of meaning, which is used to deduce and interpret tacit knowledge. The exegesis requires processes of documentation that enable insights into the matter that is exposed, to be examined and reused. We shall explore this next.

Inventory-style methods: precedent/antecedent, reuse, and superuse

I will now outline a series of inventory-style methods used in practice-led research called: *precedent/ antecedent*, *reuse* and *superuse*. They all describe the different ways in which the handling of the material, the exegesis and the existing matter can be utilised to transform the intended outcomes of the work. They are processes that mediate upon the material and then infer new use through this research that has been undertaken. The inventories-based approach to site-based practice-led research means that working with the extant and the handling of the exegesis generates a perpetual work-in-progress approach. It is one that relies on conjecture and hypothesis and is compounded by the contingencies afforded by the existing matter that is to hand.

Precedent/Antecedent

Precedent, or antecedent, is a research approach that relies on the analysis and understanding of a previous type of space, object, environment, which can be used to inform or influence the new design work. This type of work can be manifest as a *mood board*, a layout of work, of images and materials from influential projects. It is a collage of atmospheres, spaces, furniture, colours and materials which, when combined, portray a suggested overall effect. Ultimately, these methods emphasise research processes that accentuate copying, prototyping, analogy and the formulation of certain strategies with which to utilise and adapt this knowledge. This approach explores a fine line between imitation, copying and the translation of the inventory of the exemplary subjects.

Precedent/antecedent is a practice-led research inventory-based approach that continues the traditions of innovation through the interrogation of the exegesis. Precedent is often deployed to inform decisions in law, or utilised in biological sciences to describe evolutionary and analogous models of development. Antecedent, in essence, is also presumptive and is often used to describe what has gone before and what has already been undertaken. The utilisation of precedent in practice-led research allows the practitioner to develop a taxonomy or catalogue of how it has been done before. Instead, its respectability is derived from its adoption in a process that can be considered a tool for learning via that which has already been undertaken. It speaks to how Rendell described how designers will often investigate the production of work that has taken place beforehand and allow them to contextualise their work in relation to understanding who has asked the same questions before them.

Practice-led research through precedent and antecedent processes can be described as a hands-on, solution-driven activity. One where information and even answers to issues, such as structural procedures or enquiries into how certain details have been achieved, are offered up through the analysis of existing drawings, models and on-site learning. Precedents offer paradigms for new knowledge and sense through the understanding of the existing and the yet to be built. The utilisation of that learning is then translated into processes of making the new work, informed by the exegesis or understanding of the material. When both precedent and antecedent are used in the processes of the design of spaces, it means that through the understanding of existing 'typologies' and their uses, particularly through the utilisation of case studies, practice-led research is based on learning as contingent to what has already been undertaken.

In built-environment practice-led research, precedent/antecedent approaches often centre around the use of structural elements or details. These may include the technical details of junctions, such as where floors meet walls, joinery details for doors, their frames and surrounding windows. The analysis of this can provide detailed solutions to technical issues not encountered before but which are also, usually, *free* to copy. By this I mean that the structural and detail-level language of built-environment work bears no condescension towards its repeated use. Quite the opposite occurs. The quoting of details from other sources is often seen as a mark of positioning the designer's work in the milieu of others who have gone before them. Copying is considered more provocative where formal propositions have deliberately been utilised as a response to a certain context. For instance, façade copying in buildings can be considered normal practice, albeit provocative, especially in a sensitive context. For instance, on the Champs Elysees in the centre of Paris, a luxurious hotel project, and its neighbouring buildings, were transformed by copying the surrounding facades and applying them to the whole of the block. To satisfy any contentious historic preservation laws, the building facades were copied and were adapted to unify the seven facades of the block, creating a statement building in the city centre. The disjunction between the copied ornate façade, with its friezes, architraves, stringcourses and window details, and the existing levels of the interiors was profound and exemplified through a series of seemingly randomly placed windows. Through sheer audacity, the project took the logic of 'in-keeping' to a new degree of proficiency through copying. At the time, the controversy caused by the project was extraordinary. The new knowledge produced here is an amalgam of the existing, translated into the new façade as a direct copy, albeit in a different material, of the existing facades.

In another copy-method project, the centre of Schijndel, in the Netherlands, was transformed through the reappropriation of a traditional farmhouse typology, found locally, which was studied and remade into a structure that housed restaurants, shops and a wellness centre (see Figure 3.7.2). MVRDV and artist Frank van der Salm documented all of the remaining farms in the area, photographing them in detail to produce a 'typical' farm type. These images were then applied to a glass structure, itself the archetypal form of one of the buildings. The building was almost twice the size of a regular farm, monumentalising its appearance. The collage of the façade incorporated the walls, windows, roof details and even the traditional farm door, augmented to four metres high and twice its normal size. Rather than the solid masonry, timber and tiled original, its materiality was glass, rendered translucent in some places and opaque in others (see Figure 3.7.3).

Any practice-led research utilising copying means a close connection to the extant matter, such as an existing building, and to *the precedent*. In architecture, the precedent/antecedent is deployed in a formal role, and as we have seen in the two examples utilised, the learning from the 'originals' formed envelopes. In interiors, the precedent is either born out of an analysis of the qualities of the site, and what that can offer, or will be imported into the space wholesale. Like a readymade or found object, interior precedents are the site, the existing building, of practice-led research

Figure 3.7.2   Glass Farm, Schijndel, Netherlands – MVRDV.

Source: MVRDV – Daria Scagliola and Stijn Brakkee.

Figure 3.7.3 Glass Farm, Schijndel, Netherlands – MVRDV.
Source: MVRDV – Daria Scagliola and Stijn Brakkee.

which takes the form of the handling of the material with which to understand and transform. For instance, Kate Darby, with partner David Connor, was prepared to demolish an old structure on land adjacent to their listed house, to create a space for their new studio in rural Herefordshire. Existing on the site was a 300-year-old ruined barn on the verge of collapse. Because it was on the *curtilage* of the house, a law that treats adjacent structures as a listed building meant that it required the same level of protection. This ensured the demolition of the barn was not feasible. Instead, the structure had to be retained. The response to a conservation officer, who could not identify what was valuable about the building, apart from it being old, meant that the designers decided to retain the whole building. Everything was to be preserved. Figure 3.7.4 illustrates the structural elements, such as the timber frame, but also the lath and plaster, old dead vines, cobwebs and broken windows. The ruin was totally encapsulated in a new Duchamp-like ready-made black, corrugated-steel shed. This approach to practice-led-research demonstrated how precedents can be recontextualised in order to be reconfigured anew and to produce new knowledge about such architectural practices.

In essence, practice-led research through the utilisation of precedents and antecedents can alleviate the fear of the copy and can be used to disinhibit the creative processes from plagiarism, especially when appropriated wholesale.

Reuse

Reuse is the term used to characterise the appropriation of existing material into new contexts: situations to which they were often never intended to be relocated. Reuse as a practice-led approach

Figure 3.7.4 Croft Lodge Kate Darby and David Connor.
Source: Kate Darby and David Connor.

is reliant on the sensibilities of the creative practitioners and their approaches and ideas towards the redesignation of the extant matter. As Hegewald and Mitra state:

> Reuse essentially refers to using an item again. . . . It is a deliberate and selective process in which existing elements are borrowed and taken out of their former surroundings to be applied to a fresh context.[22]

Reuse methods differ from precedent/antecedent processes in that they are purely contingent and always reliant on what is found on-site. Reuse is not a new phenomenon. *Spolia*, from the Latin spoils, is an archaic term that describes the recycling of existing architectural elements by incorporating them into new buildings. In the constructed or built environment, *Spolia* refers particularly to the re-use of the elements of the classical colonnade: the column shaft, base, capital and entablature.

Figure 3.7.5   Pulpit of Torcello.

Source: Graeme Brooker.

Roman builders often constructed a colonnade in fragments, a practice that valued contingency and the potential for re-use.

It is not just the selection and edit of the fragments in spolia approaches which is important. It is also their redeployment in such a manner as to induce the viewer to understand and be cognisant with the knowledge, narrative and story of the things they are being asked to view. For instance, the Pulpit of Torcello, located in the Santa Maria Assunta Cathedral on the small Venetian island of Torcello, was formed using reused elements (see Figure 3.7.5). Rebuilt in the 9th century, the cathedral was abandoned in the 13th century as the growth of the adjacent Venice, the silting up of Torcello's canals and the onset of malaria decimated its population. The cathedral pulpit was constructed from spolia. Its steps were made from a series of reliefs that were sawn and cut to provide an edge and balustrade to the stair. The carvings were datable to the 11th and 12th centuries, and subsequent research uncovered the fact that they were dedicated to Kairos, in antiquity the symbol of time as an opportunity. The reinstatement of these fragments was not an entirely arbitrary gesture. Patricia Fortini Brown suggests that the

> restitution of the fragments to a certain degree of wholeness within the larger program calls attention to another occurring, and consummately Venetian, concern: to create a density of time within their major monuments through the employment of rediscovered relics.[23]

The steps of the Torcello pulpit were formulated using a series of leftover fragments. The remains were valued, but the carvings were handled and subsequently treated quite forcefully. They were cut to fit the steps and then edged with a reclaimed frieze detail, yet they were retained for their residual meaning: their connection to time. Spolia-based approaches, in practice-led research, are as contemporaneous as they are antiquated. New knowledge is generated through the contrast and differences between the chronologically differing elements in the approach. Ningbo Museum by Wang Shu, in China, is a building that pragmatically incorporates various tiles and bricks found on the site of the project. The reuse of the spolia of the ruined buildings was designed to comment on the rapid urbanisation of China, through the appropriation of elements of the razed village upon which the museum was constructed. Continuity is manifest through the appropriation of old into the new; on this, Wang Shu states:

> I like to build with old recycled bricks and tiles in the tradition of the region in which I live. In this way, the materials are saved, various possibilities of material application are expressed, ample and exquisite crafts are developed, and meanings of memory and time are kept.[24]

Reuse, whether through spolia or other means, is an approach that utilises its research processes through contingencies, that is, being open to incorporating new information or material through uncovering what might be found on a site, and is a very hands-on approach to practice and research. It works with all manner of found sites, objects, situations and field work.

Superuse

Superuse approaches to practice-led research emphasise closing the loop of all linear-based design thinking. Superuse is a term that goes beyond reuse to describing the production of elements, objects and buildings through intervening into their production in order to utilise them in new contexts.[25]

Superuse describes processes that reverse-engineer normative processes of design, research and also construction by utilising what is available locally and often cheaply, as *surplus*. Started in Holland by Architecten2012, *superuse* was coined as a term to phrase the practice-led research approach of understanding and then utilising materials, along with their flows of production, to realise furniture, interiors, buildings and, in the future, whole cities, such as Rotterdam, in developing circular-economic approaches to all aspects of its infrastructures. Its site or field/work methods are derived from a key strategic research device known as a *harvest map*. A zone is produced on a map of the local area by drawing a line from the centre of the site of the project. This is usually around 50 kilometres wide and provides the context from which the material for the project is to be salvaged. The map may include factories, building sites, depots – all places where waste, and production excesses, may be found and therefore utilised. The map size is never fixed, and as Jan Jongert states:

> The idea of such a map illustrates a way for systematic surveys, but it's not a panacea for all architectural projects. It's not as if you would exclude materials because they are outside the marked area. The map serves as a main scouting guideline and as a means to generate ideas.[26]

Superuse essentially denotes processes of practice-led research that emphasises the processes of handling material in the field and through research redesignating it for a new use. Finding material for the project through the utilisation of a harvest map approach is one method. It is one that the designers suggest can manifest itself in three main flows:

- One is the utilisation of production waste. This includes reworking cut-offs, leftovers and elements that cannot be recycled back into the normal flows of productions.
- The second type of superuse processes is the utilisation of materials that can be withdrawn from their production processes and reused in another context. This can include recycled elements and things sold and bought many times over: products superseded through changes in regulation, such as car parts or white goods.
- The third process of superuse is when a product has reached the end of its lifecycle. Old tyres, car parts etc. are all products that are often not recycled and end up in scrap yards or landfill.

Whatever the reason for the redesignation of materials, these processes aim to make the most of resources and cut down on the growth of landfill and waste. Superuse approaches reverse-engineer traditional methods of research by providing the answer through the extant material recovered and then requiring the designer to ask the questions of how to repurpose it. They require the designer not only to handle the material but also to immerse themselves in its various flows of production and use. The architect Jos de Kreiger states:

> Reversing the design process from giving shape to something that comes from your mind, to something that comes from the material itself. Listening to the what the material is capable of and putting that to the best use possible.[27]

The work of ROTOR, a co-operative design practice producing buildings and interiors as well as policy and economic proposals, is described as a 'Reverse-Engineering Methodology and Practice' by Alison Creba and co-founder Lionel Devlieger. Now a 20-strong practice, started in 2005, their approaches to handling material in the field, or sites of many locations, result in the disassembly of numerous objects, interiors and buildings. ROTOR's research focuses on their commission to reuse buildings, along with Rotor DC, a company that facilitates the purchase of the deconstructed products either online or via their showroom in Brussels. Practice-led research involves the disassembly of existing buildings through the analysis of the environment and the assessment of its possibilities for repurposing. For instance, in 2017 they spent three months disassembling Antwerp's historic city hall. The 1560 building was undergoing adaptation and ROTOR was asked to decommission several parts of the building that did not comply with contemporary fire standards and conflicted with the restoration of the building. As ROTOR stated, their work began with making an inventory of the building:

> After conducting a detailed inventory and evaluation of materials with high reuse values, the team carefully documented and disassembled a range of components; from parquet and limestone flooring to wall sconces and decorative mirrors.[28]

The high value of the reconstituted salvaged elements meant that their quick resale negated the business model of the whole of the salvage: a process of not just reinvesting the costs into the project but also of distributing the salvaged elements back into the community from which they were extracted. In essence, the money raised from the salvage work not only offset the costs of the project but also ensured the continuation of the life of the building back into the community of which it was a representative. The practice-led approach of disassembling, indexing, cataloguing and valuing the finds then initiated new questions around the market and how the salvaged pieces could be dispensed with. This approach is closely connected to superusers. As they state:

> The building next door is going to be demolished, you can make an inventory of all the elements that come out, measure them, weight them if necessary, map them and make them available to the public. So before it gets demolished, they can start to design with it and see what they can or want to buy from that building and put to reuse.[29]

Utilising the superuse approach as a strategy relies on the harvesting of easily available materials. This contrasts with the traditional processes of a project. It is practice-led research that relies on 'supply' as opposed to 'demand' in order to make space in new ways. In other words, a superuse project is reliant on what is available at any given time. This contingent approach often relies on efficiencies in securing materials. For instance, the HAKA office project designed by the Dutch practice Doepel Strijkers was built from the reuse of a variety of materials extracted from various building sites from around the port of Rotterdam. To facilitate this process, an inventory of materials for use in the design was developed and subsequently changed periodically. This was exemplified in the rethinking of the sourcing of a consignment of warehouse doors that could not be extracted due to the building being squatted days before demolition. Yet, an overabundance of textiles meant that the 'Urban Living Lab', a hub that would act as a catalyst for the area as a 'clean-tech activity' space, could utilise the material to create a wall constructed from eight tons of second-hand clothing. This enclosed a new auditorium. Both visual and acoustic separation was achieved through layering the clothes into a 600 mm deep timber frame, configured as a shelving unit. The wall was on wheels and could be reconfigured when needed (see Figure 3.7.6). Haka was the first office in

Figure 3.7.6 HAKA office by Doepel Strijkers Architects.
Source: Doepel Strijkers Architects – photographer – Ralph Kamena.

the Netherlands built on rigorous circular design principles, with not just reused materials but also detailed analysis on costs, CO2, even the distance travelled by the people making it. The outcome of the research demonstrated a 70% reduction of CO2, material and labour costs compared to a traditionally built interior.

All inventory approaches prioritise a closed-loop approach to practice-led research. Closed loop-ing assumes greater responsibility with material flows, design, delivery and issues around the obso-lescence and legacies of a project. A circular approach is transformative in that it asks the designer to consider the endurance of their work, that is, what life is beyond the immediate use you are putting the materials and the work to. As Jonathan Chapman states:

> We need to move away from linearity in our design thinking, and to reconnect with design on a more circular and systematic level . . . these new approaches require designers to take greater controls over material flows; closing the loop.[30]

Along with the blue economy, enacting the careful marshalling of the health of the oceans, superuse is the process of connecting diverse flows of production and products into effective closed-loop systems of circular production. For example, Blue-City in Rotterdam exemplifies this thinking. Among the dereliction of a swimming pool complex, alongside the River Maas in Rotterdam, 30 businesses created an ecosystem that shared resources and utilised each other's waste. It's a true circular economy. Through practice-led approaches to research and non-linear thinking, material flows are carefully managed throughout the systems of the buildings' inhabitants. It's a model which Rotterdam is beginning to implement. There are designers, engineers, bio-scientists, growers, cater-ers, brewers – and it's the home to superuse architecture studios. Holland has had a circular strategy in place since 2016, and because of its port, Rotterdam is attempting to significantly reduce its raw material usage. Blue-City sits in the swimming pool complex, with the various offices, meeting rooms and workshops sited among the glass dome and colourful slides of the pool complex. Its originators state that:

> Rotterzwam mushrooms . . . feed off coffee waste produced by BlueCity's Aloha restau-rant. The C02 emitted by the fungi is used by another BlueCity company Spireaux, to make Spirulina, an algae paste full of vitamins and minerals which can be used to make veggie burgers and smoothies. Once harvested, RotterZwam mushrooms are sold to Aloha and, shortly after, appear on the menu.[31]

Conclusion

Practice-led research approaches that foreground the redesignation of the existing foreground the various notions of site (discussed earlier), the working with what is found and the abilities to respond accordingly to the emergence of the unexpected. In essence, *inventories* are practice-led approaches to a research strategy that take what is already extant and the information as material for research, which in turn compacts it into and upon what is already conceived and built.

Normative linear-based processes of more-established research methods, such as the devising or the posing of a question and then contextualising and answering it, can be considered to be less effective in inventory-based practice-led research. This is because utilising an *inventory* approach promotes contingency and mediation upon the material qualities of the found. It results in an aggre-gated combination of material, process, research and design. Of course, cataloguing and taxonomis-ing have always been a part of the documentation of research, in particular, through the processes of collecting and organising information. But inventories prioritise practice-led research that primarily takes place through the auditing and collating of extant on-site material, handling it, cataloguing it and then redeploying it back into the work. In essence, *inventories* approaches in site-based practice-led work can engender ways of working that offer alternatives to conventional ways of researching.

Perhaps more importantly, two things emerge from these approaches which may be of use to the reader:

1   The retention of existing matter responds to one of the foremost challenges of design education: the discarding of orthodox models of design that prioritise the *new*. Instead, working with the existing releases the practitioner, the student, the researcher, the academic from the anxieties of authorship. Appropriating existing matter requires educators, students, practitioners to rethink the formation of their approaches. Practices of inventorising require the raising of very different questions with regards to the insertion of agents into the flows and systems of the formation, production and construction of what is already in existence, in order to recast and subsequently redesignate knowledge and practice. Changing behaviours in approaches to the sensibilities that favour working with the not new, as opposed to the unfettered new. Arguably, in a climate-emergency, these approaches must now be a priority for all educators and practitioners.

2   Perhaps more useful is that through the examining and utilising of the practices of appropriating and working with the work of others, the exploration of the political, economic, cultural forces that have formed work is essential. It is through these processes that questions of race, class, gender can be raised. Inventories approaches require the development of processes of choice and edits, judgements that require particular sensibilities with regard to what exactly is retained for continuation, how that extension of its value will be articulated and with what additions to its existing materials.

Finally, Inventories approaches prioritise what is already on-site. Having the material close at hand with which to reappropriate the existing foregrounds design processes that are contingent and reliant on what is already in situ. This requires a very different approach to the usual origins of formation in creative endeavours such as design. It impacts upon research because it raises questions about how materials, buildings, cities will incorporate the continuance of their contents through redistribution and redesignation, rather than discard and waste. This challenges traditional approaches to research through prioritising reflection upon the extant matter and how it might suggest its redesignation, and then into what systems and flows is it then redeployed. The combination of site-specific or localised approaches not just to the interior, but all site-related creative practices, united by the tacit agreement that agents engaged with the discipline are understood to be reshaping what is already with us and will ensure that inventory-based approaches can have a central role to play in practice-led research in the 21st century.

Notes

1 Editor's note: the term *practice-led* as used in this chapter is defined in the same terms as this handbook does *practice-based*. This is an example of a field such as interior design's adoption of this term as a community. This is also evident in other chapters in this handbook authored by researchers from design in general.
2 Bolt (2019).
3 Smith and Dean (2009).
4 Ingold (2013).
5 Barnhart (1988, p. 542).
6 Rendell (2013, p. 117).
7 Ibid.
8 Ibid.
9 See Brooker and Stone (2018), Brooker (2016), Scott (2007), Cramer and Breitling (2007), Littlefield and Lewis (2007), Plevoets and Van Cleempoel (2019), Stone (2019).

10 Walker (2015, p. 83).
11 Burns and Kahn (2005, p. 12).
12 Ewing et al. (2011).
13 Speed (2011, p. 63).
14 Heidegger (1962).
15 Bolt (2019, p. 30).
16 Bennet (2010).
17 See Maiklem (2018).
18 Ibid.
19 Preston (2012, p. 94); see also Brewster (2009, p. 130).
20 Carless (2013, p. 158).
21 Bolt (2019, p. 33).
22 Hegewald and Mitra (2012, p. 3).
23 Fortini-Brown (1996, p. 6).
24 Chau (2018, p. 115).
25 Van Hinte et al. (2007, p. 3).
26 Ibid.
27 Interview with Jos de Kreiger. *Architecture Now*. August 2016. https://architecturenow.co.nz/articles/archi tect-interview-jos-de-krieger/ (accessed January 23, 2020).
28 Creba and Devlieger (2019, p. 98).
29 Ibid.
30 Chapman (2017, p. 162).
31 Langley (2019).

Bibliography

Barnhart, R. (1988). *Dictionary of Etymology*. Edinburgh: Chambers Publishing.

Bennet, J. (2010). *Vibrant Matter: A Political Ecology of Things*. Durham: Duke University Press.

Bolt, B. (2019). The Magic Is in the Handling. In E. Barrett & B. Holt (Eds.), *Practice as Research: Approaches to Creative Arts Enquiry*. London: Bloomsbury Visual Arts.

Brewster, A. (2009). Beachcombing: A Fossickers Guide to Whiteness and Indigenous Sovereignty. In H. Smith & R. T. Dean (Eds.), *Practice Led Research, Research Led Practice in the Creative Arts*. Edinburgh: Edinburgh University Press.

Brooker, G. (2016). *Adaptation Strategies for Interior Architecture & Design*. London: Bloomsbury.

Brooker, G., & Stone, S. (2018). Rereadings: Interior Architecture and the Principles of Remodelling Existing Buildings. *RIBA Enterprises*, 1–2004, 2–2018.

Burns, C., & Kahn, A. (Eds.). (2005). *Site Matters- Design Concepts, Histories, and Strategies*. London: Routledge.

Carless, T. (2013). An Architectural Gleaning. In G. Cairns (Eds.), *Reinventing Architecture and Interiors*. London: Libri Publishing.

Chapman, J. (2017). Product Moments, Material Eternities. In D. Baker-Brown (Ed.), *The Reuse Atlas: A Designers Guide to the Circular Economy*. London: RIBA Publishing.

Chau, H. W. (2018). Rapid Urbanization and Wang Shu's Architecture: The Use of Spolia and Vernacular Traditions in China. In R. Crocker & K. Chiveralls (Eds.), *Subverting Consumerism: Reuse in an Accelerated World*. London: Routledge.

Cramer, J., & Breitling, S. (2007). *Architecture in Existing Fabric*. Basel: Birkhauser.

Creba, A., & Devlieger, L. (2019). Deconstructing Research: A Reverse Engineering Methodology and Practice. In D. Saunt (Eds.), *The Business of Research: Knowledge and Learning Redefined in Architectural Practice*, 3(89), May–June. New York: Wiley.

Ewing, S., McGowan, M., Speed, J., Bernie, C., & Clare, V. (Eds.). (2011). *Architecture and Field/Work*. London: Routledge.

Fortini-Brown, P. (1996). *Venice and Antiquity*. New Haven: Yale University Press.

Hegewald, J. A. B., & Mitra, S. K. (2012). *REUSE: The Art and Politics of Integration and Anxiety*. Thousand Oaks, CA: Sage.

Heidegger, M. (1962). *Being and Time*. New York: SCM Press.

Ingold, T. (2013). *Making*. London: Routledge.

Langley, E. (2019). Could London Take Inspiration from Rotterdam's Zero-waste Blue-City? *Evening Standard*.

Littlefield, D., & Lewis, S. (2007). *Architectural Voices. Listening to Old Buildings*. Chichester: John Wiley & Sons.

Maiklem, L. (2018). *Mudlarking: Lost and Found on the River Thames*. London: Bloomsbury Circus.

Plevoets, B., & Van Cleempoel, K. (2019). *Adaptive Reuse of the Built Heritage: Concepts and Cases of an Emerging Discipline*. London: Routledge.

Preston, J. (2012). A Fossick for Interior Design Pedagogies. In K. Kleinman, S. J. Merwood-Salisbury, & L. Weinthal (Eds.), *Aftertaste: Expanded Practice in Interior Design*. Princeton: Princeton Architectural Press.

Rendell, J. (2013). A Way with Words: Feminists Writing Architectural Design Research. In M. Fraser (Ed.), *Design Research in Architecture*. Surrey: Ashgate.

Scott, F. (2007). *On Altering Architecture*. London: Routledge.

Smith, H., & Dean, R. T. (2009). *Practice Led Research, Research Led Practice in the Creative Arts*. Edinburgh: Edinburgh University Press.

Speed, C. (2011). Field Work and Site: Introduction. In S. Ewing, M. McGowan, J. Speed, C. Bernie, & V. Clare (Eds.), *Architecture and Field/Work*. London: Routledge.

Stone, S. (2019). *Undoing Buildings*. London: Routledge.

Van Hinte, E., Peeren, C., & Jonger, J. (2007). *Superuse: Constructing New Architecture by Shortcutting Material Flows*. Rotterdam: 010 Publishers.

Walker, S. (2015). Demystifying Cultural Research Methods. In A. Dye & F. Samuel (Eds.), *Demystifying Architectural Research*. London: RIBA Publishing.

3.8

REFLECTIVE PRACTICE VARIANTS AND THE CREATIVE PRACTITIONER

Linda Candy

Introduction

In this chapter, the nature of reflective practice is explored and reframed, drawing on studies of creative practitioners. From this research, I have learnt that the creative process of each individual, regardless of field or discipline, involves conscious reflection in practice, as well as what I describe as 'non-reflective' processes of creativity and artistry. What is evident across all creative fields is that the making of works is in itself an 'investigation' in which the practitioner explores new ideas and experiments with materials. Such investigations run in parallel with the design and making of artefacts and installations, and there is a continuous appraisal of whether something works well or does not. These iterative processes are key to the way a practitioner creates and reflects. The process of making something can facilitate a form of 'thinking through making'[2] as the practitioner moves towards knowing how to move forward. The process can give rise to challenging questions that require answers in the immediate and longer term, depending on the kind of work undertaken. Appraising works usually involves asking questions such as: 'Does it help me understand where to go next?' 'What have I learnt from this?' Sometimes it is important to experience surprise, and some people work in ways that provoke the unexpected.

By focusing extensively on creative practice from the practitioner perspective, I came to understand that the differentiation between reflection-*in-action* and reflection-*on-action*, as Schön characterised it,[3] is too broad to account for the variety of reflective activities and situations in the creative process. In creative practice, reflective thinking occurs at different levels of granularity during and after actions take place. Several additional types of reflective thinking were identified from interviews and studies of creative practitioners; these variants, on reflection in and on action that occur throughout reflective creative practice, are the main subject of this chapter. Before that, a note on method.

Gaining insight into the innermost thinking and working practices of the creative practitioner can be done in different ways, and I have tried many of the known approaches and methods. These include interviews that facilitate the opening up of thoughts that might otherwise remain hidden; diaries recording everyday events and ideas which give a sense of immediacy that only personal narratives offer; and observations of creative work in progress in the studio and in public exhibitions, which are invaluable for what they can reveal about practitioner ways of seeing, making and reflecting. My primary sources are interviews and conversations with practitioners working in a wide variety of creative and professional fields. For the research, I used formal and informal methods of

DOI: 10.4324/9780429324154-30

gathering information. Interviews were carried out using a semi-structured method which centred around the history and nature of the practitioner's creative practice and its outcomes as well as their experience of collaboration and their awareness of reflection in practice.

The concept of reflective practice has contributed to the methodological foundations of practice-based research in a substantial way. The main subject of this chapter is reflective practice itself and additional types that were identified from the practice of creative practitioners. Reflective practice as a method for practice-based research is discussed in the lead chapter on methodology in Part III of this handbook.

Reflective practice

The term reflective practice is one familiar to researchers working in creativity, education, human–computer interaction and design. I first learnt to appreciate and apply it as a method in my own practice before going on to study it in the work of other practitioners. An understanding of reflective practice in its original form is a necessary precursor to considering what happens when we move the concept into the world of creative practitioners. If Schön's ideas are completely new to the reader, it would be a good idea to read the lead chapter to the section on methodology prior to carrying on with this chapter. For those familiar with the basic concepts, I give a brief reminder next.

Reflection within practice can be understood at its most basic as thinking about what you are doing in the moment of action or what happens as a result of your actions. On the basis of moments of reflective thinking in a particular situation, the practitioner learns what to do next time. In other words, reflection leads to greater awareness of the specific actions in hand and their consequences and any implications that require different actions. Being reflective throughout everyday practice is invaluable for learning how to be effective when faced with new situations and unexpected events. Reflective practice has benefits in increasing self-awareness, a key element of emotional intelligence and at the same time, it contributes to a better understanding of others.

Schön's contribution to our understanding of the way practice-based knowledge operates and evolves is pivotal. He argued that what was needed was a new 'epistemology of practice' which is 'implicit in the artistic, intuitive processes some practitioners do bring to situations of uncertainty, instability, uniqueness and value conflict' and that reflection in action was a 'legitimate form of professional knowing'.[4] By reflective practice, he meant the integration of thought and action within the specific situation of practice. Reflective practice involves taking actions and making judgements that are informed by the domain knowledge and wisdom of a particular professional field. Throughout Schön's writing, the words reflection and action are inextricably combined, and the concepts of reflection *in* and *on* action are key to understanding reflective practice and its challenge to the prevailing wisdom about the nature of knowledge.

In weighing up the significance of Schön's work, his theories continue to be extended in many fields of professional practice. This is particularly heightened where reflective practices have travelled furthest through schemes and courses for advancing professional competence. For Schön, the entire business of reflective practice is central to the way practitioners deal with uncertain, unique, unstable, conflicting situations, whether it is reflection before action, in the very moment of action or reflection sometime after the action. Practitioners are more likely to initiate reflection when uncertain as to how to move forward, and the mark of the highly expert professional is knowing what kind of thinking process will help. These features are exactly why such an approach lends itself to other forms of practice, including creative ones. The reader is advised to read more on Schön's thinking in the lead chapter on methodology.

In the following section, we take a look at creative practice and how reflection has similar features, but with some important differences from Schön's original concepts, which were derived from his protocol studies of professional practitioners such as management, town planning, psychotherapy and science-based professions.[5] Creative practice in the arts is influenced by situations that are different from those that typically face professional practitioners working in real-world situations under regulatory conditions governed by law. This can have implications for how reflection takes place which are situation dependent. In distinguishing between professional practice and creative practice, I do not wish to imply that professionals are not creative and creatives are not professional, and there is more to be said about the differing situations of practice for professional practitioners and creative practitioners: this issue is explored further in *The Creative Reflective Practitioner*.[6]

Creative reflective practice

In his article, 'How I Write Music', composer Nico Muhly[7] describes how he plans an 'itinerary', his word for the creative journey that he is about to embark on before any musical notes are created. Out of the process, specific questions arise that indicate the kinds of qualities he looks for in a particular piece of music, such as the dynamics or rhythmic complexity, the number of voices or instruments, which in combination determine the quality of the sound. The questions are typically:

> what are the textures and lines that form the piece's musical economy? Does it develop linearly, or vertically? Are there moments of dense saturation – the whole orchestra playing at once – and are those offset by moments of zoomed-in simplicity: a single flute, or a single viola pitted against the timpani, yards and yards away? These questions provoke moments of conscious reflection throughout the composition process.[8]

The answers he arrives at through 'conscious reflection' guide the music he makes and express his personal success criteria for assessing how well he has achieved his intended outcome. In a sense, these criteria represent expert practitioner knowledge which is usually tacit unless the reflection is articulated in the way Muhly does in his writing.

In creative practice, practitioners can develop their capacity for reflection over time and, in doing so, learn to progress their outcomes and their working methods. Throughout the process, there is appraisal of whether something seems to work or not; this may involve different levels of thinking, sometimes quick and decisive, sometimes slow and considered. Reflective practice in creative work involves many interwoven activities, as practitioners searching for understanding through making works of various forms. The process of making something facilitates thinking through making as the practitioner moves forward. Making a work and then reflecting on the process (the making and the process of reflecting-in-action) and outcomes (the thing that is made) is a pathway to understanding the practitioner's working hypotheses or 'theories-in-use', to use Argyris's term.[9] I discuss different categories of theories in my chapter, 'Theory as an Active Agent in Practice-Based Knowledge Development', in this book.

Reflection practice variants and the creative practitioner

In this section, the nature of reflective thinking in practice is examined and reframed. From my research into the thinking and making of creative practitioners, I have identified several variants of reflective practice drawing upon the views of practitioners as well as my observations of their work. This has provided me with a clearer picture of how creatives reflect through making.

Creative reflective practice is a multi-layered phenomenon that depends on the activity in hand and the particular situation of practice. How and when particular types of reflection occur depends upon the activity type, the point reached in the process and whether different actions are needed in the light of lessons learnt. Four kinds of reflection in practice were identified from studies of creative practitioners.

- Reflection-for-action
- Reflection-in-the making moment
- Reflection-on-surprise
- Reflection-at-a-distance

Reflection-for-action precedes action in the present moment as part of the intensive preparation required for certain kinds of actions. Practitioners contemplate their previous actions, thoughts and achievements in order to understand the implications of what has taken place and learn how and where to go forward. This includes reconsidering their existing works and products and reviewing relevant knowledge with a view to determining ways of proceeding. When making a physical form, a sculpture, for example, there is the roughing out on paper, testing types of materials, building models, discussion with other people and so on, during which conscious reflection takes place. There is a great deal of decision making to be done: 'I chose this material over this because it fits the particular purpose' or 'I'm going to work within this space because it is right for the piece', all of which involve deliberated thought. Past processes are brought forward into present projects and adaptations based on previous learning are incorporated.

A key element of reflection-for-action is the ability to identify the kinds of constraints that will have an impact on the anticipated work. Being faced with too many options can be paralysing, and it is a useful skill to have a way of handling the complexities and conflicts before embarking on a new work. If actions are to have the potential to move the practitioner forward, it is necessary to consider what has already been done and to assess the available options. This awareness makes it possible to reflect and learn from past outcomes and is preparation for future action.

Brigid Costello[10] is an artist working at the boundary of design and digital interaction to create participatory audience experiences. Her perspective on reflection-*for*-action highlights the need to handle constraints and reduce risk, which is a necessary element of making works for public spaces:

> Reflecting is mainly about reflecting about what the constraints of a project are. At the start of a project you are faced with an array of possible approaches and that can be paralyzing. So, when you say 'I am going to make something', you need to first make a few key decisions about practical constraints. For instance, is it going to be exhibited? How much time do I have? Do I have to send the materials overseas?[11]

Esther Rolinson[12] is an artist whose carefully structured approaches to making art involve a high degree of preparation through reflection-*for*-action. This preparedness is a precursor to intensive moments of making that require less conscious thinking. During the drawing process, when thoughts intrude, she consciously 'lets them go' as if in a meditative state. Her mind is prepared for the drawing actions in the moments to follow.[13] This is an example of reflection that takes place in the very moment of making, a category of reflection described next.

Reflection-in-the-making moment is a form of reflection-*in*-action that highlights the immediacy of certain activities involving inter-twined reflective thinking and making processes. This form of reflection can occur for relatively brief moments, often fleetingly in response to specific actions, sometimes in short breaks, sometimes brought about by external interventions or interruptions. They can be repeated frequently during a longer period of creative time. These moments consciously make space for reflection on the detail of a work in progress. It involves working with the 'material' of the situation, whether it is paint, musical notes or computer code. The scale of the activity is crucial and the timeframe embraces many pauses and breakpoints. These interruptions to the ongoing processes can happen at any point and for different reasons.

An example of *reflection-in-the making moment* is in the mixing of paints, for example as described here:

> I put the colour that I am thinking of putting on the canvass, onto the making tape so I can see that colour almost where it's about to go. It isn't perfect . . . so 'that is not going to work and I need to put a little more blue in it'. I can see it's too dark, needs to be a bit lighter.[14]

Here we see how the act of putting paint on canvas is just the first step to finding the right colour; it is a process that requires iteration because of the effect of light and surfaces on the artist's visual perception. Only by mixing the colour, applying and testing it by eye is it possible to judge what works for that painting on a specific surface, in natural light at a particular time of day. The reflection in the immediate moment involves assessing what the effect is each time a new layer of colour is applied. The colour is assessed visually but, note, not verbally, and therefore the process is not apparent to the observer. Uncovering this tacit knowledge requires the practitioner to reflect and articulate those reflections either verbally or in written words as the preceding artist was able to do.

There are several features of reflection-in-the-making moment which can occur in short turnaround moments as well as over longer periods of time The materials, the tools and the technologies are important elements that facilitate and, at the same time, shape the thinking and making process. Tim Ingold characterises 'making' as an inherently mindful activity in which things emerge from 'the correspondence of sensory awareness and material flows in a process of life'. He describes how learning takes place through a direct, responsive engagement with physical materials. For example, using wood as a material, a good craftsman requires both intellectual and physical responsiveness to create a well-crafted object.[15]

The reflection that takes place as practitioners manipulate and shape materials is integral to that process and so closely intertwined that there is often no perceptible difference between making and reflecting. Artists experience the act of drawing as a way of seeing, like 'a kind of reflective conversation' with the materials of a design situation.[16] Through drawing, the artist sees what is there, draws in relation to it, sees the result, judges its quality, learns from it and draws again. If, in this process, unintended consequences are discovered and judged to be good, this has a key role in justifying another move. An important background point is that our natural cognitive capacities may constrain the amount of prior consideration of all consequences relevant to our reflection on the result. That is why the process of iteratively drawing–seeing–drawing again is essential to the evolution of the practitioner's understanding. Reflective practice involves iterations of: *making–seeing* → *reflecting* → *making again* → *reflecting again*. The learning that goes on throughout the process advances knowledge in practice. This can help to build knowledge and foster the development of practitioner personal frameworks, for which conscious reflection is vitally important. For many practitioners, reflecting in the making moment is a significant part of their creative practice, and this extends to areas of creative activity such as improvisation.[17]

Reflection-on-surprise is a category of reflective thinking that has considerable value for creative practitioners and practice-based researchers, depending on how they respond to the unexpected. According to Schön, new knowledge can be acquired when practitioners move in a way that has a surprising outcome. He identified two ways in which the moves are surprising: 'desirable' and 'undesirable' surprise. Undesirable surprise can be negative because, having anticipated a particular result from taking an action and what occurs is not what you expected, this may suggest that your ideas are unsatisfactory ('incomplete') in some way. To address this, the practitioner is forced to come up with an explanation as to why this has happened and then use that understanding in the next action. For Schön, desirable surprise did not pose such a problem because if the outcome of action is both surprising and desirable, the practitioner has no need to re-think what has been done.[18]

Surprise in creative practice and in practice-based research comes in different guises, and there are more kinds of surprise than Schön's categories of desirable and undesirable suggest. Practitioners may respond in different ways, depending on the nature of the surprise and the context in which it is encountered, and some may engineer surprise in order to jolt them out of a familiar path into a new direction. The creative practitioner is frequently open to what a surprise might offer, even where it means having to reject and abandon things already done. When new ideas emerge as works are created, the results can give the practitioner surprises that are welcome as well as those that are not. In creative practice, those that displease can be disregarded or corrected, while the pleasing surprises offer opportunities to explore unplanned avenues. For some, the natural response might be to reject an unexpected development in the work and start over, while others may see it as an opportunity to follow the surprise in a new direction. Good surprise involves recognition and a positive response to 'go with it', either trusting one's instinct or relying on one's judgement, acquired through years of experience. A bad surprise can prompt reflection on what went wrong: 'how did that happen?' 'How can I move forward?' Some artists provoke surprises while others are more concerned to give themselves challenges in a different way, a manner that is less random but with the potential to throw them off course. The need is to come up with something that is both stimulating and satisfying at the same time as learning from rising to the challenge.

Experiencing surprise may evoke a variety of responses from practitioners. For some, provoking surprise is intentional, and using mechanisms to achieve this is part of their practice. Actions such as dropping leaves or string or paint splashes or introducing chance elements are familiar ways of bringing the unpredictable unexpected into a creative process. Marcel Duchamp and John Cage are well-known examples of artists who exploited this notion in their work. Duchamp's method of research and reflection addressed the notion of invention in what was possible. When asked which work he considered to be the most important, he cited *3 Standard Stoppages* (1913–1914), a work that used chance as an artistic medium. The idea of letting a piece of thread fall on a canvas was accidental, but 'from this accident came a carefully planned work'.[19]

Another way of provoking surprise appears in a branch of generative art that uses models of artificial life. The basic idea is that the practitioner creates software that sets a process in motion in which each step relies on some rules and events that are not known in advance. Techniques have often been based on cellular automata, a mathematical system that can be thought of as a way of modelling living processes of birth, death and evolution, themselves unpredictable processes. An example of such work is that of Dave Burraston, who created experimental music with generative processes using cellular automata and through his practice-based research contributed to new knowledge in that field.[20]

But does 'desirable' surprise have *no* place in creative practice? Stephen Scrivener sees the potential for creative thinking in the 'desirable' surprise that Schön considered unsuitable for reflection because you don't have to attend to it in any way. Responding to it positively allows the practitioner freedom to explore where it leads without reflecting on why it happened.

> I understand now that desirable surprise is a crucial aspect of practice . . . you do something and that produces an unintended outcome, but you can like it, you can find it appealing and you can just go with it, follow it.[21]

Stephen expands on this theory and his creative practice in an interview available online.[22]

When faced with an unwelcome development in a work in progress, the initial reaction might, not unnaturally, be to check to see if what was seen was really there and then wait some time before working out what to do next. How a practitioner responds to unforeseen situations is crucial to an understanding of that individual's working practice. In particular, it provides us with insights into how initial intentions are altered by engaging in a creative process that involves appraising the outcomes as they emerge from the making itself. The very experience of seeing or hearing something unexpected in a work that does not feel right has the potential to stimulate reflection about the making method itself.

In creative practice where surprise is a driving force, it extends beyond making of works. If practitioners like what happens, they can proceed along the line suggested by the pleasing outcome without having to reassess it. This can be simply a matter of following through or repeating the same process to see what happens next. It is seldom talked about when it comes to reflecting on finished artworks, and yet a practitioner's ability to learn and become better at what they do may depend upon how they respond to unexpected and surprising events arising from their actions – whether pleasing or not.

Reflection-at-a-distance is a type of reflection-*on*-action that can occur when a degree of detachment from the process is invoked or is warranted. The creative work will usually have reached a sufficiently developed state to allow for a change of space and viewpoint. There are ways in which distance can be achieved:

- The first is to change the context of the work from the practitioner's perspective.
- The second is to expose the work to other perspectives outside that of the practitioner's own experience.

In both cases, this is a way of stimulating reflection in the practitioner by breaking with the familiar existing status of the work either in progress or at completion.

The first audiences or viewers are the practitioners themselves. They are also the ones closest to the process and the outcomes. For those without ready access to willing observers, there are, nevertheless, ways of achieving distance. By placing the work in a different context, the practitioner's perspective can be altered by seeing or experiencing it in a different kind of space. For example, changing the viewing environment by taking works out of the studio to a different location, or changing the form by transferring a hand-written text to a screen and then producing paper copy. Altering the form and presentation of a work in progress can throw it into relief and reveal aspects not previously considered.

There is always a sense of excitement when talking to practitioners who are in the middle of creating new works and are seeking to encourage and provoke responses from others. It seems as if a crucial ingredient of living creative practice is never tiring of what happens when you reveal your work to the public and see it through their eyes and their behaviour. In the example that follows, the audience plays a key role in the artist's reflections on her creative practice and the installations she makes.

Julie Freeman[23] explores the relationship between science, nature and interaction with the works she makes. In her creative practice she experiments with transforming complex processes and data sets into sound compositions, objects and animations. Her focus is on the use of interactive digital technologies and scientific techniques to manipulate an audience's senses. She aims to create experiences that engender different states in those who participate, a form of *reflection-at-a-distance* that enables her to see her work in ways she has not anticipated through the way others engage with it.[24]

By placing creative outcomes in a different context and exposing them to the world, practitioners give themselves opportunities to experience them afresh and through the eyes of others. It is hard to reflect when things are too close to hand–body–mind, and the physical environment can heavily influence the experience of them. Factors such as the nature of the light or acoustics in a studio, the quality of the materials and how they are combined in different spaces and the sheer familiarity of the working environment affect one's appreciation and understanding of what is there.

The four variants of reflective practice described here were identified from different creative fields of practice in the arts. However, this does not mean they do not occur in other domains and disciplines, and may also be found in professional practice, as discussed by Michaels in this book.[25]

Non-reflective actions

Having described features of reflective practice that are arguably always conscious and rational processes, I wish to acknowledge an aspect of practice that is frequently experienced by many and is sometimes referred to as 'intuitive', sometimes as being 'in the flow', among other descriptors. Many creative practitioners, irrespective of their field, know the feeling of being immersed in the activities as if they were led by the instrument or material in use. In those kinds of situations, we might think of the scenario as 'non-reflective', where we act but are not conscious of thinking about the minutiae of the specific actions, for example when making improvised music or drama. Creative actions sometimes seem to come almost automatically from deep within, perhaps from emotional or aesthetically charged forces. Many people experience a phenomenon where they do something without being consciously aware, but the underlying reasons for this are hard to unpick. There are a number of dimensions to this. A simple statement like 'I don't think about it. I just do it' expresses a familiar experience for many when we take an action that seems to come out of nowhere, as if we are acting almost involuntarily. We have not consciously had a thought that says 'now draw these lines' but have acted spontaneously, perhaps responding to the feel of the pencil moving on paper. These moments are as if the unconscious mind takes hold and acts independently of the conscious mind. I have explored this phenomenon in more depth in other writing.[26]

Although non-reflective actions continue to elude agreed scientific explanation, they have an important place in creative practice because they are part of the felt experience of most creative practitioners. However, they represent only one part of the spectrum of actions that characterise the whole of practice. It is the rich variability of the many ways of practice in creative work that provides opportunities to learn new ways of thinking and making from those for whom it is central to their lives. This is also of great value to the practice-based researcher from whichever field, discipline or domain of expertise because of the opportunities it affords for acquiring new methods and techniques for learning how to be more reflective.

Learning from the creative reflective practitioner

Being a reflective practitioner means cultivating the many ways we can learn through experience. This can be expressed in simple terms as thinking about what you are doing or what happened as a result of your actions and then deciding what to do differently next time. Being reflective through everyday practice is essential to learning how to be effective when faced with new situations and unexpected events. We learn from the practitioners that I interviewed that living creative practice involves strong motivation, determination and an ability to manage uncertainty and take risks. Reflection in their creative practice facilitated the practitioners' investigations and their ability to move forward through the making process. In this way, reflection played a vital role in enabling them to learn from the process and its outcomes. In particular, we can learn from the variant forms of creative reflective practice that were identified. A quick resume of the main features follows.

- *Reflection-for-action* involves extensive preparation, including constraint identification and devising structured approaches before the main activity begins. It means having a fairly systematic working practice that involves contemplating previous actions, thoughts, and achievements and reassessing prior strategies and solutions. To move forward, the practitioner researcher needs to assess relevant information and identify any constraints that might influence future actions.

- *Reflection in-the-making moment* is prompted by external factors such as interruptions or, more frequently, pauses imposed by the uncertainty of what to do next. It implies simultaneous thinking and action in the immediate situation. Expressions like 'thinking on one's feet', 'thinking with the hands' and 'thinking through the body' capture some of the experiences of reflection in the making moment. In this situation, a practitioner researcher needs alert senses: the looking, listening, feeling that is indicative of heightened awareness of mind and body and requires a degree of conscious awareness that goes beyond intuitive actions and is critical to dynamic flow states. This is a state where conscious and unconscious actions seem to work in parallel.

- *Reflection-on-surprise* requires a flexible attitude to the unexpected. Positive responses to surprises imply a willingness to embrace challenges. Even where a surprise outcome is not welcome – is 'undesirable', to use Schön's term – it can prompt valuable questions. Some practitioners provoke surprise as they look for ways to identify and disrupt tacit assumptions and, in doing so, learn something new. In these ways, being open to surprise can bring rewards, and engineering surprise as a personal working method can provoke questions. Unexpected surprises can prompt reflection and lead to new directions of travel.

- *Reflection-at-a-distance* happens when the process is in a sufficiently developed state to allow for a change of space and viewpoint. It can be provoked by external factors and events. It can be achieved through detaching from the process in hand to disrupt over-familiarity with work in progress. By moving location or materials or tools, the practitioner researcher can change focus and sometimes arrive at new insights or alter the direction of travel. The same effect can also be achieved by participative practices that involve exposing one's process and outcomes to others and responding to what you learn from their feedback.

Does working creatively facilitate or encourage more reflection in practice? Or does reflective practice promote creative thinking? I believe that working creatively leads to more reflection overall because of the close interchange between making and assessing the results. Creative practitioners exhibit different kinds of reflective practices that go beyond known reflection in and on action categories. In learning from these examples, practitioners can expand their reflective repertoire.

People are often urged to learn to 'think outside the box', but how do we achieve that? One way that seems to be important is learning to take a lateral thinking perspective.[27] But how does this work in practice as distinct from learning general-purpose techniques? One way of stimulating more reflection is to deliberately adopt a tangential perspective in relation to one's familiar ground, area of expertise or everyday practice. This kind of approach can be seen in scientists and artists alike. By taking an unusual perspective, the basis for challenging existing assumptions and opening up new avenues of exploration is established in everyday practice. This is the kind of reflective ethos that may ultimately lead to larger challenges to the existing canon of knowledge.

How will we know if we are learning to be a reflective practitioner? A quick way is to recognise some key features. Here are a few hints on how to inhabit a more creative mindset:

- Embrace doubt and scepticism as a way of provoking new ideas.
- Adopt playful behaviour and take initiatives to go outside your comfort zone.
- Respond to the materials you use for making things and seek out novel techniques.
- Be open to different perspectives that stimulate your creative thinking.
- Shift your normal boundaries and refuse to rest easy in a familiar space.
- Seize unexpected opportunities and follow unfamiliar paths.
- Take courage as you search for new and fertile ground.
- Reflect and reflect and reflect again, but never forget to follow your instincts and act when the moment feels 'just right'.

And finally

Does being a reflective practitioner lead to better practice?

Many people think so. The growth of advice and guidance in, for example, the medical, nursing and legal professions is testimony to the success of the claim that being reflective in practice is key to improving professional expertise. Reflection in and on practice has been made 'official', even mandatory, as new codes of practice and guidance on how to reflect as part of continuous professional development have been introduced.

Can *creative* reflective practice be learnt? This experienced creative practitioner thinks so:

> To me the reflection is essential. It is not an added extra that makes everything better. The whole thing is a life-long quest and if you don't reflect on what you are doing and how you do it, I can't see anyway that life quest is progressing. I can't believe that any artist worth their salt doesn't spend a lot of time reflecting. I throw away works- this was a reflection on what had been done. As I moved from 50 paintings a year, and throwing away 45 to ten paintings a year, I looked back on it and found I kept the same number 5 or 6. Obviously my reflection was leading to a better prediction of what was going to work. I didn't produce any more that were any good.[28]

Human beings are thinkers and makers by nature and, given the opportunity, will seek to create and to learn from experience. As we see from the examples described in this chapter, practitioners have very variable approaches when it comes to working creatively as well as professionally. As the people who informed my writing for the *Creative Reflective Practitioner*[29] book exemplify, there are multiple ways to be a reflective practitioner. Above all, practice involves continuous learning in which the process of reflection can act as a method for learning through practice.

For further discussion on adopting reflective practice as a method and how and what to document in the practice-based research process, I refer you to the lead chapter in Part III.

Notes

1 Candy (2020).
2 Thinking through making: creative practice involves a continuous exploration of ideas, materials and tools both prior to and during the making process. The process of making something can facilitate a form of 'thinking through making' as the practitioner moves towards a clearer understanding. In his 1934 book *Art as Experience*, Dewey considers the question as to why art is so bound up with making (Dewey, J. (1934) *Art as Experience*, Penguin Books: London, pp. 48, 50): 'Art denotes a process of doing or making' and 'Man whittles, carves, sings, dances, gestures, moulds, draws and paints'. Noë asks the question of *where* do we think, and makes a case for extending the landscape from the brain into the body and the world beyond (Noë 2015: Chapter 3, pp. 27–28). As he puts it: 'Thinking is more like bridge building or dancing than it is like digestion' (p. 27).
3 Reflection-*in*-action: (Schön 1991, pp. 68–69); reflection-*on*-action (Schön 1991, p. 26).
4 Schön (1991, p. 69).
5 Schön, D. A. (1991). *The Reflective Practitioner: How Professionals Think in Action*. Aldershot: Ashgate Publishing Ltd. First published by Basic Books in 1983.
6 Candy (2020); Schön (1991), chapters 2 and 3.
7 Nico Muhly: http://nicomuhly.com/biography/ (accessed January 21, 2021).
8 Muhly (2018).
9 Argyris et al. (1985, pp. 82–83).
10 Brigid Costello: www.arts.unsw.edu.au/our-people/brigid-costello.
11 Brigid Costello: in Candy (2020), chapter 3, 'Reflective Creative Practice', pp. 54–55.
12 Esther Rolinson: www.estherrolinson.co.uk (accessed January 21, 2021).
13 Esther Rolinson: in Candy (2020), chapter 3, 'Reflective Creative Practice', pp. 58–59.

14 Ernest Edmonds: Interview, p. 5: http://lindacandy.com/CRPBOOK/cases/interviewPDFs/CRP-Edm onds.pdf.

15 Ingold (2010).

16 Schön and Wiggins (1992).

17 Candy (2020, chapter 3, pp. 57–60).

18 Schön (1991, p. 153).

19 Molderings (2010).

20 Burraston (2007).

21 Stephen Scrivener: in Candy (2020), chapter 3, 'Reflective Creative Practice', p. 69.

22 Stephen Scrivener: http://lindacandy.com/CRPBOOK/scrivener.

23 Julie Freeman: https://translatingnature.org/about/ (accessed January 23, 2021).

24 Julie Freeman: in Candy (2020, pp. 65–67).

25 Michaels (2022).

26 Candy (2020, chapter 3, pp. 60–63).

27 Lateral thinking is a term coined by Edward de Bono in 1967.

28 Ernest Edmonds: Interview, p. 6. http://lindacandy.com/CRPBOOK/cases/interviewPDFs/CRP-Edm onds.pdf.

29 Candy (2020).

Reference

Argyris, C., Putnam, R., & McLain Smith, D. (1985). *Action Science: Concepts, Methods, and Skills for Research and Intervention*. San Francisco and London: Jossey-Bass Publishers Inc., 82–83.

Burraston, D. (2007). Fundamental Insights on Complex Systems Arising from Generative Arts Practice. *Leonardo*, 40(4), 372–373.

Candy, L. (2020). *The Creative Reflective Practitioner*. London and New York: Routledge.

Dewey, J. (1934). *Art as Experience*. London: Penguin Books.

Ingold, T. (2010). The Textility of Making. *Cambridge Journal of Economics*, 34, 91–102.

Michaels, J. (2022). Interdisciplinary Perspectives on Practice-Based Research. In C. Vear (Ed.), *The Routledge International Handbook of Practice-Based Research*. London and New York: Routledge.

Molderings, H. (2010). *Duchamp and the Aesthetics of Chance*. New York: Columbia University Press.

Muhly, N. (2018). How I Write Music. *London Review of Books*, 40(20), 38–39, October 25.

Noë, A. (2015). *Strange Tools: Art and Human Nature*. New York: Hill and Wang, Farrar, Straus and Giroux.

Schön, D. A. (1991). *The Reflective Practitioner: How Professionals Think in Action*. Aldershot: Ashgate Publishing Ltd (First published by Basic Books in 1983).

Schön, D. A., & Wiggins, G. (1992). Kinds of Seeing and Their Functions in Designing. *Design Studies*, 13(2), 135–156, April.

3.9

REFLECTION IN PRACTICE

Inter-disciplinary arts collaborations in medical settings

Anna Ledgard, Sofie Layton, and Giovanni Biglino

Introduction

These are exciting times for artists, researchers, and clinicians who seek to work in collaboration. There is a growing understanding of the value of the contribution of artists to medical spheres, as indicated by findings of the All Party Parliamentary Group Arts and Health in the UK in 2017[1] and by more recent evidence published by the World Health Organisation in 2019.[2] Reflective arts practices, used commonly in education settings, can be a vital scaffold to such work. A culture of reflection is essential to arts practice and, effectively employed, is a cornerstone of any collaborative interdisciplinary arts process. Our definition of reflection in this context draws on Dewey[3] and Schön.[4] Dewey describes reflection as thinking which involves linking between aspects of thought:

> The successive portions of the reflective thought grow out of one another and support one another . . . Each phase is a step from something to something . . . The stream or flow becomes a train, chain, or thread.[5]

This chapter describes different contexts for reflective thinking, and different protagonists engaging in reflection; while reflection in creative processes may not be linear, the reflections we describe are interconnected and action is taken as a consequence of such reflection. In Dewey's words: 'Reflection thus implies that something is believed in (or disbelieved in), not on its own direct account, but through something else which stands as witness, evidence, proof, voucher, warrant; that is, as ground of belief'. So, the deep listening of an artist to the testimony of a patient gives rise to reflection, which informs the truth of the artistic response. Without such 'witness[ing]' and 'evidence' of experience, the artist's search for truth would be compromised. In Schön's writings, reflection and action are inextricably combined, and similarly the reflection we describe is not randomised thinking but is rather deliberately facilitated and framed and informs subsequent thought or action, or indeed the choice not to act or to act differently.

This chapter will explore how reflection has been used in medical settings with diverse communities – clinicians, producers, artists, carers, and patients – drawing on a body of powerful artistic practice rooted in deep participation with patients in hospital settings and resulting in high-profile public outcomes reflecting on the experience of illness from both patient and clinical perspectives.

DOI: 10.4324/9780429324154-31

Reflective practice in the field of arts and health

Making a case for participatory arts practices with partners from other disciplines such as medicine can be challenging. The field of participatory arts practice is not well theorised: much less has been written about it than other areas of the contemporary arts. While the education sector has made strong progress in defining reflective arts practices in schools,[6] patient and public engagement with the arts in health settings is a newer area which is growing fast. With it is a recognition that engaging with patients requires an arts practice which has strong values and principles of participation and integrates opportunities for reflection.

Matarasso, in his extensive work on participatory arts practice,[7] points to the importance of reflection processes for all who engage with the work on ethical grounds, with the making of art concerned with the creation of meaning 'with ethical relations between human beings and social power'.[8] Reflection matters at every point in the engagement of artists with others, and even more so in settings such as health and medicine where artists are facilitating patients and health partners, perhaps for the first time, into experiential processes of powerful and at times deeply personal creative engagement and self-discovery.[9] Reflective practice and an evaluation process representing the interests of all collaborators are key factors in ensuring quality of process in such engagement, particularly in interdisciplinary relationships where aspirations and assumptions at the outset could be very different.

A carefully managed environment for reflection should ensure that perspectives can be shared from different standpoints. Above all, reflection is at the very heart of what the artist does, whether they name it as such or not: the process of creativity is one of constant dynamic and reflexive reflection on what is known or not known and of testing this knowledge against the new in order to derive new understandings. As Candy explains, the artist uses 'tacit knowledge and insights based on experience through practice'.[10] The examples that follow weave in and out of Candy's four principles of *reflection for action*; *reflection in the making moment*; *reflection on surprise*; and *reflection at a distance*, providing practical examples to these principles, which are fully expanded in Candy's work.[11]

Principles

Key principles underpinning the practice discussed in this chapter are reported in Table 3.9.1. Keeping to these principles requires skilled coordination and a dynamic approach to reflection and forging rich complex partnerships:

> All collaborative and socially engaged arts practices require negotiation; around ethics, ownership and representation in the planning, delivery and evaluation of projects. . . . These negotiations are not a burden; they are part of a crucial dynamic, because the work is always in relationship; to others, to organisations, to spaces. Dialogue, translation, debate, barter, adjustment, insistence and appreciation are all part of these acts of engagement.[12]

This work (i.e. the end result generated through this practice) can be broadly framed as referring to the practice of narrative medicine.[13] Theoretically rooted in narratology, aesthetics, and phenomenology, narrative medicine is based on witnessing healthcare encounters with attention, representation, and affiliation. As narrative scholar Rita Charon discusses: 'we need to tell, perform, write, paint, sculpt, compose these complex experiences in order to see them, and we need readers or viewers or listeners to help us understand what we ourselves have created', further quoting John Berger on 'the act of giving form' to what is witnessed, what is received.[14] Deep attention and thoughtful representation make affiliation (between the doctor and the patient, the teller and the receiver) possible. The work is values-led, including a producing framework encompassing

Table 3.9.1 Key principles underpinning the practice discussed in this chapter.

Principle	Description in arts-and-health context
Time	Projects are long-term in nature (2–5 years) and rooted in a sustained, reflexive dialogue between artists and medical or educational partners
Participant-centred practice	Projects create entry points for a range of participants so that each can express their own life experience, illness narrative, culture, values, curiosity, and passion through artistic engagement
Facilitating role	Projects include a facilitating role (producer or relational curator) which frames the role of artist in the non-artistic realm
Interdisciplinary and collaborative art practice	Projects build on individual narratives and collectively generated ideas, irrespective of arts discipline
Relational practice	Artists offer reflexive workshop practice which seeks to be responsive and open to the narrative content, rather than having a fixed relationship to a particular arts product or discipline
Co-creation	Projects are characterised by shared planning with partners and in consultation with participants who are co-authors in the creative process wherever possible
Cross-cultural	Hospitals and schools reflect social and cultural demographics and responsive arts processes seek to represent and give voice to diversity of experience
Cross-sector	Projects engage individuals and institutions from the worlds of medicine, science, education, arts, participation
Ethicality	Ensuring trust, respect, and care for all involved
Shared culture of reflection	Projects include regular planning and review supported by institutional and producing partners in a culture of reflection and learning shared by all

institutional relationships, project management, finances, communication, and oversight of care and reflective processes.[15]

In order to further unpack these principles in practice, this chapter will examine four stages in the life of a project: 1) project conception; 2) engagement with patients; 3) translation and representation of patient narratives; 4) presentation to a public audience, drawing from four different examples.

Project conception (case study: *The Barometer of My Heart*)

In conceiving a project involving multiple partners, reflective practices are employed both to evolve trust between key players and to draw up shared objectives for the project. *The Barometer of My Heart*[16] was an exploration of men's experiences of erectile dysfunction (ED) and impotence through visual, sound, digital media, and performance. It emerged from five years of collaboration between artist Mark Storor, consultant endocrinologist Dr Leighton Seal, and producer Anna Ledgard. During the conception phases of a project, the principles of time (to build trusting relationships); co-creation (to arrive at shared objectives); and responsive engagement practices (in consultation with participants) were key.

In this case, the work was informed by encounters with men living with ED and workshops with men in male-dominated communities over a number of years. Storor was interested in the role the arts can play in illuminating subjects which are difficult to talk about, even taboo, i.e. masculinity and the loss of life potency with ageing. The work evolved through a genuine process of discovery, fuelled and informed by the people and experiences encountered along the way.

Reflective practice was essential to ensure that voices were heard and perspectives aired from different angles, but even before that phase, reflection was integral to the conception of the project.

There are several elements on which the artist or the team may reflect during project conception, and, as in the cases described next, in each case learning from reflection informs subsequent action. Elements to reflect on during project conception:

1   Previous work (or body of work), reflecting on a line of artistic enquiry begun in earlier work.

    In this case, the concern with masculinity begun in 2006 with the *Boy Child* project.[17]

2   A response to a societal need, news in the media, or a taboo in our society.

    Such as ED and its associated social and health risks.

3   Refinement of the plan for the ensuing work in response to partner or funder feedback and/or peer review, whereby fine tuning can occur without fundamentally altering the key theme(s) being explored.

4   Refinement of the project in response to funder's feedback or peer review, contributing to creating a context.

5   A response to encounters with participants or new collaborators revealing a facet of the theme(s) being explored that perhaps was not previously considered or offering different/novel tools to enrich the exploration of such theme(s).

    In this case, Storor attended a small number of consultations, and it became clear to the artist that the issue of male impotency presented a wider social challenge beyond the consultant's room. The men whom Storor met highlighted the genuine difficulty of talking about or seeking help for issues to do with sexual potency, with the result that it is often hidden behind stress, marital breakdown, or other chronic illnesses. This led to reworking the project, extending the workshops to include not only men already experiencing ED but also men who might be at risk of it, in the heart of mainstream male environments, such as military, religious, and corporate contexts. This was a subtle but important shift from the original intention, and it meant that by talking with men in both clinical and non-clinical settings, *The Barometer of My Heart* could be positioned within a wider societal frame, thus expanding its potential reach.

6   Excellent dialogue with an existing clinical partner and a meeting point between clinical and artistic priorities.

    Interests aligned in finding new ways to communicate about the physical consequences of loss of potency and particularly ED.

7   One's own experience.

    Storor is a male in middle age, not wishing to be defined by definitions of sexual potency and seeking 'a laboratory for a true collaboration between the participating men, myself and the commissioned artists, modelling aesthetic, form and content'.[18]

8   Access and permission to enter the clinical environment as artist.

    It was through careful initial reflection and tuning that Storor ultimately gained permissions to attend 64 consultations at St George's Hospital Endocrinology Clinic (London). That was the kind of immersion and engagement that the artist perceived as necessary in order to explore such a delicate theme, rooting the ensuing artistic re-presentation in lived experiences of patients representing a cross-section of the urban demographic served by the hospital and reflecting different aspects of the medical condition as well as differing cultural attitudes towards it. He also met with a number of the men outside the clinic during the conception and refinement of the project. These meetings and consultations became the core of the work, the source material for the show and devising workshops that informed it.

Engagement with patients (case study: *Under the Microscope*)

Artists who are leading engagement processes with patients in medical settings use listening and reflection as relational building blocks to establish trusting relationships. Listening and reflecting-back informs reciprocal conversations between artist, clinical partners, and participants, who can both respond to and reveal the actual experience of illness. This example will explore the principles of *relational practice*, with the artist working reflexively in a practice which is not led by a specific art-form but is responsive and open to what emerges from the relationship with participants. We will also expand on the ethical principles which underpin such engagement.

Under the Microscope was a project conceived by artist Sofie Layton in collaboration with GOSH Arts (the arts programme at Great Ormond Street Hospital). It explored the role of an artist in the hospital in relation to how children and families perceive, appropriate, and respond to medical language and medical statistics. During a one-year residency, the artist was able to access medical practitioners and researchers through the hospital as well as patients and families on the wards. The project mainly focused on children with rare diseases and young people born with congenital heart disease, including explorations of medical language and imagery and the role of imaging in today's medicine. Considering the delicate setting and vulnerable participants, a careful approach to engagement was of the essence, and deep reflection was critical to develop a process for building trust while gently and gradually engaging individuals or groups in a conversation.

Site-specific dynamics

The artist in a setting such as the one just described is faced with a complex set of relationships and spaces, and the process of working in such a deeply engaged fashion requires equally deep reflection. When approaching the participants, the artist is faced with questions pertaining to site-specific dynamics (the hospital with its schedule, alarms, meetings, emergencies, spaces) and with the necessity of attuning the process to such an environment without compromising artistic integrity. The engagement process in this case would usually involve gentle approaches, mediated by the medical team (e.g. nurse specialist, consultant, psychologist), all requiring time. The engagement may then evolve from a conversation to an activity that can be performed on the ward or in a waiting room, such as embossing or embroidery. In other settings (e.g. patients in isolation), the artist is required to gown up and wear the required protections in light of the patient's state, thus even physically immersing herself in the workings of the hospital to build the relationship, to hold that precious conversation. Where appropriate, other participants may be invited into a less medicalised space, in a creative workshop setting, creating a sense of group and community, while still requiring careful planning, including engaging relevant medical staff such as an adolescent nurse specialist (a familiar face for participants, but also trained to provide medical support if necessary), a psychologist (to observe group dynamics and ensure participants' wellbeing, identifying potential signs of distress), and a researcher (to provide the medical context alongside the creative provocations provided by the artist).

Calibration

The engagement process is carefully calibrated, with the artist working 1:1 with some participants (at the bedside or outside the hospital) and in other cases bringing together a small group of patients. Similar activities take place, but the group workshop inherently presents a different dynamic. In this case, the workshop invited young people to explore the concept of their heart through self-portraiture, creative writing, and body mapping, while also discussing medical models, bouncing between the metaphorical and the more literal/medical dimension and between two-dimensional and three-dimensional representations.[19] The workshop offered several opportunities for reflection at the end of each activity (e.g. all participants and facilitators commenting together on body maps) and in a final group discussion.

Preparation

Reflection is integral in the phase of workshop preparation e.g. how to ensure emotional safety throughout the process. For instance, how to present a group of young people with something as sensitive as models of their hearts in a group setting. Collaborations and relationships are crucial. The artist builds a relationship with the researcher creating such models and with the psychologist who is involved in the evaluation of the workshop as well as the usefulness of such models for communication purposes. There is an alignment of priorities and interests, a reciprocal exchange, and learning from each other's methodologies and languages. Relationships underpin the whole process, as the artist also is in conversation with the hospital's arts team, which facilitates and brokers the relationships and provides key contacts to initiate conversations.

Boundaries

Another important consideration pertains to the boundaries that are involved in relation not just to space and audiences but also more broadly to ethical considerations. This reflection occurs on two interlinked levels. Firstly, the very important reflection around 'what is appropriate', which the artist constantly faces when dealing with such complex material and as part of their role (when the relationship is established) of acting as a conduit, of being able to translate or re-present medical and patient worlds, their languages, their lived experiences, the distress, the joy, the resilience, the pain. Secondly, the perhaps more practical but equally important reflection around medical ethics, what approvals must be sought, processes that are in place within the hospital to ensure patients' safety, yet a space that can be surprisingly unregulated when it comes to engagement activities that do not require formal approval from a Research Ethics Committee.

Methodology

In a project of this nature, the engagement methodology is constantly tuned, from the intensive care unit to the workshop space, from the researcher's desk to the surgical theatre. Engaging patients undergoing treatment or recovering from an operation and/or their relatives can be profoundly important and empowering for the participants but poses an additional set of constraints (both practical and ethical) and requires deep reflection for the whole project team.

Supporting the artist

A final important consideration pertains to reflecting on the potential burden for the artist herself, or more broadly the individual(s) leading the engagement activities. Given the sensitive nature of the conversations that are being explored and held, it is important to consider that appropriate support is in place for the artist, in terms of supervision and opportunities to debrief and reflect on the material that has been entrusted to them as part of the exchange occurring during the engagement process.

Translation and representation of patient narratives
(case study: *For the Best*)

In participatory arts work, reflective practice is an essential tool in on-going responsiveness to participants' stories and ideas. As has already been suggested, reflection is central to the process of re-presentation of narratives of illness, a journey of translation from the engagement to the final performance. The example discussed here explores the principles of co-creation with participants as co-authors in the process. Such co-creation was possible because of the depth and strength of the relationships between schools, hospital, and arts partners, held together in a strong producing framework rooted in trust and dialogue across sectors.[20]

For the Best was a performance project conceived and directed by Mark Storor and produced by Anna Ledgard. For nine months, Storor worked with children on dialysis at Evelina London Children's Hospital and in schools to devise images and stories which would become the source material for a site-specific performance. This focused on the family experience of living with the illness of a child and took place at the Unicorn Theatre in June 2009. The show drew critical acclaim, was awarded a TMA Award, and received deeply personal responses from audiences of all ages. It was a multi-faceted three-year process with strong embedded partnerships and deep collaboration across cultural, medical, and educational realms. We will draw on the insights that have been brought to the arts-making, aesthetics, partnerships, and learning relationships in this project by Mayo,[21] Nicholson,[22] and Walsh and Ledgard.[23]

Storor is artist and researcher in all phases of the project and always works collaboratively, learning through reflection on and participation in activities with others. Theories of 'situated learning', or learning embedded in social processes and physical contexts, can be applied to much of his work.[24] Storor has a clear identity as an artist always working in participation with others: 'I only want to function as an artist when I'm in a room with people and I think they're functioning as artists too'.[25] In this practice, art is being made through all phases of a project, not only in its outcomes. In *For the Best*, narratives, images, and metaphors created by participating children were selectively woven by the artist (in conversation with the patient originators, medical collaborators, and other artists) into collective narratives of the experience of illness. With time, sensitive facilitation, and trusting relationships between schools, patients, artist, and medical professionals, different perspectives were exchanged and remoulded into a form to be shared with wider audiences.

Authenticity

Reflection includes careful consideration of the fine line between representing the authentic voice of the participant and appropriation of ideas, particularly in cases in which participants who came up with the original ideas and images would not be able to perform their own stories due to their on-going medical conditions. As outlined by Sheila Preston, 'we have a responsibility towards ensuring that the representations that are made are produced through a climate of sensitivity, dialogue, respect and willingness for reciprocity'.[26]

Close listening

Close listening, staying true to participants' voices, and maintaining the integrity of their original contributions are essential. Mayo describes how when

> working with a group of mothers of patients Storor picked up on their conversations about waiting around for hours in hospital and eating too many muffins. The next week he arrived with bags of muffins and together they created a 'muffin mother' figure, an image

of feeding and being fed, of boredom and of home, which found its way into the final performance.[27]

While Storor may play with the framing and exaggerate the scale of the imagery in the final performance, the original images, and actual voices where possible, are included, and the originators recognise their contributions in the performance, describing them as 'the same, but different'.[28]

Re-framing

The process of 're-framing' can be well exemplified by the case of an image originally created by an 8-year-old boy in a dialysis ward, and its journey over 9 months, through different iterations to representation in the final performance. This was documented by Anna Ledgard observing the following scene on the ward:

> *It is 10.00 clock in the morning. A busy renal ward in a children's hospital is going about its business. Eight children are on dialysis, sitting in large blue chairs, each of them connected, literally, by clear rubber tubing to large machines, each displaying numbers and an assortment of lights.*
>
> *I look again at what the children are doing; one is working on a GCSE practice booklet; another adjusting a drawing in an art portfolio, a younger child is reading aloud to a teacher; and across the ward a spelling test is taking place. The atmosphere is one of purposeful activity: these children are at school.*
>
> *Now 10.15. A child looks up and gasps as she sees an unfamiliar sight. A tiger peeps round the Sister's desk and peers into the ward. Another child, who is very ill today, looks at the tiger, then deliberately puts on his headphones and pulls his computer screen up to block his face. The tiger gingerly pads forward, nudging the Ward Sister's legs, gradually becoming more menacing. She taps him on the head, tries to calm him with gentle words, to no avail, then one of the children suggests that she sing to him. Quietly she hums a tune. The tiger responds, his ears twitch, he listens, inclining his head, and gently allows himself to relax. The reluctant child peeps from behind his screen, trying not to break out in a smile. The tiger goes to the bedside of a young girl, she sings with the nurse, next to her another girl joins in, as does her mother beside her. As soon as they stop singing, the tiger moves in to paw them once more. Guy, the consultant, has arrived; he watches from the doorway, then the tiger spots him and approaches, roaring menacingly. 'Waltzing Matilda, waltzing Matilda', sings the consultant, picking up the game, as the tiger dives for his ankles. The tiger rolls over playfully. The consultant turned tiger tamer whirls his stethoscope at the beast and laughs. By now the small boy has moved the computer screen aside and watches the consultant and tiger intently. The tiger swings around, waves farewell, and the reluctant boy smiles goodbye.*

The tiger in this example comes from the poem written by an 8-year-old child in hospital, undergoing multiple medical interventions and on dialysis five times a week, expressing his fury at his illness: 'I am tiger, a really angry tiger, a shouting tiger. Well-built, angry, fit. Be careful when you stretch out your hand to touch me, led by inquisitive fingers, you may hear yourself scream. I am prickly'.

Re-play

Storor brought the tiger image into a busy working ward transformed by artist Sofie Layton into a mask and costume that could be worn by an actor. For a moment, the specialists on the ward had to abandon their everyday routines and play. The children, having provided the script, were directing the action, the art offering a different dimension of reflective insight to the medical team, as:

> Although nurses and specialists have known the medical narratives of these children intimately, by participating in the process of building a shared poetics, and then later still by attending the performed work at the Unicorn Theatre, they are compelled to see their patients in a new way, perhaps reflecting on their own role within the children's more complex narratives with insight.[29]

Real-world

As Storor gathers imagery for the performance, the original scary tiger will be present. However, before then, in another subtle step in the process of building connection and embodying imagery, the tiger will appear in the school hall of a West London primary school in a series of participatory drama workshops. This was the home school of the child on the renal unit at Evelina London Children's Hospital. His name was on his class register, but he rarely attended due to his illness. Together with Storor, he had devised a set of drama workshops for his classmates at school, one of which involved the class overcoming the advances of a menacing tiger. On this Mayo wrote:

> Based on the idea of a difficult journey and informed by the images and metaphors that the boy's own reflective writing expressed, the boy developed, with Storor and others in his team, activities that used participatory drama techniques in the school.[30]

Although the children were unaware at the start of who had initiated this activity, *Out of Bounds* developed into a series of six three-hour workshops. Upon realising who had set this game of challenge, Mayo states:

> When they discovered who it was their sense of connection to and understanding of their unknown classmate in hospital grew enormously. . . . It also significantly impacted on his classmates' understanding of the realities of his life and their empathy for him. His predicament had opened up a rich stream of creative and educational activity.[31]

Experiential significance

The workshop had also served an important purpose as these primary school children would later become part of the performance, taking the role of performance 'guides' for the audiences in the performances at the Unicorn Theatre. The workshops had given them the opportunity to reflect and respond to the experiences of their friend. They understood that the tiger they had overcome in the imagined realm of the drama workshop represented the difficulties and challenges for their friend and indeed any family with children living with illness. Their role in the performance now had an experiential significance for them. They needed little rehearsal to understand the quiet accompanying role they would need to play as they gently guided adults through a performance which went to the heart of what it was to be a family torn apart by illness.

Presentation to a public audience (case study: *The Heart of the Matter*)

The Heart of the Matter was a large public engagement project running between 2016 and 2019, led by bioengineer Giovanni Biglino and artist Sofie Layton in collaboration with psychologist Jo Wray and produced by Susie Hall of GOSH Arts, Anna Ledgard, and Nicky Petto. The project explored the medical and metaphorical dimensions of the human heart through a participatory process involving several patient groups as well as clinicians and artists, culminating in a public exhibition that toured the UK in 2018. The exhibition was conceived by Layton as an immersive experience, presented in different venues, including: Great North Museum Hancock in Newcastle (March-May 2018), the Royal West of England Academy and the Centrespace Gallery in Bristol (July–August 2018), the Victoria & Albert Museum in London (September 2018), and the Copeland Gallery in London (November 2018). This example will explore the complexities behind the principle of interdisciplinary collaboration in an approach involving three cities and eight organisations, aligning institutional and artistic relationships and priorities across venues, hospitals, and patient groups, while maintaining a participant-centred approach throughout.

The artworks presented in the exhibition were conceived of and created by Sofie Layton in collaboration with a team of musicians, digital animators, textile artists, jewellers, and other makers. They ranged from screen-prints and photographs to a large digital animation, from whole spaces (e.g. a lightbox shaped as a dodecahedron and abstracting a modern cathedral but also symbolising the surgical theatre) to sculptures, and was also included as part of different installations' medical imagery and medical 3D-printed heart models. The breadth of the artworks and the media employed to represent aspects of cardiovascular anatomy and narratives of illness connected to different forms of heart disease: the profound logistical differences between venues and the sensitive nature of the content emerging from a participatory process were such that careful reflection was required throughout the planning and designing phase as well as when setting up and, in some instances, directly managing the exhibitions.

Parameters of the exhibition

Reflection is essential from the moment of envisioning the exhibition, particularly in a project where a large number of people were engaged through workshops and thus a large number of narratives emerged. In this instance it was felt important to:

- Represent all stories and images (more or less overtly) in the final piece, but attention was paid to how to achieve this without compromising either the artist's vision for the artistic outcome or the original contribution of the participants.
- Find a balance between the nature of the material (e.g. potentially confrontational or moving images around complex health conditions, pregnancy terminations, and death) and the nature of the target audience (e.g. general public and families).
- Ensure members of the public could recognise elements in the exhibition (e.g. the name of a medication, a diagnosis, an emotion) from their own medical journeys and past experiences; therefore, careful interpretation was of the essence.

The whole team came together, including the lead artist and the producers as well as technical and medical partners, to agree what kind of information should be presented, the accuracy of such information, and its quantity. The team was balancing, on the one hand, not overburdening the piece and making it too didactic, while, on the other hand, ensuring that the target audience was sufficiently guided in the interpretation of the artworks.

Invigilators and explainers

The presence of invigilators and explainers, who could support the public in experiencing the piece, required careful planning and reflection, including training of collaborators or volunteers. Training sessions were held by the lead artist and one of the producers to ensure that all explainers had sufficient insight into the nature of the work and relevant details of individual installations or artworks.

Relationships with venues

Relationships with venues were also navigated with careful reflection; the choice of the venue itself was of the utmost importance. This would be guided by:

- Budget considerations.
- Alignment of public engagement priorities of the venue with those of the project, where a public venue is involved.
- The appropriateness of the space in relation to the content.

For *The Heart of the Matter*, the public outcome was presented both in public museums and in gallery spaces for hire, posing very different sets of questions for the team. In the first case, more attention and energy were focused on relationship building and forging a strong reciprocal partnership from the outset, benefiting from the local knowledge and expertise of the venue in key areas (e.g. community engagement, exhibition technical production, communication, promotion, invigilation) while carefully maintaining open and regular communication with hospital partners and patient groups, which were valued new audiences for the venue. In the second case, technical aspects of the production relied exclusively on the project team and required outsourcing. Furthermore, presenting the same work in different spaces can have technical implications in relation to the nature of the spaces, and these choices clearly can impact on the overall curation and feel of the final piece.

Inviting the participants

Considering the work had such a strong participatory component, it was important to ensure not only that all voices were included in the final piece, but also that the work could be shared with the participants themselves. This included organising private views for workshop participants, attended by the whole team, including the lead researcher and the psychologist, to allow for careful reflection throughout the visit and a conversation at the end of the private view.[32]

Re-presentation

In the case of a project like *The Heart of the Matter*, new reflections arise when the public outcome is re-presented in different venues as part of new exhibitions. Individual pieces from the original exhibitions have been subsequently invited as part of themed exhibitions in different venues, e.g. the Villa Rot museum in Memmingen or the Royal College of Physicians in London. How to present elements of the whole piece to new audiences as part of a different whole under the leadership of a curator who has not been involved in the original project and its participatory phase? Are there ethical/cultural sensitivities in different contexts or even different countries?

Conclusion

Shedding light on the principles of effective participatory arts practices is important at a time when public engagement with the sciences is being encouraged and there is a growing understanding of the role that the arts can play in health settings.[33] Most artists gain experience through practice, and there are few practical training courses for those wanting to enter this field. Understanding the principles expanded upon in the four projects discussed is fundamental to improvements in practices in the arts in health. Such principles are founded in the values of:

- Collaboration.
- Relationship-building.
- Co-creation.
- Mutual trust and care.

Staying true to these principles is dependent on artists, partners, and producers fostering and refining the conditions which make reflective practice possible, nurturing self-awareness and critical reflection, and arguing for reflexive culture as central to their practice at every stage of project design and delivery.

Reflection is essential and absolutely integral to sensitive creative/participatory work exploring narratives of illness. As exemplified here, looking at different stages of a project using different examples (indeed, potentially interchangeable), communication is paramount for this work to be rich and meaningful and, importantly, respectful and empathetic.

The principles outlined here are applicable across health and engagement settings and art form practices. What is described is a process of fine-tuning, of calibration, of reflecting through another lens – a patient's, a collaborator's, an institutional partner's, all of them. This is ultimately why projects of this nature cannot be fixed in their expected outputs from the outset, because such iterative processes will inform the framing, the development, and the dissemination of the work.

Acknowledgements

The Barometer of My Heart was generously supported by the Wellcome Trust and public funding by the National Lottery through Arts Council England (ACE) and produced in association with Artsadmin. *Under the Microscope* was generously supported by the Wellcome Trust, the Blavatnik Family Foundation, and the National Institute of Health Research (Great Ormond Street Hospital Biomedical Research Centre). *For the Best* was generously supported by the Wellcome Trust and public funding by the National Lottery through Arts Council England (ACE) and produced in association with Artsadmin. *The Heart of the Matter* was generously supported by the Wellcome Trust, the Blavatnik Family Foundation, Above & Beyond, the National Institute of Health Research (Bristol Biomedical Research Centre), Great Ormond Street Hospital Children's Charity, and public funding by the National Lottery through Arts Council England, with thanks to Artsadmin, RapidformRCA, EngineShed Bristol, 3D Life Print, and the British Heart Foundation.

Notes

1 All-Party Parliamentary Group on Arts, Health and Wellbeing Inquiry Report (2017).
2 Fancourt and Finn (2019).
3 Dewey (1933).
4 Schon (1983).
5 Ibid.
6 Ross (1978).
7 Matarasso (2019).
8 Matarasso (2013).
9 Rogers (1961).
10 Candy (2020).
11 Ibid.
12 Mayo (2014).
13 Charon (2006).

14 Charon (2013).
15 Accepted for publication (2021): Defining the role of 'relational producer' in arts-and-health collaborations in hospitals: a reflection on catalysts and partnerships. Authors: Anna Ledgard, Susannah Hall, Sofie Layton, Nicky Petto, Jo Wray, and Giovanni Biglino. Journal: Leonardo (www.mitpressjournals.org/loi/leon).
16 The Barometer of My Heart (2015) conceived by Mark Storor and produced by Anna Ledgard in association with Artsadmin and in collaboration with St George's Hospital, Tooting and supported by the Wellcome Trust and using public funding by the National Lottery through Arts Council England http://annaledgard.com/participatory/barometer-of-my-heart-the-shy-taboo.
17 http://annaledgard.com/participatory/boy-child/.
18 Mark Storor Evaluation Statement for ACE and Wellcome Trust Evaluation Reports created in 2015.
19 Layton et al. (2016).
20 Accepted for publication (2021): Defining the role of 'relational producer' in arts-and-health collaborations in hospitals: a reflection on catalysts and partnerships. Authors: Anna Ledgard, Susannah Hall, Sofie Layton, Nicky Petto, Jo Wray, and Giovanni Biglino. Journal: Leonardo (www.mitpressjournals.org/loi/leon).
21 Mayo (2014).
22 Nicholson (2012).
23 Walsh and Ledgard (2013).
24 Lave and Wenger (1991).
25 Mayo (2014).
26 Preston (2009).
27 Mayo (2014).
28 Steele (2010).
29 Walsh and Ledgard (2013).
30 Mayo (2014).
31 Ibid.
32 Biglino, G. and Wray, J. *The Heart of the Matter Audience responses*. Personal documents.
33 Fancourt (2017).

Reference

All-Party Parliamentary Group on Arts, Health and Wellbeing Inquiry Report. (2017). *Creative Health: The Arts for Health and Wellbeing*. London, UK. https://www.culturehealthandwellbeing.org.uk/appg-inquiry/

Candy, L. (2020). *The Creative Reflective Practitioner*. London: Routledge.

Charon, R. (2006). *Narrative Medicine: Honoring the Stories of Illness*. Oxford and New York: Oxford University Press.

Charon, R. (2013). Narrative Medicine: Caring for the Sick Is a Work of Art. *Journal of the American Academy of Physician Assistants*, 26(12), 8.

Dewey, J. (1933). *How We Think: A Restatement of the Relation of Reflective Thinking to the Educative Process*. Boston, MA: D.C. Heath & Co Publishers.

Fancourt, D. (2017). *Arts in Health: Designing and Researching Interventions*. Oxford: Oxford University Press.

Fancourt, D., & Finn, S. (2019). *What Is the Evidence on the Role of the Arts in Improving Health and Well-Being? A Scoping Review*. Health Evidence Network Synthesis Report, No. 67. Copenhagen: WHO Regional Office for Europe. ISBN-13:978-92-890-5455-3.

Lave, J., & Wenger, E. (1991). *Situated Learning: Legitimate Peripheral Participation*. Cambridge: Cambridge University Press.

Layton, S., Wray, J., Leaver, L. K., Koniordou, D., Schievano, S., Taylor, A. M., & Biglino, (2016). Exploring the Uniqueness of Congenital Heart Disease: An Interdisciplinary Conversation. *Journal of Applied Arts & Health*, 7(1), 77–91.

Matarasso, F. (2013). Matarasso Reflections on Quality. *Participatory Arts: Creative Progression*, 3(3).

Matarasso, F. (2019). *A Restless Art*. London: Calouste Gulbenkian Foundation.

Mayo, S. (2014). The Artist in Collaboration: Art-Making and Partnership in the Work of Mark Storor and Anna Ledgard. In C. McAvinchey (Ed.), *Performance and Community*. London: Bloomsbury.

Nicholson, H. (2012). *Theatre, Education and Performance*. London: Palgrave Macmillan.

Preston, S. (2009). Introduction to Ethics of Representation. In T. Prentki & S. Preston (Ed.), *The Applied Theatre Reader*. London: Routledge, 65–69.

Rogers, C. (1961). *On Becoming a Person*. London: Constable.

Ross, M. (1978). *The Creative Arts*. London: Heinemann.

Schon, D. A. (1983). *The Reflective Practitioner: How Professionals Think in Action*. London: Avebury.

Steele, S. (2010). *For the Best Evaluation Report, Artsadmin*. https://www.academia.edu/865456/For_the_Best_Evaluation_report.

Walsh, A., & Ledgard, A. (2013). Re-Viewing an Arts-in-Health Process: For the Best, Research in Drama Education. *The Journal of Applied Theatre and Performance*, 18(3), 216–229. DOI:10.1080/13569783.2013.810923.

3.10

MAKING REFLECTION-IN-ACTION HAPPEN

Methods for perceptual emergence

Jennifer Seevinck

Introduction

The *reframing* that takes place during reflective practice is an emergent process. In this chapter, I discuss how it can happen and some mechanisms for implementing it. *Reframing* involves interpretation, where the practitioner makes sense of an existing phenomenon to untangle and ultimately convert that phenomenon into a new personal understanding or view of the world – a new way of seeing that phenomenon. *Reframing* can lead to emergent outcomes such as new and unexpected goals or plans for making, new findings from research or new insights about a work.

An understanding of how emergence works aids in concept generation and reflexive critique during practice and eliciting themes during data analysis of reflections or other data that has been collected. It enables the practice-based researcher or Higher Degree Researcher (HDR) to shift from practice to knowledge outcomes by assisting identification of insights and novel findings from their practice, as well as directions and themes in it. In this chapter I draw on my work in emergence1 briefly explaining it to locate it within a process of reflection-in-action.

Practice-based researchers necessarily engage with the relationship of their practice to knowledge. Understanding the tacit, reflective components of professional practice has helped address concerns around the validity of professional, practical understandings in the traditional research domain;[2] however, the practitioner's experience of insights to *reframe* and solve problems during professional engagement remains a challenge to make explicit. It is important for:

- Identifying how knowledge comes out of practice
- Supporting the creation of new research
- Understanding how the novice develops professional practice
- Continuing the self-development of professional practice

In the chapter, I describe methods of reflective practice to position emergence within a reflection-in-action cycle. These are methods for *perceptual emergence* and they have been employed by practice-based researchers to make sense of what they are doing. This has included specific instances where they have found:

- New directions within creative practice
- New structure across a literature review
- New insights into empirical data
- It has also assisted in clarifying the alignment between their practice and research

DOI: 10.4324/9780429324154-32   440

The chapter begins by identifying reflective practice concepts – *framing*, *reframing* and *frame experiments*, *moves*, *situation talk-back* and *repertoire* – as methods to reflect-in-action. I then draw on my *Taxonomy of Emergence in Interactive Art* (TEIA) to briefly review emergence and its characteristics, subsequently aligning all in an example cycle of *emergent reflection-in-action*. The chapter explains how these methods work and then shifts to focus on *mechanisms for making emergence happen within reflection-in-action*. A range of these *moves* are discussed and the insights generated from this approach for emerging new creative practices, themes in data through to identifying a PhD thesis structure, are explained.

Reflective practice and framing

In the *Reflective Practitioner*, Schön presents his studies of professionals as they engage with some of the novel and unique situations that characterise their profession. He introduces new concepts and techniques to describe how these professionals handled non-routine scenarios such as *framing*, *situation talk-back*, *moves* and *repertoires*. His text illustrates various implementations of these aspects of reflective practice that can be usefully applied as methods of reflective practice. It helps to demystify the professional practitioner's approach to handling unique and challenging types of situations, demonstrating how this can be done in ways that are experimental and considered.

For example, for a design professional and researcher, the brief to 'design a house for a poet on a rocky bluff' is considered a *situation* that is notable in its complexity and novelty – rather than formulaic it is unique, new and not an everyday problem. The designer will engage with that situation and problem to learn about it, understand and make sense of it; and this new understanding will guide the design professional's subsequent action. The process is reliant on how the designer interprets that situation and can be broadly understood as *framing* that situation. It may involve actively trying something out – a *frame experiment*. This action could then generate *situation talk-back* – where the scenario provides a form of feedback that can further inform the design practitioner's thinking, and one might say the situation is 'talking back' to the designer. Actions the designer takes are *moves* – drawing a line in the architectural sketch to add a wall, changing the building orientation, etc. These may also generate situation talk-back where the designer considers the effect of that move upon the situation. In determining their moves, a professional practitioner may also draw on their past experience. This is their *repertoire*. They can draw on this for ideas, from situations with similar patterns through to analogies, to aid how they might engage with this new situation.

It's useful to talk about these facets of reflection-in-action as discrete elements, to clarify how each may work. For the practitioner, however, the action is integral to the thinking – s/he 'does not separate thinking from doing'.[3] Rather, the implementation of a new idea is built into the enquiry. Keeping that integration in mind, it remains useful to separate out some of the stages into a possible sequence, as in Figure 3.10.1.

During reflection-in-action, a practitioner will act on their new understanding of a situation to change the situation, implementing their reflections within that action. This process can continue cyclically so that understanding and situation evolve over time. The stages repeat, as shown by the long arrows in Figure 3.10.1.

Reframing, interpretation and feedback

When a professional practitioner *reframes* a situation, they gain a new way of looking at it that is different to how they understood it previously. According to Schön, the practitioner tends to mirror the openness and unpredictability of the situation, as they allow themselves to 'experience surprise, puzzlement or confusion in a situation which he finds uncertain or unique'.[4] This open-ended approach is distinct from when someone may look at a situation to confirm a hypothesis. The latter

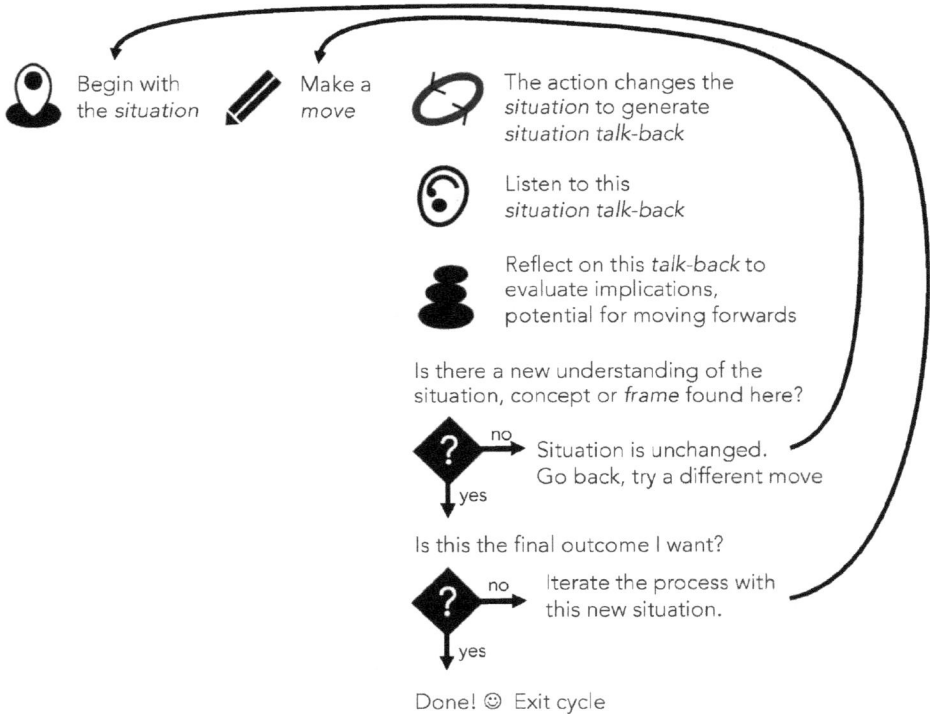

Figure 3.10.1 A sample process for implementing aspects from reflective practice and reframing a situation. Reflection-in-Action Cycle, Seevinck 2021, https://doi.org/10.6084/m9.figshare.14714886.v2.

will not *reframe* that situation – rather, it remains stuck within the preconceived realm of defined possibilities of meaning or *frames*. Instead of the verification of a hypothesis or expectation, reflection-in-action necessitates an open-minded approach.

Other key characteristics of reflection-in-action are interpretation and feedback: first, *reframing* involves being open-minded enough to be able to interpret the scenario in new ways. As Munby describes, *reframing* is seeing 'the puzzling phenomena as something else' and 'suddenly seeing the data differently'.[5] Second, during reflection-in-action one is learning and adapting their own understanding and the situation. In this process the situation is changing, and that new information is 'talking back' or 'feeding back' into the process. For example, for Schön's architect Petra, the *situation talk-back* from dealing with an unusual site helped her to *reframe* her understanding of the situation and that site. Following Figure 3.10.1, starting at the top left, she began with the design situation and then proceeded to draw into it (a *move*), resulting in a new, different drawing. As she considered this newly altered drawing (listening to the *situation talk-back* and *reflecting* on it), a new understanding of the situation came about and the situation is *reframed*. It is a looping, cyclical process with feedback from the newly created *frame* changing the situation; and this then feeding subsequent *moves* and cycles until a desirable outcome (here, an architectural design) is achieved and the cycle ends. The process relies on *reframing* via interpretation and an open-ended approach, so that the practitioner can see the situation as something else and compose a response uniquely suitable to it.

When the *reframed* understanding surprises the practitioner and is not immediately obvious to them, it can also be understood as an *emergent* outcome. The concept of emergence is debated across disciplines. The next section defines and positions it in relation to reflection-in-action.

What is emergence?

Broadly speaking, emergence occurs when a new form or concept appears that was not directly implied by the context from which it arose; and like a Gestalt, this concept is a *whole* that is more than a simple sum or grouping of its parts. Heterogeneously different to those parts and what came before, it is characteristically new, unpredictable or not immediately obvious and exists across the different levels or scales of the *whole* and parts.

This definition draws on my prior work with emergence, where I captured different disciplines' understandings of emergence in the transdisciplinary *Taxonomy of Emergence in Interactive Art* (TEIA).[16] This includes the natural sciences' perspective of emergence – for example, the self-organising, 'flocking' behaviour of birds where they do not collide but rather stay together in a seemingly structured group able to travel long distances. While this kind of emergence is possibly the more popular understanding,[7] Gestalt theory provided early, formative insight into emergence[8] which has continued to inform the approach and understanding used within design research, art and games theory. For these disciplines, the role of an observer in detecting emergence is key since they are concerned with those new understandings or insights that might be perceived and created. I class this as *perceptual emergence* within the TEIA. Another class is *physical emergence*, named after the dominant approach in the physical and natural science's emergence literature, which understands emergence as something that occurs independently of an observer.

Perceptual emergence

Perceptual emergence is concerned with the development of new insights or understandings in people. This makes it highly relevant to this chapter's exploration of mechanisms for developing practitioner and researcher insights. For example, when listening to a series of musical notes (parts), it may be possible to discern a melody (the emergent whole resulting from this combination of parts/notes – see von Ehrenfels' 1890 experiment[9]). Similarly, when looking at a drawing of two squares that overlap, one might interpret a new triangle shape at that intersection. This triangle was not explicitly drawn but is instead a new outcome, qualitatively different to the squares (the parts that make it up, Figure 3.10.2). It is considered an *emergent shape* within design research.[10]

Another form of *perceptual emergence* is a behaviour: someone playing poker can pretend to hold different cards to those they were dealt. Bluffing is not defined in the prior moves or in the rules but

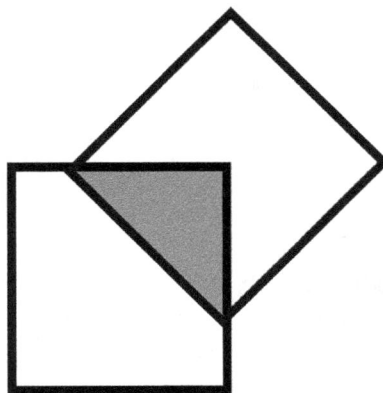

Figure 3.10.2 An emergent triangle shape can be perceived from two overlapping squares. Emergent shape, Seevinck 2021, https://doi.org/10.6084/m9.figshare.14715231.v1.

is instead a new emergent game play behaviour that developed out of poker game play.[11] In all these cases, there is a reliance on an observer to perceive the emergent melody or shape, or to subjectively make sense of a situation to enact the behaviour.

Perceptual and *physical emergence* differ in their reliance on an observer for the emergence to exist. In *perceptual emergence*, the system is expanded to include a person or observer, and they are an integral part of the unfolding emergent outcome. This expanded view provides a way of thinking about the subjective experience of emergence, enabling one to ask questions like 'how can the interpretation that occurs during perceptual emergence support *reframing*?' And 'how does *perceptual emergence* and interpretation work?' The following section draws on perception theory to answer these questions.

How does it work?

In *perceptual emergence* there are parts that are, through a process of perception and interpretation, perceived as a *whole*, and this new perception can grant new insight. Following the drawing example earlier, a designer who draws square shapes that inadvertently overlap might subsequently perceive a new triangle shape (Figure 3.10.2). This triangle is unplanned and new, different to the squares, an instance of emergence that can *reframe* the designer's understanding of her situation so that the building she is designing now gains a second story, or a third room. But how did the designer see it?

The intelligence of perception

The new triangle shape is an *emergent shape*. *Emergent shapes* are an area of research within design and psychology. They can be explained as resulting from the incoming signal to the eye, the processing of that signal and cognitive processing input from the brain. Importantly, perception has been argued to be a form of visual thinking. For example, perception theorist and psychologist Gregory maintains that perception is creatively intelligent. Our ability to perceive things such as the emergent white triangle shape in Figure 3.10.3 evidences this – 'perceptions change even though the input to the eye remains unchanged'.[12] In perceiving an ambiguity, we do, in some way, make a decision about what we see. It is a creative process.

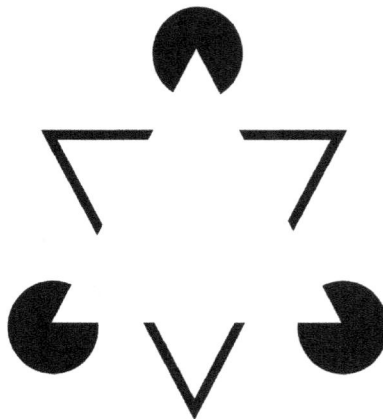

Figure 3.10.3 The Kanisza triangle (inferred white triangle) is an illusion classed as a perceptual fiction and also an emergent shape. Kanisza triangle, Seevinck 2021, https://doi.org/10.6084/m9.figshare.14715603.v1.

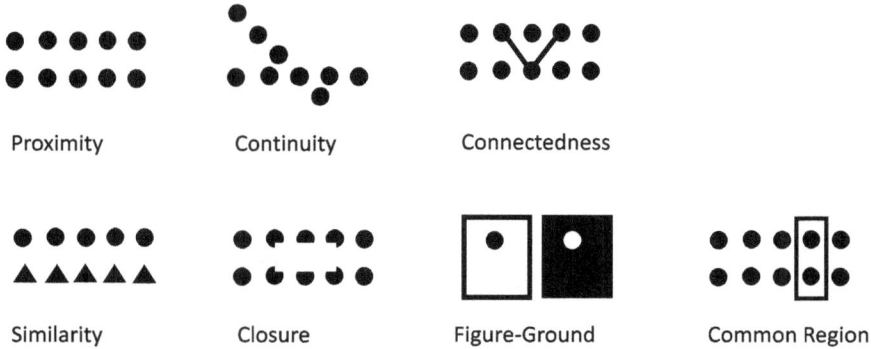

Figure 3.10.4 Principles of Gestalt. Principles of Gestalt, Seevinck 2021, https://doi.org/10.6084/m9.figshare.14715696.v2.

Illusions like the Kanisza triangle are experienced when 'a perception departs from the external world, to disagree with physical reality'.[13] The Kanisza triangle is not explicitly drawn but rather interpreted and perceived because of cognitive functions operating on the incoming signal. There is no object counterpart. Gregory explains it as resulting from the application of Gestalt perceptual processes.[14] Other visual processing studies also explain how we link up features to interpret visual forms, patterns, *emergent shapes* or illusions (see Figure 4).[15,16,17]

Human memory and experience also inform the creative processes of seeing, as it leads our interpretation and assists in recognising new outcomes. For example, seeing a dragon figure in the clouds. Here Gestalts assist one to identify elements as contiguous and part of the same figure, while top-down cognitive input guide an interpretation of the figure as a dragon: if you had no knowledge of dragons, you would not be able to recognise the likeness. Human memory and experience, along with Gestalts, can facilitate *perceptual emergence* and, in so doing, increase capacity to *reframe* during reflection-in-action.

Reflection-in-action and emergence

When the *reframed* understanding surprises the practitioner and was not immediately obvious to them, it can be considered an instance of *perceptual emergence* – new, unexpected and occurring within that person. Other literature has also identified emergence within reflective practice: Haseman and Mafe describe how the outcomes from creative practice-based research are not predictable and obvious; rather, the results 'emerge'.[18] They also emphasise the cyclic and reframing aspects to the process, explaining that the practitioner acts 'upon the requisite research material to generate new material which immediate acts back upon the practitioner who is in turn stimulated to make a subsequent response';[19] explaining that this can create a situation of 'chaos and complexity' which may 'fragment authorial control'; and describing the 'shock of recognition' that occurs when identifying a new insight as the surprising, *emergent* concept at the core of this process.[20] While Haseman and Mafe are focused on the creative practitioner, these insights into the experience of reflection and emergence also apply to other domains.

In another chapter within this book, Hübner similarly recognises that open-endedness and crystallisation of insights are critical within practice-based research and draws on emergence in the final stage of a new model for articulating practice-based research methods.[21] My focus is *how* that happens and in this chapter I identify mechanisms for effecting that emergence - by drawing on the TEIA and Schön's reflective practice.

In bringing together *perceptual emergence* and *reframing* as a method, emergence within the reflection-in-action cycle can be understood as occurring when *a new understanding, concept or frame has come about*. It occurs at the first test point in Figure 3.10.1. The emergent outcomes that can follow on from this point are shown in Figure 3.10.5. These include new knowledge outcomes, emergent practice, and a new frame for the situation.

Reflection-in-action challenges are questions for emergence

The cycle of reflective practice in Figure 3.10.5 maps out opportunities for emergence and *reframing* along the way, helping the practice-based researcher to understand a possible route for navigating the complexity of their project. Questions about how to do reflection-in-action to *reframe* work and develop new insights do, however, remain. For example: How do you detect *situation talk-back*? What *moves* might generate *situation talk-back* and *reframing* in the first place?

As discussed earlier, listening to *situation talk-back* is a process of interpretation. However, interpretation while working through the reflection-in-action cycle is something I have found novice reflective practitioners such as HDRs can struggle with. Emergence can provide answers: interpretation is a key aspect of *perceptual emergence* and there are a number of methods that can generate *perceptual emergence* to facilitate that interpretation. Some of these methods can be used as *moves* to generate *situation talk-back* and *reframing* - since, as discussed above, perceptual emergence and associated

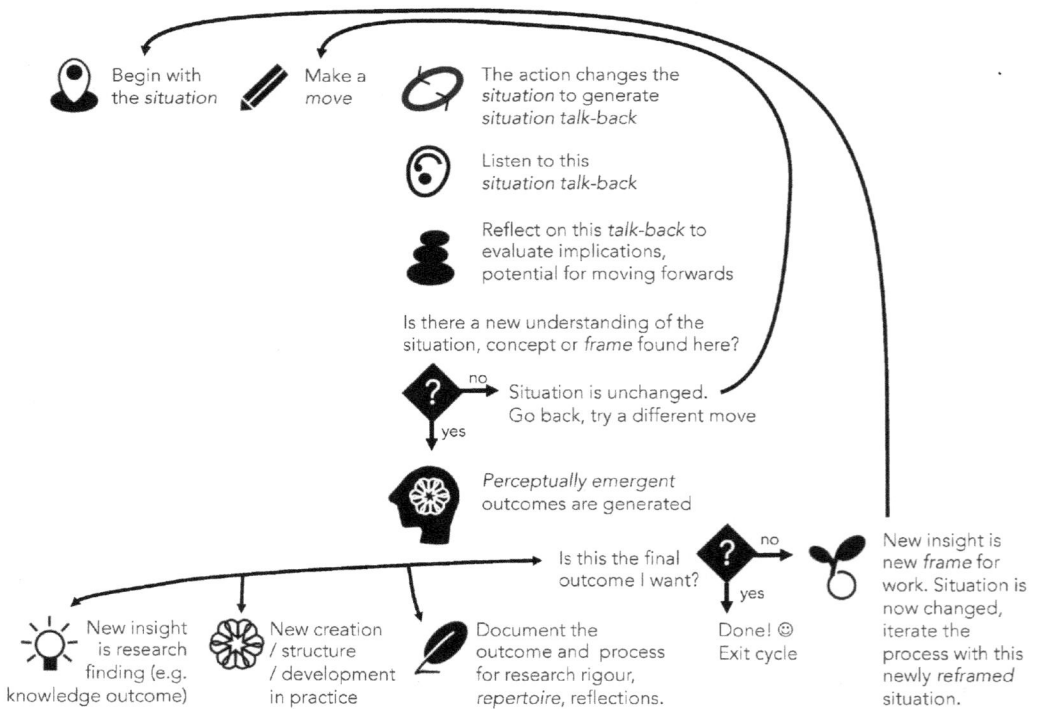

Figure 3.10.5 The process of reflection–in–action from Figure 3.10.1 is augmented to show how emergence occurs within, and as outcomes. Emergence in reflection-in-action, Seevinck 2021, https://doi.org/10.6084/m9.figshare.14714904.v1.

perceptual mechanisms do, by their very nature prompt new insights. *Perceptual emergence* therefore has the potential to scaffold reflexive thinking and working. With an emergence-focused approach, the above questions can now be *reframed* to instead ask: What mechanisms of *perceptual emergence* can generate *situation talk-back* and *reframing*?

Emergent moves for reflection-in-action

Perceptual emergence can result from cognitive processing such as Gestalts and top-down input from the brain such as past experience and knowledge. These processes assist our organisation of form and interpretation of meaning to make sense of what we see (and, arguably, of what we hear). For reflection-in-action, they provide us with ways to 'listen' (using Schön's term) to *situation talk-back* – interpreting it for *reframing* and eliciting new insights.

Perceptual emergence can also provide some mechanisms for *generating situation talk-back*. That is, methods for emergence can be used as *moves* themselves, to enact upon and change a situation, to then perceive/interpret for emergent insights and *reframing*. Figure 3.10.6 shows examples from emergence and visual thinking that can be used as *moves*. These can be inserted into the *moves* stage of the above emergent reflection-in-action cycle (Figure 3.10.5).

The *moves* illustrated in Figure 3.10.6 can facilitate insight for the PBR's practice and research. They draw on a series of situations that I have come across through my PBR work and supervision. This encompasses creative, visual practice to design interactive art installations that effect audience behaviours and dynamic visuals, drawing on interaction design methods alongside techniques from physical computing, data visualisation and augmented reality. This practice is complemented by my research enquiry to explore new possible experiences within the field of human–computer interaction and to evaluate audience experience of interactive art works; and supervision of HDRs and emerging practitioners in undergraduate design study.

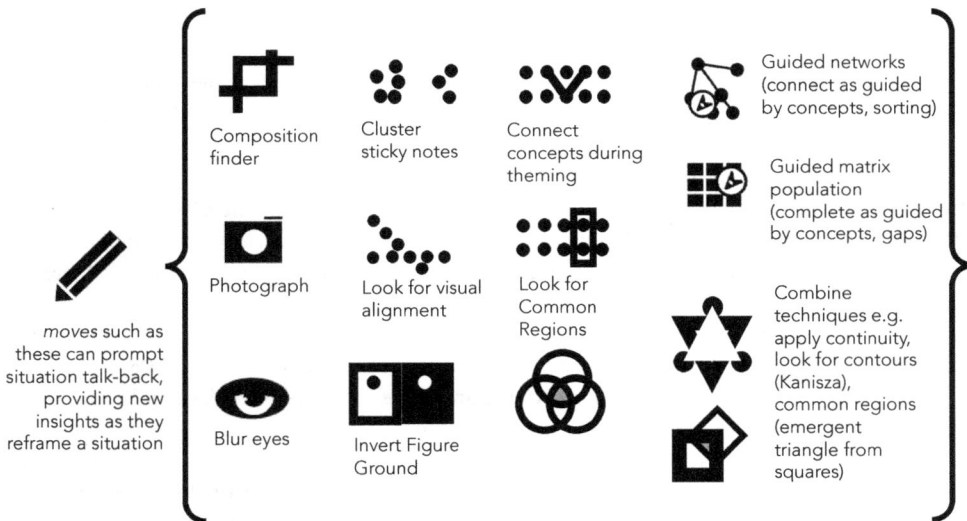

Figure 3.10.6   Moves for perceptual emergence may aid reframing and reflection-in-action. Moves for perceptual emergence and reframing, Seevinck 2021, https://doi.org/10.6084/m9.figshare.14715723.v1.

The situations discussed here focus on *reframing* for practice, such as using a composition finder, photography and playing with Gestalt principles when engaging with a creative situation. Some occurred during research activities such as data analysis – using *moves* like connecting or clustering codes during thematic analysis. Finally, some of the situations sit across practice and research components of practice-based research projects to emerge relationships and structures across these two realms. The examples shown here use visual perceptual techniques that draw on both 'bottom up' seeing, such as Gestalts of connecting, alongside 'top down' mechanisms, such as project concepts or research questions. While the emphasis here is visual – reflecting my own work and experience – the insights and *moves* for *reframing* articulated here can go beyond the visually dominated spaces. They could, for example, be adapted for non-creative and non-visual dominated situations such as nursing and surgery through to music and dance.

Moves for emergence reframe practice and research

The following situations demonstrate how mechanisms for emergence have been used as *moves* to generate *situation talk-back* and *reframing*. The first situations are practice focused, beginning with my creative practice where I used mechanisms of interpretation and visual seeing such as composition finders. These led me to emerge new concepts and understandings, which through their *reframing* of the situation, drove those creative works forward conceptually and formally.

In the next section, I discuss mechanisms for *perceptual emergence* in research situations. This begins with using emergence for data analysis, its role in thematic analysis and during the coding of research data where clustering and categorising can relate concepts. Emergence can also be used for data visualisation, and this includes methods such as embodied clustering using sticky notes, *emergent shapes* and guided mechanisms for sorting and populating a table. These draw on my research and experiences as a HDR supervisor.

Perceptual emergence in practice

The use of seeing practices and visual thinking is familiar to visual artists. Artists may study a landscape scene using visual means such as photography, cardboard composition finders, blurring the eyes and so on, in order to 'see the world in a new way'. Visual thinking processes may facilitate interpreting the feedback from a landscape to find *situation talk-back*. Visual thinking can affect *perceptual emergence* as that *situation's talk-back* and, as a result, can lead to *reframing*.

During creative practice, I consciously alter my perceptual processes to see a composition or starting situation in different ways. In creating work that is driven by natural landscapes, this has led to new ways of thinking about the given landscape (*situation*) to develop a concept and creative response. Put another way, it is an orchestrated mode of perceiving the situation that facilitates new perceptions and, in so doing, *reframes* my understanding of the situation.

Reframing has occurred for me during a site visit to a state nature park in the USA in 2006, becoming a turning point during the creation of interactive art system *+-now* (2008). Visiting the park on a late afternoon in winter meant the tree and moss shadows were dark and long, making the distinction between sky and earth (figure and ground) ambiguous. The combination of these unusual signal inputs with Gestalts led me to perceive Bald Cypress tree roots and their shadowy reflections in shallow water as two 'diamond' *emergent shapes*. This went on to *reframe* my thinking of the site and the artwork overall, to emerge a concept, or framework, for the practice.[22] Put another way, this began with a site and landscape as the *situation* (Figure 3.10.5), and I employed visual arts techniques as *moves*. These were observation, photography and the interpretation of *emergent shapes* via Gestalts (Figure 3.10.4). These led to *situation talk-back* from that site. In considering this new possible interpretation, I found a new way of approaching the work. This is a *reframed* understanding of the site and work and it emerged a new creative concept for further creative development.

Light Currents is another example. This interactive artwork was created for the foyer of the Powerhouse Theatre, an international performing arts venue on the Brisbane riverbank.[23] Here, an emergent and reflective process of *reframing* led to developing the work around the *shimmer of sunlight reflecting on the river*. As previously, I observed that riverbank and site and, by taking photographs and video I managed my perception to find new ways of looking at the site, generating *situation talk-back*. As I sought to *reframe* the site, I drew on my *repertoire* of prior work and understandings of similar sites to see if this can 'function as a precedent'.[24] That is, I sought to interpret this new site in those terms. Drawing on various aspects of perception – view finders through the camera, Gestalts to reverse figure and ground and top-down to draw on my *repertoire* (Figures 3.10.4 and 3.10.6) – I *reframed* my understanding of the site in terms of the *shimmers of light*, and this became the creative concept that drove the work visual form and dynamic behaviour.

Emergence in data analysis

The identification of new concepts or insights from data has been described as an emergent process. grounded theory methods, for example, is a qualitative data analysis process historically focused on generating new theory 'from the ground up'. Here the aim is to have 'categories emerge'[25] from the data in order to discover new theories bespoke and sensitive to that data. This is through a process of coding that data. Coding involves grouping, sorting and various ways of 'marking up' concepts to distil salient elements. It is a 'strategy that is used to find themes and patterns in qualitative data'.[26] Thematic analysis similarly employs coding to generate themes and, through these, insights. Themes are described as capturing 'something important about the data in relation to the research question, and represents some level of patterned response or meaning within the data set'.[27]

Data that is coded can take different forms. For qualitative research, data might be in the form of photos documenting practice, memos or interview transcripts or recordings. Coding typically begins by noting categories in the margins of a text (e.g. on interview transcriptions, memos or annotating on photos) that relate to what was found in that text. This is a process of interpreting the text to determine the labels. Glaser and Strauss advocate constantly comparing the codes to those already added. In so doing, the researcher develops the dimensions and understandings of a category. As explained in 'how to code' literature, at the beginning there may be many labels, but as the comparative method progresses, the activity shifts to look for how these terms relate to one another, with the aim of coalescing and developing relevant conceptual categories (for example[28]). The approach draws deeply on the text and the researcher's capacity to interpret and find structures in the texts. When this process of eliciting themes from the codes generates a new insight, it can be highly emergent.

Visualising the coding process increases the potential for generating new insights into that data, since the process of visualisation engages the practice-based researcher's perceptual processes to 'think visually' about the data. Visualising coding is also independent of the type of practice or research subject matter – it is applicable to practice-based researchers working in a range of domains. In the next section, *moves* to support visual thinking are discussed, some of which are mechanisms for *perceptual emergence* in and of themselves.

Perceptual emergence in data visualisation

A range of visual tools and techniques can support thinking about data. Data visualisation is concerned with revealing trends or patterns in the data,[29] and ways of thinking about data to visually interrogate, explore and discover things within. Visualisation works alongside analysis for the very reason that we are so well equipped anatomically to think visually[30] and indeed our visual processes have been described as cognitive processes.[31]

Visual techniques and tools include *displays* such as *networks* and *matrices*, and *moves* using emergence and Gestalt. Visual elements can be visually rearranged, grouped and sorted, and the researcher or artist can interpret new relations, compositions and structures. Miles and Huberman define a display as an 'organised, compressed assembly of information that permits conclusion drawing and action'.[32] They identify three types of display methods for qualitative data analysis, communication and exploration. The first type is *network diagrams*, to connect sequences of events or mind maps as used in brainstorming. The other types are *matrices* (tables) and *graphics*, such as Venn diagrams. Each of these are discussed next, using examples from PBR projects.

Network diagrams

Network diagrams are commonly used for connecting elements to identify and map out their relationships. They can be used to visualise processes or an unfolding series of events similar to a narrative or plot line, for example of a case study or the project as a whole. Network diagrams emphasise the relationships between the nodes or elements. These relationships between the nodes may be hierarchical, synchronous, chronological or causal.

Chronological network diagram

An example of a network diagram is shown in Figure 3.10.7. This is from a practice-based research study into audience interaction with digital interactive artwork +-*now*.[6, 33] When interacting with this work, the artwork audiences could draw into a sand-based interface to see and influence digital imagery.

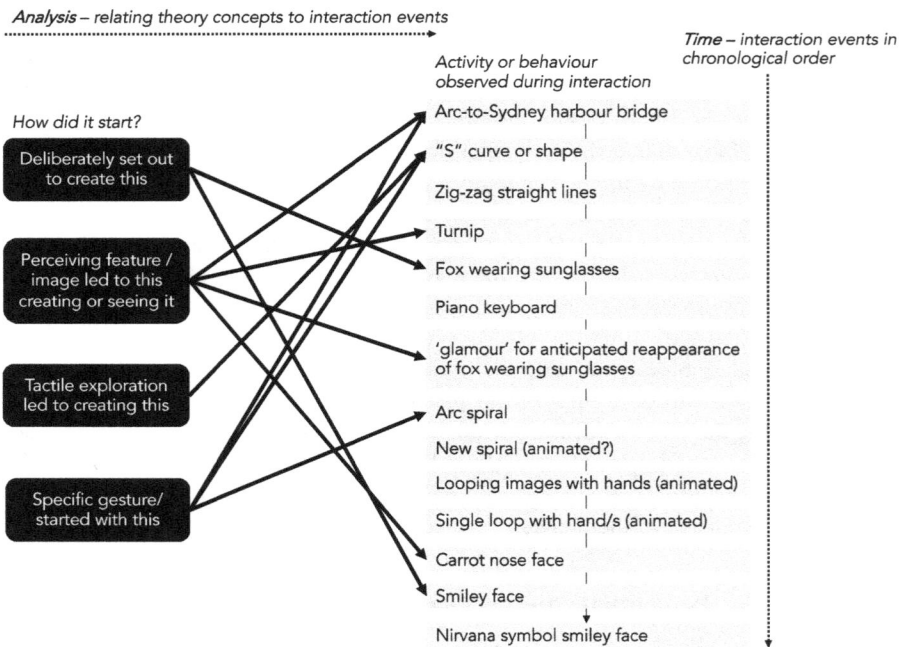

Figure 3.10.7 Network display of evaluation data aids analysis. Network display of participant interaction, Seevinck 2021, https://doi.org/10.6084/m9.figshare.14715834.v1.

450

Evaluation data was gathered for a total of 36 participants and visual network displays were created for each interaction scenario as part of the data analysis. Participant behaviours and actions with the work that were observed by the researcher and relevant to the study were noted down as discrete events, in a central column, from start to finish down the page. Salient concepts from theory were introduced and through annotations (e.g. connecting lines, colour), new categories and relations could be found.

The network diagram in Figure 3.10.7 shows how one participant's interaction with the work was broken down into this sequence of events to support analysis. For example, a theory-driven concept that is specific to this project is how participant behaviours originated: i.e. the 'starting condition'. I subsequently added the question 'how did it start' next to the observable events. It was used to interrogate each action. In this way, a theory-driven question was added to the display to help guide the interpretation of the evaluation data actions. Different types of 'starting conditions' were revealed from this and connected to the relevant events with lines.

Sorting chronologically, adding connecting lines, colouring to annotate the data in terms of concepts could all be considered *moves* for considering how the data and theory relate. Doing this process across the page facilitated a systematic as well as visual and spatial approach to interpreting the data. The network display provided a way to think about how theory concepts relate to empirical data that was spatial and graphic. This revealed new possible relations between the elements. Additionally, the display made explicit the places where any elements do not relate, revealing new insights or areas for future investigation.

Displays are useful because they facilitate specific types of visual thinking moves which can themselves lead to new findings. They also present an emergent whole: a picture of the situation.

Network maps for emerging ideas

Mind maps are another type of network diagram, often used for brainstorming. During ideation, concepts are noted down on a blank page as single words or short phrases. The effort is rapid, without self-censure, loose and free form. The relations between the concept nodes are explored through connecting lines or noting them down close to one another. Using mind maps for creative ideation or problem solving is likely familiar to the reader. Less familiar perhaps is the use of the same technique to create maps during thematic analysis.

Network maps for emerging themes

Thematic analysis maps are used to identify themes across a data corpus (for example[34]). After coding has taken place, those codes are written on a single blank page. The researcher considers the codes and looks for relationships between them, drawing connecting lines between those nodes where a relationship is found. The process is iterative and relationships may be repeatedly drawn and redrawn until meaningful groups of codes are found. These emergent groupings are themes. The perception of a theme is supported by the connecting lines, as explained by the Gestalt principle of uniform connectedness. A theme relates hierarchically to its subset of codes. Researchers may redraw the map to better communicate this hierarchical relation, such as through position or shape or size of the nodes (Figure 3.10.8).

Drawing and redrawing connections can be cumbersome rather than spontaneous. Instead, researchers have written codes on sticky notes and rearranged their positioning in close proximity or further apart, to find relations between the different concepts. Relations are represented spatially in groups and, as the Gestalt principle of proximity affords, clusters emerge and themes become explicit. This is more flexible and physically engaging than drawing and redrawing the connections. It does not, however, leave traces of prior configurations, making documentation of the explorations (e.g. photography) advisable.

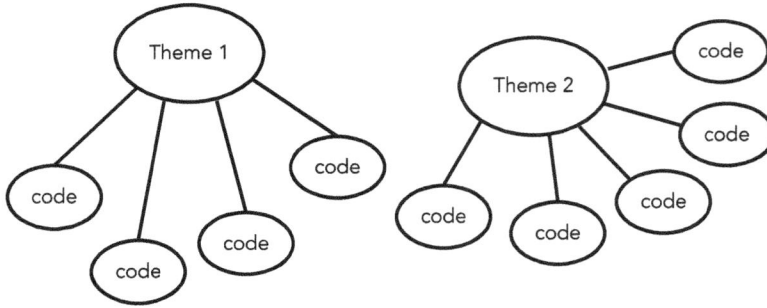

Figure 3.10.8 Network display can be used for visualising themes and their codes in thematic analysis. Network display for thematic analysis Seevinck 2021, https://doi.org/10.6084/m9.figshare.14715939.v1.

Mind mapping for a project overview

In addition to supporting thematic analysis of data, mind mapping can also support finding relationships between seemingly disparate concepts and thinking in the PhD project. To clarify PhD project direction, I have seen researchers use sticky notes to write down themes, insights and key concepts from literature review and reflective creative practice. They then re-organised these to form clusters and drew additional relationships between the clusters. Such efforts enable the researcher to see how their research and practice interrelate, and to see their project in lots of different ways. For example, they can see what categories or themes came across in both, synchronously; or track causal relationships such as where a memo on practice prompted literature searches, or those review findings sparking new concepts for practice.

Matrix diagrams

Matrix diagrams interrogate and reveal the connections between two variables. They work as an intersection of two lists, set up as rows and columns.[35] Instead of allowing for multiple connections between nodes like network diagrams do, matrices focus on the relationships between the rows and columns. This makes them less flexible than networks as they are not as free form; however, the matrix format's additional structure can also offer more guidance for visual thinking and data exploration, to emerge new understandings about that data. Matrices are also particularly useful for making gaps explicit. (Note that it's possible to increase the number of variables shown in a matrix by shading cells such as 'heat maps'.)

Matrix for background review

Matrices can be very useful for characterising a field and identifying gaps and relations across a literature or contextual review. I have used (i) key author with domain and (ii) key idea from the literature as the guiding variables in my own research (e.g. Table 3.10.1). In early research work into emergence theory, this layout assisted me in developing an understanding of the landscape of debates on the topic. The table visualised characterising, polarising and debated concepts, helping me to explore them and develop a synthesis as well as identify differences or outliers, as part of an overall picture of the topic area. I was able to see that a pattern often aligned with a view on how emergence may occur. For example, that a researcher's discipline and epistemological position (e.g. Positivist, Interpretivist) tended to link to a view of emergence as objectively observable or a subjective experience. This pattern led me to research that aspect further and develop argument around it,

Table 3.10.1 Creating a matrix display for literature review analysis can in visually identifying clustering, gaps, prompting new analysis and emerging a picture of in the field. Matrix display for literature analysis, Seevinck 2021, https://doi.org/10.6084/m9.figshare.14715987.v1.

	Characteristic 1	*Characteristic 2*	*Characteristic 3*	*Characteristic 4*
Key Author A, discipline 1				
Key Author B, discipline 2				
Key Author C, discipline 3				

and ultimately propose two primary classes in the TEIA: *physical* and *perceptual emergence*. Put another way, using this matrix display for analysis *reframed* my understanding of my practice-based research topic and led me to *emerge* a new insight – the first level of the TEIA.

A matrix for the literature review can be used in other ways as well. For example, detail can be added to each cell to create more of an overview of the project. I have also seen creative practice-based researchers who are doing a context review include thumbnail images of creative works, using rows to list these rather than authors.

Interestingly, the process of identifying characteristics and creating columns can develop over time. The activity can be like coding, where one develops understanding, criteria and columns over time through comparing characteristics from newly reviewed artworks to those previously analysed. It can reveal possible gaps and insights, because by adding new characteristics as columns in a matrix, the empty cells for previously added creative works will now prompt consideration of those works relative to the new characteristics. The result is an emerging overall picture of the topic area.

Matrix for tracking the practice-based research project

Practice-based research necessitates mapping out and keeping track of how the various practice and research efforts relate. Matrix diagrams are also useful for navigating a practice-based research project because they provide a way to map those relations, reveal any gaps in how the project is pursued and help manage project rigour.

Understanding the relations between different project aims, anticipated outcomes and the methods and evidence that will substantiate those outcomes can be quite a challenge. A visual display can support explicitly addressing each of these aspects. For example, in Table 3.10.2, the research questions are listed down the side and their corresponding outcome, method and evidence are then prompted for explicit consideration, in the rest of the table cells.

A matrix like Table 3.10.2 might not come about all at once, instead developing organically through brainstorming. During practice-based PhD studies I have seen researchers focus in on research questions by listing them across a page, then populating the space below each with post-it notes about key insights of the project to date. The use of sticky notes provides flexibility because they can move these around to order them later on. This means the ability to explore the relations, frame and reframe the connections between the aspects. Taking the time to do this helped supervisors to see how the project was answering the research questions, informed the findings discussion in the exegesis and enabled us to literally reveal any gaps of the project in responding to the research aims.

Venn and Euler diagrams

Venn and Euler diagrams draw on the Gestalt of common regions to visualise how different aspects relate. They are often used in interdisciplinary projects to show how the project addresses the gap

Table 3.10.2 An example matrix display for gaining an overview of and managing a practice-based research project. Matrix display for scoping the practice-based research project, Seevinck 2021, https://doi.org/10.6084/m9.figshare.14716662.v1.

Research question (RQ)	Outcome	Primary method to generate this outcome	Primary evidence from method
RQ1 'What is . . .'	New definition	Literature review	Literature analysis
RQ2 'What considerations for practitioner . . .'	New original practice outcome Artwork 1	Creative practice, reflective practice	Process and final work documentation: photos, video, memos
	New original practice outcome Artwork 2	Creative practice, reflective practice	Process and final work documentation: photos, video, memos
RQ3 'What considerations for context . . .'	New understanding of the contextual concerns	Naturalistic evaluation studies using observation, interview; thematic analysis	Video, photos, field notes, audio recordings

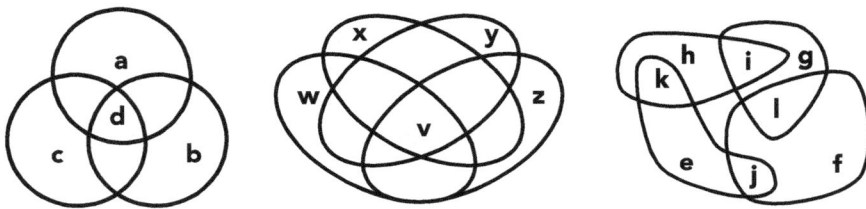

Figure 3.10.9 Common regions can lead to emergent shapes in 3 type and 4 type Venn diagrams (left, centre) or at the intersections of a Euler diagram (right). Venn and Euler diagrams, Seevinck 2021, https://doi.org/10.6084/m9.figshare.14715987.v1.

or opportunity that emerges at the intersection of the disciplines. For example, the leftmost Venn in Figure 3.10.9 shows a common region (d) across multiple project domains (a, b, c). This emerges as a new triangle shape and as new understanding.

The Venn can support guided visual analysis. This is because Venn diagrams show every possible intersection between domains, and this can prompt the researcher to consider each of these intersections between the sub-domains as a creative opportunity or for research rigour. This means the Venn can be used to question many possible domain combinations. Where there are more than three domains, a four-type Venn diagram is possible (Figure 3.10.9 centre), but as these increase, they can become complex to visualise, in which case a network or matrix might be more suitable. On the other hand, the Euler is a more flexible display option (Figure 3.10.9, right), only showing those areas of intersection that are populated. This can, however, make it more useful for early stages of ideation while trying to emerge common regions.

A practice-based researcher can start to feel lost in the relationship between their research and practice. Visual thinking such as through *network diagrams, mind maps, Venn* and *Euler* diagrams and *matrices* can assist with working through the more complex practice-based project and the relationships between these many moving parts. Creative, practice-based researchers will (necessarily) develop an artwork, letting it lead them; but how to bring it back to research? Indeed, a tension between practice and research is not uncommon in practice-based research projects broadly, and keeping the practice integrated and aligned with research can be challenging. Often, however, the relations are there; they just may not be immediately visible. As shown, making those relations explicit can be facilitated by visualisation mechanisms for *perceptual emergence*.

Conclusion

Perceptual emergence is a way of understanding the new and surprising insights and understandings that can develop during practice and research. Since *reframing* is itself emergent, insight from and techniques for facilitating emergence are highly relevant to the practice-based researcher.

The theory of *perceptual emergence* draws on areas including visual thinking, perception and Gestalts to explain and facilitate the emergence of new shapes and structures. Importantly for the practice-based researcher, this includes the processes for emerging new understandings while engaged in a practice and research cycle of reflection-in-action.

This chapter has described an example cycle of reflective practice, to explicitly identify several of Schön's reflection-in-action concepts as methods – *reflection, situation talk-back, moves* and *reframing* (Figure 3.10.1). This has been developed to describe a sample cycle of *emergent reflective practice*, positioning emergence and emergent outcomes within reflection-in-action (Figure 3.10.5). As shown, the development of *situation talk-back* is via feedback to the practitioner researcher and leads to a *reframing* of that situation. The *reframed* situation can generate new emergent outcomes from new practice (e.g. creative works) to knowledge outcomes or new project directions. And, the new *frame* is also an emergent understanding itself.

Interpretation is key to this process. The practice-based researcher must engage with an exploratory, open-minded approach, remaining receptive to ambiguity and unanticipated developments. Visual processing can play a role here, as it can be leveraged to effect *perceptual emergence* and, in turn, lead to seeing a situation differently – literally. Cultivating this open-minded approach and sensitivity to potential multiple interpretations will aid the PBR in (i) detecting feedback or *situation talk-back* to then (ii) proceed with subsequent cycles of *reframing*. Interpretation of the feedback can (iii) emerge new understanding in a researcher as well as (iv) new project relationships, themes or patterns.

There are a number of *moves* that leverage this visual processing to effect *perceptual emergence*. Composition finders such as cameras, through to data visualisation techniques such as network diagrams and matrix displays (Figure 3.10.6), can utilise a practice-based researcher's physiological and psychological capacity to prompt them to perceive structures and patterns, and lead to new, emergent insights. Examples of these *moves* for *reframing* new insights and efforts in both research and practice have been discussed, with benefit to:

- Practice – for example using *moves* that employ Gestalt perceptual techniques to *reframe* a landscape and emerge new art concepts.
- Research – for example using *moves* that employ *visual displays* to relate data analysis codes, *reframing* understanding of them and emerging themes during a thematic analysis.
- Practice-based research projects – for example using *moves* that employ *mind maps* along with *Gestalt principles* to relate concepts in each area, *reframing* understanding and emerging new project structures and relationships.

For artists who are also practice-based researchers, it may be that much of the interpretive activities identified here are familiar. This is not surprising given that artists have typically spent much of their lives studying the world to perceive it in new ways. Indeed, the creative practitioner undertaking research is arguably well equipped to identify new insights, as there is potential to leverage their already creative capacity of *perceptual emergence* within this approach to reflective practice.

Finding *moves* for *reframing* and insights may also be a broader field of research and activity for the practice-based researcher. For example, there are likely other visual art and design practices beyond the *composition finder* and *Gestalt principles* that can also prompt *reframing*. Additionally, there is potential to investigate more embodied situations for reframing. For example, in two of the situations described, paper sticky notes were used. Concepts written on those notes were positioned and repositioned spatially as the HDR made sense of their relations. Moving these notes around while thinking and looking is an interactive and embodied manipulation of the visual representation. This process of engaging physically might provide additional opportunities for feedback, interpretation and perceptual affordances, the mental models that develop across real 3D space and so on that could prompt new types of *moves* for *reframing*, emergent outcomes and emergent ways of working.

Perceptual emergence in the reflective practice cycle is a process that relies on interpretation and feedback. Cultivating this interpretive and reflective eye to engage actively and iteratively with the practice and research being conducted will strengthen both. It can lead to the emergence of new understandings and insights in practice and research situations, and like the sculptor Michelangelo Buonarotti is said to have done, to 'see the form in the stone'.

Notes

1 Seevinck (2017).
2 Schön (1983).
3 Schön (1983, p. 68-69).
4 Schön (1983, p. 68).
5 Munby (1989, p. 34).
6 Seevinck and Edmonds (2008).
7 For example, as described in Johnson (2001).
8 Wertheimer in Ellis (1938).
9 Arnheim (1954).
10 For example, see Mitchell (1993).
11 Salen and Zimmerman (2004).
12 Gregory (1997, p. 206).
13 Gregory (1997, p. 196).
14 For example, in the Kanisza triangle the eye is led without interruption (continuance principle) across black shapes to infer contours (closure principle) and 'see' the upwardly pointing white triangle. Other aspects of Gestalt are proximity and similarity – where elements that share local position or a quality such as colour or shape are understood as related (see Ellis 1938); figure-ground where one tends to dominate; and common region which can lead to emergent shapes – as in the triangle common to two squares in Figure 3.10.2.
15 Ware (2008).
16 Arnheim (1969).
17 Ellis (1938).
18 Haseman and Mafe (2009, p. 217).
19 Haseman and Mafe (2009).
20 Haseman and Mafe (2009).
21 Hübner (2022).
22 For example, see Seevinck (2013).
23 Seevinck (2015).
24 Schön (1983, p. 138).
25 Glaser and Strauss (1967, p. 23).
26 Onwuegbuzie and Combs (2010, p. 15).
27 Braun and Clarke (2006, p. 86).
28 Morse and Richards (2002).
29 For example, see Klanten et al. (2010).
30 Ware (2004).
31 Ibid.
32 Miles and Huberman (1994, p. 11).
33 Seevinck (2008).
34 Braun and Clarke (2006).
35 Miles et al. (2014, p. 105).

Bibliography

Arnheim, R. (1954). Introduction. In *Art and Visual Perception a Psychology of the Creative Eye*. Berkeley: University of California Press, i–ix.

Arnheim, R. (1969). *Visual Thinking*. Berkeley: University of California Press.

Braun, V., & Clarke, V. (2006). Using Thematic Analysis in Psychology. *Qualitative Research in Psychology*, 3(2), 77–101.

Ellis, W. D. (Ed.). (1938). *Sourcebook for Gestalt Psychology*. New York: Harcourt, Brace and Co., vol. 1.

Glaser, B. G., & Strauss, A. L. (1967). *The Discovery of Grounded Theory Strategies for Qualitative Research*. New York: Aldine Publishing Company.

Gregory, R. L. (1997). *Eye and Brain: The Psychology of Seeing*. Princeton Science Library. Princeton: Princeton University Press, 5th ed.

Haseman, B., & Mafe, D. (2009). Acquiring Know-How: Research Training for Practice-led Researchers. In H. Smith & R. T. Dean (Eds.), *Acquiring Know-How: Research Training for Practice-led Researchers*. Research Methods for the Arts and Humanities. Edinburgh: Edinburgh University Press, 229–251.

Hübner, F. (2022). The Practitioner in the Centre: A Flexible Approach to Methodology in Practice-Based Research. In C. Vear (Ed.), *Routledge Handbook to Practice-Based Research*. London: Routledge.

Johnson, S. (2001). *Emergence the Connected Lives of Ants, Brains, Cities, and Software*. New York: Scribner, 1st ed.

Klanten, R., Ehmann, S., Bourquin, N., & Tissaud, T. (Eds.). (2010). *Dataflow 2 Visualising Information in Graphic Design*. Berlin: Gestalten.

Miles, M. B., & Huberman, A. M. (1994). *Qualitative Data Analysis: An Expanded Sourcebook*. Thousand Oaks: Sage Publications, 2nd ed.

Miles, M. B., Huberman, A. M., & Saldana, J. (2014). *Qualitative Data Analysis: A Methods Sourcebook*. Thousand Oaks, CA: Sage Publications, 4th ed.

Mitchell, W. J. (1993). A Computational View of Design Creativity. In J. S. Gero & M. L. Maher (Eds.), *Modeling Creativity and Knowledge-Based Creative Design*. Hillsdale: Laurence Erlbaum Associates Inc., 25–42.

Morse, J. M., & Richards, L. (2002). *Read Me First for a User's Guide to Qualitative Methods*. Thousand Oaks, CA: Sage Publications.

Munby, H. (1989). Reflection-in-Action and Reflection-on-Action. *Education & Culture, the Journal of the John Dewey Society*, 9(1), 31–41.

Onwuegbuzie, A. J., & Combs, J. P. (2010). Emergent Data Analysis Techniques in Mixed Methods Research: A Synthesis. In *Sage Handbook of Mixed Methods in Social & Behavioral Research*. Thousand Oaks, CA: Sage Publications, Inc., 397–430.

Salen, K., & Zimmerman, E. (2004). *Rules of Play Game Design Fundamentals*. Cambridge: The MIT Press.

Schön, D. A. (1983) [1995]. *The Reflective Practitioner How Professionals Think in Action*. London: Temple Smith.

Seevinck, J. (2008). *+-now, Interactive Art Installation*. https://www.sciencedirect.com/science/article/abs/pii/S0142694X08000719.

Seevinck, J. (2013). Concepts, Water and Reflections on Practice, *Leonardo*, 46(5), 494–495.

Seevinck, J. (2015). *Light Currents, Interactive Art Installation*. https://eprints.qut.edu.au/99732/13/Seevinck DicWade-9RIGHTS.pdf.

Seevinck, J. (2017). *Emergence in Interactive Art*. Springer Series on Cultural Computing, Cham: Springer International Publishing.

Seevinck, J., & Edmonds, E. (2008). Emergence and the Art System 'Plus Minus Now'. *Design Studies*, 29(Interaction Design Special Issue), 541–555.

Ware, C. (2004). *Information Visualization Perception for Design*. Interactive Technologies. Amsterdam, The Netherlands: Morgan Kaufmann, 2nd ed.

Ware, C. (2008). *Visual Thinking for Design*. San Francisco: Interactive Technologies, Elsevier Inc., Morgan Kaufmann.

PART IV

The Practice-Based PhD

4.1

THE PRACTICE-BASED PHD

Linda Candy, Ernest Edmonds, Craig Vear

Research and practice – an introduction

As discussed in Part I of this handbook, practice-based research can be defined as a principled approach to research by means of practice, in which the research and the practice operate as interdependent and complementary processes leading to new knowledge. Research aims to seek new knowledge through a principled and original investigation, and practice aims at acting in the world, normally in a professional or creative way, for example by making something. Generating, or finding, *new knowledge* is the aim of practice-based research. However, a challenge is to identify what this new knowledge might be in relation to practice. This is particularly critical to the practice-based research PhD, which is growing in popularity in academic institutions across the world.

What is a PhD?

A PhD, or other similar higher research degree, is defined in terms of new knowledge generation. Almost all PhDs awarded through higher education institutions have similar criteria about how a PhD degree can be awarded, and in which format the submission needs to take. In general, the PhD is awarded to recognise the successful completion of a supervised programme of individual research. The results of this research are traditionally presented as a written thesis in which the PhD candidate argues the validity of their research findings. Through this argument the candidate must demonstrate:

- an independent and original contribution to knowledge
- an understanding of research methods appropriate for the study
- a critical investigation and evaluation of the topic of research

A practice-based PhD is distinguishable from other kinds of PhD because the outcomes or artefacts arising from the research process may be included as part of the thesis as defined by the rules applying to the particular institution that is hosting the candidate's studies. These outcomes take different forms, from artworks to music compositions, dance performances, architectural buildings, software, video presentations, and more, and are presented together with a written text as a multipart thesis. As such, a full understanding of the significance and context of the research can only be obtained by experience of the works created. The argument presented by the thesis, therefore, is

 DOI: 10.4324/9780429324154-34

shared between the practice-based works and the written exposition. In all other respects, a practice-based PhD has the same core characteristics as any other PhD.

In practice-based research, the results and documentation of practice can be part of the thesis. However, artefacts cannot be expected to speak for themselves in a PhD submission. For that reason, they must be accompanied by a textual exposition that describes the context, background, and methodology for the work carried out. Where artefact(s) are included in the delivered outcomes, the text must illuminate the reader about them, explaining how they are to be considered in the context of the new knowledge. Most important, it must reveal the contribution the candidate has made to knowledge and support the claim that it is both new and valid.

A written exposition arising from a practice-based research process is expected to show evidence of original scholarship, the methodology used, and material worthy of publication or of inclusion in a public exhibition or performance. The role of the written exposition is to share the understandings achieved through the research and to set out the claim, gap, aims, background, methods, and outcomes of the research. This is where the candidate shows how the work relates to the state of the art in the field and demonstrates that the knowledge generated is in new in the world.

Research questions

All research starts with a rationale for an investigation, and the practice-based PhD is no different. As the research develops this rationale may shift but must be clear and stable before the end, when the final thesis must define:

1   Specific research questions that address a gap in knowledge for the field, or clear and targeted unsolved problems within the field.
2   The context of the questions or problems to be addressed. This defines the focus of the investigation by specifying its significance, the history of research in the area, and what the gap in knowledge is.
3   The methodology used in answering the research questions. This includes a rationale for why a practice-based approach is appropriate and a justification of the specific methods employed.

This requires a focused and principled approach to PhD research, and in this respect, it is no different to any other PhD. It is in the nature of research that the process is open and things change, but at the end of the day clear aims and objectives have to be stated. It is wise to begin with a clear statement and change it if necessary, rather than work vaguely hoping for the fog to lift down the line.

A good research question pinpoints exactly what you want to find out and gives your work a clear focus and purpose. All research questions should be:

*   **Focused** on a specific problem or issue as defined by the gap in knowledge
*   **Achievable** using a clearly defined research strategy
*   **Time-bound** and feasible to answer within the duration of the project
*   **Specific** enough to investigate thoroughly and precisely
*   **Significant** enough to develop new knowledge for the field

Knowing and knowledge

There is a difference between knowing through practice and having new knowledge from research. In essence, the outcome from research is new knowledge for the betterment of the field and has direct value beyond the individual, whereas knowing within practice does not. Maintaining a principled and structured research approach is key to progressing through a PhD project, even if this approach is conducted through risky play, experimentation, and prototyping. Having a grasp of what the new knowledge is that is being sought through the investigation is part of this principled approach.

Identifying knowledge is discussed in detail in the whole of Part II, so we won't discuss it any further here; but, it is important to be clear about the difference between *knowledge* and *knowing*. Simply put:

- *knowledge* is externalised, validated, and can be archived
- *knowing* is felt within and can be demonstrated by actions

In the context of a PhD, regulations require that an independent and original contribution to knowledge (or some such wording) is demonstrated. The knowledge must be new in the world, not just to the practitioner researcher. New knowledge must be available in a form that can be shared and verified or challenged. Accepting that much of what we know is known tentatively rather than absolutely, the properties of being shareable and challengeable are more important than the absolute certain truth of the new knowledge. It is necessary to ensure that the knowledge is validated, but only in the sense that it is reasonable to believe it to be true, that we can offer good reasons to believe it. This issue is discussed in more detail in Ernest Edmonds' chapter in Part II, 'Research, Shared Knowledge and the Artefact'.[1]

New knowledge, as a presentable outcome from practice-based research, must satisfy essential criteria. First, it must advance or enhance the field through communicable and archivable representations of knowledge. Second, the knowledge must be justified by some challengeable means or another. New knowledge in practice-based research can be in (but not limited to) the following areas: the processes of practice; the forms of the outcomes of practice; the experiences of perception, emotion and cognition; the psychology of practice; the sociology of collaborative practice; the role of action and interaction between self and others in practice-based experiences; the acquisition and development of practice-based skills; or the philosophical foundations of practice-based activities. This new knowledge can emerge through practice and can bring new insights into practice.

Methodology and methods

These terms need clarification, as there is a tendency to confuse them:

- **A methodology** is the study of methods, to use its literal meaning. The second meaning, that of a set or system of methods, is commonly used.
- **A method** is a process, a technique, or a tool used in a research investigation.

These topics are discussed in greater detail in the leading chapter for Part III, 'Method' and elaborated in the other chapters of that section.

Methodology is important to understand from the outset of a PhD or any research project, practice-based or otherwise. It is an on-going process, and your methodology will need to be revisited as new insights, new goals, and new needs of your practice-based research investigation emerge.

Methods are defined, tested, systematic, and goal-orientated processes for investigation. In practice-based research, they are situated in a research strategy that is designed around practice so that there is a clear trajectory along which to conduct the research. As goals shift, perhaps after several iterations of the practice, the research strategy may need to adjust, too. The PhD candidate will need to evaluate the results from their methodology, perhaps even reassess the research questions and determine if the research design remains the best way of dealing with your research questions and your pursuit for new knowledge. The inter-relationship between the methodology and the research strategy are not fixed and are mutually dependent with on-going findings and the emerging shifts in the research needs.

Practice-based researchers need to be open to 'methodological pluralism'.[2] They can make use of a wide range of methods, investigatory tools, and techniques from a broad range of fields such as social science, humanities, technology, and scientific research. Because of this bespoke design requirement to practice-based PhDs, the research design and its specific methods can sometimes be a contribution to new knowledge and therefore part of the research output. In any case, the range of possible methods means that, unlike researchers working in tight, well-defined areas, the methodology needs to be reported explicitly. It is normally a chapter in itself in a practice-based PhD thesis.

Research design

Research design defines how and when each method is employed, how practice and research interlace. The design also needs to consider the larger design questions of ethics and data collection, protection, documentation, archiving, and evaluation of the research design and strategies, and crucially the self-evaluation of the well-being of the researcher.

Research design is important for several reasons: first, a properly defined methodology and a strategy brings about clarity in a complex research inquiry. Second, for the practitioner researcher, having an explicit design strategy and articulated methods can help improve, change, and evaluate progress through the research investigation. Third, it can make progress towards new knowledge and research insights more effective and more easily reported when it is complete. The research design is *your* project-specific pathway through *your* research investigation that gets you closer to *your* defined goal of new knowledge.

Knowledge production

Messy play, experimentation, rapid prototyping, and Lego serious play[3] are terms and strategies that are used regularly among PhD and practice-based research communities. There are a number of benefits of play-based strategies on knowledge construction that also have a profound effect upon the sense of community, shared endeavours, and well-being. Although play-based learning may seem to be isolated in the domain of early years learning, there is a surprising amount of positive evidence suggesting that play can lead to knowledge production that can be a significant contribution for the practice-based researchers, particularly in finding or clarifying research questions.

Sophy Smith, in her chapter 'Play in Research – Creating an Environment for Play Within a Doctoral Training Programme for Practice Based Research',[4] reinforces this approach and how it can 'create transformative learning environments that enable risk-taking and innovation, enhancing the creative practice of learners'. Play can also support researchers to move away from the more 'closed', traditional models of research practice, towards a more 'open' landscape.[5] It is Smith's finding over the past decade that this approach enables the 'emergence of new knowledge often across discipline areas'. Through an understanding of the dialogue and division of both playful and serious states within play, it is 'possible to move between the activity-orientated and goal-orientated mindsets that are both central to the practice-based PhD experience'.[6] As a practical guide to doctoral training, Smith offers a framework in:

- how embedding play into doctoral training can enable the development of supportive creative environments where learners are able to experiment and innovate
- importantly, how an environment conducive to risky play can be maintained within a process of monitoring and assessment

Improvisation as a strategy for play is discussed in Corey Mwamba's practitioner's voice chapter, 'Improvising as Practice/Research Method',[7] in Part V. He defines improvisation as 'a dynamic activity in which an entity attempts to make or create something, using only the resources and skills available at the time, in the present moment'. This highlights the real-time, in-vivo attunement of the researcher into the acts of practice, in which the 'presence of *resources* and *skills* highlights the fact that improvising is a social activity (with associated knowledge leading to particular skills). It is also culturally bound depending on the resources available within a society'[8]. In the context of his music-making, but applicable to any practice-based activity, he highlights the advantages of improvisation in the reflective process in this way:

1   reflection of one's own practice *requires* looking inward as well as outward to the situation
2   music making is complex, relating to intra- and extra-musical processes and events.

Improvisation as a method of play is simultaneously doing and reflecting, and critically it all happens in the here-and-now. It can be a useful method for investigating phenomena and meaning as they happens. Coupled with other methods such as video-cue recall, or stimulated recall, the researcher can 'engage critically and honestly with their own practice of improvising, and potentially create ideological or philosophical shifts within both research and practice'.[9]

Methods

Some research methods are quantitative, that is, they produce results that are numerical in form – often using statistics. Others are qualitative, producing results in the form of textual descriptions, such as interviews. Hybrid methods combine both to create a multi-faceted way of understanding the research. There is a growing body of hybrid methods in practice-based research that are gathering rigor and respect among researchers, such as autoethnography and dramaturgy (both discussed later). However, each practitioner researcher needs to determine the most appropriate methods and research design for their investigation, which may or may not be hybrid. Falk Hübner's chapter in Part III, 'Method', outlines a framework for determining this process.[10]

The goal of a hybrid approach is to use the data collection strategy from one method (e.g., reflection-in-action) in a direct conversation with another (e.g., autobiographical writing) in order to bring forth new insights and new dimensions of understanding about the practice and the research investigation. Autoethnography is one such hybrid method designed primarily in the field of practice-based research. It presents a hybrid process that concatenates two popular methods: reflective practice and autobiographical writing as a basis for socio-cultural understanding. This hybrid process aims to synthesise the reflective findings of the practitioner with the wider socio-cultural influence upon the individual, their field, and their practice at the point of doing the practice.

Iain Findlay-Walsh in his chapter, 'The Sound of My Hands Typing: Autoethnography as a Reflexive Method in Practice-Based Research',[11] states that autoethnography 'constitutes an adaptable approach to reflexive practice, embedded in the messy specificity of lived experience'. Practitioners may use writing as the main tool for expression, but they are equally use other forms of 'graphing' such as 'audio and video recording, photography, performance, choreography, song writing, scoring or crafting'. Through this, research 'becomes an iterative process of illuminating social, cultural and political conditions through an evolving, critical lens of self'.[12] The advantage of this is that it is an adaptable approach to reflection and its wider socio-cultural context. Additionally, the researchers can 'emotionally engage research audiences, involving readers in a process of locating and critically interpreting both the writer, and themselves, in the story'.[13]

Dramaturgy is another hybrid method that is discussed in more detail in Hanna Slättne's chapter, 'Navigating the Unknown – A Dramaturgical Approach'.[14] This hybrid method uses the technique of dramaturgy (from the performance field, meaning the 'structure of a story or a dramatic experience')[15] and self-reflection to map insight generation and their relationships to core aims ('the kernel') of the investigation. Slättne describes how this hybrid method is helpful 'especially for projects that are going to be in front of an audience, used by participants or consists of an experience that users need to navigate' as it can help 'excavate and identify the project kernel' and 'help explore, understand and formulate the core ideas in a responsive and dynamic process'. This mode of reflection, coupled with the route map strategy as experienced by the practitioner researcher (their dramaturgy of the research), supports the development of new ideas and insights that might be hitherto tacit or hidden. As such, this hybrid method is designed to help the researcher 'reflect on the practitioner's way of working, practice and process', at times where the practitioner's focus is inside the research.

Challenges of practice-based research

The history of practice-based research is quite short compared to many other forms, such as experimental science or astronomy. From its origins in the early 1970s, forms of practice-based research have been taking place in academic institutions, without using the exact term, as described briefly in Chapter 1, 'Practice-Based Research'.[16] From that time on, the ongoing theoretical and practical work has been documented in a series of significant publications[17] that have mapped the landscape and contributed to the ongoing development of the kind of rigorous and principled approaches represented in this handbook.

The emergence of an entirely new field of research brings with it the inevitable problems, not least of which is being represented accurately in the rules and regulations of the bodies that validate the formal outcomes of such research. Along the way towards being accepted as a valid form of research, PhD researchers and their academic advisers and those seeking funded research grants in the area have been presented with many challenges.

Andrew Johnston, in his chapter in this section, 'The Practice of Practice-Based Research: Challenges and Strategies',[18] presents five challenges and correlating strategies to deal with them at an institutional level. The challenges he outlines are:

1 *The problem of ambition: the need for resources, structure and support.* In general, the PhD candidate faces the challenge of creating works that are often 'large-scale and labour intensive' in order for them to generate something with sufficient novelty.
2 *Contributions and clarity.* PhD researchers sometimes felt a 'little dispirited or insecure because the overall contribution of their work could seem small in the context of the field and its history'.
3 *Tension between practice and research.* The question of the time and energy taken by the research part can outweigh the actual practice.
4 *Documenting the work.* The importance, and the difficulties, of effectively documenting the practice-based work that drives their project.
5 *Physical Space.* Spaces are either very difficult to arrange or are overly designed and too specific.

Johnston presents a series of strategies that can counter these challenges. They include: collaboration as an antidote to isolation;, clarifying the overall research structure early; establishing and maintaining reflective practice, and developing an ability to reflect; finding new perspectives through explicatory strategies, such as interviews; and finally slowing down and taking the time to explore.

Debbie Michaels' chapter in Part V, 'Organisational Encounters and Speculative Weavings Questioning a Body of Material',[19] reflects on what it feels like to be a PhD researcher grappling with the challenges of transposing approaches from psychoanalysis and art psychotherapy to the fine arts in an experiment with method. There are two challenges presented that relate directly to Johnston's list above: a) writing '*about*, rather than *with* and *through*, is problematic when the practice is embedded and embodied in the research; and, vice versa, with me, as practitioner researcher, at its centre', and b) the potential contribution that 'artistic processes and strategies may offer as part of a reflexive research approach, and the learning that can arise through a clash of disciplinary perspectives'. For Michaels, she approached her 'findings' as a process of discovery rather that a 'result or outcome'. Her thesis 'does not offer itself as a 'finished' work that sits *over, above*, and *apart* from what it "finds"', but it presents a 'partial, situated, view that sits *with* and *alongside* as part of an interweaving of threads in conversation, *through* which meanings and understandings may go on developing'.[20] Central to her contribution to new knowledge were her methods, which are 'revealed *through* processes of 'making' and the performance of tasks that are *on their way* to being completed'. She argues that the research value lies in the 'potential of this method to affectively (re)sensitise practitioners and researchers across arts and/in healthcare in ways that may not emerge through more traditional approaches to reflexive practice'.[21]

Building communities of practice-based researchers

Sian Vaughan's chapter, 'Community-Building for Practice-Based Doctoral Researchers: Mapping Key Dimensions for Creating Flexible Frameworks',[22] reports the results of a study of practice-based research communities and the challenges of such PhD research. She concludes that 'being part of a community can make navigating the challenges not only easier, but also more rewarding'. However, as with other forms or research, 'practice-based research itself is not homogeneous . . . and there are different disciplinary contexts as well diverse institutional contexts in which doctoral education occurs'. She presents 'dimensions, elements and activities of a framework for supporting and building community amongst practice-based doctoral researchers', but qualifies it by pointing out that there is no single framework that satisfies the needs. Rather, there is a family of frameworks that must be developed or selected to match specific contexts.

Building a practice-based research community is both supportive and timely. In the current landscape, there are increased pressures on the doctoral researcher. This can contribute to mental stress, and there is a need for well-being support from the institution. The PhD community can be a valuable lifeline. As the academic landscape responds and reflects the changing in concerns of society and industry, so too does the demands on supervisor's knowledge; it may be that senior professor's knowledge base is out of date, as such a community can help patch the gaps. Also, building a community can 'enable peer-sharing and broaden perspectives and opportunities, as well as inviting in others from different disciplines, practices and from outside higher education'.[23]

One model of how to build a practice-based research community is to consider the support structures and principles as an inter-connected *ecosystem*. Vear et al.'s chapter in this section, 'Strategies for Supporting PhD Practice-Based Research: the CTx Ecosystem',[24] outlines the strategies implemented by the Institute of Creative Technologies (IOCT) at De Montfort University in the UK. They also identify other ecosystems supporting doctoral research at institutions such as the Creativity and Cognition Studios, University of Technology, Sydney,[25] DX Arts at Washington University,[26] i-DAT at Plymouth University,[27] SensiLab at Monash University, Australia,[28] and EMPAC at Rensselaer Polytechnic Institute, Troy, New York.[29] Linda Candy and Ernest Edmonds describe the development of one approach in chapter 1.4 of this handbook.

The CTx model includes seven interlinked components of:

1   a doctoral programme (PhD) in practice-based research
2   a practice-based research doctoral training programme

3   an online 'cookbook' of methods and practices (www.pbrcookbook.com/)
4   summer school of intensive workshops and activities
5   shared space and a taught postgraduate programme
6   associated network of practitioners through *DAPPER* Digital Arts Performance Practice –
     Emerging Research network
7   creative technologists in residence programme

At the core of this *ecosystem* are the shared principles and an underlying philosophy that are valued by staff and students. This has led to several institutionally supported strategies that operate in a holistic way by providing a range of support mechanisms and structures to enhance the candidate's development. The term 'ecosystem' highlights the dynamic and inter-connected nature that grows as the researcher's practices and knowledge emerge. It is also responsive to new technologies and practices as they emerge. From this perspective, the term ecosystem does not suggest a system of organic life intertwined with and reliant on sustenance and support, but is used to define a system of research and practice within which the humans who conduct it are intertwined and benefit through sustenance and support.

Ethics and data protection

Ethics are critical and legal considerations, especially as it relates to institutionally supported practice-based research in the form of a PhD, but also for any research investigator. Ethics play a role in every research project, whether they are practice-based or not. However, acting ethically does not always mean the same as acting lawfully, despite there being some obvious overlaps. Although ethics is generally concerned with questions of moral responsibility and valued judgements, and as to what are 'right' or 'wrong' behaviours, there are also legal ethical constraints and protection of data that the researcher clearly has to address. Falk Hübner's chapter, 'Ethics Through an Empathetic Lens: A Human-Centred Approach to Ethics in Practice-based Research',[30] proposes a way of considering ethics with a 'positive' approach. He describes this as being concerned less with what 'not to do' or 'to avoid', but to 'listen to and to do right to all present human and non-human voices'. And for this to be an ongoing consideration through the whole life of the research project. Hübner points out that 'research ethics never exist in isolation, but rather in context of a particular time and culture', and significantly that they are 'intimately and inseparably entangled with methodology and the research process'.

In general, research ethics and data protection covers areas such as:

- vulnerable groups such as children or patients
- sensitive topics such as illegal behaviour or illness
- deception or carried out without the informed consent of participants
- use of confidential information or data about identifiable individuals
- processes that might cause psychological stress or more than minimal pain
- intrusive interventions (e.g., administration of drugs or vigorous physical exercise that would not be part of participants' normal lives)[31]

Despite research ethics and data protection approval generally being obtained early in a research project, there is a responsibility on the researcher to constantly 'pay attention to ethical issues' and 'stay awake' with regard to ethics and data protection, usage, dissemination, storage, and access. Furthermore, to 'act empathetically, be it in the preparatory stage, in the field or while writing up the final account'.[32]

Ethics applications for practice-based research studies

Ethics and data protection approval is important when making any kind of study involving people. All universities have formal procedures for obtaining ethics and data protection approval, and the application documentation will be different depending on the location of the research project.

The Creativity and Cognition Studios (CCS), a Sydney-based practice-based research centre, developed its own procedures through a collaborative process in which practitioner researchers wrote the material guided by the university's staff charge with ethical clearance. The documentation includes a *Standard Operating Procedures* document, setting out the practical aspects of applying ethics to research practice, and a *Code of Ethics* which details the legal codes and principles, including privacy and confidentiality; informed consent; approaches to the design, conduct, and reporting of research; professional conduct; and further sources of information about the responsible conduct of research. The CCS ethics approval process centres on a review by the community of researchers. The approach is discussed in more detail in 'The Studio and Living Laboratory Models for Practice-Based Research' by Linda Candy and Ernest Edmonds, in Part I of this handbook. Practical details are included in Ernest Edmonds' chapter in this section, 'The Practice-Based PhD: Some Practical Considerations'.

The role of the supervisor

Supervision of a practice-based research PhD can be a key factor in its success. It is obviously important that the supervisor understands the nature and principles of practice-based research alongside the specific field of study. The supervisor(s) will need to guide the candidate through methodological processes, research design, and strategy development. They will need to support the temporal rhythms that happen through a practice-based research PhD and be a critical friend through both the practice and the theoretical developments of the research. And, like all supervisors, maintain a balance between pastoral care and dialogic questioning. Specific to practice-based research is the need to support the student in navigating between the processes of practice and research.

Defining the composition of the supervisory team is especially important to get a balance of support for the individual needs of each practice-based researcher. To help guide this, the SuperProfDoc[33] research team presented a handbook that outlines different roles the supervisor can be (or are naturally). These are basically defined as:

- pastoral (e.g., emotional and spiritual support)
- directive (e.g., 'master ~ slave')
- didactic (e.g., issuing best practice solutions)
- critical friend (e.g., encouraging and support while offering honest/candid feedback)
- teacher learner (e.g., drawing out the process of learning and promoting your desire to learn)
- dialogic (e.g., presenting contrary arguments to open critical thinking)
- collegiate (e.g., supportive discussions and shared values)[34]

Because of the close relationship between the practitioner and their practice and how this can become analysed within the research process, there are other aspects of supervision that need to be considered. Following a workshop organised by the Orpheus Institute in 2020 entitled *Feed-Back, Feed -Forward: Approaches to Artistic Feedback in Doctoral Supervision*, the Royal Musical Association published a blog post based on notes taken by an attendee.[35] Although concerned with artistic-focused practice-based research doctoral supervision, there are relevant points made that can guide supervisors more generally when engaging with practice-based research. The notes are divided into three themes:

The relationship between candidate and supervisor

Dr Simon Waters[36] was the academic speaking on this subject and has experience of supervising over 50 practice-based PhDs in the field of electronic music. He highlighted the significance of discussing the proposed research before applying, as this helps to clarify the direction of the project and its viability, and also whether the applicant has the skills to conduct such a study. Crucially, it starts building trust between the supervisor(s) and the candidate. Waters also highlighted the importance of encouraging the practitioner researcher to engage with creative play as a 'mode of thinking and research'. Reinforcing the list from earlier, he also added that the supervisor needs to consciously avoid, where relevant, imposing their own personal way of doing research. The supervisor needs to be mindful that they are 'supervising the unknown', in a way that supports the journey that is 'contingent and emergent'.

Making explicit implicit and intuitive knowledge

Professor Janneke Wesseling[37] is a professor of practice and theory of research in the visual arts and talked about guiding PhD candidates through the complex process of turning intuitive and implicit knowledge into explicit forms.[38] Some of the key points made are that practice-based research can be 'driven by intuition and personal vision', and the 'importance of staying true to the hunch that begins the research, which usually arises through practice [and must be explored through practice]'. An important factor in explicating the implicit is in how the research question is phrased and focused, and that this may need to change and grow through the process as new insights emerge. The question, in this sense, becomes a productive constraint, much like a choice of material in practice. In this sense, it becomes a '"foot-hold of the mind", trusting the foothold to continuously ground the research as it develops. That this foot-holding is the condition of experimentation'.

Feed-back and feed-forward

Two academics contributed to this subject: Heloisa Amaral[39] and Vida Midgelow.[40] Amaral focused on the DasArts Feedback Method that 'encourages feedback givers to address an issue without judgement and from different positions'. This requires the feedback-receiver to formulate in advance what it is they need or would find most beneficial to the advancement of their understanding of the practice, or research that is under scrutiny. In short, it is 'learning to ask the 'right' questions in order to get the 'right' feedback back'. Midgelow's focus was on 'creative practice in/as feedback'. The focus was on giving appropriate feedback from practice-based work, rather than pure discussion or feedback of written text. She outlines a couple of principles that are worth keeping in mind when encountering a PhD candidate's practice. First, she advocates a process of feedback that is 'slow, of allowing, listening, waiting', and moving away from questions such as 'what is a thing' in favour of 'entering the experience of a thing'. She encourages supervisors and candidates to arrive at a language appropriate to the experiences of the practice, rather than naming or categorising too quickly. This could involve 'attending-to' or 'noticing what you notice' within the experience of the work. A good example of this is to start by discussing the 'miraculous moment' in the work, as perceived by the feedback-giver, as in her experiences this is not obvious to anyone else, especially the practitioner.

The practice-based PhD thesis

The thesis is an argument. This is split across the documented practice and the written exposition. To call the written section of a PhD submission the thesis and the practice something else is ultimately confusing, and we specifically use 'thesis' to mean the collection of items submitted for examination.

Practice-based research centres on thinking in, through, and with practice. This means that practitioner-knowledge can be embedded in practice, embodied through practice, or enacted with practice (refer to the knowledge-leading chapter). This means that any artefacts that practitioners create are an integral part of practice. Within PhD research, the making process provides opportunities for exploration, reflection, and evaluation. In a practice-based context, the role of the artefact is viewed as central to the research process.

This raises the question of how the outcomes of this research can be shared with the wider world. Research of a doctoral standard involves creating something novel and original that can be understood more generally, and to achieve that an accompanying text is needed. The text has a fundamental role in exposing the knowledge developed from the practice-based study. However, it is more than an explanation of any artefact that might have been produced. It is a description of all of the elements of the research. In relation to an artwork, for example, there is a

> gradual, cyclical speculation, realisation or revelation leading to momentary, contingent degrees of understanding. To this extent the text that one produces is a kind of narrative about the flux of perception-cognition-intuition. The text accounts for the iterative process that carries on until the artist decrees that the artwork is complete and available for critique, 'appreciation', interpretation, description, evaluation. All these particular practices can entail other particular texts.[41]

The practice-based doctoral research thesis arises from a structured process that is defined in university examination regulations. Knowledge arising from practice-based research is communicated in a range of outcomes: understandings about audience experience, strategies for designing engaging art systems, taxonomies of emergent behaviour and models of collaboration, to take a few examples. And, of course, there are the works themselves: the artefacts, the software, the medical practices, the compositions, the performances, the engineered bridges, the exhibitions and installations etc.

The purpose of the literature review

Sometimes called the 'state of the art review', the purpose of this is to map out the history of the field, from primary innovations to the latest advances, and to identify and outline the gap in knowledge that is addressed in the new research. This should cover the work by others which provides a basis and context for the research because it either is basing the findings to support an argument to extend something, or has identified a limitation and will be pursuing studies to address it.

The review should be drawn from primary sources: e.g., papers and books reporting results of original research, and will probably include grey literature such as catalogues, performance programmes, websites and other media used to disseminate the practice in the field. It can also outline the theoretical and philosophical structures that will be used through the thesis, if appropriate. Where the work includes a contribution that advances methodology, then the state of the art of relevant methodology will be included.

Outline for a practice-based thesis

In Ernest Edmonds' chapter in this section, 'The Practice-Based PhD: Some Practical Considerations', he proposes a few actions and issues that should be considered from the beginning. Noting that a practice-based PhD submission normally has more than one component, he argues that the text component, the written part of the thesis, needs to be thought about from the start, and he describes a draft outline chapter list. This is used to review many of the components of the work to be done and points to a way of working out a draft time-line for the PhD. The last section gives a few tips about checking the final text of the PhD before submitting it for examination. Edmonds explains how the thesis (in total) can present what the new knowledge is and suggests ways in which the text can illuminate the reader about:

- *how* it is new to the field
- *how I* went about generating the new knowledge
- *where* the new knowledge was generated in the process
- *why* this is new knowledge to the field

Starting a PhD with an initial plan for the thesis helps to structure the work. Everyone will need their own version and, inevitably, the plan will change as the research progresses; but it always provides a baseline that can be used to consider, compare, and contrast the work in discussions with supervisors and other researchers in the community.

The examination

The role of a practice-based PhD examiner is no different to the examination of any other form of PhD. There are rules and regulations implemented within each higher education institution that govern what defines a PhD and what criteria must be met in order for a degree of PhD to be awarded. The PhD candidate will make a claim to new knowledge (as part of their thesis), and then present an argument for this claim (the thesis). During the examination, the examiners will need to ensure that the PhD candidate is the original author of the thesis; that they have met the criteria set out by the institution for the award of PhD; 'pressure test' the PhD candidate's ability to defend their argument; and, if there are weaknesses in the claim or argument, assess the candidate's surrounding knowledge and ability to deal with such shortfalls.

As previously mentioned, the argument and claim of a practice-based researcher will often include the works arising from the practice. In these cases, to understand the significance and context of the new knowledge, it is necessary to see, to hear, to experience the works created rather than encounter them as illustrations in a written text. This means that as far as possible, the works should be available to the examiners alongside the written thesis and lodged in the archive for access by future researchers. Some artefacts are easier to make available and archive than others, but there are ways to address this. As an example, it is now commonplace for artists to mount an exhibition of their work to be shown at the time of the examination but beyond that period of time, the exhibition can be documented through photographs, video, and text which, while not the same experience, can nevertheless reveal much more about the contributions of the PhD than text alone.

The combination of work and exposition and the relationship of the claim and argument held across these naturally places a responsibility on both the PhD candidate and the examination team to examine the argument using all facets of the thesis: the practice-based works and the exposition. This may seem to be a self-evident statement; however, we have witnessed and still hear of examiners focusing only on the text-based exposition in the examination. This is clearly wrong where the claim of new knowledge refers directly to the work. Given the variety in practice-based PhDs and across institutions and disciplines, the specific way in which claims are made and how they might refer directly to artefacts need to be clearly articulated by the PhD candidate in the thesis. Furthermore, it would seem respectful that the examiner team also consider this when they are designing their examination strategy. This, of course, must be part of the governance process of ensuring that the institutional-specific regulations are adhered to as part of a legitimate examination.

Topics on the PhD in the chapters to follow

We have described the core elements and activities of the practice-based PhD, drawing on the material presented in the earlier sections of the handbook. In effect, we have summarised the key issues as they apply to PhD study. The emphasis has been on the distinguishing features of the practice-based PhD as well as on the particular challenges. The following chapters elaborate on a number of specific issues:

- the role of play in practice-based research
- the role that improvisation can play in practice-based research
- autoethnography as a reflexive method
- a dramaturgical approach to practice-based research
- the challenges and strategies for dealing with them
- the relationship of the researcher to the organisation
- community-building for practice-based researchers
- research ethics
- structuring and delivering practice-based research results

Notes

1 Edmonds (2022).
2 Borgdorff (2017, p. 7).
3 www.lego.com/en-us/seriousplay.
4 Smith (2022).
5 Smith and Dean (2009, p. 48).
6 Smith (2022).
7 Mwamba (2022).
8 Ibid.
9 Ibid.
10 Hübner (2022).
11 Findley-Walsh (2022).
12 Ibid.
13 Ibid.
14 Slättne (2022).
15 Ibid.
16 Candy et al. (2022).
17 See the various bibliographies in this handbook for a comprehensive list of influential publications.
18 Johnston (2022).
19 Michaels (2022).
20 Ibid.
21 Ibid.
22 Vaughan (2022).
23 Ibid.
24 Vear et al. (2022).
25 www.creativityandcognition.com/.
26 https://dxarts.washington.edu/.
27 https://i-dat.org/.
28 https://sensilab.monash.edu/.
29 https://empac.rpi.edu/.
30 Hübner (2022).
31 Denscombe (2014, p. 307).
32 Hübner (2022).
33 www.eurodoc.net/superprofdoc.
34 Ibid.

35 https://rmapracticeresearchgroup.tumblr.com/post/635862663026622464/supervising-artistic-research-doctorates.
36 https://pure.qub.ac.uk/en/persons/simon-waters.
37 www.universiteitleiden.nl/en/staffmembers/janneke-wesseling#tab-1.
38 Based on her paper 'Artistic Research at the Academy of Creative and Performing Arts, Leiden University', found at https://core.ac.uk/download/pdf/326229641.pdf#page=29.
39 https://orpheusinstituut.be/en/orpheus-research-centre/researchers/heloisa-amaral.
40 www.mdx.ac.uk/about-us/our-people/staff-directory/profile/midgelow-vida.
41 Candy (2006, p. 9).

Bibliography

Borgdorff, H. (2017). *Reasoning Through Art.* www.universiteitleiden.nl/binaries/content/assets/geesteswetenschappen/acpa/inaugural-lecture-henk-borgdorff.pdf (accessed October 6, 2019).

Candy, L. (2006). *Practice-Based Research: A Guide.* www.creativityandcognition.com/resources/PBR%20Guide-1.1-2006.pdf (accessed June 15, 2021).

Candy, L., Edmonds, E., & Vear, C. (2022). Practice-Based Research. In C. Vear (Ed.), *The Routledge Handbook of Practice Based Research.* London and New York: Routledge.

Denscombe, M. (2014). *The Good Research Guide: For Small-Scale Social Research Projects.* New York: Open University Press.

Edmonds, E. (2022). Research, Shared Knowledge and the Artefact. In C. Vear (Ed.), *The Routledge International Handbook of Practice-Based Research.* London and New York: Routledge.

Findley-Walsh, I. (2022). The Sound of My Hands Typing: Autoethnography as a Reflexive Method in Practice-Based Research. In C. Vear (Ed.), *The Routledge International Handbook of Practice-Based Research.* London and New York: Routledge.

Hübner, F. (2022). Ethics Through an Empathetic Lens: A Human-Centred Approach to Ethics in Practice-Based Research. In C. Vear (Ed.), *The Routledge International Handbook of Practice-Based Research.* London and New York: Routledge.

Johnston, A. (2022). The Practice of Practice-Based Research: Challenges and Strategies. In C. Vear (Ed.), *The Routledge International Handbook of Practice-Based Research.* London and New York: Routledge.

Michaels, D. (2022). Organisational Encounters and Speculative Weavings: Questioning a Body of Material. In C. Vear (Ed.), *The Routledge International Handbook of Practice-Based Research.* London and New York: Routledge.

Mwamba, C. (2022). Improvising as Practice/ Research Method. In C. Vear (Ed.), *The Routledge International Handbook of Practice-Based Research.* London and New York: Routledge.

Slättne, H. (2022). Navigating the Unknown – A Dramaturgical Approach. In C. Vear (Ed.), *The Routledge International Handbook of Practice-Based Research.* London and New York: Routledge.

Smith, H., & Dean, R. T. (Eds.). (2009). *Acquiring Know-How: Research Training for Practice-led Researchers.* Edinburgh: Research Methods for the Arts and Humanities, Edinburgh University Press.

Smith, S. (2022). A Play Space for Practice Based PhD Research. In C. Vear (Ed.), *The Routledge International Handbook of Practice-Based Research.* London and New York: Routledge.

Vaughan, S. (2022). Community-Building for Practice-Based Doctoral Researchers: Mapping Key Dimensions for Creating Flexible Frameworks. In C. Vear (Ed.), *The Routledge International Handbook of Practice-Based Research.* London and New York: Routledge.

Vear, C., Smith, S., & Bennett-Worth, S. L. (2022). Strategies for Supporting PhD Practice-Based Research: The CTx Ecosystem. In C. Vear (Ed.), *The Routledge International Handbook of Practice-Based Research.* London and New York: Routledge.

4.2

A PLAY SPACE FOR PRACTICE-BASED PHD RESEARCH

Sophy Smith

Introduction

This chapter will outline a number of benefits of play-based approaches to practice-based research, offering an alternative perspective on knowledge construction. This chapter will counter the assumption that play-based learning is of value only to Early Years teaching and learning by outlining how it has been used to develop and carry-out training within a Higher Education doctoral programme. The chapter will establish the value of play-based learning beyond Early Years settings and how, through play-based learning, practiced-based research doctoral candidates can move away from the more 'closed' traditional models of research practice towards a more 'open' landscape,[5] enabling the emergence of new knowledge, often across disciplines. A play-based approach to doctoral programmes can create transformative learning environments that enable risk-taking and innovation, enhancing the creative practice of learners. Nevertheless, the administrative process of doctoral study, including progress monitoring, coupled with student concerns relating to final submission, can work against any playful approach, thwarting the full realisation of potential knowledge construction. The utilisation of a more play-based approach as outlined in this chapter can support students to feel able to take risks in their research, while working within the administrative structures of doctoral programmes. The approach outlined in this chapter continues to develop, dynamically evolving in reflection of the experiences and responses of practice-based research doctoral candidates.

Through bringing together concepts including Sandseter's consideration of Reversal Theory in risk-taking in play (2010) and Kolb's learning spiral (2015), this chapter offers a play-based approach for practice-based research doctoral training providers. This framework is based on the training provision developed at the Institute of Creative Technologies at De Montfort University, which has been supporting practice-based doctoral researchers since 2012. Drawing on the experience of developing and running the Doctoral Training Programme for Practice-Based Research at the Institute of Creative Technologies at De Montfort University, and on written feedback on the programme from recent participants, this chapter outlines how the five main characteristics of play[6] have been embedded into practice and learning at PhD level. The chapter discusses how embedding play into doctoral training can enable learners to experiment and innovate, and details approaches adapted from Sandseter and Kolb to develop a supportive creative environment where a playful approach can thrive.

 DOI: 10.4324/9780429324154-35

After eight years of applying this approach at doctoral level, we are able to reflect on specific areas of impact. Two main areas of interest have been explored:

- How embedding play into doctoral training can enable the development of supportive creative environments where learners are able to experiment and innovate.
- And, importantly, how an environment conducive to risky play can be maintained within a process of monitoring and assessment.

It is important to note here that this play-based approach is not restricted to solely creative disciplines, where play may be seen as intrinsic, but extends across arts, humanities and sciences disciplines and all that fall in-between.

This chapter will be divided into three sections: firstly, the concepts and theories that underpin the play-based approach; secondly, how these theories combine to support the approach; and thirdly, the approach in practice, through the case study of the CTx Doctoral Training Programme.

The concepts and theories that underpin the play-based approach

Children doing research and researchers at play

Play is essential for children's development, building their confidence as they learn to explore, think about problems and relate to others. Children learn by leading their own play and by taking part in play that is guided by adults.[7]

Play is well established as central to the learning processes of young children[8] and is included in UK state legislation. It is also central to the Statutory Framework for Early Years Foundation Stage (2017), which states that children's learning and development 'must be implemented through planned, purposeful play'. However, the assumption seems to be that play is only important to learning until the age of 5, when children enter Key Stage 1. For example, legislation outlines how 'it is expected that the balance will gradually shift towards more activities led by adults, to help children prepare for more formal learning, ready for Year 1'.[9]

Jane Murray identifies links between children's epistemic play and the research process,[10] outlining how when leading play, children display similar behaviours to adult researchers.[11] She cites four behaviours that professional researchers specify as important to research:

> exploring, finding a solution, conceptualising and basing decisions on evidence.[12]

It is clear to see the links between these practices and those she witnessed as an Early Years teacher where she came across children:

> questioning, planning, acquiring information, analysing and interpreting, solving problems, exploring and reporting novel ideas and artefacts they had created.[13]

Murray cites Hutt et al.,[14] who describe epistemic play as

> the acquisition of knowledge and information . . . problem solving . . . exploration . . . productive, as well as focused on materials and transformations, in other words knowledge construction'.[15]

This same knowledge construction is central to the premise of research. This begs the questions:

- Could these play-based learning approaches be used to facilitate the knowledge construction by adult researchers as well as early years learners?
- How can the process of play-based learning enable the development of new knowledge for adults undertaking research?

Catharina Dyrssen[16] describes how most scientists regard play and creativity as central to their scientific work. In play, researchers must relinquish control, and this lack of control is 'a necessary part of innovation and cross-disciplinary contact and therefore not only acceptable, but also needed as an ingredient in most research processes today'.[17] In his 2011 *The Production of Knowledge in Artistic Research*,[18] Henk Borgdorff describes the methods of artistic research and reflects on the erratic nature of creative discovery, 'of which unsystematic drifting, serendipity, chance inspirations and clues form an integral part'. Exploration is as key to research as it is to play, Borgdorff reflects how 'research is more like exploration than like following a firm path'. Through a play-based learning approach, I suggest, researchers have the skills to explore with greater confidence.

The five characteristics of play

Peter Gray[19] outlines five characteristics of play – self-chosen and self-directed; intrinsically motivated; guided by mental rules; imaginative; and conducted in an active, alert, but relatively non-stressed frame of mind (p. 140). These are defined as:

Self-chosen and self-directed

Play, Gray asserts, is 'an expression of freedom',[20] where the child is led by what that they voluntarily 'want' to do, rather than what they are 'obliged' to do. Players are free to choose and direct their own action. Adults can play with children but should do so without taking control of the play. As adults are often seen as authority figures, players often feel less able to challenge the suggested direction of play.[21]

Intrinsically motivated

Play takes place for its own sake and is motivated by the activity itself. And while the play has goals, achieving the goals aren't the sole reason for the play.[22] 'The primary objective in such play is the *creation* of the object, not the *having* of the object'.[23] Referring to a University of Michigan experiment with preschool age children,[24] where a playful colouring activity was turned into a reward-based activity, Gray concluded 'It is possible to ruin play by focusing attention too strongly on rewards and outcomes. . . . When a game becomes primarily a means of proving oneself to be better than someone else . . . it becomes something other than play'.[25]

Guided by mental rules

While play is freely chosen, it has a structure which is formed through the player setting certain rules in place, 'mental concepts that often require conscious effort to keep in mind and follow'.[26] These rules, created by the player themselves, keep the player engaged with the play. Gray compares this type of player-instigated rule setting with the rules of what he calls 'formal' games, such as chess or football, with specified rules that must be adopted for the duration of the game.

Play is imaginative

Play takes place in a physical environment in the real world, uses props from the real world and is frequently based on events in the real world, yet is somehow separate from the real world.[27] Gray describes how, for example, when designing a real house, an architect will first build a 'pretend' house, imagining how it might be used and what it might look like. 'The capacity for abstract, hypothetical thinking depends on our ability to imagine situations we haven't actually experienced and to reason logically on those imagined situations'.[28]

Play is conducted in an alert, active, but non-stressed frame of mind

Gray asserts that as play is not a response to external demands; players do not feel pressure. As the player is focused on the process rather than outcome, players do not fear failure.[29] Gray cites psychologist Barbara Fredrickson's 'broaden and build theory of positive emotions'.[30] Fredrickson sets out how

> positive emotions *broaden* our perception and range of thought, which allows us to see what we didn't see before, put ideas together in new ways, experiment with new ways of behaving, and in these ways *build* our repertoire of knowledge, ideas, and skills.[31]

Conversely, negative emotions narrow out perception and focus the individual on the source of the negativity – fear, hate, judgement and failure. This activates our automatic arousal system, which enables facilitating narrow goals but disrupts creativity, learning and refection.[32]

Risky play and reversal theory

In her article 'Categorizing Risky Play – How Can We Identify Risk-Taking in Children's Play?',[33] Sandseter attempts to define 'risky' play, citing Stephenson,[34] who listed the significant elements of what makes a physical experience seem 'risky' as 'attempting something never done before; feeling on the borderline of out of control – often because of height or speed; and overcoming fear'. To this, Kaarby[35] adds 'exploration', where children undertake expeditions independently, to explore new environments.

In order to offer an enhanced understanding of the different elements of risky play, Sandseter has developed a phenomenological model of risky play, based itself on Reversal Theory as defined by Michael J. Apter.[36] Reversal Theory derives from the field of psychology – a theory of personality, motivation and emotion which focuses on the dynamic qualities of normal human experience to describe how an individual regularly reverses between psychological states, reflecting their motivational style, the meaning they attach to a given situation at a given time and the emotions they experience.[37] Reversal Theory is organised around a series of meta-motivational states, organised into four types of subjective experience, called 'domains': Means/Ends, Rules, Interaction and Orientation.[38] Each of these four domains can be experienced in two opposing ways (see Figure 4.2.1), relating to the two opposite forms of motivation.

In her study of risky play, Sandseter focuses on the Means/Ends pair – 'Telic and Paratelic' (or 'Serious and Playful'). This relates to whether an individual is motivated by achievement and future goals, or the enjoyment of process in the moment. She describes how in a serious state (telic), individuals are 'characterized by being serious-minded, goal-oriented, sensible, cautious, and arousal avoiding', compared with the playful state (paratelic), where individuals are 'characterized by being playful, activity oriented, adventurous, thrill-seeking, and arousal-seeking'.[39] In the serious state, she

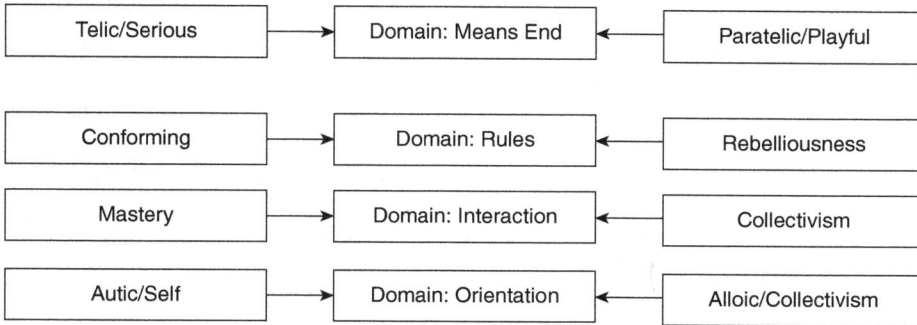

Figure 4.2.1   The 4 domains of reverse theory and their opposing styles.

Source: Adapted from Wright (2016).

continues, individuals experience high arousal as a negative emotion, and look to reduce it, while in the playful state, individuals experience high arousal as a positive emotion, and look to increase it.

The value of Sandseter's model to practice-based doctoral practice is an understanding of the relationship between play and seriousness. Practice-based doctoral candidates are often caught between these two polar states – experiencing a playful sense of adventure and thrill-seeking when engaged in innovation and experimentation, yet made cautious and arousal-avoiding by the seriousness of the academic system. In the sections that follow, we explore an approach that enables practice-based research doctoral candidates to cross between these polar states to support playful exploration within a formal academic setting.

Play and the development of new knowledge within practice-based research

The sections so far have outlined the relationship between play and research as well as concepts and strategies around risky play, including Reversal Theory. From these it is possible to draw out a number of principles that can help to support the practice-based researcher:

1   Play can have a key role in the development of new knowledge.
2   Play has a number of key characteristics, which can support the development of new knowledge.
3   Notions of risky play are particularly relevant in the development of new knowledge and, in particular, the role of meta-motivational states in Reversal Theory.

When combined, these three principles can offer a robust yet flexible approach to the playful development of new knowledge within practice-based research. In what follows, I will detail how this approach has been used in practice, as a key part of a Doctoral Training Programme for practice-based researchers.

A case study: play within the CTx practice-based research doctoral training programme

In 2013, the Institute of Creative Technologies (IOCT) at De Montfort University, developed the innovative Doctoral Training Programme in Practice-Based Research (DTP), followed by the development of the CTx programme (Creative Technologies Innovation through Doctoral Practice) in 2018.[40] These programmes have been designed to support practice-based doctoral candidates in developing and strengthening their individual practice. Key to this support is embedding play into the training programme, as a way of taking the students back-to-basics in terms of recognising and celebrating the excitement and strength of open exploration. The general findings from these programmes are that following years of formal academic education, students often arrived ill-equipped to develop ideas freely, without restriction, both individually and in groups. One particular strength of this approach is its value across discipline areas. Rather than being of relevance solely to arts-based areas, play became central to developing wider creative practice. Students participating in the DTP come from a wide range of discipline areas, including artificial intelligence, creative computing, creative writing, dance, fine art, digital art, holography, immersive technology, music composition and public art.

CTx and the characteristics of play

The structure of the CTx DTP in Practice-Based Research ensures that the key characteristics of play are embraced. Existing within a wider ethos, where risky play is embraced, the programme also includes specific training in both risky play and Reversal Theory. This ensures that students have engaged with these principles and applied them to their own practice, to give a deeper understanding of the role that risky play and Reversal Theory plays within their individual PhD journeys.

The first step was to develop an approach that embraced the five main characteristics of play outlined by Gray (see Table 4.2.1). The CTx DTP in Practice-Based Research draws on these key characteristics to develop an environment conducive to play. Work is both self-chosen and self-directed;

Table 4.2.1 CTx framework of practice, aligned with Gray's 5 characteristics of Play (2013).

Characteristic of play (Gray, 2013)	DTP component	Details
Self-Chosen/Self Directed	Playground(ed) Principles of Play Practice Cookbook	Workshops relate to individual practice and are student-led. Doctoral students are free to choose and direct their own action.
Intrinsically Motivated	Principles of Play Practice Cookbook	Doctoral students are motivated by the activities themselves rather than achieving goals. Individuals apply information to their specific experience.
Guided by Mental Rules	Principles of Play	Doctoral students set their own rules and boundaries for their practice, dependant of the aims and objectives of their research. More 'formal' rules are also in play, such as research methods, ethics procedures etc.
Imaginative	Playground(ed)	Doctoral research takes place in the real world, yet is also separate from it. To support the role of the imagination, sessions include creative formats that draw on the imagination, including workshops using Lego, drawing, haiku, rapid prototyping etc.
Conducted in an active, alert but relatively non-stressed frame of mind	Playground(ed) Principles of Play Practice Cookbook	Environment created where risk-taking and mistakes are embraced. Feelings of pressure and fear of failure are articulated and understood within Fredrickson's 'broaden and build theory of positive emotions' (2001,2003).

the quarterly whole-group sessions bring the group together to collaborate, explore and experiment in a supportive open environment. Students are able to take risks and innovate in an un-pressured environment, free from the threat of perceived failure.

Central to this play-based approach has been the development of a supportive creative environment where learners are able to experiment and innovate. In *Free to Learn*, Peter Gray describes the inhibitory effect of teacher-led learning: where students are shown a specific way to approach a problem, they will regard this as the only way. However, through a play-based approach, doctoral students explore the problem in greater detail, finding different ways to approach it and, by doing, so understand the full dimension of the problem and the 'full power of possibilities'.[41] In practice, the assurance that the result of play was valid was liberating to the PhD candidates and further validated its adoption in their research practice. As one doctoral student reflected:

> I have been encouraged to embed play in my research process. . . . Being able to 'play', rather than follow a set of strict guidelines, requirements and expectations, enabled every student to step forward with their personal views and employ their individual knowledge to contribute effectively for the accomplishment of projects.[42]

Graeme Sullivan[43] describes how knowledge creation within practice-led research occurs through 'imaginative leaps' that take place 'within an open landscape of free-range possibility rather than a closed geography of well-trodden pathways'.[44] We have found that play-based approaches can enable students to find these new knowledge pathways, with one student reflecting:

> Allowing creative play within technology-focused learning enables the student/researcher to find, and then push, boundaries that more traditional users of the technology may not encounter. The play encourages technological development by asking new questions.[45]

Another commented:

> My project is working with a relatively new technology and therefore the lack of knowledge in this area has allowed me to experiment through play. Risky play has led to processes that push the boundaries of what is possible with this technology alongside my already established movement practice. . . . Much of the early investigations I have developed within my PhD process have involved forms of risky play which have produced outcomes that would not have been possible if we had just followed the rulebook for working with this specific technology.[46]

Another reflected on how play-based approaches enable these new knowledge pathways to traverse between realms of the arts and technology:

> including the touchy, feely, emotional, poetic moments of the research journey do not exclude disseminating new technical discussions brought about by pushing boundaries in the use of existing technology. Indeed, although I deliberately approach my field of research from an artist's perspective, my contributions have been welcome at international conferences, in the form of technical papers. . . . I think that the aspect of being allowed to explore and network beyond expected boundaries is where the stardust sparkles bright.[47]

Our reflexive and responsive approach aims to create a playful environment within which new knowledge construction can take place. However, as Gray reflects, any pressure to perform well can interfere with new learning and pressure to be creative interferes with creativity.[48] Where activities involve creative though or learning a new skill, the presence of evaluation thwarts the playful state.[49] He states:

> Learning, creativity and problem solving are facilitated by anything that promotes a playful state of mind, but they are inhibited by evaluation, expectation of rewards, or anything else that destroys a playful state of mind.[50]

Play is a positive and necessary part of the PhD process, but how do we enable play for the students worried about the formal constraints of a PhD? For the CTx DTP in Practice-Based Research, the answers lie within risky play theories. By creating a safe space for more risky play, we facilitate the creative mood needed to enhance creativity[51] and the more playful state of mind needed to solve logic problems.[52]

Risky play: navigating the serious (telic) and playful (paratelic)

Risky play is a key part of the CTx ethos, and the programme includes specific workshop training in risky play and Reversal Theory as part of the Principles of Play strand (see Table 4.2.1). Through drawing on Sandseter's phenomenological model of risky play based on Reversal Theory, we have been able to enable students to retain a 'playful state of mind', play and experimentation existing alongside the rigorous PhD system, where the ultimate goal is a successful completion. In order to support this, we adapted Sandseter's figurative summary of the phenomenological structure of risky play (2010), omitting details that related specifically to children's first-hand experiences and retaining and adding those that relate to the practice-based research doctoral student (see Figure 4.2.2). For Practice-Based Research PhD candidates, both playful activity-orientated (paratelic) and serious goal orientated (telic) states are necessary for a successful PhD completion, where they are required to develop innovative creative practice while working within a strict administrative process (including numerous milestones and deadlines such as monthly progress reports, annual reviews, formal reviews and final submission). Therefore, we needed to find a mechanism whereby they can move between these two states.

The dynamic shifts across the serious and playful states, caused by either internal emotions or external events, are referred to as 'reversals',[53] and it is these dynamic shifts that allow for the student to move between the two states. The bistable nature of risky play is characterised by quick reversals between serious and playful states and contrasting emotions such as pleasure/displeasure, enjoyment/boredom and excitement/anxiety which can be experienced within a single situation.[54] Experiencing both simultaneously brings forward what Sandseter describes as 'ambivalent' emotions (2010 p. 6). By studying this framework, it is possible to understand the importance of this middle 'ambivalence' state, especially in enabling doctoral students to move across and between the serious and playful states. Rather than a bistable state, where PhD students are either in a serious goal-orientated or playful activity-orientated state, PhD students need to be confident in the ambiguous middle ground between the two. As one student reflected:

> As a practitioner who is most comfortable being in control of the artistic process and is used to working towards a final goal within a short time frame, the encouragement to allow risky play into my process supports me to find a middle ground between paratelic and telic states.[55]

For some students however, this ambiguous state is challenging. As one participant in the Doctoral Training Programme for Practice-Based Research commented:

> The fear of engaging with what I love to do has become a big mountain to climb. The access to a paratelic state of research is hard-fought. . . . I want to name the fear, the feeling

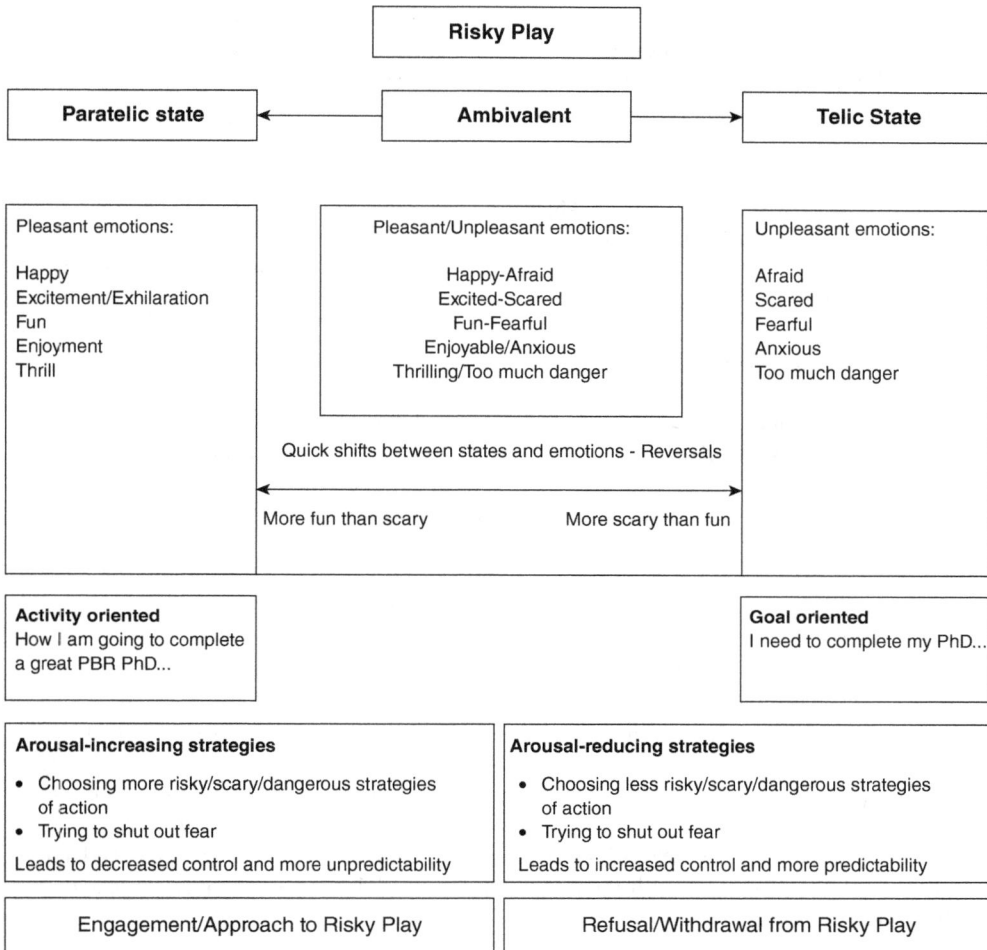

Figure 4.2.2   Risky play within PhD, based on Sandseter's figurative summary of the phenomenological structure of risky play (2010).

of insufficiency and the joy of exploring the way over the mountain, which sometimes goes
round and round and back and forth. And after all, as Tove Jansson wrote in 'The Exploits
of Moominpappa', *'For if you are not afraid, how can you be really brave?'*[56]

The tension created when practice-based researchers work within a particular system, while simultaneously needing to work against or beyond the same system, is explored in Brigid Costello's earlier chapter
concerning the rhythm of practice-based research, where she discusses how stability and instability, keeping step and breaking free can co-exist. For ideas of how a deeper awareness of the individual rhythms
of practice-based research may support a practitioner in negotiating the reversal states described earlier,
enabling them to embrace co-existence, I recommend reading Costello's chapter in depth.[57]
    A student on the DTP reflects on working within a playful paratelic state:

    While my practice was framed by existing literature and artefacts, I chose the areas I wanted
    to explore and play in/with. Extensive and well documented play resulted in experiments

which I used in my analysis. . . . There was no framework to work in and I didn't adhere to rules (not even self- imposed ones). To me playing was a way of transforming vague ideas into reality to test their viability followed by subjective evaluation. You set yourself up with tools and contexts and ideas emerge during playtime.[58]

The notion of a bistable experience created by the two polar states – serious goal-orientated and playful activity-orientated – fits with the experiences of many of our practice-based research doctoral students. On the one hand, they are keen to develop original and innovative research and relish the nature of the unknown that this brings. Equally, however, they are anxious and cautious, afraid of failing. One participant reflected:

To align my own PBR to the idea and metaphor of risky play is to imagine myself as a child on one of those 'wobble and balance' play machines whereby one foot is on the right, the other on the left and the trick is to balance between the two 'paratelic and telic' states, exploring each polarity and momentarily integrating across them – without falling off.[59]

For another, the anxiety of play, and taking the risks necessary for research, has become debilitating:

I have spent at least a month avoiding any information that is to do with my research topic, as it gives me electric shocks of guilt. On a rational level, I can see that I am terrified of engaging in the 'play and making' that defines my artistic and research journey. I am at the edge of jumping in to play, with the knowledge that I have now invested 7 years into it being successful. In additional, I have the deeper knowledge that my self-image is connected with me doing this very well, and have invested years of reading to justify my process within an academic system that has rather clumsily fitted the art school into an imagined system a PhD award. I am very aware that many people in what are seen as more traditional scientific fields of research also engage in play (experimentation) and also respond to 'hunches', without having to justify this as valid in the way that an artist within the university has to.[60]

Supporting students in a bistable learning experience

Where Reversal Theory presents a bistable experience, individuals reversing between either one state or another, we needed to look at how we could support our practice-based research doctoral students to work *across* these states. To support this, we drew upon Kolb's updated Experiential Learning Theory (ELT) of 2015. Kolb states that the aim of his Experiential Learning Theory (ELT) is to create 'a theory that helps explain how experience is transformed into learning and reliable knowledge'.[61] Kolb's updated ELT of 2015 portrays eight learning modes, grouped in dialectically related pairs:

- Concrete Experience and Abstract Conceptualisation.
- Reflective Observation and Active Experimentation.
- Assimilation and Accommodation.
- Converging and Diverging.[62]

More powerful learning emerges when these modes are used in combination,[63] and each learner combines the modes differently depending on their learning style.[64] The process ideally forms a learning spiral, where the learning recursively passes through each mode in response to the learning need, learning taking place through 'the resolution of creative tension among these 4 learning modes'.[65]

Within the CTx DTP in Practice-Based Research, we have used Kolb's ELT to support doctoral candidates to understand how they can use the theory's different learning modes to support them to move *between* the serious goal-orientated and playful activity-orientated states, necessary to move into risky play. In Figure 4.2.3, I have presented a diagram that shows how the two models (Reversal Theory and Experiential Learning Theory) can work together. In order to create this model, I have adapted McLeod's diagram of Kolb's eight learning modes,[66] to include the telic, paratelic and ambivalent states from Reversal Theory.

Importantly for PBR doctoral students, this model shows a way to move dynamically *across* serious (telic) and playful (paratelic) states, through states where feeling and doing (aligned to activity-orientated play) are combined with thinking and watching (the more serious goal-orientated focus). This is vital where the experimentation and innovation that happens during the play must also exist within more formal systems. This model supports us to develop a programme where doctoral students have the confidence to engage in risky play, while understanding where that sits in relation to the goal-driven system of a PhD. Kolb describes how the learning process is characterised by such dialectic opposition and can be filled with tension and conflict.[67] Learning, he asserts, requires abilities that are polar opposites and the resolution of these oppositions determines the level of resulting learning.[68] Through the creation of this model, and its place within the CTx DTP, we aim to offer students an approach which to resolves, or at least consolidate, such oppositions.

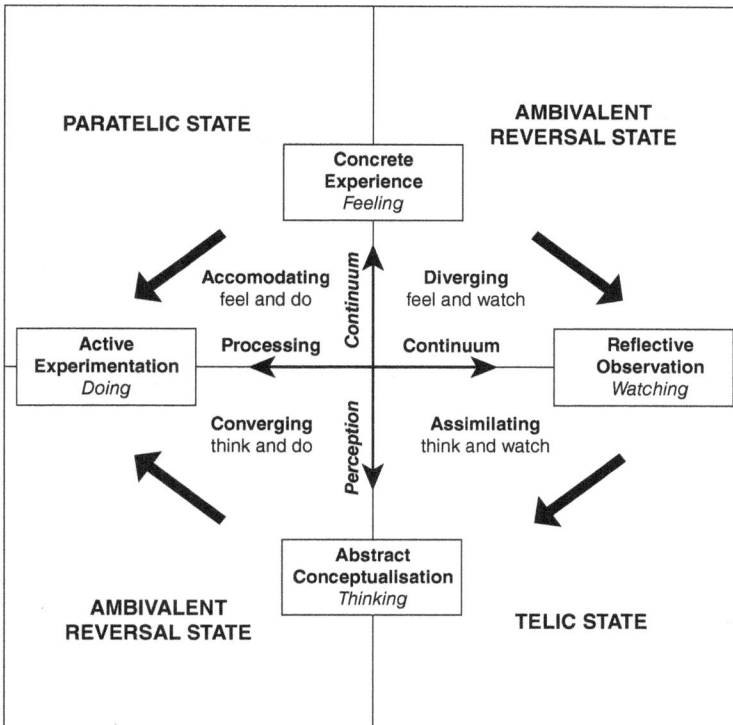

Figure 4.2.3 Notions of paratelic/telic states, applied to Kolb's learning cycle and learning styles.
Source: (McLeod (2017).

The approach of the CTx DTP has created a bridge that enables students to cross between the playful (paratelic) and serious (telic) positions, through the ambivalent reversal states which allow ambivalence to be acknowledged, welcomed and celebrated. This ambivalence is enabled by 'holding' students in these uncertain and undecided places, in a supportive, critically friendly and non-judgemental environment. Such an environment enables them to exist within Kolb's diverging state of feeling and watching, and/or the converging state of thinking and doing. By designing a system around and within the characteristics of play in a general sense, the DTP has also created a scaffold for risky play, where they feel able to take risks. All three categories of the DTP (Playground(ed), Principles of Play and the Practice Cookbook) have relevance within both the diverging (feel and watch) and converging (think and do) ambivalent states, working between and within concrete experience and reflective observation and between and within abstract conceptualisation and active experimentation. The Cookbook, foregrounding the experiences of alumni, enables students to reflect on how playful and serious approaches (and the ambivalence in between) play in each PhD journey. One PBR DTP participant reflected the importance of accepting and embracing this ambiguity:

> Sometimes this type of exploratory research felt like a seesaw and I became grounded, other times I'd fall off, bruise my ego and lose heart, wanting to hide or zoom away. However the approaches of the IOCT, both paratelic and telic, help the ability to stay with the ambiguity of not knowing. Through robust supervision and constructive peer support I gained the confidence to accept my failures, recognise new possibilities and to process my successes, emerging freer, fiercer and more focused.[69]

Interestingly, this same participant found that the risky play approach she had encountered fed directly into the way in which she worked with her research participants:

> My risky play training enabled me to support and scaffold my own participants learning, encouraging them to more confidently undertake both performative and playful risks from within their adult bodies. Specifically I created a new innovative, digital and physical arts relational framework that enabled participants to engage and create in deep and meaningful ways and encouraged them to explore and extend their own life experiences, innovating and imagining different futures.[70]

While the programme gives students the opportunity to take risks and make mistakes, they are reassured by understanding the roles that fear, rule-breaking and boundary pushing play in the PhD process, as well as valuing the roles of structure and safety. One student reflected:

> this was enlightening on so many different levels [and] gave me a new way to consider my approach to my research. . . . I valued looking at other group members work, hearing the observations being fed back . . . these events should never be under estimated to how highly valuable their contribution is to the development of researchers on their PhD journey.[71]

And another:

> Within this framework, I do feel free to play. And free to feel that play is a justifiable part of my true process. I do not believe that I could do what I do in an inspired way without it – it just becomes lumps of boring statements in a big pile of words. In a way, producing a formalised text without the spirit of exploration and without the acknowledgement of the force of play is ultimately dishonest.[72]

Students also appreciate being able to see the ambivalent state, with the quick shifts between the two states and related emotions – of feeling happy and fearful, excited and scared and feeling enjoyment and anxiety simultaneously, and recognising that as a normal part of their learning process.

For Sandseter, this ambiguity between the experiences of pleasant emotions versus unpleasant emotions were key concepts in the phenomenology of risky play,[73] allowing participants to move between the actions of engagement and withdrawal. Sandseter describes how even thinking about what could go wrong, or making a slight mistake can cause a reversal into the telic state and a withdrawal from play.[74] Students need to be enabled to recognise this shift, and by working through and to feel confident to move through the ambivalent states, work towards a paratelic state once more. An understanding of this minimises the threat of withdrawing from play completely, which is detrimental to the PhD journey. This ability to cope with anxiety, clearly articulated as a key part of risky play, is central to the success of the doctoral students on the programme. One student commented:

> The events run by the DTP have given me permission, space and time for play within my process, alleviating some of the anxiety around the doctoral process. The encouragement to share our developing processes involving play within the PlayGrounded workshops have led to a culture change and a supportive environment which embraces the unknown. For example, exploring Serious Lego play and my PhD methodology was insightful. Sharing this with other practice-based researchers allowed me to recognise that all approaches are different. This alleviated some of the anxiety around comparing my own journey to others, by understanding that every PhD experience is different.[75]

Conclusion

This chapter has offered a practical framework, grounded in psychology and learning theory, to support a playful approach to knowledge construction through doctoral practice and offering an alternative approach to knowledge construction. The value of play-based learning extends well beyond Early Years settings, as through play-based learning, practice-based research doctoral students can move away from the more 'closed' traditional models of research practice, towards a more 'open' landscape,[76] enabling the emergence of new knowledge, often across discipline areas. Through an understanding of the playful and serious states of Reversal Theory within risky play, it is possible to move between the activity-orientated and goal-orientated mindsets that are both central to the practice-based PhD experience. By combining the models of Reversal Theory and Experiential Learning Style within the CTx DTP in practice-based research, we have created a framework of practice that enables risk-taking and innovation, while valuing the necessary role of safety and structure within that play. The reflections from participating students have evidenced the value of this playful approach, which supports the development of confident PhD candidates who are prepared to take risks within their practice in order to allow the new knowledge to reveal itself through play. This approach continues to develop, dynamically evolving in reflection of the experiences and responses of practice-based research doctoral candidates.

Notes

1 Gray (2013).
2 Sandseter (2010).
3 Kolb's (2015).
4 James and Nerantzi (2018).
5 Sullivan (2010, p. 48).
6 Gray (2013).
7 Department for Education (2017, p. 9).

8 For example, Anning (2015) Moyles (2015) and Wood and Attfield (2005).
9 Statutory Framework for Early Years Foundation Stage (2017, p. 9).
10 Murray (2015, p. 108).
11 Ibid., p. 106.
12 Ibid., p. 107.
13 Ibid., p. 106.
14 Hutt et al. (1989, pp. 222–224).
15 Ibid., p. 109.
16 Dyrssen (2011, p. 238).
17 Ibid., p. 238.
18 Borgdorff (2011, p. 57).
19 Gray (2013).
20 Ibid., p. 141).
21 Ibid., p. 142.
22 Ibid., p. 143.
23 Ibid.
24 Lepper et al. (1973).
25 Gray (2013, pp. 145–5).
26 Ibid., p. 146.
27 Ibid., p. 149.
28 Ibid., p. 151.
29 Ibid., p. 152.
30 Fredrickson (2001, 2003).
31 Ibid., p. 153.
32 Ibid.
33 Sandseter (2007).
34 Stephenson (2003).
35 Kaarby (2004).
36 Apter (1981, 1982, 1984, 1989,1992, 2001), Apter et al. (1988).
37 Apter (2003).
38 Wright (2016).
39 Sandseter (2007, p. 3).
40 For further details, see Vear et al. (2022).
41 Gray (2013, p. 118).
42 Response to Student Questionnaire (2017).
43 Sullivan (2010).
44 Ibid., p. 48.
45 Response to Student Questionnaire (2017).
46 Response to Student Questionnaire (2020).
47 Ibid.
48 Gray (2013, p. 134).
49 Ibid., p. 133.
50 Ibid., p. 139.
51 Ibid., p. 136.
52 Ibid., p. 137.
53 Apter (2001, 1984).
54 Apter (1981, 2001).
55 Response to Student Questionnaire (2020).
56 Ibid.
57 Costello (2022).
58 Response to Student Questionnaire (2017).
59 Response to Student Questionnaire (2020).
60 Ibid.
61 Kolb (2015, p. xxi).
62 Ibid., p. 145.
63 Ibid., p. 101.
64 Ibid., p. 103.
65 Ibid., p. 51.

66 McLeod (2017).
67 Kolb (2015, p. 41).
68 Ibid., p. 42.
69 Response to Student Questionnaire (2020).
70 Ibid.
71 Response to Student Questionnaire (2019).
72 Response to Student Questionnaire (2020).
73 Sandseter (2010, p. 10).
74 Ibid., p. 17.
75 Ibid.
76 Smith and Dean (2011, p. 48).

Bibliography

Anning, A. (2015). Play and the Legislated Curriculum. In J. Moyles (Ed.), *The Excellence of Play*. Maidenhead: Open University Press.

Apter, M. J. (1981). On the Concept of Bistability. *International Journal of General Systems*, 6, 225–232.

Apter, M. J. (1982). *The Experience of Motivation the Theory of Psychological Reversals*. London: Academic Press.

Apter, M. J. (1984). Reversal Theory and Personality: A Review. *Journal of Research in Personality*, 18, 265–288.

Apter, M. J. (1989). *Reversal Theory: Motivation, Emotion and Personality*. London: Routledge.

Apter, M. J. (1991). The Nature, Function and Value of Play. In M. J. Apter & J. H. Kerr (Eds.), *Adult Play*. Amsterdam and Lisse: Swets & Zeitlinger B.V, 163–176.

Apter, M. J. (1992). *The Dangerous Edge: The Psychology of Excitement*. New York: The Free Press, Macmillan.

Apter, M. J. (2001). *Motivational Styles in Everyday Life a Guide to Reversal Theory*. Washington, DC: American Psychological Association.

Apter, M. J. (2003). *Reversal Theory Glossary*. http://reversaltheory.net/blog/about-the-theory/glossary/ (accessed February 20, 2010).

Apter, M. J., Kerr, J. H., & Cowles, M. P. (1988). *Progress in Reversal Theory*. Amsterdam and Oxford: North-Holland.

Biggs, M., & Karlsson, H. (2011). *The Routledge Companion to Research in the Arts*. Abingdon: Routledge.

Borgdorff, H. (2011). The Production of Knowledge in Artistic Research. In M. Biggs & H. Karlsson (Eds.), *The Routledge Companion to Research in the Arts*. Abingdon: Routledge.

Costello, B. M. (2022). Finding the Groove: The Rhythms of Practice-Based Research. In C. Vear (Ed.), *The Routledge International Handbook of Practice-Based Research*. London and New York: Routledge.

Department for Education. (2017). *Statutory Framework for Early Years Foundation Stage*. London: Department for Education.

Dyrssen, C. (2011). Navigating in Heterogeneity: Architectural Thinking and Art-Based Research. In M. Biggs & H. Karlsson (Eds.), *The Routledge Companion to Research in the Arts*. Abingdon: Routledge.

Fredrickson, B. L. (2001). The Role of Positive Emotions in Positive Psychology: The Broaden-and-Build Theory of Positive Emotions. *American Psychologist*, 56, 218–226.

Fredrickson, B. L. (2003). The Value of Positive Emotions. *American Scientist*, 91, 330–335.

Gray, P. (2013). *Free to Learn*. New York: Basic Books.

Hutt, C., Tyler, S., Hutt, C., & Christopherson, H. (1989). *Play, Exploration and Learning*. London: Routledge.

James, A., & Nerantzi, C. (Eds.). (2018). *The Power of Play in Higher Education*. London: Palgrave Macmillan.

Kaarby, K. M. E. (2004). *Children Paying in Nature*. Paper presented at the CECDE conference, Questions of Quality, Dublin Castle.

Kolb, D. A. (2015). *Experiential Learning: Experience as the Source of Learning and Development*. Upper Saddle River, NJ: Pearson Education, 2nd ed.

Lepper, M. R., Greene, D., & Nisbett, R. E. (1973). Undermining Children's Intrinsic Interest with Extrinsic Reward: A Test of the 'Overjustification' Hypothesis. *Journal of Personality and Social Psychology*, 28, 184–196.

McLeod, S. A. (2017). Kolb – Learning Styles. *Simply Psychology*, October 24. www.simplypsychology.org/learning-kolb.html (accessed February 20, 2020).

Moyles, J. (Ed.). (2015). *The Excellence of Play*. Maidenhead: Open University Press.

Murray, J. (2015). Young Children as Researchers in Play. In J. Moyles (Ed.), *The Excellence of Play*. Maidenhead: Open University Press.

Sandseter, E. B. H. (2007). Categorizing Risky Play – How Can We Identify Risk-Taking in Children's Play? *European Early Childhood Education Research Journal*, 15(2), 237–252.

Sandseter, E. B. H. (2010). It Tickles in My Tummy! – Understanding Children's Risk-Taking in Play Through Reversal Theory. *Journal of Early Childhood Research*, 8(1), 67–88.

Smith, H., & Dean, R. T. (Eds.). (2011). *Practice-led Research, Research-led Practice in the Creative Arts*. Edinburgh: Edinburgh University Press.

Stephenson, A. (2003). Physical Risk-taking: Dangerous or endangered? *Early Years*, 23(1), 35–43.

Sullivan, G. (2010). *Art Practice as Research: Inquiry in Visual Arts*. Thousand Oaks, CA: SAGE Publications, Inc., 2nd ed.

Vear, C., Smith, S., & Bennett-Worth, S. L. (2022). Strategies for Supporting PhD Practice-Based Research: The CTx Ecosystem. In C. Vear (Ed.), *The Routledge International Handbook of Practice-Based Research*. London and New York: Routledge.

Wood, E., & Attfield, J. (2005). *Play, Learning and the Early Childhood Curriculum*. London: Paul Chapman Publishing.

Wright, J. (2016). Flow Within Everyday Emotions and Motivations: A Reversal Theory Perspective. In L. Harmat, F. Ørsted Andersen, F. Ullén, J. Wright, & G. Sadlo (Eds.), *Flow Experience*. Cham: Springer. https://doi.org/10.1007/978-3-319-28634-1_13.

4.3

THE SOUND OF MY HANDS TYPING

Autoethnography as reflexive method in practice-based research

Iain Findlay-Walsh

Prelude

050820 07:13
window over window over window over window
each affords a digital possible world
I writes in this window
words appear
I farts
out into quiet
the time and space required to type this
was carefully made
I won, I stole, this silence
set the alarm, got two kids sleeping, negotiated the morning with LJ, jolted myself awake with
deadline fear
(gentrified the area, bought the house, sealed the window, drew the blind . . .)
and now here we are
in the south side of Glasgow in the quiet of a morning
crouched before an interface (laptop screen) in this little room
you're asleep
my listening surveils the flat for signs of children waking
in the past few months this place has been a site of locked down inter-familial tensions – pressures
cooking
the pandemic means that my partner, children and I
eat sleep breath live work play care teach learn here together 24/7
right now everyone is sleeping and so
I make the sound of these hands working the keyboard
as small as can be
a technique of holding the body, arching the spine, practised and refined in the very early hours of
most days
that yields now soft fluttering hammering sounding finger spasms of thought – interposed by cau-
tious halting
breathing, listening . . . rain outside the window
keyboard hand quiet room sleeping family and I
want to make a story

DOI: 10.4324/9780429324154-36

Figure 4.3.1   Dream.

Source: Iain Findlay-Walsh.

> *of the sound of hands typing*
> *and the conditions that produce the audibility of this sound as we read alone together*
> *hear and now-ing in precarious silence*

Introduction

> What if we were to create a story that would work within the handbook genre but also outside it, showing what we do as we tell about it?[1]

> . . .

> *050520 11:10*

> *Hi Iain,*

> *hope this email finds you well.*

> *I am in the process of editing the Routledge International Handbook of Practice-Based Research and am wanting to include a chapter on auto-ethnography. I wondered if you might be interested in writing such a chapter (saw your workshop invite).*

> *In general the handbook is aimed at all levels of researchers or practitioner-researchers and we are approaching it with a pan-disciplinary approach. In practice this means that the writing needs to be clear (it is not an academic discourse book), aimed at more than just PhD researchers, and that our own experiences in specific fields are exemplars that illustrate the takeaway point of each chapter.*

> *The handbook will also be politically neutral. The last thing I care about is the difference between practice-led, -as, based, etc etc. It is all the same thing: a research investigation based in/ with/ through the investigator's practice.*

> *Happy to discuss further if your interest is piqued.*

> *regards*
> *Craig*

> . . .

I'm standing near the west entrance of Queen's Park in Glasgow at around half past eleven on the morning of Tuesday, 5 May 2020, when I receive this request out of the blue while checking work emails on my phone, as my two young children, aged 6 and 9, play nearby. We are out on one of the daily walks that have become routine during the pandemic and I can hear them playing in a bush a few metres away. Schools have been closed here for nearly two months, so, like so many others globally, in this moment I am simultaneously 'at work' and caring for children who are disorientated by the sudden loss of social contact and daily routine. Making it to this park today is something of a victory, as the environment provides a sense of 'normality' and even 'fun' for them, while I get a minute to think and walk. As I read this email request, I feel a combination of excitement and dread, and begin mentally exploring initial responses . . .

"A book chapter. That's the kind of thing I should be getting asked to do at this stage in my career . . ."

"Fuck. I'm not sure I know anything about autoethnography. Yes, I am chairing a conference on it with Chris Wiley in June, but I'm concerned that he is carrying the whole thing. I'll have to write a good paper for that though, so maybe organising the event and writing that paper will get my thoughts together enough to write this thing . . ."

"Maybe this would be a good way to learn more about autoethnography and to work out once and for all whether this is something I do, or if it's just a smart-sounding term I've been using to get ahead. Whoever Craig is, they've definitely made a mistake asking me. I'm sure there are a million actual autoethnographers out there with something to say. Of course, I going to say 'yes'."

"Daddy? Daddy? I've been calling you for ages! I don't have anything to *do* here . . ."

. . .

It is now six minutes past midday on Sunday, 28 February 2021, and I am rushing to finalise this chapter, having returned to it many times in the ensuing months. I do this from a small office space in the flat in Glasgow, Scotland, where I have been living with my partner and children during the COVID-19 pandemic, in a prolonged period of alternating school closures, home-working, widespread restrictions on movement, closed parks and shops and nationwide 'lockdowns'. Accordingly, this chapter is presented as a fragmentary narrative, a story or writing and rewriting, and trying to write, and not writing, that collects and presents my efforts to discuss autoethnography as a versatile set of research methods and forms of potential use to those engaged in practice-based research.

. . .

In lieu of the more detailed introduction to the method provided in the next section, the word autoethnography can be 'unpacked' in straightforward terms to mean self (auto) culture (ethno) writing (graphy), and usually refers to research that employs critically engaged, autobiographical writing as a primary research method and presentational form (outcome). Autoethnographers write stories – scripted dramas, poems, diary entries or other narrative forms – and use the process of writing, interpreting and editing to engage in cultural inquiry, writing and rewriting the self in relation to a specific socio-cultural context. Acknowledging and tracing personal thoughts, feelings, actions and interactions that emerge during particular experiences and encounters, while considering these within wider conditions and contexts, autoethnographic researchers value and use personal experiences as instances that may yield new cultural understanding. Thus, the act of writing is practised as one way of 'reflecting-in-action',[2] and as an ongoing process of 'illumination'[3] directed towards 'self-transformation'.[4,5] In generating creatively written, layered, personal accounts that respond to cultural contexts and questions, autoethnographers often develop writing strategies that admit multiple voices of the same researcher, emerging as non-linear texts, that 'wander, twist and turn, changing direction unexpectedly'.[6] Through such methods and outcomes, autoethnographers aim to emotionally engage research audiences, involving readers in a process of locating and critically interpreting the writer, and themselves, in the story.

For practice-based researchers, autoethnography may involve critical autobiographical writing as a means of engaging in and examining relations between personal experience, the research process and socio-cultural situation. As will be discussed and shown, autoethnographic writing methods may be adapted by practitioners to include not only text-based practices, but also other forms of 'graphing', for example audio and video recording, photography, performance, choreography, song-writing, scoring or crafting. By consciously applying such methods to the examination of one's own life and culture, doing autoethnography can enable critical engagement with relations and tensions between research and lived experience, perceiving and being in the world, researcher and audience, while unearthing new questions, connections and ways of knowing, hidden desires, implicit conditions and power dynamics, as well as critical capacities and limits of one's own research and those of others. Researching becomes an iterative process of illuminating social, cultural and political conditions through an evolving, critical lens of self.

The text that follows explores my own application of autoethnography as a practice-based research method in sound art, while also discussing a range of autoethnographic practices and examples in other fields. Examples of my previous work are used to demonstrate ways in which autoethnographic methods may generate a reciprocal and fluid interplay between sound and text, practice and theory, researcher and audience, doing and meaning. Through this chapter, I discuss a range of autoethnographic methods that might be adapted and layered as elements in a practice-based research methodology. These include:

- Developing a critically engaged, reflexive practice of writing that traces, interprets and 'stories' personal experiences, interactions, emotional encounters and revelations, including those of undertaking and/or encountering practice
- Identifying specific cultural contexts, issues or dynamics that may be participated in, examined and interpreted through such a practice (i.e., a research question or aim)
- Expanding notions of autoethnographic writing (or 'graphing') to include methods specific to a range of artistic disciplines including, for example, performing, recording, scoring, crafting, drawing or other methods
- Engaging in rigorous processes of (re-)writing and (re-)editing that persistently strive to admit the local and specific, the 'here and now' of doing research, in some cases through multiple intersecting voices of the same researcher
- Generating research forms that balance personal narrative with wider context and critical framing, enfolding practice and theory rather than using one to instrumentalise the other
- Presenting research outcomes that simultaneously provoke emotional engagement and invite interpretation from readers/audiences, as engaged and engaging, direct and layered personal narratives
- Grasping the opportunity to work on one's own habits and behaviours of living, writing (about) who, why, where, how and what we are, want to be, could be and should be

. . .

When we get back to the flat, I'm excited to tell my partner LJ about the unexpected request from
Craig Vear:

"I've been asked to write a thing about autoethnography for a handbook on practice-research. It seems like pretty good timing cos I've got that conference coming up. It means I'll be writing two articles and a book chapter at the same time, though"

"Sounds like you should do it. Maybe you could speak to some live art people we know about what they do — it seems like there might be a lot of overlap."

"Yeah, totally . . . yes!"

. . .

070521 10:14

Hi Craig,

Thanks for getting in touch.

Yes, this sounds like a great project and a really useful approach (agree about the terminology turf-wars!). I'd very much like to contribute a chapter on autoethnography.

Do you have a sense of time-frame, and is there a blurb on the handbook I could take a look at?

All the best,
Iain

Autoethnography: emergence and approaches

080820 07:28 / 050820 08:05 / 100820 08:08 / 270221 13:14
This book chapter is due in 6 days, in 2 days, in 1 day
and i'm not approaching the task at all seriously
First | more | BBC | Glasg | FT How | bonn | Goog | Klays | Inbox | Soni | LIVE |
Easie | Autoe | Refle | 2 X |
alone time and the internet present too much distraction, too much temptation
is this what home-working in the post-lockdown workscape will be like?
zoming and doomscrolling remote horrors and cool music?
— — — — — — —
breathing harder now
look at all those tabs
very auto
— — — — — — —
quiet computer hum
soothes
low rumblings of neighbours through the wall
suggest connection
everyone asleep in my house

Figure 4.3.2 Right.

Source: Iain Findlay-Walsh.

rests their potential
to drift and to connect
to be cut adrift and disconnected
cars rush by too quietly to be felt
ears almost imagining
I am not here
liminally floating
on Facebook

–

none-the-less
it is truly a privilege
to work this keyboard quietly in the mornings
facing down and writing through
the fear
of being dismissed suddenly
from my job
next year

. . .

Spry argues that 'all research ultimately, pragmatically, brutally emanates from a corporeal body that exists within a socio-political context'.[7] Douglas and Carless further this by writing:

> How to tell a history? How might we write a story of what has gone before –concerning the origins and development of a research methodology we now know as autoethnography? . . . *the* history of anything does not exist – it is instead an illusion, a fiction, or a fallacy because there can be no one definitive telling of any story, history or otherwise. History, like any other story, is subject to amendment, development, alteration, expansion and change – forever re-written as new insights, stories, perspectives, contexts or understandings are uncovered. *And history, like any other story, depends on who is doing the telling.*[8]

As Douglas and Carless acknowledge in this quote, telling a history of autoethnography may seem counterintuitive, as the method is premised on a rejection of notions of knowledge as being fixed and impartial. Nonetheless, accounts of the emergence of autoethnography through the incorporation of autobiographical writing styles in social science research have been offered by Ellis and Bochner (2000),[9] Jones (2018),[10] Chang (2008),[11] Spry (2001),[12] Grant & Short & Turner (2013),[13] Denzin (2014), Douglas and Carless (2013)[14] and Gouzouasis (2020)[15] among others.

Arthur Bochner[16] has reflected upon 'the narrative turn' in (North American) social sciences during the 1980s and 1990s, which saw the development of research practices that 'pointed inquiry towards acts of meaning, focusing on the functions of stories and storytelling in creating and managing identity', and challenged traditional values of generalisability, verifiability, truthfulness and objectivity. According to Stacy Holman Jones, new calls within the social sciences to value the specific, personal, embodied experiences of practitioners emerged in response to a post-structural, post-colonial and feminist 'crisis of representation' in North American academic institutions in the late 20th century. As Jones writes, this 'crisis':

> motivated researchers to acknowledge how their own identities, lives, beliefs, feelings, and relationships influenced their approach to research and their reporting of 'findings'. This focus on representation encouraged qualitative researchers to search for more transparent, reflexive, and creative ways to do and share their research. Rather than deny or separate the researcher from the research and the personal from the relational, cultural,

and political, qualitative researchers embraced methods that recognised and used personal-cultural entanglements.[17]

While the discussions of autoethnography cited earlier often correspond to (and in some cases, directly cite) Carolyn Ellis' framing as 'research, writing, story, and method that connects the autobiographical and personal to the cultural, social, and political',[18] Tony Adams discusses the abundance of distinct applications, approaches and categorisations emerging in the late 1990s and early 2000s, exhibiting clear differences in method, aims, formats, values, and in the balance of emphases between the main constituent elements of self, culture and writing. Adams writes:

> Some autoethnographies are more analytic and social scientific . . . there are interpretive/humanistic autoethnographies that use personal experience to offer 'thick descriptions' . . . of cultural experience in order to promote understanding of these experiences. . . . There are critical autoethnographies often informed by feminist, critical-race, queer, post-colonial, indigenous and crip sensibilities. . . . There are creative, performative, and evocative autoethnographies that offer accessible, concrete and embodied accounts of personal and cultural experience[19]

During an academic conference on autoethnography and music composition (hosted by Chris Wiley and I on 17–18 June 2020), composer and researcher Soosan Lolivar cited Frantz Fanon, Audre Lorde and bell hooks as early exemplars of autoethnographic practice and ethos, with each providing models for the combining of autobiographical life-writing, socio-political analysis and emancipatory projection.[20] For Lolivar, these writing practices constitute responses and challenges to 'systemic otherness' and underscore the emergence of autoethnography through the struggles of people, cultures and identities that are marginalised.

In recent years, autoethnographic methods and forms have been adapted and applied as models for artist-researchers that seek to engage in cultural inquiry through the documenting and interpreting of their own lived experiences, including processes of making. Over the past two decades, autoethnographic studies have emerged in the fields of contemporary performance,[21] film and multimedia art,[22] music,[23] audiovisual media,[24] dance and performance, as practice-based engagements with specific socio-cultural contexts and issues. The resultant status of autoethnography as both a means of writing about doing arts-based research, and as a methodological model for creative-artistic practice, affords a fruitful reciprocity between art and writing, practice and theory, culture and self, academia and everyday life, that may be communicable through research outcomes. Through strategies of committed and persistent critical reflection, including those of emotional 'recall' and introspection, artist-researchers may trace and edit, write and re-write culturally situated experience, interpreting their own lives, selves and creative practices as sites of social, cultural and political tension.

Sonic autoethnography – methods from my practice

140221 12:04
LJ has taken the boys out
I got them up, fed and ready – we made a deal
our needs were heard mutually in the moment
now I'm at home alone on Valentine's Day with tinnitus and a whirring computer fan
immersed in a timespace that is clear
cars fading softly past the window
recently, I've been a mess

> *life seeming at once overwhelming and empty*
> *lockdown and precarious work produce an atomising pressure*
> *I've struggled to care, parent, or write*
> *but now here in this flat occurs a silence*
> *a gift (from you)*
> *which is, perhaps, enough to deliver the nourishment that can accompany the act*
> *of suspending one's self in vibration*
> *hearing feeling body emerging through open form*
> *re-embedding*
> *with / in*
> *everything*[25]
> . . .
> *Listening does not obscure but generate, we are always part of the soundscape we listen to.*[26]

My journey as an artist and researcher has developed over a 15-year period from a specific engagement with making music to a broader engagement with issues of sound and listening.

As Salomé Voegelin's writing, quoted earlier, attests, critical engagement with sonic experience may lead to new understandings of relations between lived environment, social and cultural context, self and others. Drawing upon the layered self-narratives of autoethnographers including Tami Spry, Alec Grant and Susanne Gannon, as well as practice and practice-based research examples in music and sound art by Hildegard Westerkamp, George E. Lewis, Christopher DeLaurenti, Marc Baron, Klein and others, and theories of sound and listening by writers including Kodwo Eshun, Salomé Voegelin and Eric Clarke, I have gradually developed a practice of sonic autoethnography that uses field recording, studio editing and immersive audio presentation to trace, interrogate and share the aural dynamics of specific socio-cultural contexts and situations, from the shifting perspective of a listening self/subject. The development of this approach has been focused in the related areas of field recording, critical listening and composing (audio editing, arrangement and presentation). In the following section, I summarise three projects that have emerged through this work, identifying key methods and including audio and text excerpts.

The Closing Ceremony: 'aural selfies' and immersive soundscapes

And did I forget to mention that I found a new direction and it leads back to me?[27]

The Closing Ceremony[28] is a multichannel soundscape composition for 5.1 loudspeaker array that documents the closing ceremony and concert of the 2014 'XX' Commonwealth Games in Glasgow in an effort to trace and share my lived experience of this city-wide mega-event. Through editing and layering multiple field recordings of the Games' closing ceremony and concert that captured the event from a range of listening perspectives, and generating a soundscape collage for surround sound presentation, I sought to reflexively capture, interrogate and share my situated listening perspectives and experiences in relation to the concert and to the Games.

While making this piece, I developed two key methods, which I have gone on to apply in subsequent projects, while also identifying them in the work of other practitioners.

1 The first method is the use of audio recording devices, held or worn, to trace the position and actions of my body and self in relation to others and the wider environment, capturing 'aural selfies'.[29] Conceived of as selfies, these recordings rhetorically acknowledge and foreground the presence of the recordist 'in the frame', with sonic markers of my presence, involvement in and

Figure 4.3.3 The Closing Ceremony (excerpt).

Source: Iain Findlay-Walsh.

proximity to the recording process, such as microphone 'handling noise', being clearly audible. This practice of taking 'aural selfies' is described in a subsequent journal article, as follows:

> The practice of self-consciously documenting (my own) auditory subject-position by making sound recordings has parallels with the ubiquitous cultural activity of taking 'selfies', or self-portrait photographs. While a selfie can be understood as an image of a person in a place, it can also be understood as an image of a person watching themselves taking a picture of themselves watching, in a place . . . In drawing together and layering this range of recordings, *The Closing Ceremony* can be thought of as a collage of multiple 'aural selfies'.[30]

2 The second method developed during the making of *The Closing Ceremony* can be explained in terms of a conscious engagement with the capacity of sound playback technologies (multichannel loudspeaker arrays, stereo speaker systems or headphones) to present immersive aural environments to listeners. As a means of playing back 'aural selfies', multichannel surround sound or stereo playback systems (loudspeakers or headphones) may be used to 'embed' a listener within the aural dynamics and auditory perspectives that a recordist has previously encountered and captured, occluding the listener's actual acoustic environment and temporarily imposing a new, virtual (recorded) one. The dual practices of taking and using of aural selfies to trace my listening encounters, and of using immersive audio playback systems to embed a listener within the auditory experiences and aural positionality of a recordist, were further explored through subsequent pieces composed specifically for stereo headphone playback, including the sound art release, *w/hair ph<> n mus|x*.[31]

w/hair ph<> n mus|x *and 'first-person field recording'*

> *What's this? Listening through the split spaces of experience.*
> *Brain full of place fragments and music.*
> *Auto-location and orientation. Auto-relation and alone-ness.*
> *Off-grid expanse. Virtual zero. In shops, stations and toilets.*
> *The feel of the sound of shifting in the seat looking for no-thing*[32]

Figure 4.3.4   w/hair ph<> n mus|x.

Source: Iain Findlay-Walsh.

In making *The Closing Ceremony*, I developed a specific method of sonic 'graphing' that in turn facilitated a particular approach to hearing, tracing, storying and interpreting the sonic, spatial and social dynamics of the city during the Games. An interest in capturing field recordings that trace the personal listening encounters of the recordist developed thereafter into a more deliberate practice of 'first-person field recording', and to my seeking out and studying similar methods in the work of other sound artists. In an article published in 2019, I developed a provisional definition of first-person field recording, framed as follows:

> First-person field recording is here defined as both method and material whereby environmental sound recordings are generated by a single recordist through their holding, wearing or 'being with' the microphone(s) in acts of conscious and reflexive self-documentation. In what can be understood as a departure from the standard and well-established environmental recording practice of capturing and presenting soundscapes from an anonymous, pseudo-objective perspective, first-person field recording foregrounds the presence of the recordist as an active and significant element in what can be heard. . . . First-person field recordings are thus recognisable as environmental recordings that document human proximity and personal intimacy.[33]

w/hair ph<> n mus|x[34] is an album-length collection of soundscape pieces for headphone playback, composed in 2017, that engages with and re-presents my solitary listening encounters in a range of urban and domestic environments. Various approaches to environmental recording are developed and used, including binaural recording – an approach that may be used to capture a listener's 360-degree sound environment through their wearing of microphones in the ears, as well as, in one case, swallowing and regurgitating small microphones. The release can be understood as an experiment in first-person field recording, and in embedding a headphone listener within autoethnographic soundscape collages that capture and share the intimacy of solitary listening as an aspect of contemporary urban living. After making *The Closing Ceremony* and *w/hair ph<> n mus|x*, I felt it necessary to develop an autoethnographic (text-based) writing practice to accompany the next project, more closely connecting my sonic and text-based inquiries. As such, over the past 12 months I have been engaged in field recording, soundscape composing and the writing of autoethnographic texts, as interconnected methods for examining the sonic, spatial and social dynamics of living at home with my young family during COVID-19 and under lockdown.

More no place: *listening, writing and composing as critical dwelling practices*

In the final days of March 2020, as the global COVID-19 pandemic began to significantly impact upon the practicalities of living in Scotland, I started to explore the application of autoethnographic methods as means of coming to terms with, adapting to and dealing with the sudden changes to everyday life through the pandemic, and specifically the lived realities of living with my young family under lockdown. This research journey is embodied in a range of outcomes – a series of field recordings and soundscape pieces,[35] and one text-based autoethnography.[36] Rather than summarising each of these inter-connected projects here, I will present fragments of them, as responses to the following questioning:

> What can I learn of my home and how can I participate in its transformation, as an emergent and shared site of dwelling, through a critical practice of listening?

and further,

> What possibilities arise when I conceptualise, listen through and actively participate in the relational form of my home, as co-composed music?[37]

. . .

300420 07:28 / 160221 07:51
Living here with you in this flat right now
chundering boiler longs to give up

Figure 4.3.5 Composing the field of dwelling.
Source: Iain Findlay-Walsh.

wooden bed frame creaking through fucking
brash cartoons resonating and rattling down the hall
boy racers tearing up the street for an audience of no-one
LJ on work zoom call – cold gated pro voices
low rumblings of neighbours living through the walls
computer fan hum – deep, warm
repeat kettle eruptions
murmuring radio haunting kitchen
small boys rolling around – benign screaming
T H E T O P T E N R O C K S O N G S O F A L L T I M E
prone quiet caregiving phone calls to friends
foreboding silence of the snow outside
invasive doorbell – package arrival ("is this B/2?")
bath runs full again – nurturing heat/futility portal
all together
altogether

. . .

030420 07:00
the pressures acting upon and within this space are intensified by lockdown
the kids are driving me up the wall
the inverse of social distancing is social compression
home becomes hyper-relational implosion
while the quiet streets outside fill with birdsong
and an apocalyptic politics morphs and grows somewhere far away right here under my nose

. . .

180420 06:10
if we stir the house-sound-body-machine with enough care
submit to being stirred by it, in it
your music, my music, no music, conscious music, unconscious music
unfolds in time as a phasing and layering of processes and agencies
if we try to force the elements
the whole house coughs
everything starts – out of time
chaos, ruption
(the boiler is broken)

. . .

040220 05:20
my relationship with my partner
we work to listen
with each other

. . .

100620 09.30
was this autoethnography a self-conceit?
"I artfully manage my indoor kingdom . . ."

> *". . . conserve a space of shared isolation"*
> *"straight white married guy wallows in sublime middle-class quiet enforced by shouting, hiding*
> *and withholding snacks"*
> *perhaps*
> *I mean to learn to listen better*
> *inside and outside*
> *to practice critical listening-as-dwelling*
> *becoming pedagogical*
> *in this home-frame microcosm*
> *of the social*
> *to participate in making and understanding*
> *this music*
> *of hearing ourselves*
> *in the world*[38]
> *. . .*

Taken together, the practice-based methods discussed earlier – the taking of 'aural selfies' and capturing of first-person field recordings; the editing, layering, composing and presenting of such recordings as immersive soundscape pieces; and the conscious practising of critical listening, field recording, and poetic, diaristic writing, as methods for an engaged practice of (self) critical dwelling – can be combined and applied as methods for reflexively storying, analysing, learning through and sharing culturally situated sonic experience. The application of such methods to the pursuit of a specific research aim, through a practice and process of sonic autoethnography, is one way for researchers to ground their inquiry in the culturally situated, relational and resonant sound of their own lives.

Autoethnography in practice-based research

The strategies earlier above are examples of how autoethnographic methods may be adapted and applied by practice-based researchers. Other examples of cross-disciplinary autoethnographic research include those in dance and contemporary performance,[39] audio-visual media,[40] song writing,[41] music composition[42] and crafting.[43] By conceiving of text-based and mixed-media methods as forms of 'graphing', and approaching making, editing and presenting as processes of writing and rewriting the self, of storying and interpreting personal experiences as sites of cultural inquiry, practitioners working across a diversity of disciplines may adopt autoethnography as a research method. Discussing her own autoethnographic music compositions, Lucy Hollingworth writes,

> My compositions are conceived as practice-as-research; the musical works are the research just as much as the written texts associated with them. In this sense they represent a form of *living inquiry*, they are about 'being in the world but also what constitutes our belonging in the world'. As a lifelong learner in the arts, composition became a tool for rediscovery about myself and my capacity as a researcher.[44]

Tami Spry writes of her autoethnographic performances as providing 'a space for the emancipation of the voice and body in academic discourse through breaking boundaries of stylistic form, and by reintroducing the body to the mind in the process of living research'.[45] For Spry, the admission of multiple communicative modes, or 'voices', emerging from a single researcher, allows for the integration of the 'personal, professional and political voice'.[46]

Autoethnographic methods afford specific possibilities as approaches to audiovisual research. They can be applied to the capturing and editing film, and connected with widespread and diverse existing practices and cultures of 'self-broadcast' enabled by the internet and social media, thereby generating a rich context and broad potential audience for audiovisual autoethnographies. This context informs recent examples of autoethnographic audiovisual research, including those that specifically engage with life during the COVID-19 pandemic.[47] Discussing recent Iranian film, Mazyar Lotfalian writes that autoethnographic method 'enables us to look at artistic and cultural productions across different domains (documentary, experimental art, conceptual art)' while 'self (auto) becomes an important part of the production as a vehicle for passionate engagement with society, a sort of bridge or translation device between different forms of knowledge'.[48]

At the conference on autoethnography and music in June 2020, discussed in an earlier section, composer and researcher Liz Lassiter powerfully aligned the singer and songwriter Nina Simone with autoethnographic methods and aims. Speaking of her own work, and directly quoting Simone, Lassiter stated: ' "an artist's duty is to reflect the times": to me, that is my duty'. Lassiter's framing of Nina Simone's music practice as autoethnographic encouraged me to return to performance footage of a 1976 concert by Simone that I had encountered some years previously. During the performance, Simone weaves real-time reflections on her own life and career, on race, migration and diasporic identity, as well as direct engagement and at times confrontation with the audience assembled in the auditorium, in and around a rendition of the song 'Stars', itself a critical and personal reflection upon fame and aging. True to Lassiter's claim, the performance exemplifies autoethnographic method, as an embodiment of a rigorous music performance practice grounded in and contending with entanglements between performance and identity, desire, defiance and acceptance, song as fiction and as (auto)biography. In the fluidity and ambiguity of voices and perspectives presented, it enacts and shares a layered process of understanding, of knowing, that accumulates as I work to locate both Simone and myself in the story, being variously re-positioned as protagonist, ally, enemy, and audience. I find the experience and effort of engaging with this performance emotionally demanding and impactful, as it manifests and, in turn, instigates a process of interpreting selves, culturally embedded in and speaking through the reciprocal acts of performing and listening.

and again, I don't wanna let you down and I get this feeling that the only way to tell you who I am these days is to sing a song by Janis Ian . . . and I insist upon not being one of your clowns but one of you.[49]

Postlude: defining and doing 'reflexivity'

050221 11:50
reading yesterday (Scrivener 2021 this volume; hooks 1995)
suddenly realising, remembering, knowing, and feeling, that
no 'new knowledge' of aesthetic experience and making art will be shared through this text
rather it may aspire to show and share experiences
of knowing
as stories
that live
to dig the roots of them selves
generating, inviting, becoming, meaning
in and through intra-active reading encounters
are you there?
. . .

I'm sitting in front of my computer at around 4 pm on 18 June 2020, having just hosted around seven hours of near-constant presentations and discussions on Zoom, with participants and audience

Figure 4.3.6 — in.

Source: Iain Findlay-Walsh.

contributing from across the world. I feel somewhat relieved that things have gone smoothly, and start to relax about the final task of the day, hosting a keynote presentation by Peter Gouzouasis, titled 'Autoethnography: A Reflexive Research Process'. Peter, presenting a spoken 'duet' performed live via Zoom together with his colleague, Matthew Yanko, opens by immediately bemoaning the experience of editing and writing feedback for numerous submissions for a forthcoming handbook on autoethnography and music. The keynote begins thusly:

> Peter Gouzouasis: "Hey! If I read one more paper that conflates reflective with reflexive, I don't know what I'll do . . . the experience has left me feeling . . . cognitively irritable."[50]
>
> "Oh shit", I think to myself as Peter continues. "One of those chapters is mine!"
>
> . . .

Reflexive is a term that emerges again and again in the discussion of autoethnography and in autoethnographic texts themselves, as a stated aim, goal or ideal – a kernel of the methodology. However, the word can often be ill-defined and somewhat 'slippy', while, as Gouzouasis states, the term is frequently conflated with the similar and related term, 'reflective'. If evocative autoethnography is indeed the 'exemplar of reflexivity',[51] then what exactly does the term mean in this context, and would a more precise framing and understanding of reflexivity help to further illuminate autoethnography as a useful approach to practice-based research? I will close by briefly considering, following Gouzouasis, what reflexivity might mean in this context and how practice-based researchers who choose to do autoethnography might recognise and practice it. Social work scholar Jan Fook writes:

> What is reflexivity in its simplest terms? It has been termed a 'turning back on itself'. It is an ability to locate yourself in the picture, to understand, and factor in, how what you see is influenced by your own way of seeing, and how your very presence and act of research influences the situation in which you are researching.[52]

This seems somewhat close to Donald Schön's theory of 'reflection-in-action', as in to 'turn thought back on action and on the knowing that is implicit in action'.[53] However, as Gouzouasis[54] acknowledges, Schön's reflection-in-action emphasises self-awareness 'in action' and therefore 'in the moment'. Reflexivity as it pertains to autoethnography can perhaps be better understood in terms of a longer-term, iterative, layered and staged process of writing, interpreting and rewriting the self by storying, editing, re-editing and further storying personal experience. Thus, reflexivity in autoethnographic practice moves beyond a kind of dual awareness – of simultaneously acting and at the same time reflecting upon that action – and towards a rigorous and layered practice of self-authoring. As Gouzouasis has it,

> in the reflexive process of storying the self through creative non-fictionally composed autoethnography, the writer creates new meanings and mindful reinterpretations of their actions and experiences.[55]

For Alec Grant, this process of reflexive self-authoring is ultimately directed towards self-betterment, in the sense that those who generate autoethnographies may take lessons from our stories in order to 'live the person that is storied'.[56] This goal is echoed by Stacy Holman Jones through a notion of autoethnographic 'self-transformation', with the practice affording the potential 'power to embody and to materialise the change we seek in ourselves, our lives and our worlds'.[57]

It is useful to add, as Ellis and Bocher do, that 'autoethnographers expect their readers to become deeply involved and drawn into the predicaments of their stories'.[58] Norman K. Denzin seems to reinforce this notion of iterative, reflexive self-storying as involving the interpretation and self-actualisation of both researcher and audience:

> Lives and their experiences, the telling and the told, are represented in stories which are performances. Stories are like pictures that have been painted over, and, when paint is scraped off an old picture, something new becomes visible. What is new is what was previously covered up. A life and the performances about it have the qualities of pentimento. Something new is always coming into sight, displacing what was previously certain and seen.[59]

Methods for reflexive storying, through editing, interpreting, layering and sharing culturally situated experiences become strategies for generating impactful encounters in and through which researchers and audiences write and read and re-write their emerging selves. Throughout this chapter, I have used a combination of past-tense narrative prose, poetic, diaristic writing, theoretical discussion and analysis, to present an autoethnography that is grounded in the here and now of the research process. The citing of my email correspondence with Craig Vear is another such strategy, one that grants access to remembered thoughts, feelings, motivations and tensions. Revisiting this email exchange and bringing it into the text helps me to remember, to contextualise and analyse, to feel and ultimately to better understand, the situated tensions that I experienced when first approached to write the chapter – as the consequences of a specific socio-political context of precarious working in Higher Education, and intense domestic pressures, including those of parenting and home-schooling during lockdown. Similarly, the field recording excerpts included throughout this chapter trace the sonic content and context of the research process, allowing both you and I to *listen in* to the situation, to hear and to learn through contextual specificity. Through enfolded processes of writing, interpreting, editing and presenting, as a reflexive storying, an autoethnography emerges.

Coda and conclusion

280121 07:42

Dear Craig,

I hope you are well in these strange times.

I am writing to ask if an extension of a few days may be possible for my book chapter submission for the PbR handbook. Due to the ongoing school closures, I am currently home-schooling my children (aged 6 and 9) and this has put considerable, unexpected pressure on all work and research deadlines. An additional week would be ideal, however I realise you will be trying to meet deadlines with the publisher. I am enjoying the redrafting process and look forward to submitting a much-improved version.

Please let me know your thoughts.

All the best,

Iain

. . .

280121 09:30

Hi Iain,

no problem, we are going through exactly the same. So am happy to support.

Keep me updated.

regards
craig

. . .

280221 5:45
come on
one more
day
of this
bringing this this
to some kind of timely end, in
the spring the birds are singing in the garden, in
the kitchen LJ and the boys are having breakfast, in
the city your window begins to slowly open
my computer fan hums away, agreeing
I . . .
good
day

. . .

060421 10:45

Hi Craig,

An update regarding my chapter. This will be with you early in the day tomorrow (Wed 7th).

All the best,
Iain
. . .

To conclude, autoethnography provides a framework for engaging in socio-cultural research, as the reflexive storying and sharing of culturally situated personal experience. Practice-based researchers may adapt and apply autoethnographic method to suit their research aims as well as the specific conventions of their discipline or research field, drawing upon a range of methods. These include: developing a critically engaged, reflexive practice of writing; identifying a focus for socio-cultural inquiry, a research aim or question; adapting autoethnographic writing to include other 'graphing' methods (e.g., field recording); engaging in processes of writing, editing and re-writing that admit the local and specific, examining self in relation to others; generating research forms that balance personal narrative with critical context and framing; presenting outcomes that provoke both emotional engagement and active interpretation; and grasping the opportunity to work on one's own habits and behaviours of living.

In combination, these are strategies for generating research journeys in and through which researchers and audiences may write and re-write their emerging selves in relation to others and to the world. For practice-based researchers pursuing new understanding, through processes, work and encounters that facilitate new ways of perceiving, seeing, feeling, knowing, and being in the world, autoethnography constitutes an adaptable approach to reflexive practice, embedded in the messy specificity of lived experience.

Notes

1 Ellis and Bochner (2000, p. 736).
2 Schon (1983).
3 Lorde (2017).
4 Jones (2018, p. 10).
5 Grant et al. (2013, p. 2).
6 Ibid.
7 Spry (2016, p. 37).
8 Douglas and Carless (2013, p. 84).
9 Ellis and Bochner (2000).
10 Jones (2018).
11 Chang (2008).
12 Spry (2001).
13 Grant et al. (2013).
14 Douglas and Carless (2013).
15 Gouzouasis (2020).
16 Bochner (2012).
17 Jones in Adams et al. (2015, p. 22).
18 Ellis (2004, p. xix).
19 Adams (2017, p. 63).
20 E.g., Fanon (1952/1986), Lorde (1977/2017), Hooks (1994, 1995).
21 Spry (2001).
22 Lotfalian (2013).
23 Carless (2018), Wiley (2019).

24 Szilagyi (2020), Juritz (2021).
25 Findlay-Walsh (2021).
26 Voegelin (2014, p. 24).
27 Kylie Minogue (2000).
28 Findlay-Walsh (2015).
29 Findlay-Walsh (2018, p. 124).
30 Findlay-Walsh (2018).
31 Findlay-Walsh (2019b).
32 Press release for w/hair ph<> n mus|x, Findlay-Walsh (2019b).
33 Findlay-Walsh (2019a).
34 Findlay-Walsh (2019b).
35 Including one now-published sound art release – Findlay-Walsh (2020).
36 Findlay-Walsh (2021 in review).
37 Ibid.
38 Ibid.
39 E.g., Spry (2001).
40 Szilagyi (2020), Juritz (2021).
41 Carless (2018).
42 Hollingworth (2017).
43 Kouhia (2015).
44 Hollingworth (2017, p. 153).
45 Spry (2001, pp. 719–720).
46 Ibid., p. 721.
47 Szilagyi (2020), Juritz (2021).
48 Lotfalian (2013, p. 127).
49 Verbatim speech from Nina Simone.
50 Gouzouasis (2020).
51 Ibid.
52 Fook (2015).
53 Schön (1983, p. 50).
54 Gouzouasis (2020).
55 Ibid.
56 Grant et al. (2013).
57 Ibid. (p. 10).
58 Ellis and Bochner (2016).
59 Denzin (2014, p. 2).

Bibliography

Adams, T. (2017). Autoethnographic Responsibilities. *International Review of Qualitative Research*, 10(1), 62–66, Spring.

Adams, T., Ellis, C., & Jones, S. H. (2015). *Autoethnography: Understanding Qualitative Research*. Oxford: Oxford University Press.

Autoethnography. (2021). https://en.wikipedia.org/wiki/Autoethnography (accessed February 28, 2021).

Bochner, A. (2012). On First-Person Narrative Scholarship: Autoethnography as Acts of Meaning. *Narrative Inquiry*, 22(1), 155–164.

Carless, D. (2018). 'Throughness': A Story About Songwriting as Auto/Ethnography. *Qualitative Inquiry*, 24(3), 227–232.

Chang, H. (2008). *Autoethnography as Method*. Walnut Creek, CA: Left Coast Press.

Denzin, N. K. (2014). Assumptions of the Method. In *Interpretive Autoethnography*. Thousand Oaks: Sage.

Douglas, K., & Carless, D. (2013). A History of Autoethnographic Inquiry. In S. H. Jones & T. E. Adams, & C. Ellis (Eds.), *Handbook of Autoethnography*. London: Routledge, 84–107.

Ellis, C. (2004). *The Ethnographic I: A Methodological Novel About Auto-Ethnography*. Walnut Creek: AltaMira Press.

Ellis, C., & Bochner, A. (2000). Autoethnography, Personal Narrative, Reflexivity: Researcher as Subject. In K. Denzin & Y. Linkin (Eds.), *Handbook of Qualitative Research*. Thousand Oaks, CA: Sage Publications, 2nd ed., 733–769.

Ellis, C., & Bochner, A. (2016). *Evocative Autoethnography: Writing Lives and Telling Stories*. New York: Routledge.

Fanon, F. (1952/1986). *Black Skin White Masks*. London: Pluto Press.

Findlay-Walsh, I. (2015). *The Closing Ceremony* (excerpt – stereo reduction). Digital. https://qrgo.page.link/EQeBS.

Findlay-Walsh, I. (2018). Sonic Autoethnographies: Personal Listening as Compositional Context. In *Organised Sound*. Cambridge: Cambridge University Press, 23(1).

Findlay-Walsh, I. (2019a). Hearing How It Feels to Listen: Perception, Embodiment and First-Person Field Recording. In *Organised Sound*. Cambridge: Cambridge University Press, 24(1).

Findlay-Walsh, I. (2019b). *w/hair ph<> n mus|x*. Entr'acte. https://qrgo.page.link/qe1ds.

Findlay-Walsh, I. (2020). *More No Place*. Digital. Outlet Archival. https://outletarchival.bandcamp.com/album/more-no-place.

Findlay-Walsh, I. (2021). Composing the Field of Dwelling: An Autoethnography on Listening in the Home. *Journal of Sonic Studies*. In review.

Fook, J. (2015). *Chapter 26 Reflective Practice and Critical Reflection*. https://practicelearning.info/pluginfile.php/317/mod_data/content/2740/Extract%20-%20Lishman%20-%20chapter%2026.pdf

Gouzouasis, P. (2020). Autoethnography: A Reflexive Research Process. (Keynote address). In *The Autoethnography of Composition and the Composition of Autoethnography*. Glasgow: University of Glasgow, June 17.

Grant, A., Short, N., & Turner, L. (2013). Introduction: Storying Life and Lives. In A. Grant, N. Short, & L. Turner (Eds.), *Contemporary British Autoethnography*. Rotterdam: Sense Publishers, 1–17.

Hollingworth, L. (2017). *Out of the Snowstorm, an Owl*. https://soundcloud.com/lucy-holling-worth/out-of-the-snowstorm-an-owl.

hooks, bell. (1994). *Teaching to Transgress: Education as the Practice of Freedom*. New York: Routledge.

hooks, bell. (1995). An Aesthetic of Blackness: Strange and Oppositional. *Lenox Avenue: A Journal of Interarts Inquiry*, 1, 65–72.

Jones, S. H. (2018). Creative Cultures/Creative Selves: Critical Autoethnography, Performance and Pedagogy. In S. H. Jones & M. Pruyn (Eds.), *Creative Cultures/Creative Selves: Critical Autoethnography, Performance and Pedagogy*. London: Palgrave Macmillan, 3–21.

Juritz, G. (2021). Dionysus Dies: An Autoethnographic Exploration of Tensions in the Anthropocene, Rejoicing in the Abject and Embracing the Perpetual Post-Digital Mess. *Sonic Scope: New Approaches to Audiovisual Culture*. https://doi.org/10.21428/66f840a4.cab8311d.

Kouhia, A. (2015). *Crafts in My Life – A Short Film*. www.youtube.com/watch?v=x02fmvg5yuo.

Lorde, A. (2017) [1977]. Poetry Is Not a Luxury. In *The Master's Tools Will Never Dismantle The Master's House*. New York: Penguin Modern.

Lotfalian, M. (2013). Aestheticized Politics, Visual Culture, and Emergent Forms of Digital Practice. *International Journal of Communication*, 7, 1371–1390.

Minogue, K. (2000). *Spinning Around*. Parlophone. www.youtube.com/watch?v=t1DWBKk5xHQ.

Schön, D. A. (1983). *The Reflective Practitioner: How Professionals Think in Action*. London: Avebury.

Spry, T. (2001). Performing Autoethnography: An Embodied Methodological Praxis. *Qualitative Inquiry*, 7(6), 706–732.

Spry, T. (2016). *Autoethnography and the Other: Unsettling Power Through Utopian Performatives*. New York: Routledge.

Szilagyi, P. (2020). *"(no) Fear?!" – Work in Progress*. https://digitalschoolofautoethnography.tumblr.com/post/622728433292673025/peter-szilagyi-no-fear-work-in-progress.

Voegelin, S. (2014). *Sonic Possible Worlds*. London: Bloomsbury Academic.

Wiley, C. (2019). Autoethnography, Autobiography, and Creative Art as Academic Research in Music Studies: A Fugal Ethnodrama. *Action, Criticism, and Theory for Music Education*, 18(2), 73–115.

4.4

NAVIGATING THE UNKNOWN
A dramaturgical approach

Hanna Slättne, with illustrations by Stéphanie Heckman

Introduction

This chapter takes the approach and methods of the dramaturg from the field of theatre and performance and applies it to practice-based research. It will give you an overview of dramaturgy and the dramaturgical process. It will explain how the dramaturg's focus is to support an artistic team in their journey of discovery, research, practice and decision-making. The methods presented here have been extracted from my experiences as a working dramaturg for those undertaking a practice-based research project. I hope practitioners across a varied range of fields will recognise and relate to this approach.

This chapter will focus on two different aspects of the dramaturg's method.

- The first is the process of asking questions in order to

    - Help excavate and identify the project kernel (see definition later) that is unique to the researcher and their way of working.
    - Help explore, understand and formulate the core ideas in a responsive and dynamic process

- The second is to

    - Reflect on the practitioner way of working, practice and process.
    - Note where the practitioners focus is directed and what this means for the process.

The chapter is suited to those bringing practical experience and tacit knowledge[1] into a more rigorous research process. For example, those undertaking an academic practice-based research project who might feel a tension between differentiating their practice and research processes (see Figure 4.4.1). The chapter is based on my experience of working as a dramaturg for over 20 years in the industry.

Throughout the chapter I have added reflection points with the hope that they will help you make links between what is discussed in the chapter and your own practice, experience and field. The tool kit in part 2 of this chapter focuses on dramaturgical questions with the aim to support you in your particular practice-based research project.

DOI: 10.4324/9780429324154-37

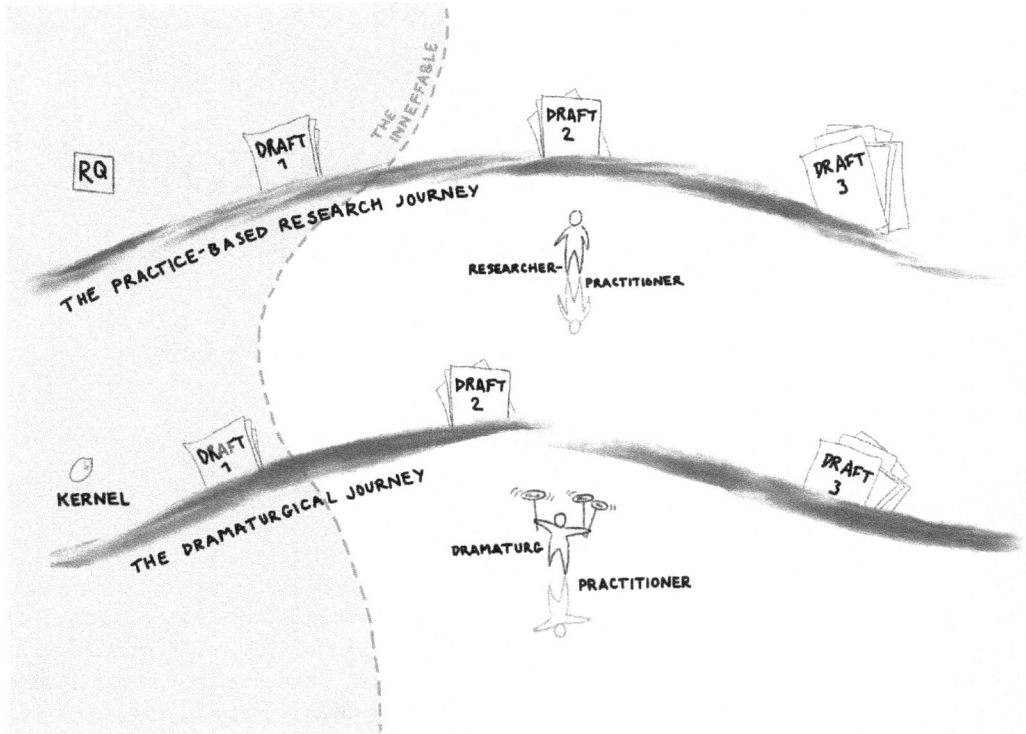

Figure 4.4.1 This chapter will look at the differences and similarities between the journey of a practice process and that of practice-based research.

Source: Illustration by Stéphanie Heckman.

Part 1: the dramaturgical process

What is dramaturgy?

Dramaturgy is defined in a variety of ways, but in essence, the term refers to the structure of a story or a dramatic experience.[2] The UK-based Dramaturgs' Network provides the following definition: 'the **dramaturgy** of any performance piece is the dynamic structure built from its various components. . . . It is also how those components relate to the experience of the work as a whole'.[3] From my vantage point as a theatre practitioner, dramaturgy is more complex and fluid than that and not exclusive to the performance arts, which is often suggested in dictionary definitions. For anyone interested in reading more about the practice and theory of dramaturgy, there are many excellent books which also provides practice examples.[4] In this section, I will set out my definitions of dramaturgy, doing dramaturgy and the dramaturgical ideas.

In the performing arts, the active components making up the dramaturgy of a performance include text, voice, sound, set, space, costume, music, lights, actors, physicality AND the structure of the experience. With more theatre practitioners working outside of traditional conventions focusing on immersion and embodiment, dramaturgy can also include how the piece affects the audience's body and mind. However, dramaturgy is not exclusive to the theatre and story arts; an expanded use might refer to the dramaturgy in a computer game, or the flow of experience in social media exchanges, or the dramaturgy of a trial, the dramaturgy of urban planning or the dramaturgy of a health or social care encounter. For the purposes of this chapter, and its context of practice-based research, I will aim to include both the complexity of what it means and other contexts in my definitions for this chapter. Therefore, my definition of dramaturgy for this chapter is:

Definition

Dramaturgy *is the way meanings are engendered in a piece of work; the components, their placement, interaction and prominence AND the structure or shape of how that piece is experienced from first encounter, by its audience, until its impact stops resonating.*

In examples from outside of the arts, for example, in the context of a patient/doctor encounter, the dramaturgy is the combination of the feel and look of the consultation room, the light, furniture and seating arrangement, the staff's body language, their voice and duration of the visit. Alternatively, the dramaturgy of a computer program encompasses the experience starting with the computer–human interface, its design and structure and how it accommodates first time users as well as more skilled user in terms of functions, workflows, usability and integration into the user's lifestyle. The process of thinking of, designing and structuring the user, audience or participant's experience of the project product or output is akin to what the dramaturg does.

Doing dramaturgy

Doing dramaturgy is a practice, a way of thinking and shaping both the process and the composition of a project or experience. It is the practice method of the dramaturg. The main tool of the dramaturg's method is observation, listening and to ask specific and probing questions of the artist, the ideas and their way of working. Everyone in the creative team involved in a theatre production are working with and shaping its dramaturgy, i.e. doing dramaturgy. It is, therefore, not essential to have a dedicated dramaturg in a process. However, having one person whose focus is on facilitating the different aspects and phases of the dramaturgical process, to keep track of changes and support the creative team's dramaturgical decisions, makes for a more efficient process. Doing dramaturgy requires a different skillset from directing, writing, designing or acting, and if a theatre artist is doing their own dramaturgy, it requires them to shift between the two skillsets.

As a practicing dramaturg across many art-forms, I define the act of doing dramaturgy across a project process as:

Definition

Doing dramaturgy *is to map the complex multiple origins of artistic instincts and ideas, through a process of excavating the ineffable.* **Doing dramaturgy** is to help identify and formulate the dramaturgical ideas, the kernel, of the project and to build a scaffold or skeleton around which to layer and craft materials. *It is to guide collaborative sense-making.* **Doing dramaturgy** is to shape the journey into which others are invited to share a particular response to the world. It is to contribute new resonances, understandings and connections through offering shape, sensations and experiences.

As discussed earlier, a production in the performance arts has its own specific dramaturgy, through how it is put together, structured and performed. In order to arrive at this dramaturgy, the company has engaged in the practice and the dramaturg has focused on 'doing dramaturgy'. The organising principle behind their decisions are the dramaturgical ideas.

Dramaturgical ideas

Artistic processes, as many others, involve collaborations across many disciplines and are at times messy and intense. The success is dependent on a joint understanding of *the dramaturgical ideas* at the heart of the project. The ideas, or starting points of artistic processes, might at first be ineffable and stem from complex instincts or deep-seated emotional responses. The dramaturgical method constructs a framework with which to effectively nurture, explore and eventually communicate these ineffable notions to a wider team of collaborators and then an audience.

The dramaturg's contribution to finding this joint understanding and artistic vision for a collaborative process is through a process of asking dramaturgical questions. In the early parts of the process, we gently probe, open up areas of questioning, reflect back on the artistic process and gently facilitate the exploration of the initially ineffable notions, the kernels of the ideas. The process from kernel to dramaturgical ideas takes time. Performance artists arrive at the dramaturgical ideas in very different ways, through practice, collaboration or reflection. It often involves a process of generating and exploring materials.

Throughout this process, the dramaturg's focus is on listening, observing and reflecting on different ways to express the kernel into the joint understanding of the dramaturgical ideas and the production. The dramaturgical ideas allow the wider creative team to respond to the projects' unique and specific nature through their own practice, skill and experience. I have been in many production and rehearsal processes, where the collaborators did not take the time to arrive at a joint understanding of the dramaturgical ideas, and it inevitably will show in the final production. With a joint understanding of the dramaturgical ideas, the dramaturg can support the collaborators and the process to respond dynamically to the specificity of the kernel. Together, the artistic team work to convey the dramaturgical ideas to an audience.

My definition of the kernel and the dramaturgical ideas:

Definition

The dramaturgical ideas *are the ideas at the heart of the project, emerging from the kernel from which the project grows.* **The kernel** *holds the artistic intuition, the gut feelings, the reason to create, the unique experience that the creative practitioner wants to share through the work.*

Both a rigorous dramaturgical process, and the shape of the final project emanate from the kernel. In a poetic sense, like the botanical kernels of plants, the growth of a kernel is dependent on the soil within which it grows, the nutrients and care given and the time and space it is given to grow. The dramaturg suggests ways to add nourishment to the process, the artists and the ideas. A dramaturgical approach should be flexible and agile. It should never be a rigid formula or set in advance. Throughout the process, discoveries will change the dramaturgical ideas and changes in the dramaturgical ideas will definitely change the process and the final output.

Reflection points 1

Throughout the chapter, I will be prompting you to make connections with the text and how I am referring to the role of the dramaturg, doing dramaturgy and the dramaturgy of a piece of work in the context of your own practice. These *reflection points* are intended to draw your attention to the ways you think and talk about your process and practice with colleagues and peers. They will hopefully help you identify opportunities in applying some of the dramaturgical thinking to your own area of expertise.

- *Is there an equivalent of the role of the dramaturg within your practice area? Someone whose role it is to help safeguard, nurture and protect core values and ideas of projects? i.e. the dramaturgical ideas.*
- *How do you respond to a hunch in your practice?*
- *How do you think about, talk about and communicate this hunch, the seed of an idea, before it is fully formed to a wider team?*
- *What is the equivalent of 'doing dramaturgy' in your practice and research?*

Dramaturgy as a method of composition and analysis

The dramaturg's methods of keeping track of how the ideas and the process evolves over time is a helpful approach for the practitioner researcher. In particular, the dramaturg's methods of noting how the composition of a performance evolves around the dramaturgical ideas; the reasons why certain aspects are foregrounded and accentuated at different times. These notes provide a roadmap of the process with clear links between the dramaturgical ideas and the compositional decisions. These are a useful source of data in a research project. Having a good record of the specific intentions and ambitions behind the dramaturgical ideas allows a means through which you can compare those intentions with how the project evolves and eventually how it is received and understood as it meets an audience.

Definition

Dramaturgy as a method of practice is the structuring of an experience, story or system. It achieves this by means of foregrounding, accentuating and drawing attention to particular components, instructions and information in a specific order. The order of the components and the way they affect each other are crafted to reveal the meanings that the practitioner wants to convey to those experiencing, reading or navigating the work.

This method can also be used in reverse as a *method of analysing* an experience, story or system created by others. For example, these questions could be applied as part of the reflective or evaluation processes of a project:

- What do the accentuated information, emotions and information make you feel and think and do?
- What does it mean to you?
- Is it a successful experience? In what way is it interesting, impactful, beautiful, thought provoking, efficient, helpful, smooth and/or transportative or whatever criteria are important to your research.
- From the choices the creator of the piece presents to you, what can you deduce about what they set out to achieve?
- How successful were they in achieving this?
- What are the underlying organising ideas? i.e. their dramaturgical ideas.

In both these methods – method of creating and method of analysing – the importance is in understanding the connection between the underlying dramaturgical ideas and how they are conveyed through the work. Of course, as a theatre practitioner, you cannot know how different audiences will receive your work. The aim is to give your project an internal coherence, a strong skeleton. Through this the beneficiaries (your audience or your research community) have a better chance of inferring a sense that there is thought, care and craft in what you are asking them to engage with and to reflect on the dramaturgical ideas that are resonating throughout the work.

Reflection points 2

The following reflection points are meant to prompt a deeper exploration of how you work and how you assess your own and others' work in the context of this chapter. During your practice-research, you will return to this exploration in different ways.

- *What are your methods within in your practice? Both practical and conceptual?*
- *How do you assess the success of your own projects? How do you gauge what others make of it or take from it? How does that make you reflect on your methods and ideas?*
- *How do you assess the success of other's projects? What do you look for? How do you identify your own and others' underlying assumptions?*
- *How does your way of thinking about your methods inform what you do next? i.e. how does it enable you to see and understand problems?*

The dramaturg in the artistic process

Another useful aspect for the practitioner researcher is how the dramaturg brings the audiences' presence into the process. We are sometimes referred to as a first audience, or the audience advocate in the process. The dramaturgical approach can be a way to consider the audience or user throughout the process without losing the freedom required in the creative process.

For us involved in making artistic experiences, the aim of the arts is to take our audience on an extraordinary journey, to speak to many different types of experiences and minds, to delight and challenge. Art can open up difficult questions, emotions and complex experiences and present them in ways that makes them resonate with its audience without telling them how and what to think and feel. The route to achieving this is intricate. It's full of wrong turnings, bad ideas, getting side-tracked, too many voices, not enough voices, too many ideas, not enough practice, too little introspection or too much self-indulgence. Or a very common problem, not allowing every part in the process the time it needs.

In my experience, being a dramaturg is not too different from being a researcher: both bring rigour to a process and both look wide and deep to make connections and to create new meanings and new knowledge. In 2016, when addressing fellow theatre practitioners, I spoke about how to keep myself creatively fit as a dramaturg by keeping actively curious.

- Keeping curious about the minutiae of how people in our team think, communicate and create and how this particular rehearsal room and process work.
- Keeping curious about human beings, cultures, social structures, the restraints, the limitations and the possibilities, so we can ask the right questions at the right time.
- Keeping curious about our art form; where it has been and where it is striving, so that our beast (the work) is in dialogue with its own.

- Keeping curious about the immediate world outside our rehearsal room as well as the wider and wider and wider context. Always taking one step further back, allowing the perspective to change and bringing into the bigger picture the smallest intimate moments.
- Keeping curious about what our peers are doing, what the scientists are doing, what the environmentalists, the philosophers and the neuroscientists are doing so that our beast can speak many languages and communicate on many levels.[5]

The same applies to practice-based researcher. In the speech I used the analogy of the dramaturgy being the skeleton (see Figure 4.4.2), and the performance the beast created by the artists and the process around it.

The following quote from the same paper helps to articulate this skeleton illustration. I am sure it resonates with anyone who makes work that is released into a public arena. Thinking about the audience experience early and throughout the process is a means to avoid an 'unwieldy beast'.

> All of us, in that team know that if we get the proportions between the skeleton and the body mass wrong, if we have too much bulk, or the bodily adornments are attached in the wrong places, or we ignore a really important, but maybe slightly inconspicuous bone. . . . Well, we all know what happens then. We have all been in the company of Beasts that fall over, fall apart or wobble their unwieldy way across our stages![6]

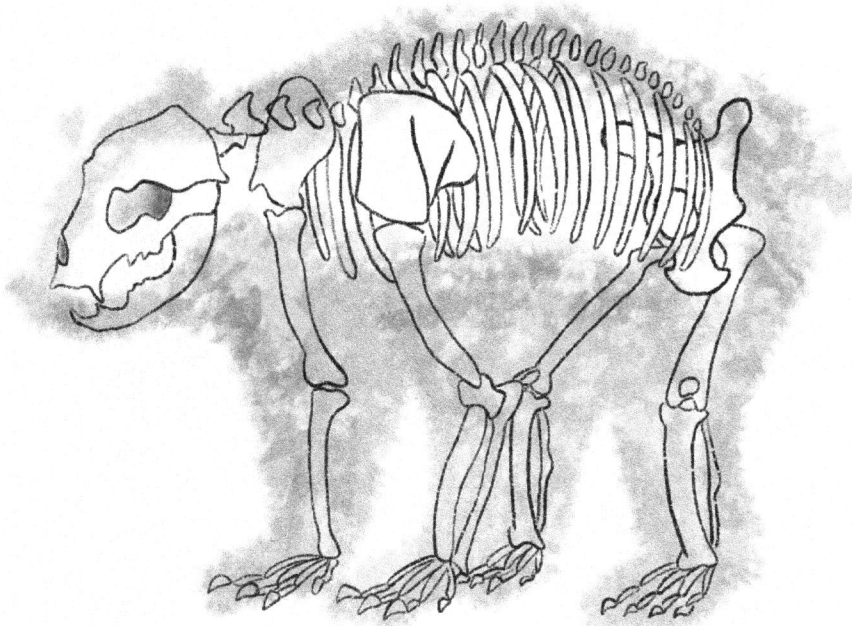

Figure 4.4.2 A healthy beast (project) has a strong, well-balanced, reliable skeleton. That is, a set of core ideas that are well tested and nurtured.

Source: Illustration by Stéphanie Heckman.

Reflection points 3

Each field and practice area finds its own metaphors and language to describe complex processes. The following reflection points ask you to reflect on the equivalences from your field of the metaphors and practices of the dramaturg.

- *Within your practice, what would be the equivalent to the skeleton?*
- *How do you reflect on and develop the internal logic of your project without dictating what it might end up being and looking like at the end of a journey of innovation and exploration?*
- *How do you challenge your own established ways of thinking and working? To keep 'fit' within your field?*
- *When in your process do you consider the experience of the end user, the beneficiary or audience?*
- *How does a user encounter your work, engage with it, and how does it affect them? i.e. the dramaturgical audience journeys.*

Being a practitioner and being a researcher

Your practice is something you know how to do without thinking about it, and sometimes thinking about it can get in the way. However, as practitioner- researchers this is part of the job. For me, practice consists of many different strands of activity, areas of focus and types of decisions. A good way to describe what this feels like is as a circus plate spinner. They used to mesmerise me as a child, how they brilliantly pay attention to all the different plates spinning high up on sticks all around them. Noticing in the corner of their eye when one is starting to slow down. Dashing over, giving it a spin, while already planning which wobbling plate to spin next while also being aware of the audience. The job is to give all the plates attention and make sure none of them slow down and crash to the ground. Figure 4.4.3 illustrates the dramaturg's three key plates; the kernel quest, the artistic process and the audience experience.

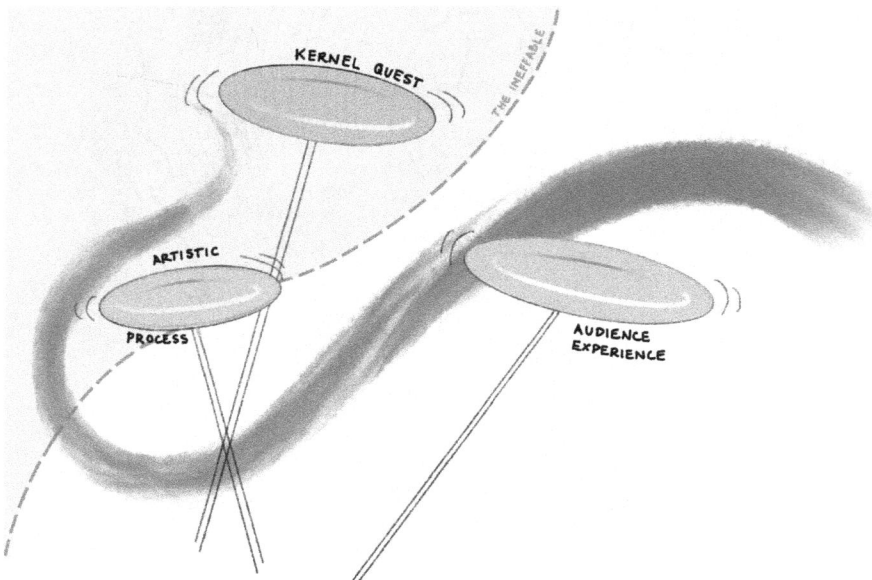

Figure 4.4.3   In any process you keep many plates spinning that dynamically inform and change each other: the quest for the question, the needs of the process and shaping the project outcome.

Source: Illustration by Stéphanie Heckman.

Table 4.4.1 In practice-based research you have additional plates to spin which require a different approach and focus.

Practice plates	Research plates
PLATE 1 **THE DRAMATURGICAL IDEAS** The quest for your kernel and how to formulate it.	**RESEARCH QUESTION** The quest to find the right research question and to delineate your area of expertise.
PLATE 2 **THE DRAMATURGICAL PROCESS** To reflect on what your practice and process is and on what the process needs in light of the specific, yet ever-evolving, ideas of the project kernel. What materials you need and how to generate them. What participants, skillsets and collaborators to engage and when. What kind of R&D and testing does the project need?	**THE RESEARCH PROCESS** Your practice and research methods and the process of reflecting on them through your evolving research question and research practice. Identify your position in the field. Formulate new knowledge. Map the field: theory and practice. Decide what theorists you will engage with and why.
PLATE 3 **SHAPING/AUDIENCE JOURNEY** Exploring options of how the world encounters your kernel through its structure and placement. How to achieve the effects and affects you want for the audience/participant/ user? The mechanics of how it works and how you got there, are most likely hidden. The experience and its impact are what matter.	**THE THESIS** Explain your position in your identified context, demonstrate your knowledge of that context and from there explain your findings. You are mapping out the journey using evidence from your practice and literature to explain it. It needs to be clear, overt and in a particular format.

Source: Illustration by Stéphanie Heckman.

These three separate yet interweaving areas dynamically inform and change each other. Identifying the interweaving strands in your practice can help to identify what additional plates you will need to spin in your practice-based research and what they involve; see Table 4.4.1. When building a PhD portfolio, for example, you might go through several practice-based processes, each with its own spinning plates, while the process of reflection on each project is part of single PhD journey.

Reflection points 4

Reflection points 4 together with Table 4.4.1 are intended to help you reflect on what your 'plates' are, how you move from one to the other, how you communicate ideas and what the difference is in your everyday practice and in your practice-based research.

- *What are the key 'plates' you need to keep spinning during your practice?*
- *What additional 'plates' will you need to spin for your research journey? And in the context of your practice research, how different or similar are they to your practice 'plates'?*
- *What additional collaborators, practical and theoretical, do you need to bring into the research project to situate your practice within your field?*
- *Is there a difference between how you respond to an idea through your practice and how you approach it as a research question?*

Part 2: a toolkit of dramaturgical questions

A dramaturg's approach to finding the kernel in practice-based research

In this section, I will share two sets of tools in the form of questions that are useful to the practitioner researcher. Questions sets 1–4 are aimed at the very beginning of the process and question set 5 is to help you reflect on your practice.

As your project proceeds, ideas and research question will change. How they change should reflect what is important to you i.e. what is contained in the kernel. These first set of questions will help you, the practitioner researcher, to find your unique connection to a research question or kernel of your research idea.

It might be helpful to think about your project as a tree (see Figure 4.4.4). The question in section 1 will focus on what interactions and exchanges created the kernel in the first place? The questions in section 2 asks why the kernel is starting to grow now, at this point, and what you can learn about it by looking at its environment. Section 3 asks what nourishments the kernel is picking up from the particular soil it is growing in, that is you, the researcher. The kernel is growing inside you. Finally, in section 4, the focus is on what can you learn from what your visions and ambition are of the kernel as it grows into a full-grown tree.

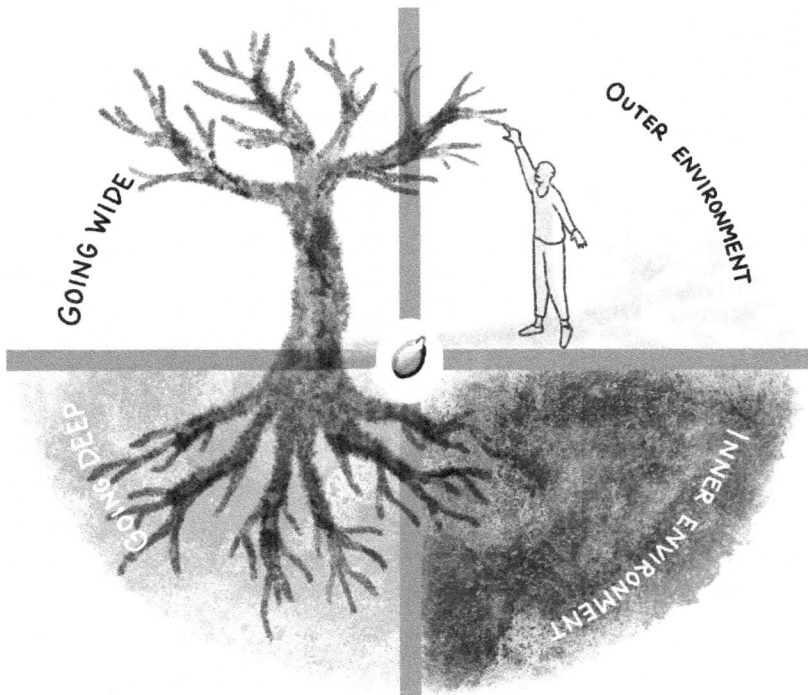

Figure 4.4.4   The origin of ideas are complex. Understanding its many different influences, will make your project and research unique and specific.

Source: Illustration by Stéphanie Heckman.

The fifth and final set of questions will help you reflect on your practice. It will ask questions aimed at making you aware of how you direct your focus and attention. This awareness can make it easier to adapt your way of working to respond to the research.

Toolkit: the kernel

WHAT IS THE STARTING POINT OF THE PROJECT?

The first set of questions is helpful during the very early stages of your practice research. Their focus is to help you reflect on the status of your research question: the starting point, your relationship to it and who else contributes towards its formulation.

Is it your own idea?

- Is this your own idea? I.e. are you the sole holder of the dramaturgical ideas?
- What are the different influences that has shaped it?
- How long have you been thinking about it?
- How well formulated do you feel it is?

If the idea originates with someone else:

- Are you contributing to someone else's area?
- If so, what are your own connection, interests and/or pre-occupations within this area?

If the ideas emerging from practice:

- Are you expecting it to arise from practice or a collaborative process?
- What are your pre-occupations, habits, processes that might steer what will emerge?
- Are you a leader or follower in a collaboration?
- Do you easily get attached to early ideas or do you leave commitment to an idea as late as possible?

All ideas:

- What state is it in? Is it an instinct, a gut feel, a hunch, an image, a more fully formed idea or a hypothesis? That, is what is your actual starting point?

WHY NOW? WHY HERE?

These second set of questions looks at the outer environment within which you are conducting your research. How do the immediate opportunities and challenges shape your project?

- Why do you want to do your research project now? How is the timing relevant to you and your experience?
- And why here? Why this particular geographical location?
- Who is your audience?
  - How do you want to engage with them?
  - What do you want to contribute to their lives?
- How does your project reflect on or resonate with the contemporary society locally, nationally and internationally?

The third set of questions focuses on what you bring as a researcher that is unique to you; your inner environment. Reflecting on these answers can help you challenge how you do things and what other viewpoints, values and expertise you might want to add to the project.

- How do you work?
- What is your default approach?
- What are the implications for your project of how you work?
- What are your pre-occupations?
- What are your values?
- What are your own personal connections to this research and why? What are the aspects you care about most? What aspects motivates you most?

HOW DO YOU ENVISAGE OR IMAGINE THE FINISHED PROJECT?

The fourth set of questions is designed to reveal aspects of your own assumptions and desires from how you imagine your project being received when finished. You can ask yourself these questions at different points in the process and reflect on how your answers change. With the caveat that you, of course, do not know at the beginning where the project might end up.

- What do you want it to achieve?
- How do you want it to be received and experienced?
- What impact on the participant/audience/user do you want it to have?
- What impact on your life can you envisage of the finished project?

A dramaturg's approach to reflecting on ways of working

In addition to asking questions about what is important to the creative practitioner within the core set of ideas, the dramaturg closely observes how the artist works in order to identify what their process is. As every practitioner works differently and every process is unique, the dramaturg needs to adjust their input accordingly.

With this approach, I will give you some tools to consider where your focus is and how that might affect how you arrive at your project's kernel and/or how you shape the final output. This approach can be helpful if you are taking your normal process into a new environment of practice-based research which will require you to direct your focus in additional and new ways to what comes natural to you.

For simplicity, I will suggest following four types of focus. There are most likely other types of focus; however, this basic format is sufficient to give you another way to think about your practice.

Process focus: Practice and the generation of materials, as a means to find the kernel and a starting point. Through the accumulated experience and expertise of the practitioners, in collaboration with others or on their own. The practitioner's focus is on, and in, the process of creation – the now, the flow, the exchange. The dramaturgical focus is to reflect on what is created through the practice and uncover its meanings or functions.

Kernel focus: Creation as a means to express and share something that is important to us, something ineffable. The dramaturgical focus is initially on that excavating and formulating the deep-rooted idea, feeling, or situation. Through their practice, the practitioner carefully generates, explores and crafts materials and decides how to direct the attention of their audience, beneficiaries or participants through the experience to reflect the kernel.

Applied focus: The practitioner has an outcome in mind or are working towards a brief. The dramaturgical motivation is to ensure the project meets the brief from a user's perspective. The practitioner's focus is to explore how different options and means engender the specific affects or effects of the brief, for the beneficiary engaging with the project.

Academic research focus: The practitioner researcher's aim is to communicate research findings, data and insights from the process and practice to fellow academics. The dramaturgical focus is on the research context, the researcher's particular vantage point within it and the story of the research. The researcher directs the reader's attention to the topography of the context, the route they took and the significant junctions on the way to their findings, arguments and new knowledge.

Toolkit: focus and process

WHERE IS THE FOCUS OF INTEREST FOR YOU IN THE RESEARCH?

This fifth and final set of questions as part of the dramaturgical toolkit are designed to be used mid-practice flow. It requires self-reflection at different points in your practice process. It is also helpful to observe where your focus is directed and when, how and why it changes.

- In the experience of the practice?
- In the methodology of a specific process? Or the methods of a general one?
- In the artwork or output on its own or in its reception?
- Is it in a system, in a formula, or an artefact?
- Is it in how it impacts on others? Through an exchange, interaction, an experience?
- Where is your expertise and practice in relation to the focus? This is relevant in interdisciplinary projects.
- What is your role within the practice and process? Within a collaboration in relation to where the focus of interest is?

The overall challenge for the practice-based researcher is to oscillate (spin plates) between the positions of the four foci (outlined previously), especially between practitioner role and the researcher role. Your practice, which is so familiar to you, can be hard to decode; and the research context can seem complex and overwhelming and difficult to relate to your practice. As there is flow in your practice, it is good to listen for and find a rhythm and a means to go between the practice process, the kernel and/or audience experience and your research process of mapping and contextualising the theory. Ultimately, it is good to be aware of how your shift in focus feels, and what their respective aims are. In essence, how to keep spinning all the plates.

There are most likely questions related to your field and specialism that you can add to this toolkit.

Reflection points 5

The final reflection points aim to direct your attention to how you feel and think about your practice and research at this point, in the context of reading this chapter.

- *How does your way of working and who you are as a researcher feed into the decisions you take? And how you shape your final project?*
- *What kind of focus does your practice-based research require and how is it different from your practice?*
- *What mental shifts and new skills do you need to acquire and apply transitioning from practice to practice-based research?*

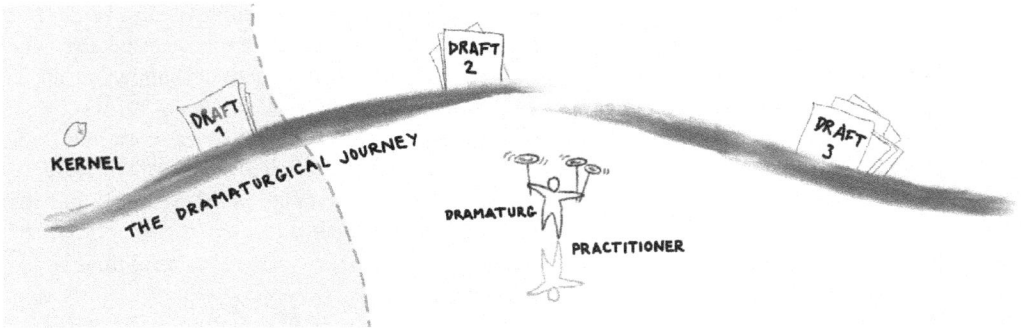

Figure 4.4.5   Helping other's find their questions and core ideas is a good way to reflect on your own project and process.

Source: Illustration by Stéphanie Heckman.

Being a dramaturg to yourself and others

The questions in the previous section prompted you to be your own dramaturg by asking yourself questions. Questions that can help you identify what is important to and unique for you in your research, your ideas and your unique way of working. The next step of being your own dramaturg is to use that understanding in thinking about how you shape your practice-research process (see Figure 4.4.5) through different iterations and drafts. What does the project need next? What do I as a practitioner researcher need next? What collaborators would I benefit from working with? How can I approach my practice in new ways? How can I challenge my way of working? How might that impact on the process and the findings?

You can also learn a lot about your own project and process when you support fellow practitioner researchers with their projects. Trying out the role of dramaturg from the outside rather than the inside. When you take on that role and start asking questions or your colleagues project, it is really important to remember to reassure them that they don't need to be able to formulate the full idea from the start. It is a quest, over time.

Help them through really listening to them: listen out for metaphors, any made-up words or doodles representing the ineffable feeling. The deep-seated complexity of ideas that they might not have words for yet. What is underneath? What is unique and personal to them? Allow time for the ideas to gestate and grow. Your colleague will potentially formulate their kernel in many different ways while exploring what the project means to them. You can listen, reflect back and gently probe the different versions. See how their narrative changes over time. It's a process of nurturing and protecting. Looking after their kernel in a world that demands researchers to express ideas in words and arguments, sometimes too soon.

Remember:[7]

- You are not telling them what to do. Or what to think. Be curious about how and why they think as they do. What is underneath? Reflect back and ask questions.
- Avoid suggestions and solutions unless they are suggestions of further exploration based on what you hear, see and feel when they speak.
- That their research question can only ever be fully understood from within their vantage point, so your job is to help them understand their vantage point so that they can explain it to others in their team or beyond.

Further down the line of the process when drafts, pilots and work-in-progress start to appear, continue to listen, observe and report back what you see, feel and how you interpret what they are presenting. Together, discuss if it does what they say that they want it to do. The aim is to highlight gaps in their story and areas that might need clarification. It's their job to find a solution. Reflect on the kernel, on what the process needs next and what the experience will be like for the audience. Make the process bespoke to the person/s that are developing the ideas.

Tensions and warnings

In this last section, I want to acknowledge the tension some practitioner researchers experience when they take their practice into a new arena such as academia and its new demand on how to direct and split their focus. For some practitioners, this might be a big shift. Working as a full-time dramaturg in a new writing theatre company, I was regularly sent plays written during a master's or PhD project. As an experienced reader of plays I could tell, just by reading them, that this was their origin. These plays very often needed to be rewritten for the professional stage. The tension between normal practice and practice within an academic process were, generally, evident in the play itself. That tension is even more present in production processes that are part of practice-based academic projects, where the artist's gut feeling, artistic ambition and sensibilities are vying for attention with their research focus. Practice flows are interrupted, and contextual ideas become both part of the practice and the research and it can be hard to keep them separate.

To some, new to practice-based research, it might feel as if the academic process infringes on the artistic as they are so similar, and this is something to look out for. You are, after all, venturing on an additional dramaturgical journey of telling the story of your research, with a different set of collaborators in the form of academic writing, data collection, evidencing and referencing. For example, my approach to this chapter. I could have gone down a more academic route; however, that would have led to a very different chapter.[8] It would have taken my focus away from really thinking through how my practice as a dramaturg and my tools can be applied across a wide range of practice-based research processes. Consequently, my dramaturgical approach will resonate with some people but not everyone. For some, it might be helpful to have a separate dramaturgical journey for the practice question and one for the research question; for others, they are one.

The most important aspect is to formulate for yourself what the dramaturgical ideas are for you, not in general, but for you specifically, as that will guide you into how you shape the process and the practice output regardless of how you think about the practice within the research. The method is intended to help you identify the moment in the process when the ineffable notions of the kernel pivots (see Figure 4.4.6) into a clearer view of your research question.

Conclusions

This chapter has outlined how the dramaturgical process and the dramaturg's way of working can provide helpful tools for a practitioner researcher. It provided definitions of dramaturgy, doing dramaturgy and the dramaturgical ideas, definitions that come from my practice as a dramaturg. The chapter looked at how three different aspects of a project – formulating your research question, planning the process and shaping the final output – continually feed into each other and that mapping this dynamic interaction provides process data for the practitioner researcher. The chapter noted that dramaturgy is a helpful method in any practice-based work, especially for projects that are going to be in front of an audience, used by participants or consist of an experience that users need to navigate. It outlined how a method of dramaturgical composition is useful to achieve the desired outcomes, meanings and affects that truly reflect the dramaturgical ideas, the kernel of your project.

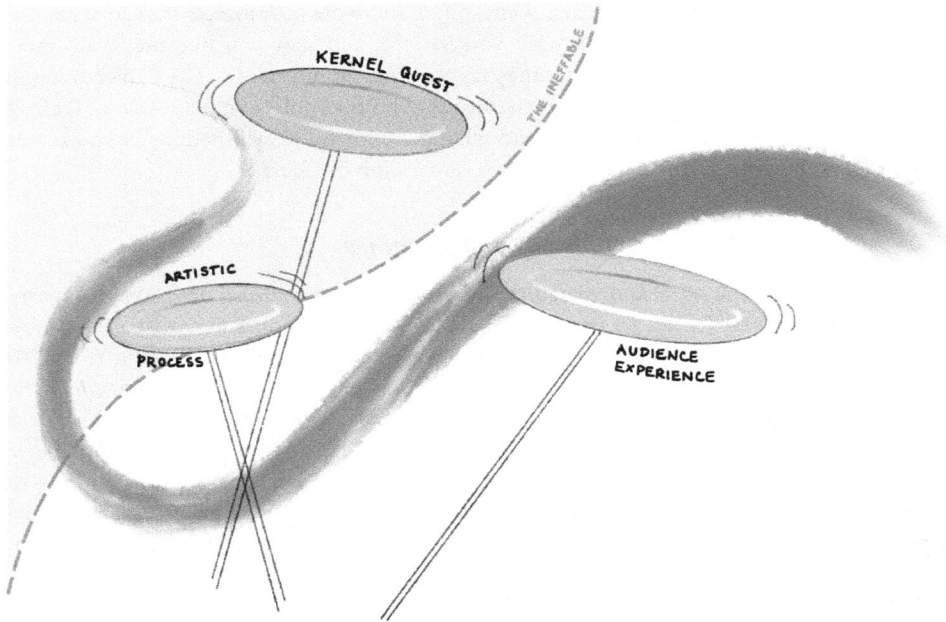

Figure 4.4.6 The dramaturgical approach to navigating the unknown helps you find the pivot point of the research question as it becomes more specific to you and your practice.

Source: Illustration by Stéphanie Heckman.

In part 2, tools and methods for reflecting on your ideas, research question and project were outlined. It proposed that the dramaturg's method of reflecting on where the practitioner's focus is directed is a helpful tool when a practitioner is adjusting to practice-based research. Furthermore, it suggested that the dramaturgical process supports a researcher in creating a road map during the practice-based research that charts the evolution of the core ideas and the process: providing the researcher with points of reference for when, how and why a project evolves. The chapter concluded by acknowledging the tension and challenges in moving from practice to practice-based research and provided advice for practitioners in how to be their own dramaturg as well as supporting peers and colleagues in their practice-based research.

Notes

1 Editor's note: tacit knowledge is discussed in the leading chapter of Part IV, but in essence refers to the type of knowledge that is difficult to explain or formalise in writing and/or as verbal instructions.
2 Dramaturgy is defined by the Oxford English Dictionary as 'the theory and practice of dramatic composition', and by Wikipedia as 'the study of dramatic composition and the representation of the main elements of drama on the stage'.
3 Dramaturgs' Network, Resources, available at www.dramaturgy.co.uk/resources (accessed January 30, 2021).
4 Some books that I return to in my practice are *Dramaturgy in the Making:– A User's Guide for Theatre Practitioners*, by Katalin Trencsényi (Bloomsbury 2015), *Dramaturgy in Motion: At Work on Dance and Movement Performance*, by Katherine Profeta (University of Wisconsin Press 2015) and *Performing Dramaturgy*, by Fiona Graham (Playmarket 2017). There is also *The Routledge Companion to Dramaturgy*, edited by Magda Romanska (Routledge 2016). which covers many different aspects of dramaturgical research.
5 For more information, visit www.dramaturgy.co.uk/dramaturgy-papers-hanna-slattne.

6 Hanna Slattne – The Art of Creating a Skeleton, 2016. Available online www.dramaturgy.co.uk/dramaturgy-papers-hanna-slattne (accessed February 2, 2021).
7 It might be helpful to explore different ways of giving feedback to fellow researchers and practitioners. Liz Lerman's Critical Response Process, initially developed in dance practice, is a very interesting place to start (website: https://lizlerman.com/critical-response-process/).
8 As an arts practitioner, I have soaked up the ideas, practices and innovations by hundreds of people around me. Through productions, through books, plays, conferences and talks or from collaborators contributions in one of the many, many processes I have been in. You will see remnants of these within my chapter; however, as an arts practitioner, unlike as an academic, I have not kept a track record of exactly what idea came from where. That is the freedom of artistic practice: your focus is on the artwork in the making, it's a forward looking, searching, probing focus about the next project. Having undertaken to do a practice-based research PhD, I am painfully aware of how the wider reference treasure trove from my years of experience is missing. However, the suggested reading in the note 4 are resources and colleagues I value that have been helpful to me in my work and therefore in this chapter.

4.5

THE PRACTICE OF PRACTICE-BASED RESEARCH

Challenges and strategies

Andrew Johnston

Introduction

This chapter presents findings from a study examining the experiences of recent practice-based PhD graduates from the Creativity and Cognition Studios (CCS) at the University of Technology Sydney (UTS). CCS has a long history in practice-based research (PBR), having its origins in a group initially established at Loughborough University by Ernest Edmonds and Linda Candy in 1992. In 2003, Edmonds and Candy moved the group to UTS, where it continued to grow and develop.[1] In 2010, Sam Ferguson and I became CCS co-directors while Edmonds became a founding director and Candy took an adjunct professor role. In 2017, I established a research group at the UTS Animal Logic Academy that applies the same basic approach to PBR but in the broad context of animation, visual effects, and the creation of interactive audio-visual experiences.

In preparing this chapter, I primarily draw on interviews conducted with a range of PBR PhD researchers and graduates, and with three experienced PBR supervisors. I have also drawn on my own experiences as a practice-based PhD researcher at CCS (graduating in 2009) and subsequently as a supervisor of seven practice-based PhDs to successful completion. (I am currently supervising an additional nine PBR PhD researchers at various stages of completion.)

My experience has been that despite the long history of PBR, and numerous publications describing its characteristics and underlying philosophies, PhD researchers can still become confused and disheartened at certain stages of their candidature. With this chapter, I wish to provide PBR PhD researchers, and their supervisors, with some well-grounded, practical insights and strategies that they may use in order to complete their PBR projects and realise the benefits that a practice-orientation can bring.

I present these insights as a series of challenges that PBR PhDs often face, and strategies to help. Throughout I have tried to preserve the voice of interviewees, and provide concrete examples that relate the text to the practice and research of the participants. I should stress that the findings I present here are very much based on the experiences of practice-based researchers at the Creativity and Cognition Studios, and some of our close collaborators. While I hope they are broadly applicable and useful, the limited, local context that gave rise to them should be kept in mind.

DOI: 10.4324/9780429324154-38

Background

In 2009, Linda Candy conducted a series of interviews with current PBR PhD researchers (Candy 2009). The study explored their understanding of PBR and considered in particular their reasons for undertaking PhD research, the relationships between practice and research, the role of the artefact, research methods and reflections, and advice to prospective researchers. Candy presented a range of insights into the PBR experience, documenting the concerns and dilemmas faced, as well as the benefits of PBR and the PhD more broadly. Dilemmas identified included lack of clarity around the role of creative work in the PhD and how it related to the written thesis, a lack of commonly agreed research methods, and reconciling analytical and creative thinking. Advice for new PBR PhDs was solicited from interviewees, and responses 'emphasised the importance of originality and individuality, and the dangers of adopting the views of others slavishly'.[2]

In this chapter, I revisit the PBR researcher experience ten years down the track to present the insights of the most recent generation, with an emphasis on identifying the challenges they faced and the strategies they employed to address them. For additional perspective, the experiences of three experienced PBR supervisors is also considered. In some ways, the observations I present here are an extension of an approach to PBR, documented in several earlier publications, which was based on my own personal PBR projects.[3] As I have subsequently worked with others on a broader range of PBR projects, it seems timely to update this work and identify some practical insights that are relevant for a new generation of practice-based researchers. While the focus is on practice-based research in a university context, the practical orientation of the participants in this study is readily apparent and will, I hope, resonate and have relevance for practitioners beyond academia. In addition, while interviewed participants were all doing higher degrees (PhDs, master's by research), the methods they applied – and the challenges they faced in making them work – are equally applicable to undergraduate students wishing to integrate creativity, practice, and theory.

Method

A series of interviews with recent practice-based PhD graduates was conducted. Six graduates, who had all completed their degrees since 2016, participated. Two current PhDs, both approximately half way through their degree, were also interviewed, along with three experienced supervisors, who had supervised numerous practice-based PhDs to completion. The interviews were semi-structured, with a series of deliberately broad questions used as a starting point for discussion. The questions were largely based on the themes considered by Candy in 2009 and focused on participants' understanding of PBR, and their experiences attempting to apply it. A particular focus was identifying the challenges they faced and the strategies they applied to address them. For supervisors, the interview questions explored similar themes but emphasised their experiences with their students and the strategies they had employed to guide and support practice-based research projects. All questions are listed in the appendix at the conclusion of this chapter. Each interview lasted approximately 60 minutes. Interviews were transcribed and analysed using coding and memoing methods drawn from Grounded Theory.[4]

Findings

Analysis of the interviews led to the identification of several areas of concern that have been broken down here into two main sections focusing on the challenges of PBR, and the strategies that participants employed and/or recommended to address these challenges.

Challenges

Interviewees identified a range of challenges faced by PBR PhD researchers. PBR provides some particular challenges beyond those experienced by PhD researchers using more traditional methods. Here I present five key challenges:

1   The problem of ambition: the need for resources, structure, and support
2   Contributions and clarity
3   Tension between practice and research
4   Documenting the work
5   Physical space

The problem of ambition: need for resources, structure, and support

The challenge of making genuinely innovative and interesting use of technology in creative areas can be significant. PBR PhDs face the significant challenge of having to create large-scale and, often, labour-intensive creative work which is sufficiently novel to generate worthwhile research outcomes while also undertaking in-depth, sustained reflections, gathering and analysing research data, and writing up a thesis.

Those interview participants involved in animation and interactive dance/theatre were part of larger teams. The credits for PhD researcher Simon Rippingale's *Jasper* animation project (Figure 4.5.1), for example, listed more than 30 participating artists and had a budget of nearly AU$500,000. Likewise, the *Creature* interactive dance/theatre work, which featured the large-scale interactive projections created as part of Andrew Bluff's PhD work (Figure 4.5.2), had a cast of five professional performers, a director, choreographer, script writer, dramaturg, costume designer, lighting designer, composer, production manager as well as many other 'behind the scenes' staff. The systems that were developed had to be reliable and robust enough to stand up to the rigours of touring.

Figure 4.5.1   Still from 'Jasper', directed by Simon Rippingale. Used with permission.

The scale of these works, and the associated larger audiences, lends credibility and a level of complexity and depth to the research – but also presents major challenges and risks to the PBR PhD researchers. For Simon, the scale of a professional production is necessary; without this, he feels that he would not be researching practice with relevance to the field within which he works:

> I'm quite ambitious about the sorts of animated images I want to make. And they sort of require movie production structure. I can't see how you do it without that. . . . For me, the value of what I'm trying to do comes through seeing through a production. . . . Whereas otherwise, I think I'd be frustrated if I was just making a few shots. . . . I'd be struggling to see the value of it.

Andrew Bluff highlighted the considerable organisational challenges that were associated with the production of *Creature*:

> Yeah, there's the practice of making the systems I did. There's a practice of actually using the system to make artworks. But there's also a considerable amount of effort in just cre-ating opportunities to actually do those artworks. So . . . the logistics of producing a show and organizing venues and, you know, making the concepts for a show, and like all those . . . [Funding?] Funding, yeah, getting a team together, managing that team. [There were] all sorts of things like that, which I didn't really have a lot of experience with.

In Andrew's case, the fact that he was collaborating with a theatre company that had produced numerous shows over many years meant that he was able to effectively outsource many of these issues, freeing him up to primarily focus on the development of new interactive systems for the per-formance, but this is of course not the case for many, perhaps most, PBR PhDs. Where the creative practice has minimal funding, PBR researchers must necessarily scale back their creative work, while ensuring it is sufficiently interesting or innovative to generate worthwhile insights.

A related issue is the somewhat slippery concept of quality. It requires significant effort for the higher-level concepts that form the intellectual core of a PBR project's contribution to knowledge to be meaningfully represented and explored through creative works. If the concepts are to be per-ceived, recognised, and respected, then there has to be a baseline of quality in the creative work – and clear links between the concepts and the work itself. Alon Ilsar, who created the gesture-based musical instrument the *AirSticks*, talks about a cycle between developing higher-level concepts and then working to craft these ideas into quality performances. Reflection then leads to further refine-ment of theories, concepts and – importantly for performers – the development of increased virtuos-ity. Conveying the quality of the work to thesis examiners was important for Alon and drove him to ensure performances were adequately documented. (See the following for more on documentation.)

In summary, the PBR projects of several interviewees required significant resources – and there-fore funding – and made extremely high demands on the PhD researchers themselves. Securing funding, and managing the associated contracts, production schedules, etc. is a significant task which, in the case of projects referenced here, required close collaboration between PhD researcher, supervisor, and external organisations.

Contributions and clarity

There was recognition of the need to consider the contributions of PBR over a long time period – and in a broader context. As with more traditional scientific research, it can take time for a consen-sus to emerge and genuine insights to percolate through and become broadly accepted in the field. Paradigm shifts – small or large – require more than one or two studies. PhD researchers sometimes

Figure 4.5.2 Performance of 'Creature: Dot & The Kangaroo' by Stalker Theatre, featuring interactive graphics by Andrew Bluff.

Source: Stalker Theatre. Photo: Darren Thomas, used with permission.

felt a little dispirited or insecure because the overall contribution of their work could seem small in the context of the field and its history.

Particularly when presenting the work to people outside the field, there was a sense that beyond the visible, creative outcomes, the broader value of the work could be challenging to convey. The claims made for individual projects are by necessity going to be limited, somewhat contingent, and specific to the creative works undertaken. For example, one PhD researcher commented:

> Sometimes it's hard to find the value of . . . Maybe one person does to sort of screw around for three years and then write a paper. But maybe it's like when 20 people do that . . . you do get a lot of value.

Perhaps due to this insecurity about the quality – or the simplicity – of the work, there is a risk that PhD researchers might be tempted to overly complicate the work – or to make larger claims for significance which are not really backed up by the work itself. For example, a PhD supervisor observed:

> There is a tendency for some people to talk about their practice in very kind of grandiose, highfalutin' kind of terms that are very difficult to relate to anything about the practice itself.

Practice-based researchers must therefore work to convey with clarity and confidence the insights that they have laboured to produce, while avoiding a tendency to oversell them. The benefit of grounding theoretical contributions in practice is that claims that are made can be backed up with references to the creative works that gave rise to the them. Quality documentation of these works is therefore critical (see 'Documenting the Work' later in this chapter), as this, along with other forms of data such as interviews or user experience metrics, provides the foundation for claims and defines the limits of theoretical contributions.

Tension between practice and research

The question of the time and energy taken by the research part of PBR – as opposed to the practice – was mentioned by several participants. While the centrality of moving between inspiration, ideas, practical implementation, and reflection was not questioned, the degree to which a PBR PhD requires formalisation of the process and expression of the findings in written form was a point of contention for some. All of the participants were already high-level practitioners in their field going into their PhD and could be said to have a 'practice orientation'. They recognised the value of taking the time required to more formally research previous works by others, gather data, and/or reflect on what had been learned through their own creative work, but sometimes the *amount* of time this took could be problematic.

This sometimes manifested as a feeling that what was done formally as part of the PhD was merely a laborious extension of what the artist would have done informally anyway. One PhD graduate said:

> So I have to be more rigorous for the PhD, so I go and interview people and I have to use some kind of methodology to kind of tease out information from that. And the process of codifying that information and doing memos, etc is . . . I found it to be just a formalised version of what I would just do naturally.

Another source of tension was the feeling of having to switch modes between artist and researcher. The artist/researcher was aware, while creating new work, that they should be capturing elements of the process – evolving sketches, creative decisions, and technical limitations, for example – but could feel that the need to attend to this was diverting their attention from the work itself. This can be particularly challenging if the work involves collaboration with other artists who may not necessarily be interested in the bigger themes being explored in the research. On this a PBR researcher and supervisor said:

> [It can be challenging] for me personally when I'm having to switch roles: am I the artist or am I the researcher? . . . There's different imperatives around what I need to do, and how I want to be in those different roles. So sometimes they can be at odds. . . . And if I'm collaborating with others as well, then maybe other people are not so invested in the research side of it.

Supervisors suggested that recognising and acknowledging this tension early is important:

> that reconciliation between the research remit . . . and keeping on with your practice – it has an obvious tension. And I think that ought to be confronted early on. In our world – in the practice-based research world – we have to talk about it. . . . We have to say, 'okay, it's going to be difficult at first.' They [practice-based researchers] will learn that having a 'research head' on you is as important as your 'practice head'. And that you really need to be able to reconcile those things if you feel they are in conflict. And the best PhD students learn how to do that.

This practice–research tension was also acknowledged by PBR PhDs in Candy's 2009 study:

> the dangers to practice it [practice-based research] presented were not inconsiderable and achieving a balance between research and practice is difficult.[5]

This is not to say that interviewees did not recognise the value of the additional rigour required to complete a practice-based PhD. In many cases they acknowledged that this did indeed lead to insights about their practice and the broader implications of their work that they would have otherwise missed. The section on 'Strategies' provides some examples.

Documenting the work

Several interviewees mentioned the importance, and the difficulties, of effectively documenting the creative work that drove their PBR project. For example, a PhD graduate discussed:

> In terms of the PhD for me, it was crucial that the examiners understood the quality and the physicality of it . . . [Interviewer: So you just couldn't convey what it was without video?] No – impossible.

They felt their credibility hinged on being able to *show* others what they had achieved and demonstrate that what had been claimed did in fact occur. Partly this was because they had themselves not been convinced by written descriptions of creative work that was not accompanied by convincing documentation. A PhD graduate highlighted that:

> When I'm reading papers, research papers and things like that. If they're talking about all this great stuff they did and they don't have good documentation on it, I have a big question mark over it.

Documentation also supports and encourages reflection (see 'Establish and Maintain Reflective Processes' later in this chapter). The presence of good documentation of a performance, for example, provides a powerful trigger for reflection – and makes it easier to demonstrate the links between reflective writings in the thesis and the performances that gave rise to them. Without the documentation, there is a chance the details of the process, the goals of the artists, and the nuances of performance will be lost. On this, a PhD researcher and supervisor mentioned:

> If we didn't stop and document straightaway, we lost so much recall of just what happened then.

The practicalities of documenting the work can be detrimental to the creative project, or at least distracting for practitioners. Setting up cameras, microphones, etc. takes a certain amount of time and energy and the amount of ongoing attention they require can vary greatly depending on the physical environment within which the work takes place. To meaningfully capture a full day of rehearsal in a large performance space, for example, is a considerable challenge. At the extreme, 'gold-standard' documentation will make production demands akin to that of a documentary film crew. Even if this level of comprehensiveness is not necessary, it may well be worth allocating documentation tasks to someone other than the researcher and their creative collaborators – if resources permit. For example, one PBR researcher/ supervisor observed:

> [Having to document makes us] possibly . . . less flexible, less able to immerse in like maybe the creative process in a way. Because then I have to be slightly detached. [I need to think] 'Oh, do I need to record this? I do want to record this session and I do need to make sure certain things are in place so that we get the data'.

In support of this, a PhD graduate mentioned:

> a lot of these works were, you know, public-facing works of considerable size. [So, my collaborators arranged] professional photographers to take . . . video and photos, and they all got included in the PhD thesis and they are quite integral to explaining what it was that I was even on about – what I did.

In some cases, the documentation can have a greater reach and impact than the original work. For example, the video of the first performance of *Encoded* (discussed earlier), the live performance of which was attended by perhaps 400 people in total, has now been viewed nearly 15,000 times. In this sense, the documentation is another channel through which to disseminate the work – and it becomes another work of art in itself. One interviewee sees the documentation as a creative act in itself, in that it goes beyond simply documenting what occurred and often can influence the kinds of performances that she crafts:

> [It's] very important to do good documentation. But also not just documentation . . . in the sense where you go, 'Okay, I'm just going to photograph this or record it', or whatever. But think[ing] up artistic ways of documenting, I think is important, because it can actually further new types of work.

Given its importance to many of the practice-based researchers we interviewed, the challenge of making documentation of the artworks available in the longer term is a critical one. Most recent graduates included video documentation with their thesis by providing links to online video repositories. The obvious risk is that in the longer term these repositories may cease to exist and the videos become unavailable. Our experience has been that university library systems are not yet as effective at archiving these forms of documentation – a situation that to the best of our knowledge remains fundamentally unchanged since Candy's 2009 study, which also identified this issue.

Physical space

Space requirements for PBR were in one sense considered simple – and largely similar to those of any studio-based work – but in practice surprisingly challenging to arrange. Flexibility was key. Spaces which were overly defined, or designed with very specific practices in mind, were undesirable and seen as often counter-productive. A PhD graduate reflected:

> Where I am at the moment in the university, we work in spaces that are already defined: 'okay, that's the meeting room. That's BFA that's MFA. That's the NFA hub'. And I never really felt really comfortable in any of these spaces because they've been defined.

While the environment should be pleasant enough that people want to work there and feel comfortable hosting visitors, the priority should be providing flexibility and detailed control over the space. The trend towards automatic lighting, air conditioning systems, and motorised window blinds in new constructions, for example, often works against the kind of detailed control over lighting and sound that many creative works require. On this the graduate said:

> I say it jokingly, but . . . it's being able to control your environment. Being able to make noise when you need to make noise. It's being able to have darkness and use light when you want to use light.

At CCS we have been fortunate to be part of a large program of university development that enabled us to be very active participants in the architecture, design, and fitout of our research spaces over several years. Our practical advice for others who are in this situation is to be tenacious and hold firm against equipment (and the design consultants it often seems to enslave) to insist on as much manual control over the environment as possible.

Strategies

Having outlined a range of challenges for practice-based researchers, we now present a series of strategies that interviewees identified as being effective.

Collaboration as an antidote to isolation

Some feelings of isolation or solitude were mentioned by several participants, particularly and most obviously by those who lived a significant distance away from UTS and therefore spent less time physically at the university. All the artists in this study collaborated with others to a significant degree to create the artworks that were presented in their thesis. This was seen as a way of getting external perspectives, providing creative challenges and also setting up helpful constraints around the work. As discussed previously, in some cases, collaboration was also necessary in order to deliver creative work of the required scale and complexity. A PBR graduate and supervisor reflected that:

> [the loneliness] can also be compounded when you're doing the kind of PBR that I was doing – which was very reflective, and very introspective at first – it can compound that cycle of not feeling like you're part of a larger community. I think in work where you are constantly working with collaborators and [conducting] interviews and working off collaborative projects and using that practice to drive theories then you're constantly working with a lot of people and that helps things.

It is the case that the PBR projects described by the interviewees are almost universally collaborative. While finding collaborators with compatible skills and availability is not always straightforward, there is strong consensus that it is worth the effort.

Clarify the overall research structure early

Taking time to work through the larger picture of how the practice and the research relate to one another early in the PhD process was mentioned by one participant as an important strategy:

> I get students that I've been working with to think about methodology early, and to think about how the creative practice is going to be positioned [and] if they're going to do reflective practice, getting them to formalize it – and actually write early.

This was considered more important if the project was based primarily on individual reflection rather than feedback from users or other practitioners via interviews or other sources of external data. A PhD graduate highlighted:

> Getting that sort of momentum is sort of important. And to get that early . . . to have the understanding of: what is this PhD? And what does it entail? To have the research methods to understand all of that early on, because it is a lot to grasp in the beginning, these

concepts, as an artist, because it's so foreign somehow. . . . But once you know what all these things are, you can apply them within your practice, and they actually support the practice.

Having said this, the structure and methods of the PhD are likely to evolve throughout the project. For some participants, a certain amount of relatively less-constrained exploration in the early phases of the project was seen as an essential part of finding their own path. In theory, having the method mapped out early would result in faster completion, but this is not always realistic or even desirable. A PhD graduate reflected:

I spent a couple of years at the beginning of it kind of swanning around [and] not really getting a clear picture of what direction [I] would take. But in my case, I think that was necessary because I was new to the whole research area. But if I was to do it again [I would try] to set up a lot of it before I even enrolled . . . almost have that mapped out before I started and I think then I'd be able to do it more quickly.

In summary, while flexibility and freedom to explore are important in the early stages of practice-based PhDs, it is of course necessary to identify the high-level structure of the project and how the components of practice, research studies, and reflection relate and provide structure for the overall contribution of the thesis. Establishing this 'big picture' structure early is considered beneficial, even if plans subsequently change.

Establish and maintain reflective processes

Establishing and maintaining effective reflective processes is an important strategy for the PhD researcher. However, the messiness of the creative/research process may mean the overall structure of the project only becomes fully apparent in retrospect. One PhD graduate mentioned:

I'm pretty fidgety and want to try a lot of things. I think a lot of these things overlap and are messy at the time, but when you step back you can create your own timeline that makes a bit more sense.

In addition, while Schön[6] provides a useful conceptual framework for reflection (and Candy's recent publication helps provide concrete advice),[7] some PhD researchers mentioned that they had difficulty finding clear guidelines for how to apply these concepts in a practice-based PhD. A PhD graduate and supervisor reflected that:

I used the ideas about reflective practices and model but there wasn't like a 'how-to' that I that I used. So that I found a bit challenging.

The formats used to capture reflections varied. What is important is capturing the evolving creative process and the practitioner's thoughts about them as the work develops. One artist-researcher, for example, whose PhD involves the creation of innovative animation techniques, uses a (digital) folder of material within which he stores a wide range of material, including sketches and video content. As he does this, he has in mind a developing narrative of the project that he is likely to present to an audience at a later date. For example:

I like thinking about some sort of 'behind the scenes' keynote presentation. . . . Some sort of document, a folder of stuff that I throw things into. And some of it is video content, some of it is images, some of it is drawings and sketches. Some of it is writing . . . a little bit of interviews as well.

As this material accumulates, there are certain events that cause him to step back and take a broader view of the project and where it is heading. This helps him see how the many sketches, ideas, videos, etc might fit into a larger, more coherent picture in the near future:

Like if someone says, 'can you come and show us your work?' I do quite enjoy that process of . . . putting that into a slideshow, audio visual presentation. And for me, as well as the writing, that's an interesting part of the process because [I'm forced to consider] how do I say this in 20 minutes with pictures? . . . Making that presentation is for me always pretty useful.

Interviewees adopted a range of different reflective practices (see 'Develop the Ability to Reflect' later in this chapter), largely depending on the particular characteristics of their individual projects and their personal preferences. The key point is that reflective practices are consciously selected and applied as early as possible.

Finding new perspectives through interviews

The artist-researchers were continually looking for new perspectives on their work and ideas for how to move it forward. Insights were sought via a range of means, from informal conversations with their supervisor and peers, with audience members after a performance or exhibition, and from the broader field through paper submissions and presentations.

In addition to these channels of feedback, it was interesting to hear that all eight of the current and recent PhD graduate participants conducted formal or semi-formal interviews during their candidature, usually with experts in the field. This was even the case for those artist-researchers whose method was primarily based around individual reflection. In several of these cases, the interviews were not formally analysed and written up in the thesis – even if this was the intention at the time. Where this was the case, however, the interviews were still perceived as a very valuable part of the process, as they provided the artist-researcher with insights and perspectives that they had not previously considered. Taking the time, therefore, to prepare and conduct a range of interviews thoroughly and carefully with people who are in a position to provide insights (usually fellow artists in the cases examined here) is a strategy that may help move a PBR research project forward.

The interviews that were discussed by participants in this case went beyond casual conversations after a performance or exhibition; the questions were pre-prepared, interviews scheduled, and usually audio or video recorded. As such, the discussions were more in depth and considered than the more frequent (and also very valuable) after-show chats. A couple of PhD graduates reflected that:

I did get a lot out of that interview. That's quite true. And maybe it helped me to structure some of my own thoughts, having spoken in depth with him about that one particular piece.

I did carry out a lot of interviews actually. I think that really helps . . . to get the insights from different people in different disciplines. Those conversations are so crucial.

Slow down – take time to explore

Given that PBR PhD researchers are likely to already be expert in their area of practice, it is important they remain open and willing to explore areas they are less familiar with. One PBR supervisor mentioned:

> It's only too easy, if you're an expert, to think you know everything because in a certain domain you do, probably, but just to the side there's something else and you just don't know anything about it. Some people are naturally very open and kind of listen and think. And some people not. These people can succeed but they're hard work to supervise, because they basically don't listen to you!

Equally, it is important they are open to re-examining and sharing their creative process. This is not something that is necessarily required of outstanding practitioners; there is no need for great musicians or visual artists, for example, to be open about their creative processes – although many are, of course. However, given the purpose of a practice-based research program is to develop, explore, and share new approaches to creative practice, PhD researchers who are fixed in their opinions and/or narrow in their views are unlikely to succeed.

Perhaps equally important is the willingness to share the artworks that are created. Being open about the process is not meaningful if the artworks themselves are not exhibited in some way for people to experience and, potentially, comment upon. In many, perhaps most, cases artist-researchers will seek feedback from audience members and/or experts in the field in order to get new perspectives on the work and ideas for future developments. Sometimes this feedback is more formally documented, analysed, and used in the thesis, but often it is used simply as a source of new ideas and inspiration. A PBR supervisor commented that:

> One of the characteristics of creative practitioners is that they're constantly challenging themselves. So it's only a step forward, really, if you put your work out in the world. . . . And in a sense, that's a way of challenging yourself. Putting work out . . . is one of [the ways] of stimulating reflection. It's a way of them looking at things differently, but it's also an invitation to other people to feedback. . . . And I think a reflective practitioner will be doing that kind of thing as a matter of course.

A key theme from the interviews was the need for openness and willingness to adapt, at times radically, in response to audience and domain expert feedback.

Develop the ability to reflect

Developing the ability to reflect is a critical strategy for the PBR researcher. On this Linda Candy wrote: 'How to document reflective practice and use it effectively is a skill that has to be learnt and practical advice is useful'.[8] Practitioner notebooks, journals, sketches, etc. are valuable sources for reflection and can be kept for this purpose. It is not necessary for them to become part of the data gathered formally and presented in the thesis, but without them the reflective process risks becoming detached from the reality of the practice – and the resulting reflections may turn into post-hoc rationalisations, or misrepresentations, of the process. A PBR supervisor highlighted that:

> the challenge is in kind of being slightly more rigorous about how that stuff [journals, etc] is kept and making sure the right notes about the things that fail [are kept], especially – even more than the things that succeed, because often that's what you learn most [from].

While the ability to effectively and honestly reflect on the practice is seen as a critical skill, the interviewees were cautious against being overly formal or formulaic about reflections. The three PBR supervisors mentioned:

> I'm very keen that they have them and use them to help them reflect upon what's been going on. But I'm against formalizing the reflective process. Because I think that misses the point in a way.

> I mean you pretty much have to use reflection on action for some of the stuff. That's all that's possible. But if the distance is too great, and then you kind of make it up as if it was a more organized thinking process than it really was. Then, although it kind of looks better in a kind of formal research sense, it may be more invalid, because it's too far away from what actually happened.

> My main trick is to try to persuade the student to be more open and honest [about their process, successes and failures] I try to persuade the student that it's perfectly okay to be quite straightforward about this and, you know, just be honest about it.

Additionally, the temptation to over rationalise, or construct post-hoc explanations that are not grounded in reality, can be a result of lack of confidence or insecurity. One PBR supervisor mentioned:

> helping people to feel confident about their practice, and what they think about and in their practice, must be an element of supervision I think for such work. Because, you know, everyone has a problem there: 'maybe what I'm doing is a load of rubbish'.

The supervisor therefore has a role in ensuring that the practice is valued in itself, and not reducing it to a by-product of the research process. In practical terms, this often means simply taking an interest in the work itself. With few exceptions, the participants in this study all had well-established creative practices before commencing the PhD, and thus their ability to produce quality work was not in doubt. The supervisor was rarely required to assist with the work itself – even if they were able to. In practice, therefore, supervisors spend most of their time helping with methods and thesis writing. To counterbalance this, it can be important to demonstrate that the creative work is of equal, or greater, importance.

Conclusion

In this chapter, the experiences of current PBR PhD researchers, recent graduates, and experienced supervisors, examined through a series of interviews, have been presented. Some consistently challenging areas for PBR PhDs emerged, along with a series of strategies which were found useful. The intention is that these will be of use to others working in similar areas. The research group within which these artist-researchers worked has been operating at the University of Technology Sydney since 2002 – and indeed, for many years prior to that in the UK – and so these PhD researchers are participating in, and building upon, a considerable body of knowledge and experience in practice-based research. For those working in situations where PBR is relatively new or where the prevailing research culture is more traditionally 'scientific', it is hoped the insights presented here will be of practical assistance and encouragement.

Notes

1 For further information read Candy and Edmonds (2022).
2 Candy (2009).
3 Johnston (2011, 2014, 2015).
4 Glaser (1992), Glaser and Strauss (1967).
5 Candy (2009, p. 15).
6 Schön (1983).
7 Candy (2020, Chapter 7).
8 Candy (2020, p. 241).

Bibliography

Candy, L. (2009). *Practice-Based Research and the PhD: A Study.* http://research.it.uts.edu.au/creative/linda/PBR/pdfs/CCS-PBR%20Main%20Report.pdf.

Candy, L. (2020). *The Creative Reflective Practitioner.* London and New York: Routledge.

Candy, L., & Edmonds, E. (2022). The Studio and Living Laboratory Models for Practice-Based Research. In C. Vear (Ed.), *The Routledge International Handbook of Practice-Based Research.* London and New York: Routledge.

Glaser, B. G. (1992). *Basics of Grounded Theory Analysis.* Mill Valley, CA: Sociology Press.

Glaser, B. G., & Strauss, A. L. (1967). *The Discovery of Grounded Theory: Strategies for Qualitative Research.* New York: Aldine de Gruyter.

Johnston, A. (2011). Beyond Evaluation: Linking Practice and Theory in New Musical Interface Design. *Proceedings of the International Conference on New Interfaces for Musical Expression*, 280–283.

Johnston, A. (2014). Keeping Research in Tune with Practice. *Interactive Experience in the Digital Age: Evaluating New Art Practice*, 49–62, Springer.

Johnston, A. (2015). Conceptualising Interaction in Live Performance: Reflections on Encoded. *Proceedings of the International Workshop on Movement and Computing*, 60–67.

Schön, D. A. (1983). *The Reflective Practitioner: How Professionals Think in Action.* New York: Basic Books.

Appendix

Interview questions (PhD graduates/researchers)

What led you to undertake a PhD?

What does 'practice-based research' mean in the context of what you did for your PhD?

What were the advantages of PBR for you?

What were the disadvantages of PBR?

What challenges did you face in your PhD work?

What strategies did you use to address these?

What is the relationship between practice and theory (or practice and 'research') in your work?

If you were to supervise a PBR project, how would you approach it? What specifically would you do? What would you avoid?

What factors affected your choice of university/supervisor?

What would you do differently if you were to start over?

What approaches did you contemplate (or can you contemplate now) other than PBR?

What are you doing now? (recent graduates)/What would you like to do when you graduate? (current students)

What research methods did you apply/are you applying? (e.g. interviews, observation, questionnaires . . .)

What is/was the role of reflection in your PhD?

What is/was the role (if any) of documentation (e.g. video/audio recordings, photographs, etc) in your PBR? Did/Will you include these with your thesis?

Interview questions (supervisors)

What led you to PBR?

What does PBR mean for you – in terms of the PhDs you supervise now or have previously supervised?

What are the advantages/disadvantages of PBR you have observed in your experience supervising PhDs?

Have you supervised PhDs which were not practice-based?

What are the biggest challenges you face as a PBR supervisor?

What strategies do you use to address these?

What do you see as the role of reflection in PBR?

What have you observed in relation to reflection in the PhDs you have supervised? Are there common pitfalls?

What are the characteristics of a successful PBR PhD researcher in your experience?

Are there particular coaching/supervisory strategies that you have found effective (or ineffective)?

Are there traps you see inexperienced supervisors falling into?

What makes an effective PBR environment for students in your experience?

PBR is 'directed towards enlightening and enhancing practice'. How do you feel we do in this regard? What is the difference between PBR that does 'enlighten and enhance' and PBR that doesn't?

What role does the artefact play in 'impact' of PBR? In your experience, does most impact arise from writing or from the artwork? (or both?)

4.6

COMMUNITY-BUILDING FOR PRACTICE-BASED DOCTORAL RESEARCHERS

Mapping key dimensions for creating flexible frameworks

Sian Vaughan

Introduction

Undertaking a practice-based doctorate can be challenging. There is clear need for community for practice-based doctoral researchers and distinct benefits that can result from being part of a community.[1] However, as with other forms of research, practice-based research itself is not homogeneous (and nor should it be), and there are different disciplinary contexts as well as diverse institutional contexts in which doctoral education occurs. The activities and elements that build and sustain doctoral communities for practice-based researchers will need to respond to and reflect particular contexts. As argued elsewhere in this volume, such research communities also need to be fluid and enable agency; a sense of community cannot be successfully imposed.

While this chapter has arisen from both my own experience supporting doctoral researchers in Art and Design and my research into doctoral communities across Art and Design, Performance, Music, Digital Arts and the Humanities internationally, my aspiration is that the ideas and questions it prompts can be beneficial whatever the particular flavour and discipline of practice-based research context. My intention here is not to provide a blueprint or template for practice-based doctoral research communities, but rather to offer possibilities and points of departure for institutions and individuals.

I begin by briefly examining some of the reasons why higher education institutions should build and support communities for practice-based researchers undertaking doctorates. Initially, you need to know what you hope to achieve from supporting and/or belonging to a community, as that will inform your selection of appropriate types of community-building activities. However, this is not a set of recipes from which outcomes can be predicted if instructions are followed. Doctoral community is a community of practice[2] – in fact, often several interlocking communities of practice, each involving a common focus but diverse individuals and all the complex power and social dynamics these can involve. The equally amazing and frustrating thing about research communities, particularly in practice-based research, is that they retain an element of unpredictability.

I outline the dimensions and elements that I believe should be considered and that can be mixed in building and supporting community for practice-based doctoral researchers. These dimensions are not exclusive facets of community, but should be conceived of as entangled and fluid. The relative

DOI: 10.4324/9780429324154-39   544

importance and priority attached to each will depend on the possibilities and particularities of your context. Then, taking each core dimension in turn, I will use them as lenses to outline types of activities and structures that could form and encourage community. Each thematic section includes some key questions that are not merely rhetorical but also offer important points for consideration in supporting practice-based doctoral communities in different institutional contexts. I also offer suggestions for practice-based doctoral researchers as to how they might individually engage with and benefit from particular types of community-building activities.

Reasons for building and supporting community in doctoral education

Arguably, there has never been a greater need for community support in doctoral research. For example:

- In a doctoral landscape in which the expectations of professional development activity have increased alongside the precarity and competitiveness of the academic job market, the numbers of doctoral candidates has grown significantly worldwide.
- Recent literature has drawn attention to the mental stresses faced by doctoral researchers, and to the need for higher education institutions to support their well-being and professional development alongside their contribution to knowledge made through independent research and articulated as a thesis.
- The doctoral landscape and academic job markets are evolving, and supervisors whose own lived experience may quickly become outdated should not and cannot be the only source of preparation and intelligence that practice-based doctoral researchers have access to.
- Building community can enable peer-sharing and broaden perspectives and opportunities, as well as invite in others from different disciplines, practices and outside higher education.

It is not just about joining an intellectual community; for all doctoral researchers a sense of collegiality, and community has been shown to significantly benefit the doctoral experience.[3] As a higher-level qualification and an induction into the processes of academic research, undertaking a doctorate should be a complex, stimulating and challenging endeavour. Importantly:

> Institutions need to find ways to support PGRs to disconnect the 'healthy stress' related to the intellectual challenge of undertaking a doctorate from other stresses that have a negative impact on wellbeing and mental health.[4]

Recent literature has drawn attention to the mental stresses faced by doctoral researchers in all disciplines.[5] For example, a much-cited large-scale European study concluded in 2017 that almost a third of doctoral researchers were at risk of having or developing a common psychiatric disorder.[6] *The Wellbeing Thesis*, an online resource developed by the University of Derby and King's College London in collaboration with the Student Minds charity, is an initiative that demonstrates growing recognition that higher education institutions need to pay greater attention to supporting the mental health and wellbeing of their doctoral researchers and the academic staff that support them.

Alongside the inherent intellectual challenge, and the supervisory relationship which is the focus of a elsewhere in this volume, Mackie and Bates identified a wide range of other factors that potentially impact on the mental health of doctoral researchers, including a lack of transparency and unclear expectations in university doctoral processes.[7] This can be particularly challenging for practice-based doctoral researchers as most university doctoral processes still assume that a doctorate is articulated in a text,[8] leading to additional anxieties around perceived issues of hierarchy and legitimacy.

There are significant benefits in interdisciplinary and cross-disciplinary communities of practice-based researchers, as approaches and knowledges are shared, translated and appropriated. Challenges can be exacerbated by traditional models of doctoral education that focus on supervision and the research project, rather than considering more holistically a doctoral experience. Such traditional models can be reinforced by the implicit assumption that because our doctoral researchers are often highly regarded creative professionals, the majority with a master's degree and viewed as experienced students, there is less need for attention to psycho-social support for the doctoral becoming of practice-based researchers.

Research has also shown that doctoral researchers are actively seeking support mechanisms and a sense of community to support their wellbeing.[9] The growth in popularity of Twitter hashtags and accounts such as #phdchat and #phdlife, and of blogs such as Inger Mewburn's *ThesisWhisperer*, testify to the need felt by doctoral researchers to share their doctoral experiences and challenges and to feel a sense of belonging. Informal support networks are important for doctoral researchers.[10] These informal support networks can include family, friends and peers within and outside academia and have been characterised as part of a rich penumbra of the doctoral landscape.[11] Recognition of such a penumbra on the peripheries of more formal doctoral education does not, however, absolve institutions and academics who supervise practice-based doctoral research from needing to consider how they support their doctoral researchers as communities.

The dimensions and elements of community: mapping a framework

There may not be a singular 'Framework' that can be applied to build and support doctoral community for practice-based researchers; there are, however, key dimensions of community that I argue should form part of any such doctoral community. These key dimensions of doctoral community are the:

1   discursive
2   active
3   social
4   virtual

In particular for practice-based research, there are particular needs around provision for sharing and engaging in practice, for active elements of community for doing as well as thinking and socialising.

These are not necessarily discrete and separate dimensions: an event can be both discursive and social, and an activity can promote doing, thinking, discussing and socialising. A space, whether studio or office, can enable making, thinking, doing and being as well as informal social connections that create the sense of belonging that differentiates community from simply gathering. These dimensions need to be considered as entangled, fluid and responsive – think more of constellations coming together at different times and places to create and sustain community rather than a static map or structure. Similarly, doctoral community, while a useful generalisation as a term, in an institution will actually comprise multiple communities and should not be conceived of as homogenous or singular.

It is also important to recognise other significant aspects that orbit and influence these four key dimensions of doctoral community. For example:

* There will be times when a distinct focus on *disciplinarity* will be appropriate; at others, the focus and community itself may well be distinctly *cross-disciplinary* or *interdisciplinary*.
* Institutions have a responsibility to support doctoral researcher communities and provision the environment, concurrent opportunities and support for the *agency* of doctoral researchers

themselves is as important. Practice-based doctoral researchers are experienced practitioners and deserve respect as such alongside opportunities to learn to be a researcher. There is more sustainability and engagement with community where there is a sense of common endeavour and ownership; self-directed community activities promote this.

- Alongside this, the *position of staff* is a further significant orbiting aspect for consideration. There are multiple potential roles for supervisors and academics in relation to doctoral community, and it is also important that practice-based researchers are welcomed into the research communities of their schools, departments and institutions.
- The *inclusivity* of research communities needs to be kept in mind to avoid hierarchical positioning and the worst of academic competitiveness, territorialism, othering and disenfranchisement.
- Finally, I want to draw attention to the significance of *serendipity* in practice-based research communities. Community works when it is valued by participants and when it is enjoyable. We cannot know for certain what will work or precisely how, and we should allow flexibility and space for unknowing and the unexpected.

In summary, the key dimensions of doctoral community for practice-based researchers are the **discursive**, **active**, **social** and **virtual**. A local framework for community must consider all and make space for *disciplinarity*, the *cross-disciplinary* and *interdisciplinarity*, *agency*, *inclusivity* and *serendipity*. Taking each core dimension in turn, I will use them as lenses to outline types of activities and structures that could form and encourage community among practice-based doctoral researchers, with key questions from an academic and institutional perspective as well as tips directly aimed at doctoral researchers. Inevitably, there are overlaps and repetitions as an activity can encompass more than one dimension.

Dimension 1: discursive communities

A crucial dimension of community for doctoral researchers is the opportunity to discuss research with others. For practice-based researchers, this is particularly important in enhancing confidence and further legitimising practice-based research within institutions. It is also important that activities that promote discussion and critique of research also facilitate engagement with practice and not just with the more standard academic formats. Table 4.6.1 identifies some of the elements and activities that can build discussion in communities that are critical but supportive.

Table 4.6.1 Activities that can encourage discursive community.

Discursive community building activities	
Seminars	Academic papers, work-in-progress presentations, invited speakers (academics and industry professionals), panel discussions, Crit Clubs
Reading groups	Disciplinary and/or thematic, peer-review groups
Writing activities	Peer-review groups
Academic skills training	Referencing and data management systems, research ethics, networking skills, conference participation, academic publishing, abstract writing, understanding peer-review
Workshops	Creative and technical skills-based, project-based, collaborative activities, work-in-progress demonstrations, rumbles, mudpits, sandpits
Events	Festivals, conferences, performances, showcases, exhibitions, symposia
Space	Group studio space, office space, social spaces, studio-takeovers
Social media platforms	Blogs, Facebook and WhatsApp groups

In reality, almost any community activity that brings researchers together can facilitate creative and critical conversations. In a context of growing numbers of practice-based researchers graduating with doctorates and growth in the precarity and competitiveness of the academic job market across most disciplines, it is important that doctoral programmes prepare candidates for multiple career routes. Skills workshops and professional development activities obviously address this need, and they can simultaneously be framed and critiqued in relation to practice-based research discourses.

When developing the discursive dimension of a community, it would be helpful to consider these key questions:

Key questions when thinking about the discursive dimensions of community

- What discursive formats for events can encourage respect for difference and exploratory conversations about practice-based research?
- How can we balance embedding practice-based doctoral researchers in active research cultures in our institutions while also providing spaces for them to share experiences of learning to be researchers?
- What is the role of academic staff in supporting practice-based doctoral researchers as a community, and how is this supported by the institution?
- How can we be honest about the positives and negatives of academic careers, while enabling a supportive environment for developing our doctoral researchers?

If you are a doctoral researcher and you intend to go to an academic event or an event that focuses on academic careers, here are some tips that have proven to be beneficial:

- If practical, don't just go to events that seem to have an obvious link with the subject of your own research. There are always links to be made with another's work, even if it just provides reassurance and confirmation that you don't want or need to do something a different way!
- Try not to view participating in events as time away from your 'research proper'; the discussions and networking are as important for your development and will move your thinking forward.
- Think about how you can find out what events are running in your department or school, but also elsewhere across your institution and beyond. What conversations do you want to be part of, and how can you seek them out?
- Work at finding and building your networks, and at bridging professional and higher education communities. It takes time but does open up opportunities.
- Don't see an academic role as the pinnacle of post-doctorate achievement; it isn't.
- Seek opportunities to understand the reality of academic life and the range of professional and hybrid roles with higher education.

Dimension 2: active practice-based communities

It seems self-evident to state that community for practice-based doctoral researchers needs to facilitate and activate the sharing and doing of practice, the bringing of significant knowledge from the direct execution of the relevant practice. With the wide range of practices that can be encompassed in practice-based research, the precise facilities and forms of practice-sharing will vary. Table 4.6.2

Table 4.6.2 Activities and elements that can encourage active community.

Active community elements	
Workshops	Creative and technical skills-based, project-based, collaborative activities, work-in-progress demonstrations, rumbles, mudpits, sandpits
Space	Group studio space, office space, social spaces, studio-takeovers
Events	Festivals, performances, showcases, masterclasses
Academic skills training	Referencing and data management systems, abstract writing, understanding and undertaking peer-review, teaching in HE
Writing activities	Writing retreats

illustrates some of the elements that can support community building through doing, through activity and the processes of practice, not just the written articulation of research.

Practice is individual and can be collaborative. Practice-based research is centred on the individual practitioner and their knowledge and skills with their practice; practice-based research also stretches and develops practice. There are technical and professional skills that practice-based doctoral researchers may need to acquire, and adopting cohort or community-based models for skills training can aid the sense of community and belonging. Within a doctoral community, there will be a variety of skill sets that can be shared, particularly if the doctoral community is interdisciplinary and cross-disciplinary. The provision of space to enable practice-based researchers to carry out, share and disseminate their research needs consideration, even if institutionally problematic. Even the potentially less creative practices of learning referencing or additional forms of software can encourage community learning and a sense of shared challenges and frustrations. Writing retreats often include the use and testing of different writing strategies live in the moment, with the sharing of experiences, strategies and tips facilitating the sense of collective challenges and possible solutions. Writing is, after all, a practice in itself with many forms.

When developing active space dimensions for community building, it would be helpful to consider these key questions:

Key questions for thinking about active spaces for community building

- How can we create spaces for community among practice-based doctoral researchers that are playful and creative?
- What opportunities might there be for doctoral researchers to share in the doing of practice?
- Can the different activities in practice-based research be co-located?
- How can we create inclusive communities that enable and include doctoral researchers who cannot be as physically present due to part-time status, caring responsibilities and/or health conditions?

If you are a doctoral researcher, here are some tips about how to get the best out of shared spaces and the sharing of skills, even across disciplines:

- Not everyone uses office, studio or making space in the same way. Think about what you need and want, and about how you can play your individual part in helping to create an ethos within a space that is inclusive and supportive, playful but respectful.

- Think about the skills that you have that might be useful to your peers; these might be disciplinary, professional, technical or even from other previous work roles. What can you offer to share?
- Identify skills or techniques that might be useful, interesting or just fun to learn or try out yourself. Who do you know in your community who might help you learn?
- If there is a particular workshop or skill that you think you need, speak up, as there may well be others who would also benefit. Talk to your peers, your supervisors and doctoral education leads as they can signpost to opportunities. If it doesn't exist already, help them make it happen.

Dimension 3: social communities

The social aspects of community and the sense of belonging are crucial in addressing the recognised mental health issues that doctoral researchers may face. Yet the social aspect of doctoral communities can be the most fragile, as the community membership is in constant flux. While all community events and activities that bring doctoral researchers into contact with one another will have a social dimension, there are particular types of activity that can encourage the ethos of peer-support and psycho-social support, as Table 4.6.3 suggests.

The informal social interactions within a community can also address the additional concerns that practice-based doctoral researchers may have concerning creative identity and the legitimacy of practice-based research. Shared experiences can help dissipate concerns regarding hierarchical positioning and lack of institutional recognition, though of course checks and balances may be required to ensure that such sharing is supportive and constructive rather than fanning the flames of anxiety and othering. Enabling the agency of doctoral researchers to define and create their own community activities can assist with engagement and create the sense of ownership that facilitates belonging and peer-support.

When developing the social community dimension, it is important to consider these key questions, especially if you are supporting doctoral researcher-led initiatives or they are dealing with psycho-social support for doctoral researchers:

Key questions for thinking about supporting doctoral researcher-led initiatives and psycho-social support through community

- How can we balance intellectual debates, professional training and social support in practice-based doctoral communities?
- How can we balance enabling individual success without engendering competitiveness, rivalry and antagonism?
- How might peer-support and peer-mentoring be enabled across disciplines and without segregating practice-based research as other?
- What opportunities are there to co-create doctoral community and co-design support for practice-based doctoral researchers?
- How do we advocate to create budgets to fund doctoral researcher-led initiatives, and to provision the environment not least through catering?
- How do we balance enabling the agency of doctoral researchers while providing support and guidance as staff?

Table 4.6.3 Activities and elements that can encourage social community.

Social community activities and elements	
Well-being and social events	Breakfast and lunch clubs, coffee-and-chat sessions, mindfulness events, meditation classes, dance classes, end-of-term parties
Mentoring	Peer-mentoring, buddy system, pastoral groups
Space	Group studio space, office space, social spaces, studio-takeovers
Social media platforms	Websites, online profiles, blogs, Facebook and WhatsApp groups
Events	Festivals, conferences, performances, showcases, exhibitions, symposia
Writing activities	Shut-up-and-write groups
Academic skills training	Networking skills events

If you are a doctoral researcher preparing and developing community events, here are some tips to help you maximise these opportunities:

- There are various doctoral support communities on social media (twitter, Facebook, Instagram, WhatsApp etc.) that you can join and participate in. Does your department have student-led and/or doctoral-researcher-led groups; if not, could you start one?
- Create informal reading groups if there a few of you interested in a particular subject or theory. Similarly, Crit Clubs can be formed to share practice.
- Something as simple as a regular 'coffee and chat' arrangement can be a great starting point.
- Be kind. Territorialism, extreme competitiveness and rivalry are rarely valued inside or outside the academy.
- Try blending physical and virtual participation in community activities yourself; this will help you understand how best to enable virtual or offsite engagement by others.

Dimension 4: virtual and blended communities

Our collective recent experiences of the Covid-19 pandemic and the lockdowns which prevented physical gathering and face-to-face educational activity mean that the consideration of virtual and online community activity to support doctoral education has been imperative. Blending on-campus events with virtual spaces and online activities also helps to address inclusivity issues to support doctoral researchers who are part-time, live at some distance from campus, have caring responsibilities and/or disabilities; groups who have historically been less able to access on campus provision and thus disenfranchised. With a little creativity, almost every activity and element of community introduced so far can be moved online, as Table 4.6.4 shows.

The benefits of online events can to some degree outweigh the lack of informal real-life social interactions. The affordances of platforms such as MSTeams, Zoom and Skype with screensharing, collaborative whiteboards, file-sharing and chat functions can enable rich conversations and interactions. While there can be accessibility issues in terms of bandwidth, technology, health and/or neurodivergence with screen interactions, online events which do not require travel can be more convenient for many. The online world also makes national and international connections more feasible without either the financial resource or carbon footprint of travel (not that the digital is without its own carbon footprint). The challenge is in enabling community that exists across on-campus and virtual spaces.

Here are some key questions to consider when developing this virtual dimension:

Table 4.6.4 Online community activities.

Suggested online community activities	
Online seminars	Online seminars, invited speakers (academics and industry professionals), panel discussions Especially the use of automatic breakout rooms
Online reading groups	Disciplinary and/or thematic, peer-review groups
Online writing activities	Writing retreats, shut-up-and-write groups, peer-review groups
Academic skills training	Online interactive workshops
Online workshops	Creative and technical skills-based, project-based, collaborative activities, work-in-progress demonstrations, rumbles, mudpits, sandpits
Online events	Conferences, live-streamed performances, virtual exhibitions, symposia
Online well-being and social events	Breakfast and lunch clubs, coffee-and-chat sessions, mindfulness events, meditation classes, dance classes, cocktail/mocktail evenings
Social media platforms	Websites, online profiles, blogs, Facebook and WhatsApp groups

---

Key questions for thinking about online activities for community building

- How can a balance and boundaries between professional and private lives be maintained in online environments (for staff and for doctoral researchers)?
- What professional development in online pedagogy and technical support is available to assist staff with moving activities online?
- Are there ways to blend encourage community across online and on-campus activities both asynchronously and as blended events?

As a doctoral researcher engaging with online and virtual events, here are a few tips to help you maximise the experience and opportunity for further development:

- When identifying key words for literature searching, think also of potential hashtags to follow on social media platforms.
- Participating in online events can help you build networks and links beyond your own institution to include national and international communities. How will you find these events and online discussion fora?
- Always double check the time differences for international events.
- Reflect on your own online profile: does it represent you as both a practitioner and a researcher?

Some concluding thoughts

Communities are built and supported through the provision of spaces and discursive activities as well as making and doing opportunities for engaging in and sharing practice. The sociality of community can be encouraged through thinking about how to provision the environment, and the importance of coffee cannot be underestimated.[12] Community cannot be imposed or forced by staff; doctoral researchers need a reason to come together. The role for academic staff here is complex, needing both to be part of the community and also to acknowledge the difference in positionality to handle the power dynamics sensitively; fostering and sustaining community, while enabling the agency of doctoral researchers rather than leading or imposing. Securing resources, whether financial or spatial, for doctoral research can be difficult. Also, we should not underestimate the issues in recognising such academic citizenship and in supporting and enabling staff to have the time and energy to in turn support doctoral researchers as communities as well as individual supervisees.

I have here presented dimensions, elements and activities of a framework for supporting and building community among practice-based doctoral researchers that is not a 'Framework'. The individual nuances of each practice-based research project, the shift in individual needs across the lifespan of their doctoral research and differing disciplinary and institutional contexts mean that a singular 'Framework' would be inappropriate and ultimately ineffective. We need a parallel plurality in the models and methods that we use to build and support community. Different activities will meet different community needs and be more appropriate to some contexts than others. The modes of community for practice-based doctoral researchers need to be fluid and to evolve alongside our changing doctoral populations. As Daučíková states, we need to:

> Develop the research milieu, building upon specific local (albeit 'globalised-local') resources, traditions and issues rather than presuming that there is an international norm or ready-made ideal model of research milieu to which we can all adhere. Consciously consider the challenge of maintaining an openness to new insights and the radical unpredictability of research practices.[13]

There are also still significant challenges in how such community participation is enabled for part-time doctoral researchers and those with caring or other responsibilities and conditions that restrict their physical access to higher education. Virtual and blended community activity is part of a response but not a panacea and challenging in the contexts of material practices. Within an institution or particular context, any framework for encouraging and supporting community for practice-based doctoral researchers needs to be diverse, fluid, inclusive and ultimately enable passion for practice and for research to flourish.

Notes

1 Elsewhere in this volume I explore the diverse lived experiences of academics who have tried to build and support community for practice-based doctoral researchers in their own institutions, revealing the benefits of community but also the challenges in creating it;– please see Vaughan (2022).
2 Lave and Wenger (1991).
3 Wisker et al. (2007).
4 Metcalfe et al. (2018, p. 30).
5 Schmidt and Hansson (2018).
6 Levecque et al. (2017).
7 Mackie and Bates (2019, p. 567).
8 Vaughan (2021).
9 Mantai (2019), McAlpine and Amundsen (2009).
10 Mantai (2019), Sweitzer (2009).
11 Wisker et al. (2017).
12 Boultwood et al. (2015).
13 Daučíková (2013, p. 71).

Bibliography

Boultwood, A., Taylor, J., & Vaughan, S. (2015). The Importance of Coffee: Peer Mentoring to Support PGRs and ECRs in Art & Design. *Vitae Occasional Papers Volume 2: Research Careers and Cultures, CRAC*, 15–20.
Daučíková, A. (2013). Developing Third-Cycle Artistic Research Education. In M. Wilson & S. van Ruiten (Eds.), *Share Handbook for Artistic Research Education*. Amsterdam: Elia, 65–71.

Lave, J., & Wenger, E. (1991). *Situated Learning: Legitimate Peripheral Participation.* Cambridge: University of Cambridge Press.

Levecque, K., Anseel, F., De Beuckelaer, A., Van der Heyden, J., & Gisle, L. (2017). Work Organization and Mental Health Problems in PhD Students. *Research Policy*, 46(4), 868–879.

Mackie, S. A., & Bates, G. W. (2019). Contribution of the Doctoral Education Environment to PhD Candidates' Mental Health Problems: A Scoping Review. *Higher Education Research & Development*, 38(3), 565–578.

Mantai, L. (2019). A Source of Sanity: The Role of Social Support for Doctoral Candidates' Belonging and Becoming. *International Journal of Doctoral Studies*, 14, 367–382.

McAlpine, L., & Amundsen, C. (2009). Identity and Agency: Pleasures and Collegiality Among the Challenges of the Doctoral Journey. *Studies in Continuing Education*, 31(2), 109–125.

Metcalfe, J., Wilson, S., & Levecque, K. (2018). *Exploring Wellbeing and Mental Health and Associated Support Services for Postgraduate Researchers.* Cambridge: Vitae.

Schmidt, S., & Hansson, E. (2018). Doctoral Students' Well-Being: A Literature Review. *International Journal of Qualitative Studies on Health and Wellbeing*, 13(1), 1508171.

Sweitzer, V. B. (2009). Towards a Theory of Doctoral Student Professional Identity Development: A Developmental Networks Approach. *The Journal of Higher Education*, 80(1), 1–33.

Vaughan, S. (2021). Practice Submissions – Are Doctoral Regulations and Policies Responding to the Needs of Creative Practice? *Research in Post-Compulsory Education*, forthcoming.

Vaughan, S. (2022). Understanding Doctoral Communities in Practice-Based Research. In C. Vear (Ed.), *The Routledge International Handbook of Practice-Based Research.* London and New York: Routledge.

Wisker, G., Robinson, G., & Bengtsen, S. E. (2017). Penumbra: Doctoral Support as Drama: From the 'Lightside' to the 'Darkside': From Front of House to Trapdoors and Recesses. *Innovations in Education and Teaching International*, 54(6), 527–538.

Wisker, G., Robinson, G., & Shacham, M. (2007). Postgraduate Research Success: Communities of Practice Involving Cohorts, Guardian Supervisors and Online Communities. *Innovations in Education and Teaching International*, 44(3), 301–320.

4.7

STRATEGIES FOR SUPPORTING PHD PRACTICE-BASED RESEARCH

The CTx ecosystem

Craig Vear, Sophy Smith, Stacie Lee Bennett-Worth

Introduction

CTx is a branch of the Institute of Creative Technologies (IOCT) at De Montfort University dedicated to the support of doctoral training, specifically in creative technologies. The staff in the IOCT share principles and an underlying philosophy when supervising and supporting doctoral researchers, most of whom are practice-based. This has led to several institutionally supported strategies that operate in a holistic way by providing a range of support mechanisms and structures that enhance the candidate's development. This has developed into an eco-system (defined below) which is built from seven different components that support doctoral research (detailed below). CTx is not an isolated initiative and has similarities with other ecosystems supporting doctoral research at institutions, such as the Creativity and Cognition Studios, University of Technology, Sydney,[1] DX Arts at Washington University,[2] i-DAT at Plymouth University,[3] SensiLab at Monash University, Australia,[4] and EMPAC at Rensselaer Polytechnic Institute, Troy, New York.[5]

The CTx ecosystem consists of seven interlinked components. These are:

1   A doctoral programme (PhD) in practice-based research
2   A practice-based research doctoral training programme
3   An online 'cookbook' of methods and practices
4   Summer school of intensive workshops and activities
5   Shared space and a taught postgraduate programme
6   Associated network of practitioners through *DAPPER* (Digital Arts Performance Practice – Emerging Research network)
7   Creative Technologists in Residence programme

Part 1: the nature of CTx

Ecosystem meaning

The term ecosystem has been adopted here to highlight the inter-twined relationships of the practice and knowledge generation of a community of researchers within, and associated with, the

   DOI: 10.4324/9780429324154-40

IOCT. Sharing the flow of research activity and practices and cross-synthesising the development and outcomes of such practices can be a positive and enriching factor in the well-being of the IOCT researchers. This community includes PhD candidates, supervisors, research professors, research fellows, masters students, professionals-in-residence, and academic leadership.

The term 'ecosystem' highlights its dynamic and inter-connected nature that grows as the researcher's practices and knowledge emerges. It is also responsive to new technologies and practices as they emerge. The use of the term was inspired and informed by Tim Ingold's writing about *entanglements* and *meshwork*, in which he states:

> I shall show that the pathways or trajectories along which improvisatory practice unfolds are not connections, nor do they describe relations between one thing and another. They are rather lines along which things continually come into being. Thus when I speak of the entanglement of things I mean this literally and precisely: not a network of connections but a meshwork of interwoven lines of growth and movement.[6]

From this perspective, the term ecosystem does not suggest a system of organic life intertwined and reliant for sustenance and support, but is used to define a system of research and practice within which the humans who conduct it are intertwined and **benefit through sustenance and support.**

Underlying philosophy

The philosophy underpinning the ecosystem of CTx is built upon three elements (explained next): first, we positively encourage and endorse creative play, messy exploration, and rapid prototyping; second, CTx aspires to operate on a flat structure and management, avoiding a top-down approach, instead leading/inspiring from bottom-up approaches of making and doing; third, Sister Corita Kent's *Ten Rules for Students and Teachers*[7] operates as inspiration for the underlying principles of CTx ecosystem.

These elements are understood as:

1 **Creative play and rapid prototyping.** This prioritises playful research and messy practice[8] approaches that highlight the importance of manifesting practice-based ideas early in the domain in which they belong: namely practice. This can circumvent any meta-physical mistakes by studying such outcomes in the wrong domain, such as brainstorming. In fact, we have adopted the term *body-storming* to draw attention to this. For example, an idea about practice in dance research should be realised in dance through *body-storming* before it can truly be evaluated, reflected upon, and felt. Our experience is that assessing such practice-based ideas as an intellectual discussion (or brainstorming) will only get the researcher so far, and may even waste time by mis-calculating or biasing certain essential aspects of the research.

To aid the rapid generation of ideas, Vear introduced a RIPA system based on these principles. RIPA stands for:

R: Rapid Generation = quick gain, small steps, user experience first
I: Iterative = unpick experience, rebuild with inductive/deductive solution, alpha test ready for participant
P: Performative = authentic environment for real-world realisation
A: Agile = 'quick and dirty' approaches prioritising the results from the practice-based research as indicators for next stage development.

RIPA is an iterative process based on four guiding frameworks: a) the agile manifesto,[9] which prioritises a versatile development process allowing new knowledge and parameters to influence the direction of the research; b) the Scrum method, which is an agile framework for developing ideas using small, time-bound, and mini-goal orientated processes; this supports the agile manifesto's aims; c) Wizard-of-Oz testing, in which complex systems believed to operate autonomously (such as computer interactivity) are first developed using unseen human operators (such as the Wizard hidden behind the curtain in the 1939 film). This way of experimenting has two main advantages: first, it is quick and cheap and gets to the heart of user-experience, thereby supporting the agile manifesto, and second, it allows the researchers to become, or embody, the role of the interactive software system, offering them deeper insights from inside the performative situation; and d) is body-storming (mentioned earlier).

2 **Flat structure and management.** This is inspired by Shackleton's leadership[10] of his crew while being stranded in Antarctica for 19 months, insofar that CTx values the role that each member plays in the well-being of the community, and the importance of each member to the sharing and enrichment of knowledge. CTx is led by example and feeds the ecosystem through motivation and ideas from all the members. Qualities such as these that follow are also encouraged and implemented thereby enriching the ecosystem through hands-on inspiration:

- Have a clear mission focus
- Improvise when needed
- Maintain a sense of emotional and social intelligence by expecting/welcoming change
- Be persistent and resilient
- Be organised and manage vital details
- Freely share knowledge with CTx and the wider community
- Learn from mistakes

3 **Sister Corita Kent's *Ten Rules for Students and Teachers*.** This operates as inspiration for the CTx ecosystem. These operate more as guiding principles than rules, or as questions to pose to oneself while researching. Many of the IOCT researchers were introduced to these while studying the American composer John Cage, who is usually attributed wrongly as the author of these rules. Nonetheless, they operate as a value-driven set of guidelines, as opposed to the normal formal rules imposed by institutional management. The rules are:[11]

a *RULE ONE: Find a place you trust, and then try trusting it for a while.*

b *RULE TWO: General duties of a student: pull everything out of your teacher; pull everything out of your fellow students.*

c *RULE THREE: General duties of a teacher: pull everything out of your students.*

d *RULE FOUR: Consider everything an experiment.*

e *RULE FIVE: Be self-disciplined: this means finding someone wise or smart and choosing to follow them. To be disciplined is to follow in a good way. To be self-disciplined is to follow in a better way.*

f *RULE SIX: Nothing is a mistake. There's no win and no fail, there's only make.*

g *RULE SEVEN: The only rule is work. If you work, it will lead to something. It's the people who do all of the work all of the time who eventually catch on to things.*

h *RULE EIGHT: Don't try to create and analyse at the same time. They're different processes.*

i *RULE NINE: Be happy whenever you can manage it. Enjoy yourself. It's lighter than you think.*

j *RULE TEN: 'We're breaking all the rules. Even our own rules. And how do we do that? By leaving plenty of room for X quantities' (John Cage).*

k *HINTS: Always be around. Come or go to everything. Always go to classes. Read anything you can get your hands on. Look at movies carefully, often. Save everything. It might come in handy later.*

Part 2: the eco-system

In this section, we will outline each of the components of the eco-system separately and include an evaluation from the student body. After this, we offer a discussion on how these are inter-linked in the day-to-day running of the IOCT, followed by a conclusion.

A doctoral programme (PhD) in practice-based research

The doctoral programme (PhD) in practice-based research (hereafter doctoral programme) is one of the central components in the CTx eco-system. It was validated by De Montfort University's (DMU) Doctoral College in 2017, with the first full cohort submitting their theses in 2021. The rationale for designing and developing this programme was to intervene with an observed problem: it was noticed by the Doctoral College that there was an increasing pattern of late submissions from practice-based researchers, and IOCT was challenged with finding a solution.

Almost all PhD candidates in the IOCT subscribe to the programme and follow the path of study. The aim of the doctoral programme is to create opportunities for artists, performers, sculptors, video makers, photographers, programmers, digital media creators, software developers, game designers, professionals, and others to discover and share formal knowledge by thinking in, through, and with their practice. Creating new works (in the broadest sense) is at the centre of all activities. The commitment from the PhD candidates is on maintaining this programme so that it supports their individual progress and needs, rather than imposing a dogmatic one-size-fits-all programme. As such, the doctoral programme approaches a PhD investigation using a principled, methodical modular approach. This involves designing and implementing a series of sub-projects that identify key research questions within each study and are time bound with strategic objectives, milestones, and deliverables. The benefit of this approach is that key outcomes and new knowledge can be evaluated at specific points in the development of the research, which in turn are prioritised for further examination in following phases.

The programme starts at the point of first contact with the prospective PhD researcher. They are required to complete an application form (see Appendix 1), in addition to the institutional requirements of the Doctoral College. This CTx application operates as:

1   A starting point for a conversation with the IOCT supervisors so they can guide and strengthen the applicant's project
2   As a means with which to start the process of thinking about practice-based research in a principled and focused way

Once accepted into the university as a PhD candidate, the doctoral programme starts immediately with a focus on practice and continues to build the stepping stones of new knowledge and insights through, with, and in their practice. The first project will typically take the form of a rapid prototype of the central hypothesis. This is a way of bringing the intellectual and theoretical stuff from inside-the-mind out into the domain where it belongs (their practice). A minor project is typically 2–6 months duration and concentrates on a specific task, method, question, practice mode, or prototype system. These are essential to breaking down and formulating novel processes and new insights towards a larger or composite problem. A major project typically lasts 6–12 months and deals with more complex problems or investigates the inter-relationships of smaller component elements that have been developed as minor projects. The major and minor projects are broadly defined and will need to be interpreted by the PhD candidates and their supervisorial team. The priority is that these projects are defined and refined iteratively through conversations with a) the supervisors and b) the ongoing evaluation of insights that inform the needs of the next steps of the research investigation.

At the start of the doctoral programme, each PhD candidate will complete a learning contract (see Appendix 2). This is a 'live' document and will be updated as new knowledge and insights gained from the research inform the next steps. It operates as a constant focus for the research questions and ensures that any ideas that emerge from the investigation of these questions are taken on board and examined; developments of the initial aims and objectives must be acknowledged, valued, and assessed through their added value or detriment.

At the end of the research programme the PhD candidates complete an *Abstract Generator* (see Appendix 4), which scaffolds the construction of a clear and focused abstract for their PhD thesis. As with all these strategies, this is a starting point for a conversation and a framework for clarity, for both their own thinking and the logic of their argument.

Major/minor project design: route map approach

Each PhD research project on the programme is designed and developed using a series of minor and major projects (see Appendix 3). They also form part of the annual and mid-year review process. These act as a route map for the research projects; as an evaluation framework for the exposition of knowledge; and as a conversation point with the supervisors. The design process is typically:

1 Conceptualisation and proposition design based on rationale (from preceding PbR findings) and context (from ongoing State of the Art Review)
2 Define research questions, aims, issues, problems, and objectives
3 Determine research strategy, methods, work plan, timeline, milestones, and deliverables; funding, dissemination, and documentation
4 Propose agile, iterative, experimental, linear of open process
5 Collate data/evidence/artefact(s)
6 Decoding against initial or emergent questions/issues/problems
7 Preliminary conclusion
8 Further problems/issues/challenges arising from project
9 Documentation, dissemination, scrutiny, sharing

Review points and deliverables

Next is a typical timeline for review points, deliverables, and milestones. The specific documents required for each review point will be negotiated between the supervisors and the candidate. Practice is an important document in this process. The main aim of these review points is to maximise on the research insights, protect against damaging creep, and support a containable, SMART approach to research project development. The goal with such a process is to ensure that the research is agile to the ongoing findings of the research project, so that any presumptions/intentions by the PhD candidate are questioned and evaluated regularly.

This is a typical timeline for full-time study; the part-time version would double the expected delivery points e.g. end of years 2, 4, 6.

Year 1

Mid-year review:

Learning contract
Evaluation of initial project (first prototype)
Outline draft of State of the Art Review

End of year review:

Complete draft of State of the Art Review
Evaluation of projects from Y1
Design of project route-maps for Y2

Year 2

Mid-year review:

Evaluation of projects from Y2
Design of project route maps for next projects
Draft methodology chapter

End of year review:

Draft exposition for each completed PbR projects this year
Draft introduction and preliminary studies chapters
Draft methodology chapter

Year 3

Mid-year review:

Draft 'New Studies' chapter
Evaluation from projects from Y3
Design of project route map for final round of projects (consolidation)

End of year review:

Draft exposition for each completed PbR projects this year
Draft 'Results' chapter and preliminary conclusions

Value to the PhD candidates[12]

Overall, the Doctoral Practice-Based Research Programme is designed to be flexible; but equally meant to develop a systematic and principled approach to practice-based research offering us (the PhD candidates) a frame that prioritises regular reachable goals and breaks down the PhD study into component parts. The programme suggests that each component is designed with clear objectives, methods, milestones, and deliverables and should be regarded as a framework for success, providing the PhD candidate with a sense of completion after each major and minor project. This route allows the researcher to evaluate new insights at appropriate points and make informed decisions towards the next steps of the research. Another positive outcome from this approach is that the PhD candidate gets a sense of achievement, with each project providing a step towards their research goals, helping to avoid mid-term lulls and the sometimes-negative sense of 'where is this going?' Above all, this process offers the PhD candidate an experience in the rigorous process of systematic research design that underpins post-doc academic research and scholarship.

A practice-based research doctoral training programme

IOCT and the Doctoral College at De Montfort University (DMU) conduct a practice-based research doctoral training programme (hereafter DTP) to support practice-based research across all four faculties of the university. The DTP was developed in 2012 and rolled out across the university from 2013.

DMU has been recognised since the 1970s across the globe for pioneering practice-based research, including the development of high-quality doctoral training programmes in this field. Practice-based study at DMU involves a broad category of research that can include designing or making objects, the staging of performances or events, and the documentation of artefacts. The definition of an artefact is broad, ranging from a physical entity, such as a painting, a novel, or a computer program, to a transient entity such as a performance. The experience and expertise of academics associated with the training programme range across a breadth of practice-based disciplines, including digital arts, creative writing, music, computer game design, and performance.

The DTP is designed to meet the needs of PhD candidates at different stages of their PhD journey. Over the past decade of delivering the DTP, the IOCT have developed a model where the practice-based researchers meet four times a year for a day-long session, supported by two intensive residential sessions. This gives the PhD candidates a focal point for their project timelines and maximises the general benefit of these events. Previously, the programme included monthly events, but it was found that the focus of such events wasn't always relevant to the needs of most candidates at those particular points in time (of the academic year and their progression of their PhD research).

The current model is a year-long programme that is structured around three key components – *Playground(ed)*, *Principles of Play*, and *Practice Cookbook*.

1   *Playground(ed)* sessions run as practical workshops, where PhD candidates investigate and interrogate aspects of their practice through a range of creative activities, including Lego, drawing, haiku writing, and rapid prototyping. These sessions offer the participants different perspectives on their doctoral practice, leading to new reflections and realisations. In addition, PhD candidates have the opportunity to share their developing practice in a scaffolded, critically supportive environment and are encouraged to use these as an opportunity to gain feedback or alternative perspectives from the wider group.

2   *The Principles of Play* sessions are participatory workshops, where PhD candidates bring their own experience to reflect on the 4 Ps:

   a   Process (research methodologies in practice)
   b   People (the range of ways that other people may be involved in the practice-based research)
   c   Product (thinking about the practise of practice and how their creative artefacts will be presented)
   d   Proof (collecting and collating documentation).
       These sessions include input from practice-based research alumni who reflect on how these principles manifested themselves in their own doctoral journeys.

3   *Practice Cookbook* (see later in this chapter). These are sessions that foreground the experiences of alumni who have been awarded PhDs, to demonstrate the variety of approaches that can lead to a successful submission. Rather than present a 'tidy' experience, speakers are encouraged to present an honest and accurate 'warts and all' picture of their doctoral experience. As well as including their experiences of the 4 Ps covered in the *Principles of Play* session, speakers are encouraged to discuss other equally important aspects such as the supervisory relationship, work–life balance, coping strategies etc. This component is supported by an online resource,

the *Practice-Based Research Cookbook*, which offers further insights into a specific practice-based PhD thesis, through interviews, documentation, and completed thesis from alumni who have successfully completed.

The workshops are designed to encourage peer support and learning, across what can be up to an 8-year journey. In *Playground(ed)* sessions, work is shared from those just starting out their PhD journey, to those preparing for their viva. Participants are encouraged to share works-in-progress in an atmosphere of openness and generosity, giving participants a sense that their work has value, and the confidence to show work that is unfinished. Workshops within the *Principles of Play* sessions are based on loose themes, for example 'research design', 'documentation', 'the role of the audience', 'journal writing', and 'presenting your artefact'. Activities are task-based differentiated by outcome, allowing meaningful engagement across all experience levels. This offers a rich resource where PhD candidates reflect upon their own experiences in relation to those of others, learn new techniques and tips, and are able to discover strategies from a range of disciplines and approaches.

Value to the PhD candidates

The practice-based research doctoral training programme (DTP) is a system for navigating the PhD process, offering tools, methods. and approaches to help us progress and achieve meaningful milestones at a pace that suits our research. The DTP gives the candidate access to multi-modal workshops and interdisciplinary perspectives that help them to develop the appropriate skills necessary to test, iterate, and share work within a non-judgemental, critically engaged environment. The programme presents opportunities for PhD candidates to connect with ideas outside of their field in a way that brings a new focus to their projects, through the proposition of alternative methods, new modes of practice, and critical reflection. The programme offers space for peer-to-peer discussion and conversation among researchers at varying stages of the process, presenting possibilities for the PhD candidate to extend their support system and research community. The DTP has included monthly meetups on the request of candidates who enjoy the regular anchor points throughout the year; however, these had irregular attendance and a stronger core group of researchers seemed to be formed when the meetups were less frequent but more focused.

For distance-learners, these semi-regular meet-ups, for face-to-face conversation and contact with the DTP research community, have proved extremely valuable, particularly in enabling them to feel connected and supported in what could otherwise be a potentially isolated process. These in-person meetups are also supported by regular activity in online spaces such as the CTx Facebook group and online webinars. The combination of synchronous workshops and events paired with asynchronous activities offered digitally forms a sound basis for any PhD candidate to develop new skills and form meaningful relationships, both of which provide a great framework of support throughout the peaks and troughs of the PhD research process.

Candidates have reported that they found the peer-support and possibility of a testing ground to prototype new ideas an invaluable tool to their progression and well-being within the process. As more candidates begin to explore the possibilities offered by the programme, more opportunities are presented for the cohort to generate additional strategies that could provide more bespoke testing grounds specific to certain subject areas. This tends to occur quite informally within the present structure; however, current candidates have reported that more formal developments of these kind of micro-communities (such as music-focused sessions) would contribute to their own development and help them locate themselves within their specific field. The DTP community has been described as 'extremely welcoming, proactive and forward thinking', with candidates expressing how the programme helped offer a sense of place in the academic world.

An online 'cookbook' of methods and practices

As part of the doctoral training programme (DTP), CTx developed an openly available web resource, the *Practice-Based Research Cookbook* (*PbR Cookbook* for short). It was designed and populated 2017–19 and is being augmented on an on-going basis. Here successful PhD candidates talk about their journey to thesis submission, including the relationship between practice and theory, telling the story and the presentation of practice. This resource addresses one of the main difficulties faced by practice-based research doctoral candidates: that no two submissions look the same. We have, over the past 16 years, found that each one differs in their reflection on the different artefacts submitted and the different way that the new knowledge is demonstrated/exposed in the written text. This can cause much stress and confusion among PhD candidates who are unsure how to present their work and feel unable to make informed decisions about how best to demonstrate their original contribution to knowledge. While practice-based research PhDs have been in existence of the last couple of decades, it is still not easy for PhD candidates to access successfully completed theses. Practice-based research submissions are notoriously difficult to access and PhD candidates are desperate to see how others have approach the challenge of artefact/thesis presentation. The *PbR Cookbook* enhances current research training programmes through the fulfilment of this need. The website URL is http://pbrcookbook.com.

The *PbR Cookbook* web resource takes the form of a web site that offers real case studies from successful practice-based research, presented in a way to break down the mystique that surrounds these submissions. The resource is structured into three sections:

- *The Main Course*, including abstracts, full thesis, and links to the artefacts
- *Recipes*, including interviews with the candidates reflecting on their own journeys towards thesis submission, the viva experience, and beyond
- *Ingredients*, where interviews are broken down into key themes, including the development of a methodology, documentation of the creative process, reflective practice, the relationship between practice and theory, how to 'tell the story', and the presentation of practice

Value to the PhD candidates

The *PbR Cookbook* is an online resource for practice-based research doctoral candidates that shares the stories of past successful PhD candidates as they reflect on their PhD journey. The interviews provide insight into the broad ranging possibilities for, and unique forms of, practice-based research PhD projects and give us an inside glimpse into the very personal experiences of life as a PhD researcher at DMU. Featured alumni said that they found being part of the cookbook a positive experience, allowing them to rediscover their PhD process with a new focus. Current candidates felt that the invitation to preserve the legacy of such projects in this way was a positive indicator of a larger support network that doesn't simply cease after submission.

Overall, the *PbR Cookbook* is a useful companion for PhD candidates to refer back to throughout their studies, offering audio-visual content as well as written commentaries and links to other resources that can be revisited when needed. Candidates have stated that the resource has provided guidance through challenging moments of their studies and they particularly enjoyed the celebration of non-formulaic approaches to practice-based PhD studies and outcomes. Others have said they used the cookbook to promote the department and share the work of successful candidates to friends and colleagues interested in joining the programme. The cookbook is currently in its first iteration; however, as the resource continues to grow, users have suggested that audio quality particularly could be improved for ease of listening and interviews could potentially be formatted into 'podcasts' as well as being available as a video.

Summer school of intensive workshops and activities

CTx rolls out a series of intensive workshops and activities across the year. These can take the form of a creative technologist-in-residence (see later in this chapter) leading a technical skill-based activity for ½ day; the *DAPPer* network (see later in this chapter) leading a full day workshop on a theme; or an ad-hoc rapid-prototyping unit exploring emerging research opportunities.

Central to the academic year's activities is the intensive two-day workshop. The first was launched in 2018 but recently have been postponed due to COVID restrictions. The priority of this workshop is to share practice-based research, including practice, thinking, and writing with other practice-based doctoral candidates at DMU and across the region. The activities across the two days include talks, discussion groups, workshops, and practice sharing and features a keynote lecture from internationally prominent figures in practice-based research, discussion groups about the nature of practice-based research such as the role of the artefact in knowledge generation, and workshops around important approaches and novel methods such as dramaturgy, auto-ethnography, play, or ethics.

Crucially, the intensive has been designed to share, discuss, test, and evaluate practice-based research in a critically supportive and enhancing environment. It is scheduled at a point in the academic year that offers a focal point and milestone for deliverables associated with the major and minor project structure of the doctoral programme (PhD) in practice-based research (discussed above).

Value to the PhD candidates

The balance between new approaches and practice-sharing allowed the CTx PhD community time to reflect on our research and find new methods and approaches to understanding both the process and possible outcomes of their practice-based research. All candidates sharing their work have commented on how the workshops are 'rich, inspiring and stimulating',[13] and how they are a useful opportunity to user test 'first iteration and to gather data which can be analysed to develop the project further'.

Being immersed in an intensive workshop of this design also changes the nature of the community's inter-relationships beyond those forged through the day-to-day collegiate-ness of the CTx ecosystem and the shared lab spaces. It seems to bring a deeper, richer sense of the humans behind the research activities as individuals spend a couple of days together. This length of time together means that members of the community can be more engaged and focused, supporting a generosity and curiousness about each other's work and working processes. The combination of sharing work, data collection, and learning new methods has been valuable to the PhD candidates and helps us consider refinements to their data collection process. It also exposes alternative ways that questions can be posed in the future in order to gather rich data and complement the experiential work.

Postgraduate taught programme and a shared learning/maker space

The Creative Technologies MA/MSc was established in 2006. It is central to the IOCT and the only taught programme offered. The master's programmes were set up as innovative and transformational initiatives that cross traditional subject disciplines and enable students to work at the convergence of technology and creative practice. This mirrored the organisation of the IOCT, which was established as a pan-disciplinary unit spanning across the whole university, deliberately side-stepping faculty-focused bias and silo mentality. Students may be technologists with creative dimensions, artists working with technologies, designers with programming skills, or a combination of other traditional disciplines. As such, the course is designed to give all these types of students the vision to learn and be in tune with contemporary creative technologies. These MA/MSc students experience a multi-disciplinary approach to learning and research, bringing together eScience, digital arts and design, and digital humanities in a way that aims to cross traditional disciplines and boundaries, encouraging innovation and developing new modes of collaboration.

Crucially to the CTx ecosystem MA students, PhD candidates, and Institute staff all work within a shared space. This is the dedicated central teaching space, which also operates as a research lab for all members of CTx and IOCT. There are also satellite lab spaces (such as the multi-parametric tracking lab) that are available to CTx researchers. The IOCT lab, however, is in regular use for a range of activities and is the gathering space and 'hang-out' for all members. When taught classes are scheduled, PhD candidates may remain (but must respect the need of the learners); similarly, lunch meetings can occur around practical experiments.

It is a non-hierarchical, inclusive space that welcomes the community. No priority is given to any one programme or individual needs. In fact, collaboration, help, assistance, communication, nurturing, and intertwining of interests are positively encouraged. PhD candidates will also teach on the master's programme, and master's students will assist with PhD research.

Value to the PhD candidates

Within this environment, the richness of an individual's knowledge and practice can be seen to be truly valuable for all who engage. This, in turn, encourages self-development of one's own projects and a keenness to share this with others. The intertwining of the ecosystem also enables students to transition from taught to research provision and to consider their Creative Technologies MA/MSc projects as potential PhD propositions as they gain a deeper understanding of the PhD journey and the holistic experience. Overall, it encourages deeper levels of peer support and critical friendship, which adds depth to MA studies and can avoid the width-over-depth prioritisation common at this level.

Associated network of practitioners through DAPPer Digital Arts Performance Practice – Emerging Research network

DAPPer is a practice-based research network and incubator space where people working in all areas of digital performance come together to capture, share, discuss, experiment, and develop work and ideas relating to digital art and performance. The membership includes professional practitioners, technologists, academics, organisations, and PhD researchers. It is DAPPer's contention that while many individuals work within their own specialist area or sector, innovation occurs when we have the opportunity to collaborate and cooperate with others. DAPPer aims to provide a regular meeting space to focus on and interrogate the range of inter/transdisciplinary approaches specifically from the perspective of artistic process and practice; and a main component of this is knowledge exchange. A key presence in this knowledge exchange process are the PhD candidates from CTx, and importantly, this knowledge exchange is bi-directional. Not only are CTx candidates exposed to cutting edge R&D within external cultural and commercial organisations, but CTx candidates also have the opportunity to directly showcase their research, demonstrating to them the value of their research to sectors beyond academia.

Knowledge transfer events such as SIVE (Storytelling in Virtual Environments) in 2018 and Risky Play (an R&D prototyping space for external organisations) in 2019/20 placed CTx candidates at the centre of knowledge exchange. Through participation in these initiatives, candidates were able to appreciate the value of the research process and witness first-hand how, through collaborations with other researchers, professionals and organisations are able to develop truly groundbreaking work that would not be possible in primarily commercial environments. For example, Paul Mowbray, Creative Director for the award-winning NSC creative immersive storytelling studios, commented how the events enabled him 'to open [his] mind to other ways to generate and develop ideas for immersive projects', and Luke Richie identified that what the Philharmonia required was 'time and space to mess around, make mistakes and try things out'. Relationships forged through DAPPer have led to a number of CTx candidates being directly involved in live funded research projects, providing them with vital experience as early career researchers as well as increasing the impact of their own research.

Value to the PhD candidates

CTx PhD candidates gain an understanding of research, especially with technological innovation as a focus within a wider context of the creative industries. They also gain an insight into the professionalism and priorities of creative technology practitioners, while exploring the relevance of their research and practice within the broader world of the creative industries. This can in turn lead to a potential increase in our understanding of impact. The benefits for us are in the development of relationships and collaborations across sectors, which has led to successful explorations of interests of benefit to both parties and to a vital exchange of knowledge between sectors. Candidates reported that the connection with DAPPer's range of artists, technologists, and practitioners was extremely helpful for networking and for engaging more widely within the field. Events were described as inspirational and seen to offer a sense of affirmation of the validity of practice-based research and its application in the 'real' world.

Creative Technologists in Residence programme

As an extension to the DAPPer activities, CTx designed the *Creative Technologists in Residence* scheme, which offers creative technologists an opportunity for a concentrated period of original research and development supported by the Institute. The first residents started in 2019, with activities moved online following COVID restrictions.

The term 'creative technologists' encompasses anyone making, building, designing, developing in a creative way through or with technology. This can include professional artists, architects, coders, dancers, designers, engineers, hackers, musicians, sculptors etc. across many disciplines and interdisciplinary practices. The creative technologists are given the status of research fellow in IOCT and have access to the facilities and lab-spaces. This includes access to technical support, staff consultations, and the body of student and post-grad researchers. The scheme is designed for creative technologists to be autonomous with an original project, but to have an opportunity to engage with the culture and community of IOCT.

The roles of the creative technologists are to:

- Conduct a programme of original research and development that advances IOCT's vision and pushes cultural horizons
- Contribute to the research environment of the IOCT through engagement with the taught and research programmes
- Commit to developing a follow-on funding bid involving the IOCT in some way (e.g. PhD scholarship, Arts Council grant, research fellowship)
- Contribute to an IOCT ongoing research project investigating how creative technologists make practice-based work with creative technology

The first fellowships were awarded to Indira Knight[14] (a creative technologist and developer with a background in 2D and 3D animation, XR, emerging technologies, and data visualisation) and Ben Neal[15] (creative technologist and digital producer specialising in VR and performance at the Psicon

Lab). Since he began his creative technologist role, Ben has become involved in a number of funded research projects at the IOCT.

Value to the PhD candidates

Aside from being involved as participants in the creative technologists' research programme, the CTx community gains two specific benefits from their presence: first, we are generally invited in as consultants to the creative technologists' projects, which gives the PhD candidates, and some invited master's students, confidence in the value of their research. Individual research can, for the most part, be a lonely solo pursuit, not really appreciating its true value until the viva when someone from outside their community is able to study their thesis.[16] Working alongside professional creative technologists, who may have a PhD or not, offers a feeling of relevance to the world outside CTx and DMU. Secondly, the working processes and attitudes of professional creative technologists inside the research institute highlight the importance and relevancy of the CTx philosophy and the cruciality of the major-minor project design of the doctoral programme (PhD) in practice-based research. This give us strength of conviction that these processes and methods are relevant to professional and post-doc work, and generally lead to successful outcomes.

How do the different parts of the ecosystem work together?

In an ecosystem, individual organisms work together to form a 'bubble of life', each, either directly or indirectly, dependent on the other. The CTx 'bubble of life' is discussed next and relies on the seven interlinked components introduced earlier. The ecosystem has developed a shared community space across physical, virtual, and conceptual realms, which has two key characteristics:

- Firstly, the way in which the components support interaction across four inter-relationships:

  - Practice ~ reflection
  - Individual ~ community
  - Internal ~ external
  - Individual research ~ wider sector

- Secondly, the phased approach to component delivery

The first characteristic enables participants to position themselves and their research on a flexible continuum across practice and reflection (e.g. the PhD programme and the DTP), the individual and the community (PB Cookbook and Shared Space), internal and external focus (summer schools and Creative Technologist in Residence), and individual research and the wider sector (DAPPer).

The second characteristic of the ecosystem means delivery of the components is designed around overlapping time frames with the different components allowing the PhD candidate to access support that relates to their needs at any particular point within their PhD journey. As such, each component runs over a different timescale, with some working across the four years, others working annually, and some looping over a matter of months. As the seven components run simultaneously but unsynchronised, this results in a phasing effect, where the relationship between the seven components is constantly shifting. If a standard four-year PhD journey was used as an example, the doctoral programme itself would roll out over four years. Within the ecosystem, this is supported by the doctoral training programme, which is designed on a year-long pattern, with content changing annually. Summer schools occur across July and August, with the Creative Technologists in Residence being active from February to July. Both the PBR Cookbook and the Shared Space are available throughout the four years, with external and internal research projects operating within.

DAPPer initiatives are dotted over the four years, with perhaps 2–3 opportunities annually. This approach supports the PhD candidate to find a pattern of engagement that suits their individual progress, while enabling them to feel part of a wider community.

Conclusion

We have observed that the ecosystem approach enriches practice-based research for the doctoral candidate. This approach values knowledge and people regardless of status or appointment. It presents models of practice-based research across a range of sectors and applications and values the common principles that bind these as research. The sharing of core philosophies throughout the team (faculty, fellows, master's students, PhD candidates etc.) leads to a culture of growth and positiveness, ownership and authority of one's own knowledge rather than a siloed mentality that pervades some of the graduate school industry. While the organisers of CTx acknowledge that there is always room for improvement, the responses from the PhD cohort are positive and encouraging. Critically, they say that they feel part of a much larger flow of knowledge generation through practice and creative technology, and 'plugged into' a world outside the invisible walls of academia.

Notes

1 www.creativityandcognition.com/.
2 https://dxarts.washington.edu/.
3 https://i-dat.org/.
4 https://sensilab.monash.edu/.
5 https://empac.rpi.edu/.
6 Ingold (2010).
7 Kent and Steward (2008).
8 See Smith (2022).
9 Further information can be found at https://agilemanifesto.org/principles.html (accessed February 1, 2021).
10 Craig Vear spent three months working as composer-in-residence with the British Antarctic Survey over the Austral summer of 2003/4. Here he experienced the Shackleton way and the influence of a flat management structure.
11 Kent and Steward (2008).
12 Each element of the eco-system is concluded by a *Value to the PhD Candidate* section. This is a compilation of anonymised responses from an open questionnaire that was distributed across the cohort and alumni of IOCT PhD researchers. The responses were collated by Stacie Lee Bennet-Worth, who is currently in her final year of PhD research in CTx, and included in this chapter. No staff members edited these or manipulated the results.
13 Taken from informal feedback from 2018.
14 http://indiraknight.com.
15 www.psiconlab.co.uk.
16 A comparable feeling can be gained by attending and presenting at academic conferences.

Bibliography

Ingold, T. (2010). *Bringing Things to Life: Creative Entanglements in a World of Materials*. University of Aberdeen, July. http://eprints.ncrm.ac.uk/1306/1/0510_creative_entanglements.pdf (accessed February 1, 2021).
Kent, C., & Steward, J. (2008). *Learning by Heart: Teachings to Free the Creative Spirit*. New York: Allworth Press.
Smith, S. (2022). A Play Space for Practice Based PhD Research. In C. Vear (Ed.), *The Routledge International Handbook of Practice-Based Research*. London and New York: Routledge.

Appendix 1
CTX DOCTORAL (PHD) PRACTICE-BASED RESEARCH PROGRAMME

Framework for developing research proposal

Breaking down your research idea

1 Your aim – *I seek to find new knowledge/ insights/ etc in* . . . (simple statement of your intent)
I seek to find new knowledge/ insights/ in. . . . .

2 Problem/gap definition (state of the art) – *this is the gap in knowledge, which I aim to address.*

3 Problem/gap solution – *having identified this gap, this is how I intend to deal with it.*

4 Preparedness – *this is why I am the perfect person to conduct this.*

5 Your hypothesis and research questions – *having identified the gap, and a solution, this is my hypothesis.*

6 Objectives – *these are my stepping stones to deliver my hypothesis* (bullet point the key objectives).

7 Methodology – *how do you intend to solve this problem, and why is this the most suitable solution?*

8 Method – *what tools and techniques will you employ for seeking this new knowledge.*

9 Analytical/evaluation frameworks – *how are you going to turn the data/evidence/practice/stuff from your method into knowledge. Why is this the best way?*

10 What do you intend to produce as your practice-based outputs – *what will be in your box when you submit?*

11 What is your proposed OUTLINE timeline?

12 Academic and scientific reach/ impact. Who benefits 3-D perspective: a) subject significance, b) reach/ influence into other disciplines, c) impact into PESTLES (politics, economic, societal, technology, legislative, environmental, scale) – *give it a go; try to identify some beneficiaries from each of the 3-D.*

Appendix 2
CTX DOCTORAL (PHD) PRACTICE-BASED RESEARCH PROGRAMME

Learning contract

'One Goal; Different Routes' (http://superprofdoc.eu)

Date completed

Details

Name of PhD Candidate:

Provisional Title of PhD: Supervisors:

Research Goals

Aim – what knowledge to you seek in this PhD study? SINGLE AIM!

Objectives – what do you need to do in order to reach your aim?

Outcomes – what do you aim to create through the course of your PhD? Do you envisage a box of media? Website? What size will your thesis be (50% @ 40,000 words?) How does it help articulate your aim and objectives?

Implementation

Timeline – what is your timetable to achieve your outputs/objectives in order to reach your aim (including documentation and critical evaluation/ refection)?

Milestones – what milestones do you need in order to critically move your project forward? When do these happen?

Management

Supervisors' Engagement – when do you need to see your supervisors in order to most efficiently and effectively manage your project with their support/ expertise? How do these meetings support your timeline? Who reaches out to whom?

Supervisors Role – each expert supervisor can have different roles in this project. What would you like to propose these are? How does this enable you to manage their allotted time as expert Supervisor on your project?

Mode of Contact – face-to-face; written; Skype. Which is best to maximise their allotted time, and your management of your project?

Style of Contact – how would you like the expert–supervisor to support you, and how does this link with your project support structure/ route plan?

- Pastoral (e.g. emotional and spiritual support)
- Directive (e.g. 'master ~ slave')
- Didactic (e.g. issuing best practice solutions)
- Critical Friend (e.g. encouraging and support while offering honest/candid feedback)
- Teacher Learner (e.g. drawing out the process of learning and promoting your desire to learn)
- Dialogic (e.g. presenting contrary arguments to open critical thinking)
- Collegiate (e.g. supportive discussions and shared values)

Supervisor's section

End of Year 1 checklist (FT; PT=end of Year 2):

- Complete draft of Lit Review chapter
- Written evaluation for each major/minor projects from Y1
- Design of project route-maps for Y2 (one per project)

End of Year 2 checklist (FT; PT=end of Year 4):

- Draft exposition for each completed PbR projects this year
- Draft introduction and preliminary studies chapters
- Draft methodology chapter
- Design of project route-maps for Y3 (one per project)

End of Year 3 checklist (FT; PT=end of Year 6):

- Draft exposition for each completed PbR projects this year
- Draft 'Results' chapter and preliminary conclusions

Supervisor's Signature

Please confirm that you

- have been consulted and agree with this learning contract, and
- have discussed the outline work and route plans for the major/minor projects for this coming year, and
- can confirm that the work from the previous year is up-to-date or on-track (see checklists)

Signature:
Date:

Appendix 3
CTX DOCTORAL (PHD) PRACTICE-BASED RESEARCH PROGRAMME

Route plan for major/minor PbR projects

- Pre-project planning
- One per project
- Placing the expert and the expertise in the driving seat of their research. Further info at http://superprofdoc.eu
- 'I'd rather a roadmap for success than a crash report any day of the week'

Project goals

Aim – what knowledge to you seek in this project? (linked to learning contract details). SINGLE AIM!

Objectives – what do you need to do in order to reach your project aim?

Outcomes – what do you aim to create through the course of your project, and how does it help articulate your aim and objectives?

Implementation

Timeline – what is your timetable to achieve my outputs, objectives in order to reach your aim (including documentation and critical evaluation/ refection)

Milestones – what milestones to you need in order to critically move your project forward? When do these happen?

Deliverables – what do you need to deliver in order to be on track?

Management

Supervisor Engagement – when do you need to see your supervisor in order to most efficiently and effectively manage your project with their support/expertise? How do these meetings support your timeline?

Supervisor Role – each expert supervisor can have different roles in this project. What would you like to propose these are? How does this enable you to manage their allotted time as expert supervisor on your project?

Mode of Contact – face-to-face; written; Skype. Which is best to maximise their allotted time, and your management of your project?

Style of Contact – How would you like the expert–supervisor to support you, and how does this link with your project support structure/ route plan?

- Pastoral (e.g. emotional and spiritual support)
- Directive (e.g. 'master ~ slave')
- Didactic (e.g. issuing best practice solutions)
- Critical Friend (e.g. encouraging and support while offering honest/candid feedback)
- Teacher Learner (e.g. drawing out the process of learning and promoting your desire to learn)
- Dialogic (e.g. presenting contrary arguments to open critical thinking)
- Collegiate (e.g. supportive discussions and shared values)

Contingency – what are your plans if things don't go to plan?

Ethics – do you need to seek ethical approval?

Appendix 4
CTX DOCTORAL (PHD) PRACTICE-BASED RESEARCH PROGRAMME

Framework for developing your thesis abstract

Part 1: breaking down your research claim

1 Your claim to new knowledge. (Simple statement of your intent)
My claim to new knowledge is . . .

---

2 Problem/gap definition (state of the art/ literature review). *This is the gap in knowledge, which I have addressed.*
Or . . . *These are the boundaries of known knowledge, which defines a gap which I have solved.* Be specific; make a clear statement about what the gap is (was).

---

3 A short statement about your hypothesis – *having identified the gap, this was my hypothesis (pathway to a solution)*

---

4 Research questions. *List your research questions that framed your research*

---

5 Problem/gap solution – *having identified this gap, and outlined research questions, this is my solution.* (Describe your framework/ theory, philosophy, model, portfolio etc.)

---

6 Foundation work – *describe earlier work that you have done that provided a foundation to this PhD, or your significant background experience.*

Breaking down your Method *and research design*

7 Methodology — *describe the methods you choose to use to solve this problem, and why were these the most suitable solution?*

8 My method — *describe how you applied these tools and techniques in your overall research design. This was my method.*

9 Analytical/evaluation frameworks — *how did you turn the data/evidence/practice/stuff from your method into knowledge. Why was this the best way?*

10 Objectives and design — *what were your key stepping stones to conducting your research.* (Bullet point the key objectives and structure.)

11 What was your role through your research design?

Breaking down your findings

12 Outline the key studies that you conducted to support your new knowledge generation process. *These are the studies that I conducted that directly informed the generation of new knowledge.*

13 What were the conclusions that you got from these? *The critical and novel results from these studies were . . .*

14 List your conclusions and how you got there. *Together, the results from the above helped me conclude/ evaluate (give value to) my claim.*

15 What did you produce as your practice-based outputs — *what is in your box when you submit?*

16 Academic and scientific reach/ impact. Who benefits 3-D perspective:

a Subject significance
b Reach/influence into other disciplines
c Impact into PESTLES (politics, economic, societal, technology, legislative, environmental, scale) — *give it a go, try to identify some beneficiaries from each of the 3-D.*

17 Future studies. *Now that I have completed my PhD research, this is what the next steps of development could be . . .*

4.8

ETHICS THROUGH AN EMPATHETIC LENS

A human-centred approach to ethics in practice-based research

Falk Hübner

Introduction

At the outset, it is important to define the specific focus of this chapter. Ethics is broadly concerned with questions of value and moral, distinguishing between what is good and bad, what is right and wrong. This includes several sub-categories such as meta-ethics, normative ethics and applied ethics. The focus of this chapter is largely in the field of normative ethics, concerning 'questions such as: What kinds of actions are right or wrong? . . . What is the basic matter of moral concern? And what are the fundamental or basic moral truths?'[1] However, the focus here is not so much a philosophical one, but rather to do with the behaviour of researchers predominantly in relation to other people, or to other human and non-human entities in general.[2] This involves questions such as: How are we behaving in relation to others, in an ethical way? How can we develop 'ethical antennas'[3] and, as Raymond Madden puts it, take care of 'being ethical in the world'?[4] As one possible answer to such questions I propose an ethical stance that is more empathetic and positive, rather than seeing ethics as a hindrance, by doing right to all present.

In the field of research ethics, this chapter is centred on five behaviours as suggested by social scientist Uwe Flick:[5] *being pushy*, *being ignorant*, *being accurate*, *being fair* and *being confidential*. These points will help frame some of the difficulties that emerge through the often hybrid professional identity of the practitioner researcher. This hybridity of role means she often has to fulfil even more roles in relation to clients, other parties or contexts, alongside the research and the practice. The chapter aims to raise awareness on a number of ethical issues relevant to a variety of settings, with the aim of motivating the individual researcher-practitioner to take initiative to actively 'transpose' these issues and steps to her own context.

All practitioner researchers should obey and follow ethical guidelines. However, it is important to understand that these guidelines and ethical protocols are particularly strong in the context of academic and institutional research situations, where they are an absolute requirement. Due to the multidisciplinary nature of this publication, which aims to include all kinds of practice-based research, it is challenging to provide a detailed guide to research ethics that can directly be applied to a specific discipline.

Traditionally, research ethics are a particularly important aspect of fields such as medical research and the social sciences, where patients or participants 'should be protected from researchers who might be tempted to use any means available to advance the state of knowledge on a given topic'.[6] Essentially, the interests of the people who are involved in the research must be protected, and it must

DOI: 10.4324/9780429324154-41

be ensured that no one is potentially harmed by the inquiry. Next to minimising the risk of harm, general principles include protecting anonymity and confidentiality, obtaining informed consent, avoiding deceptive practices and giving participants the right to withdraw from a project.

Adherence to this requirement is commonly achieved at universities or other research institutions by a procedure in which research proposals need to be first approved by an ethics committee. These are often complex and lengthy procedures, sometimes taking weeks or even months before approval is given to begin the research, and mostly involve assessing the primary data collection (surveys, observations or interviews). Social researcher Martyn Denscombe offers several possible scenarios in which it would become necessary for researchers to be assessed by an ethics committee. These include:

- vulnerable groups such as children or patients
- sensitive topics such as illegal behaviour or illness
- deception or carried out without the informed consent of participants
- use of confidential information about identifiable individuals
- processes that might cause psychological stress or more than minimal pain
- intrusive interventions (e.g. administration of drugs or vigorous physical exercise that would not be part of participants' normal lives)[7]

Costley and Fulton mention a number of key ethical principles as outlined by the Economic and Social Research Council:

- Research should aim to maximise the benefit for individuals and society and minimise risk and harm.
- The rights and dignity of individuals and groups should be respected at all times.
- Wherever possible, participation should be voluntary, consensual, and appropriately informed.
- Research should be conducted with integrity and transparency.
- Lines of responsibility and accountability should be clearly defined.[8]

These key principles are certainly subject to the discussions in ethics committees. A good example of how such lengthy institutional ethics procedures can be organised more effectively is the work of the Creativity and Cognition Studios (CCS) at the University of Technology Sydney (UTS). The CCS research group has developed a process based on an application to the ethics committee in order to get a 'programme clearance'.[9] This application thoroughly describes the group's kinds of research activities (such as interviews, focus groups or surveys), the topics under study and the processes that are followed, such as how participants are made to be non-identifiable or the template for informed consent. Essentially this concerns transparency, informed consent and the right for participants to withdraw from participation at any point, from data collection to publication. The ethics committee has approved the group's standard operating procedures and methods, and from there the individual project applications can be sent to an internal committee in the group. These applications are much shorter (as they fall under the overall framework of the group), which expedites this process from three to four months to a couple of days, with a standard form of just a few pages. Anything that is more complex or more controversial than these standard procedures still needs to go through the normal ethics application procedure.[10]

As necessary as ethics committees are, their approval of research projects can never guarantee that it, and all of its operations, are carried out in an ethical way. Ethnographer Raymond Madden remarks that:

> while these formal approval processes are indeed about ethics at some level, they are also about managing 'risk' and avoiding the commissioning institution becoming liable to legal action as a consequence of the behaviours or research practices of an employed researcher.[11]

As important as the avoidance of institutional risk is, I agree with Madden that most ethical choices are made and carried out quite performatively in the moment of the here and now, 'on the ground' of the research-in-action and by the researchers themselves.

Regarding ethics committees, there are two further points that are worth discussing:

First, research ethics never exist in isolation, but rather in context of a particular time and culture. As Helen Kara (2018) argues, comparing the ethics of the Euro-Western research paradigm next to the ethics of the Indigenous research paradigm can show that 'research ethics' is not a single or a universal approach in either paradigm'.[12] An even stronger argument, as I will elaborate later, is that ethics are intimately and inseparably entangled with methodology and the research process. Or, as physicist and philosopher Karen Barad puts it, 'questions of ethics and justice are always already threaded through the very fabric of the world. . . . Epistemology, ontology, and ethics are inseparable'.[13]

Second, despite the ethics approval having an operative function in many research projects and institutions, the view represented here is that the researcher constantly needs to pay attention to ethical issues and 'stay awake' with regard to ethics and act empathetically, be it in the preparatory stage, in the field or while writing up the final account. This includes aspects of taking time, care and paying attention to all human and non-human entities[14] in a research project. Kara simply calls this 'the real world', in which 'ethical research requires an ongoing and active engagement with people and the environment around us'.[15] In the same spirit, this chapter takes a quite concrete approach, which is to raise a number of thoughts by the practitioner researcher, to consider before, during and after carrying out a research inquiry.

One could assume that research ethics are of smaller importance in practice-based research than in other research strands, as much practice-based research is predominantly centred on the practice of the practitioners themselves. Furthermore, they are at the same time the ones who carry out the research and write about it. However, other humans and non-human entities are quite quickly and regularly involved in one or another way: for example, people working in the context of inquiry or acting as participants; the spaces, organisations and institutions in which the researcher works; or, as Cesar Baio and Lucy Solomon explain in their chapter, technology, other species or non-living organisms.[16]

From this ethical perspective, it is important to understand that as soon as research happens in and through the researcher's practice, it is likely that she is close to others in this field of practice. Therefore 'one must consider the ethical dimensions of being close to others'.[17] This includes dimensions of uncertainties, vulnerabilities, discomfort, shame or embarrassment, which need to be noticed, addressed or resolved.[18]

Additional ethical challenges can emerge exactly because it is one's own practice that is subject to investigation and therefore 'at stake'. As practitioners, our own practice means a lot to us and we are potentially ready to make sacrifices for it – as artists are practicing, rehearsing and editing their work for countless hours in order to create what they want to create – without looking at office hours. This is not a problem as such, in ethical sense, but we need to be aware of potential over-enthusiasm that affects others, such as participants or collaborators, who are not as deeply involved (see 'Being pushy' later).

There are a few questions that emerge from the hybridity of roles: What can be possible ethical issues that emerge from such hybrid roles? In which ways do practitioner researchers need to stay aware and awake regarding their role as insiders? And what kinds of behaviours are helpful to deal with these ethical issues? Furthermore, as the practitioner enters her own context in the role as researcher, power relationships to others – who might be superiors, colleagues or even friends of the researcher – need to be managed and sometimes re-negotiated in cases where previous relationships existed.[19] Another challenge is making one's own position as insider-researcher transparent and specific in writing and reporting, but also to participants and colleagues during the research itself.[20]

Preparatory matters

In some cases, a 'Just do it!' approach is adopted is 'as being the best way of going into the field, finding something new there and developing interesting knowledge from it.[. . . However, we should think about how to prepare our research(ers) carefully for working in the field'.[21] Flick observes the tendency of not wanting to overthink, but to sort out any emerging issues 'in the moment'. He argues from a sociological perspective, but I have observed this approach present in some practice-based research as well. To avoid this scenario (whether intentional or not), the researcher needs to ask herself:

- Do I really know what I am doing, and how to do it – also in potentially unexpected situations?
- To which degree am I certain about which people will be affected by me and my work, and in which way?
- Am I prepared for all kinds of people or events I might encounter in the field, and how to deal with them?

Such ethical considerations need to be made throughout an entire project, but they start in the preparatory and planning phase (including writing proposals). Here most if not all decisions about collaborators are made, and one should decide about a clear approach to acting in the field, guiding participants, using and interpreting information. Additionally, the researcher needs to be aware that they 'typically enter fields in positions of relative power to those of the participants',[22] and prepare for such. This is not to suggest that the researcher is able to oversee all situations and ethical potentialities from the start at all times and in each and every project. But, as Costley and Fulton remark, this should not mean 'that researchers should blindly attempt things, but rather to learn from and use learning experiences to shape the iterative ethical development of the research'.[23]

For better or worse: five behaviours

In order to develop an awareness for ethical issues and to develop an ethically aware way of acting while carrying out research, I opted to follow a set of five behaviours, as argued by sociologist Uwe Flick.[24] These five behaviours (next) signify kinds of behaviours towards other people; some of these behaviours are meant to be pursued, others to be avoided. Overall, these behaviours outline an empathetic, ethically sound way of acting of the practitioner researcher.

Being pushy

Practice, education and research are all dialogues, consisting of asking people for access: to spaces, information, biographies, stories, observable processes. It is therefore important to 'develop a feeling of the limits of our participants, . . . when we should stop insisting'.[25] This means respecting borders of privacy and intimacy and being aware of (and grateful for) the time participants take in order to join in with one's research. Integrity also means staying in the areas of the other person's life to which we are invited.

Being ignorant

Contrary to behaviour 1, there are situations in which participants provide more information than they were asked for, or offer unforeseen information and stories. Flick argues that 'in this context it is again the balance between working with the participant in a very focused way and taking him or her seriously in what they reflect about the issue beyond what we expected'.[26] When such situations arise, researchers have a responsibility to follow these emergent stories with integrity, deciding how to manage the participants' urgency to share in the most appropriate and context-related way.

Being accurate

This behaviour addresses accuracy while evaluating collected data or experiences. This involved reading and re-reading the collected material, in order to understand it fully. At the same time, this involves considering the most appropriate methods and instruments for analysis, as different kinds of data, experiences, learning and knowing ask for different kinds of analysing and reflecting on them.

Being fair

It is important to avoid interpretation of data in a way that comes along with a devaluation of people or objects. On the whole, researchers need to respect people's intentions and agencies while interpreting their practices or statements. This means being as non-judgmental and neutral as possible, particularly in the case of conflicts between data, such as contrary experiences by participants of an experiment.

Another risk is over-generalisations, which might result in developing certain types of people, or patterns of behaviour. Essentially, the researcher must stay faithful to the multiplicity of individuals, rather than induce any kind of collective identity through generalisation. A common strategy, if possible, is to go back to participants and to review, possibly in a debriefing, the correctness of certain interpretations.

Being confidential

Confidentiality implies that private data will not be reported, in order to take care for anonymity and privacy. A widespread technique is to anonymise or pseudonymise information immediately. This is to ensure that there is no risk of making people or places recognisable. However, this might be difficult in particular contexts of small and easily recognisable groups, or situations in which participants want or *need* to be credited (I will refer to a case study later).

Guidelines for planning human-centred ethics in practice-based research

This section outlines some guidance to the researcher regarding a human-centred approach to ethics. This is certainly not meant as a set of fixed rules but rather a strategy for internalising thoughts on which the practitioner researcher can fall back and think about potential issues. It is quite literally meant as a list that can be reviewed repeatedly during the various phases of a project. It is important to remember that these steps are relevant in all phases of a project, from initial planning and preparation, to carrying out the research, field work and the final delivery and eventually closing discussions.

The aim is to provide a process for the development of an 'integrated' behaviour or habit, to constantly be intuitively aware of these aspects. This approach is supported by action researcher and Professor Emerita in Learning and Leadership Judi Marshall's notion of 'holding things lightly'.[27] As such, this process is not intended to be a repetitious full ethical scan, but rather a light process of going through the list of questions, to take a few minutes to review and revisit one's ethical choices. Furthermore, these questions are not intended as a linear step-by-step plan, but as a series of questions the researcher needs to ask herself:

- In which way does my/our work involve others?
- What is the background and context in which I am/we are working, concerning people, spaces, ethnicities or other groups, objects and processes?

- Which knowledge from other professions/professionals might be necessary or helpful to consider?
- Which people are affected by my/our work, and in which possible way(s)?
- How do I/we make sure that human and non-human entities (people, spaces and objects)[28] are not affected in a negative way? What are possible strategies and concrete measures to protect them?
- How can I/we ensure not to behave pushily or ignorantly?
- How do I/we ensure I/we act accurately, fairly and confidentially?
- Have I/we made sure that all participants know what they are participating in, including possible consequences, and have they agreed to do so?
- In which way might this work have impact on me/us as practitioner researcher-person(s)? (Eventually work with a (more experienced) colleague or supervisor.)

Following this process, however, does not guarantee that a research trajectory will run without any ethical issues, challenges or difficult choices. Various publications on ethics suggest different ways to work with ethical challenges, ranging from very rigid to rather loosely. Madden, for example, suggests a three-step order of 'ethical priorities', based on 'issues of doing what is right by one's participants . . ., doing what is right by oneself . . ., and doing what is right by the discipline'.[29] Madden's reference point is the discipline of ethnography, but his argument is valid for other disciplines as well.

Examples

The following examples illustrate different ways in which a human-centred approach to ethics can be adopted. They also highlight how ethical agreements can work within a specific discipline, even implicitly, without necessarily writing or talking about them specifically.

Concerning anonymisation and ethical paradox

The first example highlights how emerging ethical questions might be affecting a project during its process,[30] in this case the anonymisation of participants. This case study concerns the work and creative process of a group of artists in a transdisciplinary research project between medicine, neuroscience and interdisciplinary arts. Nine artists from different artistic disciplines work in a collaboration with a group of eight young people between 8 and 28 years old.

The aim of the project, 'If you are not here, where are you?' (IYANTWAY), was to find a language for the often fearful and misunderstood experiences that children have during absence seizures. Science and art are connected in an experiment that aims to make the invisible experience visible, audible, experienceable. Alone or in duos, the artists worked with the youngsters on artistic utterances (music, paintings, interactive installations and so on) that match with the experience before, during or after a seizure, and which tried to do justice to the multimodal nature of the experiences. The participating neurologists were mainly present to inform the artists at the beginning of the project, as well as during presentations. The outcomes provide the young people with alternative ways to communicate with the world about their disease, in image, sound and experience rather than by language alone.

This project sparked ethical questions on various layers. First of all, we are working with young people, some of them children who suffer from an illness that is not openly shared in many situations or contexts (such as the classroom). Furthermore, the children were filmed in a variety of situations, and this footage has been used for a research film; in the final phase of the project, all material may potentially be used for the final documentary (of which our research forms just one part). All of these activities took place with the consent of either the participants or, in the case of younger participants, their parents. However, the issues described here are not so much concerning the legality of the use of the material, but the ethical discussion that arises out of this.

A crucial argument of the project, and observation made by all artist researchers in the project, was a considerably heightened amount of ownership on the side of the participants that emerged during the process of creating the artistic works, up to the point of co-creatorship. The shared ownership was not intended or planned from the outset, but emerged through the work itself. Through this emerging co-creatorship, the ethical question of anonymising or not became an issue. From a traditional ethical perspective, patients and participants as a vulnerable group should always be anonymised. However, from a perspective of being artistically ethical, it could be argued that was unethical to anonymise co-authors of a work.

The final decision was to 'fully acknowledge the ownership and artistry of the participants, maintaining their names, their stories, and the connection between both'.[31] This is an affirmation of the artistic argument, but at the same time does not defy traditional aspects of research ethics. Indeed, my final argument in this case study is that through making this decision about an ethical question, we not only prevent harm (one of the traditional values of research ethics) but also use this discussion for the opposite: to empower the participants.

The ethics of telling other people's stories

Writing about other people is an aspect of many forms of research, including practice-based forms. Anna Derrig,[32] researcher and tutor on life writing and ethics, offers a number of thoughts on how different forms of writing spark a number of moral and ethical questions, questions of responsibility and agency. These can range from scholarly and non-scholarly forms to interview reports, memoirs, and visual and non-textual forms such as photographs or moving images. She presents these questions, which can help draw the researcher's attention to a number of issues:

- What are any possible negative consequences for anyone participating in a research (or writing) project?
- How does telling someone else's story and using someone else's material and life define this person's future?
- In how far and to what degree do participants actually have the capacity to give consent and can oversee all possible consequences?
- To what degree can participants actually be *fully* informed about what is being done to them? 'Can . . . any of us truly consent to our story being told?'[33]
- In the final phase of a project or prior to publication: Has the researcher or writer gone back and checked with everybody?
- Does anonymisation actually help, even if the context of community is very small? What are possible alternatives to keep participants save?
- Just using other people is not good: What's in it for them?

In her lecture *Other People's Stories*, Derrig explains why ethical considerations in life writing are so important: 'Our stories are our most precious possessions. They are our identity. Getting them wrong, misusing them can cause real hurt and harm'.[34] The issue of consent is one of the most interesting and central questions concerning the ethics of life writing for Derrig. This resonates with many discussions on research ethics, as Costley and Fulton also remark:

> The consent of participants must be gained and information may need to be made accessible, thereby ensuring that participants can provide informed consent. This is paramount and should never be viewed as an extra step to be taken, but as an integral part of the research process.[35]

However, regardless of the researcher's effort to provide informed consent, the question remains of how far people have the capacity to actually give consent, especially in the time of the internet. Also, as Derrig asks, 'Can . . . any of us truly consent to our story being told?'[36] As with other aspects of ethics, the responsibility lies on the side of the researcher;[37] she needs to ask herself, 'How do I manage my obligations to truth and to the person concerned?'[38] As Jodie Taylor shows in her research of 'intimate insiderness', consent alone is never a guarantee to behave ethically:

> Looking back over my interview transcriptions, in each one I see occasions where I have inserted '[off the record]'. . . . not . . . because my informant explicitly said so, but because I understood implicitly that what they were telling me here was not as a researcher but as a friend and therefore – it felt to me – unethical to transcribe this statement for future analysis.[39]

It is of crucial importance to understand that the decision to exclude specific parts of the conversation and hold them 'off the record' is not due to regulations or to matters of consent on the side of the interviewees, but the result of careful and empathic consideration by the researcher. Taylor refers to settings in which her interviewees are also friends, but the argument still applies to situations of practice-based research.

A long list for the visual arts

One of the points several authors have made before is that ethics is an area that is usually not clear cut. By that I infer that ethics does not have rules as strict as the law and presents many grey areas in which it is the researcher or the practitioner who is responsible for her own ethical behaviour and decisions. In this respect, it is important to note that ethical principles are not necessarily the same in different areas or fields; different disciplines and professions have different views on ethics, and ethical practices. As Costley and Fulton put it, it is 'important to emphasise that many professions have a code of practice, and that this is something which should shape and guide our actions'.[40]

Karen Atkinson offers a perspective on ethics from the perspective of the (visual) arts in her text *Ethics for Artists*,[41] where she discusses a number of attitudes, behaviours and principles. Through this, she argues that these are either important to follow or necessary to avoid. Atkinson refers to the visual arts and the core practices of making work, presenting the work in places such as galleries and selling it – as artwork; so it is not research ethics she is concerned with. Nevertheless, I think it is interesting to look at this example in the context of what might be called 'disciplinary ethics', or 'discipline-specific ethics', which defines an ethical code of conduct specific to a certain discipline. Atkinson is not alone in thinking that such a code in the visual arts world is necessary and important, as Patricia Maloney and Kara Q. Smith also point out in their 2015 column that 'no real shared code exists . . . [f]or practitioners in the contemporary art world'.[42] They clearly argue for the necessity of such a code, in particular with regard to fair practices in the industry.

Atkinson lists a number of ethical behaviours specific to the context of the visual arts world. Some of these points might sound entirely natural and self-evident, and therefore possibly spark the question of why include them at all in such a list. But the fact that Atkinson mentions them already suggests that they might not all be as obvious as they seem, and that what happens in daily practice is often at risk of violating these 'ethics for artists'. Another aspect to consider is that none of these are strict rules in the visual arts world, but rather implicit; and they are usually not written up in a way that Atkinson has done, but rather learned through (bad) experience or taught by mentors, teachers or supervisors in the course of a (young) artist's education and/or career. Her list includes the following relevant points:

- Treat colleagues with respect.
- Don't keep information to yourself; share it.

- Don't be selfish.
- Don't tread on other artists' spaces.[43]
- Follow the agreements you make – all of them.
- Don't steal other people's ideas.
- Give back.
- Do what you say you are going to do.
- Be professional.
- Avoid deception at all times.
- Take care for the safety of your audience.
- Be thankful to those who support you, regardless of the nature of their support.
- Be conscious and thoughtful when you ask others for recommendation letters.

It is important to realise that in an art practice, ethical issues of stolen ownership, behaviour towards other artists or organisations are not always recognisable as easily as one might think. This is true especially in cases of power structures and differences in authority, such as when, as in an example Atkinson offers, artists with great reputation copy work of students, unknown and potentially even unaware that copying has taken place. In this sense I think that Atkinson's list offers a helpful compass to young(er) artists entering the field, and at the same time good reminders for more experienced professionals – especially because many of her points are transferrable to other artistic disciplines.

Towards a 'positive' approach

As the previous three examples have illustrated, often when ethics in research are discussed, a substantial part of the conversation (and the literature) suggests what kind of things 'not to do' or 'to avoid'. Parts in this chapter until this point have done so too: for example, the five behaviours from Flick include such examples, as we should avoid being pushy, being ignorant, devaluating or over-generalising.

To be clear, I don't mean to generally discredit such 'stances of avoidance', as certain behaviours and situations definitely need to be avoided and to be cared about. However, at the same time I can understand some practitioner researchers who experience thinking about ethics as a hindrance. This might happen when they feel they are requested to take a high number of precautions that seem disproportionate for the situations in a research project. I think this front-loaded sense of hindrance shouldn't be necessary, and ethics should rather stay in our continuous thinking and acting as an *integrated* moral compass. The advantage of this approach can make us more caring about other humans and non-humans, about our surroundings and our world at large, in a responsible and positive sense.

So, towards the end of this chapter, I like to stress a view or a perspective on ethical behaviour that, in my view, is slightly more empathetic, positive and possibly poetic, rather than seeing ethics as a hindrance. In short, what I am suggesting is to:

listen to and to do right to all present human and non-human voices.[44]

The idea of using *voices* as a metaphor builds on the work of playwright and research professor Nirav Christophe and his work about co-creation in the performing arts, and the different (internal and external) voices in complex situations of collaboration.[45]

Different kinds of questions might be evoked by thinking about ethics in this way that nonetheless support the core (institutional) process. For example, how can we work in a way that does justice to all human and non-human voices that are there? If we think about the voices involved in a project or in a certain situation, experiment or conversation, which ones might run the risk of being overthrown by others? Which voices have the risk of becoming marginal and thus need to be taken care of more than others? Where are our own potential blind spots in terms of ethics?

In the case of the project IYANTWAY mentioned earlier, this notion of listening to the present voices – particularly the voices of the participants, their stories and creative impulses – clearly led to a heightened ownership of the participants of the project, up to true co-creatorship in the creation of the artistic works.[46] The voices present in this kind of work were diverse: the voice of the patients/participants, their parents (in some cases), the spaces in which our conversations took place, the various materials the participants shared with us as artists, our own voices and professional experience.

Karen Atkinson's list can also be seen from this perspective, with a multitude of different voices to pay attention to, such as colleagues, agreements made, other people's ideas, private or public property, audiences or one's own promises. Lastly, the initial quote by Anna Derrig in the section on telling other people's stories accentuates the same point: treating others with respect as a decent human being. Looking back on the example of life writing, it is not so much the question of what to *avoid* in writing, and what to be cautious *not to do*, but rather to tell the stories of other persons with respect and empathy – which can be both inspiring and spark creative writing in its own right.

I do not mean to suggest that these points essentially change the argument for ethics in relation to other literature, or that the necessity for ethics or an ethical stance change, and neither do the concrete choices regarding ethics change necessarily. My point here is a change of the researcher's inner attitude and perspective towards ethics. That is, not as an obligation or hindrance, but rather as a behaviour of empathy, of thinking about and being aware of others.

I am also sure that many texts, narratives and conversations on ethics are meant to be like this or some such similar way, but my point here is that they can be not perceived in this way. In relation to this, I close this chapter with a final quote from Jodie Taylor.[47] She reminds us that regardless of the information on ethics and various examples, which hopefully help the practitioner researcher to make informed choices and decisions, it remains important to

> not only think but also feel our way empathetically in the field.

Notes

1 Copp (2007, p. 19).
2 See Hübner (2022).
3 van Zilfhout and Wouters (2017).
4 Madden (2017, p. 89).
5 Flick (2018).
6 Denscombe (2014, p. 306).
7 Ibid., p. 307.
8 Costley and Fulton (2019, p. 78).
9 This system is explained in more detail in Candy and Edmonds (2022).
10 I have to thank Andrew Johnston for sharing information on this work and process. It needs to be noted that this process mainly works because the activities of the group are at a relatively low risk, ethically speaking; in most cases, this simply involves participants talking about their experiences while engaging with technology, artworks or installations. Another element to keep in mind is that this process is based on trust. The institution's ethics committee provides trust to the research group, which on their side respond to this trust with integrity, responsibility and care.
11 Madden (2017, p. 82).
12 Kara (2018, p. 1).
13 Barad in Dolphijn and Van der Tuin (2012, p. 69).
14 Regarding the notion of human and non-human entities, see Hübner (2022).
15 Kara (2018, p. 1).
16 Solomon and Baio (2022).
17 Madden (2017, p. 81).

18 Ibid., p. 88.
19 Costley and Fulton (2019, p. 87).
20 In order to do this responsibly, Jodie Taylor has developed a practice in which she makes transcripts and writing drafts accessible for informants, participants and others involved, prior to submission or publication. This enables the people involved to review and to react on what has been written about them (Taylor 2011, pp. 16–17).
21 Flick (2018, p. 86).
22 Madden (2017, p. 77).
23 Costley and Fulton (2019, p. 86).
24 Flick (2018, pp. 90–92).
25 Ibid., p. 90.
26 Ibid.
27 See Marshall (2016, pp. 19, 31, 54).
28 See Hübner (2022) for a more in-depth discussion of entities as an element of crafting research methods.
29 Madden (2017, p. 87).
30 I have written about this case study in greater length in Hübner (2017a).
31 Hübner (2017a, pp. 212–213).
32 My sincere thanks go to Anna Derrig for having an in-depth conversation with me about the ethics of life writing and sharing her thoughts for the purpose of this publication. The ideas and issues mentioned here are largely based on this conversation and a lecture Derrig gave on the BBC (Derrig 2020), without which this section would not have been possible.
33 Derrig (2016).
34 Ibid.
35 Costley and Fulton (2019, p. 88).
36 Derrig (2016).
37 It is also interesting in this context to look at the etymology of 'consent' itself. The term comes from Latin and combines the syllables 'con', meaning 'with, together'; and 'sent', originating in 'sentire', meaning 'to feel'. The two meanings combine into 'to feel with', which resonates with the notion of empathy and responsibility of the researcher.
38 Derrig (2016).
39 Taylor (2011, p. 14).
40 Costley and Fulton (2019, p. 86).
41 Atkinson (2017 [2011]).
42 Maloney and Smith (2015).
43 This point refers to 'invading' other artists' spaces in order to present one's own work, or invite curators to one's own studio, rather than paying attention to the others' work and leaving the space for them.
44 The notion of 'listening' is to be understood not only in terms of sounds, but rather in a multimodal sense: listening as paying attention with all the senses.
45 See Christophe (2017). Such voices can be the voices of collaborators, participants or commissioning parties, but also inner voices such as the voice of one's own imagination, experience, inner critic or reflection.
46 For a more in-depth account on the aspect of heightened ownership in this project, see Hübner (2017b).
47 Taylor (2011, p. 19).

Bibliography

Atkinson, K. (2017 [2011]). *Ethics for Artists*. www.huffingtonpost.com/karen-atkinson/ethics-for- artists_b_826053.html.
Candy, L., & Edmonds, E. (2022). The Studio and Living Laboratory Models for Practice-Based Research. In C. Vear (Ed.), *The Routledge International Handbook of Practice-Based Research*. London and New York: Routledge.
Christophe, N. (2017). The Art Is in the Encounter. In H. Dörr & F. Hübner (Eds.), *If You Are Not There, Where Are You? Mapping the Experience of Absence Seizures Through Art*. Amsterdam: International Theatre & Film Books, 88–105.
Copp, D. (2007). Introduction: Metaethics and Normative Ethics. In D. Copp (Ed.), *The Oxford Handbook of Ethical Theory*. Oxford: Oxford University Press, 3–35.
Costley, C., & Fulton, J. (2019). *Methodologies for Practice Research: Approaches for Professional Doctorates*. London: Sage.

Denscombe, M. (2014). *The Good Research Guide: For Small-Scale Social Research Projects*. New York: Open University Press.

Derrig, A. (2016). *Other People's Stories* (lecture). www.bbc.co.uk/programmes/b07z4djm (accessed December 11, 2020).

Dolphijn, R., & Van der Tuin, I. (2012). *New Materialism: Interviews & Cartographies*. Ann Arbor, MI: Open Humanities Press.

Flick, U. (2018). *Designing Qualitative Research*. London: Sage.

Hübner, F. (2017a). 'I Care for Them'. On Artistic Ethics in Research. In H. Dörr & F. Hübner (Eds.), *If You Are Not There, Where Are You? Mapping the Experience of Absence Seizures Through Art*. Amsterdam: International Theatre & Film Books, 202–215.

Hübner, F. (2017b). Is It Mine or Yours? On the Shift of Ownership Between Artists and Participants. In H. Dörr & F. Hübner (Eds.), *If You Are Not There, Where Are You? Mapping the Experience of Absence Seizures Through Art*. Amsterdam: International Theatre & Film Books, 70–87.

Hübner, F. (2022). The Practitioner in the Centre: A Flexible Approach to Methodology in Practice-Based Research. In C. Vear (Ed.), *Routledge Handbook to Practice-Based Research*. London: Routledge.

Kara, H. (2018). *Research Ethics in the Real World*. Bristol: Policy Press.

Madden, R. (2017). *Being Ethnographic: A Guide to the Theory and Practice of Ethnography*. London: Sage.

Maloney, P., & Smith, K. Q. (2015). Is It Possible to Create a Code of Ethics for the Arts? *Art Practical*, June 25. www.artpractical.com/column/is-it-possible-to-create-a-code-of-ethics-for-the-arts/ (accessed September 4, 2020).

Marshall, J. (2016). *First Person Action Research. Living Life as Inquiry*. London: Sage.

Solomon, L., & Baio, C. (2022). Thinking Together Through Practice and Research: Collaborations Across Living and Nonliving Systems. In C. Vear (Ed.), *The Routledge International Handbook of Practice-Based Research*. London and New York: Routledge.

Taylor, J. (2011). The Intimate Insider: Negotiating the Ethics of Friendship When Doing Insider Research. *Qualitative Research*, 11(1), 3–22.

van Zilfhout, P., & Wouters, E. (2017). Ethische Antennes [Ethical Antennas]. In E. Wouters & S. Aarts (Eds.), *Ethiek van praktijkgericht onderzoek. Zonder ethiek is het al moeilijk genoeg (Ethics of Practice-Based Research: It's Already Difficult Enough Without Ethics)*. Houten: Bohn Stafleu van Loghum.

4.9

THE PRACTICE-BASED PHD

Some practical considerations

Ernest Edmonds

Introduction

This chapter covers ground that is discussed throughout the handbook but brings things together in a way that is focused on the practice-based PhD and, in particular, being organised so as to make a successful final submission. Hence, much of what is said repeats points that are made elsewhere. However, the chapter should help the practice-based PhD researcher focus attention on certain key issues in planning and presenting the work.

Getting organised

As with all PhDs, a practice-based PhD has to make a new contribution to knowledge. So, in the end you need to demonstrate that what you have done has not been done before, and also show what your claim to knowledge is. The knowledge can be in very many different forms and can be of many different types, but it must be able to be shared, reviewed and understood.

Before starting, it is helpful to read up about practice-based research, starting with this handbook and, for example, with Candy and Edmonds' 2018 paper.[1] As soon as possible, it will be important to work with your supervisor(s) or adviser(s) to identify specific training courses on offer that can help you develop as a researcher. In any case, such courses may well be a required part of your study programme. Before you start, however, you need to develop a plausible topic and approach. Basically, it is important to be clear about your initial expectations in relation to the research problem, its context, the research methods to be employed and the likely outcomes. Your *initial proposal* would do well to cover those points. Remember that the idea is to do research: you find things out as you progress and so plans change. But it is important to start with a plausible problem and plan, even though it is likely to evolve, contract, expand or even change all together.

The core of any practice-based PhD will, of course, be the practice – in whatever discipline is under study. Almost always, the researcher will be well into that practice before they start the PhD and often will already be expert or, at the very least, competent. In a sense, then, the practice is not the problem. The issue is to conduct research about, in, with or through that practice. The first thing is to ensure that a good record is being kept of the work being done. Most practitioners keep some form of logbook or diary, and this will be a vital tool in the PhD. My advice is to ensure that all thoughts are recorded, including notes about what went wrong or was rejected and what actions

DOI: 10.4324/9780429324154-42

you took as a result. Full documentation of the practice is important to keep and be available for you to consult throughout the research.

Starting with your *initial proposal*, develop a plan of action for year one. Every case will be different, but normally by the end of the year you should have:

1   A refined version of the initial proposal (a draft contribution to chapter one).
2   A first draft of the state-of-the-art review – this confirms that the research problem/issue has not been dealt with before. The project addresses a gap in our knowledge.
3   A first draft of the methodology chapter, identifying the expected research methods to be employed. This also enables you to have a first go at a plan for the rest of the work.
4   Examples from your work of documented practice – which may be thought of as initial explorations into the ideas, even if it is too soon to make the work formally contribute to the research.

Saving the writing up for later is always a dangerous approach. Draft chapters should be written as the research progresses, starting as suggested in year one.

It is worth thinking early on about the fact that at some point the thesis, the argument, has to be shown to be convincing. This will need to be done throughout the thesis, and in many different ways, but should be outlined in the methodology chapter. The best way to start is to look at the claim that you hope to make. For example, if the claim is that a new art form has been found that increases audience awareness of poverty, then some kind of survey or interview-based study might confirm that. If the claim is that a new software painting system has been invented that can draw circles faster than any previous system, then a series of timed tests would show that. On the other hand, a philosophical claim would be justified by a very tight linguistic argument and would be discussed in the methodology chapter.

The contents of a practice-based thesis

In a practice-based PhD it is often necessary to submit more than a written thesis. Where the regulations allow it, supplementary material such as music, films or the documentation of sculptures or performances may be included. Indeed, the examination may involve the viewing of an exhibition or performance, for example. It is important to understand that the full thesis, text, artefacts, documentation etc., should be seen as an integrated entity rather than as a collection of unconnected, or loosely connected, elements. A possible chapter structure for the written component is given in "Thesis Structure" later in this chapter. The discussion in this section can be used to help plan the contents of each of your thesis chapters.

The next section goes through the likely components of a practice-based PhD thesis, summarising the key points to consider. First of all, I discuss the supplementary material and then the various components that are delivered as text.

Supplementary material

The outcomes from a practice-based research project often include an artefact, in some sense or another, as has been discussed in many chapters of the handbook. The distinguishing feature of the regulations that normally govern a practice-based PhD is that they allow "supplementary material" to be included in the thesis. Such material is often left largely undefined but limited in that it must be possible to archive it together with the written text. The examiners may be given the privilege of experiencing the material in its natural form where it cannot be archived, such as a live performance, provided that suitable documentation is provided. An increasingly important special case is the demonstration. Sometimes, a written description of a new way of doing something may be hard to compose, while it is easy to demonstrate it. In such cases, a video of the demonstration might be able to be justified as an archivable representation of that knowledge. Looking up on YouTube how to fix something round the house is, after all, a common activity today.

We can see that the supplementary material can be in many forms but may be a crucial part of the thesis. Some thought needs to go into the presentation of any artefacts that are important to the research. A film, for example, can be submitted in a form that can be stored with the written component. It might be stored in a digital form on a DVD or USB stick, perhaps, and bound in with the volume that constitutes the written submission. While the technology is certainly ever changing, even a DVD can still be read today and, in the future, its contents can always be transferred to another medium, even if it requires specialist facilities. However, files stored on the internet, a website for example, cannot normally count as adequately archived because, as most people know to their cost, servers may close down, and files are deleted or moved to unknown locations. A copy of the relevant files needs to be stored as part of the thesis if the content is important.

As mentioned earlier, while the examiners might be able to view a performance or an interactive sculpture, for example, neither will be easily archived in the library. Hence, the PhD candidate will need to provide documentation, in some form or another, that can be archived. Careful thought has to be given to ensure that the documentation reveals the significant features that support the claims made in the thesis.

There is another concern to consider. The reader of the thesis will need to know how to "view" the supplementary material, the delivered artefact(s), in the context of the research outcomes and claim. It is not sufficient to simply provide an artefact and assume that its research significance will be understood by all, over time. Somewhere in the written component, the researcher needs to explain how to approach the artefact, how to view, listen to or otherwise experience it. What about it is new? How does that generalise beyond this specific case? How does the artefact illuminate the argument, the new knowledge?

An example of text dealing with some of these matters is this text from Sarah Moss' PhD,[2] which investigated the experience of presence in interactive art systems using eye tracking. She bound a hard drive into the submitted thesis. This is a brief excerpt from her description:

> The external hard drive accompanying this thesis contains a number of items that assist the reader in their understanding of the artist, the art system and the participants' experiences of the art system. . . . folder . . . contains all the media files required to preview the system in an interactive format. Other media clips on this drive illustrate how the eye-tracking device facilitated an interactive experience for participants. This folder contains an application specifically devised in order to demonstrate the visual media contents and create a visual display that imitates the form of the system.

Now consider the various text components of a thesis.

Overview

The written thesis should start with an overview of the research. Four key elements need to be briefly described. Of course, this can only be completed once the work is done, at the end of the research. However, it is good practice to write a version early on, describing the expectations and hopes that you have. Things will change and grow and so these statements will evolve, but it helps the planning and focus the research if you have draft versions form the start.

The problem

This is a concise statement of the research questions or issues that the project, and the thesis, addresses. Taking an example from Sarah Moss again, she posed three questions:

1   How can previous research in presence inform the design of an art system?
2   What influence does eye-tracking technology have on the experience of presence-generating art systems?
3   What are the factors influencing an individual's ability to experience presence-generating art systems?

It is important that the questions are posed in a way that indicates what would count as an answer, and the answer will need to be a contribution to knowledge: new knowledge, as discussed in Part II of this handbook. The first of these questions will be answered by a description of at least one art system concerned with presence, together with an explanation of how the research has been employed in its design. For the answer to be valuable, one would expect that a generalised, validated description would be given and that it would be shown to be new in the world. The other two questions would each be answered by lists of factors, accompanied by evidence of the claimed influences, again in the form of new knowledge.

The context

This covers the main work that has been done in the past. What is the gap in knowledge, the opportunity, and what is the significance of the research issue? The state-of-the-art report, discussed below, elaborates on this.

The method

This indicates the approach to solving the problem (experimental, practice-based, analytic etc.). Within the overview, this is basically an abstract of the methodology discussion covered later.

The outcomes

Here the key contribution(s) to knowledge are concisely described. They are the things that arise from the work that are new and shown to advance understanding or practice both nationally and internationally. The value of these outcomes will be to one or more communities (composers, computer scientists, artists, theoreticians etc.), and it is important to be clear who they are. Outcomes are particularly hard to predict at the start, and hence the initial statement might only identify the kind of things that are expected, such as a new theory or a new method. In the end, however, the outcomes must be the answers to the questions posed under "The Problem." Sometimes, as a project comes to its conclusion, the researcher finds that the outcomes are answers to slightly different questions to those posed at the start, or even very different ones. I will come to this problem later in the section on "Results."

State-of-the-art review

This section presents the results of a survey of the area(s) of study. It should be a critical review in the context of the stated research question and related issues. This chapter answers questions such as: Who is doing what? Who has done what? Who first did it or published it? The survey is taken from published papers, research monographs, catalogues etc. It must be based on and refer to primary sources, not textbooks or other such reports on the work of others. It is to be expected that this chapter identifies a new structured view of the field of study. Its conclusion is that there is a gap – something not yet covered by anyone. That gap is what was given as "the problem" above.

It is sometimes possible to offer a new view of the field, so that the survey becomes a research contribution in its own right. Then, as well as being part of the thesis, it might be publishable in the research literature. In Dave Burraston's PhD, he wrote such a survey which was published in *Digital Creativity* and is often cited.[3]

Methodology

This is a key part of the thesis. It provides a description and justification of the research methods used. Normally, the methods will be selected from known and proven examples. In special cases, the development of a new method may be a key part of the research, but then this will have been described in the overview and reviewed as part of the state-of-the-art. A full section of this handbook is devoted to this issue. It is a particularly important element of a practice-based PhD thesis because there is no single standard way of conducting such research. Where there is, then the methodology might be taken for granted, but here, where there is considerable choice, you need to clearly tell the reader where you stand and what you have done in order to obtain the new knowledge presented.

As an example, consider Lizzie Muller's PhD,[4] in which she begins her discussion of methodology by stating:

> To formulate a methodological approach to this research I have drawn from two main established sources. Donald Schön's account of the way that knowledge is produced through practice provides an epistemological starting point, and his description of "reflective practice" is the basis for my approach. . . . Steven Scrivener has built on Schön's work to develop a structured process of documentation and reflection surrounding creative practice. . . . I have based the process of my research on his detailed account of the shape and structure of a creative-production PhD. In this chapter I describe the aspects of reflective creative practice, as described by Schön and Scrivener, which are most important to this thesis. I begin by describing the context of the practice-led approach and discussing examples of similar projects in the field of curating.

And concludes with this brief summary:

> The creative curatorial practice underpinning the project is process based and collaborative. It exists as a crafted set of situations, discussions and interventions which enable exchanges between people (audiences and artists) that feed into emerging artworks through public exhibition. Since this work has already taken place, it can only be accessed through documentation. My practice contributes to the production of artworks made by the artists through the case-studies. These artworks are a form of evidence of the curatorial work which took place. Video and photographic documentation of the artworks as they appeared in their several iterations during the course of the project demonstrate the trajectory and impact of my curatorial work. The nature and quality of the experiences of the audiences for each exhibition is evidence of the applied experiential research which made these experiences possible. Quotations from audience's descriptions of their experience, as

well as photographs of the artworks therefore appear throughout the text, and full transcripts and video documentation are included as appendices.

As part of her methodology section, Lizzie Muller also provided a table that showed which method was used at which stage and for what purpose. A clear report of the research process and the justification for using the methods employed is important, and this example indicates one way in which it can be presented.

Foundation work

This optional section is a chance to describe earlier work done by the practice-based PhD researcher (possibly with others) that provides a foundation or significant background to the PhD. It may be helpful to revisit and reassess earlier work in the light of the research focus of the PhD, particularly if it is not easily available in the public domain. This will not be needed in some PhDs, where the work is from a fresh start, but where it is appropriate it is normally quite easy to write. The researcher is simply reporting on the earlier practice that put them in a position to start the study. This section, if used, is not a biographical overview but is a description of the relevant work done by the researcher.

New studies

The core of the thesis is a description of the new studies/software/artworks and the process of production. It answers the questions: What has been done? How was it achieved? What was the rationale? This can, for example, be a report on the design and execution of a set of experiments or the development of an innovative software system or the making of innovative art works. In a practice-based PhD an artwork, for example, can be presented for examination. If this is the case, this chapter will explain what is important and novel about the new work, as discussed earlier under "Supplementary Material."

While this is the crucial section of the thesis, there are as many variations of it as there are projects and so little can be said in general, except that it must be clearly and unambiguously reported. Ideally, the reader should learn enough about what was done to repeat it or go through a very similar process using the same methods.

Results

What came out of the studies? What is the new knowledge? How do we know that it has been validated in some sense? The research leads to a claim that the posed questions have been solved by delivering new knowledge. The evaluation of the new software/artwork or the analysis of the results or processes of the new studies, as described in the previous section, may have led to these results. The results must be presented and shown justified by the reported studies. This section shows that the outcomes, as promised in the overview, have been achieved. More significantly, it shows that the problem has been solved, the questions have been answered.

I mentioned earlier that the answers delivered in the end can be to different questions than those posed when the project began. This need not be a problem. It is important to remember that the thesis is, in total, an argument describing what is known, and why, at the conclusion. It is quite reasonable to change the questions to match the work actually done. In fact, it is unhelpful and confusing to give as the questions ones that were never answered. A poet who starts to write a sonnet but ends up using free verse is not expected to call the result a sonnet. It is the same for the researcher, who may of course also be a poet.

Conclusions

This chapter revisits the initial overview and explicitly relates it to the results described earlier. A discussion can now be provided that puts a wider perspective on the results and discusses the implications of them for other broader areas and domains. Future work and outstanding questions are normally also discussed in this section.

Bibliography

Use a standard reference format, such as Harvard, and be careful to check each entry. It is tempting to presume that a software system will ensure a perfect reference list, but that all depends on exactly how each entry was stored. There is no substitute for a line-by-line check. Depending on the regulations, it is common to include your relevant published papers within the thesis.

Ethics

All research that involves people in the collection or analysis of data is subject to ethical considerations. Some of these considerations are subject to legal constraints, which vary from country to country and state to state. Others may be controlled by procedures determined by individual institutions. Beyond the constraints, one would expect every researcher to conduct themselves in an ethical way and to be open about how they did that. An important aspect is the consideration of data management. How data is stored and accessed as well as how long it is kept are important issues to be clear about.

An early stage in designing your research is to prepare a research proposal that outlines the methodology and identifies any ethical issues. Researchers must follow the law and university rules on ethics and, in particular, obtain the informed consent of persons participating in the research before it begins. Particular requirements apply when dealing with children or vulnerable people in general.

The normal process, having identified the ethical issues, is to plan how they will be dealt with and to submit research methodology, including the ethics plan, for formal approval within the relevant university. Once approved, then that aspect of the research can begin. A description of one approach to specifying and reviewing ethical considerations is given in "Ethics Procedures."

Reading the final draft of the thesis

In this section, I offer a few tips for your final check before you confirm that you have finished, and that the written component of the thesis is ready to be examined. You need to read the final draft as if you were an examiner.

Top-level structure

Look particularly at the introduction and conclusions. A PhD thesis is an argument that soundly and persuasively convinces the reader of the thesis proposed: the claim made. You need to confirm that:

- You have clearly explained the problem to which you have found a solution. What is the problem and what is that solution? You can always phrase your thesis around a problem. For example, if you have seen an aesthetic opportunity (X) that nobody else has shown how to address, then the problem is "How can opportunity X be realised?"
- You have clearly stated your assumptions and demonstrated that they are correct or, at least, reasonable.

- You have shown that you are aware of any counter-arguments to your thesis and convinced us that they are false.
- You have validated the claims that you make.
- You have clearly scoped your thesis, explaining when it does not apply as well as when it does (you have probably not solved global warming, but you might have discovered something that will help).

The argument

Close reading of the whole thesis is very important at this stage. Most people work better from paper, rather than a screen, when reading in this way. This is not only looking for spelling errors, missing references and so on. It is also looking at the strength of the argument. The most important thing is to see what questions your text might bring to the reader's mind and then to check that you have provided answers. For example:

- Have you made a claim without giving an argument? "Interactive art works must be constantly changing." Why?
- Have you failed to provide a reference? "Plato thought . . ." Where and when did he say so? Are you just quoting a textbook? Might the author have misquoted Plato?
- Have you considered recent research relating to old quotes that you rely on? What does recent philosophical scholarship have to say about Plato's thoughts?
- Will the reader be convinced that your source is reliable? "David Beckham said that the origins of the earth are . . .," "Stephen Hawking said that, when playing football, you must always . . ."
- Have you generalised? "Sailing on Sydney Harbour is very peaceful." Is that true during a yacht race?
- Have you taken a black and white view? "X is not always true: hence, it is always false."
- Have you argued against the person rather than their position? "The Prime Minister has made bad decisions about immigration, so we cannot agree with his tax reforms."
- If the submission includes a USB stick or DVD, for example, that you claim represents new practice-based knowledge, does the thesis explain how and why that is the case?

Remember that, although you wish to be persuasive, the thesis is a matter of argument, not rhetoric. People are often persuaded by false or weak arguments that, for example, attack the person rather than the issue (just read the newspapers or follow a political campaign), but such approaches have no place in a PhD thesis. To repeat the most important thing: see what questions your text might bring to the reader's mind and then check that you have provided answers.

Thesis structure

This section provides a thesis chapter outline. It is a starting point for the plan of the written element of a practice-based PhD submission. As with all plans, it will probably change and, in any case, will need to be modified to fit individual needs. The discussion in "The Contents of a Practice-Based Thesis" section of this chapter can be used to expand the ideas about what each chapter might contain.

It is a good idea to think about this structure early in a PhD project, for several reasons. Firstly, it is valuable to start writing from the outset – not leaving the writing to be a concluding task. Secondly, planning your work, navigating between practice, research and writing, can be partly facilitated by having a draft chapter plan to hand.

C1 Introduction

Four key elements are briefly described in this section.

i   *The Problem*
    The research question or issue that the thesis addresses.
ii *The Context*
    The main work that has been done in the past.
iii *The Method*
    The approach to solving the problem.
iv *The Outcomes*
    The key contribution(s) to knowledge.

C2 State-of-the-art review

The results of the survey of the area(s) of study. A critical review in the historical and current context of the research.

C3 Methodology

A description and justification of the research methods used.

C4 Foundation work

This optional chapter describes relevant earlier work done by the researcher.

C5 New studies

The core of the thesis is a description of the research done. If an artwork, for example, is presented as part of the thesis, this chapter will explain what is important and novel about it.

C6 Results

The new studies will have led to certain results. The chapter describes them and shows that they answer the problem posed in C1.

C7 Conclusions

This chapter revisits C1 and explicitly relates it to the results described in C6. Future work and outstanding questions are normally also discussed.

C8 References

Formatted in a standard reference format.

Ethics procedures

The procedures used by the Creativity and Cognition Studios (CCS) at UTS[5] provide an example of how ethics considerations can be managed. The university has given CCS the power to give ethics approval within certain constraints:

> All research investigations that involve human subjects and participants need to apply for ethics approval.
>
> Creativity and Cognition Studios (CCS) has been approved to administer research investigation projects, conducted using a specific set of methods recognised together as a methodology appropriate to the CCS field of investigation. Each project must be approved internally at a CCS Ethics meeting. The process is not a chore but an essential step in planning research investigations and providing access to the experience and expertise represented in the wider CCS research program.
>
> If the research methodology proposed in an application is not covered by the CCS Generic Ethics, then a full application will need to be submitted to the UTS Human Research Ethics Committee.
>
> As part of the methodology, issues of Informed Consent, Privacy and Confidentiality are central to ethical behaviour. All researchers are required to read and understand the information and conditions described in this and associated documents forming part of the CCS Generic Ethics Approval.

The process is as follows:

> The first stage in seeking Ethics Approval is to follow the pro-forma 2-page application form. . . . The form is designed to cover the essential information required by a CCS Ethics meeting to describe the specific investigation planned involving human participants.
>
> The form seeks the following information: Title of Project; Aims of Research; Description of Methodology; Significance of the Research; Number of participants and justification of numbers; Selection/exclusion criteria for participants; Children; Procedures to be used; Time commitment for participants; Location of research; Consent procedures; Additional Risks; Strategies to cope with additional risks; Any other issues.
>
> The investigation approved may extend over a period of time, but any alteration made to gathering or using the data that received final approval from the CCS Ethics meeting, will require a fresh application. (The approved investigation may be only one of several different approaches taken within an overall research project.)
>
> Even if a research group has not been granted power to approve ethics proposals, following the CCS method is still valuable as it enables group discussion about projects, from which the individual researcher and the community can learn.

Conclusion

In any research project, PhD or otherwise, the results have to be reported. This chapter has reviewed the core elements of what such a report might contain and so provides a checklist that any researcher can use to ensure that they have followed the advice contained in the handbook as they report their results.

Notes

1 Candy and Edmonds (2018).
2 Moss (2011).
3 Burraston and Edmonds (2005).
4 Muller (2008).
5 Candy, L., & Edmonds, E. A. (2022). *The Studio and Living Laboratory Models for Practice-Based Research*. This volume.

Bibliography

Burraston, D., & Edmonds, E. A. (2005). Cellular Automata in Generative Music and Sonic Art: A Historical and Technical Review. *Digital Creativity*, 16(3), 165–185.

Candy, L., & Edmonds, E. A. (2018). Practice-Based Research in the Creative Arts: Foundations and Futures from the Front Line. *Leonardo*, 51(1), 63–69.

Moss, S. L. (2011). *Presence-Generating Arts Systems*. PhD thesis. Sydney: University of Technology.

Muller, E. (2008). *The Experience of Interactive Art: A Curatorial Study*. PhD thesis. Sydney: University of Technology.

PART V

Practitioner Voices

5.1

PRACTITIONER VOICES

Craig Vear

The final section of this handbook presents a series of practitioner's voice chapters from a range of fields and disciplines. The authors have written a short piece that centres on a practice-based research project they have undertaken, whether part of a PhD or another kind of research project. The focus is on their understanding of the work as it relates to both practice and research, and presents lessons learnt in a way that speaks outwardly to the diverse readership of the handbook. To do this, they concentrate on personal reflections of the experiences and what they learnt through doing their projects. As such, they are not specific to a single case study in a single field, but speak outwardly to all practitioners. Some decided to focus on the challenges they faced, or the pitfalls and problems, conflicts and solutions they needed to negotiate. Others discuss outcomes such as frameworks and models that contribute to the kind of knowledge or methods that other practitioner researchers might be able to use to advance their practice-based research. Others present discursive reflections on the processes and methods of their projects, including an account of how any creative works were made and appraised or evaluated. Overall, these chapters present a rich insight into the processes and research practice of a diverse range of practice-based researchers in order to bring back into the practice the knowledge that has been discussed in the previous sections of handbook.

The chapters are sequenced in alphabetical order, and discuss the following topics:

1 **Alice Charlotte Bell:** A New Framework for Enabling Deep Relational Encounter Through Participatory Practice-Based Research.

   In this chapter, Alice discusses the generation of a Participatory Practice-Based Research (PartPb) framework through which deep relational encounters between practitioners and participants can be enabled. It provides a scaffold for generating opportunities for sustained one-to-one encounters between creative practitioners operating in performance, arts, and health contexts with participant-subjects. This chapter incorporates a discussion of the ethical and artistic challenges faced in the realisation of her PartPb research and signposts the new knowledge and insights gained.

2 **Oliver Bown:** Risk, Creative Spaces and Creative Identity in Creative Technologies Research.

   Ollie draws on over ten years of practice-based research that applies emerging technologies, algorithms, and code as a creative medium in the areas of music performance, music composition, and media art installation. He considers the competing demands of making things work technically and artistically, and the nature of collaborative work. This is considered through his

 DOI: 10.4324/9780429324154-44

own recent research and that of a current PhD student as case studies in how this works in an academic context.

3 **Øyvind Brandtsegg and Alexandra Murray-Leslie:** FEEDBACK: Vibrotactile Materials Informing Artistic Practice.

Øyvind and Alexandra discuss their collaborative project 'FEEDBACK: Vibrotactile Materials Informing Artistic Practice'. Through this they outline new dialogic spaces between practitioners' diverse and, at times, opposite processes, and how this informed a polymorphic art practice in asynchronous ways. Their work is based on research into engineering natural and artificial materials which are developed in tandem with musical and technological devices. This chapter outlines their project and reflects on aesthetic and practical lessons learned through this multi-disciplinary collaboration.

4 **Ben Carey:** Co-evolving Research and Practice – _derivations and the Performer-Developer.

Ben discusses the approach taken to practice-based research in his doctoral project entitled _derivations and the Performer-Developer: Co-Evolving Digital Artefacts and Human–Machine Performance Practices. His chapter details the approach to reflective practice taken in this creative-production research project, reflects upon the benefits and challenges of this methodological approach and outlines the theoretical findings that emerged from cycles of practice and reflection in the research process.

5 **Maria Chatzichristodoulou**: Publishing Practice Research: Reflections of an Editor.

Maria's chapter reflects on the publication of practice research from her perspective as Editor-in-Chief of the International Journal of Performance Arts and Digital Media (IJPADM). It offers a different perspective to the practice research debate, which can be relevant to researchers themselves as well as other academic journals, publishers, and research publication outlets. The discussion is framed by her own experience as a practitioner researcher.

6 **Balandino Di Donato:** From a PhD to Assisting BioMusic Research.

Balandino describes his practice and experience in transitioning from being a PhD in Music Technology candidate to a research assistant on a large-scale funded project. The practice of these works was driven by computer science approaches applied in a musical context. Challenges imposed by the research and the timeline were significant, but at the same time, they fostered musical creativity and contribution to knowledge. Different aspects of practice-based research are described using his observations and experience.

7 **Kerry Francksen:** The Curious Nature of Negotiating Studio-based Practice in PhD Research: Intimate Bodies and Technologies.

Kerry's chapter considers some of the important activities involved in negotiating practice in PhD research. She contemplates the often complex, changeable, and multi-layered processes of negotiating practice-based research, and discusses key topics such as methodology and knowledge as practice. Specifically, she reflects on some of the practicalities of exploring practice via a studio-based investigation and highlights a number of key discoveries that he encountered during this process. And how this ultimately helped to define the emerging thesis and ultimately enabled her to explore the production of knowledge via practice. As such, some of the discoveries made pose interesting questions for the practice-based researcher.

8 **Petra Gemeinboeck and Rob Saunders:** Encounters at the Fringe: A Relational Approach to Human–Robot Interaction.

Petra and Rob focus on their long-term collaborative practice, which aims to contribute to the public imaginary of our sociotechnical future by exploring how machines creatively and aesthetically participate in social encounters. In this chapter, they begin by briefly describing their long-term art-science collaboration and how methodological development and knowledge production are deeply entangled in their practice. Additionally, they give insights into their underlying core concepts that are constitutive to and materially enacted by their methodology of embodied, performative inquiry, as well as the new insights and understandings they produce.

9 **Pearl John:** The Impact of Public Engagement with Research on a Holographic Practice-Based Study.

Pearl introduces her field of practice as an artist working with holography and lenticular imaging. Drawing from her experience and her own development as a researcher, she discusses lessons learnt during her PhD research journey, aiming to answer the questions: In what ways is the *Vitae* Researcher Development Framework (RDF) beneficial to the research process? And how could supervisors make better use of the framework to assist researchers with their development?'

10 **Gail Kenning:** Project-based Participatory Practice and Research: Reflections on Being 'In the Field'.

Gail's chapter focuses on an arts health research project. The aims of the project were to understand how older adults felt about where they lived, use arts-based approaches for data collection, and create a digital artwork for exhibition in the community. She outlines how art and design engagement approaches can become methods for use in research, opening up new possibilities, and suggests her practice as an artist and designer prepared her for engaging in this research.

11 **Sofie Layton:** Bearing Witness – The Artist Within the Medical Landscape: Reflections on a Participatory and Personal Research by Practice.

Sofie discusses her practice as an artist, which is rooted in participatory arts within the medical context, and reflects on the key research elements within her practice. These include the participatory process, collaboration, and the artistic filtering of the narratives and metaphors which emerge through the participatory research process as a final artwork. Through this she highlights how, within the arts, and arts/health, and scientific fields of transdisciplinary research, the partnership with the institution and the development of individual relationships, which may become a collaboration, are essential aspects of good research practice.

12 **Debbie Michaels:** Organisational Encounters and Speculative Weavings: Questioning a Body of Material.

Debbie reflects on an interdisciplinary project in which she transposes approaches from psychoanalysis and art psychotherapy to the fine arts in an experiment with method. Conceptualised as a 'speculative weaving' in three transpositions, her research follows the intertwining dialogues and entanglements that emerge as she traversed institutional boundaries in healthcare and academia. Through this new understanding emerges *in/through* the moving, (re)assembling, handling, (re)configuring, and sharing of diverse practices and material, the interweaving of dialogues, and the negotiation of tensions and resistances encountered at the borders between domains.

13 **Corey Mwamba:** Improvising as Practice/Research Method.

Corey's chapter outlines his work as a musician, using vibraphone and audio software in both improvised and prepared settings. His research engages with the relationship between the vibraphone and the improviser, and he discusses how that relationship projects the idea of a personal sound in jazz and related forms. From this, he advocates using improvising as a method for investigation in practice-based research, and how the practitioner researcher can use improvising to form new research questions and insights.

14 **Fabrizio Augusto Poltronieri:** Dreaming of Utopian Cities: Art, Technology, Creative AI, and New Knowledge.

In his chapter, Fabrizio discusses some of the questions that intrigue him as researcher and artist working with practice-based research. He outlines how his pragmatic, *in vivo* experience has showed him the importance of working with prototypes and small-scale projects using a methodology developed mainly from art. He applies this to a practical project involving Creative AI, *ArchXtonic*.

15 **Deborah Turnbull Tillman:** Curating Interactive Art as a Practice-Based Researcher: An Enquiry Into the Role of Autoethnography and Reflective Practice.

Deborah discusses her investigation into the role of reflective practice and autoethnography in curatorial practice. It examines two case studies, the first focusing on the methods articulated in a curatorial case study as part of a practice-based PhD on curating interactive art; the second focusing on an enquiry into methods acquired academically being useful professionally.

16 **Marloeke van der Vlugt:** Please Touch!

Marloeke's chapter outlines her field of practice, before describing and presenting the variety of methods that she deploys in her research. She zooms in on the case study *Thresholds of Touch*, a performative experiment based on an inter-disciplinary collaboration between a composer/researcher, a sociologist, and an artist/researcher. Through this, she shares the collaborative methodology between social science and artistic research, and what it contributed to researching *touch* from her perspective on practice-based research. She outlines the power relations between disciplines, methods, and forms of expression/knowledge and reflects on how different documentation strategies would have foregrounded other experiences, insights, and/or knowledge.

5.2

A NEW FRAMEWORK FOR ENABLING DEEP RELATIONAL ENCOUNTER THROUGH PARTICIPATORY PRACTICE-BASED RESEARCH

Alice Charlotte Bell

Research design

My Participatory Practice-Based Research (PartPb) methodology is built from the transdisciplinary fields of arts and health. Consequently, I incorporate both exploratory approaches to producing artistic artefacts and to facilitating participants ethically. The main research strategy used within my PartPb is Action-Research, formulated by Kurt Lewin,[1] combined with iterative in-vivo reflection-in-action and in-vitro reflection-on-action as outlined by Donald Schon.[2] I also utilise the established self-reflexive methods of Ray Holland[3] and ethnographic methods of Carolyn Ellis et al.[4] Additionally, I extend a form of Interpretive Phenomenological Analysis (hereon IPA), useful for its focus on relationships and the idiographic concerns of lived experience, as expounded by Smith, Flowers and Larkin.[5]

Within my PartPb research processes, I also interweave relational materials influenced by psychoanalytic and feminist theory. Key concepts include from Bracha Ettinger,[6] thinking m/otherwise, carrying and co-habitation, all of which incorporate a sense of co-mingling within artefact generation as researcher and participant. I also make use of psychoanalyst Donald Winnicott's[7] concerns with holding, becoming, transitional objects and phenomena whereby a researcher provides a 'good-enough' caregiving environment within which participants can explore and relate deeply within a PartPb project world.

My new framework consequently provides a guide to the associated behavioural values and ethics required of a PartPb researcher. These include propensities such as balance, vulnerability, empathy, generosity, support, resilience, maturity, confidence and humility. Future PartPb researchers may arrive experientially ready with these qualities; if not, they can cultivate such capacities further through their own formative projects, professional and personal development.

Alongside these relational approaches, my framework employs PbR artistic strategies and utilises multimodal art forms: human–computer, digital, performative, screen and sculptural as well as the arts psychotherapeutic. In PartPb approaches, questions continually arise during the process of making between practitioner researcher, practitioner-participants and artwork, with all resultant new knowledge generated from inside the participatory practice itself. Because of the inclusion of human participants as both co-creators and subjects within PartPb artefact invention, their experiences

DOI: 10.4324/9780429324154-45

also inform practice outcomes. Outputs therefore include practitioner procedural, operational and behavioural guidelines, as well as artistic outputs, digital dialogues, performance encounters, screen narratives and interactive sculptural objects.

PartPb framework overview

As an output itself, my new PartPb framework has both an outer PbR construction informed by Linda Candy's and Ernest Edmonds' PbR Trajectory,[8] to include elements of Practice, Evaluation and Theory and an inner Gestalt core, stemming from psychotherapeutic Gestalt theory informed by Fritz Perl's.[9] At its centre are the researcher, participants and resultant artworks. In Figure 5.2.1, the background 3D Möbius image represents a PartPb artefact world, here the project 'Transformational Encounters: Touch, Traction, Transform', (hereon *TETTT*). The multiple narrow arrows indicate the movements a PartPb researcher traverses iteratively within the outer PbR trajectory through elements of Practice, Evaluation and Theory. The single wide dotted arrow denotes the Gestalt Cycle of Experience that a participant operating within a PartPb process also navigates.

When engaging human subjects within a PartPb project, various challenges are inevitably faced. The design of my new framework assessed and addressed in-action such events in application to the *TETTT* project. The main issue that arose through *TETTT* was how to manage participants within the Practice element, whereby their involvement as co-creators meant that I could not practically and ethically just carry on making autonomously when participants stalled creatively or psychologically. To gain clarity on this problem I revisited Candy and Edmonds' PbR model[10] and started to modify it to suit my needs first practically, by introducing three newly

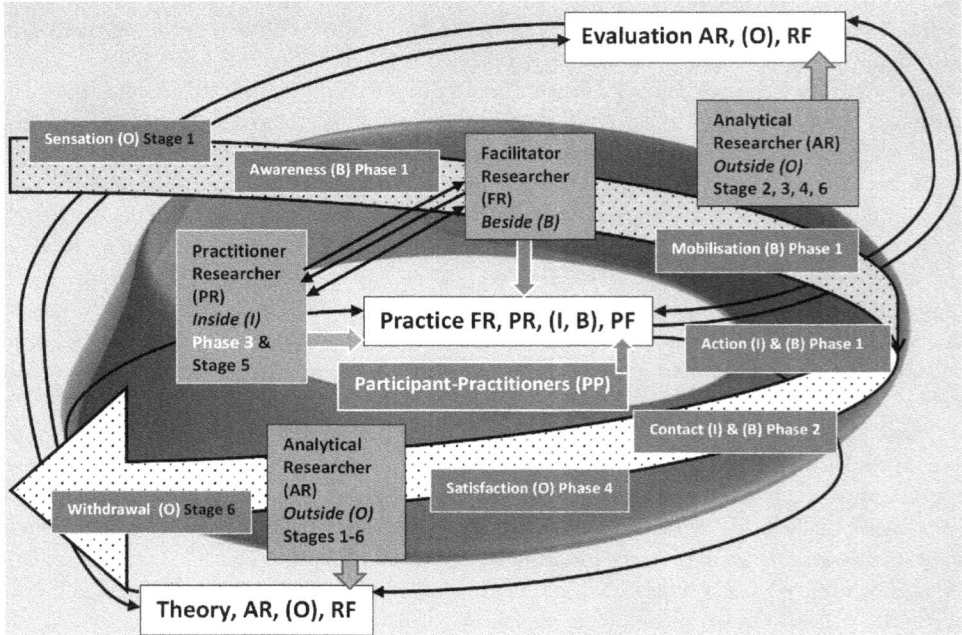

Figure 5.2.1 My new PartPb framework for enabling deep relational encounter. (Möbius Image contained beneath my own diagram licensed under 'File: Möbius strip 3D red.png' by BojanV03 under CC BY-SA 4.0).

Source: Alice Charlotte Bell.

defined specific researcher positions. These are Analytic-Researcher (AR), Practioner-Researcher (PR) and Facilitator-Researcher (FR) denoted on Figure 5.2.1 to define my operational role at each point of PartPb making. The Analytical-Researcher (AR) combines both Theoretical and Evaluative determinants; the Practioner-Researcher (PR) and Facilitator-Researcher (FR) both operate in the Practice element.

Reaching these definitions allowed me as AR to self-reflexively regain and maintain overall project overview even when interpersonal-psychological aspects impacted the progression of the research. As FR, it also allowed me to concurrently intermingle with my Participant-Practitioners (PP) within the Practice element, searching for new creative solutions to re-empower participant agency. Evaluatively, my three new researcher positions operated Outside (O), Beside (B) or Inside (I) participatory artefact generation. The Theoretical and Evaluative elements predominately happened Outside (O) participatory artefact making and provided necessary reflective and analytical researcher opportunity. These I came to term as Researcher-Facing (RF) Stages. All Practice elements I referred to as Participant-Facing (PF) Phases, and these were conducted with and for participants and later audiences. The location of these Stages and Phases are also mapped on Figure 5.2.1 in relation to the *TETTT* project extrapolated later in this chapter.

Having compartmentalised the theoretical and evaluative components of my PartPb research, the real challenge was now how to stay in relationship with my participants within the Practice element comfortably and without halting the overall process. This was challenging, at times overwhelming, and asked for patience, resilience and trust. In these moments I experienced the process of artefact generation as a morphing field of multimodal objects and subjects in intra- and interpsychological[11] interplay. I found that trying to transition from any one of my new researcher positions (AR, PR, FR) to another, could at times be met with points of agitation between multiple states. This is shown on Figure 5.2.1, occurring mainly between the positions of Practitioner Researcher (PR) operating Inside (I) and Facilitator-Researcher (FR) operating Beside (B) the participatory Practice element. Such disturbances often manifested around a particular participant concern that became stuck within the system.

To address this, I had to devise a secure means of navigating such moments of uncertainty, to find a way as FR that I could remain responsive to and co-responsible for participants' emergent needs but without losing my own footing. As such, I added an inner psychological-phenomenological core within my framework based on Perls' Gestalt Cycle of Experience.[12] In Figure 5.2.1, the seven phrases in white text articulate the inner states a participant can encounter within, which can either facilitate creative flow or be met with resistance. In Gestalt terms this is through a positive experience, or its opposite:

- sensation/desensitisation
- awareness/defection
- mobilisation/introjection
- action/projection
- contact/retroflection
- satisfaction/egotism
- withdrawal/confluence[13]

While there is not space to cover these experiential states in full, it needs to be noted that there are also many other inter and intrapsychological[14] elements at play within the PartPb world of artefact generation. In this sense, the artefact is in reality a complex field of human and artistic components in metarelational flux. To try and express this, I came to term the outer PbR scaffold and inner Gestalt core of my new PartPb framework as a form of *feeling architecture*, within which participants are held and activated both psychologically and creatively. My phrase draws upon both the architectural and phenomenological concepts of Juhani Pallasmaa[15] and Raymond Williams' 1970s concept of 'structures of feeling'[16] as well as my own observations.

TETTT overview

TETTT was a year-long project made by myself in engagement with 12 participants who prototyped my new PartPb framework-in-action. My new PartPb contains six RF Stages and four PF Phases and allows participants to engage, slow down and create in deep and meaningful ways. In providing for a scaffolded, supported and sustained relational encounter with a practitioner researcher, participants can explore, narrate, and perform identity through multimodal artistic means. Specifically:

- Phase 1: *Digital Dialogues* sees a process of sharing stories digitally and provides an alternative online space within which researcher and participant can relate intimately. In *TETTT*, the Phase 1 section consisted of three 7-day sets of researcher multimodal prompts themed *Touch*, *Traction* and *Transform*. The subject matter for each prompt is planned thematically in advance. The prompts are also shaped throughout the participatory process customised to each project.

- Phase 2: *Performative Encounters* enables participants to perform aspects of self playfully. It also fulfils a need for a new form of transformative face-to-face, one-to-one encounter to be provided that is primarily participant-subject, and not researcher-practitioner, led. These live performance encounters are also filmed by the researcher in specific locations with precise props and participant objects. The content and format of the events are informed by the analysis of participants' Phase 1 data in Stage 2, and also allow for spontaneous improvisation within the encountering moment. The type of encounter will continue to vary from project to project, researcher to participant, as it is customised to the specific content stimulated and shared.

- Phase 3: *Screen Narratives* embodies, carries and intensifies participants' digital and performed storied material through into screen media. Here the researcher edits the films of Phase 2 *Performative Encounters*, interweaving footage with Phase 1 content and themes as defined in Stage 2. The application of a feminist ethnographic and somatic approach is foregrounded to give a sense of being beside participants, and not speaking for them.

- Phase 4: *Relational Artworks* addresses the need for the placement of participant stories, performances and screen media within art objects that invites subsequent embodied responses from audiences. This is a public exhibition of large-scale interactive sculptural objects containing Phase 3: *Screen Narratives*, able to hold audience members' entire physical bodies. This phase integrates all findings from Phases 1–3. In this phase, participants are also referred to as participant-audience, providing them with an opportunity to reflect on their journey within the project. The 4 PF Phases of the framework hold, carry and intensify meaning cumulatively across virtual and physical spaces. Such expressions of Self serve to reach beyond normative social, cultural, personal and artistic boundaries.

The 4 Phases of practical artefact generation are also aligned to inner Gestalt facilitation processes as mapped on Figure 5.2.1. The first seven days of *Touch* are designed to bring participants into 'Awareness'[17] with their inner material. The subsequent seven days of *Traction* take participants into 'Mobilisation',[18] a form of active recognition with their disclosures. The last seven days of *Transform* bring them into 'Action',[19] a process of wanting to dynamically do something with their findings. In Phase 2, participants enter a period of full 'contact',[20] with distilled inner material within their *Performative Encounter*. The Phase 3 *Screen Narratives* presents their performance back to them, augmented by practitioner observations. These films also allow for a sharing of participants' experiences with others in edited form. The final Phase 4 exhibition enables participants to see the full impact of their co-creations on Self and Other and feel 'Satisfaction',[21] before a slow 'Withdrawal'[22] from the project ethically. However, my new PartPb framework is scalable and agile and can be reduced or expanded by future PartPb researchers in application to their own projects. The Phase 1 themed sets of *Prompts* could be called any number of things in application to future project interpretations, as long as they have the desired effect of activating initial awareness, mobilisation and action within project participants.

Experiential learning

Despite careful planning, the inclusion of participants in PbR brings with it unforeseen surprises that cannot necessarily be anticipated. These events, needs and revelations enable what Candy terms as 'reflection-on-surprise',[23] unexpected moments within the practice that pushes the PartPb knowledge forward. Such moments will differ between cohorts and projects. This demands agility, patience and confidence on behalf of the PartPb researcher to negotiate emergent challenges or desires sensitively.

In *TETTT*, the surfacing needs of participants in-action following Phase 1, necessitated the introduction of extra sub-phases within *TETTT* that involved collective and not just one-to-one participant-to-researcher encounters. I have not detailed these additional Phases on Figure 5.2.1; however, they did add complexity to the project. While this brought with it a certain richness, it also introduced supplementary layers of intricacy demanding both a practical and energetic aspect. In *TETTT*, I endeavoured to respond to such requests as opportunity for a potential revelation of new participatory knowledge. In self-support, I selected and oscillated between my AR, PR and FR positions to both acknowledge participants as co-creators of the PartPb artefacts and give them increased agency. However, I also maintained an overall aesthetic organisation, gently guiding the direction of the project inclusively.

Navigating such nuances led me to conclude that my new PartPb framework could be either expanded to accommodate (or omit additional sub-phases), according to how much intimate one-on-one or collective relational experience future PartPb researchers chose to reveal to other participants, particularly in Phase 1. Phase 1 is the most multifaceted of the 4 Phases. It sees a complex interrelated exchange of 21 multimodal and themed *Prompts* delivered digitally to all participants by the researcher. In turn, all participants return a *Response*. The researcher then makes sensitive and non-judgemental observations which I termed as *Noticings*. *TETTT* started with 21 participants, and by the end of Phase 1 it still had 16 subjects fully engaging. I had not anticipated so many participants sustaining Phase 1. This proved stressful in terms of time, energy and quality, particularly because the format of Phase 1 is the most important in terms of setting up the future success of subsequent Phases 2–4.

With 21 *Prompts* going out, 16 *Responses* coming back and 16 individual *Noticings*, (excluding *Group Noticings* I made in addition within *TETTT*, that in turn served to fuel the additional sub-phases 1a and 1b); this amounted to 693 exchanges in 21 days. Experientially, this put pressure on maintaining quality, calm and sensitive attentiveness, which are so integral to building bespoke and meaningful relational encounters with participants through PartPb projects. Eventual findings therefore point to a ratio of 8:1 participant to researcher as an ideal balance for a 12-month project. However, my new PartPb framework could also be fulfilled simultaneously by several researchers working concurrently or collectively with different groups of participants across various thematic projects. Indeed, it could also be extended, reduced and revisited through Phases and Stages to add new knowledge to this new field of PartPb research.

Conclusion

This chapter discussed the generation of my new participatory practice-based framework in relation to the project *TETTT*. I specifically explained how my framework is constructed to support both researcher and participants within its *feeling architecture*. I articulated how my design comprises both a robust outer creative PbR scaffold interwoven with an inner Gestalt core. I described that this can be used to support and stimulate sustained one-to-one relational encounters between creative practitioners with participant-subjects across multimodal art forms. I gave a brief overview of the 4 Phases and 6 Stages contained within my framework and brought detail to Phase 1 to indicate both its importance, complexity and relationship between artistic provocations and psychotherapeutic facilitation. I defined my three different researcher positions and behaviours in relation to being Inside, Beside and Outside participatory artefact generation. There isn't space here to illustrate all the detailed inner workings of Phases 2–4, or each analytical and evaluative Stages 1–6 but I have noted these are designed (and do) distil, traverse and deepen relational encounter across multimodal artforms. I also discussed the ethical and artistic challenges that I overcame in designing this new framework for PartPb research. I conclude that my original framework can enable future practitioners to generate meaningful and innovative new participatory projects and in doing so, deliver robust, expressive and continuing knowledge to my new field of PartPb research.

Notes

1 Adelman (1993).
2 Schon (1983).
3 Holland (1999).
4 Ellis et al. (2010).
5 Smith et al. (2009).
6 Ettinger (1995/2001).
7 Winnicott (1971/2001).
8 Candy and Edmonds (2010).
9 Perls et al. (1997).
10 Candy and Edmonds (2010).
11 Vygotsky (1978).
12 Perls et al. (1997).
13 Ibid.
14 Vygotsky (1978).
15 Pallasmaa (2012).
16 Franks (2014, p. 10).
17 Perls et al. (1997).
18 Ibid.
19 Ibid.
20 Ibid.
21 Ibid.
22 Ibid.
23 Candy (2019).

Bibliography

Adelman, C. (1993). Kurt Lewin and the Origins of Action Research. *Educational Action Research*, 1(1), 7–24. DOI:10.1080/0965079930010102 (accessed January 4, 2021).

Candy, L. (2019). *The Creative Reflective Practitioner*. London: Routledge.

Candy, L., & Edmonds, E. (2010). Relating Theory, Practice and Evaluation in Practitioner Research. *Leonardo*, 43(5), 470–476.

Ellis, C., Adams, T., & Bochner, A. (2010). Autoethnography: An Overview. *Forum Qualitative Sozialforschung/ Forum: Qualitative Social Research*, 12(1), Art. 10. http://nbn-resolving.de/urn:nbn:de:0114-fqs1101108 (accessed July 3, 2020).

Ettinger, B. (1995) [2001]. *The Matrixial Gaze*. Leeds: Leeds University Press.

Franks, A. (2014). Drama and the Representation of Affect – Structures of Feeling and Signs of Learning. *Research in Drama Education: The Journal of Applied Theatre and Performance*, 19(2), 195–207.

Holland, R. (1999). Reflexivity. *Human Relations*, 52, 463–484. doi:10.1177/001872679905200403 (accessed March 9, 2021).

Pallasmaa, J. (2012). *The Eyes of the Skin: Architecture and the Senses*. Hoboken: Wiley, 3rd ed.

Perls, F., Hefferline, R., & Goodman, P. (1997). *Gestalt Therapy Excitement and Growth in the Human Personality*. London: Souvenir Press.

Schön, D. A. (1983). *The Reflective Practitioner – How Professionals Think in Action*. New York: Basic Books.

Smith, J., Flowers, P., & Larkin, M. (2009). *Interpretative Phenomenological Analysis*. London: Sage Publications.

Vygotsky, L. S. (1978). *Mind in Society: The Development of Higher Psychological Processes*. Cambridge, MA: Harvard University Press.

Winnicott, D. W. (1971) [2001]. *Playing and Reality*. London and New York: Routledge.

RISK, CREATIVE SPACES AND CREATIVE IDENTITY IN CREATIVE TECHNOLOGIES RESEARCH

Or why it's okay for academic creative technology outputs to look scrappy and be buggy

Oliver Bown

"Brakes, don't talk to me about brakes! Anyone can make a car go slow! It takes a genius to make a car go fast!!" From "The Race", by Peter Ustinov.

Research and practice background

I have worked for over 15 years as what I often describe as a creative technologist. I make both new creative works using technology and new technologies for creating creative works. In creative technologies practice, these are interrelated activities.[1] Primarily, my work contributes to knowledge through algorithms, interface ideas, other design ideas and creative practice methods and concepts that are ultimately aimed at benefiting other people in their work. Two obvious and different ways this can be manifest are:

i   new ways of doing things that are sufficiently well documented and explained as to be useful to others
ii   new tangible tools, such as a new piece of software, that can *actually* be put to use by other people

For the former, it is possible to work as an experimental artist who reflexively examines and reports on their process using practice-based research methods. Engaging in inquiry-driven artistic practice can be enough to constitute significant technological experimentation leading to new insights, and sharing one's process through such experimentation can reveal new ways of doing things that benefit others in the field. Striving to explore the creative possibilities of technology, manifest through one's own personal creative outcomes, makes for a valuable contribution to the field when successful. Note, however, that not everyone who experimentally applies technology to creative outcomes does so in order to then abstract and share principles of creative technologies practice; many do so, for example, to critique or philosophically query the technology's implications.[2] But that has generally been *my* focus.

DOI: 10.4324/9780429324154-46

Creating new tools involves at least one significant extra step beyond the work involved in creating artworks themselves: the building of a *functional* system, not just a working prototype, is a distinct extra tranche of work. From a design perspective, making tools also involves additional work in terms of ensuring the tool is usable for its users (possibly starting with oneself), for example through well-designed UIs. Working and usable tools depend on good design methods focusing on developing appropriate systems for their intended users. And yet these two activities (creative function and usable tools) frequently overlap and are sometimes indistinguishable, with artists producing tools for their own use primarily, with the sharing of these resources being a natural extension of their work.[3] In other cases, the researcher is not really pitching themselves as an artist, but rather as a producer of tools, and yet nevertheless takes on the work of exploring and demonstrating a system's creative possibilities themselves (as a temporary or placeholder artist).

Practice-based and design approaches can also be structured in a natural progression from experimentation to product. Such a progression roughly plays out as follows:

- create an artwork that explores some technological idea
- assuming success, reflect and report on how it was made (the creative practice, as well as the technology, its affordances and interaction possibilities)
- consider what software might support the application of this process
- prototype the software
- iterate the software through further development and user research

With so much to do, such research requires highly pragmatic time management. Several of the preceding steps may be beyond the scope of an academic research agenda. In particular:

i   Producing creative works may be more or less "prototypey" (versus "high production values" work). In the context of the creative possibilities of technology, one may wish to focus more effort on the creative exploration than on the production of a finished artwork. To achieve the goal of discovering or communicating new ways to work effectively, the researcher is arguably wasting time by putting effort into polishing a work.

ii   Likewise, producing software may be more or less "complete" from an end-user perspective. Again, the time it takes to make software work, being bug-free and with great UX, is arguably work that shouldn't take up research time.

The ambiguity of the term "creative technologies" itself suits this situation. It can mean both "being creative with technology" or "making technology for creative people".[4]

Figure 5.3.1 summarises how one might step through a creative technologies project with different outcomes in mind.[5] At the end of each path lies something that would be considered a knowledge outcome, i.e., something that could be published as an academic paper. This is an incomplete representation; there are likely to be many other pathways, but as a categorisation of ways to achieve knowledge outcomes it covers the key methodological ideas. What this diagram does not represent, and what is critical to the discussion here, is the relative effort that one may need (or wish) to invest in each of these areas. Developing and reflecting on a single proof of concept may consume an entire project.

Cases

From personal experience, I consider two concepts that go further in framing the challenges of the pragmatic creative technology researcher. The first considers the question of time commitment and risk when undertaking specific activities, in relation to expected outcomes. I will look at how

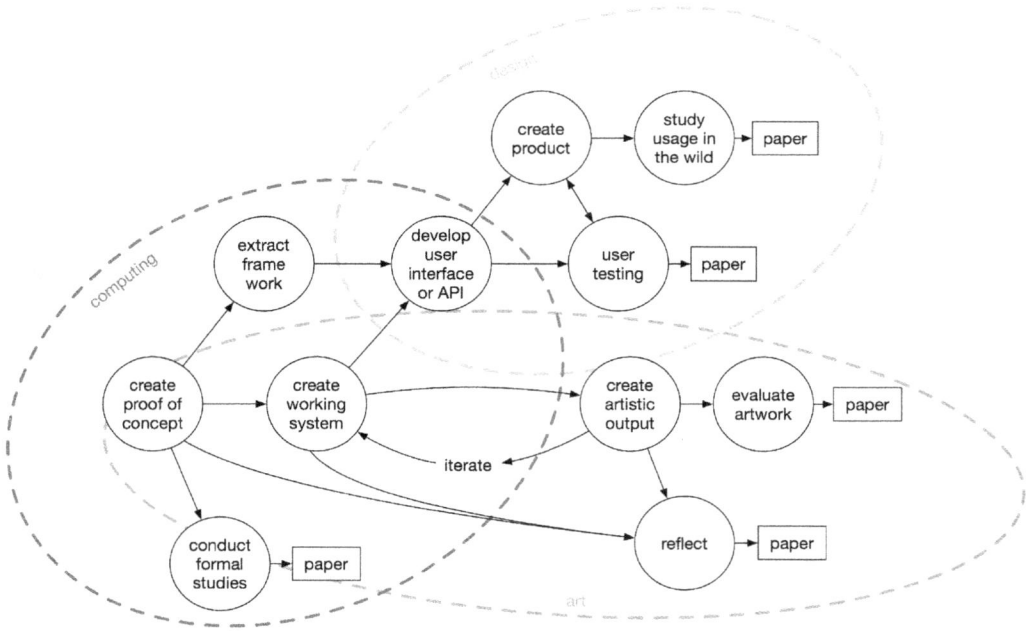

Figure 5.3.1 Different pathways leading from an initial creative technologies idea (demonstrated in a proof of concept) towards different knowledge outcomes that can be published as academic papers.

a practitioner handles the risk of an idea not working out at all, and the more pragmatic risk of unexpected barriers to success. The second considers how much effort goes into setting up a creative space, through prior technical work and design thinking. I will look at examples of building in creative freedom into a tool. I discuss each in turn with specific reference to my own work and that of my PhD student, Steffan Ianigro. In relation to these two points, I will also consider the issue of expertise and the development of an artistic identity and look at the very long timescale over which this happens, and how this influences one's strategies.

The question of time commitment and risk

My work *Zamyatin* for live improvised performance used evolutionary computing to discover novel real-time performance behaviours that could be incorporated into a hand-coded responsive, generative performance system.[6] The system was developed over multiple years beginning around 2010, following a previous concept iteration (previously I had used evolved neural networks, a specific non-linear system design,[7] and for *Zamyatin* I developed my own non-linear system architecture based on decision trees and feedback).[8]

As *Zamyatin's* developer and only user, I coded the system in an Integrated Development Environment (Eclipse) but also needed to interact with the system in multiple ways: running evolutionary algorithms to search for evolved behaviours, auditioning behaviours, interacting with the system in real time and tweaking configuration parameters (both in real time and offline). Coding time could be spent in a vast number of ways: implementing better evolutionary algorithms (which might improve the search process), implementing better evolutionary fitness functions, implementing a better real-time performance system, building better graphical user interface tools, building other analytical tools to help understand how the algorithms were behaving. In addition, coding time competed with time devoted to other activities such as auditioning, preparing test recordings, preparing sample banks and other raw musical material and rehearsing.[9]

Accordingly, making polished user interfaces, and even polished works and recordings, took away time from exploring the system from an academic standpoint, even though these things might have other benefits such as improved promotion of the work. Conversely, conducting formal user studies could have been effective, but only if the system was well-developed enough and also conceptually clear enough to other potential users, which required significant development from the system I used in my own work.

A significant risk of creative technology research, which can also be found throughout experimental creative practice work and technology innovation, is the unpredictability of outcomes. This is fundamental to creative work, which by broad consensus of definition is work for which there is no prior known path to an outcome. Creative search has been defined as blind[10] and inherently undirected and heuristic.[11] As with many creative technology ideas, the existential risk with *Zamyatin* was not coming up with any worthwhile behaviour at all. The development of a new evolvable structure based on decision trees and feedback was highly speculative. Thus, it was essential to develop the work in an agile way, first attempting to develop any evidence that there was something there worthwhile, in as short a time as possible.

An agile methodology has emerged as one of the leading solutions to risk management in technology development. Development occurs in short bursts with frequent review, with a readiness to revise design goals and plans. In an agile approach to the initial ideation and experimentation in creative technologies research, there is some potential to adapt goals on the go. I found that my evolved decision tree approach produced interesting-enough behaviour to justify its use, but as expected, its complexity was hard to control or steer to specific goals via evolution. I therefore focused more on the creative practice of hand-coding behaviours that responded to the decision tree output, which gave me a form of meta control, and constituted a kind of slow collaboration with this complex system. That shaped the final form of the system and the kind of research that was involved.

In his PhD research, Steffan Ianigro has continued to work with similar systems, picking up my original evolved neural network research.[12] Ianigro's focus has been much more on the user interface possibilities, for which he has made an interactive web page where users can search through large populations of evolved behaviours, and an Ableton Live device, which allows users to immediately incorporate these dynamic non-linear behaviours into their creative work. To use this, users must deal with some very complex concepts and behaviours, but via a simple interface that "black boxes" much of this complexity. Ianigro has produced his own suite of practice-based compositional works using these tools but is also in the process of gathering user feedback from the fully functional tools he has shared online. For both of us, there are many more systematic experiments we could attempt in order to find better underlying system behaviours, and much more we could have done to make improved, working user interfaces, but through practical compositional and design decisions, we have navigated the trade-off between time commitments.

Effort and setting up a creative space

The second of my principles concerns the creative space, and the ways in which it is possible to set yourself up more or less easily to creatively explore. Consider how the knobs, sliders and buttons of a good synthesiser stimulate exploration, or how good modular frameworks, as in modular synthesiser design, can enable more high-level exploration. Such work can be grounded in a powerful body of research into creative systems design, creative technologies practice and the design of productive tools.[13]

Working in code, one is faced with an infinite space of possibilities, but only some of those possibilities are ready to hand. Some time spent coding a useful framework for oneself to use might alter what is to hand, thereby enabling the potential for greater exploration. In *Zamyatin*, I went to great lengths to ensure that my system had good modular design, in particular because in this system I combined evolved behaviours (clearly defined as a modular component) with hand-coded musical structures for each performance. I needed to ensure that I could easily prototype those hand-coded structures and connect them into the larger system interchangeably.

In Ianigro's case, part of the research was explicitly about providing an interface for rapid search: a website that one could explore for different behaviours (generated using a technique called novelty search, which promotes a diversity of generated outcomes), with the ability to embed those behaviours into a DAW for immediate use. Such search-enabling interfaces will normally promote search across some dimensions at the expense of others: it is about enabling not an exhaustive search space but a productive one. In Ianigro's current implementation of the interface, the beginner user can start with a set of six pre-sets that represent a diverse set of curated behaviours. This very small, curated set did not encourage more detailed search of the space of system behaviours, but did help focus search on how one applied the systems into a composition, much as I found with my coded behaviours.

Reflection

In Figure 5.3.1, I presented a diagram of different paths through a creative technology research process from an initial proof-of-concept system. This shows how one may arrive at different knowledge outcomes, which I highlight in Figure 5.3.2 for each of the projects discussed.

Navigating any creative technologies project with this high-level map helps identify the possible exit points where knowledge outcomes can be achieved. Just like a real map, which may not show the terrain or traffic jams, what the map does not show is the time and effort that might be required (or desired) to invest into each of these stages, which will vary radically depending on the immediate challenges presented for the current task. Adopting an agile approach to navigating this map means re-evaluating where one is at regularly and seeing what steps are needed to arrive at any knowledge outcome. Importantly, while complete working software, tested in the hands of users, or polished creative works presented in high profile public areas, are highly desirable outcomes; there are many other effective knowledge outcomes along the way that may serve the wider community to produce these things. Any of these stages might be vast multi-year projects or even unsolvable problems that need to be aborted. Conversely, third-party tools, prior research or examples might already exist that act as platforms for researchers to start their paths at different stages (e.g., design researchers conducting in-the-wild research into existing products). Some journeys through this space may be rapid, while others are stuck on one node for a long period. Creative technology researchers with limited time must plan accordingly how they get to the best knowledge outcomes given the circumstances in which they find themselves.

Occasionally system designs arise that open up vast spaces of creative exploration, usually by taking an existing working concept and introducing an interface to that which is fast and agile. For example, the Runway program for interacting with machine learning (ML) models has made it very easy for non-programmer artists to explore creative applications of machine learning. The opening of such creative spaces offers both important knowledge outcomes that emerge from the processes just described and also new creative ecosystems that enable entirely new genres of creative technologies practice. Students at my school of art and design with no programming background, for example, are creatively exploring the generative power of ML tools via Runway, which lets them navigate new pathways to knowledge outcomes.

Likewise, creative technologists are constantly setting up their own spaces of exploration, through programming libraries and interfaces or configuring datasets and tools. This can be critical groundwork to support their own artistic creativity, or can be an end-in-itself, as a design outcome seeding a new wave of creative possibilities. It is important to identify these critical innovations in supporting creative work as a specific goal in creative technologies work, whether they are entirely there for the

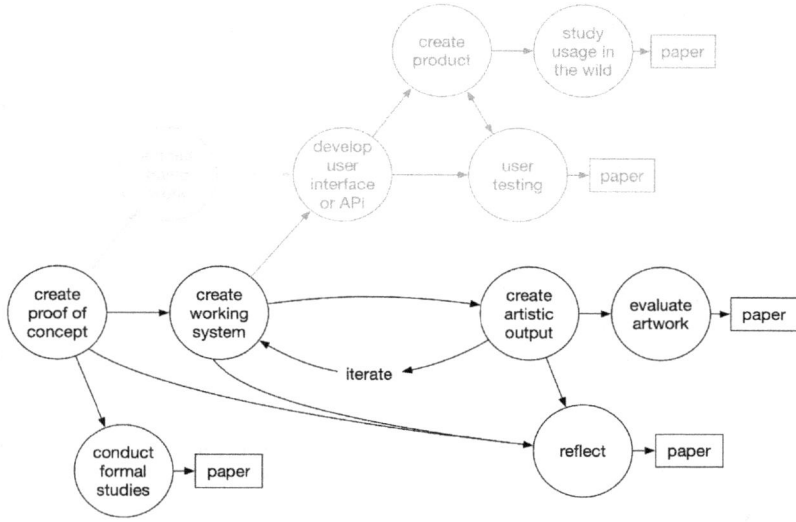

Bown

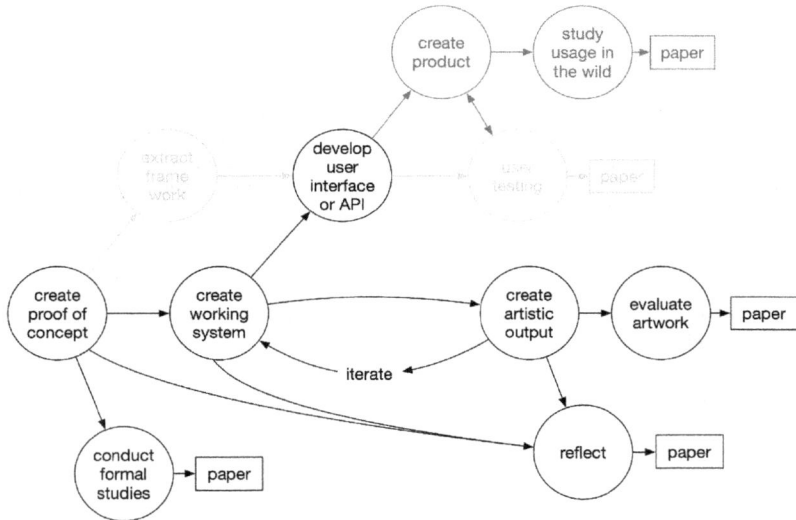

Ianigro

Figure 5.3.2   Two different research projects shown as paths through possible outcomes (highlighted for core outputs and peripheral outputs).

individual researcher to use themselves to improve their own creative outcomes, or as key landmarks in the development of effective creative tools for others. Practitioners should become familiar with thinking about the setup of these creative spaces, whether building tools for others or using tools themselves; they are critical foci of creative technologies research and they may turn out to be the critical outcome of a research project.

The preceding discussion relates to individual projects that one might think of as having a finite duration (such as a PhD project). A more holistic view of creative technologies practice recognises the long-term investment of individuals as they embark down these paths of exploration.

A practitioner may spend many years gradually moving between nodes in one of these search spaces, developing expertise and honing a creative practice. They do so as part of a community of practice, bringing the baggage of their prior investments in time and technology (what skills, programming languages and tools they know and have ready to hand) and their embeddedness in a social network (what work they are exposed to, what status the aspire to, where they can get help and what they are known for by others). Mostly, this longer-term embeddedness remains invisible from the standpoint of everyday practice, but as with the opening of creative spaces, this should be a constant source of reflection; often this is simply to know one's strengths and the scope of one's effective work, but occasionally it is because the true innovation lies in the (possibly blind) leap into a different way of working that will define one's research direction for a long time to come. Many academics find themselves at this juncture at the beginning of a PhD where they are deciding on a path for their project, but also at the end, where their process of reflection may define their future career trajectory. In this sense, enabling such reflection should also be a crucial part of a PhD training.

In these comments, I hope the above topics define useful ways to help others frame their project planning. This is perhaps particularly useful to PhD students and other research projects with limited resources and ambitious innovative goals. Sometimes you want to make something go fast and can add the brakes later.

Notes

1 Edmonds et al. (2005).
2 For one example among many, see Gemeinboeck and Saunders (2013).
3 A great example in creative technologies is the work of Rebecca Fiebrink, which exists in an ecosystem of practice and technology research. See the discussion in Fiebrink et al. (2010), for example.
4 It could also refer to the technology itself being creative, but we should draw the line and reserve the common terms "creative AI", "generative systems" and "computational creativity" for this area.
5 This figure was inspired by Smith and Dean's iterative cyclic web of practice-led research and research-led practice (Smith 2009).
6 Bown (2011, 2015, 2018).
7 Bown and Lexer (2006).
8 Bown (2011).
9 Further to all of this, one must decide to work on a project such as this rather than write a paper or a grant application.
10 Simonton (2011), Stanley and Lehman (2015).
11 Amabile (1983).
12 Ianigro and Bown (2016, 2018, 2019).
13 E.g., Blackwell and Green (2003), Shneiderman et al. (2006), Biskjaer and Halskov (2014), Resnick and Robinson (2017).

Bibliography

Amabile, T. M. (1983). The Social Psychology of Creativity: A Componential Conceptualization. *Journal of Personality and Social Psychology*, 45(2), 357.

Biskjaer, M., & Halskov, K. (2014). Decisive Constraints as a Creative Resource in Interaction Design. *Digital Creativity*, 25(1), 27–61.

Blackwell, A., & Green, T. (2003). Notational Systems – The Cognitive Dimensions of Notations Framework. In *HCI Models, Theories, and Frameworks: Toward an Interdisciplinary Science*. San Francisco: Morgan Kaufmann.

Bown, O. (2011). Experiments in Modular Design for the Creative Composition of Live Algorithms. *Computer Music Journal*, 35(3), 73–85.

Bown, O. (2015). *Player Responses to a Live Algorithm: Conceptualising Computational Creativity Without Recourse to Human Comparisons?* ICCC, 126–133. https://www.researchgate.net/publication/313236755_Experience_Driven_Design_of_Creative_Systems.

Bown, O. (2018). Performer Interaction and Expectation with Live Algorithms: Experiences with Zamyatin. *Digital Creativity*, 29(1), 37–50.

Bown, O., & Lexer, S. (2006). Continuous-Time Recurrent Neural Networks for Generative and Interactive Musical Performance. In *Workshops on Applications of Evolutionary Computation*. Berlin, Heidelberg: Springer, 652–663.

Edmonds, E. A., Weakley, A., Candy, L., Fell, M., Knott, R., & Pauletto, S. (2005). The Studio as Laboratory: Combining Creative Practice and Digital Technology Research. *International Journal of Human-Computer Studies*, 63(4–5), 452–481.

Fiebrink, R., Trueman, D., Britt, N. C., Nagai, M., Kaczmarek, K., Early, M., Daniel, M. R., Hege, A., & Cook, P. R. (2010). *Toward Understanding Human-Computer Interaction in Composing the Instrument*. ICMC. https://quod.lib.umich.edu/i/icmc/bbp2372.2010.027/-toward-understanding-human-computer-interaction-in-composing?view=image.

Gemeinboeck, P., & Saunders, R. (2013). *Creative Machine Performance: Computational Creativity and Robotic Art*. ICCC, 215–219. https://www.semanticscholar.org/paper/Creative-Machine-Performance%3A-Computational-and-Art-Gemeinboeck-Saunders/d94ebdb7fc45fbae2d4d719314f07836bed5b7c1.

Ianigro, S., & Bown, O. (2016). Plecto: A Low-Level Interactive Genetic Algorithm for the Evolution of Audio. In *International Conference on Computational Intelligence in Music, Sound, Art and Design*. Cham: Springer, 63–78.

Ianigro, S., & Bown, O. (2018). *Exploring Continuous Time Recurrent Neural Networks Through Novelty Search*. NIME, 108–113. https://www.semanticscholar.org/paper/Exploring-Continuous-Time-Recurrent-Neural-Networks-Ianigro-Bown/5d957aaeb6fe0364f8c4a5e5790809baa1b2341f.

Ianigro, S., & Bown, O. (2019). Exploring Transfer Functions in Evolved CTRNNs for Music Generation. In *International Conference on Computational Intelligence in Music, Sound, Art and Design (Part of EvoStar)*. Cham: Springer, 234–248.

Resnick, M., & Robinson, K. (2017). *Lifelong Kindergarten: Cultivating Creativity Through Projects, Passion, Peers, and Play*. Cambridge: MIT Press.

Shneiderman, B., Fischer, G., Czerwinski, M., Resnick, M., Myers, B., Candy, L., Edmonds, E., Eisenberg, M., Giaccardi, E., Hewett, T., & Jennings, P. (2006). Creativity Support Tools: Report from a US National Science Foundation Sponsored Workshop. *International Journal of Human-Computer Interaction*, 20(2), 61–77.

Simonton, D. K. (2011). Creativity and Discovery as Blind Variation: Campbell's (1960) BVSR Model After the Half-Century Mark. *Review of General Psychology*, 15(2), 158–174.

Smith, H. (Ed.). (2009). *Practice-Led Research, Research-Led Practice in the Creative Arts*. Edinburgh: Edinburgh University Press.

Stanley, K. O., & Lehman, J. (2015). *Why Greatness Cannot Be Planned: The Myth of the Objective*. Cham: Springer.

5.4

FEEDBACK

Vibrotactile materials informing artistic practice

Øyvind Brandtsegg and Alexandra Murray-Leslie

Background and context

The use of feedback in artistic practice has a long tradition, as seen in the sonic performance works of Alvin Lucier, Elaine Radigue and in the *Relational Aesthetics* of Nicolas Bourriaud,[1] understood as art based on human relations and their social contexts. A thorough description and analysis of feedback in musical instruments can be found in Sanfilippo and Valle.[2] The making, playing and conceptualising of feedback instruments is discussed by Eldridge, Keifer, Overholt and Ulfarson.[3]

The current project fosters feedback as a sound-generating mechanism, as a tool for the analysis of resonant properties of vibrating objects and how they inform processes and ideas embedded in artistic practice and theatrical performance.

This feedback-oriented method can be seen as a practical counterpart to scientific analysis and modelling. Both methods offer specific perspectives on how we can understand material properties. The project made use of polycarbonate plates and bioplastics as resonating objects. The polycarbonate plates were chosen because of the resonant properties of the material. It has a sound that is similar to wood, while also being much more consistent and thus easier to model. The fact that it looks like the now all-too-familiar pandemic screen was something we became aware of during the performative experimentation.

Bioplastics have been around for quite some time. The first known bioplastic, Poly Hydroxy Butyrate (PHB), was discovered by Maurice Lemoigne in 1926. They are not widely used for a variety of reasons, including poor mechanical performance, high production cost and traditional plastics industry resistance.[4] However, there is a renewed interest in bioplastics due, for example, to climate change concerns and the accumulation of plastic in oceans and landfills. Initiatives such as the *European Green Deal*[5] have played an important role, as have the objectives of the New European Bauhaus, which Ursula von der Leyen called the "renovation wave",[6] suggesting that it is "not only about looking into existing building stock. It is the start of a forward-looking process to match sustainability with style".[7]

The current project was done by a cross-disciplinary group of researchers at the Norwegian University of Science and Technology. The principal investigators were the authors, Professor Alexandra Murray-Leslie from the Academy of Fine Art and Professor Øyvind Brandtsegg from the Department of Music. The team was expanded with Associate Professor Kaspar Lasn from the Department of Mechanical and Industrial Engineering, Associate Professor Nina Katrine Haarsaker from

DOI: 10.4324/9780429324154-47

the Department of Architecture and Technology and María Azucena Gutiérrez González and Jussi Evertson from NTNU Oceans Research with collaborating students: Michelle Rassmussen, Thea Fahrstad Blindheimsvik, Weronika Wojtczak and Andréas Méndez. Water cutting of the polycarbonate plates was conducted by Børge Holen. The performative vocabulary was developed in interaction with the performers and the costume designers taking part in the productions being made during the project. This contribution was prominent for a performance in April 2021 in Trondheim, for which we wish to especially credit the contributions of Amalia Wiatr Lewis, Jordan Sand, Melissa E. Logan, Mona Krogstad, Mette Rasmussen, Joel Hynsjö, Faezeh Valadan Zoej, Shiva Sherveh and Dinu Boddiciu.

Methods

The research method is based on an artistic search within the potential served by the materials and techniques. One could call it *explorative and experimental prototyping*; an organic way of developing artistic practice, based on a selected set of known and unknown materials. Still, it is in key respects different from developing an artwork based on known materials, techniques and controlled outcomes (for example, where musical instruments and performance elements are defined from the beginning). In our case, the instruments and techniques were developed and refined during the artistic search, which therefore also informed the themes, costumes and dramaturgy of the performance. We wanted to investigate how we could develop musical instruments using audio feedback as the main sound generator, and in the process reveal some inner characteristics of the materials through which the feedback loop is created. This process was then set in dialogue with the process of actually making music and building a performance dramaturgy with these instruments. As part of the process, a new microphone technique was developed in which the audio pickup is mounted on the performer's finger. This allowed performative exploration of resonance modes by pointing and touching different sections of the vibrating body. The choice of material to use as a resonating body was also part of the search, and we experimented with various metals, plastics, wood, ceramics, concrete, glass and self-made bioplastics.

We are aware that the terms *exploratory prototyping* and *experimental prototyping* are used in other fields of design; see for example Mayhew and Dearnley.[8] We use the terms slightly differently, as our prototypes also end up as part of the final output (if they work well). In this respect, our approach might be closer to "evolutionary prototyping".[9] However, many aspects of our prototyping process are less orderly than what is called for in a commercial product design process. For us, the exploratory, experimental, evolutionary and validation prototyping steps are intermingled.

This difference also informs the evaluation of prototypes. In the artistic search we find that we have no exacting set of quality criteria, and the quality assessment is done through artistic intuition and experimental iterative processes. The intuition based on what we expect might create something interesting to perform with. Richness of interaction and aspects of controllability balanced with the material's own agency are some of the factors that come into play. The process spans the relation between intuition, expectation and performance.

In the case of creating the polymer plastic plate musical instruments, we set out to have something that is controllable enough that one can produce the same result (sound) repeatedly at will, while also having some "lively", slightly uncontrollable inherent characteristics. Such exploratory prototyping methods led to the introduction of a material that is, in many ways, an opposite. The bioplastic plates allowed us to explore the agency and unknown factors of this unpredictable material (where the natural material itself led the process of making, shaping, playing and the sound palette of the instrument). Put simply, its physical form and sound are unrepeatable. The trade-off between controllability/uncontrollability and liveliness is also balanced by the potential for the materials and techniques to generate sounds in a musically useful range of frequencies. The visual element of performative action on the instrument also contributes to artistic choices, as a visually expressive performative action, and can enhance the delivery of performative intent.

Musical instruments

Our cross-disciplinary research project develops novel musical instruments by utilising vibrational modes of plate-like structures. The sound is generated and amplified in a feedback loop using the vibrational modes of instrument plates which, in turn, reflect the stiffness properties of plate materials.[10] A set of artistic designs for polymer plate instruments in our project are shown in Figure 5.4.1. Some of their design features – localised perforated areas, and "keys" and "strings" made by slits – enable decoupled local vibrations. This allows the performer to steer and get better control over the semi-chaotic sounds such plates generate. Other design features are purely of symbolic/artistic value. The plates were cut using a water cutter. The designs proved to be a challenge to cut, due to the large number of perforations. For each single perforation, the waterjet must restart the flow of water and sand, leading to a small risk of clogging the nozzle. With several hundred perforations, this small risk leads to a certainty of clogging. To fix a clogged nozzle, the operator must disassemble, clean and reassemble.

The bioplastics plates are biodegradable and compostable plastics made from non-depletable, renewable seaweed sources (a wild species of brown algae growing in underwater kelp forests off the coast of the Norwegian island of Frøya was harvested for this project). The sound generating mechanism and playing technique of the bioplastic plates is the same as described for the polymer plates. The material almost seemed to take on a mind of its own during the making; idiosyncrasies such as air bubbles and irregularities occurred through the drying process, creating internal waves, forming abrupt changes across the plate (a three-dimensionality in the surface as can be seen in Figure 5.4.2 on the sides and corners of the plate particularly). The uneven surface affects the density and consistency of the vibrational areas in an unpredictable manner, partly due to the air trapped inside air bubbles.

New imaginaries: cross-pollination with bioplastics project

The initial work of exploring vibrational modes with feedback was done with just polycarbonate plates. The addition of the bioplastic instruments was somewhat opportunistic, as Alex

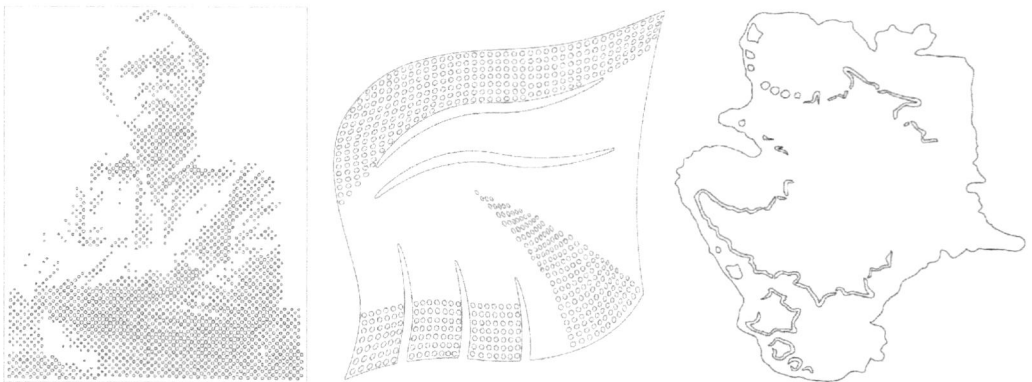

Figure 5.4.1 Artistic designs for water cut polymer plates. "Alvin" (left) "The Eye" (centre) and "Mandelbrot" (right).

Source: Øyvind Brandtsegg.

Figure 5.4.2 Kelp alginate bioplastic feedback instrument "A New Continent", 2021.

Source: Photographer Jordan Sand.

Murray-Leslie was already working on them in another project; they would contrast the formal and manufactured aesthetic of the clear plastic plates, and also reflect on the environmental aspects of these objects. Because this was not a planned step in the initial project, it offers a clear example of how the peculiar qualities and behaviours of the material in general and in the specific can serve as a conceptual and ethical "mechanism". Another instance might have produced fundamentally different results and therefore led to a fundamentally different trajectory. Thus, this ethics-inspired aesthetic feedback during the creative process led to new observations and ideas developing organically around linear and circular aesthetics (circular aesthetics being a feedback loop in itself).

We observed that the two instruments embody different models of production, with different aesthetics flowing from them: on the one hand, a linear aesthetic, on the other a circular aesthetic:[11]

- Linear aesthetics – which we could associate with the plastic polymer plates – describes artefacts produced from materials that are extracted, processed and usually thrown away after a limited, often single use. Broadly speaking, this approach is emblematic of the artificial.
- Circular aesthetics, in contrast, are afforded via an experimental process led by environmental and ecological factors, in which the material is seen as having agency and shapes itself; the outcomes are sustainable and far more representative of the natural.

Bioplastics have a circular aesthetic quality: imperfect and experimental in substance and form. We discovered their peculiar qualities are embedded in the artefacts of that experimentality in process, product and its performance. Through the experiences of working with the two materials in a rehearsal studio, while developing the performance *The Things We Need Feedback* at TKM Gråmølna, where the respective materials collided sonically and aesthetically, acting as an unexpected dramaturgical metaphor for the performance to travel along. In effect, the performance reflected pressing societal questions of the future need for materials that are more sustainable, reliable and universal than ever with globalised standards, in order to be able to trust these materials. The performance brought together these contradictions, highlighting radically different aspects of a material, its use in a new instrument and the way it sounds and is played.

During the rehearsals and performance, we found the bioplastic plates perhaps harder to control, with less sonic variety, and that the object retains its experimentality. Its "inconsistent" and "unpredictable" qualities deviate from the uniform vibrational modes found in the polymer plates. We found the challenge, then, was to learn to work (and collaborate) around two materials that are opposed in more than just their specific material features: one predictable and controllable, the other idiosyncratic, unpredictable and organic. The circularity of the feedback systems meets the circular aesthetics, creating a double loop-within-loop with exponential complexity.

The bioplastic plates forged a meaningful connection with nature, informed by kelp's intrinsic properties. They were developed in parallel and as a response to the polymer plates leading to an experiment with opposites, artificial and natural, as a way to draw a comparison of how the differences might inform the sonic, aesthetic and performative, reflecting timely ideas around transition design, compelled by a sense of urgency and society's need to shift away from linear modes of production and consumption.

How to approach music: a play of productive opposites

As with the polymer plates and the bioplastics, the way we approached the music making was like a call and response. This play of productive opposites between practitioners first needed to be made visible; only then could we consider and appreciate them. It was this complex process that allowed us to build a meaningful set of instruments and to design their actuation based on what we had learned from each other's opposite approaches. At this point, it makes sense to review the PI's disciplinary contexts.

Øyvind Brandtsegg comes from a trained background in music composition, performance and music technology. In contrast, Alexandra Murray-Leslie comes from a DIY art band Chicks on Speed; her early experiences with music and musical instruments were not entirely successful, and this informed how she would later approach audio-visual performance. Alex's more performative approach, as exemplified in the *High Heeled Shoe Guitar* and other OBJEKTINSTRUMENTS made by Chicks on Speed,[12] approaches instruments and music-making objects as performative props: instruments of staged sonic expression. Whereas Øyvind is concerned about the precision of touch required to produce specific sonic results and how to shape the dramaturgy of the shape of sound over time.

These two approaches are at times mirrored in the musical/material side of the collaboration feeding back into each other. Even though Øyvind's practice is already experimental, in the sense that there are no rules, Alex would easily break any conventions remaining. Neither approach is right or wrong. The tension between the approaches, however, guides and nourishes the development of the musical instruments, how they are performed and the unique kind of aesthetics generated. We have learned a great deal from one another's respective way of thinking about music, how to understand it, how it is made and how it can be visually performed. We discovered that, at root, even sophisticated and seemingly technical descriptions of these aspects encode (and sometimes mask) fundamental questions about comfort zones. The challenge, then, involves finding constructive ways to recognise and articulate these differences. In jam sessions and rehearsals in the pre-production phase, the meta-task of critically and constructively engaging with each other's positions serves as a kind of calibration and balancing feedback, informing a new and better collective artistic language.

Reflection in material

Due to the complexity of creating numerical models of the physical objects and their relation to the feedback loop, we rely heavily on practical experimentation to develop both the instruments and the playing technique. Playing these instruments feels like chasing a moving target, the instrument always escaping rationalisation, forcing an intuitive approach. As a trained musician, it is common to expect precise control over pitch. Still, being trained to play by ear in melodic improvisation seems to function as a guide in the exploration, even though one cannot always control if the next pitch is going up or down from the current one being played.

Even though we cannot always control pitch deliberately, playing the instrument is not a random action. It is rather a feeling of accepting what is given, going from there, playing a part *together* with the instrument. As Eldridge et al.[13] describe, it feels as having a *dialogue* with the instrument. Performers interviewed in this study also reported disappointment whenever the instrument was extended with processing mechanisms that would enhance specific pitch control, so it seems we are in good company when being fascinated and intrigued by the wildness of these instruments. This complexity is multiplied when several performers play the same surface simultaneously, as we experienced when Amalia Wiatr-Lewis and Øyvind Brandtsegg played as a duo on one single polycarbonate screen (at the *OnlyConnect* performance). The actions of one performer would in this case change the potential resonances available to the other performer, enabling a dual chaos-attractor search for a common language of expression.

Polymorphism

The project embraces the notion of polymorphous practice between artists and engineers to build new common languages of trust. Something we can build on as a sort of toolbox to help others navigate cross-disciplinary collaboration. The working processes taught us to stay open, to listen to each other's areas of expertise and to blur the boundaries of our practices. We also saw how ideas can take form in different states, a transformational process over time of creating new meanings and understandings. Although it is far more time consuming early on, it later becomes a creative and a lucrative practice to cross-disciplinarily wander outside one's comfort zone.

The project reflects, and in some ways advances, the urgency of the need to work together cutting across disciplines and areas of expertise, to find solutions to societal challenges from a finished and controlled world, but that will change; in 50 years, people will be using bioplastics widely, and this side of the project hopes to promote this positive poetics.

The numerical and the experiential

The analysis of expected vibrational modes of plates was done with appropriate numerical tools as briefly described earlier. The resulting natural frequencies, when activating vibrational modes with the feedback device, deviated significantly from the modelling predictions. Some of these deviations can be explained by including the feedback delay (the latency of the feedback circuit), and also allowing for constructive or destructive interference due to phase cancellations. Still, a significant portion of the variances in the feedback frequencies did not have analytical counterparts in our models. The numerical modelling was very useful in order to create a rough map of potentials, and we used this to refine our plate shape designs, and also to guide the sonic exploration onto areas of plates where certain variations could be found.

We have yet to understand the exact nature of these feedback instruments, which is part of what makes them still interesting to explore. Even though we see deviations in the exact frequencies between the numerical and the artistic methods of exploration, we can also see a sort of musical interplay between the two different approaches. One of call and response – estimation and evaluation – calculation and intuition.

Discovering realms of exception: a new circular aesthetics was created which will be interesting to explore in future development of instruments. Artistic experiments with bioplastics expose more humanistic, less technocratic, more culturalist approaches to doing and making; and these, in turn, provide additional and maybe better ways to discover what is needed where. The bioplastic work serves as a kind of encouragement to artists to think ethically about what they produce, its lifetime as artefacts in this world, ones that have an enduring and more respectful partnership *with* nature.

Notes

1 Bourriaud (1998).
2 Sanfilippo and Valle (2013).
3 Eldridge et al. (2021).
4 Philp et al. (2012).
5 von der Leyen (2021).
6 Ibid.
7 Ibid.
8 Mayhew and Dearnley (1987).
9 Ibid.
10 E.g., Lasn et al. (2015a, 2015b).
11 A Manifesto on Circular Aesthetics can be explored here: https://medium.com/@flrnschndr/circular-aesthetics-b47e8e3b29c.
12 For further reading about the Chicks on Speed OBJEKTINSTRUMENTS (self-made musical instruments): "Chicks on Speed: Don't Art, Fashion, Music" book authored by Alexandra Murray-Leslie, Melissa E. Logan and Judith Winter, Booth-Clibborn Editions, 2010.
13 Eldridge et al. (2021).

Bibliography

Bourriaud, N. (1998). *Relational Aesthetics*. Paris: Les Presse Du Reel.
Eldridge, A., Kiefer, C., Overholt, D., & Ulfarsson, H. (2021). *Self-Resonating Vibrotactile Feedback Instruments ||: Making, Playing, Conceptualising :||*. Feedback Musicianship Network. https://feedback-musicianship.pub-pub.org/pub/kl8m5o5y.
Lasn, K., Echtermeyer, A. T., Klauson, A., Chati, F., & Décultot, D. (2015a). Comparison of Laminate Stiffness as Measured by Three Experimental Methods. *Polymer Testing*, 44, 143–152. https://doi.org/10.1016/j.polymertesting.2015.04.006.

Lasn, K., Echtermeyer, A. T., Klauson, A., Chati, F., & Décultot, D. (2015b). An Experimental Study on the Effects of Matrix Cracking to the Stiffness of Glass/Epoxy Cross Plied Laminates. *Composites Part B: Engineering*, 80, 260–268. https://doi.org/10.1016/j.compositesb.2015.06.005.

Mayhew, P. J., & Dearnley, P. (1987). An Alternative Prototyping Classification. *The Computer Journal*, 30, 481–484. doi:10.1093/comjnl/30.6.481.

Philp, J., Bartsev, A., Ritchie, R., Baucher, M. A., & Guy, K. (2012). Bioplastics Science from a Policy Vantage Point. *New Biotechnology*, 30. doi:10.1016/j.nbt.2012.11.021.

Sanfilippo, D., & Valle, A. (2013). Feedback Systems: An Analytical Framework. *Computer Music Journal*, 37(2), 12–27.

von der Leyen, U. (2021). *A New European Bauhaus: Op-ed by Ursula von der Leyen*. https://ec.europa.eu/commission/presscorner/detail/en/AC_20_1916. As seen in "Notes on the New European Bauhaus". https://link.medium.com/g2yfVFFkrgb (accessed May 20, 2021).

5.5

CO-EVOLVING RESEARCH AND PRACTICE – _DERIVATIONS AND THE PERFORMER-DEVELOPER

Benjamin Carey

Introduction

I am a composer, performer and practice-based researcher in the area of electronic music based in Sydney, Australia. I make work on the modular synthesiser, develop bespoke music software and create audio-visual works. My research and practice are concerned with musical interactivity, generativity and the delicate dance between human and machine agencies in composition and performance. Since beginning my PhD in 2010, I have been researching and practising in the area of interactive musical performance. This area of creative work is concerned with the development and arrangement of software and hardware environments that promote exploratory interactivity between musicians and technology. In my work I see the development of such environments as akin to the creation of a non-linear, interactive musical score, one whose affordances and constraints are revealed in the real-time act of performance. I am interested in complexity and surprise in both composition and performance, and my artistic output is the result of designed interactions between semi-autonomous musical systems and human agency – most often my own.

Practice-based research (PbR) has been at the core of my work as a researcher since I began my PhD. In my work I generate both artistic and conceptual ideas through doing, 'in the plane of practice', as Andrew Pickering once wrote.[1] While I often have broad artistic and research concerns coming into a project, it is through practice that I surface researchable problems which I then interrogate further through both my artistic practice and considered reflective practice. This is a cyclical process in which artistic ideas and outputs, theoretical concerns and methodological approaches emerge through practice, and subsequent interrogation of that practice. I have found this approach to research a fruitful way of deeply understanding my own approach to art making, which in turn helps move my work forward. In addition, by acknowledging, analysing and communicating tacit practitioner knowledge throughout the life of a project, it also provides opportunity to contextualise one's practice with broader theoretical concerns, to develop novel methodological approaches to artistic research and to develop new understandings of existing and burgeoning practices.

_derivations and the performer-developer

I developed this approach to PbR throughout my doctoral project, entitled *_derivations and the Performer-Developer: Co-Evolving Digital Artefacts and Human-Machine Performance Practices*.[2] This

DOI: 10.4324/9780429324154-48

project centred upon the development and use of a bespoke software environment developed for human–machine improvisation, eventually titled *_derivations* (see Figure 5.5.1). As a creative arte-fact, *_derivations* exists somewhere along a continuum between a software tool, a musical work and an artificial performance partner. The software was developed to listen to, record and analyse the musical input of an improviser, and to perform spontaneously with the human performer live on stage, using material derived from the musician's past performance/s. The system derives its sonic vocabulary entirely from recordings and analyses of a performer as they play, allowing it to make contextual decisions about its contribution to a performance based on timbral similarity between its growing database of material and the current input of the human performer.[3,4]

By the point of publication of my doctoral thesis, *_derivations* had evolved into a stabilised soft-ware environment used in my personal artistic practice, as well as the practice of other improvis-ing musicians. The software was distributed freely online in 2013 and has since been performed internationally, and its interactions with musicians had been documented in an album release entitled *_derivations: human–machine improvisations*.[5] While the software was eventually presented as a stabilised technological artefact, complete with a dedicated website, software documentation and tutorials, its genesis was messy and unpredictable. The software was not designed in the traditional sense but represented the end point of a chain of design decisions that evolved through interac-tions between software development, performative testing and live performance with developing code. This project is an example of what Scrivener has defined as a 'creative-production' research project, whereby technological development is foregrounded as a creative practice and an integral part of the research methodology. This is in contrast to a 'problem-solving' project, where the development of a technological artefact seeks to solve a well-defined problem, or to improve an existing artefact.[6] As an artist-researcher, what was of most interest to me in this work was the interactions between development, testing and use in performance, and how these interactions continually surfaced researchable theoretical and conceptual concerns in the area of interactive musical performance.

Figure 5.5.1 Joshua Hyde rehearsing with *_derivations* at IRCAM, Paris, 2012.

Source: Benjamin Carey.

Three central chapters of my PhD thesis charted the development of my programming and the associated performance practice over the life of the project, forming what I have termed a 'narrative of development'.[7] Through analysis of data, including written research memos, archived software code and audio recordings, this narrative of development traced the evolution of my practice, opening up the development and creation process that led to _derivations' stabilisation as a distributable software artefact. Importantly, these core chapters made methodical use of Schön's concept of 'reflection-on-action' in identifying, correlating and making sense of theoretical ideas that were emerging throughout my work.[8] In the first instance, this process surfaced what Scrivener has defined as 'issues, concerns and interests' emerging from my developing artistic work,[9] identifying research themes and giving voice to the tacit knowledge embedded in my programming practice. Secondly, by surfacing and discussing these through sustained reflective practice, I identified three key areas of interest that required further theoretical discussion. Three reflective essays formed the final chapter of the thesis, contextualising my artistic work through an engagement with research in science and technology studies and musicology. These essays explored themes of machine agency and authorship, human–machine symbiosis and musical interpretation.[10]

I embarked upon this project with an interest in musical interactivity and improvisation, nascent programming skills and an openness to expanding my creative practice. At the time I began my PhD, I was interested in furthering my creative practice by investigating the design and use of interactive musical systems. This trajectory would draw upon prevailing interests and knowledge but would also require me to learn new skills and to place myself in a position of creative vulnerability. Developing software was a relatively new part of my practice, and I used the first part of my PhD to develop my technical skills, while investigating the musical implications of designing interactive musical systems.

Developing a methodology – programming as creative practice

Throughout this project, I grappled most with the concept of methodology. Embarking upon my literature research, I encountered numerous research projects in the area of New Interfaces for Musical Expression, Computer Music and Sound and Music Computing in which new technologies were created and justified as a solution to a problem in the domain of research and practice. In addition, many such projects based their research contributions on the novelty and utility of their technical contributions to the field. These might include the development of a new musical interface, a novel synthesis algorithm, an improved networking architecture etc. I noted that the research methodology of many such projects often followed a formula appropriate for a design discipline: define the problem area, design a technological solution, test the solution in the problem domain and then evaluate and refine the technology. Often, artistic outputs generated with these artefacts were ancillary, and the development process itself is not foregrounded.

In the early stages of my PhD, knowing the areas I wanted to explore through practice, I acknowledged that technological development would be required to achieve my aims. However, as a musician and not a trained software developer, the technology I was developing could not be characterised as a generalisable solution to a well-defined research problem. I was programming bespoke software, often designed for my own creative needs, and therefore lacked any meaningful claim to generalisability. I found it hard, therefore, to see how my approach to development could fit within such a research paradigm. After reviewing the literature, I was fascinated by the many and varied approaches to algorithm and system design taken by those in the field. The burgeoning fields of 'Musical Metacreation' and 'Live Algorithms for Music' were replete with cutting-edge algorithmic approaches to musical generativity and interactivity.[11] While I took great interest in the novelty and technical sophistication of such approaches, I found that the most interesting examples were those that foregrounded artistic outputs, emphasised the collaborative nature of the work or dived deep into the theoretical implications of the technology.

In my work, I was interested in exploring the musical implications of interactive music systems, as well as understanding the design of such systems as a creative practice. Throughout my project, I instinctively practised software development as a form of creative exploration; I was not concerned with solving a research question through the development of generalisable software. Instead, I began by defining creative problems for myself through the development process and sought low-level, programmable solutions to my creative ideas. This exploration through software code was always tightly linked to performance-based testing and iterative refinement. As a saxophonist, interacting with these pieces of software through sound, what I termed 'performative testing' was essential to developing an understanding of their interactive potential. Throughout the project, I came to understand that the cyclical process of development, testing and refinement existed as the principal site for generating and interrogating artistic concepts in my research. My hybrid practice of software development and performance became a space for problem-setting and creative exploration. Defining myself as a 'performer-developer', I used the practice of interactive system development to problematise my existing practice and to co-evolve new practices alongside the development of bespoke software artefacts.

Reflective practice

In order to surface and interrogate the 'issues, concerns and interests' driving my practice, regular reflective practice became an essential method. While developing software, testing and performing, I archived digital research data generated from my practice (code and recordings), and took regular 'reflections on action' in the form of dated research memos. These research memos surfaced tacit knowledge emerging from practice, allowing me to identify and expand upon emergent research themes in the written thesis. Through this approach to reflection on action, I acknowledged the agency of software code and algorithms in the process of developing and interacting with my software, and in the development of new forms of performance practice. I also began to understand more about the kind of software developer I was becoming, and the entanglement I was experiencing between software development and the emerging theoretical and conceptual concerns of the project.

This learning experience had implications for both framing and justifying my research methodology and the eventual contributions of the project as a whole. As Gray and Malins have stated elsewhere, artistic research projects often embrace novel and hybrid methodologies that are uniquely suited to the specific nature of the research project in question.[12] Through reflective practice, I became aware of the emergent nature of research methodology in my work, understanding this as a natural part of a practice-based research project. In my project, I saw value in presenting an expanded methodology chapter in the written thesis, outlining the relationship between software development and testing, reflective practice and theoretical engagement. Throughout the project, I contextualised my work using McLean and Wiggins' concept of 'bricolage programming' and Pickering's conception of the 'mangle of practice'.[13] I argued that the unique performer-developer context necessitated a 'mangling' of ideas and practices, and an exploratory form of software development that was responsive to feedback from the software's developing agency.

Theoretical insights

Moving from the 'narrative of development' to my findings chapter, I used reflective practice to connect developing understandings of practice to broader research themes. In one sustained essay, I reflected upon the relationship between development and use in my practice as a performer-developer, turning to numerous theorists in science and technology studies to help solidify my developing ideas in this area. In developing my software, I was interested in provoking surprise in performance with a human improviser, and the concept of machine agency in both development and use of *_deriva-tions* surfaced as core research concern.[14] As the developer and the primary user of the software, I was interested in my unique position of the performer-developer in this practice. How could one program a system to elicit surprising interactions, when this user has themselves programmed its internal behaviours? While my system did not make use of machine learning or AI approaches to musical generativity, the complex interactions between deterministic and aleatoric processes in my code provoked surprise in my own interactions with the software. Given how intimately I knew the detail of my codebase, I theorised that my interactions with *_derivations* required a 'suspension of disbelief' in performance and testing. In my reflections on this topic, I considered Akrich's concept of an artefact's 'script',[15] the notion of 'black-boxing' from actor-network theory[16] and Hamman's analysis of an artefact's *episteme*.[17] This research led to an understanding that, in my practice as a performer-developer, I was engaged in a form of 'intentional black-boxing', whereby the details of the internal components receded from my awareness, in order to interact with the system as a black box. As Akrich has noted regarding technical objects: 'as they become indispensable, objects also have to efface themselves'.[18]

In another sustained reflection, reflective practice facilitated a theorisation of the relationship between improvisation and interpretation in the performance practice of human–machine improvisation. Once established as a stabilised technical artefact, *_derivations'* fluid role as a performance partner, software tool and musical work raised questions about the nature of musical improvisation in this human–machine context. In this essay, it was argued that performative engagement with interactive performance systems, as distinct from human–human improvised contexts, could be characterised as a form of musical interpretation due to the programmed nature of the software system. Regardless of the sophistication of a system's algorithms, such systems represent the musical proclivities and interactive intentions of a human programmer, and as such performers engage with such systems in a similar way to that of a graphic or textual score. This essay emerged from several reflective memos written throughout my PhD, in which I problematised the concepts of improvisation, authorship and machine agency in the practice of human–machine improvisation.

Discussion

These sustained reflections rounded off an approach to PbR that I believe shows the potential for reflective practice to connect to larger themes beyond one's own practice. Having developed a new software artefact and an associated performance practice, I was now able to reflect upon the outcomes of my work in some depth, enabling me to connect these emergent ideas with literature in the field of science and technology studies, and musicology. The outcomes of these reflections provided novel insights into the practice of a performer-developer, as well as the benefits of reflective practice in my discipline area. Alongside the developed software artefact, associated recordings and the novel methodological approach, these reflections became part of the core research contributions of my PhD. With respect to methodology, by the conclusion of the project I had solidified a research methodology that was faithful to the way in which I develop my artistic work. Given the interdisciplinary nature of my research and practice, technological development will always be an integral part of any research project I am engaged with. However, I learned throughout this project that there is value in opening up and reflecting upon development as a creative practice. As Scrivener has observed, in problem-solving projects, the twists and turns of the 'problem-setting' phase of a research project are often not written up in a final thesis document.[19] However, where software development is exploratory, this phase of the project is rich and generative. By reflecting upon and communicating the tacit knowledge emerging from creative explorations in software, I was able to make connections between aspects of my work that I was previously unaware of. In turn, this learning spurred on deeper theoretical discussions that connected my work to much broader bodies of research and practice.

Notes

1 Pickering (1995, p. 20).
2 Carey (2016a).
3 A performance with _derivations can be viewed here: www.youtube.com/watch?v=GHxHumlCZOQ.
4 Documentation videos on the software can be viewed here: https://vimeo.com/showcase/3291705.
5 Carey (2014).
6 Scrivener (2000).
7 Carey (2016a).
8 Schön (1983).
9 Scrivener (2000, p. 6).
10 Two of these essays have since been published separately as Carey (2016b) and Carey (2019).
11 Young and Blackwell (2013), Pasquier et al. (2017).
12 Gray and Malins (2004).
13 McLean and Wiggins (2010), Pickering (1995).
14 This theme of surprise and machine agency with _derivations is discussed in depth in an interview with Linda Candy, appearing in Candy (2020, pp. 201–203, 226–230).
15 Akrich (1992).
16 Latour (1990), Akrich (1992).
17 Hamman (1999).
18 Akrich (1992, p. 221).
19 Scrivener (2000, p. 8).

Bibliography

Akrich, M. (1992). The Description of Technical Objects. In W. Bijker (Ed.), *Shaping Technology/Building Society: Studies in Sociotechnical Change (Inside Technology)*. Cambridge: The MIT Press, 205–224.

Candy, L. (2020). *The Creative Reflective Practitioner*. London and New York: Routledge.

Carey, B. (2014). *_derivations: Human-Machine Improvisations*. Bandcamp: Sound Recording.

Carey, B. (2016a). Artefact 'Scripts' and the Performer-Developer. *Leonardo*, 49(1), 74–75.

Carey, B. (2016b). *_derivations and the Performer Developer: Co-Evolving Digital Artefacts and Human-Machine Performance Practices*. PhD thesis, University of Technology. https://opus.lib.uts.edu.au/handle/10453/43452 (accessed May 10, 2021).

Carey, B. (2019). Musical Interpretation in Improvised Human-Machine Performance. *Sound Scripts*, 6(1), 1–9.

Gray, C., & Malins, J. (2004). *Visualizing Research: A Guide to the Research Process in Art and Design*. Aldershot: Ashgate Publishing.

Hamman, M. (1999). From Symbol to Semiotic: Representation, Signification, and the Composition of Music Interaction. *Journal of New Music Research*, 28(2), 90–104.

Latour, B. (1990). Technology Is Society Made Durable. *The Sociological Review*, 38(S1), 103–131.

McLean, A., & Wiggins, G. (2010). *Bricolage Programming in the Creative Arts*. Paper presented to the 22nd Psychology of Programming Interest Group, Madrid.

Pasquier, P., Eigenfeldt, A., Bown, O., & Dubnov, S. (2017). An Introduction to Musical Metacreation. In *Computers in Entertainment*. New York: Association for Computing Machinery, vol. 14.

Pickering, A. (1995). *The Mangle of Practice: Time, Agency, and Science*. Chicago: University of Chicago Press.

Schön, D. A. (1983). *The Reflective Practitioner: How Professionals Think in Action*. Aldershot: Ashgate Publishing, 1995 ed.

Scrivener, S. (2000). Reflection in and on Action and Practice in Creative-Production Doctoral Projects in Art and Design. *Working Papers in Art and Design*, 1.

Young, M., & Blackwell, T. (2013). Live Algorithms for Music: Can Computers Be Improvisers? In B. Piekut & G. E. Lewis (Eds.), *The Oxford Handbook of Critical Improvisation Studies*. Oxford: Oxford University Press, vol. 2.

5.6

PUBLISHING PRACTICE RESEARCH

Reflections of an editor

Maria Chatzichristodoulou

Introduction

Practice Research (insert any favoured term, such as practice-as-research, practice-based research, practice-led research and so on) is a methodological approach to academic enquiry that leads to non-conventional types of creative research outputs such as performances, compositions, artworks and exhibitions, rather than to conventional text-based academic papers and books. Practice Research is important in validating different types of knowledge and insights – particularly knowledge(s) and insights that have not had a 'place at the table' of academic enquiry previously, because they were deemed as out of scope, unreliable, irrelevant or subjective. This particularly applies to tacit, embodied, situated and material forms of knowledge and enquiry which, as Barbara Bolt has argued,[2] 'ha[ve] the potential to be generalised so that [they] set wobbling the existing paradigms operating in a discipline'. Those types of knowledge, which result from processes of performing and making, bring a richness of immeasurable value to the context of academic enquiry. Combined with critical and historical approaches, those new (to academia) types of knowledge offer us the opportunity to re-envisage and reorient academic endeavour in the arts and humanities, so as to include a wider range of voices, approaches and insights. Practice Research approaches open up our way of 'thinking about knowing' to allow for processes of making-as-thinking and of thinking-through-making. They, alone, have the ability to ground research in the arts into the materialities and situated processes of the substrates that carry it. Because, as we know, matter does matter.

Personal journey

In 2002, I registered as a PhD student at Goldsmiths University of London, to undertake a practice as research PhD in theatre and performance. Coming with a practitioner background of performer and media arts curator, I wanted to embark upon a process of critical enquiry led by practical considerations in studying and developing innovative forms of theatre that experimented with digital technologies, such as multi-user virtual environments and online platforms; and in offering such artistic endeavours for public consumption, exploring effective ways of curating digital performance art. I was the first student to undertake a practice as research PhD at the Drama Department of Goldsmiths College. Unfortunately, I felt unsupported and misunderstood by my main research supervisor. The academic context in the department was not fertile for either Practice Research or

DOI: 10.4324/9780429324154-49

interdisciplinary investigations – and mine was both. I am sure that the particular department and supervisor represented more the rule than the exception at the time.

Luckily, a new academic context was being created at Goldsmiths, and in 2005–6 (after several miserable years . . .) I found my intellectual home at Goldsmiths Digital Studios: an interdisciplinary centre for postgraduate studies that was based between Computer Science and Art, bringing together artistic and scientific approaches and the combined expertise of the two visionary academics that led the endeavour: Janis Jefferies, Professor of Visual Arts[3] and Robert Zimmer, Professor of Computing. Digital Studios was truly interdisciplinary in engaging with academic staff and students across many disciplines at Goldsmiths other than Art and Computing: Music, Drama, Cultural Studies, Media and Communications, Design and Education, among others. This interdisciplinary approach led to a sense of openness and inclusivity. Interdisciplinarity requires that we challenge assumptions based in established disciplinary frameworks and engage with different vocabularies. We need to rethink what we think we know and are invited to look at the world and our research from a different perspective – a previously unoccupied position. I suggest that interdisciplinary environments offer fertile ground for Practice Research due to their inherent openness, the fact that they bring together people who carry different understandings, speak different vocabularies and bring different value systems into the enquiry. Practice Research in the arts is no different to a Computer Scientist than Art Historical research. Indeed, its material substrates and, often, tangible or otherwise concrete outcomes, might make Practice Research more accessible to researchers from non-arts/ humanities disciplinary frameworks.

The Digital Studios, and Professor Janis Jefferies as my main PhD supervisor, offered the mental and creative space to experiment with Practice Research and interdisciplinarity. This was a fertile environment that nurtured work which challenged formats and conventions, experimented with processes, blurred boundaries, spoke to environments outside academia, engaged with communities and social systems and enquired into the nature of academic enquiry itself, questioning what constitutes valid academic research into the context of a doctoral programme.

This personal narrative, I hope, evidences the need for the right environment to nurture Practice Research. Though academic contexts have now shifted – in the UK at least – to accommodate and even invite practice research approaches, this type of investigation requires a different type of attention; is more resource-hungry; is riskier in terms of outcomes, as it does not follow prescribed steps; often relies on external conditions and partners (for example, it may require funding, facilities and/or technical support to showcase the work; it may need collaborators); and its robustness is still a matter of debate (and, frequently, disagreement) among academic environments, as it challenges or even undoes centuries-old traditions modelled on positivist scientific methodologies, which are deeply distrustful of artistic processes that are deemed unscientific and unreliable.

Editor journey

I have offered my personal journey in Practice Research as a context to my current practice as Editor-in-Chief of the *International Journal of Performance Arts and Digital Media (IJPADM)* – an academic journal which, since 2017 when I took on this role, actively invites Practice Research contributions (see Figure 5.6.1). *IJPADM* invites robust research contributions, which can be conventional research articles or anything else that we can – or cannot yet – imagine as a research output, all under the category of 'Documents'. Documents can be: practitioner reports, reflexive accounts, interviews, creative writing contributions, video/photo/multimedia essays, manifestos – or can take any other format. The journal is open to receiving and reviewing any manifestation of robust research process.

Publishing non-conventional contributions in the context of an academic journal does not come without its challenges. Here I try to identify some of the key challenges that myself and IJPADM's

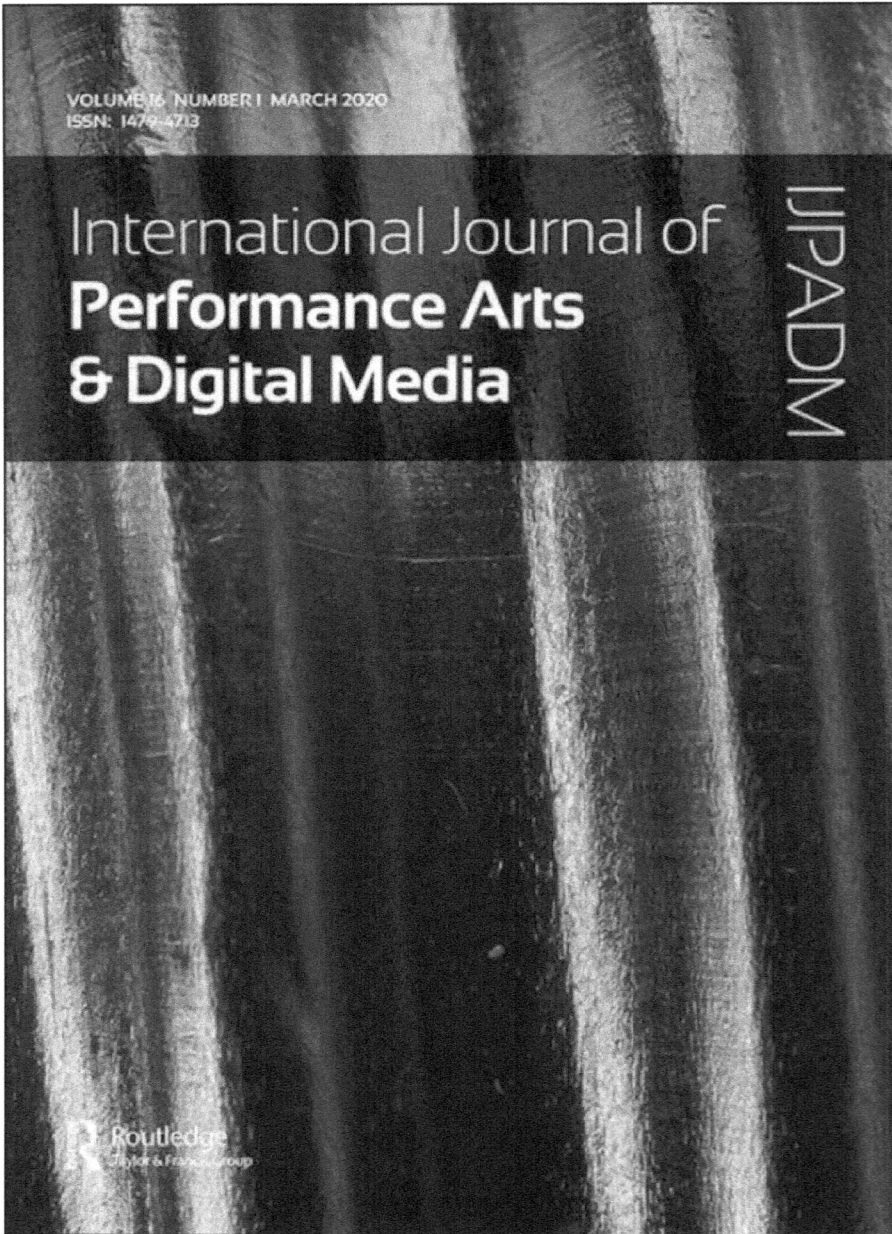

Figure 5.6.1   International Journal for Performance Arts and Digital Media. IJPADM cover artwork @Nick Hunt
from series: 'In Which the Quotidian Becomes Heroic'.

team of Associate Editors (Sarah Bay-Cheng, Kevin Brown, Nick Hunt, Peter Kuling, Elise Morrison and Toni Sant) have encountered in our process of attracting and publishing Practice Research
contributions.

Firstly, getting the word out there is a challenge in itself. Making practitioner-academics and
critical/reflective practitioners aware that they are able to submit a different range of formats and

types of outputs to the journal, which do not have to abide by the established academic conventions of a research article, and raising their confidence about the value of such work, requires a communications campaign – such that no academic publisher has the resources to undertake. Indeed, raising confidence in processes and outcomes of Practice Research might be our biggest challenge yet, as I found from the continuous contradiction of Practitioner Researchers on the one hand lamenting the fact that, in their view, their work is not being accepted as of equal value; and on the other, submitting work that self-censors by imitating the format, structure and language of a conventional academic article, shying away from exploring and experimenting with its own means of expression. The fact that Practitioner Researchers need to have their confidence raised in the context of publishing and presenting their work within standard academic outlets is hardly a surprise, given the fact that those practices were denied the status of 'proper' research in the recent past. The question perhaps is, what concerted effort do academic publishers in arts and humanities need to make in order to shift the culture of academic publishing towards one that is welcoming of Practice Research outputs on their own terms – rather than a culture that attempts to shape-shift those outputs into more familiar and convenient, but often unsuitable, formats. This does not come without a risk in quality and reputation for some academic journals, as experimentation carries an inherent risk of 'failure' of some sort. However, it is essential that publishers and journal editors take this risk if they want to be part of future developments in arts and humanities research, making for a more inclusive, diverse and richer culture.

Once a Practice Research article is submitted to the journal, we are faced with a number of other challenges in relation to how this article is handled, starting with the peer review process. A Practice Research output would need to be flagged up and assigned to peer reviewers who have the knowledge and understanding to review the submission – again – on its own terms, rather than approaching it with the expectations they would have of an academic article. The peer review of Practice Research requires particular care and attention; rejecting Practice Research submissions because they do not perform well against the standard assessment criteria used for conventional academic outputs, or returning derogatory feedback that belittles the work and requires it to conform to standards not inherent to its particular, practical or applied, research dimension, is damaging to the individual researcher(s) involved but also the academic arts and humanities research community as a whole. However, a well-meaning peer reviewer of Practice Research might also require guidance in relation to the criteria that is appropriate to apply to such outputs. Certain research criteria, for example 'originality', might be more suitable for judging Practice Research outputs – or, at least, easier to apply – compared to others, such as 'rigour'. Indeed, following Biggs and Büchler,[4] I would argue that we need to reconceptualise 'rigour' as a criterion for Practice Research more widely; not only in relation to its publication standards but also in how we understand and evaluate the robustness of a Practice Research process of enquiry without being constrained by positivist, science-led methodological approaches that are either unsuitable or 'not the most important aspect to be judged'[5] in the context of Practice Research enquiry.

Other, practical challenges in publishing Practice Research, include issues to do with copy editing and producing a document which experiments with form, such as a multimedia essay or a text-based document that requires particular formatting attention. It is fair to say that submissions which require individual, tailored attention, are not built into the business models of academic publishers who, increasingly, resort to AI applications for the copy editing and post-production of academic journals. Another concern is the fact that most academic journals, including *IJPADM*, are published both online and as a hard copy. Subscribers receive hard copies of the journal and public libraries store them. However, the presentation of a multimedia piece in a hard copy version is incomplete and reductive. In the case of *IJPADM* (as applies to most journals), a video is substituted in the hard copy, by a number of still images and the web link where the video can be accessed. This means that the reader has to go online and type the link into a browser to access the full content – but how many readers will do so? Publishers will need to consider a range of more sophisticated options for hybrid publishing practices going forward as this will become more common. It is possible that hard copies of academic journals will gradually eclipse due to economic and environmental concerns, as well as concerns to do with content, format and presentation. This, despite the dismay of those readers who appreciate the physical artefact of a printed journal issue, could benefit multimedia, hybrid or photographic Practice Research submissions as it will entail consistent presentation throughout. Finally, the archiving and longevity of multimedia content is also a concern; whereas hard copies will be preserved in libraries and archives, digital content is more unstable and technological changes can render it obsolete.

I believe that engaging with Practice Research is hugely beneficial to *IJPADM*, despite the challenges presented above. Through the process of engaging with practitioner researchers in accommodating, peer reviewing and publishing their Practice Research outputs we, as an academic community, have an opportunity to contribute to shaping the field of academic enquiry in Arts and Humanities research internationally. Publishing Practice Research means validating[6] as equal different types of knowledge and understanding, promoting an 'enriched and expanded epistemic culture'[7] and contributing to debates about the nature of academic research and the place of practice in it. Including practitioner researchers, their particular methods and the types of knowledge they generate in our *IJPADM* community, and the wider academic community, means including a more diverse range of voices in academic discourse. Those are not just the voices of practitioner researchers, who oftentimes have been excluded from academic discourses about practice (even when it is their own), but also the voices of other people engaged in arts practice as or through research enquiry, leading to what Sally Mackey has termed 'polyphonic conversations'.[8] Those polyphonic conversations are hugely important in bringing different approaches, perspectives, positions and experiences to bear on our learnings from arts and humanities research, often resulting in the 'dislocation of knowledge itself'.[9] Other benefits of publishing Practice Research include: experimentation with form and content, exploring different ways of presenting and communicating academic knowledge; reconsidering the stylistic and literary conventions of academic writing which make it hard-to-access, introspective and ineligible to wider, non-academic audiences; innovating in novel, hybrid, media-rich forms of academic publishing; widening the reach of academic enquiry to engage with different audiences and communities of knowledge and practice; making for a richer, more engaging and stimulating reader experience.

Conclusion

The epistemic shift towards Practice Research in Arts and Humanities disciplines offers an opportunity to reconsider the nature and function of academic publishing, particularly in the context of academic journals. The range of different types and formats of submissions a journal such as *IJPADM* can expect to receive requires that editors apply extra care and consideration in handling those submissions with respect to their particular nature; and that publishers consider novel, hybrid ways of presenting Practice Research-generated content, which suit the particular needs of such content – rather than forcing it into the format of a conventional text-based academic article. Given society's increasing concern about our carbon footprint on the planet, perhaps it is time academic publishers eliminated the use of hard copies, at least for those journals that include Practice Research or other rich media submissions. Though it is understood that digital platforms also have a carbon footprint, journals are already online – printing and posting hard copy artifacts around the globe places an additional strain on the environment.

Overall, it is clear that Practice Research, as a method and ideology, is here to stay and to disrupt academic conventions and discourses longer term. Finding ways to support Practice Research through productive peer review exchanges and to publish it as a valid mode of knowledge production is, thus, crucial.

Notes

1 See: www.tandfonline.com/toc/rpdm20/current.
2 Bolt (2007).
3 For Jefferies' perspective on the *Digital Studios*, read Jefferies (2022) in this handbook.
4 Biggs and Büchler (2007).
5 Ibid., p. 68.
6 Though I don't believe researchers themselves need to have their research validated, this is certainly valuable within an institutional or systemic context.
7 Borgdorff (2012, p. 31).
8 Mackey (2016).
9 Kershaw et al. (2011, p. 84).

Bibliography

Biggs, M. A. R., & Büchler, D. (2007). Rigor and Practice-based Research. *Design Issues*, 23(3), 62–69.

Bolt, B. (2007). The Magic Is in Handling. In E. J. Barrett & B. R. Bolt (Eds.), *Practice as Research: Approaches to Creative Arts Enquiry*. London: Tauris, 27–34.

Borgdorff, H. (2012). *The Conflict of the Faculties: Perspectives on Artistic Research and Academia*. Leiden: Leiden University Press.

Jefferies, J. (2022). Research Doctorates in the Arts – A Goldsmith's Perspective. In C. Vear (Ed.), *The Routledge International Handbook of Practice-Based Research*. London and New York: Routledge.

Kershaw, B., Miller, L., Whalley, J. B., Lee, R., & Pollard, N. (2011). Practice as Research: Transdisciplinary Innovation in Action. In B. Bershaw & H. Nicholson (Eds.), *Research Methods in Theatre and Performance*. Edinburgh: Edinburgh University Press, 63–85.

Mackey, S. (2016). Applied Theatre and Practice as Research: Polyphonic Conversations. *Research in Drama Education: The Journal of Applied Theatre and Performance*, 21(4), 478–491.

5.7

FROM A PHD TO ASSISTING BIOMUSIC RESEARCH

Balandino Di Donato

Introduction

One of the possible steps that one can take after a PhD is becoming a Postdoctoral Research Assistant. In this chapter, I will describe my experience in transitioning from being a PhD student to a research assistant within a larger team of academic and industry partners. These reflections were made by looking back about two years after I left the position of research assistant. Thus, this chapter is primarily of interest for young researchers, but also informative to the broader community.

The research context in which I worked was the multidisciplinary field of electronic music performance at the intersection of music and Human–Computer Interaction. I conducted my PhD studies at Royal Birmingham Conservatoire, Birmingham City University, and later I was a research assistant on the BioMusic project[1] led by Professor Atau Tanaka and hosted at Goldsmiths, University of London. Broadly speaking, in both situations, the research regarded the realisation of bio-musical instruments and the design of embodied interactions to perform live electronic music. Bio-musical instruments are digital instruments that use biological and kinaesthetic body data to generate and process sound during performance. In these works, new technologies were developed combining electromyography (EMG) with Inertial Measurement Unit (IMU) sensors to detect body movements and custom authored software to control electronic audio signal processes through bodily gestures. Practice was guided by two methodologies: User-Centred Design (UCD) and Rapid Prototyping. These were used for the design of gesture–sound interactions, and realisation of software and hardware solutions. This chapter will primarily focus on my own experience and practice as PhD student and research assistant, and not on research outcomes. Please refer to the literature referenced within this text to know more about the research work itself.

Working with musicians at the centre of the design

In the early 2000s, researchers adopted Human–Computer Interaction (HCI) approaches to create and formally evaluate interfaces for musical expression.[2] One well-established HCI approach is User-Centred Design (UCD). UCD focuses on human and context-specific aspects when designing new technologies.[3] It fosters an iterative pluralistic approach that not only looks at solving singular tasks using a product, but also considers the context in which it will be used, the social interactions that it might enhance or disrupt, the way we interact with these products taking into account bodily actions and emotions when perceiving their feedback or achieving the objectives. The UCD method also includes the usability of a product. It provides guidelines to make the product and its features visible and understandable, to foresee errors and to provide ways for fixing potential mistakes.[4]

The UCD method was adapted here to be applied in a musical context, and it centred around participatory workshops, rehearsals and performances.[5] These were designed to leverage practitioner knowledge in performance and to understand their use of the software and hardware, and bring these into the design.[6] Qualitative data about participants' experience while using the proposed technologies were collected and later analysed adopting Open and Axial Coding Methods.[7] The introduction of performers' practice and experience supported the investigation of the benefits of our software and the potential issues in a contextualised situation. Here, UCD supported the gathering of vital information regarding interaction design decisions, the workflow and technical computing aspects. By observing participants' musical practice using technological solutions, relevant information emerged that answered the research questions. For example, the degree of adaptability of both the musician and the system through using the body as musical instrument, or highlighting a feature of the system so to stimulate their creativity. Conversely, this process also gave insights into the musicians' adaptation to an instrument, or if it needlessly impaired the musical performance.

Although this approach allowed the gathering of essential evidence about musicians' practice, it was tedious and complex to implement. Finding the right balance of stimulating participants during the workshop without biasing their feedback represented a challenge. Technical aspects were another issue. We required several cameras and microphones to capture the workshop and the experience of each participant, which also might be an element of distraction and influence for the participant. To minimise the risks of biasing, data procedures were rehearsed, practiced and evaluated from both the researcher and participant perspective. Practicing and experiencing the data collection was of immense value, especially as a young researcher, when many issues are not fully understood until the theory is put into practice.

Time was another critical aspect I had to consider. During my PhD, I worked over an extended period of four years. I had room for exploration, diverging and converging towards the aims of my research multiple times, with a certain freedom of error. By contrast, the BioMusic project ran for a considerably shorter period of 18 months. It required a higher degree of focus and necessitated following a systematic research plan. In this period, I had to respond to my colleagues, but foremost to the funder organisation, in my case the European Research Council, which ultimately is funded by tax payers, the public. From my experience, iterative approaches to practice research are immensely informative and help gain experience on looking at an issue from different angles and finding that gap in the knowledge. However, it can easily lead to a never-ending research process; thus, it is essential to carefully plan each iteration and make sure they are correctly executed.

Prototyping with musicians, academics and industry

As part of the BioMusic project, we built the EAVI EMG board[8] (see Figure 5.7.1). The board is the result of a rapid prototyping iterative process. We, the BioMusic research team at Goldsmiths, University of London and Rebel Technology,[9] an audio technology company based in London (United Kingdom), worked closely toward the prototyping of the EAVI EMG board.

The transition from working on my own during my PhD studies, to working in a team on a research project designed by somebody else, did not happen at the moment when I received my PhD award, nor when I signed the research assistant contract. In my experience, this shift was characterised by daily progressive adaptation to others' way of working and thinking until the end of the project. Maintaining my own identity as a young researcher, sometimes with the fear of making mistakes, and the pressure of contributing to the team and producing research worth taxpayers' money was overwhelming sometimes. Yet, this feeling went away through practice. The act of realising an artefact and the creative process behind had positive impact on the research itself, as well as my own

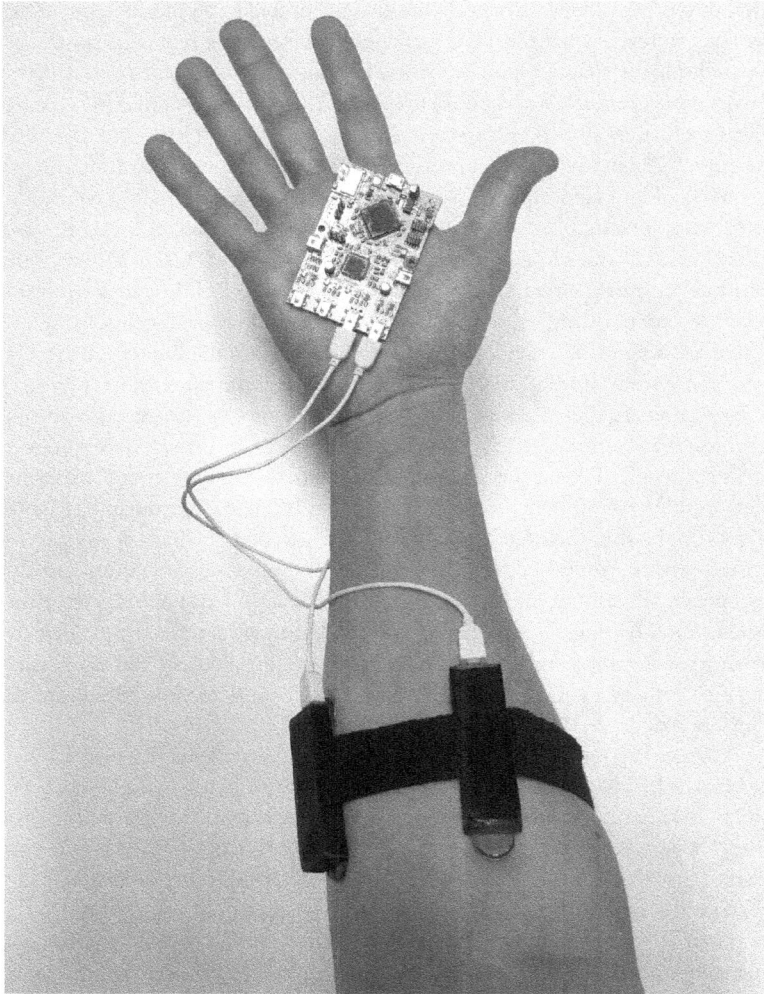

Figure 5.7.1 EAVI EMG board.

Source: Adapted from Di Donato et al. (2019).

wellbeing, through a sense of reward when looking and "holding the research between my hands". The impact that methodologies have on our wellbeing is another important issue in the life of a young researcher, which often is forgotten.[10] From my experience, the sense of reward that practice-based research provides is beneficial to both the research work itself and the researchers' wellbeing.

Drawing on the experience of working with musicians and dancers in performative settings, we individuated a series of musicians' needs and technical requirements for our prototype. Afterwards, we moved into the software and hardware implementation, which was realised through an iterative process where academic and industrial teams worked closely, constantly sharing ideas, designs and feedback on each other's work.

The board and software were then tested through musicians' use of the board in a real-world application, composition, rehearsal and/or performance. The band Chicks on Speed[11] performed using the EAVI EMG board at Muzeum Susch (Susch, Switzerland) on 28 December 2019. Before the performance, together with Dr Federico Visi (researcher and musician on the BioMusic project), the band went through a one-day experimentation and research period. They looked at ways to use creatively the board during the performance through musical and technological exploration, and at a custom wearable solution, so that the board would become part of their performance clothing and not a functional device only. To design head movement–sound processing interactions and sound control through shoulder and neck isometric activity, Alex Murray-Leslie placed the board (with IMU sensor) on her head by clipping it onto the hair with a hair clip, and the electrodes attached to the interested muscle groups (see Figure 5.7.2). To track the leg movements and muscle activity, Krõõt Juurak adapted the EAVI EMG board within her tights (see Figure 5.7.3). As in the previous case, this solution was found by looking at both functional and aesthetic aspects. The tights served as a band for the electrodes and the board, so to keep it in place for the whole duration of the performance. At the same time, the wiring and the web-alike disposition of the electrodes blended well with the performer's tights.

Figure 5.7.2   Alex Murray-Leslie with EMG board held in between hair.

Source: Photo credits Silke Brie 2019.

Figure 5.7.3 Krõõt Juurak with EMG board on her left leg and held by the tights.

Source: Photo credits Dr Federico Visi 2019.

Through the practice of using of the prototype, we gathered that the EAVI EMG board fosters new ways of interaction with sound, and its flexibility in use permitted it to be customised and adapted to visual aspects of the performance, in which technology has a strong stage presence (see the performance's video).[12] If not through practice, it would have been very difficult to collect this knowledge. However, working with "artists creating experiences that are then deployed and studied in the 'wild' of public performance" poses several challenges, such as balancing artistic and research interests, building relationships from small ideas to bigger projects and finally ethical issues.[13] Practice-based research can generate substantial research outcomes, yet it can also function as a catalyser for getting off the tangent and gathering a large volume of information that is not relevant to answering research questions. Diverse aspects that fall outside the scope of the research goals can emerge. This is incredibly valuable to highlight innovative aspects of the research, to think outside the box and to foresee future research directions. However, it might be also a counterproductive effect on the timeline of the current research. As also highlighted earlier, time is a critical aspect for research outputs delivery.

In relation to the two periods, practice in the design and prototyping process was the common denominator. The research itself, timeline, relationships with other researchers and industry partners, workflow and other aspects were very different; yet, putting theory into practice and collecting data from real-world applications was the shared procedure between the two periods.

UCD and Rapid Prototyping supported the musical, scientific and engineering practice alike. In my work, the two fields of Music and Computer Science blended seamlessly through practice, yet issues arose when conducting a formal evaluation of the work as a whole. Computer Science benefits from rigorous and quantitative evaluation methods that are difficult to apply in practice-based music research. This is because of the nature of the context in which practice happens. Specifically, unanswered questions were, how rigorous was the evaluation of a creative process, which is wild and prone to diverging and converging at a random pace? How objective are the data? How quantifiable are the data? Stepping into the shoes of the musician instead, different questions arise, such as: how is it possible to evaluate the artistic practice through methods that aim to quantify a hardly quantifiable process, and that leave little room for artistic exploration? What is the contribution of a rigorous evaluation to the artistic work? Is a rigorous approach to artistic practice suited to identify the contribution to knowledge? Will the artwork benefit from it? Ultimately, in my experience, in the evaluation phase of the research, the technological and scientific work served as a tool in support of the artistic practice.

From PhD candidate to Research Assistant: learned lessons

In the transition from a PhD candidate to a research assistant, I have experienced different challenges of my practice that have contributed to the research. This section lists the primary observations that I made during this period:

- Research practice is the common denominator in this transition. Although, with different dynamics, objectives and timeline, practice is what fuses artistic, scientific and industry research.
- Research practice gives room for realising a diverse range of research outputs: publications, performances, software and hardware. Thus, to provide the community with tangible artefacts as well as knowledge.
- Research practice can support one's wellbeing while working long hours. Practice is conducted through different methodologies and activities that can help keep engaging research daily. The diversity of activities to carry out can be very stimulating but at the same time more challenging.
- Practice is complex to deliver and evaluate because of the many variables that render the research into a continuous divergent process. It provides rich findings, but these can be hard to frame and sometimes out-of-scope to answer research questions.
- Time is a critical factor in research practice. Time management becomes more complicated when stepping up the academic ladder. Deadlines become tighter and deliverables more complex. Publishing, teaching and delivering other activities while carrying out the PhD research is a precious experience in the transition towards a research assistant.
- Research practice relies on cross-sector collaborations. These became more and more important in the transition towards research assistant and later career stages.
- Greater responsibilities and exciting challenges bring higher expectations when becoming a research assistant. These can lead to additional pressure at the early stages of the career as a researcher and practitioner. Working under pressure, approaching responsibilities

methodologically and managing expectations are essential skills to learn while a PhD candidate; and practice is a great school for this.

* Good relationships with your colleagues are fundamental to relieving some of this pressure and focus on research. It applies in different working environments, yet I feel that relationships are critical to fostering the interaction of minds, thoughts and research practices.

Conclusions

Transitioning from my doctoral studies to my post as research assistant has been a continuous process during which practice-based research has played an important role under many aspects. This chapter summarised some of the methodologies and outputs of my research practice, as well as observations made during this period. Practice-based research is an approach that definitely had a strong positive impact on my progression to a research assistant position. While it posed several challenges along the way, they were nevertheless ones that pushed the research practice forward.

Acknowledgements

The presented work was funded by the H2020-EU.1.1. – EXCELLENT SCIENCE – European Research Council (ERC) programme, Grant agreement ID: 789825.

I would like to thank all contributors to the BioMusic project, Prof Atau Tanaka, Dr Michael Zbyszyński, Dr Federico Visi, Martin Klang, Geert Roks, Chicks on Speed and Muzeum Susch.

Notes

1 https://cordis.europa.eu/project/id/789825.
2 Orio et al. (2001); Tsandilas et al. (2009).
3 Norman and Draper (1986).
4 Norman (1988).
5 Di Donato et al. (2020), Tanaka et al. (2019), Zbyszyński et al. (2020).
6 Correia and Tanaka (2017).
7 Corbin and Strauss (2014).
8 Di Donato et al. (2019).
9 www.rebeltech.org/.
10 Schmidt and Hansson (2018).
11 http://chicksonspeed.com/.
12 https://vimeo.com/387881683.
13 Benford et al. (2013).

Bibliography

Benford, S. et al. (2013). Performance-Led Research in the Wild. *ACM Transactions on Computer-Human Interaction*, 20(3), 22, July.

Corbin, J., & Strauss, A. L. (2014). *Basics of Qualitative Research: Techniques and Procedures for Developing Grounded Theory*. Thousand Oaks: Sage Publications, Inc, 4th ed.

Correia, N., & Tanaka, A. (2017). *AVUI: Designing a Toolkit for Audiovisual Interfaces*. Proceedings of the 2017 CHI Conference on Human Factors in Computing Systems (CHI'17), Association for Computing Machinery, New York, 1093–1104.

Di Donato, B., Dooley, J., & Coccioli, L. (2020). HarpCI, Empowering Performers to Control and Transform Harp Sounds in Live Performance. *Contemporary Music Review*, 38(6), 667–686, January 8.

Di Donato, B., Zbyszyński, M., Tanaka, A., & Klang, M. (2019). *EAVI EMG Board*. International Conference on New Interfaces for Musical Expression, Porto Alegre, Brazil.

Norman, D. A. (1988). *The Psychology of Everyday Things*. New York: Basic Books, vol. 5.

Norman, D. A., & Draper, S. (1986). *User Centered System Design; New Perspectives on Human-Computer Interaction*. New York: Erlbaum Associates Inc.

Orio, N., Schnell, N., & Wanderley, M. (2001). *Input Devices for Musical Expression: Borrowing Tools from HCI*. Proceedings of the International Conference on New Interfaces for Musical Expression, Seattle, WA, 15–18.

Schmidt, M., & Hansson, E. (2018). Doctoral Students' Well-Being: A Literature Review. *International Journal of Qualitative Studies on Health and Well-Being*, 13(1), August 13.

Tanaka, A., Di Donato, B., Zbyszyński, M., & Roks, G. (2019). *Designing Gestures for Continuous Sonic Interaction*. Proceedings of the International Conference on New Interfaces for Musical Expression (NIME), Porto Allegre, 180–185.

Tsandilas, T., Letondal, C., & Mackay, W. (2009). *Musink: Composing Music Through Augmented Drawing*. Proceedings of the SIGCHI Conference on Human Factors in Computing Systems, Boston, MA, 819–828.

Zbyszyński, M., Di Donato, B., Visi, F., & Tanaka, A. (2020). *Gesture-Timbre Space: Multidimensional Feature Mapping Using Machine Learning & Concatenative Synthesis*. Proceedings of 14th International Symposium on Computer Music Multidisciplinary Research (CMMR), Springer, Marseille, France.

5.8

THE CURIOUS NATURE OF NEGOTIATING STUDIO-BASED PRACTICE IN PHD RESEARCH

Intimate bodies and technologies

Kerry Francksen

Introduction: negotiating the terrain of practice: why a PhD?

I work as an artist primarily engaged in the medium of dance as a choreographer, dancer, and educator. I am fascinated by the exceptional capabilities dance has to convey and articulate human nature through embodied expression. My passion over the last 20 years has been to explore the unique knowledge a dancer has as both a maker and a performer. A dancer's valuable insights can be tricky to identify because of the indefinable and often ineffable nature of movement. I have therefore been examining how a dancer's embodied inventiveness and understanding can both be realised and ultimately understood. To do this, I have been seeking new situations in which to inspire movement making. Over the last ten years, I have been engaged with the exciting developments in digital dance performance. My work focuses on the dancer's unique understanding of this new medium and, importantly, champions the rich perceptual and embodied knowledge she has of moving within media-rich environments (for example, see Figure 5.8.1). This is both the subject and originality of my PhD research.

My exploration into *Intimate Bodies* comprised a largely qualitative and empirical study. In terms of my practice-based research process, embodied experience was crucial for the knowledge I uncovered. My study was therefore conducted as an original investigation into the dancer's insights and appreciation of perceiving and generating movement from inside the embodied experience. It was through an ongoing and principled enquiry into the practice itself that enabled me to explore and test such insights. Without the practice, my discoveries would not have arisen. Essentially, it was searching into the what, how, and why of moving that took me on an interesting journey into PhD research.

Importantly, I recognise Candy and Edmonds' suggestion that 'an important distinction between personal practitioner research and doctoral practice-based research is the form that the knowledge generated takes'.[1] For me, the work I was doing was not personal practice research because the knowledge and insights gained were centred on the dancers' embodied experiences, and my study was largely based on the continuous discovery of new movement potentials. And so, rather than working towards a commission, or focusing on making an end product (i.e. a dance piece), my motivation was based on exploring an emerging set of behaviours, which could reveal a unique way of approaching movement within digital dance performance from which others could benefit.

DOI: 10.4324/9780429324154-51

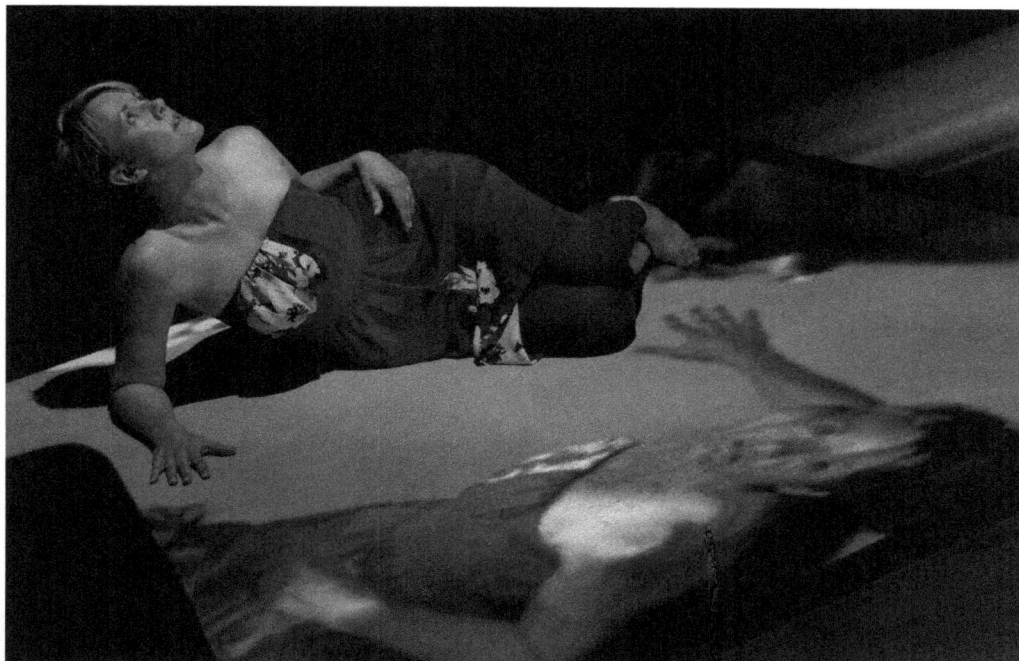

Figure 5.8.1   Shift digital dance (2011).
Source: Photograph: Kerry Francksen.

I began to discuss each work as an encounter within a generative system. For that reason, I did not create fixed works as a predefined sequence of movements; rather, I created an ongoing practical enquiry into working in real–time within this system. These were distilled into three moments when the work was shared with an audience. Sharing the work with an audience, in effect, concentrated those explorations into what might be called a performance. However, throughout the process, I was very clear that these works were a measure of the ongoing journey and each one was a constituent feature of the other. Consequently, the written submission became an analysis and document of the overall process or methodology. My submission includes images from the three works and was interspersed with dancer testimonials and significant moments of interaction. There was no video documentation included in the submitted thesis (purposely); however, the examiners were given video recordings of some of the ongoing practical investigations, and crucially they were asked to experience a version of the work live.

So, as a practising artist, why undertake research this way?

Nelson makes a case[2] for arts practice as knowledge-producing, which in and of itself articulates a research inquiry.[3] In my case, the dancer was presented with a challenge: *How might the integration and influence of digital media within her performance environment inspire her to generate and open up to new movement potentials?* Because my intrigue was based on the very act of making movement, I needed to find a basis for enabling the dancers to engage in a fluid and open-ended process.

That is not to say that such an enquiry could not have been explored in a purely artistic context. On the contrary, such an exploration could just as easily be undertaken outside of the academy. However, I realised very quickly that this project needed a different approach and that my usual artistic and choreographic methods were too reliable for extending the dancers' decision-making processes in this scenario. On reflection, it was framing the dancing through academic scrutiny that helped me to explore a process that was based on discovery rather than adhering to the conventions of [my] dance making per se. Those challenges I encountered enabled me to unpick my understanding of some of the landmark activities of movement-making in both a deep and sustained way. For example, I began to explore a number of situations, some of which were designed to integrate the mediated image into the dancer's physical and perceptual environment (see Figures 5.8.2 and 5.8.3). In such situations, her responses to the emerging real-time digital media meant that her perceptual awareness and subsequent management of the movement changed. This, alongside other interactive environments I designed to heighten and extend her responses further, helped me to develop an effective framework for generating new forms of embodied knowledge within a generative and dynamic system.

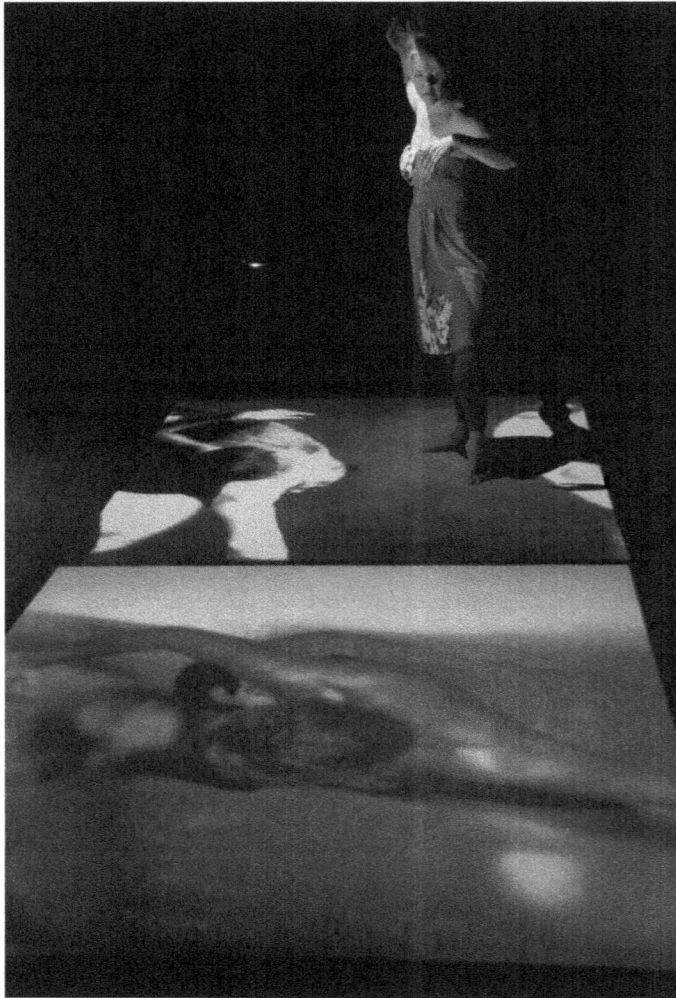

Figure 5.8.2 Shift digital dance (2011).

Source: Photograph: Kerry Francksen.

Figure 5.8.3   Shift digital dance (2011).
Source: Photograph: Kerry Francksen.

All of the research questions that arose came from the experiences of moving that were explored during my practice-based research. Essentially, my PhD sought to discover how moving in media-rich environments might afford the dancer with new and/or alternative choices for perceiving and making movement in such situations.[4] The dancer's abilities to activate movement as part of a living process meant that the methods I used to understand and assess the developing research were constantly shifting and being redefined. This is what sat at the heart of all my work – not the pursuit of creating an artwork but trying to unpick the nature of the practice itself. In the end, the PhD became its own generative system, and my originality of knowledge was the method itself.

Consequently, the interconnections between the mediums of dance, image, sound, and the concentration on academic enquiry called for approaches that were multi-modal and theoretically varied. By its very nature, the creative process did not follow a linear logic and the integration of conceptual ideas and philosophical readings emerged as part of the developing practical explorations.

For that reason, I set about designing an appropriate testing ground for embodied discovery, which meant that the research needs were driven by both the physical and conceptual insights of the dancers. The research process itself had to provide the dancer with a challenge: how to encounter technology not as prosthesis[5] but as a characteristic and qualitative function of her evolving movements. In the end, the very nature of my research helped to define the overall shape of the PhD and the studio environment itself became an important breeding ground for change.

Journeys through the mire: establishing methods and creating positive frameworks for practice

Given my focus was on the continual discovery and transformation of movement, my approach needed a level of flexibility and openness. This meant that the research process was dependent on establishing methods that could inspire the dancers to move differently. Significantly, the experience of moving was fundamental to the construction, evaluation, and development of any new knowledge, and this played out most meaningfully within a practical setting (for me this included a number of settings: a video-recording studio – see Figure 5.8.4 – my kitchen, a basement, a variety of corridors, and any other corner I could commandeer to set up a screen, a camera, and a projector – guerrilla-style studio set-ups were a must!).

As Matthew Reason describes,

> The particular forms of knowing that can be generated through arts practice are those of embodied, tacit and material knowledge, where discovery happens through the action of arts making, and in reflection in and upon that action. Located within action, the particular claim of practice-based research is that if offers not just a different way of doing things than more traditional research methodologies but rather, and more importantly, access to different forms of knowledge.[67]

The usefulness of having to establish what new and original knowledge might be as the dancers embarked on a journey of discovery became instrumental in helping me to define my approach and is what led me towards a practice-based research enquiry. Most notably, the degree of uncertainty and serendipity that my enquiry necessitated, did, as Smith and Dean remark, challenge the idea of 'knowledge as being an understood given'.[8] The idea that knowledge, as they propose, 'take[s] many forms and occur[s] at various different levels of precision and stability'[9] became a way for me to articulate the different forms of knowing through movement practice.

Understanding the nature of my research was a massive clue for defining the method. It was tricky trying to narrow down or define those methods I should follow. In many ways, these struggles for definition were important for making me realise that my PhD's fundamental features (i.e. the somatic and perceptual insights of the dancers) would become instructive of the methods I needed to use – much like being able to let go of my usual methods for choreographing described earlier.

Similarly, reading, exploring, and analysing existing methods related to this particular field of research was helpful; but in my experience, it was paying attention to the needs of the practice, or more excitingly the problems[10] inherent in the practice, which enabled me to develop an approach that was more meaningful to me. In essence, I allowed the work to tell me what the appropriate methodological response should be. This was at once both scary and exhilarating!

This realisation supported the prospect for my developing practice-based research to be unstable, unpredictable, and somewhat unsystematic in its presentation of originality. Once I realised this, I was able to begin to conceive of the practice in a new and dynamic way. For any artist, in any discipline, I believe that searching for the character and essence of an individual's work is key to

Figure 5.8.4   Betwixt & Between (2013).
Source: Photograph: Michael Huxley.

defining the right approach. Some of the most exciting moments of my research were delving into the characteristic features and qualities of what I do and why I do it.

In my thesis, I described my process as an iterative cycle of

moving, responding, reflecting, programming, and experiencing,

in which the practical and theoretical insights gathered became interchangeable. This draws on Candy and Edmonds[11] and Nelson,[12] who discuss variously the reciprocal processes of practical, reflective, and conceptual methods in practice-based research. My creative process was by no means orderly. The nature of the research required the dancers to constantly change and adapt to respond to the experiential insights being gathered in the studio. Thus, my working methods developed from establishing a systematic approach, rather than following a codified method. However, that was due to the practical characteristics of the research, which became instructive of a self-generating process. This was practical, theoretical, and philosophical in nature.

In many ways, grappling with what my method should be in the end made me realise that I could not fit my research into any tried and tested methods. There was no off-the-shelf approach that I could follow. Consequently, the knowledge that arose comprised a method for understanding and subsequently generating new movement potentials within mediated environments. Something I learnt through my process was that only the work itself – practicing in the studio – could reveal the specific method. While a challenging and somewhat scary thing to do, I would argue that there is a case for encouraging any practice-based researcher to remain unfixed in their pursuit of the right method (if there is such a thing?). In many ways, one of my 'ah-ha' moments was realising that I needed to let the work lead me.

Trying to articulate such a continuously changing and slippery process can prove tricky, particularly when it comes to presenting your findings at, for example, a formal or annual review. The need for constant re-invention, therefore, became a key factor for my analysis and evaluation, and this was written into the formal documentation throughout the PhD. Being clear that the research was based on rigorous inquiry was of course important, and this helped me to structure my thinking. However, not only did I apply academic scrutiny to help explain and validate the practical outcomes, but I also found that my theoretical and philosophical engagement became a mechanism for allowing the practice to be unravelled. In accordance with Candy's chapter in this handbook,[13] the trajectory of my research was dependent on what I consider to be a type of theory-in-action, whereby the dancers were able to identify, act upon, and ultimately effect a change within an emerging and generative system. This change came about through the reciprocal process of practicing, theorising, and evaluating. In effect, this process became a way to un-pick my understanding. I became more interested in finding ideas or theories that problematised what I was doing over and above trying to identify a logical method of practice, although of course contextualising the work within current paradigms was important.

Unravelling practice

Once I was able to trust that the method was in the process of revealing itself, I was able to reach another turning point: managing the relationship between theory and practice. Rather than thinking about the writing as a means of explanation, my theoretical and philosophical application helped me to challenge what was arising, and importantly helped me to manage a process that placed me in a counterintuitive position of actively wanting to become less familiar with what I was doing. My initial feelings of wanting to demonstrate the success of the practice gave way to a somewhat alarming but exhilarating process of placing myself in situations that could inspire a sense of the unknown. Consequently, by recognising a need for spontaneity and risk, I began to find different ways to define the research imperatives.

As a consequence, certain theories became instructive for the arising practice, and in equal measure, the practical enquiries helped to drive the theoretical questioning. Crucially, the performance works I submitted as part of my PhD became instruments for uncovering a number of key insights concerning the dancers' perceptual and experiential responses. Notably, my PhD was submitted as a portfolio of works accompanied by a written document that provided a textual analysis of the significant insights gained throughout the research process.

Another tricky aspect of my work was how best to document what was a living and evolving process. It was extremely important that all of my works were experienced. Due to the nature of the knowledge building, those individuals who shared the work with me (both collaborators and audience alike) also became instructive of the emerging behaviours. Significantly, the experiential encounters of all who experienced the work became affective[14] within the system. This meant that traditional forms of documentation did not always capture the important nuances of my practice or the embodied experiences generated throughout the system. This is why I used a combination of video and photographic documentation, reflective journals, audience interviews and performer/audience feedback. It was also very important that the examiners were able to experience my work for themselves via live performance events (which also presented its own set of challenges, i.e. having to perform directly before one's viva voce examination – something I do not recommend).

Breeding grounds for practice

The key to the knowledge I uncovered was the dancer's awareness of how to 'be' in such environments. The explicit focus on embodied practice was therefore crucial for developing such a perspective, and particular methods for creating an encounter with visual and aural materials continued to shape the portfolio of works. While my thesis does not propose a step-by-step guide to making what I termed 'live-digital' dance, it does offer a purposeful framework for re-considering the difficult relationships between bodies and mediated images (see Figure 5.8.5 for an illustration of how these live-digital encounters materialised). What is more, such a process brought about methods that, in and of themselves, established a non-binary relationship between human dancer and the mediated presences. This was what ultimately answered the main question of the thesis: *how to challenge the dominance of the digital*, and which allowed me to present an original methodology to the field.

Figure 5.8.5   Betwixt & Between (2013).
Source: Photograph: Michael Huxley.

For any artist, the decision to undertake PhD research is fraught with many challenges and wonderful possibilities. On reflection, my own experiences point towards allowing the practice – or otherwise said the doing of the practice[15] – to guide you through the academic rigor of PhD research. For me, it was fore-fronting the experiential encounters and charting the exciting happenstances of live and digital materiality that helped me to unpick some of the fundamental building blocks of my own artistic practice. Gathering knowledge through embodied, practical, and philosophical exploration helped me to define and expand the knowledge I discovered *in* practice. Significantly, it was the practice itself that formed the basis for new knowledge. As I have described, this journey could only have taken place in the breeding ground of practice, within a studio setting. In the end, it was the curious nature of having to negotiate what Nelson describes as 'the making visible of an intelligence which nevertheless remains fundamentally located in embodied knowing'[16] which inspired an ongoing method for practice that I am still pursuing to this day.

Notes

1 Candy and Edmonds (2018, p. 66).
2 As an artist engaged in embodied practice, I found Robin Nelson's discussion of 'doing-knowing' particularly useful. In his book *Practice as Research in the Arts: Principles, Protocols, Pedagogies, Resistances* (2013), he develops the idea that different modes of knowledge production can be achieved through a multi-modal research enquiry. Via his model titled, 'Modes of Knowing: Multi-mode Epistemological Model for PaR', Nelson illustrates how 'doing-knowing' can encompass different 'modes of knowing (tacit, embodied-cognition, performative' (Nelson 2013, p. 38). Significant to my research was how new knowledge can be gained through perception and experience.
3 Nelson (2013).
4 Please see my chapter 'The Implications of Technology in Dance: A Dancer's Perspective of Moving in Media-Rich Environments' (2018). Kerry Francksen. In Sarah Whatley, Rosamaria K. Cisneros and Amalia Sabiescu (Eds.), *Digital Echoes. Spaces for Intangible and Performance-based Cultural Heritage*, for a more detailed description.
5 I refer here to Erin Manning's notion of 'prosthesis' in her article 'Prosthetics Making Sense: Dancing the Technogenetic Body'.
6 Reason (2012, p. 195).
7 For a similar reading of dance as a form of knowledge, see Ann Pakes 'Knowing Through Dancemaking: Choreography, Practical Knowledge and Practice-as-Research', in *Contemporary Choreography: A Critical Reader* (2009).
8 Smith and Dean (2009, p. 2).
9 Ibid., p. 4.
10 I use the term problem here to represent the positive opportunities in rethinking the relationships between choreography and performance described by Bojana Cvejić. She states, 'Choreography doesn't merely precede performance as the creative process that then culminates in an event, nor can it be reduced to a technical, craft orientated definition. . . . The making continues to operate in the performing in the sense that its problems persist and give rise to different solutions in the performing of, attending to, and also thinking beyond the spatio-temporal event of the performance' (Cvejić 2015, p. 14).
11 Candy and Edmonds (2018).
12 Nelson (2013).
13 Candy (2022).
14 I use the term affective in the context of Erin Manning's discussion of an ecology of forces. She states, 'Affect, understood along the lines of Whitehead's concept of feeling, is a transductive force that propels being to become across the phases of its individuation. Affect is of milieu. . . . Affect activates the very connectibility of experience' (Manning (2013, p. 26). Such theories therefore became useful as a means to understand the experiential encounters of the work, especially in terms of the dancers' connectivity within an evolving situation. Again, the nature of the work directed me towards particular theories/concepts that were meaningful to the practice.
15 Nelson (2013).
16 Ibid., p. 40.

Bibliography

Candy, L. (2022). Theory as an Active Agent in Practice based Knowledge Development. In C. Vear (Ed.), *The Routledge International Handbook of Practice-Based Research*. London and New York: Routledge.

Candy, L., & Edmonds, E. (2018). Practice-Based Research in the Creative Arts: Foundations and Futures from the Front Line. *Leonardo*, 51, 63–69.

Cvejić, B. (2015). *Choreographing Problems: Expressive Concepts in European Contemporary Dance and Performance*. Hampshire: Palgrave Macmillan.

Manning, E. (2013). *Always More Than One: Individuation's Dance*. Durham and London: Duke University Press.

Nelson, R. (2013). *Practice-as-Research in the Arts: Principles, Protocols. Pedagogies, Resistances*. Hampshire: Palgrave Macmillan.

Pakes, A. (2009). Knowing Through Dance-Making: Choreography, Practical Knowledge and Practice-as-Research. In J. Butterworth & L. Wildschut (Eds.), *Contemporary Choreography: A Critical Reader*. Abingdon: Routledge, 10–22.

Reason, M. (2012). Part IV: Artistic Enquiries: Kinesthetic Empathy and Practice-Based Research. Introduction. In D. Reynolds & M. Reason (Eds.), *Kinesthetic Empathy in Creative and Cultural Practices*. Bristol: Intellect, 195–197.

Smith, H., & Dean, R. T. (2009). *Practice-led Research, Research-led Practice in the Creative Arts*. Edinburgh: Edinburgh University Press.

5.9

ENCOUNTERS AT THE FRINGE

A relational approach to human–robot interaction

Petra Gemeinboeck and Rob Saunders

Doing art and science collaboratively – diffractively

Our art–science collaboration began 15 years ago. While at first it was perhaps a typical configuration of a lead artist working with an expert programmer/engineer to realise their projects, it soon evolved into a more horizontal, non-hierarchical partnership, where ideas, goals and approaches intertwine and extend and transform each other. While any interdisciplinary collaboration requires significant effort to develop a shared language to navigate and negotiate different disciplinary meanings and expectations,[1] over the years we have immersed ourselves in each other's ways of thinking and doing – allowing for new, essentially transdisciplinary meanings, expectations and relations to emerge. At its best, our practice is about what Born and Barry[2] refer to as 'effecting ontological change in both the object(s) of research, and the relations between research subjects and objects'.[3]

At the same time, we still maintain distinct identities, motivations and objectives that drive each project, arising from, drawing on and reaching back into our respective fields of practice. Petra's field of practice is experimental art with a focus on situated reconfigurings of so-called disruptive technologies that explore critical questions of embodiment and agency. Prior to working with Rob, her works comprised site-specific, interactive installations that sometimes also involved performance elements, staging dynamically evolving human–nonhuman (e.g., electronic) encounters. Rob's field of research is Computational Creativity, a subfield of Artificial Intelligence concerned with the computational modelling and study of creativity. Prior to our long-term collaboration, Rob's work focused on the computational modelling of curiosity in artificial creative agents, and the study of social and cultural situatedness in simulated societies.

Practice-based research provides us with a rich investigatory playground to join forces and experiment with ideas, concepts and processes in material, embodied ways. Our collaborative projects span experimental arts, robotics/AI, performance and new materialism – bringing these fields into a unique conjunction through a performative, embodied and process-focused practice of inquiry that often draws on concepts from radical embodied cognition. We align ourselves with the emergent field of creative robotics, which explores human–robot configurations, building on the history of robotic and cybernetic art,[4] albeit pushing its boundaries to include feminist practices and methods from computational creativity by exploring posthuman notions of machine performance.

DOI: 10.4324/9780429324154-52

The Machine Movement Lab project

Our collaborative projects explore the performative potential of nonhuman agents by evoking emergent dramaturgies of encounter with creative but distinctly non-humanlike machine performers. Our current Machine Movement Lab (MML) project, which began in 2015, extends this practice but focuses in on relation-making with machines and investigates how they become social and creative participants in the situation of encounter. From the start, we were interested in exploring the social potential of abstract robotic artefacts. How can we relate to such machines without relying on mimicking human features and behaviours and replicating human social protocols? How can we instead embrace and aesthetically exploit their machinic embodiment and its unique, non-humanlike capacities? This apparent conundrum of seeking the relational, while not only recognising but also exploiting the differences between humans and robots, became the creative engine of this project that has shaped our conceptual framework and methodological development.

To approach this puzzle, we recruited collaborators from dramaturgy, choreography and dance to explore how movement and its situated, connection and meaning-making potential could open up an embodied inquiry beyond the subject–object divide. Rather than looking at movement as a mode of navigating a terrain or expressing a predefined narrative, we aimed at investigating its capacity for shaping the dynamics of the encounter and for transforming the artefact and its relational possibilities in the process. We began with a series of workshops in which we asked dancers to negotiate the kinetic relation-making potential of a variety of simple shapes and materials (see Figure 5.9.1). Early experiments revealed that, in order to foreground the generative and affective potential of movement, we needed to aesthetically put to work the contraposition of abstract, plain appearance and dynamic, delicate movement qualities. To our surprise, we found that the familiar, humble cardboard box, when inhabited and activated by a dancer, was rendered strange and transformed into something else, more than an object through the dynamics of movement. To avoid possible humanoid proportions, we began to probe deeper into the relational potential of the regular, omni-directional cube shape, deployed as a purpose-built, wearable costume for a dancer (see Figure 5.9.1).

We refer to our approach as Performative Body Mapping (PBM). In PBM, the costume becomes a mapping instrument for harnessing dancers' kinaesthetic expertise to bodily probe into the unique material embodiment and performative potential of a machinelike artefact. The dance performer makes use of the costume to empathically immerse themselves and kinaesthetically feel into and move with this 'other body'. We then capture the movements resulting from this dancer-cube entanglement to inform machine learning processes for a robot, which resembles the shape and size of the costume (discussed in more detail next). So far, we have built two robotic prototypes, called *Cube Performer #1* and *Cube Performer #2* (see Figure 5.9.1). To us, the cube performer is part of the embodied proposition and 'experiment in progress' for rethinking and reimagining human–machine relationships put forward by MML.

In 2019, we were able to secure more funding for situating PBM in the sociocultural milieu of the dance studio. The aim is to develop an extended performance-making practice as a relational, embodied research tool to explore human–machine relationships and probe into and reimagine subject–object boundaries.

MML in practice: entangling the bodily, material and data-driven

Methodological development and knowledge production are deeply entangled in our practice, producing and propelling each other through an ongoing webbing of interknottings and feedback loops. Experimenting with movement and how it dynamically 'bodies-forth'[5] and generates new relations and meanings in the process is our central 'mode of *doing* research',[6] connecting all of our processes.

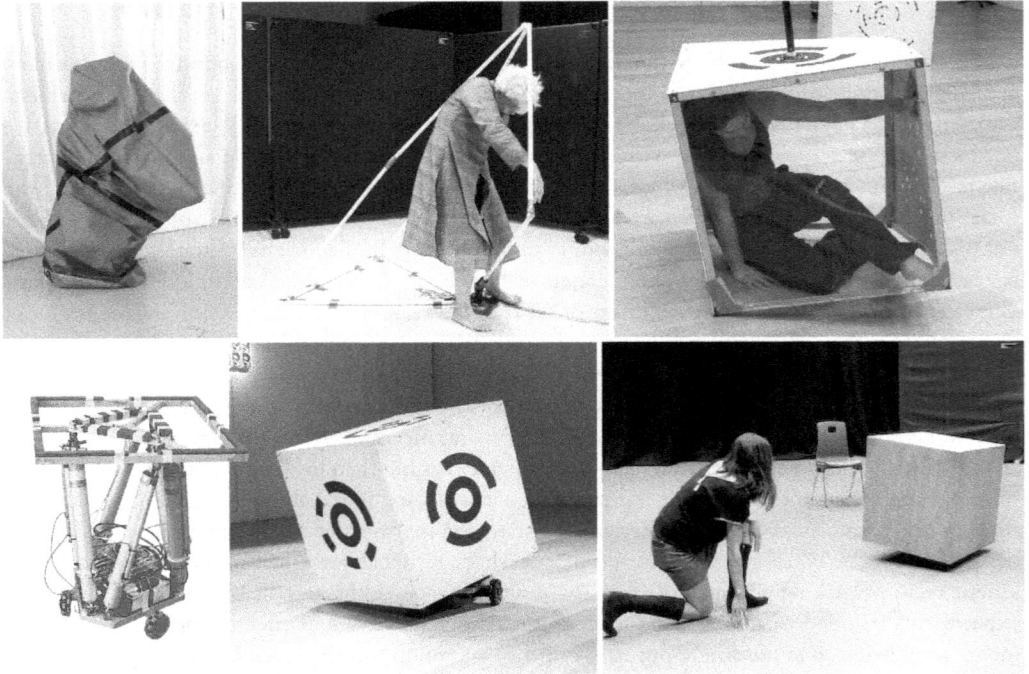

Figure 5.9.1 Snapshots from our PBM processes and outcomes, including the early form-finding stage (top left), tetrahedron-shaped costume activated by Tess De Quincey (top centre), a cube-shaped costume inhabited by Audrey Rochette (top right), the cube performer's mechanical frame (bottom left), *Cube Performer #1* at the Games and Performing Arts Festival, UK, 2018 (bottom centre), and a participant encountering *Cube Performer #2*, UNSW, Sydney, 2019 (bottom right).

Source: Petra Gemeinboeck.

Our PBM approach, for instance, brokers dancers' bodily negotiations with nonhuman artefacts and their material affordances, as well as movement analysis with the mechanical design and machine learning. This produces a movement-language specific to the artefact's embodiment, which patterns the cube performers' relational capacities.

One of our core concerns when developing the methodological frame for PBM was to carefully attend to the multiplicity of human and nonhuman collaborators and to give plenty of space to the emergence of possible relations, as well as productive sparrings. Treusch argues that 'situated co-engineering is about adhering to interference patterns as they emerge',[7] which, as we will further explore, also means attending to and conspiring with nonhuman co-agencies. With this relatively open structure for inquiry, where process has primacy over the final product,[8] any limits we encounter, whether arising from the dancers' bodily negotiations, material affordances or technical limitations, become enabling constraints that channel and steer our process.

In PBM (see Figure 5.9.2), our performance-based inquiries of non-humanlike relation-making shape the choreographic scores that, in turn, shape the movement dynamics that the dancer performs with the costume. The dancer's performances are moulded by the costume's shape, weight and size, as well as the spatial affordances they produce when put in motion. Machine learning requirements, e.g., data sets of variations of the same movement phrase performed in different qualities, not only structure but also aesthetically frame the sequentially unfolding choreograph scores. And, vice versa, the robot's design and its machine learning system are equally shaped by our embodied enquiries. For example, the mechanical engineering followed our movement experiments rather than the other way round, and our machine learning approach foregrounds the embodied, embedded nature of the costume's motion data in relation to the robot's own machinic embodiment.[9]

This interplay of bodily, material and data-driven processes also produces the intersection points for analysing the performative outcomes of our methodology, which, in turn, becomes further source material for our inquiry (see Figure 5.9.2). The performative, affective potential of recorded kinetic traces (motion data) is examined in relation to the choreographers' intent using a Laban Movement Analysis lens to develop descriptors for the machine learner. Studies of participants encountering the cube performer for the first time examine how they express their bodily resonances with its movements, which we correlate with dancers' accounts. These different correlations allow us to study how embodied empathy translates from the dancers' entangled encounters with the costume to experiential encounters with the robotic artefact.[10] Data recordings of participants' encounters, in turn, provide relational patterns for simulating the robot's perceptual world to develop its improvisational machine learning.

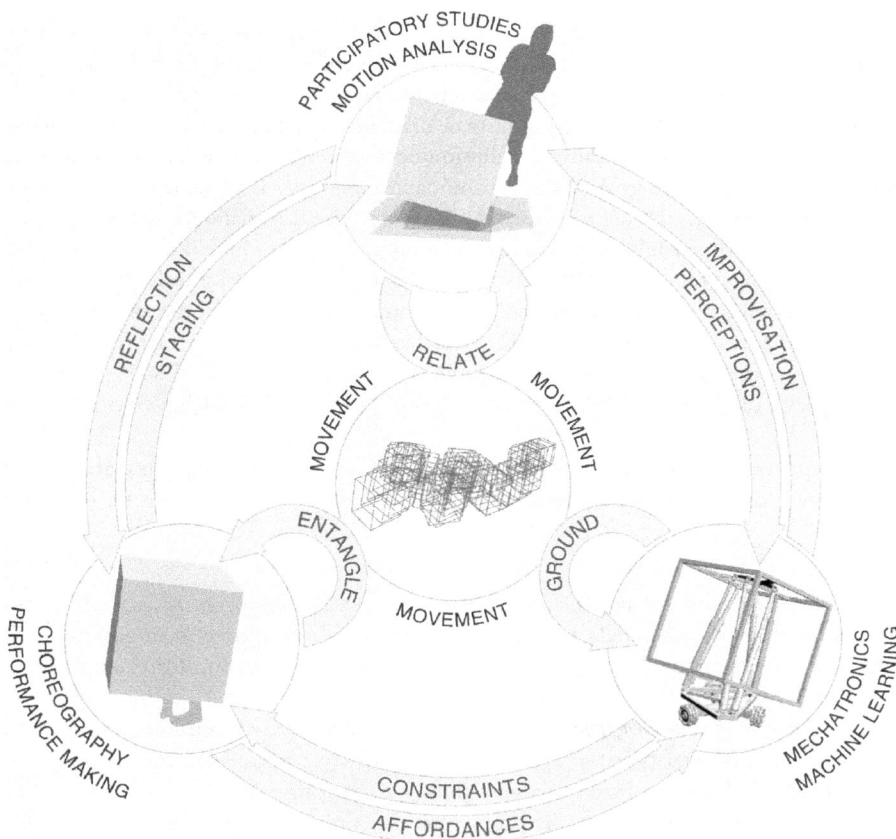

Figure 5.9.2   Diagram showing the interconnectedness of core PBM processes and the centrality of movement-making as research.

Source: Petra Gemeinboeck and Rob Saunders.

MML in practice: constitutive concepts for a posthuman dramaturgy

Our project aims to challenge the binary thinking that still dominates robotics practices by fore-grounding the emergence of agencies, meanings and, with it, subject and object boundaries through material, performative relations. Human–robot interaction practices are commonly characterised through a series of man-made constructs and beliefs that define the boundaries of human and machine based on universalist, binary categories, as if they were naturally given or historically and culturally neutral.[11] In practice, these fixed hierarchical boundaries often serve as blinkers, 'shielding' us from the active, participatory role that nonhuman constituents, capacities and constellations have in co-shaping our design processes. Eschewing these engrained boundaries and hierarchies, includ-ing a static understanding of subjects and objects, and unanchoring the machine from stereotypical, often gendered role-based or task-driven narratives, complicates the design process. We previously referred to this as moving beyond the mirror (a world remade in our own image) and straight through the looking glass.[12]

Navigating a world beyond the humancentric mirror plane, we consider our practice more akin to an ongoing negotiation of dynamic processes and collaboration with materials (explored in more detail later in this chapter). By negotiation we specifically refer to notions of 'giving space' to the material, relational dynamics that unfold in the interactional situation, as part of the design process as well as in the process of audiences encountering the robot. Hence, we can no longer speak of designing a human–machine relationship or even designing for a successful interaction scenario; rather, we design with potential dynamics and becoming agents, human and nonhuman, that together participate in the enactment of a relationship, unique to a particular situation. This is more akin a notion of emergent, posthuman dramaturgy, where we set the stage for human and nonhuman participants, yet without the illusion or desire of being able to control what unfolds in the encounter. Instead, the goal of our posthuman dramaturgy is to socioculturally situate the nonhuman participant, i.e., the cube performer, based on the relational dynamics of movement to facilitate meaning-making while at the same time embracing its machinelike otherness. This, in a nutshell, is how PBM attempts to creatively tackle the aforementioned puzzle. 'Giving space', to us, then means that we attend to the specificities of material capacities and movement qualities and the kinds of affordances and entangled relations they can produce in a particular situation and how they contribute to the enactment of meanings. Meaning making in PBM, with its focus on the generative capacity of movement, relates to what Froese and Fuchs have referred to as 'intra-bodily resonances'.[13]

Core concepts or principles that drive our process and, at the same time, are materially enacted as part of it draw on the relational focus of new materialist practices and can be summarised as following:

- *A robot's social agency is not an individual property but, instead, emerges in the interactional dynamics.* Our approach draws on Barad's posthuman account,[14] where agency is an enactment, arising from intra-actions rather than something that can be given to a robot, for example humanlike features.[15]
- *Empathy is a form of embodied, relational mattering and not confined to human matters.* In PBM, the enactment of social agency is tightly coupled with how we make meaning as part of a bodily attunement with the other, beginning with the dancer getting entangled with the robot's shape, in the form of the customs, to bodily probe into its meaning-making potential through move-ment. It is this ongoing attunement that gives raise to intra-bodily resonances, without the need for symmetry (e.g., the other looking like us or having similar cognitive capacities).[16]
- *A relational approach to human–machine relationships challenges hierarchical dualisms.* PBM materially investigates how humans and sociotechnical artefacts constitute each other.[17] A performative,

relational view of a robot's sociality thus opens up a horizontal ethics of human–machine relations.[18]

MML in practice: a material collaboration

As aforementioned, our relational approach relies on attending to and collaborating with nonhuman materials and the affordances and agencies they coenact. As we aim for invoking notions of embodied empathy with nonhuman agents, we also need to practice material, posthuman empathy as part of the investigatory process, which heavily relies on forming alliances with nonhuman participants and their capacities. From a new materialist perspective, our relationships with artefacts are material, not only in terms of how we make use of material capacities but also with regards to their material effects – the meanings enacted through our relations literally matter. They are situated and specific, not only with regards to the interactional encounter but also the sociohistorical networks they are embedded in.[19] We thus position ourselves and our practice in the middle of the human–nonhuman encounter, negotiating and designing with these relationships, rather than positioning ourselves above (e.g., by attempting to define and control them) or outside (e.g., by designing for them as if we could detach ourselves or the artefact from the process). Our process of making new relations with nonhuman agents thus favours performativity, an ongoing emergent, posthuman dramaturgy of becoming, coupling and bodying, over representationalism.

Needless to say, this material collaboration, which requires us to remain open to any unforeseen 'inputs' and direction changes from the nonhuman participants involved, can be a struggle and, at times, quite a humbling affair. Nonhuman participants include, for instance, the costumes' materials that, over the years, have comprised varying composites of cardboard, folded paper, elastic and nonelastic textiles, plastic panelling, aluminium extrusions, plywood and plexiglass sheeting, steel bolts, nuts and gaffa tape. Entangled with human bodies and literally invited to dance, their equally varying material responses, compliances and transformations have, sometimes dramatically, shaped the dancers' bodily experience, not only with regard to what they can do with them but also how they need to reconfigure their own bodies to negotiate this close encounter.[20] Importantly, these performative-material relations we develop along all stages of the design process later co-shape the relational possibilities and effects, when the robot encounters other, human participants.[21]

The building of robots relies on complex material alliances that are forged from digital processes, data materials, sensors and material-mechanical composites to enact material performances across the apparent physical and digital divide. In MML, we attend to the performative potential of the machinic embodiment to develop a learned motor controller that is grounded in this specific, nonhuman embodiment and its unique capacities. Our goal for the cube performer is that it does not simply replicate the movements captured from the dancer–costume entanglement but learns to improvise: moving in machine-like ways, albeit grounded in the aesthetically and socioculturally coded biases and constraints that are specific to our PBM motion data. It is this material negotiation of biases that are specific to the mechanical and electronic processes and sociocultural scaffolds that shape the robot's unique movement qualities and allow the cube performer to participate in the meaning-making of the encounter.

One major challenge arising from our PBM process is to do with the significantly different time scales of the material collaborations involved. The delay between our experimental processes in the dance studio and the technical development and validation processes can be significant. While this inevitably compromises the rich and necessary knowledge exchange between experimental and developmental processes and the embodied, situated insights they offer,[22] we also found that it structures and impacts our technical development in ways that foreground the primacy of embodied, situated knowledge. Put differently, these timing differences prohibit us from possibly pre-shaping relationships based on assumptions we may otherwise make in the development process; instead, they allow us to mould technical specifications (e.g., regarding the robot's movement or perceptual capacities) according to the relations that emerge in the embodied-material encounters, whether in the dancer–costume entanglement or with audiences in exhibitions and studies.[23]

Concluding remarks

Connecting back to the discussion of our interdisciplinary practice, it is worth noting that our goals and outcomes are co-shaped by the institutional, funding-oriented milieu it is embedded in. They are, however, equally shaped by our practice-based, relational approach and motivation to effect ontological changes[24] beyond the bounds of our practice. Attending to nonhuman agential 'matter-ings'[25] in embodied, relational ways that criss-cross academic domains as well as tacit and performa-tive practices challenges the dominant modus operandi where knowledge production is confined to disciplinary boundaries and knowing can be separated into scholarly or practiced-embodied forms.[26] In line with that, it is critical that we engage a wide range of audiences, from the publics to different disciplinary research communities. Catering to diverse audiences, institutional settings and disci-plines then requires a continuous process of reframing and resituating our work, not dissimilar to a series of site-specific re-performances of our practice.

One of the most notable differences between our prior experimental artistic practice[27] and our current practice-based research project is that MML started with a question. By asking how we can empathically relate to machines in ways that embrace their unique machinic otherness, we set ourselves on a path of open experimentation from the outset. In contrast, our artistic projects often began with a more or less formed idea for a particular experience or relationship, which we then sought to materialise. While notions of experience and emerging relationships remain key to MML, they play a more significant role in defining our process, rather than providing a goal to be realised. We found that seeking to reimagine human–machine encounters in bodily, performative ways involves a seemingly endless series of relational encounters that carve out our path of discovery. As we, sometimes quite literally feel our way forward in this process, attending to movements and matterings, and how they shape relational affordances is, in our experience, the only means of dis-covering new relations. Hence, a major insight for us was that we cannot design *for* encounters but that we design *with* them and the many material, performative and relational possibilities they open up. This, we believe, allows for gradually shaping a space of relational possibilities from which new encounters can emerge. By 'designing with encounters', we refer back to our practice of ongoing material collaboration and how it promotes an emergent, posthuman dramaturgy of becoming. Our practice thus contrasts notions of design that submit materials, bodies and relations to a predefined, singular version of encountering. Favouring performativity over representationalism, we found, grounds our process and inventions in specific materialities and situated constellations of bodies, materials and meanings. At the same time, it opens up a space of possibilities for new relations to be generated by unfixing any pre-defined boundaries, relations and ideas. In our practice, which seeks the relational by embracing difference, it seems befitting to seek emergent possibilities by attending to the specificities of our entanglements.

Acknowledgements

We would like to thank our collaborators Marie-Claude Poulin (kondition pluriel, CA; Univer-sity of Applied Arts Vienna, AT), Roos van Berkel (Eindhoven University of Technology, NL), Rochelle Haley (University of New South Wales, AU), and Audrey Rochette (CA), Maaike Bleeker (Utrecht University, NL), Tess de Quincey (De Quincey Co, Sydney, AU), Linda Luke (AU) and Kirsten Packham (AU).

The project discussed in this chapter has been partly supported by: the Australian Government through the Australian Research Council (DP160104706); the EU Framework Program (FP7) European Research Area Chairs Scheme project (621403); and the Austrian Government through the Austrian Science Fund (FWF, AR545).

Notes

1 See Papastergiadis (2004).
2 Born and Barry (2010).
3 Born and Barry (2010, p. 105).
4 Gemeinboeck (2017).
5 Manning and Massumi (2014, p. 39).
6 Hunter (2018).
7 Treusch (2021, p. 63).
8 See Ingold (2010).
9 See Gemeinboeck (2021).
10 Gemeinboeck and Saunders (2018).
11 See Nakamura (2003).
12 Gemeinboeck and Saunders (2021).
13 Froese and Fuchs (2012, p. 212).
14 Barad (2007).
15 See Alac (2015), Gemeinboeck (2021).
16 See also Despret (2013).
17 Suchman (2007).
18 See also Damiano and Dumouchel (2020).
19 See Suchman (2007).
20 See also Noland (2009).
21 See Gemeinboeck (2021).
22 Gemeinboeck (2021).
23 See Gemeinboeck and Saunders (2021).
24 Born and Barry (2010).
25 See Barad (2007).
26 See Guimarães et al. (2019).
27 E.g., see Gemeinboeck and Saunders (2013).

Bibliography

Alač, M. (2015). Social Robots: Things or Agents? *AI & Society*, 31(1), 519–535.
Barad, K. (2007). *Meeting the Universe Halfway: Quantum Physics and the Entanglement of Matter and Meaning.* Durham, NC: Duke University Press.
Born, G., & Barry, A. (2010). Art-Science. *Journal of Cultural Economy*, 3(1), 103–119. DOI:10.1080/17530351003617610.
Damiano, L., & Dumouchel, P. (2020). Emotions in Relation: Epistemological and Ethical Scaffolding for Mixed Human-Robot Social Ecologies. *Humana.Mente Journal of Philosophical Studies*, 13(37), 181–206. www.humanamente.eu/index.php/HM/article/view/321 (accessed April 20, 2021).
Despret, V. (2013). Responding Bodies and Partial Affinities in Human-Animal Worlds. *Theory, Culture & Society*, 30(7–8), 51–76. DOI:10.1177/0263276413496852.
Froese, T., & Fuchs, T. (2012). The Extended Body: A Case Study in the Neuro-Phenomenology of Social Interaction. *Phenomenology and the Cognitive Sciences*, 11, 205–236. DOI:10.1007/s11097-012-9254-2.
Gemeinboeck, P. (2017). Creative Robotics: Introduction. Creative Robotics: Rethinking Human Machine Configurations. *The Fibreculture Journal*, 28. DOI:10.15307/fcj.28; https://twentyeight.fibreculturejournal.org/2016/12/22/fcj-203-introduction-creative-robotics-rethinking-human-machine-configurations/ (accessed April 23, 2021).
Gemeinboeck, P. (2021). The Aesthetics of Encounter: A Relational-Performative Design Approach to Human-Robot Interaction. *Frontiers in Robotics and AI*, 7. DOI:10.3389/frobt.2020.577900; www.frontiersin.org/article/10.3389/frobt.2020.577900 (accessed April 23, 2021).

Gemeinboeck, P., & Saunders, R. (2013). Inventing Cultural Machines. In A. Dong, J. Conomos, & B. Buckley (Eds.), *Ecologies of Invention*. Sydney: University of Sydney Press, 37–46.

Gemeinboeck, P., & Saunders, R. (2018). Human-Robot Kinesthetics: Mediating Kinesthetic Experience for Designing Affective Non-Humanlike Social Robots. In *Proceedings of the 2018 IEEE RO-MAN: The 27th IEEE International Conference on Robot and Human Interactive Communication*. New York, NY: IEEE, 571–576. DOI:10.1109/ROMAN.2018.8525596.

Gemeinboeck, P., & Saunders, R. (2021). Moving Beyond the Mirror: Relational, Performative Meaning-Making in Human-Robot Communication. AI & Society, Springer. DOI:10.1007/s00146-021-01212-1; link.springer.com/article/10.1007/s00146-021-01212-1 (accessed October 9, 2021).

Guimarães, M. H., Pohl, C., Bina, O., & Varanda, M. (2019). Who Is Doing Interand Transdisciplinary Research, and Why? An Empirical Study of Motivations, Attitudes, Skills, and Behaviours. *Futures*, 112. DOI: 10.1016/j.futures.2019.102441; www.sciencedirect.com/science/article/pii/S001632871830483X (accessed April 23, 2021).

Hunter, V. (2018). Dance. *New Materialism Almanac*. https://newmaterialism.eu/almanac/d/dance.html (accessed April 23, 2021).

Ingold, T. (2010). Bringing Things to Life: Creative Entanglements in a World of Materials. *ESRC National Centre for Research Methods, NCRM Working Paper Series*, 5(10), 2–14. http://eprints.ncrm.ac.uk/1306/1/0510_creative_entanglements.pdf (accessed April 15, 2021).

Manning, E., & Massumi, B. (2014). *Thought in the Act: Passages in the Ecology of Experience*. Minneapolis, MN: University of Minnesota Press.

Nakamura, N. (2003). Prospects for a Materialist Informatics: An Interview with Donna Haraway, Electronic Book Review. Reprinted in M. Bousquet & K. Wills (Eds.), *The Politics of Information: The Electronic Mediation of Social Change*. Boulder, CO: Alt-X Press, 154–168.

Noland, C. (2009). Coping and Choreography. In *Proceedings of the Digital Arts and Culture 2009 (DAC 09)*, UC Irvine. https://escholarship.org/uc/item/0gq729xq (accessed April 21, 2021).

Papastergiadis, N. (2004). The Ethics of Collaboration. In C. Green (Ed.), *Australian Culture Now*. Melbourne: National Gallery of Victoria, 57–61.

Suchman, L. (2007). *Human-Machine Reconfigurations: Plans and Situated Actions* (2nd ed.). Cambridge, MA: Cambridge University Press.

Treusch, P. (2021). *Robotic Knitting Re-Crafting Human-Robot Collaboration Through Careful Coboting*. https://www.degruyter.com/document/doi/10.14361/9783839452035/html

5.10

THE IMPACT OF PUBLIC ENGAGEMENT WITH RESEARCH ON A HOLOGRAPHIC PRACTICE-BASED STUDY

Pearl John

Introduction

Vitae is a charitable organisation describing itself as: 'the global leader in supporting the professional development of researchers, experienced in working with institutions as they strive for research excellence, innovation and impact'.[1] The RDF outlines researcher competence in four domains:

1) Domain A: *Knowledge and intellectual abilities:* The knowledge, intellectual abilities, and techniques to do research.
2) Domain B: *Personal effectiveness:* This includes the personal qualities and approach needed to be an effective researcher.
3) Domain C: *Research governance and organisation*: Knowledge of the professional standards and requirements to do research.
4) Domain D: *Engagement, influence, and impact:* includes the knowledge and skills to work with others and ensure the wider impact of research.

The last domain of *Engagement* has provided an essential element to my research process. Public engagement is a valuable pathway to impact for researchers and can be used to provide impact case studies for the Research Excellence Framework (REF).[2] Public engagement, as defined by the National Coordinating Centre for Public Engagement (NCCPE), describes 'the myriad of ways in which the activity and benefits of higher education and research can be shared with the public. Engagement is by definition a two-way process, involving interaction and listening, with the goal of generating mutual benefit'.[3] The NCCPE supports universities and research institutes in their engagement with the public and is funded by Research Councils UK, the Higher Education Funding Councils, and the Wellcome Trust. The REF is a process of expert review, undertaken by the UK's main funding bodies, and is carried out by expert panels made up of senior academics, international members, and research users. The review examines how university research funding can be allocated more efficiently so that universities can focus on carrying out world-leading research.

 DOI: 10.4324/9780429324154-53

Art and holography

Holography[4] has been central to my art practice over the last 35 years, and through it I have developed my own visual, critical, and practical vocabulary. I have a Fine Art and English Literature BA Combined Honours degree from the University of Exeter, specialising in time-based – or 4D – media (including film-making and installation work) and I have an MA in Holography from the Royal College of Art.[5] I undertook a PhD at De Montfort University as I knew that it would help me maintain a focus on my practice and give me access to valuable equipment and resources, as well as a research community that would otherwise be unavailable to me. However, my postgraduate studies were part-time, during which I continued to work as a science communicator. I am a Photonics outreach specialist and work as the Public Engagement Leader in Physics and Astronomy and Future Photonics Manufacturing Hub at the at the University of Southampton. Working in public engagement and being involved with the production of REF Impact Case Studies as part of my job informed my practice as an artist and as a researcher.

As an artist working with holography, I am part of a small but dedicated international community estimated by the historian Sean Johnston in *Holographic Visions: A History of New Science*[6] to be of less than 300 people, a majority of who are women. Making holograms requires a great deal of equipment: an engineering table; optical equipment (mirrors, lenses); a laser; and photographic developing facilities, and is a highly technical process.[7] The equipment and knowledge required for making holograms is relevant to the field of Photonics: the science of light, typically located in University Optical Engineering and Physics Departments. Historically, artists have worked with scientists and engineers to learn their craft or undergone an informal apprenticeship with a professional holographer. A recent publication by Andrew Pepper, *Holography: A Critical Debate within Contemporary Visual Culture*,[8] explored contemporary artists' concerns, which included: the limits of holography in documenting life; the nature of reality and illusion; concerns of installation and display, with artist Melissa Crenshaw describing Holograms as 'Virtual Sculptures' requiring specific installation; holography and the art market; and holography in Permanent Collections. Pepper's own paper discussed using the art gallery as a location for research-informed practice and critical reflection and my own research using the evaluation method – 'A Silent Researcher Critique'.[9]

A brief description of my research

My research project was entitled *Temporal and Spatial Coherence: Chronological and Affective Narrative within Holographic and Lenticular Space*.[10] My thesis argued that the Z and X axes of lenticular and holographic space can be used to store images chronologically, providing an audience with a new experience with affective and authentic impact. I created a new element to the lenticular and holographic artform, as my contribution to knowledge. My research presented my family's archival material dating back to the 1800s: photographs, film, text and objects – in a sequential order within the Z and X axes of holographic space, creating an animated four-dimensional (4D) family album in which my ancestors receded into holographic space and members of the current generation floated in front of the surface of the media. I evaluated different audience's experiences of the artworks I produced through surveys, observations, and a 'silent research critique' with experts, providing evidence of the research study's contribution to knowledge. While all the audiences had new experiences in interacting with the artworks and were moved by what they saw (physically and emotionally), only a group of experts in art and holography were able to identify and comprehend the novel conceptual use of the Z-axis of holographic space.

Methodology

Practice-based research in my field and at De Montfort University necessitated the completion of a body of artwork and a 40,000-word dissertation. The research process involved making artwork (holograms and lenticular images), exhibiting the work and evaluating audience responses to the work, documenting the research to assist with self-reflection (sketch books, lab books, weblog), informal audience observation, discussion with peers, and a critique by experts. I also wrote and published papers to disseminate research findings at the following international conferences and symposia: *HoloExpo2011* at the National Academy of Science, Belarus (2011); the *International Symposia on Display Holography (ISDH2015)* at ITMO University, St. Petersburg (2015); and at the *ISDH2018*, University of Aveiro, Portugal. The body of work culminated in the creation of a digital animated holographic artwork which answered my research questions. The image was animated by the movement of an audience member/participant in the interactive artwork. A still from the work, built using *Maxon Cinema 4D*, a 3D animation software program, is shown in Figure 5.10.1. (A video of the artwork produced as part of my research can be found on my website: www.pearljohn.com.)

My study had three distinct phases within it which engaged three different types of audiences. Phase 1 consisted of the production of a series of time-based lenticular prints, exhibited to 11,000 visitors at the Royal Society Summer Exhibition in 2012. To see the 4D images, viewers had to move laterally, or to-and-fro, in front of the lenticulars to make the images move. During the exhibit I became frustrated at the limited amount of audience feedback I was able to obtain in that environment because of my work with public engagement. Specifically, I was only able to collect the number of audience members (or impact reach), but was unable to determine the impact significance of the research on its audience through informal observation and discussion with audience members. During this phase of the research, I grew to believe that it was important for me to ascertain the cognitive and affective impact of the artwork on the audience, who were also participants in the interactive artwork. I had been influenced by the work of my supervisor, Professor Ernest Edmonds:

> It is impossible to directly observe the inner feelings of the audience . . . being able to explore the 'Interaction space' involves some form of evaluation with audience cooperation.[11]

Figure 5.10.1   Sketch for Digital Hologram Production 'Passing Time, Distant Memory', 2018.
Source: Pearl John.

In Phase 2, I created a body of analogue holograms from family archival material in which objects were sunk into the depth of the holographic images. The artworks were exhibited in a Solo show entitled *The Virtual Artist*,[12] attended by approximately 500 people, at the University of Southampton in 2015. The work was evaluated by gathering feedback from members of the public who attended the exhibition, via a paper-based survey, along with an analysis of a discussion by a small focus group of artists who assisted with a 'silent researcher critique' session in which the experts asked me questions which I did not answer. This was a technique adapted from one utilised in art educational settings. Andrew Pepper describes it as a 'robust pedagogic approach tested in contemporary fine art degree teaching 'the show and Listen Seminar (or silent student critique)'.[13]

In Phase 3 of the research, I produced a digital animated hologram which was shown at the *Time and Space* Group exhibition at the City Museum Aveiro, during the International Symposium on Display Holography 2018, Portugal. The work was evaluated by nine international experts in art and holography at the Museum during a silent researcher critique session. The questions asked were transcribed and analysed to assist with the evaluation to determine whether my research questions had been answered.

Lessons learned

During the research process, I learned a number of lessons about professional and personal development, and in hindsight, these were described by the Vitae Researcher Development Framework (RDF), which I was only vaguely aware of at the time. This section will outline the lessons learned and make recommendations – both for the research student and supervisor – on how to use the framework to support the research journey.

The Framework exists to help researchers evaluate and plan their professional development and as a guide for managers and supervisors of researchers to support the development of researchers There were two domains described in the RDF which I believe were underutilised by other researchers and supervisors: Domain B: *Personal Effectiveness*, which includes the personal qualities and approach needed to be an effective researcher, and Domain D: *Engagement*, which related to public engagement with research. My experience of supervisory support was that Domains A and C were extremely well supported and that B and D were under-utilised during tutorials and annual reviews.

Domain B: Personal Effectiveness

My research journey was not an easy one personally for a couple of reasons: 1) my mother sadly underwent two open heart surgeries and died during the course of my postgraduate studies, necessitating a pausing and then an extension to my studies, and 2) having to carry the burden of the imposter syndrome, which I will discuss in more detail.

Imposter syndrome was first described in the article 'The Impostor Phenomenon in High Achieving Women: Dynamics and Therapeutic Intervention'.[14] Two academics, Bothello and Roulet, who experienced the condition, described feeling 'a growing sense of anxiety and self-doubt about the legitimacy of our profession and our position within it'.[15] The syndrome is generally misunderstood to be an individual – private – problem of faulty self-esteem; however, it is now thought more likely to be the result of a toxic environment within higher education.[16] Abu-Lughod describes situating the affective landscape of imposterism in a socio-political context, exploring intersections of class, gender, race and ethnicity, disability, sexuality, and factors including caring responsibilities, being of the first familial generation to enter HE. Abu-Lughod recommends analysing feelings of imposterism to determine a 'diagnostic of power'.[17]

When reviewing my own situation in terms of its socio-political context, I became aware of at least three internal burdens I carried which slowed down my research progress and limited my outputs:

1) Being of the first generation of women to undertake a master's degree. While my paternal grandmother had undertaken an BA in Fine Art, she once told me (when displeased about my appearance) that it was more important for a woman to have neat hair than a degree, which said something about the historical value of women's education to society which I may have internalised.

2) The societal belief that artists' work is not of equal importance to that of engineers and scientists – Sean Johnston quotes Harriet Casdin-Silver, a pioneer artist working with holography, who was 'concerned about the "Second class citizenship of the artists in this supposed union of scientists and artists"'.[18] It is customary, at least in physics, for PhD researchers to be fully funded, whereas all of the artists in my research group were self-funded (although I was extremely grateful for a contribution toward my fees by the School of Physics and Astronomy at the University of Southampton, who also approved my spending 10% of my time on my research). So, a hierarchy of worth exists between self-funded female artists and funded male scientists and engineers (there are of course exceptions).

3) Being a part-time, mature PhD student, which left me somewhat isolated.

Considering these different aspects, the toxic environment which created the burden of my imposter syndrome became entirely visible. The research journey is naturally challenging and the more reading I did as a researcher, the more I became aware of my own ignorance. However, as the research process relies on self-doubt as a method for seeking weaknesses in my own understanding, personal reflection as well as a great deal of practical activity is required in which to create new knowledge. Imposter syndrome interferes with the confidence that it takes to continue in the research process in the face of self-doubt and can exacerbate procrastination, making the research journey at times feel thoroughly unpleasant; in short, it slowed me down.

On reflection, what may well have supported me further was a supervisor and I using the RDF framework to provide me with evidence of my situation and progress. Domain B provides subdomains and descriptors and I would have benefited from using the framework as a diagnostic tool, particularly in the areas B1 (Personal Qualities), B2 (Self-management) and B3 (Professional and Career Development) to see at what stage – or Phase – I was situated. I believe that these would have helped provide evidence to show that my lack of self-confidence was inappropriate.

Notwithstanding, the following activities and interventions assisted in combating the syndrome:

1) Public engagement activities which increased my confidence in my skill and my field of research.
2) Publishing papers and giving presentations at international conferences.
3) Supervisor encouragement.
4) Co-working: to tackle the isolation of being a part-time researcher working at a distance from my research group I now use the free website *Focusmate*[19] which pairs strangers up for virtual coworking sessions; this would be useful for distance learners.
5) Funding from an external body.

Domain D: engagement

Delivering outreach and public engagement activities teaches the researcher valuable skills which contribute to the research process.[20]

In my capacity as the Public Engagement Leader in Physics and Astronomy and the Future Photonics Manufacturing Hub at the University of Southampton, I was principal investigator on two European-funded Photonics Outreach grants, Photonics4All and **PHABLABS** 4.0. These two grants provided funding which enabled me to make holograms with the public and with schoolchildren, engaging people with my own research and photonics outreach more generally. I delivered holography and photonics workshops in schools and colleges and gave talks to teachers and members of the public, reaching roughly 18,000 school children and members of the public via a travelling laser light show. In these I introduced the field of Photonics (the science of light), describing myself as an artist, and introduced my holography PhD research. I also managed and delivered Photonics outreach and holography workshops with another 4,500 people during the period I was a PhD researcher. This work formed a substantial contribution to a potential REF impact case study as a result of the extensive evaluation of the activities which determined levels of participant enjoyment, changes of attitude toward Photonics as a result of the interventions, and a measurement of the learning objectives met.

The benefits of public engagement work to the research, for myself and other researchers,[21] included the following:

1) An increase in confidence in presenting – which in turn benefited my ability to teach undergraduate students, deliver papers at conferences, and ultimately defend my thesis in my viva.
2) Verbal and written communication skills: useful during grant writing, explaining my area of research in my dissertation and in grant writing when communicating with non-specialists.
3) Identifying weaknesses in my own understanding of my research area. During presentations and question and answer sessions after talking with audiences, I became more aware of areas of weakness in my own understanding.
4) Gaining experience in event management, budget setting, and financial management.

Unfortunately, it has been my experience as a Public Engagement Leader that some supervisors have been reluctant to allow their postgraduate students to spend time on public engagement and outreach activities, as it appears to take them away from their research. However, within the context of the development of professional and personal development outlined in Domain D of the RDF (and in my own experiences), this work is important, providing the student with valuable skills and experiences for their career progression, inside or outside of academia.

Conclusion

I have discussed my field of research as an artist working with holography and lenticular imaging, describing my practice-based research project as one which involved participants whose experience of the interactive artworks was evaluated. The study benefited from engaging the public, providing me with valuable feedback which helped to develop my work, including the crafting of research questions and methods. I also undertook a programme of public engagement and outreach, as recognised by the Vitae Researcher Development Framework's Domain D, providing me with valuable research and professional skills. I discussed some of the difficulties I had with self-confidence during my journey and have argued that researchers may be empowered by their supervisors referring to the different phases of Domain B: Personal Effectiveness in the RDF during the research journey. I will continue to use the RDF as my career in academia continues, helping me set long-term career goals at annual appraisal meetings. I hope one day to supervise PhD students myself and will explore students' annual progress using the context of the four domains, while supporting their practice-based research work.

Notes

1 VITAE (n.d.).
2 A review by the NCCPE suggested that nearly half of the submitted case studies for the REF in 2014 made some mention of public engagement as a route to the claimed impacts. See the NCCPE website for the full report. Available on-line: www.publicengagement.ac.uk/about-engagement/current-policy-landscape/public-engagement-and-ref.
3 NCCPE (n.d.).
4 Holograms and lenticular prints are rarely used as fine art media and require a brief introduction; both media combine multiple views of objects which the eye combines to create three-dimensional (3D) images, and both media can also be used to appear to animate objects, in which case they can be described as four-dimensional (4D), as they include the dimension of time.
5 The Royal College of Art Holography Unit closed in 1994 and is described in Johnston's *Holographic Visions* (Johnston 2006, pp. 307–308) and was documented in Holography Unit Exhibition in 2012 and available from: http://thegluefactory.org/holography-unit/.
6 Johnston (2006).
7 Making digital holograms requires 3D software skills and access to a specialist print facility.
8 Pepper (2019).
9 John (2019).
10 John (2018).
11 Edmonds (2010, p. 2).
12 The Virtual Artist exhibition was documented in a short film available online: www.youtube.com/watch?v=53bAKW6-7Eg.
13 Pepper (2019, p. 30).
14 Clance and Imes (1978).
15 Bothello and Roulet (2018).
16 Abu-Lughod (1990), Breeze (2018).
17 Abu-Lughod (1990).
18 Johnston (2006, p. 303).
19 Focusmate.com.
20 Posner (2017).
21 Posner (2017).

Bibliography

Abu-Lughod, L. (1990). The Romance of Resistance: Tracing Transformations of Power Through Bedouin Women. *American Ethnologist*, 17, 41–55.

Bothello, J., & Roulet, T. (2018). The Imposter Syndrome, or the Mis-Representation of Self in Academic Life. *Journal of Management Studies*, 56 (accessed March 20, 2021).

Breeze, M. (2018). Imposter Syndrome as a Public Feeling. In Y. Taylor & K. Lahad (Eds.), *Feeling Academic in the Neoliberal University. Palgrave Studies in Gender and Education*. Cham: Palgrave Macmillan.

Clance, P., & Imes, D. (1978). The Impostor Phenomenon in High Achieving Women: Dynamics and Therapeutic Intervention. *Psychology and Psychotherapy: Theory, Research and Practice*, 15, 241.

Edmonds, E. (2010). The Art of Interaction. In *'Create 10' Proceedings of the 2010 International Conference on the Interaction Design, UK June 30–July 2*. Swindon: British Computing Society, 5–10. www.bcs.org/upload/pdf/ewic_create10_keynote3.pdf (accessed April 4, 2021).

John, P. (2018). *Temporal and Spatial Coherence: Chronological and Affective Narrative within Holographic and Lenticular Space*. Unpublished Ph.D. dissertation. De Montfort University. www.dora.dmu.ac.uk/handle/2086/18127 (accessed April 1, 2021).

John, P. (2019). The Silent Researcher Critique: A New Method for Obtaining a Critical Response to a Holographic Artwork. In A. Pepper (Ed.), *Holography: A Critical Debate Within Contemporary Visual Culture*. Basel: MDPI, 38–49 (A Reprint from *Arts* 2019, 8(3), 117. www.mdpi.com/2076-0752/8/3/117 (accessed April 1, 2021).

Johnston, S. (2006). *Holographic Visions: A History of New Science*. Oxford: Oxford University Press, 309.

National Coordinating Centre for Public Engagement. (n.d.). www.publicengagement.ac.uk/about-engagement/what-public-engagement (accessed April 4, 2021).

Pepper, A. (2019). Holography: *A Critical Debate Within Contemporary Visual Culture*. Basel: MDPI, 38–49 (A Reprint from *Arts* 2019, 8(3), 117). www.mdpi.com/2076-0752/8/3/117 (accessed April 1, 2021).

Posner, M. (2017). *Optical Integrated Circuits for Large-Scale Quantum Networks*. Doctoral thesis. University of Southampton. http://eprints.soton.ac.uk/id/eprint/417392 (accessed April 1, 2021).

VITAE. (n.d.). www.vitae.ac.uk/researchers-professional-development/about-the-vitae-researcher-development-framework (accessed March 20, 2021).

5.11

PROJECT-BASED PARTICIPATORY PRACTICE AND RESEARCH

Reflections on being 'in the field'

Gail Kenning

Introduction: the practitioner researcher background

My background is as an art practitioner coming into research through a practice-based Doctor of Philosophy (PhD) programme at the University of New South Wales, Australia. Undertaking a PhD expanded my approach to art practice and consolidated my interest in research. However, it was the projects undertaken after completing the PhD that changed my practice; those projects introduced writing as a practice, collaboration as a way of thinking, and encouraged socially engaged practices. This involved getting out of the studio and doing research and practice 'in the field' with communities, aged care facilities, libraries, and galleries. This prompted questions about the artists' roles, who the artists were, audience diversity, and the extent to which 'mainstream' artworks addressed or marginalised specific audience needs and wants.

My field of practice

My practice and research are located at the nexus of what at various times have been considered four distinct, but related, or overlapping disciplines: art, design, health (clinical/medical), and wellbeing (engaging with individual and social factors of health). For the sake of brevity and clarity, however, I refer to myself as an arts health practitioner and researcher. This is a term in common use in which 'art' relates to creativity and encompasses design and 'health' may refer to clinical and medical health or wellbeing in a social context.

My work is primarily project-based, inter-disciplinary (and often transdisciplinary), collaborative, participatory, and socially engaged.[1] I explore how creative activity and arts and design engagement impact health and wellbeing, how creative activity and engagement become data, and how creative activity and engagement projects inform the disciplines of art, design, health, and wellbeing. After discussing one of these projects, I will explore how the factors justed noted present ongoing challenges and misunderstandings and become complexities for practitioners and researchers working in the arts health space.

An arts health project case study

The *Woollahra Visualisation Emotion Experience* (WEVE)[2] was a collaborative project between myself, working as artist and researcher, artist Warren Coleman, and older adults in the Woollahra Municipal

                    DOI: 10.4324/9780429324154-54

Council Area. Funded by the Council, the aims were to bring together older adults and gain insights into their health and wellbeing by exploring emotions and feelings related to where they live, through a creative experience, eventuating in an artwork for exhibition. Working with specifically designed workbooks, in a workshop environment, participants engaged in art and design activities which explored emotions, community, and home. The workbooks supported discussion about the topics being explored and prompted a creative activity, followed by a debrief discussion. For example, a workshop exploring the concept of home began by exploring Eastern and Western philosophies of home, going from and coming home, stillness and home as place or a position of comfort, the childhood home, and what home is now. Participants were introduced to poets, writers, artists, and contemporary social commentators.

After exploring Bachelard's *Poetics of Space*,[3] participants were invited to make a plan of their childhood home and to include details on rooms, gardens, and streets (Figure 5.11.1). This drawing

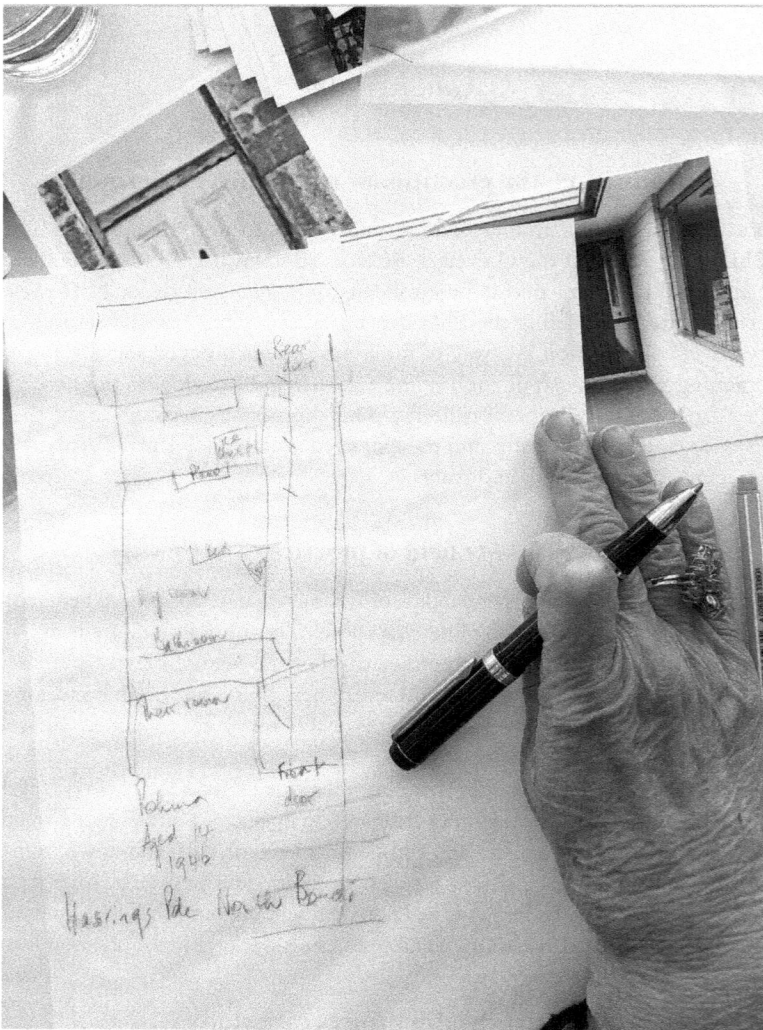

Figure 5.11.1   Participants were invited to make a plan of their childhood home.
Source: Photo courtesy Author.

was then overlaid with drawings on tracing paper of furniture and possessions. A third layer was added relating to favourite places in the home, associations with people, and feelings, thoughts, and emotions encapsulated in the rooms and possessions. In another workshop exploring everyday emotions, Plutchick's *Wheel of Emotions*[4] was used as a starting point. This provided participants with access to a broader vocabulary and language to describe emotions they experienced. They began to explore shapes and colours by making drawing about emotions.

After each art/design engagement session, participants were invited to comment on the experience, offer feedback on what they experienced, and complete a short questionnaire. All workshops were audio and video recorded, the drawings made by participants were photographed, and workbooks were collected for post-event analysis. The recordings were analysed for themes and frequency and intensity of engagement. A database was set up consisting of keywords, key quotes, drawn images, and textual imagery created from the writings and spoken words of participants.

One of the outputs from the project was a synchronised multi-screen installation (Figure 5.11.2). Two separate installations of five and ten screens accessed data from the database via Wi-Fi. The data, categorised according to various themes, issues, and emotions, was divided into chapters for onscreen viewing. Each chapter contained a range of images made from participants' writing and drawings. Snippets from stories, key quotes, and words related to the themes explored in the analysis simultaneously ran across each of the screens with accompanying soundscapes. The work was exhibited in the community from which the stories came via installations in three libraries in the eastern suburbs of Sydney. It is now being developed for a more permanent exhibition.

Figure 5.11.2   The project eventuated in a synchronised multi-screen installation.

Source: Photo courtesy Author.

Project context

This project, and most of my work currently, is practice-based research in an arts health context. It is both participatory practice and is participatory research 'to gain new knowledge by means of practice and the outcomes of that practice'.[5] Projects operating in the context of arts health practice and research are often subject to what we might call a 'doubling' effect whereby the 'practice' may engage with and inform both an arts practice and health practice and arts research and health research. This can cause confusion and complexities in the evaluation or assessment of a project and in deciding the success of a project.[6] This will be addressed in more detail in my reflections on the project outlined.

Personal reflections on experiences and learning related to this project

In this section, I will address some of the challenges, pitfalls, problems, conflicts, and fixes negotiated throughout the project and revisiting the challenges set out in *my field of practice*.

Art/Design

In my own practice, differentiating between art and design projects is of limited interest. My personal interest is in what these projects *do*, how they engage with people, and what happens as a result. However, I am aware from training in a Fine Art school that for some, identifying what is art and what is design is important. Furthermore, funders often require further clarification with regard to whether a project is an art project or a community/health project. When engaging with communities 'in the field', the use of language can be key to getting people involved or not. We began the WEVE project by engaging with social groups and communities in the eastern suburbs of Sydney and inviting them to take part in this art project.

Many of the potential participants were 'gallery goers' and had pre-defined ideas about what an art project might look like and how they would engage – expecting to be engaging in drawing or painting and contributing to a mural and were put off by the intersection of art and health/wellbeing. Conversely, some potential participants were put off by the idea of engaging in art. Suggesting they were not 'arty', 'can't draw', or were 'not the artist in the family'. Sometimes a simple shift of emphasis from an art project to a design project and describing the engagement through a design lens was enough to trigger a positive response, particularly for the those who 'didn't see the point of art' but 'reckon [they] have a good eye for design'.

Discussion about how to define the project occurred often. As the participants on the WEVE project were all well educated and primarily motivated to learn more, they were provided with an overview on shifts in art and design practices and how the emergence of socially engaged practices and 'the sematic turn'[7] in design embraced participatory and reciprocal approaches aimed at connecting people.

Health and wellbeing and research

In the WEVE project, as with many other projects, the discussion of health and wellbeing in the context of art leads to assumptions that the project was to be art therapy. For the WEVE project, this was one of the greatest misconceptions to be overcome and was off-putting for many potential participants. This was compounded by the labelling of the project as arts health *research*. Potential participants had assumptions about health and wellbeing research, expecting that it would involve clinical and medical tests carried out by clinicians, and that they would need to declare their medical conditions. Their prior experience or knowledge of research was primarily related to random control trials (RCTs) and medical *interventions*. As a result, we were questioned about reducing variables, use of placebos, control groups, representative samples (of participants), and objective measures.

For many, wellbeing was not a familiar term, and they equated it with esoteric mindful and meditation practices, seeing they had little connection to health or mental health. Furthermore, many potential participants could not see how art and design projects that focused on creative activities could constitute research. While these questions and challenges were not necessarily unexpected, the investment of time and energy needed to overcome some deeply entrenched views had been underestimated. Discussion of the role of art and design practice and how they related to health and research was ongoing throughout the project, meaning we needed to reiterate the nature of the art/design health and wellbeing.

Trans, inter, multi, cross-disciplinary experiences

Participants on the WEVE project were surprised at how the project traversed disciplines. Many understood the concept of disciplines. However, their experiences were primarily of having been schooled in the arts or sciences; few had experience of stepping outside of their disciplinary paradigm. The workshop engagements were made up of arts engagement activities, design activities, psychology, philosophy, and involved discussion of mental health and wellbeing from clinical and social viewpoints. Participants began to enjoy looking at the same concept (for example, *Home*) from very different perspectives. The interactive synchronous, multi-screen media artwork created from the data generated in the workshops required a further disciplinary input in the form of human–computer interaction (HCI) design and interactive media skills. The cross- and inter-disciplinary nature of the project highlighted the wide range of skills sets and approaches needed when engaging in practices that operate across disciplines and paradigms, such as art/science, arts/health projects. As this was the first cross-disciplinary project that many of the participants had engaged with, they were keen to see other projects similar in approach and were surprised by the widespread nature this way of working.[8]

Collaboratory, participatory, socially engaged practices

The WEVE project, like many others using collaborative and participatory approaches, called for the project leads to have extended skillsets as they worked as artist, designer, facilitator, manager, arbitrator, and communicator. Like other participatory projects that attempt to challenge hierarchical organisation, the project leads/artist invited participants to facilitate some of the engagement activities or offered to share the role. Participants in the WEVE project were surprised by the attempt to reduce the hierarchy in the workshops, with some perceiving it as a reallocation of workloads. So, while there was some curiosity about this approach, for the most part there remained an expectation among the older adults that they would be taught, given information, or would learn techniques or skills

Artist/designer and audience

With regard to my reflection on my practice and the WEVE project, the final area I would like to address is in relation to the role of the artist and audience. In 'Collaboratory, Participatory, Socially Engaged Practices', I commented on the advanced range of skillsets needed to work in this space. When working with any audience in a participatory way, the wants and needs of all involved may change. Working with diverse or marginalised participants or people with access needs reminds us that addressing these shifts can be crucial for project outcomes. For example, the participants engaged on the WEVE project were highly educated, engaged, and socially active, and for the most part, they thrived on new information and ideas. However, their desire to fully understand new approaches in art, design, health, wellbeing, and research meant that workshops needed to be adapted to provide more in-depth background material, time for discussion and to accommodate, sometimes, dissenting views. More time needed to be allocated to the project to compensate for this.

Key take-aways

Working on a participatory project across disciplines and with wide varieties of audiences, with very different abilities, in different types of spaces and locations, has provided insights into some of the key factors that can lead to successful outcomes. This has given rise to some key take-aways for ongoing projects.

An important aspect when engaging with participants from communities is to develop a **reciprocal** approach, particularly when engaging in art health research. It ensures that the participants get something out of the project. With the WEVE project, we as researchers were collecting data; as artists we were co-creating an artwork. Some participants wanted to be involved and engage in the production of the artwork, while for others the project was an opportunity to meet people and connect, to be engaged in a shared experience, to learn new things, or simply to get out of the house.

Arts and design engagement approaches can be used in a variety of ways to engage people in talking about what is important to them. The central tenet is in the use of objects, images, prototypes as catalysts for associations and conversations.[9] This is a way of facilitating discussion and communication and can be used with those who are highly articulate and people with communication difficulties. In the WEVE project, using this approach enabled workshops to begin simply and evolve into highly complex discussions.

Investment of **time for recruitment** can be overlooked. Calling for participants in both art projects and is often carried out through advertisement and flyers; we recognised we needed to meet people face-to-face or through online communication channels to build confidence in us as artist researchers and in the project.

We recognised in the WEVE project, and in many others, that the need for **iterative** engagement ultimately demands **flexibility.** Using an iterative approach requires ongoing reassessment of the project in light of new findings as the project progresses and time for people to adapt. This means that the project needs to have the flexibility to embrace change.

Concluding comments

My practice as an artist was great preparation for the work that I do now in research. The need to be flexible and find new ways of doing things were constants. Being reflexive about the creative journey in making artefacts meant using an iterative approach, always asking questions and adapting accordingly. Changes were sometimes small and ongoing and often required more intense change. For example, recognising that an installation that had worked so well in one space did not work in another calling for significant change. At times, change was needed in shifting theoretical frameworks for the work being carried out. All of this was great preparation for both considered change and thinking 'on the fly' as well as responding to change while not compromising quality.

The WEVE project involved recruiting older participants from a specific community in Sydney to take part. There were challenges in communicating the aims of the project – because there were multiple aims. There was confusion because of project crossed disciplines; there was push back to the idea of engaging in an art project. However, the project was successful in terms of reaching an older population and collecting data on their wellbeing, in providing a space and place for shared creative experiences, and in the production of a media artwork for public viewing. Many of the participants who initially were sceptical and critical became the most avid participants and have gone to engage in follow-up projects.

Acknowledgement

This research was funded in part by the Australian Government's Australian Research Council. Scientia Professor Jill Bennett, lead of fEEL (felt Experience and Empathy Lab), UNSW is the recipient of an Australian Laureate Fellowship FL170100131 and by Woollahra Municipal Council

Notes

1 Stock and Burton (2011).
2 Further information can be found at https://gailkenning.wordpress.com/2020/05/16/weve-it-feels-like-home-2/.
3 Bachelard and Jolas (1994).
4 Ibid.
5 Candy (2006).
6 Bishop (2012).
7 Bang and Vossoughi (2016), Bishop (2012), Bourriaud (1998), Thompson (2012), Kenning (2020), Krippendorff (2006), Wildevuur et al. (2013).
8 Stock and Burton (2011).
9 Kenning (2022).

Bibliography

Bachelard, G., & Jolas, M. (1994). *The Poetics of Space*. Boston: Beacon Press.
Bang, M., & Vossoughi, S. (2016). Participatory Design Research and Educational Justice: Studying Learning and Relations Within Social Change Making. *Cognition and Instruction*, 34(3), 173–193. https://doi.org/10.1080/07370008.2016.1181879.
Bishop, C. (2012). *Artificial Hells: Participatory Art and the Politics of Spectatorship*. London and New York: Verso Books.
Bourriaud, N. (1998). *Relational Aesthetics*. Dijon: Les Presses du Reel.
Candy, L. (2006). *Practice Based Research: A Guide*. www.creativityandcognition.com/resources/PBR%20Guide-1.1-2006.pdf.
Kenning, G. (2020). Reciprocal Design. In R. Brankaert & G. Kenning (Eds.), *HCI: Design in the Context of Dementia*. London: Springer.
Kenning, G. (2022). Arts Engagement: Experiences to Support Wellbeing and Collect Data to Inform Understanding of Lived Experience. In P. Crawford & P. Kadetz (Eds.), *Palgrave Encyclopedia of the Health Humanities*. London: Palgrave Macmillan.
Krippendorff, K. (2006). *The Semantic Turn: A New Foundation for Design*. Boca Raton: CRC, Taylor & Francis.
Stock, P., & Burton, R. J. F. (2011). Defining Terms for Integrated (Multi-Inter-Trans-Disciplinary) Sustainability Research. *Sustainability*, 3, 1090–1113. doi:10.3390/su3081090.
Thompson, N. (2012). *Living as Form: Socially Engaged Art from 1991–2011*. Cambridge: The MIT Press, 1st ed.
Wildevuur, S., Dijk, D. V., Hammmer-Jakobsen, T., Bjerre, M., Ayvari, A., & Lund, J. (2013). *Connect: Design for an Empathetic Society*. Amsterdam: BIS Publishers.

5.12

BEARING WITNESS – THE ARTIST WITHIN THE MEDICAL LANDSCAPE

Reflections on a participatory and personal research by practice

Sofie Layton

The participatory gathering and making process

As an artist with acknowledged experience in participatory practices, I have a profound interest in the medical landscape and how medical concepts are perceived and re-appropriated by patients and society. My research practice is an investigation into the concepts and narratives associated with illness and the sick body. Recent work has focused on the heart and specifically questions how women understand their own and their babies' internal bodies when captured through medical imaging techniques. The art practice explores the discourse surrounding the mother-and-child dyad and questions whether imaging data, such as ultrasound, can be used in new ways outside the clinical setting.

My artistic practice is collaborative and interdisciplinary. Working in the medical landscape has required me to develop a rigorous understanding of the scientific language held alongside the patient experience honouring the seriousness of illness, while ensuring that the artistic practice maintains its own authority. Working within a medical setting, it is possible that the artist can become subservient to the medic or scientist and is not seen as having authority in her own right. The best collaborations happen when each specialism respects and is interested in what the other person's professionalism brings to this interface. My practice requires a bearing witness to the narratives of others as well as my own. I hold and witness the narratives of those I work with, alongside my own experiences of sickness. My practice-based research is into how an artist can bring a new articulation and methodology to this space, allowing patient stories, conducted through my own experiential reality, to be materialised through narratives and medical imagery. Working with texts such as Anne Boyer's *The Undying* and Susan Sontag's *Illness as Metaphor* – alongside my own research by practice, I seek to investigate the metaphors and experiences of the parent and patient in the medical setting. My work intersects an arts and health approach where the creative workshop process has a benefit to a person's wellbeing; however, it is also a research process and these encounters are what generates the content of the artworks.

My practice can be broken down into a series of different approaches:

- Devising appropriate creative workshop activities which can be done at the bedside on a hospital ward or in the clinical setting. Equally it could extend to group workshops in a non-clinical

DOI: 10.4324/9780429324154-55 684

space. I often begin the research process by integrating creative writing, drawing, embossing, and body mapping exercises through which questions can be explored. These workshops can be one-to-one or in small groups as published in the *Journal of Applied Arts and Humanities*.[1]

• Collaboration is central to my practice, working with scientists and clinicians to explore how data, language, and medical imagery can be used within the participatory process. I have been able to bring 3D images developed for scientific research purposes into the artistic discourse with the patient/parent and this allows for a new metaphorical language to develop around the medical image.

• The filtering of the workshop narratives becomes the creative starting point, a spore of an idea that necessitates being realised as an artwork. The need to make it tangible through a hands-on approach is the beginning of the art making process. Depending on the commission, this may develop as a design that becomes an installation, created through a series of collaborations with sound designers and different fabricators.

• The final presentation of the artwork is an essential part of my practice as the work becomes a reflection of the patient's journey through their own particular illness. Held within the artistic vision, it opens up the dialogue to a wider audience who are then able to reflect on this from a universal perspective.

I will contextualise my research by practice by using an example of working with a particular mother of a child with a serous congenital heart condition which began as part of 'Under the Microscope' (UTM), an artist residency at Great Ormond Street Hospital London (2015–2016). The working method that developed from the residency led to the development of a large public engagement project which culminated in the national touring exhibition *The Heart of the Matter*[2] (2016–2018). This project explored the medical, experiential, and poetical dimensions of the heart. In this chapter, I will look at my socially engaged participative practice within the medical landscape – the collection and translation of these narratives – through the artwork *Sacred* 2018 (Figure 5.12.1; an immersive lightbox installation with a soundscape) created as part of *The Heart of the Matter* exhibition. I will also look at my reflexive process and the investigation into my own heart data as part of my research using *Excavations*, an autobiographical exploration into the metonymies of the heart.

What distinguishes my research practice from a more traditional participative process within an arts-in-health context, is that often the artist's own narrative is not intentionally positioned within the work. However, I believe that the acknowledging of one's own narrative, even if it is not directly foregrounded – due to ethical and/or caring concerns around participants wellbeing and clinical staff – is central to my method. In this practice I, the artist researcher, also have to undergo another inquiry, which is a form of filtering and the translation of others' as well as my own narrative, to create an artwork which bears witness to the complexities of these stories but is authenticated through my own experiential understanding of the sick body.

I will illustrate this process and my research by practice by exploring the role of the artist as witness within the clinical setting, working with and translating patients' narratives and the medical image through a haptic engagement with 3D objects and the making of artworks. This is an investigation into how medical data can be transformed and reimagined to explore the emotive reality of disease and loss.

Additionally, I will discuss the process of making *Excavations*, the artwork mentioned previously, created through a research process, which was the literal excavation of my own cardiac Magnetic Resonance Imagery data, manufactured using a robotic arm as part of a Computer Numerical Control process. The process of making an artwork which used a robotic arm to cut away layers to reveal the internal chambers of my heart was a witnessing of my own maternal relationship to the emotional complexities of caring for a sick child within a medical context. By acknowledging

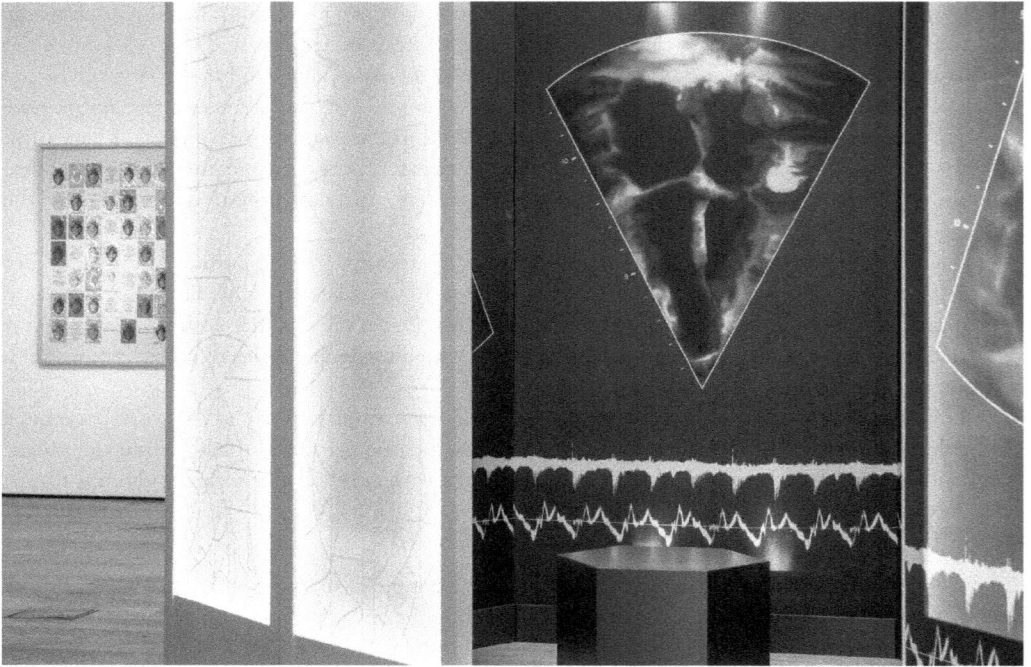

Figure 5.12.1   Sacred (installation detail) 2018 – Sofie Layton.

Source: Photo Stephen King.

and making concrete my own narrative, which was not an emotionally comfortable experience, it allowed me to authentically reflect on the parents' position in the clinical setting.

Reflecting on the participative engagement process, bearing witness, and art making

My UTM residency at Great Ormond Street Hospital studied how patients and families understand disease and whether the artist can bring something to this translation process. I began working on the cardiac ward and met a mother of a 3-month-old baby with a complicated congenital heart problem. We talked while embossing medical heart images that had been printed onto thin aluminium foil. Through a mark-making process, the aluminum can be patterned and worked into producing a beautiful votive offering. These became a collective artwork gathered through individual workshop encounters with patients and families as part of the residency. For me, this haptic engagement process allows for a different engagement process. The process is repetitive, and participants often describe it as being relaxing and therapeutic: a by-product of the making process.

During this residency, I met Dr Giovanni Biglino, bioengineer, who was researching the 3D printing of patients' heart data to create 3D heart forms. I used one of these heart forms, reduced in scale, and cast it in bronze as a way of materialising the preciousness of the heart as an object that could be held. The mother of the young baby had seen the bronze heart that I had made and asked if we could print her daughter's heart. Dr Biglino facilitated this and six weeks later in the intensive care unit, I was able to show the mother her daughter's heart form. As I placed the model in her hands, she looked down at this white powder 3D print and said, 'I feel like I am holding snow in my hands'; a few moments later, the nurse on duty looked at the heart model and explained the morphology of the heart form, and in this moment the poetics of the form became medical once again. This encounter around the boundary object, whether it is a medical 3D print or an artwork, is a key element within my research practice and demonstrates the translation that happens between the artist and participant and the new narrative understanding that emerged from this experiential encounter.

As part of the development for *The Heart of the Matter* and the creation of the installation *Sacred*, I worked with two mothers who both had young children with congenital heart problems. Working with creative writing and body mapping exercises, which are part of my workshop approach, I asked each mother in separate 1:1 workshops to describe their daughters as a building. One mother said that her daughter would be the Sagrada Familia in Barcelona because she was still under construction; the other said she would be an old, refurbished church. This reflection illustrates how profound the lived reality of their daughters' congenital heart problems was and that they would have ongoing medical interventions for a significant part of their early lives.

As an artist, the realisation that both of these parents had this extraordinary vision of their young children as these iconic buildings was quite phenomenal. I realised that I had a responsibility within the context of the exhibition to find a visual language with which to translate these narratives. This became *Sacred*, a light box installation created as a dodecagon with external printed stained-glass windows using heart cells as the basis for the design of the frosted glass patterns. Internally, a series of six silk panels of echo cardiogram images of the heart were first produced as silk paintings, which were then enlarged to two metres and digitally printed. Visually, I explored the inside and outside reality of the invisibility of a heart condition: on the outside of the cathedral, one is aware that there is a pattern on the stained-glass windows, but it's not till you're inside the church and you're looking out that you are aware of the light and the complexity and the magic of that internal space. The installation also became a sacred space where the audience could enter and listen to one of the mothers talking about the life choice that she made to continue with her pregnancy. The artwork was potentially challenging as it explored difficult life and death decisions which posed an element of risk to the mother whose story was used and audience members who might find the subject challenging. The presentation of the parent and surgeon's story within the sound piece also allowed the audience to experience the complexities of a child's journey with a serious heart condition from both sides. The artwork itself becomes a representation of the complexities of this interconnected story. Creating an artwork like this also became an exploration of the medical imagery associated with congenital heart disease.

Alongside this, I was examining the notion of the sanctity of the operating theatre, the surgical space, and the enormous responsibilities of the cardiac surgeon to perform surgery on these babies' hearts. As part of my residency, I attended a couple of heart operations to understand the reality of that aspect of the patient journey and to experience the medical space, which has its own sanctity. There is a strange fascination in being able to access these very intimate spaces and have these encounters with the illness narrative; however, I am aware that it is also a privilege that has been earned through a practice of care and respect for the patients and staff working within that environment. My research by practice develops through a process of collaboration and trust that is built through this interdisciplinary process. I do not take this work and having access to people when they are vulnerable lightly; it is a delicate and ethically complex landscape to navigate. The artist could be seen to be exploiting people's personal and often sensitive narratives in order to make an artwork. It is therefore important to ensure that the participants are consulted before anyone's narratives and particularly any voice recordings are used within the artwork. As an artist, I am constantly reflecting on what the most appropriate way of conveying these stories is.

For me, within the arts and health/arts and scientific fields of transdisciplinary intersection, the partnership with the institution and the development of relationships which may become collaborations is an essential aspect of a good research practice. The relationships are built through a mutual respect for what the scientist and arts can bring to the patient space. I believe it is essential that the artist develops an understanding of the science and medical issues in order to gain the respect of the medical partners. The weight of this can be quite arduous; however, there is an imperative that the work is underpinned by a respect for medical information and knowledge: otherwise, the art practice becomes insubstantial. In the same way, the parent or patient becomes medicalised and assimilates the medical language and terminology in order to empower and equip themselves in the translation of the medicalised world that they now inhabit. From an artist's perspective, the really meaningful interdisciplinary collaborations occur when the scientist also endeavours to understand the possibilities that the arts contribute to this conversation. It is not just a means to illustrate the science or patient's narrative but brings an equally rigorous method and language with which to approach and research the complexity of disease. The art process will not save someone's life, as an emergency operation does, but it may help articulate the complexities of living with a disease for both the patient and the clinician/scientist. The feedback in response to my UTM residency and exhibition at GOSH in 2016 illustrated the way that my artistic practice brought a new dimension to this landscape. Patients felt that their stories were being heard and articulated through the artworks and the clinicians found that the reality of the medical science alongside the patient's narrative was being 'beautifully described'. Narrative medicine is an established field of medical humanities developed by the eminent scholar Rita Charon, but the articulation of the medical science, in particular medical imagery and data, alongside the patient's narrative, is a new area of research.

This bearing witness and understanding the complexity of the medical conditions that the patients live with, the medical setting, and how an artist holds and represents these narratives in a way that is appropriate to all of the individuals involved is a central tool in my research practice. As an artist, I act as a conduit for these conversations, but it goes beyond the holding and representing of others' stories. My own practice is contextualised through an autobiographical understanding of the complexity of the medical journey underpinned by my own lived experience of this landscape, which enhances the emotional insight and the reinterpretation of these narratives researched within this environment. As part of my research method, I found it important to also acknowledge my own narrative and my own negotiation of the mother–child relationship within a medical context.

Working as an artist on one's own archival narrative has a different focus but requires a similar attention to detail and rigorous practice. It is potentially individually more challenging, as it is one's own story and needs to be navigated as carefully and empathetically as one would the engagement within the medical framework with patients. However, perhaps as artists we are not always so mindful of ourselves. In order to do the work in the medical setting, I originally set up a professional support framework as I acknowledged that it would be challenging working in this context but that my lived experience would also enable me to assess the appropriateness of certain ways of working. It is important not to disclose one's own narrative within a participative practice as the work is not about one's own experience but rather an ability to hold and reinterpret others' narratives; at points this became quite complex as the boundaries between professionalism and the connections with participants became more personal. One has an ethical responsibility to protect the participants and ensure their wellbeing within this process.

As part of my reflective practice, I undertook an MRes at the Royal College of Art, where I created the artwork *Excavations* using Walter Benjamin's text *Excavation as Memory* as a way to interrogate my own autobiographical narrative. Using my MRI heart data, I translated it as sculpture using a seven axis Kuka robotic arm, literally excavating my own heart form, exploring the idea of absence through the void space. I worked collaboratively with a technician, Steve Bunn, to develop the best way of manufacturing the piece. I explored scale, enlarging it by 400 per cent. The manufacturing and witnessing of the process became part of my own reflective process. Working with the haptic materialisation of making an empty physical space allowed me to interrogate the complexities of sickness and loss through a manufacturing process. As part of this reflective approach, I also filmed and created a series of written reflections of my own excavation process.[3]

This collaboration and questioning of the artist's narrative and the responsibility of the artist to the collective participatory narrative, as realised as part of a socially engaged practice, is something that I'm constantly adjusting and working with. My practice-based research process needs constant revision and adjustment; it is something that is reflexive, not fixed – it has to be a changing organic process, with a method that can be returned to, and yet it constantly requestions the art-making and collection process. In the same way that a doctor should approach each consultation afresh but build on the knowledge of the patient's history, the artist has a set of tools and experiences which are nuanced and bring a new way of looking at patients' narratives as within my own practiced-based research practice.

My collaborative practice allows these complicated and emotive realities to be shared with a team of artist and scientists. The work is orchestrated by myself through my interactions but brought to life through a chorus of others who support and add to the collective finished work. The making of the artwork, the imaging and designing of the final output, and the presentation of these ideas is another essential part of my research by practice as it is the interface with a public that may or not identify with the landscape of the sick and more particularly the specific narrative that is being articulated through an artwork. Within my practice, I research the question of how to articulate the complexities of illness and the patients' reaction to being ill. This is represented to a public audience in order to disseminate this learning.

The artistic decision making, both in how to convey the essence of an idea along with the making and crafting of an idea into an installation, a screen print, sculpture, film etc., for me is another part of this journey. Here I remove myself from the clinical space and withdraw to the artist studio and gather my artistic collaborators around me. In the same way that the surgeon who leads an operation holds the overview of the procedure and is supported by a team of specialists, I also draw on a team of artists with their own specialisms to realise some of the final artworks. I present a design, or sound recording, and work with a team to extrapolate the best way of editing and representing the learning from the engagement process. This ongoing collaboration between the participative parent/patient, the doctor/scientist, and the artist/artistic teams creates an ongoing reflexive non-hierarchical research process.

Reflection on the context and process of engagement

My practice is constantly evolving and builds on the previous learning and knowledge that was acquired. The collaborative process is central to my work. Without the generosity of collaborators like Dr Biglino and his expertise on the medical heart, I would not have had my own MRI scan of my heart, and without that data I would not have conceived *Excavations*. This constant review and appraisal of process, the rebuilding and reconnection with the sick body, and an exploration of the methods of how artistically to articulate these narratives are what underpins the work. Creating work that helps to articulate the complexities of living with a disease for both the patient and clinician is an incredibly sensitive process and a fragile space to inhabit. However, the learning and the cross-boundary work that happens in this setting through collaboration and the experience of working with people in a participative way goes beyond the individual artist's own lived experience and has a potential benefit for the way in which medical narratives are communicated in the future.

At the heart of this research by practice is the trusting of the not knowing and learning through personal interactions with participants along with the transdisciplinary experimentations which may help to create the form of the artwork. As part of my collaborative process, I am also privileged to have worked with some extraordinary artists and makers who support my artistic practice and help me to realise something new and unique through my research process.

Notes

1 Layton et al. (2016).
2 www.insidetheheart.org.
3 https://thepolyphony.org/2020/10/15/heart-excavations/.

Reference

Layton, S., Wray, J., Leaver, L. K., Koniordou, D., Schievano, S., Taylor, A. M., & Biglino, G. (2016). Exploring the Uniqueness of Congenital Heart Disease: An Interdisciplinary Conversation. *Journal of Applied Arts & Health*, 7(1), 77–91.

5.13

ORGANISATIONAL ENCOUNTERS AND SPECULATIVE WEAVINGS

Questioning a body of material

Debbie Michaels

Introduction

I write this chapter while tussling with (re)organising material in preparation for the submission of my doctoral project. However, to write *about*, rather than *with* and *through*, is problematic when the practice is embedded and embodied in the research; and, vice versa, with me, as practitioner researcher, at its centre. So, I start by just beginning to write.

By way of personal introduction, I do not have a formal art training, but start my professional career in 1981 as a commercial interior designer. Some years later, prompted by the aftermath of illness and life-changing surgery, I shifted direction to follow a developing interest in psychotherapy, psychoanalytic thinking, and artmaking, pursuing a second career in art psychotherapy. Entering art academia in 2015, I hover on the edges of retirement, but am driven by a desire to make use of my experience – to *draw on it* as a resource and *draw it out* further through my research. Still, inevitably, I come 'lugging this great heavy sack of stuff', as writer Ursula Le Guin might say – a repertoire of life experiences, personal, social, and cultural understandings, and ways of doing things that will undoubtedly affect how I approach the situation.[1] I *am* and can only be *here* because I have been *there*, and done and experienced *that*.

In brief, my enquiry gathers threads from psychoanalysis, art psychotherapy, and the arts to (re) examine the psycho-social role of reflexive artmaking in honing sensitivity to the affective dimensions of human situations and experience. It is motivated by a concern that increasing systemisation and scrutiny of the human services, with corresponding expectations of efficiency, speed, and busyness, means that there is little time for slow, meaningful reflection that enlivens rather than deadens emotional sensitivities. Paradoxically, this makes paying attention to how we affect and are affected even more important.

Conceptualised as a 'speculative weaving' *in/through* three transpositions (discussed later in this chapter), my research follows the intertwining dialogues and entanglements as I traverse institutional boundaries in healthcare and academia, *unmaking*, *making*, and *remaking* a body of work. Put simply, I combine approaches from different disciplinary cultures and practices in an experiment with method concerned with art as a means of enquiry into organisational processes. Embracing 'experiment' as an experiential process of feeling my way forward and into a situation in which I am intimately implicated, the emphasis is on 'learning *through* experience', where 'experience' is understood as undergoing an encounter – being impressed by something in the sense of feeling, sensing,

 DOI: 10.4324/9780429324154-56

and imagining, as well as practical contact with and observation of events. The implication is that the world may reveal itself through fluctuations and movements in the situation I set up as artist-researcher, that I am 'part of – *affecting and affected by* – the research process, and that the situation can answer back and contribute to this interaction'.[2] Through this process I have come to a deeper understanding, of not only the events and situations that I experience, but also the process *through* which this understanding develops.

Personal reflections

Crossing boundaries

Transposition I – 'unmaking'

In *Inherently Interdisciplinary* Clive Cazeaux suggests that it is 'not simply the case that we leave one way of shaping experience and move to another; it is the tension between the two that is decisive for our purpose'.[3] A prime motivation for situating my research in the arts is that it might offer different frames and lenses through which to experience and observe organisational processes (my own included) without the constraints of clinical practice. Still, how art is 'framed', where it is situated, and where maker and viewer are positioned in relation to it affects how it is understood and valued, and may be very different when viewed from art academia, art psychotherapy, or other perspectives. Crossing disciplinary boundaries into art academia from the world of art psychotherapy, I feel like an intruder in a foreign land. Without familiar codes and conventions, I find myself personally and professionally exposed, and in a state of conflict and confusion about where the boundaries are. However, engaging with the creative opportunities offered in this unfamiliar setting, I begin to explore and document the dialogues that emerge as I reorientate myself.

For the *Testing Testing* project in 2016, I stage an encounter, (re)imagining and fictionalising a certain situation in the context of another through *unmaking, remaking*, and *exhibiting* an 'art therapy object' in an art research setting (see Figure 5.13.1).[4] Although unable to articulate what I am doing at the time, working *through* the dramatisation I get a *feel* for what it means to open myself to a different way of learning as, like the object, I find myself caught *between* what I have been and what I might yet become – in an unravelled, vulnerable state, and under intense scrutiny. While developing a capacity to stay with or 'contain' disturbing and ambivalent feelings is part of my training and practice as art psychotherapist, the research situation is unfamiliar and unsettling. Drawing on prior experience and learning as well as the support of my supervisors and others is invaluable in helping me to negotiate the uncertainties and emotional turmoil.[5]

Transposition II – 'making'

In Transposition II, I cross boundaries again, transferring, modifying, and repurposing approaches from previous practices and setting up further 'rules' as guiding structures for my research. Negotiating a twelve-week placement in a healthcare setting, I assemble frames through which to observe and experience the organisation and myself in it. Specifically, I expand on a model of observing organisations described in the psychoanalytic literature which emphasises the subjective experience of the participant-observer and their ordinary human capacity to 'pick up vibrations', or intuitively tune into the atmosphere of a situation.[6] Adapting the model for this research context, I draw in threads from art psychotherapy which involve the use of artmaking as a reflexive space for broadening an understanding of a situation through allowing it to find an 'echo' in one's own inner life.[7] Like my art psychotherapy practice, I use my personality, sensory, and emotional sensitivity, as an apparatus for receiving and processing subjective information, 'lending' my body to the endeavour

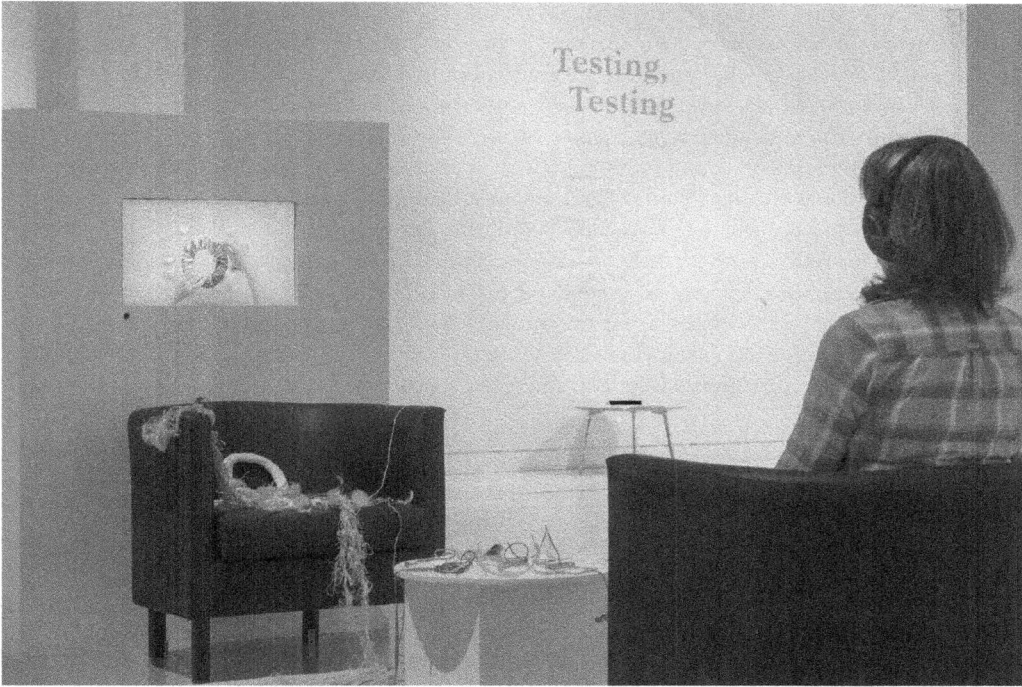

Figure 5.13.1 'Be│tween', multi-media installation, in 'Testing Testing', Sheffield Institute of Art, 2016. Video, audio, and other material, 340×150×150. Duration: 50 + 10 minutes (loop).

as I pay close attention to how I *affect* and *am affected by* it.[8] Of course, this is not a clinical situation and situating my research in the arts shifts the emphasis from art *as* or *in* therapy to artistic experience as a primary mode of enquiry. Still, as wanted or not, feelings and bodily reactions are an inevitable part of research and, as Patricia Townsend argues, just as the analyst listens for the potential meaning of the patient's communications through the resonances of her own internal responses, 'so the artist listens to the communications of the developing artwork through their effects on her'.[9]

Experiencing the healthcare setting and myself therein – once a week for one hour a week, at a regular time and place – I transfer the observational frame from organisation to studio, moving attention to the site of artmaking, and to my *making* process as a multi-layered response to the research situation.[10]

Within this framework, processes of making become 'virtual' spaces *in/through* which experiences, situations, relations, feelings, and thoughts may be evoked and provoked, as well as explored, tested, and (re)enacted.[11] For the duration of the twelve weeks, a range of recording devices document my process; however, the emphasis is not about obtaining an accurate or complete record of events. Rather, their presence as containing 'bodies' offers further material for critical reflection and the development of artwork, *with/through* which I might present an account of my sensitivity with all its inconsistencies and flaws.[12]

Entering the organisation as artist-researcher is a strange situation.[13] While not intended as such, my observing presence on a weekly basis is a provocation which challenges convention. For example, it raises questions about who and what I am, what I am doing there, if I am 'doing' nothing, or just come each week to watch the telly. In the studio, work emerges without prior conceptualisation through using 'stuff' I have to hand, including my emotional sensitivity. However, disciplinary tensions between fine art and art psychotherapy erupt as I present early 'raw' material at a seminar several weeks into the placement, and pressures to 'disrupt' my process and break with familiar conventions and languages unsettle me again. Of course, I cannot know how things may have proceeded otherwise; however, while resisting *felt* pressures (internal and external) to modify my practice in order to 'fit in', the encounter nonetheless provokes a turning point through the effect it has on me. Staying with the turmoil, and troubling affect, rather than avoiding it, I endeavour to slow the pace, working *with* and *through* the tensions, entanglements, and dilemmas I encounter during the making process, and the feelings evoked as I weave through different institutional spaces.[14] Gradually, the emphasis moves from something 'made' to the gestural, repetitive, and constructed nature of the work and the performance of some *thing* in the making – an intertwining of 'undergoings' and 'goings on', through which I access the thinking.[15] As I am pressed to notice and *feel into* the complexities of the situation more acutely my attention is drawn to 'making' as a space for imaginative encounter and performative enactment, and to the speculative, entangled, affective nature of the research process.

Transposition III – 'remaking'

Although intimately implicated in its making, as a material 'body' with its own anatomy separate from mine, the work in the studio is more than just a projection. That it has a life of its own becomes evident when I cross boundaries again in Transposition III. Moving the residual 'body' of work out of the studio, I *remake* it as I (re)situate and (re)present it, initially in the healthcare organisation, and then in other settings that bridge art, healthcare, and academia. Challenging more conventional ways of presenting research, the psychosocial presence of the 'body', and its associated parts, is amplified as different audiences respond to it. For example, situated where I had sat experiencing the healthcare setting, the material body provokes a range of responses from 'just a load of materials' to 'reaching out to something that is difficult to grasp'.[16] At a conference, delegates are invited to engage with the material body and 'the voice of its making' in a confined space which evokes (among other things) thoughts of 'someone in distress – trying to escape a situation' (see Figure. 5.13.2).[17] Sharing my sensitivity across disciplines, I test prior conceptions and open a space for dialogue which, Leavy suggests, is vital to the negotiation of meaning through the incorporation of multiple perspectives.[18] Whether through interest and engagement, indifference, or dismissal and devaluation, the audience is thus implicated in the meaning-making process. This deepens my understanding of the human situation through the way that both I, and others, *affect* and are *affected*, drawing attention to hospitality and an ethics of responsibility, attention, and care *for/of* the body. The embodied act of moving, handling, (re)situating, and (re)presenting the 'body' of work draws attention to the nature of different sites and offers an insight into how the art is *working*, as affective understanding insists on being *unmade* and *remade* with each 're' iteration and performance of it.

Learning through *experience*

The work of *making as reflexive practice* is integral to the development of my research and the design of my project. It is the work of my research which, like the work of meaning-making, is a slow, messy, uncertain business that does not happen in a linear, orderly fashion, but through the complex interplay of different elements spanning internal and external worlds. Although each transposition foregrounds a primary gesture, each is intricately interwoven with the others in a reflexive conversation that continually loops back and over as I gradually feel my way into the situation moving towards 'knowing how to move forward'.[19] Feelings, thoughts, and insights emerge at different times and from all directions as material 'made' in the past is 'remade' through touching the stuff of new

Figure 5.13.2　'Is this just my projection'. Multimedia installation and conference intervention, 'Double Agency', Design4Health 2018, Sheffield Hallam University.

situations in the present – the conceptualisation emerging through the ongoing work of *unmaking*, *making*, and *remaking* the material of the situation. Sites of making and research therefore extend to the performance of *writing* which, as an *affect-laden* process, *in* and *of* itself, generates reflexivity through the work of its construction.

There seems, therefore, to be a fundamental difficulty in saying this is the art and that is the thesis – to separate out the practice from the research as is required by the academic institution. Not unlike the question that brings me to the research, presenting the work becomes a matter of how to organise material in a way that keeps the practice alive and meets the necessary standards, while preserving rather than resolving tension in meaning-making; when I approach the entangled 'body' in the studio to move it, I cannot simply dismantle and separate its parts dispassionately without damaging its integrity. Organising my practice submission to sit in parallel with my thesis, I therefore write in a manner that reflects the process of its emergence; assembling threads and fragments in a multi-layered narrative which seeks to unfold practice and theory as it emerges. As with the work of *art*, the work of *writing* has an organising and structuring function but resists fixed or static meaning.

Findings

The demands of a PhD to specify a contribution to knowledge sets up an artificial frame around an ongoing, evolving process – focusing the attention but limiting the view. Arguing for the situated nature of knowledge, Haraway suggests that 'only partial perspective promises objective vision' and that 'the self is partial in all its guises, never finished, whole, simply there and original; it is always constructed and stitched together imperfectly'.[20] My thesis does not offer itself as a 'finished' work that sits *over*, *above*, and *apart* from what it 'finds'. Rather, it presents a partial, situated view that sits *with* and *alongside* as part of an interweaving of threads in conversation, *through* which meanings

695

Transposition I – 'Unmaking'

'Unmake' previous practices to
open spaces for learning *through*
experience.

Transposition II – 'Making'

'Make' a place for myself in art academia.
Assemble frames through which to observe and
experience a healthcare setting and myself *in* it.

Transposition III – 'Remaking'

'Remake' the residual *'body'* of work.
Amplify its psycho-social presence through
involving others in meaning-making.

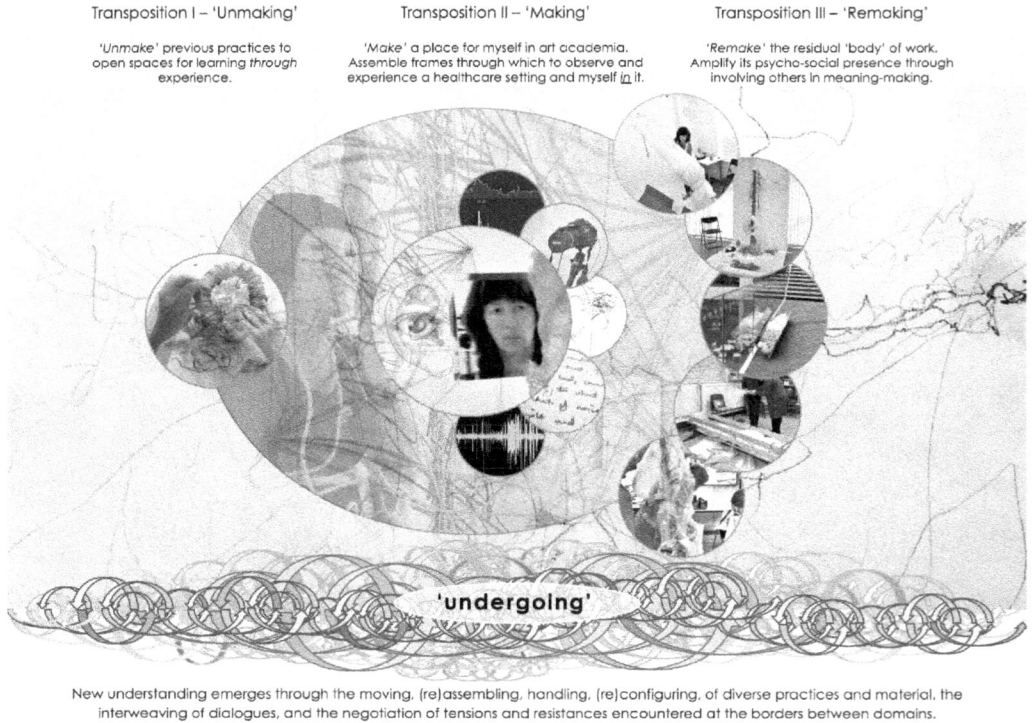

'undergoing'

New understanding emerges through the moving, (re)assembling, handling, (re)configuring, of diverse practices and material, the
interweaving of dialogues, and the negotiation of tensions and resistances encountered at the borders between domains.

Figure 5.13.3 Unmaking, making, and remaking: a speculative weaving in three transpositions.

and understandings may go on developing. I therefore approach 'finding' as a *process* of discovery rather than result or outcome. Although I gain insights about the organisational situation, my main contribution to knowledge lies with 'method' – what may be revealed *through* processes of 'making' and the performance of tasks that are *on their way* to being completed. The diagram in Figure 5.13.3 encapsulates my research process as it emerges and takes on meaning within the constraints of the research situation. While documentation and 'things' made along the way act as residual evidence that something has taken place, the emphasis moves to 'making as undergoing', a process I conceptualise as 'speculative weaving'. Implicit in this is 'time' and a capacity to endure and sustain the slow, ongoing, messy, material, affective, and psychological 'work' bound into *unmaking, making,* and *remaking,* from which insights are gained. New understanding emerges *in/through* the work of moving, (re)assembling, handling, and (re)configuring diverse practices and material, the interweaving of dialogues, and the negotiation of tensions and resistances encountered at the borders between domains. As researcher-practitioner, the contents of the great heavy sack I carry are reorganised and transformed *in/through* the 'making' process, opening a potential space for the introduction of new concepts to make sense of the research situation and my experience of it.[21] More broadly, as a site in/through which one may be pressed to notice and feel more acutely, the research value lies in the potential of this method to affectively (re)sensitise practitioners and researchers across arts and/in healthcare in ways that may not emerge through more traditional approaches to reflexive practice.

Notes

1 Le Guin (2019, p. 35).
2 Knudsen and Stage (2015, p. 5).
3 Cazeaux (2008, p. 128).
4 Testing, Testing is a project that aimed to explore the process of artistic production as research methodology. Initiated and produced by practice-based PhD researchers in the fine art subject area at Sheffield Hallam University, the project took the form of an exhibition at SIA Gallery, a symposium event, and two publications. Further details can be found at http://testingtesting.org.uk.
5 Artist and researcher Jean Carabine offers some useful thoughts for learning to work with not-knowing and uncertainty in artistic practice (Carabine 2013).
6 An adaptation and application of psychoanalytic thought outside the consulting room into a social and cultural context, the model was originally designed as a training exercise for psychotherapists and psychoanalysts working in organisations, aimed at honing intuitive sensitivity to human situations and experiences. (See Hinshelwood and Skogstad 2000; Hinshelwood 2013.)
7 In the profession this is known as 'response art', which aims, among other things – through dialogue and processes of shared reflection – to open, build, and deepen attunement and understanding in an empathic cycle. (See Nash 2020; Fish 2012.)
8 I refer here to the psychoanalytic concept of 'countertransference'. While psychoanalysis or psychotherapeutic practice is not generally thought of as research, social and qualitative researchers have become increasingly interested in application of psychoanalytic ideas in the research setting, particularly the use of 'countertransference' (see Holmes 2014). Akin to empathy, 'countertransference' is a rather clunky psychoanalytic term for what Robert Hinshelwood describes as 'the most human of all human characteristics and functions . . . the essence of the live connection between human beings' in relationship. While its application must be approached with caution, Hinshelwood's description reflects more contemporary ideas of transference and countertransference as a complex entanglement that emerges increasingly as a form of narrative enacted in the analytic setting – through an attention to process, an emotional sensitivity, and a capacity to pick up 'vibrations' that, he argues, is part of human nature (Hinshelwood 2016).
9 Townsend (2019, p. 92).
10 The model becomes 'one hour observation in the organisation + one hour making in the studio' at a regular time and place with the aim of transferring the intensity of experience from organisational setting to studio without interruption. The studio is situated away from the organisation, in a space adjacent to, but separate from, my art psychotherapy practice room.
11 I refer here to 'virtual' in the sense described by Donald Schön, which may be understood in psychoanalytic terms as the 'transference'. In such a virtual world of experiential, performative engagement, Schön suggests that it may become 'possible to slow down phenomena which would ordinarily be lost to reflection' (Schön 1983, p. 160). The implication is that if emergent material can become an object of shared enquiry – brought into the present so it can be attended to – then puzzling, uncomfortable thoughts and feelings, as well as inconsistencies and confusions, may become sources of surprise and wonder, rather than triggers for action.
12 Documentation includes written notes of observations, reflections between observational sessions, audiovisual and photographic documentation of studio-based happenings. I also employ Fitbit tracking technology to record aspects of my experience that are outside my conscious awareness, such as: journeys travelled, heart rate, and speed.
13 The observational placement was hosted by a multidisciplinary NHS service providing day rehabilitation for adults living in the community with conditions such as stroke and Parkinson's disease. I visited for one hour a week, at a regular time each week, and observed from the same place in a communal area. Here, patients gather, are offered tea, coffee, and lunch, and await treatment from a range of health professionals including nurses, support workers, physiotherapists, occupational therapists, and speech and language therapists.
14 During the summer following the twelve-week observation I apply for a three-month suspension of study in order to allow myself time to digest and process the intensity of the experience more fully before presenting my work for PhD confirmation.
15 'Thing' is understood here in the manner described by anthropologist Tim Ingold who, following philosopher Martin Heidegger, describes it as a place where several 'goings on become entwined' (Ingold 2010, p. 4).
16 Extracts from anonymous staff responses to the installation of the material 'body' in the healthcare setting, entitled 'Interrupting the Flow', 6 August 2018.

17 I refer to the 'Double Agency' intervention delivered as part of the Design4Health 2018 conference hosted by Sheffield Hallam University, and alongside the launch of the Critical Arts in Health Network (CAHN) and the 'Double Agency' publication (Smizz and Walters 2018). The intervention comprised a series of sequential one-on-one material and dialogical 'encounters' with four researchers – all health practitioners and artist/designers – working with creative methods to look critically at aspects of the healthcare system. Delegates who booked an 'encounter' were invited to sit with me, the 'body', and the 'voice of its making' for ten minutes in a small tutorial room – to move around it, touch its threads, move its parts, ask questions, express thoughts, and feelings in response, or just to sit quietly. The 'voice of its making' refers to a layered audio recording of the twelve hours of studio making (during which time the 'body' is constructed), compressed into one hour. This work was made specifically for the conference.

18 Leavy (2009, p. 18).

19 Candy (2019, p. 51).

20 Haraway (1988, pp. 583, 586).

21 Cazeaux (2008).

Bibliography

Candy, L. (2019). *The Creative Reflective Practitioner: Research Through Making and Practice*. London: Routledge.

Carabine, J. (2013). Creativity, Art and Learning: A Psycho-Social Exploration of Uncertainty. *International Journal of Art & Design Education*, 32(1), 33–43.

Cazeaux, C. (2008). Inherently Interdisciplinary: Four Perspectives on Practice-Based Research. *Journal of Visual Art Practice*, 7(2), 107–132.

Fish, B. J. (2012). Response Art: The Art of the Art Therapist. *Art Therapy*, 29(3), 138–143.

Haraway, D. (1988). Situated Knowledges: The Science Question in Feminism and the Privilege of Partial Perspective. *Feminist Studies*, 14(3), 575–599.

Hinshelwood, R. D. (2013). Observing Anxiety: A Psychoanalytic Training Method for Understanding Organisations. In S. Long (Ed.), *Socioanalytic Methods: Discovering the Hidden in Organisations and Social Systems*. London: Karnac Books, 47–66.

Hinshelwood, R. D. (2016). *Countertransference and Alive Moments: Help or Hindrance*. London: Process Press.

Hinshelwood, R. D., & Skogstad, W. (2000). *Observing Organisations: Anxiety, Defence and Culture in Health Care*. London: Routledge.

Holmes, J. (2014). Countertransference in Qualitative Research: A Critical Appraisal. *Qualitative Research*, 14(2), 166–183.

Ingold, T. (2010). *Bringing Things Back to Life: Creative Entanglements in a World of Materials*. NCRM Working Paper. Realities/Morgan Centre. http://eprints.ncrm.ac.uk/1306/1/0510_creative_entanglements.pdf.

Knudsen, B. T., & Stage, C. (2015). *Affective Methodologies: Developing Cultural Research Strategies for the Study of Affect*. Basingstoke: Palgrave Macmillan.

Leavy, P. (2009). *Method Meets Art: Arts-Based Research Practice*. New York: Guilford Press.

Le Guin, U. K. (2019). *The Carrier Bag Theory of Fiction*. London: Ignota, 165–170.

Nash, G. (2020). Response Art in Art Therapy Practice and Research with a Focus on Reflect Piece Imagery. *International Journal of Art Therapy*, 25(1), 39–48.

Schön, D. A. (1983). *The Reflective Practitioner: How Professionals Think in Action*. New York: Basic Books.

Smizz, S., & Walters, J. (2018). *Double Agency*. Sheffield: Independent Publishing Network. https://doubleagencypublications.tumblr.com/.

Townsend, P. (2019). *Creative States of Mind: Psychoanalysis and the Artist's Process*. London: Routledge.

5.14

IMPROVISING AS PRACTICE/ RESEARCH METHOD

Corey Mwamba

Introduction

As a musician, I predominantly use the vibraphone and audio software, either in real time or preparation. I frame my field of work in two ways: 1) in the domain of genre, I frame my field as *jazz+* – essentially "jazz (and more, related)"; 2) primarily, I am an improviser. I define improvising as:

> *a dynamic activity in which an entity attempts to make or create something, using only the resources and skills available at the time, in the present moment.*

The presence of *resources* and *skills* highlights that improvising is a social activity (with associated knowledges leading to particular skills) and also culturally bound (depending on the resources available within a society). In my case, improvising is a music-making method; and my improvising is strongly related to my connection with jazz+, even though I have a strong interest in other forms of improvising.

My research interest has been focused on the critical relationship between the vibraphone and the improviser, and the development of personal sound in jazz+. The US clarinettist and saxophonist Sidney Bechet suggested that, by working with various modulations of sound, a musician can develop a personal mode of expression that is intelligible within jazz+ practices. Whitney Balliett recounts Bechet's words to a student:

> See how many ways you can play that note – growl it, smear it, flat it, sharp it, do anything you want to it. That's how you express your feelings in this music. It's like talking.[1]

This focus on varying a single note recalls what the writer Nathaniel Mackey describes as "othering'" within Africana artistic practices. Mackey suggests that the practice of creating variations translates the static into the dynamic.[2] Exploring a dynamic process such as music requires a model of action research. Kurt Lewin first coined the term action research in 1946; and his initial spiral of research stages (planning, acting, observing, reflecting) has been adopted, critiqued, and modified by branches of the social sciences, particularly in education. The observing and reflecting stages are commonly grounded in social science methods of qualitative analysis. In the social sciences, this grounding is seen as necessary, to avoid the criticisms of action research being inward-looking and inappropriately complex when compared to the situation in which it is used.[3] But for inquiry rooted in the arts and humanities (and specifically music), those criticisms are in fact strengths, as:

1   reflection of one's own practice *requires* looking inward as well as outward to the situation
2   music making is complex, relating to intra- and extra-musical processes and events

The various methods of improvising are very similar to models of action research. The processes of improvising are simultaneously *active* (planning, doing) and *inquisitive* (observing and reflecting through consideration of the social, cultural, sonic, and physical environs in which music is made). What is different is the temporality. The planning cycle of improvising can occur both before and during the act of music-making; and the observation and reflection stages can occur during and afterwards. The act of listening (a core musical skill) is also an action research cycle; listening is an action in, and an observation of, a sonic space. But these processes deviate from the classical model of action research; the planning stage can be convolved within reflection. While improvising, a musician *decides* by and from *doing (or not doing)*. In musical listening, the listener *deduces* by and from *describing*. These actions – doing, deducing, describing, and deciding – are interrelated in improvising (see Figure 5.14.1).

What cements improvising as a potential mode of investigation is how reflection (deducing from what has been done or described) is treated after the music has been made. In the active/inquisitive process of improvising, reflection is not a stage (as with Lewin and later models of social science action research) but a required mode of action throughout. The objects of concern during the improviser's reflection are generally observable, and include:

1   **the sonics made by the self:** this observation relates to reflection that centres on the physical operations of music-making, and our understanding of the sounds within improvising
2   **the relationship between the sonics made by the self and the environs:** this reflection is concerned with how the music the improviser is making interacts with everything that is happening in the world. This includes the music other people are making, as well as any audience, recording devices, the space in which music is being made, and societal factors
3   **the relationship between the self and its environs:** this concern relates to how the improviser as a person situates themselves within various social setting, chronologies, and cultures (the band, the audience, music business dealings, etc.).

Each of these concerns positions improvising as a potential method for autoethnography.[4] Ethnography and autoethnography study the social subject in particular settings[5] – describing "personal experience of cultural experience".[6] In the case of ethnography, the researcher chooses the subject or subjects and particular setting in which they survey the subject, but it is a necessary step for the ethnographer to reside within the setting. By being in the field, the ethnographer or autoethnographer can elicit cultural knowledge, investigate patterns of social interaction, and critique or analyse cultures and societies.[7] But it is important that the ethnographic descriptions declare and account

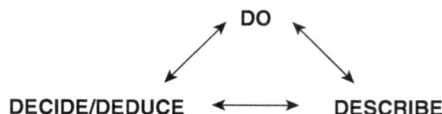

Figure 5.14.1 The improvising/research cycle.

Source: The Author.

for the value assumptions of the ethnographer in the field.[8] This is a pressing issue for the musician/researcher. As an improviser, I am intimately familiar with their own practice. But that familiarity could lead to missing important aspects of my practice that were important in a research context. There is also the danger of misremembering details and controlling the narrative to present a desired result. It is thus necessary to prepare for the research situation, by finding out about oneself through other means than relying on one's own memory.

To do this, I set up what Jesse D. Ruskin and Timothy Rice would term an indirect encounter[9] between myself as musician and myself as the researcher. Indirect encounter is a method from stimulated recall, which examines historical materials about the individual to develop a picture of that person's social and ideological world. From my beginnings in learning music as a teenager, I kept notes of theories, sketches, compositions, and thoughts on a variety of papers and notebooks. These *practice ephemera* span a time period from 1995 to the present. I also have kept many letters and notes of conversations in that time period. By analysing these texts, I sought to extract information about the development of my practice and map out the trajectory of my formative experiences. This "map" gave a sense of the various environments in which I encountered music making and performance (or dissemination). From there I began to shape connections between those environments and what I term different *socialries*:[10] collections of social groups, modes, and behaviours that are bound to or associated with a particular social or cultural practice.

After this, there needs to be some consideration about the practical sessions. In setting up the recording sessions, it was essential that I prepared the recording space in the same way each time, that I noted the general environs and my own physical and emotional state. The questions I intended to explore while improvising were always set beforehand. Then, as well as recording the sound or video, I ensured that my emotional/mental state about the music that has just been made was documented as soon after the recording as possible. I did this through a personal journal, which was separate from my research journal. Then, after two or three days, I would sit with the recording and my personal journal, and listen through to the recording twice. While the recording was playing, I would consider what was happening in the sonic space while I was improvising. This was based on what I can hear or see, as well as what I had written in both the research journal and the personal journal. These considerations were then documented in the research journal. One of the listening sessions documented how I was feeling at the time; the other documented what can be heard at the time.

These preparations enabled me as the researcher to gain some critical distance from me as a musician. Simultaneously, it shifted my position as a musician to one where I was responding to the research and changing my practical-theoretical approaches within the music. What follows is an example of how these methods are used in context.

In context

(s)kin is a collection of 13 digital releases recorded throughout 2018.[11] The music served various functions within my research: as research output that has public impact with an immediate reach; as a philosophical inquiry into the performer–instrument relationship, and as autoethnographic data. The idea of doing *(s)kin* had arisen from my thinking about how my electro-acoustic work had diminished over time. The avenues available for me to work within the realm of contemporary electro-acoustic music were closed off: in summary, this was due to my geographical location, my race, my educational background (I did not come through a conservatoire or music college), and my lack of connection with others within that scene. I resolved to regain that aspect of my practice. I wanted to reflect the personal facets of my life in meaningful and musical ways; and I also needed a musical way of thinking through my research on the vibraphone. *(s)kin* was the method through

Figure 5.14.2 A self-portrait from my practice space.

Source: The Author.

which I felt I could achieve all these things. The series began in January 2018 and finished in the December of that year (see Figure 5.14.2).

The recording of the music for *(s)kin* would follow a procedure. I would enter my garage, which I have converted into a working space for music, in the mornings: usually between 9 a.m. and 10 a.m. I would then improvise, and record as I did so. I would aim to make five pieces each time, although sometimes I made more. I relied on single takes; any "mistakes" were left in. I would also write text and record speech; these supplemented the music. The output from this process was quick, with most albums being started and completed in a day; I would always stop working in the studio in the evenings, especially in the winter months. This process created 4 hours and 45 minutes of music.

One of the sessions produced *(de)t(e)r(min)a(t)i(o)ns*. It was recorded on 10 April 2018[12] and is a piece about journeys (trains), points of change (node), and continuance (time). "Trains", "node", and "time" can be rearranged to spell "determinations". At this point, I had returned from a life-changing trip to Norway; I was in the throes of building a tool that more accurately represented the British and Irish jazz and improvised music scenes;[13] and the pianist Cecil Taylor had died four days before.[14] It was also an investigation into the quality of "heaviness" produced by the vibraphone while improvising using soft mallets. Soft mallets are generally considered to have less articulation, as the contact sound between the mallet and the bar is lower in amplitude than with other mallets. In addition to this, there is strong presence of the fundamental frequency of the note with fewer upper frequencies (called harmonics), and this is described as "soft". Soft mallets are associated with quiet playing, or lightness. I wanted to test the limits of that quietness and find ways of activating the vibraphone so that it had a heavier presence. Below are excerpts from the reviews. The researcher persona spots questions (marked Q) and potential insights (RI) while listening and reading.

As can be seen in Table 5.14.1, although the musician voice and the researcher voice have slightly different concerns, there are points of dialogue, honesty about mistakes (and what improvising does

Table 5.14.1 Self-directed musician–research review of "(de)t(e)r(min)a(t)i(o)ns".

Description	The musician	The Researcher
The microphone is already in place. I go to my desk, to the computer. I set up the digital audio workstation to record. I arm the track, and then carry the wireless keyboard with me to the instrument. I then press record. I pick up four soft mallets, and begin with a small figure that evolves to: **g_5 $♭♭b_4$ g_5 $a♭$ b_5 –, $a♭$ g_5 g_5 $b♭$ $a♭_5$** (0:00–0:16).	This was the third "trains" piece. I was still processing everything from the last couple of weeks. Let's do something that shows a contrast.	Initial inquiry: I am mapping the bar. The bar has spaces. I am thinking about weight; the quality of "heaviness" in my voice using these mallets; negating the connection between quietness and softness.
My ear is drawn to the traffic from the ring road outside my house (0:15).		
As I strike the b_5 bar near – but not at – the nodal point, I quickly shift my foot to release the pedal. The damper bar flicks upwards and activates the other bars. In that moment, my foot pushes down on the pedal again (0:17).	Got it. That's heavy. Need to move forward with this.	Here multiple sounds with one stroke of the bar.
I move the figure towards b_5 again, this time without the pedal motion; then exactly on the nodal point; then slightly away again (0:18–0:21).		I can still hear the frame; and the contact between the mallet and the bar. The sound still has the higher harmonics of the note, but the contact sound is slightly louder. How do I describe this?
As I strike the **b_5** bar near – but not at – the nodal point. I quickly shift my foot to release the pedal. The damper bar flicks upwards and activates the other bars. In that moment, my foot pushes down on the pedal again (0:22).	This is a choir/escape. *The frame shudders; the singular tone is accompanied by a low thud and is surrounded by the ghosts of siblings.*	
I move the figure towards **b_5** again, varying each time. At one point while attempting *choir/escape*, I strike **c_6** at the same time (0:35). I repeat this, this time avoiding activating the other bars and insuring that **b_5** is louder than **c_6** (0:37).	I'm not sure I like the frame sound with the scream. That's a mistake.	Q: Is "weight" being represented through the frame, or contact between the bar and the mallet? Or is it both? The *scream* doesn't sound "heavy" when the frame is activated.

(Continued)

Table 5.14.1 (Continued)

Description	The musician	The Researcher
	My ear is drawn to a siren approaching (0:45).	
I step back; and then approach again. I run up in fourths to ab_4 striking near the centre of each bar (0:58) and play a figure that ends with an accented f_5 and an unaccented f_5 (1:00). I follow this with two blues related motivic fast runs downwards ensuring to strike near the nodal point of each bar (1:01–1:03); and then an upwards blues figure to a *choral escaped a₄* (1:05), finally moving to a *choral/escaped a₆* (1:07).	Just let it go. You can't do anything about the road. Clean the slate.	Is "weight" being represented through the frame or contact between the bar and the mallet? Let's try with faster playing. Start with a "soft/quiet" run and vary from there. RI: *I need to ensure the sound of the frame is balanced just slightly quieter than the contact sound of the mallet, and strike near – but not on – the nodal point.*
a_5 e_5 $a\#_5$ $c\#_6$ $a\#_5$ g_5 f_5 $f\#_5$ $d\#_5$ $g\#_5$ b_5 e_5 (1:08).	Yeah. Okay, let's move on.	Explore weight at the higher register.
	Wait, what is that?	
	There is a beeping sound coming from outside (1:08–1:11).	
		I was distracted here.
And then on the nodal point a_5 (1 09).		
I begin to vary the motif, chopping and repeating it into smaller phrases (1:12–1:52).		
I intersperse the variations with a chord formed by striking the bars at the centre (1:52).	I'm staying on this.	Q: Am I really hearing "heaviness"? Play a four-note chord that sounds full. Play a thin chord.
I then form chords using the shafts of the mallets (1:56–2:01).		The thin chord sounded "lighter". Maybe explore that more.
I continue to vary the motif (2:02–2:15). I then interrupt this with shaft-played chords (2:16–2:20).	I'm focusing on the shape of the line, and the sound, not the pitches. The sound of the whole vibraphone.	Checking that thinness again. Q: is "light" the right word? Is "heavy"? Q: what "sound"?
The motif variation becomes shorter in length, leading to a final sustained c_4 (2:34), underpinned by a quieter ab_4.	I'm winding down here.	Q: How do we describe "the sound of the whole vibraphone?"

Source: The Author.

with those mistakes), and a sharing of the musician's terminology (*thin; full; scream*), and insight into the physical operation that create "heaviness". But at the end, I question whether the word "heavy" is indeed the right terminology. The insight could have been revealed by trying to carry out the operation repeatedly; but then the labelling (*choir/escape*) may not have been brought forward as the action would have been separated from a musical context. In addition, the eventual lack of resolution (*how do we describe the sound of the whole vibraphone?*) shows how improvising can generate research and musical questions in ways that operational repetition cannot.

Conclusion

Improvising is a useful method within practice research, as it provides a way of investigating phenomena and meaning within the dynamic context of practice. In combination with a stimulated recall approach, the practice researcher can engage critically and honestly with their own practice of improvising and potentially create ideological or philosophical shifts within both research and practice. Improvising can engage with the sonics made by the self; the relationship between the sonics made by the self and the environs; and the relationship between the self and its environs, and lead to new ways of thinking about practice in all these areas.

Notes

1 Balliett (1983).
2 Mackey (1992).
3 Robson (1995, pp. 440–441).
4 Editor's note: for a more detailed account of autoethnography, refer to Iain Findley-Walsh's chapter "The Sound of My Hands Typing: Autoethnography as a Reflexive Method in Practice-Based Research" in Part IV of this Handbook.
5 Hammersley and Atkinson (1983).
6 Ellis et al. (2010).
7 Hammersley and Atkinson (1983, pp. 1–6).
8 Hammersley (1992, chap. 1).
9 Ruskin and Rice (2012).
10 Here I am using this term to differentiate from *sociality,* a term that already does a lot of work in critical and sociological theory.
11 These can be found at https://coreymwamba.bandcamp.com/.
12 "(de)t(e)r(min)a(t)i(o)ns, by Corey Mwamba", *Corey Mwamba* https://coreymwamba.bandcamp.com/track/de-t-e-r-min-a-t-i-o-ns.
13 Corey Mwamba and Tom Ward, "THE RHIZOME", 2018 www.coreymwamba.co.uk/resources/rhizome/ (accessed June 4, 2018).
14 Marilyn Crispell, "Shared Post from Marilyn Crispell", *Facebook*, 2018 www.facebook.com/coreymwamba/posts/10155651591228198 (accessed April 9, 2021).

Bibliography

Balliett, W. (1983). *Jelly Roll, Jabbo, and Fats: 19 Portraits in Jazz*. New York: Oxford University Press.
Ellis, C., Adams, T. E., & Bochner, A. P. (2010). Autoethnography: An Overview. *Forum Qualitative Sozialforschung/Forum: Qualitative Social Research*, 12(1). www.qualitative-research.net/index.php/fqs/article/view/1589.
Hammersley, M. (1992). *What's Wrong with Ethnography? Methodological Explorations*. London and New York: Routledge.
Hammersley, M., & Atkinson, P. (1983). *Ethnography: Principles in Practice*. London and New York: Tavistock.
Mackey, M. (1992). Other: From Noun to Verb. *Representations*, 39, 51–70. https://doi.org/10.2307/2928594.
Robson, C. (1995). *Real World Research: A Resource for Social Scientists and Practitioner-Researchers*. Oxford: Blackwell, 1, published and reprinted.
Ruskin, J. D., & Rice, T. (2012). The Individual in Musical Ethnography. *Ethnomusicology*, 56(2), 299, 311. https://doi.org/10.5406/ethnomusicology.56.2.0299.

5.15

DREAMING OF UTOPIAN CITIES

Art, technology, Creative AI, and new knowledge

Fabrizio Augusto Poltronieri

Introduction

I want to start this chapter by saying that my relationship with technology – and more specifically, computers – is something inseparable from my life. Long before computers were practically ubiquitous, I already had a computer at home. I remember when my father arrived home with a mysterious box when I was around 8 years old. Inside this box, something even more mysterious, a Brazilian clone of the famous ZX Spectrum, a British 8-bit personal computer. Before that, I had owned a few video games, including a Magnavox Odyssey and an Intellivision, but nothing compared to the mystery of that little black box, which, when connected to a television set, displayed only a flashing number "1" and a letter "K", with no hint of what to do.

At that time, computers were accompanied not by assembly or user guides but by programming manuals. Having a computer meant learning how to program. So, the day after the computer arrived, I started to dedicate my free time out of school to study the manual, which taught programming in a language prevalent in the 1980s, called BASIC (Beginners' All-purpose Symbolic Instruction Code).

This was the beginning of an activity that had a profound impact on me, since programming became not only a hobby but an activity that I hardly spend a day without exercising, even if only for five minutes. From the knowledge acquired with BASIC, I started to try new languages. Access to better computers was challenging and extremely expensive in Brazil in the 1980s, as the importation of electronic devices was not allowed, and the local industry was decades behind. Today I program in various languages and consider programming to be an extremely creative activity.

Although I was also lucky enough to grow up in a house with a large library, with books in various areas and subjects, many about art, it took me a while to realise computers' real creative potential. After graduating in mathematics, which awakened my interest in mathematics as a creative field, closer to poetry than to engineering, I studied graphic design and started exploring the possibilities of merging the two areas, programming and design, which led me to an interdisciplinary Master's in Education, Art and History of Culture and my PhD in Semiotics, with a thesis on the role of chance in computer art. It was during my PhD that I came into contact with practice-based research. Or rather, it was during my PhD that I managed to systematically formulate something that was already part of my history because the desire to learn and discover by doing is something that has accompanied me since childhood, whether programming computers, developing new algorithms or mathematical explanations for phenomena, or seeking new forms of expression through art.

DOI: 10.4324/9780429324154-58

I consider my artistic practice to be computational art, specifically Creative AI, which allows me to explore aesthetic hypotheses and many philosophical problems. I see practice-based research as a play of constant interchange between the questions that form my academic research, my artistic practice, and a set of philosophical theories that I have been exploring over the years.

I do not believe in isolated research, unrelated to the external world and the social and scientific context in which we live. I believe in research that dialogues with my own doing and with already existing authors and theories. It is the kind of research which exerts an influence on me while I exert an impact on a theoretical framework. The project this chapter is about, *ArchXtonic*, is an example of this vision, since it was born from a series of philosophical concerns, materialised in practical research through art.

The project's starting point is one of the theoretical ideas of the Czech-Brazilian philosopher Vilém Flusser (1920–1991). For him, after the invention of photography, a new era, called post-historical, began. Post-history is mainly marked by the transition from the realm of machines to the age of apparatuses. In general terms, apparatuses are not machines because they are dedicated to transforming the world through plays of symbolic permutations – which is precisely what computers do.[1] Another characteristic of apparatuses, according to Flusser, is that they are always ready to attack. A photographic camera, for example, frames the world in its way, seducing the photographer to click its button. As "fierce beings", the apparatuses always frame our perception, making the play against them seductive and exciting. How many photos of the Eiffel Tower have we seen? How many of these pictures are different, not resembling a framing provided by the camera, which seduced its user to freeze that instant, already repeated exhaustively by other apparatuses and photographers? Faced with the fact that many of the images produced using cameras are more the apparatuses' vision of the scene than a new image, I decided to explore these limits in this project that started in 2015.

During this time, I travelled to many countries, including China, Brazil, the United States, Canada, Germany, Portugal, Spain, the UK, and France. On these travels, and in my daily life, I always carried the same camera, a Blackmagic Pocket Cinema Camera. My methodology was always the same, consisting of leaving the camera filming at random, capturing whatever the apparatus would frame on its own. My question was to discover what kind of aesthetics could emerge from such seemingly banal images and how such images could serve as a creative source to feed my Creative AI research, following what I believe practice-based research should be, as developed in the next section.

The theoretical side: science, art, technology, and practice-based research

As an artist-researcher, something that was not very clear at the beginning of my career was how these two "categories" – art practice and scientific research – intertwined and informed each other. I have always believed in the power of science, maybe an inheritance from my time devoted to mathematics. However, there is something in art that always made me discover new things. I had this feeling this is because art is an activity that makes substantial changes in the world, and this feeling has been the leading thread of my research. I can say that the biggest lesson I have learned throughout all these years working with practice-based research, and especially with *ArchXtonic*, is that the balance between the artistic and the scientific spheres is delicate. Still, art offers a unique path to rationality, to unveil the new.

What interests me in practice-based research is how new methodologies can be developed to generate new knowledge, leading to question the very nature of the human spirit from a semiotic perspective. This is because such spirit is characterised by possessing a scientific, pragmatic, ontological mind that learns from experience. Experience is something difficult to be catalogued or placed in hierarchies, in separated boxes. For this reason, current scientific knowledge has renounced the concept of single truth in favour of openness to complexity. There is still those opposed to this idea, and for them, practice-based research is probably something that does not make much sense because they see absolute truth as the supreme good to be pursued.

What does it mean to say that science has renounced the truth in favour of complexity? It certainly does not mean that truth is not complex or that it is no longer relevant. It means that the most advanced science doubts things – including itself – more than in any other time. To build new knowledge is to produce doubts, and this is what I do in my projects: doubts that lead to projects that create temporary conclusions, leading to novel doubts that trigger new projects.

We live on the surface of concrete materiality. But traditional science and technology sought to bring the material level to stagnation through extreme mechanisation driven by modernity. Science is an abstract numerical plane that is materialised through its technological applications. In the early and intermediate phases of capitalism, technological phenomena were predominantly available in durable goods. Today the focus of advanced capitalism is increasingly on communication technologies and services. The problem is that technology must seem to be infallible to be sold. There can be no doubt.

It is then up to the ones on the top of the scientific pyramid to doubt. Doubt that must be distilled in ever faster and more efficient processes to be transformed into technology. This is the post-industrial reality, moved by processes that transform doubt (abstract thought) into technology to be massively consumed in the form of durable goods or, more interestingly, in communicational services. These communicational services deal with certainty. They are abstractions that have been refined for the masses. Thus, my research aims to reverse this idea, bringing uncertainty as a central and driving point, accentuating the complexity of practical experience through art. And here falls my personal reflection on what I consider art.

In my view, art is an activity aimed at creating fictions, and this idea has been informing all my research. Art is the territory of the most radical nominalism, owing nothing to reality. Nothing prevents art in its task of creating fictions. It is the field of playful games, of impossible combinations, of the refusal of deterministic rationality. It is where doubt presents itself in all its splendour and beauty. Its fictions are created through a method of thought that rests on the uneasiness of chance.

There is a method in art, as there is a method in all activity performed by the mind. What differs the method of art from those observed in other fields? In the first place, it is a method whose target is in uninterrupted movement. How to reach a target that presents itself in such a way? How to calculate the position of something whose motion cannot be calculated? Only experimentation in its broadest sense can give some solace to these questions. Anything that the experimental method of art hits can be considered a valid target. It is up to an experienced researcher to decide whether this target is good enough or not. If it is not, the methodological bidding of possibilities should start again. One shoots in the direction of new targets. New goals are sought. New horizons are projected. It is an open method.

The method of art is given by the practice that is concerned with the essence of discovery. This method is, therefore, a play. It needs to be played with absolute philosophical and methodological seriousness by the researcher and the artist. Artistic practice cuts out possibilities from an infinite and continuous flow and transforms them into phenomena. In this process, the most important things are the ideas that emerge free and indeterminate. This process goes to selecting, cutting out elements from a great repository of possibilities to make the research object be born and grow. This also aligns with the ideas of one of the forerunners of what we might call a method for practice-based research. The English reverend, statistician, and philosopher Thomas Bayes (1701–1761), famous for his solution to a problem of inverse probability, ran counter to the deeply held conviction that modern science requires objectivity and precision. His method is a measure of belief, showing that we can learn even from missing and inadequate data, from approximation, and from ignorance.[2]

Figure 5.15.1 A diagram depicting my methodological process.

Source: Fabrizio Augusto Poltronieri, 2021.

I now turn to the practical implications of *ArchXtonic*, which involves the methodology (see Figure 5.15.1) I have described.

The practical side: combining creative AI, images, and collages

After the long period of capturing images, I found myself with many hard drives filled with video files collected over the years. My first challenge was to devise an automated procedure to process at least some of this material. I was careful enough to organise the files by location and date, making the process much easier. It is essential to highlight this because when dealing with a massive amount of data in a long-term research project, it is necessary to pre-plan at least the basic procedures. Otherwise, the risk of wasting time and documents is significant, and this is a lesson I have learned from previous endeavours. During the years I dedicated to filming these places, my ambition was to create an artificial intelligence algorithm that could generate new real-time narratives from video fragments. A kind of automatically generated video collage, using computer code as the vehicle to express this visual language.

Moreover, my goal was not to use entire frames of the videos but to cut out each frame's various elements through masks. This way, I would have a large database storing the information contained in each frame. For example, the tenth video's first frame has a couple walking until they leave the scene at frame number 348 of the same video. The first frame of this video also has a building, a tree, and a car moving until frame number 158, and so on.

I imagine that even someone who has no idea how many files and information this represents can visualise the arduous amount of work required to extract valuable data from this collection. Even with prior planning and my experience working with creative programming, artificial intelligence, and databases, I began to get an idea of the size of the challenge only in my first attempts to extract the information I needed using just one video as a test.

In 2015, when I started to think about which computer vision system I could use for such a task, the situation was not very encouraging. Many available technologies were not mature enough, and I did not have access to the necessary equipment. Even with such setbacks, I sought creative solutions to start the experiments regarding object identification in images, optimising processes and reducing the time needed to process each frame. I did several experiments with a well-known computer vision library called OpenCV,[3] which gave me confidence that the process was feasible, even if the results obtained at that time did not completely match what I expected.

This methodological posture has always been part of how I work, dividing the project into small steps and performing the validation through prototypes. These prototypes do not need to be completely functional or well elaborated. They serve to validate ideas and help to foresee challenges and new problems, helping to analyse evidence, change my mind as I get new information, and make rational decisions in the face of uncertainty. By updating my initial belief with objective new information, I get a new and improved belief, learning from experience. In fact, it is a logic for reasoning about the broad spectrum of our experience that lies in the grey areas between truth and uncertainty. We often have information about only a small part of what we wonder about. According to the pragmatic doctrine, we all want to predict something based on our past experiences. However, all the pragmatists warn us that we must change our beliefs as we acquire new information.[4]

As my practice is based on creative technologies, it has been vital to me over the decades to organise a set of libraries, frameworks, and algorithms to write programs quickly and prototype various situations. In most cases, I use a programming language called Python,[5] which allows rapid development of prototypes, tests, and the final product.

During the period of image collection, I followed closely the progress made in an area of artificial intelligence called "Image Segmentation", which is the process of dividing a digital image into multiple segments, intending to simplify or modify the representation of an image into something more meaningful and easier to analyse. What it does is to segment the image into its instances, i.e., into categories previously trained in the artificial intelligence algorithm. The aim is to obtain masks around the various objects identified in each image (Figure 5.15.2).

With a functional small database, I started to develop an artificial intelligence algorithm capable of selecting different objects from different video files and, in real-time, generate a new narrative, a new video, with these elements. A kind of video collage obtained with what I call "apparatus memories", which are these images automatically obtained through a professional video camera, automatically selected utilising my image segmentation and tracking algorithm, and, finally, regrouped also automatically by this second program, which generates surreal images, as shown in Figure 5.15.3. These images look like utopian (or many times dystopian) cities from a not very distant future or past. As a science fiction fan, Philip K. Dick's books and the idea of devices or androids dreaming of utopian or dystopian landscapes, cities, or realities have always significantly influenced my imagination.

Conclusion

The process of video analysis is still ongoing, as it is time-consuming. Still, I have collected enough information to generate endless hours of new video narratives without repeating elements. However,

Figure 5.15.2 An example of Image Segmentation. The original frame (left) and its segmentation (right).
Source: Fabrizio Augusto Poltronieri, 2021.

Figure 5.15.3   A frame generated by *ArchXtonic* (2021).

Source: Fabrizio Augusto Poltronieri, 2021.

I have no control over the process since the computer, as anticipated by Flusser (2000), is the one who controls the whole narrative.

In its current development stage, *ArchXtonic*'s[6] algorithm does not have a sufficiently advanced semantic insight to make accurate decisions regarding the position of the various elements that form a scene. This means that the algorithm knows people, cars, trees, buildings, etc. because each item in the database contains this information. However, for these elements to interact with each other coherently, a new and more advanced level of semantic relations would need to be built, establishing interrelationships between all types of components. While I am happy with how the algorithm makes its aesthetic choices, through a procedure using a neural network trained to build urban scenarios, a more elaborate semantic layer would open up new horizons for research and could be used for the dynamic and automated construction of scenarios for video games or virtual reality environments. The development of this kind of intelligence is my next goal for this long project.

Ultimately, *ArchXtonic* is about producing new knowledge through artificial intelligence with a methodology based on artistic practice, establishing interfaces with science, as epistemology, science, and art cannot be dealt with only in their specific domains.

Observing science from the art perspective, we notice in science a fictional aspect that remains ignored. Only through the construction of new, non-compartmentalised horizons, free of the barriers of linearity, more complex relationships reveal themselves. Science seeks the truth, but imagination plays a central role in the process of building new knowledge. Science aims to reveal truths, but it also invents them.

Since art knows the mechanisms of invention, it is no secret how much invention is involved in science. This is the problem of science from the point of view of art – the concealment of truth when science tries to unravel it. From the point of view of science, the problem of art is also clear, because art to science is fiction. The two are mirrors facing each other, reflecting each other. For me, the great challenge of works like *ArchXtonic* is to reveal a creative methodology that benefits from the rigour and weaknesses of both areas.

In a world that is becoming more complex, I can state that the most important lesson practice-based research has taught me is that there are scientific methods to embrace a reality in constant movement. In my specific case, the fictional model of art enables me to see new insights, which are put into practice and foster further theoretical discussions. The combination of practice-based research, art, and technology provides me with the ideal scenario for playful experimentation combined with scientific rigour.

Notes

1 For a deeper discussion regarding Vilém Flusser's theories and what constitutes apparatuses, see Poltronieri (2014).
2 A good source of information on Bayes is Mcgrayne (2012).
3 https://opencv.org/.
4 See Menand (1997), James (1995), and Peirce (1986).
5 www.python.org/.
6 *ArchXtonic* will be shown to the public for the first time at the UK pavilion at World Expo 2021 in Dubai.

Bibliography

Flusser, V. (2000). *Towards a Philosophy of Photography*. London: Reaktion Books.
James, W. (1995). *Pragmatism*. Mineola: Dover Publications.
Mcgrayne, S. (2012). *The Theory That Would Not Die: How Bayes' Rule Cracked the Enigma Code, Hunted Down Russian Submarines, and Emerged Triumphant from Two Centuries of Controversy*. New Haven: Yale University Press.
Menand, L. (1997). *Pragmatism: A Reader*. New York: Vintage.
Peirce, C. S. (1986). *Philosophical Writings*. Mineola: Dover Publications.
Poltronieri, F. (2014). Communicology, Apparatus, and Post-History: Vilém Flusser's Concepts Applied to Video games and Gamification. In M. Fuchs, S. Fizek, P. Ruffino, & N. Schrape (Eds.), *Rethinking Gamification*. Lüneburg: Meson Press.

5.16

CURATING INTERACTIVE ART AS A PRACTICE-BASED RESEARCHER

An enquiry into the role of autoethnography and reflective practice

Deborah Turnbull Tillman

Introduction

This chapter examines the methodology applied to an enquiry on how reflective practice and autoethnography can be useful to creative practitioners working in academic ways in professional settings. It focuses on what I discovered about being an 'independent curator-as-producer' working across the medium of interactive art and compares it to other creative practitioners working with higher degree by research methods in professional settings. This chapter contains the preliminary criteria for utilising an autoethnographic and reflective approach (my *Criteria for Curating Interactive Artworks*). The key learning outcomes from this chapter are insights gleaned from reflections and testimonials on my methodology from creative practitioners working in art galleries and as independent curators.

My *Criteria for Curating Interactive Artworks* are:

1   **Accessibility and inclusivity:** In order for the artwork to be available to most people who approach it, whether the point of engagement or the content they are engaging, it would be best practice to have a legible and easy-to-use or intuitive interface, while not simplifying the story/message/other communication in order to do so.

2   **Plurality of voices and mediums for diversity in stories and storytellers** is important in navigating an interdisciplinary medium with practitioners from different fields coming together on one or a series of related projects.

3   **Authoritative and contributory:** For those in directorial, managerial, or marketing positions (including the generation of social media content), you want to be sure the information you are grouping together and how you communicate it is strong and contributes to an existing dialogue.

4   **Relational and representative:** A context and core message/critique/communication is important, especially when showing prototypes or works-in-progress.

5   **Absent presence and present absence:** This refers to the ability of the curator-as producer to be present but removed as work unfolds around them, accepting that they are unable to control outcomes of experimental practice while also observing, reflecting, and documenting the process analytically.

    DOI: 10.4324/9780429324154-59

6   **Tacit and physical knowledge:** These are knowledges internalised and accessible to a practitioner from having performed and thought through the process previous.[1]

Autoethnography and reflective practice: backgrounds and definitions

The PhD case study: ISEA2015: disruption

Here I took an autoethnographical and reflective approach to data collection and analysis. This data revealed reflective practice models for moving through the production phases of a festival over distributed sites. The framework criteria (laid out earlier) that emerged advises creative practitioners, curators, producers, artists, and technologists on what they might provide to their collaborators, namely clients and audiences.

Autoethnographic research is the study of reflexively writing about oneself. It is divergent of the traditionally discipline-based qualitative methods based on the research of others. Social scientist Carolyn Ellis, credited with being the originator and a key developer of autoethnography within her study of qualitative research, has noted several 'traditional criteria' for 'good autoethnography'. This was described in her book, *The Ethnographic I: A Methodological Novel about Autoethnography.* Ellis first borrows from multiple practitioners to create an informed definition then relying on Laurel Richardson's approach to any social experiment via the below checklist:

1   **Substantive contribution:** Does the piece contribute to our understanding of social life?
2   **Aesthetic merit:** Does this piece succeed aesthetically? Is the text artistically shaped, satisfyingly complex, and not boring?
3   **Reflexivity:** How did the author come to write this text? How has the author's subjectivity been both a producer and a product of this text?
4   **Impactfulness:** Does this affect me emotionally and/or intellectually? Does it generate new questions or move me to action?
5   **Expresses a reality:** Does this text embody a fleshed-out sense of lived experience?

Later Ellis, in collaboration with Tony Adams and Stacy Jones,[2] developed a four-stage process for the evaluation of autoethnographic work that encompasses 'descriptive, prescriptive, practical and theoretical goals'. As such, any good autoethnographical study should:

* Make contributions to knowledge.
* Value the personal and experiential.
* Demonstrate the power, craft, and responsibilities of stories and storytelling.
* Take a relationally responsible approach to research practice and representation.

An **Appreciative System** was also articulated in this foundational case study as a way to gauge one's own creative and professional progress through a sustained cycle of reflection-in/on-action. This tool for gleaning knowledge from theory via process originates with Donald Schön and his seminal text *The Reflective Practitioner,*[3] but has been used by a variety of education-based practitioners who modelled their process, most relevant to my own process being Lewin and Kolb (learning focused), Cook and Graham and Muller (curatorially focused). Ways of engaging your Appreciative System are note taking, reflecting, articulating what works, setting aside what works, learning while doing, and modelling these processes for others, usually for education purposes. Linda Candy also refers to this as 'a reflective repertoire'.[4]

What this chapter aims to articulate is how Ellis et al.'s criteria and my own hold up in contemporary curatorial practice in professional settings.

The expanded case study: recent curatorial fieldwork

My criteria related to interactive art differs to Adams et al.'s[5] list in that mine has an emphasis on a) iterative process that informs an Appreciative System, b) evaluation of data within practice-based research, and c) the value of the practitioner/researcher to apply theory to in-situ research methods. The iterative process allows for testing similar methods more than once, and the evaluation of data is usually more interpretive and analytical then merely tallied or opinion-based. Autoethnography takes into account the theory and methods of many professions.[6] As such, it needs to be flexible and adaptive in its approach to analysis. This is linked to an array of creative practice in Iain Findlay-Walsh's chapter communicated across 'a combination of past-tense narrative prose, poetic, diaristic writing [and] theoretic discussion and analysis to present an autoethnography that is founded in the here and now of the research process'.[7] Pragmatically rather than poetically, I too move across past texts and analyse them in relation to my current practice. I attempt to situate what I have done and do as a reflective practice curator-as-producer within a range of practitioners working in and around curatorial practice. Findlay-Walsh also references Ellis, Jones, and Bochner, more specifically regarding the role that autoethnography can play in extending practice through reflexivity 'by actively mov[ing] beyond a kind of dual awareness – of simultaneously acting and at the same time reflecting upon that action'.[8] Of particular interest and relevance is a connection between Jones and Alec Grant and the potential for this dual awareness to trigger a kind of 'self-transformation . . . ultimately directed towards self-betterment', through lived experience.[9] What I discovered in the field in relation to these ideas and an autoethnographic/reflective approach was very interesting.

Results and general applicability for professionals: testimonials from the field

In setting up this fieldwork, I extended Ellis' methodology in terms of bringing together different authors and practitioners to expand this part of my criteria. I utilised interviews for this in accordance with Linda Candy's position that 'interviews can facilitate the opening up of thoughts that might otherwise remain hidden'.[10] I used a 360-degree approach to gleaning knowledge from those working in the field by targeting an emerging curator (Subject A), an independent curator at a similar level to myself, but focused on a different specialism (Subject B), and a gallery employee responsible for public programs in a university gallery (Subject C).[11] All Subjects have worked with art, science, and technology modes within museum and gallery settings. I prepared the Subjects by first providing them with the relevant criteria from the original case study (see earlier). I also included an explanation of my curatorial research and the questions that would frame the interview in relation to thinking through the above criteria.

Subject A (Australia)

Synopsis

Subject A is an emerging curator practitioner, educated at a master's level, and has an artist's understanding of materials. As she is open to new ways of understanding through *Third Space* curating, she easily relates to the *Visual Matrix* and art-science approach to curating art.[12] Regarding the question of using criteria in general as useful, she finds it double-sided: for example, a 'best practice' approach can feel quite strict, but it does offer a measure of comparison to others and provides context, which she finds a useful consideration. Her reflections on my curatorial *Criteria* are captured in Table 5.16.1.

Table 5.16.1 Qualitative data for Subject A.

My curatorial criterion	Qualitative data	Insight
Accessibility and inclusivity	Familiar as an enabling communication to the general public about 'higher learning' subject matter	
Plurality of voices and mediums for diversity in stories and storytellers	This criterion could be a subset of inclusivity	
Authoritative and contributory	Aspires to	Part of a more mature curatorial reflective practice
Relational and representative	More drawn to the relational aspect as this is the building and maintaining (sustaining) of professional relationships	Important in the emerging stage of any professional practice to have mentors and peers to connect with, check in with, and reach out to
		Happens through social media followed up with face-to-face meetings
Absent presence and present absence	Aspires to	Part of a more mature curatorial reflective practice
Tacit and physical knowledge	Confident in her abilities as a trained artist with a physical knowledge of the arts Welcomes artists' opinions on broad briefs for this reason	Knowledge in both these areas creates confidence to experiment with more sophisticated themes and mediums (like art/science)
	Also confident with the tacit knowledge through installation experience	
Engaging her Appreciative System (including autoethnography)	Utilises systematic tools such as note-taking and reflecting on those notes regarding developing her practice	This reflective way of digging through acquired knowledge transfers theory to tacit knowledge
	Creating and continuing conversations around themes and practice past the exhibition floor	Community building – tapping into distributed knowledge and experience
	Spends reflective time mocking up and reviewing practice-based worksheets for different modes of curatorial practice: i.e. applications, budgets, and equipment spreadsheets	
		Criteria can be double-sided: best practice approach can feel quite strict, but the different scenarios provide comparison and context

Insights

Subject A demonstrates a type of curatorial practitioner that is emerging and would find guidance in the form of curatorial criteria useful. Here we find that some of my criteria are familiar and some are points to aim for but take longer to develop over a sustained curatorial career. I find her feedback useful as she represents a certain part of my own curatorial history. I have been where she is before and can recognise what she articulates as needs and her insights from thinking about my curatorial criteria resonate with why I began this research in the first place: to aid emerging practitioners. Her feedback serves as the control, something I have experienced that she validates upon reflecting on my criteria.

Subject B (UK and Australia)

Synopsis

By way of contrast, Subject B is a mid-career independent curator, PhD educated, and practice-based with institutional and academic experience. She is also using the *Visual Matrix* audience evaluation method in a *Third Space* curatorial context.[13] With her specialist curatorial themes including health, wellbeing, and technology, where the predominant dialogue is psychology, she understands (post-PhD) that through a curatorial process, and engagement with relevant subject matter, the site of an exhibition can become a space for psychological transformation that she facilitates. She believes the aesthetic experience is an appropriate method for this experience, which is presented through technological engagement. In short, the medium (technology-based art) represents the subject matter (mental health) in a realistic enough manner to leave the audience transformed in some meaningful way. She calls this applied research methodology *The Knowledge Continuum*.[14] Her comments on my curatorial criteria in relation her curatorial own practice are outlined in Table 5.16.2.

Insights

Here we find that the more research we can do in an exhibition space, the more it opens up the space, and the more it impacts the work being done around the experience being had by the audience. In thinking about Jones' and Grant's ideas around autoethnography having the potential for self-transformation and betterment through lived experience, and in considering the curator in the role of practicing auto-ethnographer, there does seem to be potential through the exhibition as the site of audience engagement for this transformation to occur. For this transformation to result in better exhibition experiences for the audiences is compelling qualitative data, indeed. With Subject B's research, we can turn theory into understanding through action research, particularly regarding the sociological activity taking place in the exhibition space between artists and audiences and as facilitated by curators. In analysing the insights from Subject B's interview, I learned that there are curators other than myself looking at applied curatorial research methodologies for technology-based art experiences. Where Subject B was clear the criteria wouldn't work in every exhibition, she could see herself using them as a starting point to make transformational experiences more relatable by assigning language and structure to them. Reflecting on these aids in the ideation or list-making portion of curatorial process would be useful organisationally, where the transfer of tacit to physical knowledge might be the most important criterion as it underpins iterative process. To curators working alongside artists practicing iteratively, this proves very useful. Subject B recognised this confirms what our community voiced more broadly in the PhD case study. This feedback and activity serve as the extension from what is known about curatorial research practice. As demonstrated with Subject A confirming a sort of experimental knowledge control, Subject B expands from there, experimenting successfully with this knowledge in a live gallery space to create, in her own words, a kind of continuum.

Subject C (Australia)

Synopsis

Subject C was a public engagement and programs officer for an Australian university art gallery and interested in a preliminary framework or set of processes for self-care in a production role.

Table 5.16.2 Qualitative data for Subject B.

My curatorial criterion	Qualitative data	Insight
Accessibility and inclusivity	Pondered collective exhibitions over individual ones – a curator can offer contextual and logistical experience where an inclusive and accessible exhibition provides a platform for data collection from a wider ranging audience	An iterative exhibition process highlights the impactfulness of a PBR-trained curator on the way the final artwork is generated
Plurality of voices and mediums for diversity in stories and storytellers	According to her specialism, the dominant voice tends to be the psychiatric community – as an interdisciplinary art/science curator can bring alternate voices and differing collaborators together when making an exhibition	Curators can be facilitators for multiple disciplines explored through cultural platforms
Authoritative and contributory	Can sometimes be negative Has felt some pressure around selecting topics and artists that are popular rather than risking trying something new	There is a need to strike a balance with an authoritative voice and a contributory one
Relational and representative	Relates more to the representative criterion In this there is the understanding of a need to set up expectations for the reality of the exhibition space, especially when experimenting with technology The audience needs to know what to expect (i.e. living lab vs finished gallery exhibition)	Careful consideration of the audience in relation to the medium
Absent presence and present absence	No comment	n/a
Tacit and physical knowledge	Thinks that prompts, cues, and reflective tools are useful to production-based creative professionals	There isn't time scheduled for this in most production-based roles
	Believes criteria can be useful in context specific and adaptable formats – has done this herself for different curatorial contexts	Modifiable criteria allowed for an expanded curatorial process
		Expandable criteria allow non-specialists to adapt my criteria to their own organisational tools such as lists and ideation If there was time built into a production-based role for reflecting on the conversion of tacit to physical knowledge, this might be the most important
Engaging an Appreciative System (including autoethnography)	A want to give audiences more and better experiences	
	Having a curator engage an Appreciative System toolkit impacts how the audience experiences an exhibition Her research calls it *The Knowledge Production Continuum*, a very reciprocal process	About developing a collective unconscious between the curator, the audience, and the artist which is focused on how they make experiences together It is very porous, not directional or hierarchical

Table 5.16.3 Qualitative data for Subject C.

My curatorial criterion	Qualitative data	Insight
Accessibility and inclusivity	Imperative (Subject came from community arts organisations so this is at the core of her practice)	Consideration and planning around what community and the different access points to that community can't exist without criteria addressing it
Tacit and physical knowledge	Successful events are very important particularly around exhibitions There isn't a research method attached to this success, more of a loose trajectory plotted with understood tasks that need to be completed	More about care to ensure successful outcomes
	Roadblocks can crop up at any point in the process and 'success' in relation to a programmed event can slip through your fingers	Strong networks are imperative to success, if you don't reach the right person in the 'schedule–evaluate–revise' production process, your event will be compromised, as will the exhibition
Engaging an Appreciative System (including autoethnography)	Wouldn't use the tools of an Appreciative System in a conscious way. More a way to centre herself in the task at hand within the programs she's working on	Writing or thinking doesn't come from the producer, but from reflecting on the communities they're engaging with in their programs, more a part of daily tasks than a separate process to go through

A connected and conscious practitioner working in a creative setting, she didn't always feel there was time or space within her role to be creative due to the production demands of the role. Her comments on my curatorial criteria are in Table 5.16.3. I have only listed the criteria she commented on.

Insights

Subject C reiterates that she is interested in this research because she currently doesn't possess a vocabulary to express or discuss herself in relation to the production needs of her role with her institution in a straightforward way. She stipulates that the considerations for an artist or curator and their audiences are different to that of a producer. She believes that a lot of what a public program or engagements officer might do in practice is an interpretation of the works, or of the management of the curator or audience's interpretation of the works. It's not as direct as what a curator might do, or the audience might be directed to do, but it might be relational or thematic to those groups and their modes of participation. Having viewed these preliminary criteria, she sees a way forward in terms of a care-based labour framework around which discussions can be tabled and boundaries can be affirmed around emotional labour. As this wasn't available to her in this role, Subject C has since left this position. If Subject A data served as this experiment's control, and Subject B's data as the continuum, I would posit that Subject C's data shows a variant between what is required in terms of translating tacit and physical knowledge and how that is not only understood but carried forward as embodied knowledge: something to include in an Appreciative System.

Reflections and lessons

The practice-based research methods of reflexivity and autoethnography became useful in living, recording, recalling, and re-iterating my experience of a 'curator-as-producer' role. When the research that took place in the PhD case study resulting in *My Criteria for Curating Interactive Art* is applied to contemporary curatorial contexts, it appears that similar methodologies were already in use within a collaborative community of practice. Interestingly, they aren't often articulated (spoken or written) due, first, to a need for the development of language and frameworks to articulate what methodologies are required in which roles; and, second, there also appears to be a requirement to schedule time for reflection and auto-ethnography into time-based production schedules based on physical but requiring emotional or care-based labour. Where many of my criteria seemed familiar to each Subject as we reviewed them, they hadn't actually had time to create a platform from which they could consider the language and frameworks or labour models necessary to their job roles.

Where I had always thought of curating as a reciprocal process of taking knowledge in, embodying it, and then returning it in a new form in a new way (i.e. considering artworks, developing a curatorial theme, and then presenting the artwork in context to others in a curatorial platform), I found elements of what Subject B articulated as *The Knowledge Continuum* in the feedback from Subjects A and C. This speaks to how I think of curating as a constant back and forth, where you look back at what you have done, sit with what you are doing, and then look forward at what you're going to do based on what you've done, all the while contemplating what worked and what didn't. This also fits well with what Subject A articulates as a way forward with her emerging curatorial practice, and what Subject C articulates as a model of emotional or care-based labour necessary to successful public and educational programs.

Conclusion

In conclusion, the proposed key learning outcomes from this chapter are insights gleaned from reflections and testimonials on my own methodology from creative practitioners working in art galleries and as independent curators. Through interviewing three such practitioners, this chapter aimed to articulate how Ellis et al.'s criteria and my own hold up in contemporary curatorial practice in professional settings. What I discovered in the field in relation to these ideas and an autoethnographic/reflective approach was insightful. Where Ellis et al., articulate the *characteristics* of ethnographic/reflective practice, Jones, Grant, and (in this text) Findlay-Walsh are able to articulate the *understanding* found in auto ethnographic/reflective practice through reflecting on the lived experience of their practice.

In my own criteria, this transition would happen between criteria 5 (absent presence and present absence) and 6 (tacit and physical knowledge), where the state of duality achieved in one's curatorial process (reflection-in-action) transforms into tacit and physical knowledge (an Appreciative System) by reflecting on lived experience. This was brought out in contemporary curatorial practice by introducing language and criteria around autoethnographic/reflective practice with producer/curators by using interviews as an investigative research tool. Through academic rigour, these interviews became a formal enquiry.

This enquiry both extends the PhD case study by testing the usefulness of academically acquired knowledge in professional scenarios. It also suggests that applied theory written and reflected on through lived experience can result in tacit knowledge acquisition. The qualitative analysis and ensuing insights from the interview subjects are depicted in Tables 5.16.1, 5.16.2, and 5.16.3. The *Criteria for Reflection and Autoethnography* established as part of my PhD appear to have use as starting points in planning and as frameworks for considering these activities in professional production-based environment. They need to remain adaptable for contextual viability, but this small sample size displays a welcome platform at emerging, established, and adjacent creative practitioner levels. What we have learnt here is that from this starting point we can extend knowledge and understanding around practice-based research curating through autoethnographic and reflective tools by plotting the time and space to do so in the production process and articulating this through writing reflectively about one's lived experience. Adjustments to one's practice may need to be made along the way but ultimately, hopefully, and optimistically to the betterment of the discipline overall.

Notes

1 Turnbull Tillman (2019).
2 Adams et al. (2015).
3 Schön (1983).
4 Candy (2022, p. 11).
5 Adams et al. (2015).
6 Bocher (2014).
7 Findlay-Walsh (2022, p. 20).
8 Ibid.
9 Jones (2018), Grant et al. (2013).
10 Candy (2022).
11 The Ethics Approval Code for this research project, under which the interviews took place, is HC190710 and is titled: *Independent Curating and Practice-Based Research: Reflection and Auto-ethnography.*
12 Muller (2015).
13 Bartlett (2019a).
14 Bartlett (2019b).

Bibliography

Adams, T. E., Jones, S. H., & Ellis, C. (2015). *Autoethnography: Understanding Qualitative Research.* New York: Oxford University Press.

Bartlett, V. (2019a). Digital Design and Time on Device; How Aesthetic Experience Can Help to Illuminate the Psychological Impact of Living in a Digital Culture. *Digital Creativity*, 30(3), 177–195. DOI:10.1080/1 4626268.2019.163789.

Bartlett, V. (2019b). Psychosocial Curating: A Theory and Practice of Exhibition-Making at the Intersection Between Health and Aesthetics. *Medical Humanities.* DOI:10.1136/medhum2019-011694.

Bocher, A. P. (2014). *Coming to Narrative: A Personal History of Paradigm Change in the Human Sciences.* Walnut Creek, CA: Left Coast Press.

Candy, L. (2022). Reflective Practice Variants and the Creative Practitioner. In C. Vear (Ed.), *The Routledge International Handbook of Practice-Based Research.* London and New York: Routledge.

Findley-Walsh, I. (2022). The Sound of My Hands Typing: Autoethnography as a Reflexive Method in Practice-Based Research. In C. Vear (Ed.), *The Routledge International Handbook of Practice-Based Research.* London and New York: Routledge.

Grant, A., Short, N., & Turner, L. (2013). Introduction: Storying Life and Lives. In A. Grant, N. Short, & L. Turner (Eds.), *Contemporary British Autoethnography.* Rotterdam: Sense Publishers, 1–17.

Jones, S. H. (2018). Creative Cultures/Creative Selves: Critical Autoethnography, Performance and Pedagogy. In S. H. Jones & M. Pruyn (Eds.), *Creative Cultures/Creative Selves: Critical Autoethnography, Performance and Pedagogy.* London: Palgrave Macmillan, 3–21.

Muller, L. et al. (2015). *Understanding Third Space: Evaluating Art-Science Collaboration.* ISEA 2015: Proceedings of the 21st International Symposium on Electronic Art, Vancouver, Canada. ISBN:978-1-910172-00-1.

Personal Communications, Interview Questions Emailed to Subjects Dated, August 27, 2019 {Subjects A & B}, and January 17, 2020 {Subject C}.

Schön, D. A. (1983). *The Reflective Practitioner: How Professionals Think in Action.* New York: Basic Books.

Turnbull Tillman, D. (2019). Past the Museum Floor: Criteria for curating experience. Book chapter in T. Giannini & J. P. Bowen (Eds.), *Museums and Digital Culture: New Perspectives and Research.* Springer Series on Cultural Computing. Libri: Springer, May, 115–145.

5.17

PLEASE TOUCH!

Marloeke van der Vlugt

The aesthetics of touch: researching the tactile sense in and through art

My area of research revolves around artistic strategies that activate tactility (the experiential sensation of touch(ing)). This examines tactility through imagination or in the real, as a prominent component within the aesthetic process that locates itself at the crossroad of performance, scenography and visual arts.

In and through the creation and presentation of tactile-centred performative installations, I explore how to interact with bodies, organisms, objects and materials in a speculative manner, to explore how the different components of such a composition relate to each other rather than to reaffirm what is already known. One of my motivations for making these time-based artistic propositions is to raise awareness of our *reciprocal* nature of being in the world. I share the view of philosopher Jane Bennett, who states that all materialities have their own form of agency and have a dynamic relationship with their surroundings.[1] To experience this open world view, I employ performance strategies, like multi-sensorial activation, spatial arrangement or guided tours, to encourage the audience to deliberately oscillate within the durational interval that a tactile experience can evoke. My installations provide time and space for me, as a maker, and for an audience/group of participants to explore the events created by the reciprocal relation between our bodies and other materialities, in order to become sensitive to a world populated by animate things rather than passive objects and to open up for the emergence of new voices, claims or rights.

In my current PhD project, I have formulated the following questions:

1   What artistic strategies can be used to activate the sense of touch within the aesthetic process that locates itself at the crossroad of performance, scenography and visual arts?
2   What formal qualities of a relational artistic event and/or artefact (e.g. materials or dramaturgy) can enact the *reciprocal nature of touching* within artistic creation and reception processes?
3   How might these experiential, ephemeral processes that assist artists/participants/the public in exploring their tactile interaction with the world be documented and disseminated?

I research these questions in and through my interdisciplinary artistic practice. I alternate between methods: creating and presenting artworks; the study of literature and theory; the observing, documenting, reflecting and disseminating of the creation and reception processes of these works through writing and filming; and the teaching of and co-creating with students and fellow artists. As such,

DOI: 10.4324/9780429324154-60   722

the roles of artist, researcher and teacher are entangled and a part of the overarching creation process. I do not follow a linear preconceived path; some methods are implicated within each other, and the variety of methods are practiced in a cyclical and iterative manner. Some are used in parallel, others in sequence, and each step informs the next step and vice versa. As such, the constant reformulation and/or reframing of my methods is an important part of my research output.

Fieldnote[2]

I am making one knot after another in the silk: binding, stopping, stroking the cloth to locate a new spot that is flexible, pinching the material, pressing the fingers to put the thread around the silk stack, loosening or pulling the knot etc. The material influences the process, I have to 'listen' to the material in order to continue.

The actual knotting evokes a heightened awareness of the alternation between the active exploratory action of touching and passive touching through the stimulation of skin by the material.

This alternation of touches and the diversity in qualities of touches (hard, soft, sharp, short, long) make it an immersive experience and ignite my body to perform a range of physical sensations combined with emotions, memories and imaginations.

The inner sensation of moving my body in time and space becomes highly influenced by the action itself and the materials I am working with, especially the size of the cloth. While knotting, I start to relate the passing of time to the amount, size and distance between the knots I (still need to) make. The macro space around me disappears, the experiential space surrounding my body (pericutaneous) takes over. For instance, when I try to make little knots close to each other, I need to bend over and bring the silk close to my eyes, making micro movements with my fingers, evoking the sensation of being trapped in a small, narrow and low corridor. However, when the cloth is big, just unfolding it or throwing it out in the air and draping it over my body, expands my sensation of space, giving it a vast, infinite quality.

As my full attention, both mental and physical, is hyper-focused on the haptic interaction with the cloth, I somehow incorporate these formal but flexible qualities of the material. It makes me aware of the relationality of our experiences and the performative qualities of the tactile sense.

As this fieldnote demonstrates, within my practice-based research I treat the body as an area of investigation in its own right. In other words, I execute 'embodied research', as formulated by theorist Ben Spatz:[3] I take my own body as a tool to create with and to reflect on. I consider this method as an experimental and embodied pathway into researching the interrelatedness of feeling and sensation and of the tactile, aural and the visual senses. I regard the body itself as 'a process of intersecting forces (affects) and spatio-temporal variables (connections)'.[4]

Given the goal of my research is to critically explore tactility within the realm of art making and reception, it is implicit that other people are necessary to differentiate and speculate about its outcomes.[5] Doing and sharing research within a live set-up is closely connected to Brad Haseman's concept of Performative Research.[6] Performative Research manifests itself by 'doing artistic interventions, whereby makers, researchers and audience meet and exchange, and can experience new collaborative forms, share and experiment together'.[7] As part of my research strategy, I unfold and extend the concept of Performative Research through an iterative trajectory of practical set-ups and through the confrontation with actual theories on performativity. With the term performative, I build on Judith Butler's concept of performativity,[8] in which we *perform* our identities, which are realised through physical and repetitive acts. New Materialism builds on this foundation, underscoring the co-constitution of material and discursive productions of reality, accentuating how (in)animate objects perform their agency.[9] Physicist and philosopher Karen Barad specifies this construct of performance by reworking the idea of causality, stating that 'distinct agencies do not precede but rather emerge through their intra-action'.[10] This concept points to the entanglement between set-up (apparatus) and research outcome, and questions the possibility of 'archiving' experiential processes while the impact on the participant/maker of the tactile interaction within an artwork is performed in the moment and on the spot.

Fieldnote[11]

Teaching a class with performance students on the hierarchy between the senses, we examined our physical reactions while watching other students caressing, pinching, folding and stretching a networked, conductive cloth. Their manual handling of this cloth triggered pre-recorded audio files, together evoking a complete narrative. Just watching them performing a variety of touches produced a strong tactile sensation within my body. It evoked the physical memory of wearing a navigation belt: I felt the tiny motors vibrating around my waist – they made me feel like flying and I wanted to dance. I realised that just the visual observation of a tactile encounter between material and bodies made me enact a strong, intimate and sensual sensation of touching. Talking afterwards about these ephemeral, relational experiences was hard to do as descriptions tend to 'search for' meaning, and become merely representations of the experiences. We started to try diverse documentation methods, like recording videos and watching them immediately afterwards, staying in the same space, or discussing, drawing and moving in space while the performance was going on. Like this we assembled data to compare the impact of changing the composition of components (for instance a different sequence or quality) on the durational experience that was performed.

Through my practice-based research, I explore manners of documentation and dissemination such as experiential and performative set-ups that facilitate the participants to overcome the sort of inter-action in which subject and object are seen as two separate unities and surrender to what Barad calls 'intra-action'; where knowledge is fluid and constantly changing in the moment.[12] The whole process of conception until and beyond the interaction with the audience is, according to Barad,[13] how matter starts to 'matter'. In other words, as we touch someone or something, the thoughts that are acted out are created in a complex collaboration between the object itself, the situation and the performative moment. As such, when the participants are physically present in the setup, the situation will have the ability to build its own unknown reality.

Case study: *Thresholds of Touch*

After attending the conference *Hold me now* at the Stedelijk Museum Amsterdam in April 2018, I became acquainted with social scientist Carey Jewitt, researcher/director of the *In-Touch* project, University College London. This became the start of a long-term, ongoing collaboration between Carey, the composer and researcher Falk Hübner and myself that, until this day, amounted to the performative experiment *Thresholds of Touch* (January 2020) and a research paper (Jewitt, van der Vlugt, and Hübner, 2021).

The collaboration started off with a sincere and mutual interest in each other's practices. Carey expressed her interest in being involved in an artistic project from the start. I visited the lab in London, acquainted myself with *In-Touch*'s projects and research methods, based on multimodality and sensory ethnography, talked to the other team members, learnt a lot about the history and politics concerning touch and visited an exhibition the lab was involved in. I walked through the exhibition, touching the artefacts and other visitors, and attempted to verbalise the multi-sensorial embodied experiences that were evoked within the gallery space. I experienced their manner of assembling the reactions of the audience: a combination of 'live' and video-recorded observations of and interviews with the visitors, always aware and critically questioning how their position as researchers influenced the outcomes. As an artist/researcher, I inwardly focused on my sensorial interaction with the materialities that were present, for example:

Please touch!

Fieldnote[14]

Questions to myself: What materialities invite, evoke or require tactility to keep my attention? What imaginary or real interactions with materialities over time make me curious and open, enacts a mystery, shifts something inside of me?

I am always searching for an aesthetic experience that evokes me to oscillate between a prelinguistic being-in-the-moment and my thoughts, immersing me into an internal space, a space-in-between, that informs and resonates with my 'making practice'. This search was hardly noticeable in my actions and I did not attempt to verbalise my embodied reactions. The sort of knowledge I get out of these experiences can be best described as ontological: poetic, mysterious and unfathomable.[15] It becomes a source of inspiration as it gives me direction, although undefined and ephemeral, in my (subconsciously) aimed-for qualities of a new artistic creation.

Here a difference between Carey and myself became clear: as a social scientist, her goal with the project is to critically explore and discuss methods and concepts around touch in relation to socio-logical concerns. The *In-Touch* project aims to generate, capture and interrogate touch moments through felt empathetic route into participants' bodily known practices and imaginaries in search of a better understanding of its social and affective character.[16] While these are important goals, my attention differs; first and foremost, I move to a state of 'unknowing' as a silent, passive reception[17] as I relate my experiences, framed by the topics I am researching, partly to the re-articulation of my research questions, and partly to working with materialities within the process of creation.

For one and a half years, the three of us exchanged information, notes and ideas through and within digital and physical spaces. Carey visited our studios in Utrecht and Amsterdam. During these meetings, we each moved in and out of our roles as researcher, artist, performer/assistant, participant, and at times it was hard to delineate such clear lines between us. For the live meetings, I would assemble materials and create artefacts for Falk and Carey to experience. We slowed down and took time to question the components of the events that were created in the moment. Our own bodies became a starting point for touchy explorations and we acquired a vocabulary and insights in/ through our joint explorations. We shifted roles and at times involved our professional biographies: I created knitted structures of electric wire and Falk experimented with sound files that would intensify the experience of touching. Carey contributed social theory and concepts of touch, offering real-time socially orientated analytical commentary. This shifting between roles became a central strategy for our individual and collaborative explorations and inspirations.

Returning to our places, we continued to research separately. In these periods, I developed several artefacts that were presented at four semi-public spaces: a gallery space, the theatre academy, my research group and the presentation space of an artist initiative. These periods and presentations were important, as they gave space and time for my research activities to position themselves within the artistic research field. I arrived at a few conclusions that needed to become experienceable in the artistic artefacts:

Reflective diary[18]

To activate tactile 'reading' I will work with materials and techniques that implicate 'vitality' and have a 'dynamic form'.[19] As a result, the objects:

- *activate the Tactile eye,[20] meaning watching the (shape of the) material activates the sensation of touching, for instance through its resemblance to skin or landscapes*
- *present traces and marks (reveal the process and a fabricator).*
- *lengthen and intensify the haptic exploration while the material reacts to skin-to-skin contact, either through changing its shape or through its conductive properties that trigger audio files.*

> As a result, to get to 'know' the object, only seeing is not enough. It's not possible to predict how
> the backside looks, without turning it around, to understand what is inside without pressing it or to
> sense its temperature nor predict its weight without picking it up. The objects require haptic explora-
> tion, as this is indispensable for effective decoding and mobilising imagination.

In order to disseminate these temporarily research outcomes, I set myself a few rules and limited myself to three techniques/crafts and materials:

1   The first technique is the craft of Bandhani, which is a type of tie-dye textile, where cloth is tightly tied into many small knots that form a design. The cloth is subsequently dyed, and the thread later removed to leave a circular design on the cloth. The structure strongly invites the fabric to be touched with the top of the fingers, an action reminiscent of reading a story written in braille.
2   The second technique is the use of casting with urethane foam. Working with this process is not entirely predictable as the invisible chemical process is a part of what creates the final shape, and this is influenced by temperature, stirring, movement, colour pigment, elements added within the foam, surface it is poured on etc. As a result, these foams can create a variety of tactile quali-ties to 'read'. The reciprocal nature of these objects (i.e. the tactile responsiveness) makes them very appealing to grab: they immediately return your squeeze as they try to get back into their original shapes (see Figures 5.17.1 and 5.17.2).
3   The third technique is knitting electric wire, as it connects both fear and intimacy in relation to touch. Fear related to touching anything that has current running through it and intimacy triggered by the physical act of knitting, think of the homely, feminine and intimate qualities knitted surfaces still evoke. The capacitive sensing qualities of the wire itself and the possibility of connecting the wire to a variety of sensors made it possible to translate the quality, duration or amount of touching of the wire with one's hands or body into sounds, words, images etc.

Experiment

In the autumn of 2019, Carey, Falk and I began to develop a collaborative event: the performative experiment *Thresholds of Touch*. Carey secured a venue: the Bloomsbury theatre studio in London, that – although affiliated with the UCL – seemed to position the work within an artistic context.

Fieldnote[21]

> From the moment the venue was brought into consideration, the fluid, productive roles we performed so
> far slightly shifted. My professional background, director/scenographer, pushed itself to the foreground.
> I went into production modus.

We devised the experiment as a diptych: the first part was an interactive performance, partly based on my artistic research findings so far, including some didactic and dramaturgical insights I assembled as a teacher. Within the structure of the whole event, it came to function as a 'tactile preparation chamber', while the second part was organised as an exploratory workshop that extended the themes of the performance and was designed to assemble data on the embodied experiences of the partici-pants. As Carey put it: 'the workshop aims to generate accounts and stories of touch and interrogate our tactile imaginations of digital touch communication'.

In this second part, the participants were invited to reflect on, explore and document their felt experiences of the various propositions. The participants were asked to use *their* skills and forms of expression, based on the materials available e.g. notebooks, post-cards, storyboards and video. For

Figure 5.17.1 and 5.17.2   Tactile Objects _ Series 1.

Source: Photo: Marloeke van der Vlugt.

documenting the whole experiment, video recordings, observations, work- and fieldnotes were made by the members of the Lab and a few members of the audience.

Reflective diary[22]

It feels like part one, the performance experiment, has become instrumental to the collection of data in part two. Are the sculptures I created regarded as research objects or do they also evoke an aesthetic experience? Can they enact both? Did the objects transfer knowledge that I embodied within the creation process? How to critically discuss the tension between theatre space and research environment? How is this translated in the invitation or the publicity? There are expectations related to the words research, experiment, theatre, performance. If the audience become the performers themselves how does this impact their relation to materialities?

In production modus, creating the script, the scenography, and performing myself, it was impossible to slow down and find manners to position the first part, the performance, as a dissemination AND as a 'performative research environment' that would inform my practice-based research, the question: how to 'document' the events taking place in **that** *moment, was pushed to the background.*

Only in retrospect have I realised that during this period my roles of artist and researcher were in conflict. This was realised through the tensions of presenting the performative experiment within the framework of our collaborative method, with its time pressure, contradictory spatial expectations and the ethical regulations to comply with. These tensions led to the first part – the performance – being only video-recorded once and, related to technical difficulties, not for the complete duration. As I was also 'performing', I could not simultaneously take on the role of observer/experiencer/researcher. As a result, the aesthetic experience and relational knowledge that was enacted in the moment was solely documented from the perspective of data assembling that could assist in reconstructing intimate 'stories', rather than from an artistic perspective. As a result, the documentation failed to give insight in the relationality between the components of the experience itself nor instigated me to articulate new artistic questions. This failing became a lesson for the future collaboration. However, the outcome of the reflective part was inspiring in various other ways, as Carey formulated:

> the framework provided researchers and participants novel ways to get in touch with touch, it helped them to attune to touch, offered a platform and vocabulary for exploring touch, and generated insights on the felt, sensorial, and affective dimensions of touch. This helped to bring touch out from beneath the other senses in the aesthetic process, and to hold it up to critical reflective address and exploration.[23]

Conclusion

The practice-based research project *Thresholds of Touch* became a multi-disciplinary, experimental method, blurring the boundaries between artistic and sociological research by offering an innovative possibility to research and disseminate the multimodal experience of touching. The performance experiment heightened participants' sensitivity, awareness and reflection of touch. The notebooks provided a reflexive space.

A performative research environment like this can provide the objects/performance/space to investigate 'ephemeral and performative experiences' that get enacted in the moment, as it invites the audience, participants and researchers to explore not what the tactile experience *means* but what the tactile sense can *do*. However, it is crucial to consider how to document these experiences. As an option, I would suggest video-recording the whole performance and immediately afterwards watching/listening to the recordings with the participants, especially paying attention to the potentiality of their bodies, as even the slightest movement, the tensing of a body part, might indicate a shift in perspective.

Our collaboration continues. The team of *In-Touch* is analysing the collected data combining two inductive data-driven approaches: multimodality and sensory ethnography. They will work across the assembled research materials, participant notebooks, the researcher fieldnotes and the video recordings to make sense of the participant touch experiences. In parallel, as an artist/researcher, I aim to open up the analysis and broaden the collected data (the responses of the audience) by taking a (specific part of) the collected data – the notebooks – and explore notions of affect through my '*felt responses to them*', rather than consciously reflecting on, articulating or categorising them. The ontological knowledge related to this embodied research will inform the development of a new artistic work. This artistic method will contribute to the analysis by articulating insights concerning tacit knowledge and by documenting artistic strategies that evoke the emergence of experiential and material forms.

Acknowledgements

The collaborative case study *Thresholds of Touch* was undertaken with support from the InTouch project, a European Research Council Consolidator Award (Award Number: 681489).

Notes

1 Bennett (2009).
2 Taken from a personal document written on 26 May 2020.
3 Spatz (2017).
4 Braidotti (2002, p. 21).
5 Goodman (2018).
6 Haseman (2006).
7 Haseman (2006, pp. 98–106).
8 Butler (1988).
9 Bennett (2009.)
10 Barad (2007, p. 33).
11 Taken from a personal document written on 10 October 2019.
12 Barad (2003, pp. 801–831).
13 Barad (2003).
14 Taken from a personal document written on 2 July 2018.
15 Visse et al. (2019, p. 2).
16 Taken from personal fieldnote (2019).
17 Ibid (p. 6).
18 Taken from a personal document written on 28 October 2019.
19 Massumi (2008).
20 Sobchack (2004).
21 Taken from a personal document written 30 January 2020.
22 Ibid.
23 Jewitt et al. (2021).

Bibliography

Barad, K. (2003). Posthumanist Performativity: Toward an Understanding of How Matter Comes to Matter. *Signs: Journal of Women in Culture and Society*, 28(3), 801–831.
Barad, K. (2007). *Meeting the Universe Halfway: Quantum Physics and the Entanglement of Matter and Meaning*. Durham and London: Duke University Press, 33.
Bennett, J. (2009). *Vibrant Matter*. John Hope Franklin Center. Durham, NC: Duke University Press.
Braidotti, R. (2002). *Metamorphoses: Towards a Materialist Theory of Becoming*. Cornwall: MPG Books.
Butler, J. (1988). Performative Acts and Gender Constitution: An Essay in Phenomenology and Feminist Theory. *Theatre Journal*, 4, 40.
Goodman, A. (2018). *Gathering Ecologies Thinking Beyond Interactivity*. London: Open Humanities Press.
Haseman, B. (2006). A Manifesto for Performative Research. *Media International Australia Incorporating Culture and Policy Theme Issue 'Practice-led Research'*, 118, 98–116.
Jewitt, C., Vlugt, M. van der, & Hübner, F. (2021). *Sensoria: An Exploratory Interdisciplinary Framework for Researching Multimodal & Sensory Experiences*. Methodological Innovations. DOI: 10.1177/20597991211051446

Massumi, B. (2008). The Thinking-Feeling of What Happens: A Semblance of a Conversation. *Inflexions Online Journal*, 1(1).

Sobchack, V. (2004). *Carnal Thoughts*. London: University of California Press.

Spatz, B. (2017). Embodied Research: A Methodology. *Liminalities: A Journal of Performance Studies*, 13(2).

Visse, M., Hansen, F., & Leget, C. (2019). The Unsayable in Arts-Based Research: On the Praxis of Life Itself. *International Journal of Qualitative Methods*, 18, 1–13.

INDEX

4'33" (Cage) 296

Abu-Lughod, L. 672
academics: criteria for possible PhD program 174; for success 173
academisation of creativity and morphogenesis of practice-based researcher 60–73; chapter summary 6; concluding thoughts 72; demise of Art College 61–63; ecology allowing practice-based research to thrive 65–72; event/space to perform 67–69; instruments/artefacts to perform 69–72; introduction 60–61; network activities 66–67; research in curation and conservation 63–65
acquaintance, knowledge by 246
action research 208
active agent, meta-materiality as 258–260
active communities 548–550; activities that encourage 548–549; key questions when thinking about 549; tips for doctoral researcher 549–550
active reviewing cycle 48
activist, in cognitive processing model 230, 234, 236, 237
activities, in Crafting Methods 331
actors in curatorial projects, other 273–275
Adams, L. 125–126
Adams, T. 497, 714, 715
Advanced Telecommunications Research (ATR) 77
aesthetic knowledge 228, 233, 234, 237
affective knowledge 227–228, 235
agency, in practice-based doctoral research communities 546–547
Agius, J. 351
AHRC 33, 110
AHRC ICT Methods Network 148
AirSticks 101–103
Akrich, M. 634
Albers, J. 296

Amaral, H. 470
ambition, in practice of practice-based research 530–531
analysis: dramaturgy 515–516; emergence in 449; mapping practitioner knowledge 222–223; new knowledge 222–223; *see also* data analysis
analytical-researcher (AR) 212, 609
Anderson, J. R. 245
Andre, C. 362
anonymisation 583–584
antecedent 403–405
apparatuses 707
appreciative systems 267–275; chapter summary 11; expanding 273–275; introduction 267; work of 267–269
Apter, M. J. 478
Archer, B. 32, 299
ArchXtonic 707–712
argument, in practice-based thesis 597
Argyris, C. 48, 210, 293, 417
Arlander, A. 333
art: artmaking as hypothesis testing 389–393; collaborative practice, in reflective practice 428; computational technology and 146–149; design and, differentiating between 680; holography and 670; making 686–689; object (*see* art object does not embody a form of knowledge); practical value of 288–289; practice, methods used in 142; practice-based research in, origins of 32–34; process, dramaturg in 516–517; research and practice beyond domain of 383–385; research doctorates in 139–153; scholarship of 169; science collaboration and 660; tactile sense in and through, researching 722–724; theoretical side of 707–709; as transient experiment 387–389
Art College 60–65, 68
artifacts: to perform 69–72; PhD research and 87–88; in regulatory codes of practice 88;

731

For Product Safety Concerns and Information please contact our EU
representative GPSR@taylorandfrancis.com
Taylor & Francis Verlag GmbH, Kaufingerstraße 24, 80331 München, Germany

www.ingramcontent.com/pod-product-compliance
Lightning Source LLC
Chambersburg PA
CBHW081208220326
41598CB00037B/6708